D0138186

Applied Animal Nutrition

FEEDS AND FEEDING

Third Edition

Peter R. Cheeke

Department of Animal Sciences
Oregon State University

Upper Saddle River, New Jersey 07458

Library of Congress Cataloging-in-Publication Data

Cheeke, Peter R.
Applied animal nutrition : feeds and feeding / Peter R. Cheeke.—3rd ed.
p. cm.
Includes bibliographical references (p.)
ISBN 0-13-113331-4
1. Feeds. 2. Animal feeding. 3. Animal nutrition. I. Title.

SF95.C463 2005
636.08'5—dc22 2004044341

Executive Editor: Debbie Yarnell
Editorial Assistant: Jonathan Tenthoff
Marketing Manager: Jimmy Stephens
Marketing Assistant: Melissa Ursburn
Production Editor: Carrie Rausch, Carlisle Publishers Services
Managing Editor: Mary Carnis
Director of Production and Manufacturing: Bruce Johnson
Marketing Assistant: Melissa Orsborn
Production Liaison: Janice Stangel
Manufacturing Manager: Ilene Sanford
Manufacturing Buyer: Cathleen Petersen
Creative Director: Cheryl Asherman
Design Coordinator: Christopher Weigand
Cover Design: Kevin Kall
Cover Illustration/Photo: Golden Labrador puppies, courtesy of Getty Images, Inc.-Image Bank; Horse and colt, courtesy of Getty Images, Inc.-Photodisc, Farmer with a laptop beside a pigpen: courtesy of PhotoEdit, photographer: Mark Richards; Lambs, courtesy of American Lamb Council; Holstein cows, courtesy of Corbis/Bettmann, photographer: Richard Hamilton Smith.
Composition/Full-Service Project Management: Carlisle Publishers Services
Printer/Binder: Courier Westford

Pearson Education LTD.
Pearson Education Singapore, Pte. Ltd
Pearson Education, Canada, Ltd
Pearson Education-Japan
Pearson Education Australia PTY, Limited
Pearson Education North Asia Ltd
Pearson Educación de Mexico, S. A. de C. V.
Pearson Education Malaysia, Pte. Ltd

10 9 8 7 6 5 4 3 2
ISBN: 0-13-113331-4

CONTENTS

PART IV FEEDING AND NUTRITION OF LIVESTOCK AND SPECIALTY ANIMALS 317

CONTRIBUTORS

H. G. Bateman II, Agricultural Research Center, Louisiana State University, Baton Rouge, Louisiana.

Timothy DelCurto, Eastern Oregon Agricultural Research Center, Oregon State University, Union, Oregon

Ellen S. Dierenfeld, St. Louis Zoo, St. Louis, Missouri

George C. Fahey, Jr., Department of Animal Science, University of Illinois at Urbana-Champaign.

Larry A. Lawrence, Kentucky Equine Research, Inc. Versailles Kentucky

David R. Ledoux, Department of Animal Science, University of Missouri, Columbia, Missouri

Anders Skrede, Agricultural University of Norway AS, Norway

Julie K. Spears, Department of Animal Science, University of Illinois at Urbana-Champaign.

Robert A. Swick, American Soybean Association Singapore.

James M. Thompson, Department of Animal Sciences, Oregon State University, Corvallis, Oregon.

Trygve L. Veum, Department of Animal Science, University of Missouri, Columbia, Missouri.

PREFACE

The objective of this book is to describe the properties of feedstuffs used in the feeding of domestic animals and to provide information on feeding practices for a variety of domestic and exotic animal species. It is intended that the book be suitable as a text for undergraduate students, at both university and junior college levels. It should also be useful to professional animal nutritionists, extension agents, veterinarians, and livestock producers.

Recognizing this diverse audience, I have endeavored to make the presentation of subjects technically complete but not technically difficult. The writing style is one that should help accomplish this goal. Detailed discussion of metabolic pathways and use of chemical structures and terminology have been avoided. My philosophy is that these topics are better presented in a book on principles of nutrition, rather than in one on feeds and feeding. The literature citations are recent and serve as an entry to the literature for those who need more detailed information. Where possible, literature citations are from major journals that should be available at most university and college libraries. References are cited extensively in the text to introduce and/or reinforce the concept of documenting statements with verifiable data from peer-reviewed scientific journals.

Chapters dealing with the nutrition and feeding of the various livestock species have been contributed by leading specialists in each area. Although for the first two editions I was the sole author of the entire text, the complexity of contemporary animal nutrition has necessitated the input of specialists. I have endeavored to integrate their contributions so as to maintain a uniform writing style.

Animal agriculture is undergoing rapid industrialization and globalization, providing new challenges and opportunities. More than ever before, students should have some exposure to the implications of the global economy on their field of study. Various contentious issues impact animal feeding and nutrition. The emergence of BSE or "mad cow disease" in Great Britain has focused worldwide attention on the use of meat meal as a feed for livestock. *E. coli* contamination of beef has drawn public attention to the use of chicken excreta as an animal feedstuff. Harvesting fish and krill (Antarctic shrimp) for use as animal feed has raised environmental concerns. These types of contemporary issues, which impact animal feeding, are discussed throughout the book.

Awareness of environmental and ecological effects of livestock production is critical to animal science students if they are to be adequately prepared for coping with the challenges of the twenty-first century. These challenges include the continually increasing human population coupled with rising expectations for an improved diet, the emergence of a global economy, increased public concerns about exploitation of natural resources and environmental degradation, and continuing food safety concerns, many involving animal products (e.g., microbial contamination, BSE). Air, water, and soil pollution associated with intensive animal production is becoming a major issue. Aspects of "ecological nutrition" with nutrient management used to minimize environmental impacts of animal production are discussed. Thus, I have endeavored to discuss animal feedstuffs within the larger framework of societal values and concerns. In keeping with the public mood and my own biases, the treatment of contemporary issues has an environmentally friendly slant. We should recognize that many students today have a sincere and very strong interest in preserving the environment and are turned off by a strictly economic "least cost, most efficient" bottom line.

Another unique feature of this book is my attempt to bridge the gap I perceive between animal scientists and agronomists. Increasingly, students do not have a farm background, and they may have little practical knowledge of different crops and forages. They may know about alfalfa meal as an ingredient but have no idea of what sort of plant alfalfa is, what clover looks like, and so on. This sort of background knowledge has been taken for granted in the past. Consequently, I have endeavored to provide more agronomic information about the sources of feedstuffs than might usually be the case.

A series of questions and study guides follows each chapter. These instructional materials are intended to aid students in comprehending the text. Some of the questions "pull things together" so that students are encouraged to think and develop a rational answer. In some cases, the answer per se is not in the text, but students should be able to synthesize a reasonable answer from the information given. In other cases, there is no "right" answer; these questions are intended to make students think critically about issues and formulate opinions, using the background gained from the text. They should provide a good starting point for classroom discussions.

The treatment leans more heavily toward the discussion of properties of feedstuffs, with a less detailed consideration of feeding practices. The objective of Part II (Applied Nutrition) is to provide the student with some background on the feeding requirements of each of the major species. For specialists and professional nutritionists, more detailed sources, such as the NRC publications, should be consulted. The wide coverage of species, including fish, zoo animals, and wildlife, is probably unique in a text of this type. Consideration is given to minor species to a greater extent than might be warranted from an economic viewpoint. My reasoning is that it is much easier for the reader to find information in other sources on dairy cattle than on ostriches, for example. This book will provide basic information and access to the literature for most species that students and practitioners of animal and poultry nutrition are likely to encounter.

Later chapters build on earlier ones. Topics such as essential amino acids, omega–3 fatty acids, and so on are introduced in the early chapters and encountered again later. It is intended that appreciation for different nutrients and feedstuffs will be enhanced when they are encountered several times in different contexts. I hope that by using this approach, students can learn and gain increased understanding without necessarily realizing it at the time. It is intended that there be several layers of depth so that, according to individual needs, students can achieve different levels of understanding as is appropriate.

I have found that a very useful teaching aid to accompany this text is a student subscription to *Feedstuffs* magazine. Inexpensive student subscriptions for a term or semester are available. I provide a weekly assignment from the current issue of *Feedstuffs.* Throughout the term, many of the topics covered in class are encountered in *Feedstuffs.* I draw attention to these by a selection of questions from each issue. This gives students an appreciation that what we discuss in class really has "real-world" application and reinforces material presented in lectures and the text.

Many thanks to students and colleagues who have made suggestions and comments, many of which have been incorporated into the new edition including Leland S. Shapiro, L.A. Pierce College; H.G. Bateman, Louisianna State University.

Peter R. Cheeke

INTERNET RESOURCES

AGRICULTURE SUPERSITE

This site is a free online resource center for students and instructors in the field of Agriculture. Located at http://www.prenhall.com/agsite, this site contains numerous resources for students including additional study questions, job search links, photo galleries, PowerPoint slides, *The New York Times* eThemes archive, and other agriculture-related links.

On this supersite, instructors will find a complete listing of Prentice Hall's agriculture texts, as well as instructor supplements that are available for immediate download. Please contact your Prentice Hall sales representative for password information.

THE NEW YORK TIMES eTHEMES OF THE TIMES FOR AGRICULTURE AND *THE NEW YORK TIMES* eTHEMES OF THE TIMES FOR AGRIBUSINESS

Taken directly from the pages of *The New York Times,* these carefully edited collections of articles offer students insight into the hottest issues facing the industry today. These free supplements can be accessed by logging onto the Agriculture Supersite at: http://www.prenhall.com/agsite.

AGRIBOOKS: A CUSTOM PUBLISHING PROGRAM FOR AGRICULTURE

Just can't find the textbook that fits your class? Here is your chance to create your own ideal book by mixing and matching chapters from Prentice Hall's agriculture textbooks. Up to 20% of your custom book can be your own writing or come from outside sources. Visit us at: http://www.prenhall.com/ agribooks.

Yellow corn is the major feed grain used for swine, poultry, and cattle production in much of the world. It is called *dent corn* because of the indentation formed at the tip of the kernel when the starch shrinks during drying. Corn is high yielding, palatable, and has a higher energy content than most other feed grains. (Joe Munroe/Photo Researchers, Inc.)

PART I

FEEDSTUFFS AND THEIR PROPERTIES

This part is intended to develop an appreciation of the comparative value of feed ingredients and to illustrate the synergism among crop production, sustainable agriculture, and livestock production. Many feedstuffs are agricultural by-products; livestock have an important role in human welfare in converting crop residues and by-products to high-quality food products.

Part Objectives

1. To understand the use of the following feedstuffs in animal feeding:
    - concentrate energy feeds: grains
    - other concentrate energy feeds
    - protein supplements
    - forages
    - mineral and vitamin supplements
    - feed additives
2. To understand feedstuffs in terms of the following:
    - agronomic features
    - nutritional value
    - nutrient deficiencies
    - processing needed
    - deleterious factors
    - use in animal feeding

CHAPTER 1

Introduction to Nutrients, Digestive Physiology, and Feed Analysis

Objectives

1. To discuss required nutrients and their functions
2. To compare the major types of digestive tract physiology and relate differences to animals' abilities to digest and utilize various feedstuffs
3. To introduce the common methods of feed evaluation:
   microscopy
   proximate analysis
   digestibility
   energy determination
4. To identify sources of information on feedstuffs

The objective of this book is to provide a comprehensive but readily understood treatment of the characteristics of feedstuffs and their use in animal feeding. Feeds will be discussed in terms of their agronomic features, nutritive value, special properties, processing needs, suitability for particular kinds of livestock, the presence of deleterious or toxic factors, and any other considerations that relate to their utility as feeds. The role of livestock and their feedstuffs should be considered in terms of the entire agricultural scene; thus, it is desirable for animal science students to have some appreciation of agronomic features of crop and forage plants and how the utilization of crops and by-products as feeds relates to the integration of livestock production into the entire agricultural complex.

Since the dawn of agriculture, livestock production has been an integral component of all farming systems. Livestock production promotes soil fertility by necessitating crop rotation with forages, particularly nitrogen-fixing leguminous forages, and by the direct contribution of manure as a fertilizer and soil amendment. Livestock convert crop residues, wastes, and by-products into useful commodities. Animal production contributes to rural stability and an enduring agriculture. Monoculture production of crops in modern American agriculture has been viewed by some as soil mining, with high short-term yields at the expense of soil erosion and loss of the rural fabric. Cultivation of lands best suited for grass production led to the infamous "dust bowl" of the 1930s. Crop rotations involving forages for animal production can markedly reduce soil erosion problems.

This book is intended to be useful to students in animal science and related fields, to farmers and ranchers, and to people in the feed industry. Chemical terminology has been kept to a minimum, consistent with the needs of the intended users.

Nutrition is a very broad discipline, and it is difficult to cover it all at once. Therefore, the nutrient categories are simply introduced in this chapter; further elaboration will be provided when needed. For example, when discussing corn as a feed, its contributions of energy and deficiencies of essential amino acids must be considered. The concept of lysine as a critical amino acid will be discussed at that point, rather than in a separate chapter on protein and amino acid metabolism. It is hoped that this approach will be more interesting to the student, by showing the direct relevance of the information. Each chapter builds upon information provided in earlier chapters.

NUTRIENT CATEGORIES

A **nutrient** is a dietary essential for one or more species of animals. Not all animals require the same **nutrients.** For example, ruminants (cattle, sheep, goats, etc.) have quite simple nutrient requirements in comparison to nonruminant (monogastric) species, such as swine, poultry, and humans. All of the known nutrients are in one of the following categories: protein, carbohydrates, lipids (fats and oils), minerals, vitamins, and water.

Protein

Protein is composed of substances called **amino acids.** There are several hundred known amino acids in plants. However, only 20 amino acids make up animal proteins. Of these, 10 can be formed in the tissues, whereas the others must be provided in the diet. The **essential amino acids** must be provided in the diet (or in the case of ruminants, are synthesized by microbes in the gut). Those that can be synthesized in animal tissues are referred to as **nonessential amino acids** because they are nonessential to the diet. They generally are not of concern in ration formulation. The essential and nonessential amino acids in animal nutrition are as follows.

Essential Amino Acids		*Nonessential Amino Acids*	
Arginine	Methionine	Alanine	Glycine*
Histidine	Phenylalanine	Aspartic acid	Proline
Isoleucine	Threonine	Cysteine	Serine*
Leucine	Tryptophan	Cystine	Tyrosine
Lysine	Valine	Glutamic acid	

*Under some conditions, these are dietary requirements for poultry.

All amino acids contain nitrogen as part of the amino group; therefore, proteins contain nitrogen. In general, proteins contain about 16 percent nitrogen. The protein content of feeds can be measured by determining the nitrogen content and multiplying by the factor of 6.25. **Crude protein** is defined as nitrogen $\times$ 6.25 (16 g of N come from 100 g of protein; therefore 1 g of N is associated with $100/16 = 6.25$ g of protein).

Carbohydrates

Carbohydrates are the products of photosynthesis in plants. With the appropriate enzymes, plant tissues convert carbon dioxide and water to sugars, using solar energy trapped by the chlorophyll in chloroplasts. The basic units of carbohydrate structure are sugars such as glucose. These are called simple sugars or monosaccharides. More complex carbohydrates, such as cellulose and starch, are composed of large numbers of simple sugar molecules joined together.

The main carbohydrates in feeds are starch, cellulose, and hemicelluloses. Starch is a readily digested carbohydrate stored in plant seeds. Cereal grains and some roots and tubers (e.g., potatoes, cassava) are high in starch. Cellulose and hemicelluloses are major components of plant fiber. They are important constituents of fibrous feeds (such as roughages) and various agricultural by-products (milling by-products, hulls, etc.).

Lipids

Lipids are the substances in plant and animal tissue that are soluble in organic solvents like ether. The lipid content of feeds is measured by extraction with ether; thus it is called the **ether extract.** The principal lipids of importance in animal nutrition are fats and oils. **Fats** are usually of animal origin (tallow, lard), whereas **oils** are from plants (vegetable oils) and marine animals (e.g., cod liver oil). Fats and oils are composed of glycerol and three fatty acids and are often referred to as **triglycerides** (*triacylglycerols,* in newer terminology). The fatty acids are of two types: saturated and unsaturated. **Saturated fatty acids** have no double bonds; thus they are saturated with hydrogen. **Unsaturated fatty acids** have one or more double bonds, which means they are capable of taking up hydrogen (they are unsaturated with respect to hydrogen). Unsaturated fatty acids may be monounsaturated (one double bond) or polyunsaturated (two or more double bonds). **Hydrogenation** is the process of converting an unsaturated fatty acid to a more saturated one by the addition of hydrogen. For example, corn oil is converted to margarine by hydrogenation. In ruminant animals, hydrogenation of unsaturated fatty acids occurs in the rumen, with the result that the body fat of ruminants contains saturated fatty acids. The properties of a triglyceride are determined by the fatty acids of which it is composed. Fats have a higher proportion of saturated fatty acids than oils, which contain predominantly unsaturated fatty acids. The degree of unsaturation affects the melting point of the lipid, accounting for the fact that those with saturated fatty acids are usually solid at room temperature (e.g., tallow), whereas those with unsaturated fatty acids (e.g., corn oil) are usually liquids. An exception is coconut oil, which is saturated but is a liquid at room temperature, because it contains a high proportion of short-chain fatty acids (containing fewer carbons than is typical in most fatty acids). The shorter the carbon chain of a fatty acid, the lower its melting point (i.e., the greater its tendency to be a liquid).

Minerals

Various mineral elements are dietary essentials for animals. Some of them are required in relatively large quantities. These are termed the **macroelements** or "macrominerals." Others are needed in very small (trace) amounts and are referred to as the **trace elements** or "microelements." Macrominerals generally function as components of the tissue structure (e.g., bone), whereas trace elements function as activators or cofactors of enzymes. The macro- and micro-elements are as follows.

Macroelements	*Trace Elements*
Calcium	Manganese
Phosphorus	Zinc
Sodium	Iron
Potassium	Copper
Chlorine	Molybdenum
Magnesium	Selenium
Sulfur	Iodine
	Cobalt
	Chromium

In addition, several other elements such as silicon, nickel, arsenic, vanadium, and tin may be essential, but they are not of practical importance because a deficiency can only be demonstrated with highly purified diets.

Vitamins

A **vitamin** is defined as follows: (1) it is an organic nutrient (distinct from the other organic nutrients—carbohydrates, lipids, and proteins); (2) it is required in extremely small quantities in the diet; (3) it is essential for normal metabolism; (4) when it is absent from the diet or not present in adequate quantity, a specific deficiency symptom develops; and (5) it cannot normally be synthesized in the animal's body and therefore is a dietary essential. There are a few exceptions to the last characteristic of a vitamin: vitamin D can be synthesized by animals exposed to sunlight; niacin can be synthesized in some animals from the amino acid tryptophan. Vitamin C (ascorbic acid) can be synthesized by most animals except primates, guinea pigs, and some exotic species such as the fruit-eating bat and the red-vented bulbul bird. (On an evolutionary basis, animal species whose diets typically contain fruit may lose the enzymes necessary for synthesis of vitamin C; because it is in their diet, they do not need to make it. Vitamin C is a simple molecule derived from glucose.)

Vitamins, like minerals, can be classified into two groups: fat soluble and water soluble.

Fat Soluble	*Water Soluble*	
Vitamin A	Thiamin (B_1)	Nicotinic acid (niacin)
Vitamin D	Riboflavin (B_2)	Folic acid (folacin)
Vitamin E	Pyridoxine (B_6)	Biotin
Vitamin K	Cyanocobalamin (B_{12})	Choline
	Pantothenic acid	Vitamin C (ascorbic acid)

Fat-soluble vitamins are stored in fatty tissues of the body and are poorly excreted. Thus, a long period of time on a deficient diet is needed for a deficiency of fat-soluble vitamins to occur. In contrast, **water-soluble vitamins** (except vitamin B_{12}) are readily excreted in the urine. If they are not provided in the diet, they rapidly become deficient because they are poorly stored in the body. Vitamin B_{12} is efficiently stored by the liver, unlike the other water-soluble vitamins. With the exception of vitamin C, the water-soluble vitamins are also referred to as the **B-complex vitamins.** There are a number of **bogus vitamins** (e.g., vitamin F, vitamin H, vitamin P, etc.). These names are coined by entrepreneurs selling dietary supplements. To be a vitamin, a substance must meet the five criteria listed in this section.

Water

Water is regarded by many nutritionists as the most important nutrient. Water does not totally fit the definition of a nutrient because it is not required in the diet but is usually consumed separately.

NUTRIENT FUNCTIONS

Structural Role

Some nutrients function primarily in making up the structure of the animal body. Muscle tissue (meat) is composed largely of protein and water. Proteins are an integral part of all cells as components of cell membranes. Fats have a structural role, with protein, in making up the lipoproteins that are found in cell membranes. Proteins are the main constituents of connective tissues, such as arteries, veins, ligaments, and tendons. Bone is composed of a cartilaginous protein matrix that becomes mineralized with calcium and phosphorus. Carbohydrates are components of glycoproteins, which are constituents of connective tissue (cell walls, collagen, bone matrix). Thus, nutrients with major structural roles include protein, calcium, phosphorus, and, to a lesser extent, lipids and carbohydrates.

Sources of Energy

All animals require sources of energy. Energy is used for locomotion and thermoregulation (regulation of body temperature). The main energy requirement is somewhat more difficult to understand. This is the biochemical energy required for maintenance and growth of living tissue. Animal cells are thermodynamically unstable and require the constant expenditure of energy to maintain them in the living state. Thus, there is a large requirement of energy for tissue maintenance. In addition, all of the reactions involved in the formation of new tissue require an input of energy. For example, muscle proteins are synthesized from absorbed amino acids. The amino acids do not react together spontaneously; chemical energy is required to form the bonds between them that join them together to form a protein.

Energy in feeds and animal metabolism is expressed in calories, a measure of heat energy. A **calorie** is defined as the amount of heat required to raise the temperature of 1 gram of water by 1 degree centigrade. The calorie is usually not a practical term for use in animal nutrition; generally the kilocalorie (kcal) and megacalorie (Mcal) are used. These are equal to 1,000 and 1,000,000 calories, respectively. In many countries the **joule** is used instead of the calorie; a calorie is equal to 4.184 joules.

The main nutrient categories that provide animals with energy are carbohydrates and lipids. **Cellular metabolism** of animals is basically the reverse of photosynthesis. Glucose is metabolized in a series of biochemical reactions, during which chemical energy is released as adenosine triphosphate (ATP), which in turn is used as the fuel for driving other reactions (e.g., protein synthesis). In addition to ATP production, a major part of the energy of carbohydrates is given off as heat. The overall reaction of cellular metabolism is

$$\underset{\text{glucose}}{C_6H_{12}O_6} + 6O_2 + \underset{\substack{\text{(adenosine}\\\text{diphosphate)}}}{ADP} \xrightarrow[\text{enzymes}]{\text{cellular}} ATP + \text{heat} + 6CO_2 + 6H_2O$$

Although protein can be metabolized to yield energy, this is generally undesirable because it is a more expensive source of calories than are carbohydrates and lipids. In ration formulation, diets are balanced to try to minimize the use of protein for energy.

Regulatory Functions

Most of the required nutrients function in the regulation of cellular metabolism. Sodium, potassium, and chlorine function in fluid balance to maintain concentrations and concentration gradients of dissolved substances in optimal conditions for metabolic reactions. Vitamins and most minerals function as **cofactors** or activators of enzymes, which catalyze the thousands of chemical reactions that occur in animal tissues. Deficiencies of these nutrients result in impairment of the specific metabolic reactions for which they are required, causing development of characteristic deficiency symptoms. For example, vitamin A functions in enzymatic reactions in vision; thus, a vitamin A deficiency causes blindness.

NUTRIENT REQUIREMENTS

The purpose of providing feedstuffs and formulating diets is to provide animals with the nutrients they require. Energy is the major nutritional need. At least 80 percent of the total feed intake of most animals consists of sources of calories. Feed intake is regulated according to energy need; **animals eat to satisfy their energy requirements.**

A much smaller portion of the total diet is protein. The protein requirement varies with animal species, stage of growth, and type of production, but generally is less than 20 percent of the diet. For example, for mature beef cows, the protein requirement is about 6 percent of the diet; for a growing pig, 13–15 percent; and for a mature ewe, 9 percent.

The requirement for minerals can be met in approximately 3 to 4 percent of the diet. Salt is added at a level of 0.25 or 0.5 percent; calcium and phosphorus requirements for most species are in the 0.5 to 1 percent range, and the requirements for trace elements are met with less than 0.5 percent of the total diet. Vitamins are needed in trace amounts, much less than 1 percent of the diet.

Thus a "typical" diet for livestock will contain 10 to 20 percent protein, 80 to 90 percent energy-yielding nutrients, 3 to 4 percent minerals, and a trace of vitamins. The cost of nutrient sources will vary in a similar manner. Thus most of the cost of diet ingredients is for energy sources and, to a lesser extent, protein supplements.

DIGESTIVE TRACT PHYSIOLOGY

The nutritional requirements of animals and their ability to utilize feedstuffs are greatly dependent on their digestive tract anatomy and physiology. It is readily apparent, for example, that poultry, humans, and swine could not survive on a diet of hay, which will support a beef cow, a sheep, or a horse.

Livestock can be classified into three groups according to their digestive tracts: (1) simple nonruminants, (2) ruminants, and (3) nonruminant herbivores.

Simple Nonruminants (Monogastrics)

The **simple nonruminants** have a pouchlike, noncompartmentalized stomach and do not depend much upon microbial digestion in any part of the gut. They accomplish their own work of digestion by digestive enzymes secreted into the gastrointestinal tract (GIT). They are often referred to as *monogastric animals,* but that term is not totally satisfactory because all animals have only one stomach, although in some (e.g., ruminants) it is subdivided into compartments. Examples of simple nonruminant species include swine, poultry, dogs, cats, rats, and humans. The basic features of the digestive tract of the simple nonruminant are shown in Fig. 1.1. The functions of each major segment of the GIT will be briefly described.

The **stomach** is a muscular organ that functions by storing ingested feed and metering the ingesta into the small intestine in amounts that intestinal digestion can accommodate. Churning activity causes mixing of the ingested material with stomach acid and enzymes. Hydrochloric acid (HCl) is secreted into the stomach, resulting in a low pH of 1.5–2.5. This low pH is bactericidal, killing most of the bacteria ingested with the feed. The strong acidity also accomplishes some digestive activity, particularly in the hydrolysis of proteins.

The major site of digestion and absorption in simple nonruminants is the **small intestine,** consisting of three segments: the duodenum, jejunum, and ileum. Digestion is accomplished by enzymes secreted by the animal into the small intestine. A major source of digestive enzymes is the **pancreas gland.** It produces enzymes that digest proteins (trypsin, chymotrypsin), carbohydrates (amylase), and lipids (lipase). In addition, pancreatic secretions contain buffers (e.g., bicarbonates) to neutralize stomach acid. Pancreatic enzymes, secreted into the duodenum via the pancreatic duct, result in a partial digestion of the large protein, carbohydrate, and fat molecules.

Digestion is completed by enzymes in the **brush border** or intestinal mucosa (Fig. 1.2, 1.3). The mucosa is made up of projections, called **villi,** which in turn have minute projections called **microvilli.** At the tips of the microvilli are filamentous webs, the **glycocalyx.** The microvilli and glycocalyx constitute the brush border. Enzymes secreted here complete the process of breaking down proteins, carbohydrates, and lipids into amino acids, sugars, and fatty acids, respectively. Nutrients released into the brush border are absorbed into the microvilli, and then into the villi and circulatory system. Certain antinutritional factors (ANFs) in feeds may cause damage to the brush border, such as lectins in beans (see Fig. 4.4). The blood supply from the intestine travels directly to the liver, where modification of absorbed nutrients may occur before they enter the general circulation and are transported to other tissues. Also, toxins absorbed from the intestine may be removed from the blood and detoxified in the liver. An additional

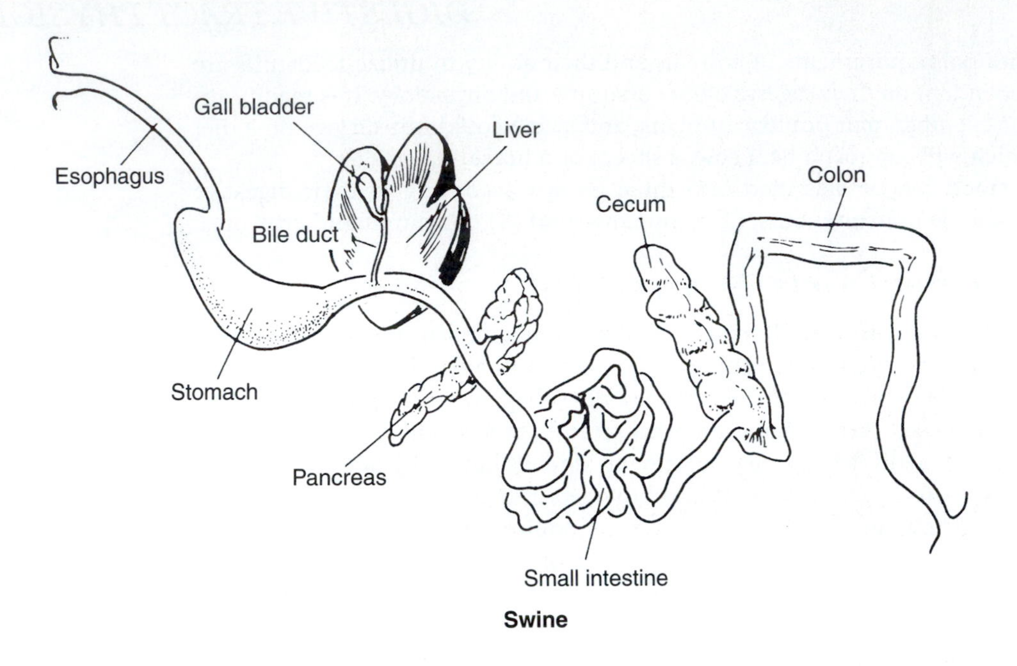

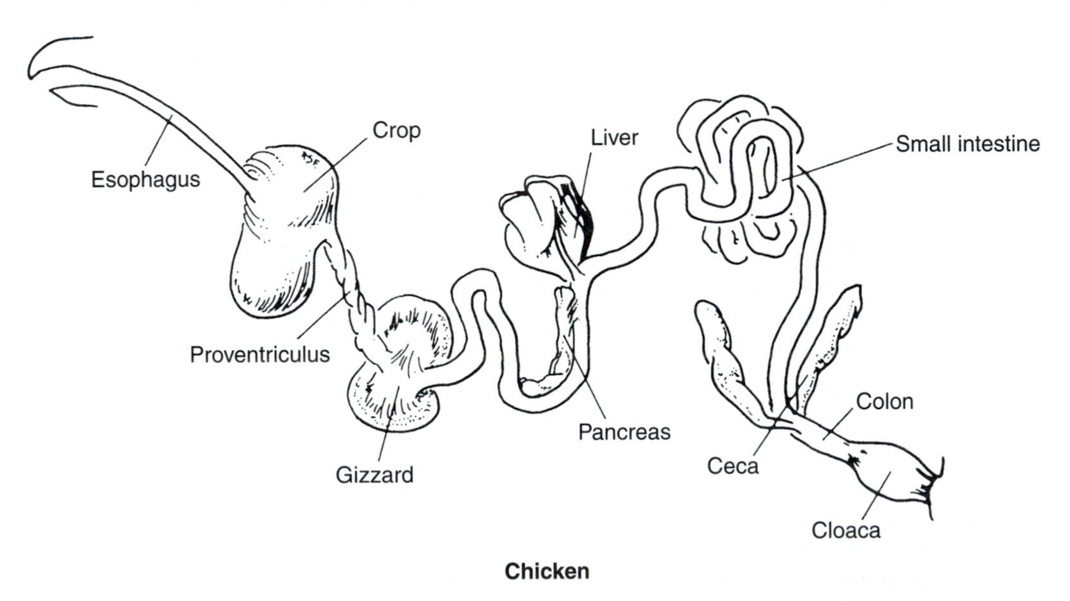

FIGURE 1.1 The general features of the simple nonruminant digestive tract. Note the simple, noncompartmentalized stomach and the differences in the foregut of the mammalian type (swine) and avian species (chicken).

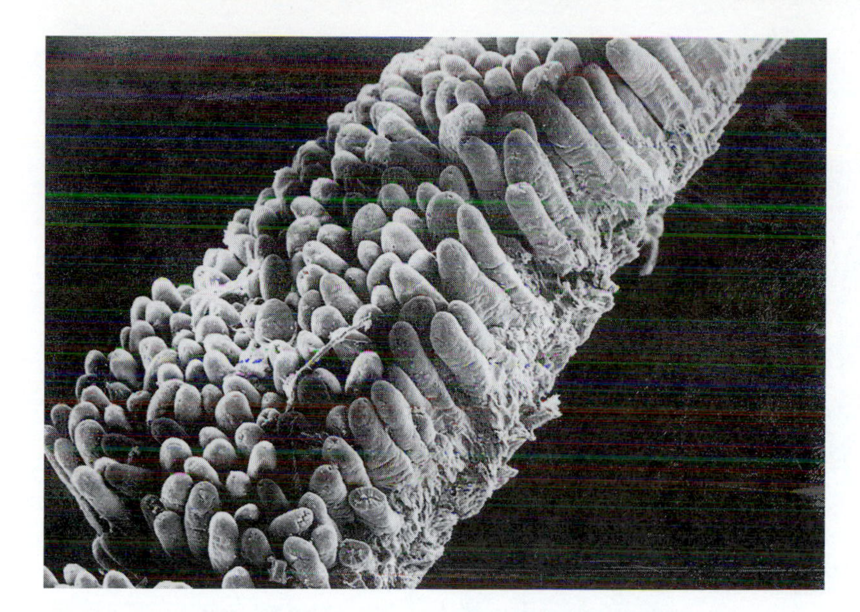

FIGURE 1.2 An electron microscope view of a section of the intestinal mucosa of a turkey showing the villi. With higher magnification, it would be apparent that each villus is also covered with small projections, called *microvilli.* The villi and microvilli greatly increase the surface area available for nutrient absorption. (Courtesy of R. A. Swick.)

role of the liver is that it produces bile, which is secreted into the duodenum via the bile duct. Bile is essential for the digestion and absorption of fat.

The major site of nutrient absorption in simple nonruminants is the jejunum. Most of the nutrients that will be utilized are absorbed by the time material passes from the ileum to the colon.

The stomach and small intestine are referred to as the **foregut.** The hindgut is composed of the cecum, colon, and rectum. The primary functions of the **hindgut** in simple nonruminants are the absorption of water, some bacterial fermentation, and formation of the feces. The contents of the small intestine are quite fluid, whereas the feces are normally much lower in water content due to fluid absorption in the colon. Bacterial growth in the hindgut results in synthesis of B-complex vitamins. This can be of nutritional significance in animals that consume their feces (e.g., rats), a practice known as **coprophagy,** and when fecal contamination of feed occurs (e.g., poultry raised on deep litter rather than on wire, or pigs on solid floors rather than slotted floors).

Although most simple nonruminants share these general characteristics, there are species differences in specific aspects of GIT anatomy and physiology. For example, strict **carnivores** (e.g., mink, cats) tend to have a short intestinal tract, and a rapid transit time of ingesta through the GIT, because of the high digestibility of their meat-based diet. **Omnivores** such as swine tend to have a long small intestine and a somewhat enlarged hindgut with a much more significant microbial population and some fiber digestion. In avian species (birds), the foregut is modified from the mammalian system, with the crop, proventriculus, and gizzard replacing the simple stomach and teeth (Fig. 1.2). In birds, ingested food enters the crop, a temporary storage site, where moistening of the food occurs. The **proventriculus** is the true stomach, where hydrochloric acid and digestive enzymes are secreted. The **gizzard,** a muscular organ with a tough lining, performs the functions of mammalian teeth. The churning action of the gizzard and the grit (small stones) it contains grind feed into small particles.

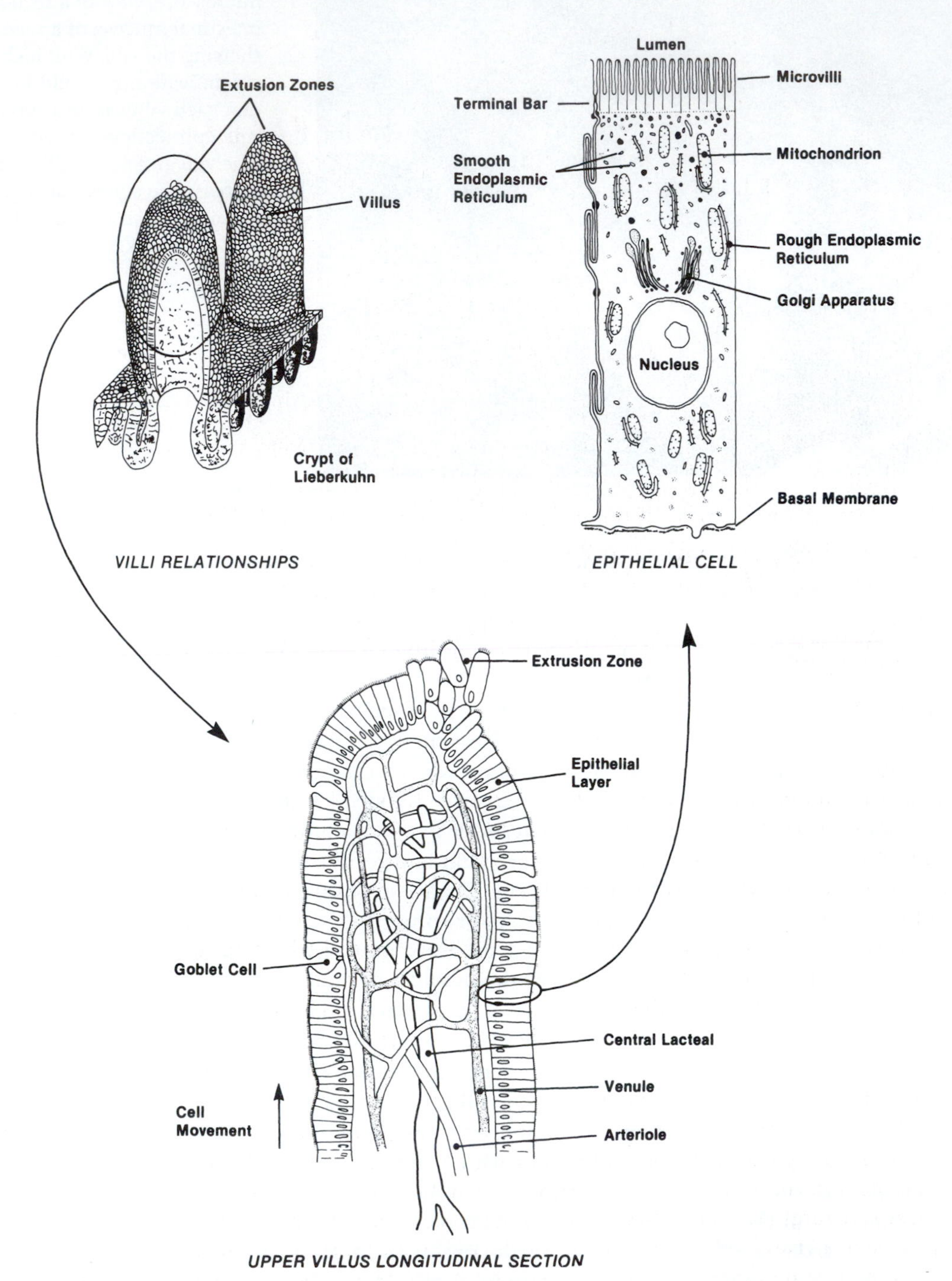

FIGURE 1.3 Swine intestinal villi. The illustration shows a three-dimensional appearance (top left), a longitudinal section of a villus with enterocytes (epithelial cells) (bottom), and an expanded view of an enterocyte (top right). The glycocalyx is not illustrated; it is a filamentous web connected to the microvilli. (Courtesy of Edwin T. Moran, Jr.)

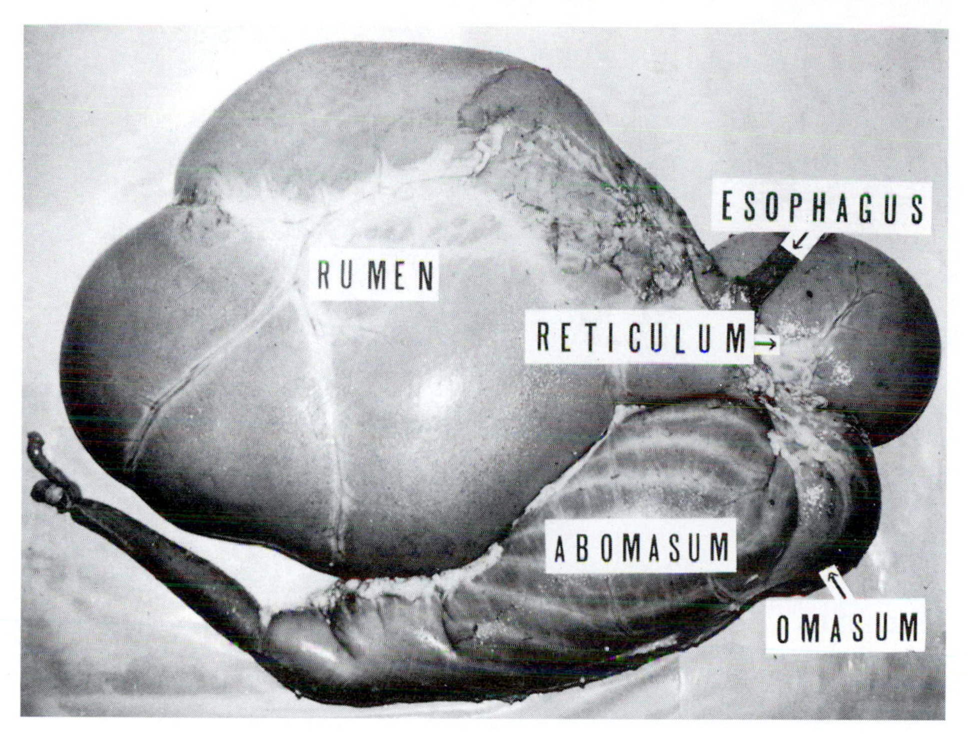

FIGURE 1.4 A dried preparation of a ruminant stomach that shows the locations of each of the compartments.

Ruminants

A ruminant animal differs from a simple nonruminant in two major respects. It has a large, compartmentalized stomach (Fig. 1.4) and much of the work of digestion is accomplished by microbes that inhabit the large stomach, rather than by enzymes the animal produces itself. Rumen microbes are described in Appendix C (p. 588).

The largest segment of the ruminant stomach is the **rumen,** which functions as a fermentation vat. It contains an immense microbial population of bacteria, protozoa, fungi, and yeasts that ferment the ingested feed. The major sources of absorbed energy in ruminants are fermentation end products, known as **volatile fatty acids (VFA).** Fermentation produces large quantities of gases, mainly carbon dioxide (CO_2) and methane (CH_4). These gases are removed by the process of **eructation** (belching). The breakdown of feeds, particularly roughages, into smaller particles to facilitate fermentation is accomplished by the process of rumination. A bolus of rumen ingesta is formed by muscles at the base of the esophagus, regurgitated, and chewed. **Rumination,** or "chewing the cud," is a characteristic of all ruminants. The rumen wall is lined with projections called **papillae** that function in nutrient absorption in much the same way as the villi in the small intestine (Fig. 1.5).

The second segment of the ruminant stomach is the **reticulum,** an extension of the rumen. The reticulum and rumen are often considered to be one

FIGURE 1.5 The papillae are projections of the rumen wall that function in absorption of nutrients from the rumen.

compartment, the reticulorumen. The reticulum (Fig. 1.6) has honeycomb-shaped internal projections that trap foreign materials (hardware) such as nails, wire, and stones that are inadvertently consumed. These materials accumulate in the reticulum, thus preventing puncture of the intestines. If the gut is punctured, bacteria can enter the abdominal cavity, causing infections and liver abscesses. This is known as **hardware disease.** In animals that will be exposed to hardware, such as dairy cattle fed wire-tied alfalfa hay, a magnet can be administered orally. It lodges in the reticulum, and traps and holds metallic objects such as nails and wire.

Movement of the ingesta from the reticulorumen to the abomasum, the true gastric stomach, is regulated by flow through the **omasum.** The omasum is a small compartment containing membranous leaves that act as a sieve or filter. Ingested feed must be degraded to small particles in the rumen before it can pass through the omasal leaves. Thus the omasum functions in retention of feed in the rumen for maximum fermentation efficiency.

The **abomasum** is the site of secretion of gastric juices such as hydrochloric acid and proteolytic enzymes. The acidity kills rumen microbes, as the rumen contents flow through the abomasum. This allows digestion of microbial protein, which serves as a major source of amino acids for the ruminant animal.

The anatomy and functions of the small intestine and hindgut of the ruminant are similar to those described for the simple nonruminant.

The ruminant stomach has profound implications in animal nutrition and feeding. There are a number of advantages conferred to the animal possessing the ruminant stomach, as well as some disadvantages. A major advantage of fermentation is that it allows the utilization of fibrous feeds. No mammal produces cellulase, the enzyme that degrades cellulose. Cellulose is a major constituent of plant fiber and is considered to be the most abundant organic compound on earth. Rumen mi-

FIGURE 1.6 The reticulum lining has a honeycomb appearance.

crobes secrete **cellulase,** which breaks the chemical bonds linking glucose units together in the cellulose molecule. Thus, ruminants can utilize roughages, forages, and other fibrous feeds, in contrast to simple nonruminants, which cannot.

Another important advantage conferred by rumen fermentation is the ability of rumen microbes to synthesize amino acids and proteins from ammonia. A major source of protein for the ruminant animal is the digestion of microbial proteins. Thus ruminants can be fed poor-quality proteins and even just sources of nitrogen (nonprotein nitrogen, e.g., urea); these are upgraded by rumen fermentation to **microbial protein.** In addition to providing the animal with a major portion of its energy and protein needs, fermentation is also important in providing **vitamins.** All of the water-soluble vitamins and vitamin K are synthesized by rumen microbes, so the animal has a dietary requirement only for vitamins A, D, and E. In fact, for ruminants that graze green forage and that are exposed to sunlight, there is no need for vitamin supplementation at all. Thus the protein, energy, and vitamin needs of ruminants can be satisfied with a much simpler diet than is the case with monogastrics.

There are some nutritional disadvantages associated with rumen fermentation. Fermentation of high-energy feeds such as cereal grains yields less digestible energy to the animal than when the grain is digested directly in the small intestine. Energy is lost during fermentation, including through heat production by microbes (heat of fermentation) and through production of rumen gases (such as methane). **Methane** production results in a loss of carbon, which otherwise could have been oxidized in the animal's own metabolic processes. Similarly, there are other losses associated with the rumen metabolism of protein. The conversion of dietary protein to microbial protein frequently results in a reduction of its biological value. Ammonia produced in rumen digestion of protein may be absorbed and excreted in the urine, representing a loss of dietary protein. However, on balance, the ruminant stomach confers a tremendous advantage in the utilization of low-quality feedstuffs and is responsible for the special niche that ruminants occupy in world food production.

Nonruminant Herbivores

Nonruminant herbivores have digestive tract modifications to facilitate microbial fermentation, with many of the same functions performed as in the rumen. Intermediate between simple nonruminants and ruminants in their digestive physiology and nutritional requirements, nonruminant herbivores share similarities with each. They are herbivorous, consuming forages and other vegetation. Domestic nonruminant herbivores include the horse, rabbit, and guinea pig; examples of wild species are the zebra, elephant, and hippopotamus. Nonruminant herbivores can be divided into three groups: the foregut fermentors, the colon fermentors, and the cecal fermentors.

Foregut Fermentors A number of nonruminant animals have a compartmentalized stomach, with one or more pouches where microbial fermentation occurs, separate from the gastric stomach. Examples of animals with a segmented forestomach include some of the Australian marsupials (such as some species of kangaroo and wallaby), the hippopotamus, peccary, colobus monkey, and tree sloth. The latter two are tree dwellers whose diet consists primarily of tree leaves. The colobus monkey has a four-chamber stomach, and can ferment herbage with an efficiency approaching that of ruminants. There are no important domesticated species of nonruminant **foregut fermentors.** There is one known avian species with foregut fermentation: the hoatzin. This is a neotropical bird whose primary food source is tree foliage (see Chapter 22).

Colon Fermentors Equids (horses, donkeys, zebras) are examples of colon fermentors. The enlarged colon is a site of microbial fermentation. Readily digested nutrients such as high-quality protein, starch, and lipids are digested in the small intestine, as they are in the simple nonruminants. Material that resists breakdown in the small intestine (autodigestion), primarily fiber, enters the hindgut and is subject to fermentation by microbes similar to those in the rumen. Compared to rumen fermentation, hindgut digestion is somewhat less efficient because of its anatomical location. In the rumen, microbes get the first opportunity to digest feed, so they have a rich substrate containing sugars, proteins, minerals, and so on. In the hindgut, the microbes get what the animal cannot digest, so there is a much less favorable nutritional environment for them. For this reason, fiber digestion is less efficient in a horse than in a cow, as can be appreciated by comparing the physical appearance of horse versus cow manure. Much fibrous material with the cellular structure of grass still present is seen in horse manure, but is absent in cattle feces.

Another disadvantage of colon fermentation is that the products of microbial metabolism, such as bacterial protein, are less available to the host animal than is the case with ruminants. In ruminants, the products of fermentation are subject to the animal's digestive enzymes in the small intestine, whereas with hindgut fermentation, they are not. Although fiber digestion is less efficient in horses than in ruminants, it is at first paradoxical that equids are better adapted to low-quality, high-fiber diets than are ruminants. For example, in Africa, it has been noted that **zebra** can live on low-quality forages on which antelope and other ruminants cannot survive (Janis, 1976). This is because in ruminants, the flow rate of ingesta through the stomach is regulated by the omasum. Material cannot leave the rumen until it has been degraded to a small particle size by the rumen microbes. With very low-quality fibrous feeds, the breakdown of fiber in the rumen is so slow that the rumen becomes impacted with indigestible fiber, and feed intake is restricted. With hindgut fermentors, such restriction does not occur, and, with a high intake of low-quality roughages, they are able to obtain sufficient nutrients to survive. Thus hindgut digestion is a superior digestive strategy for very low-quality forages. In Africa, nonruminant herbivores, such as the zebra, selectively consume the lowest quality forage, whereas ruminants feed on the more digestible plant parts, and the mixed herds of nonruminant herbivores and ruminants utilize the total forage production very efficiently. Perhaps the competitive ability of nonruminant herbivores can be further appreciated by considering the other (besides Africa) major grasslands of the world. Producers of domestic ruminants (mainly cattle and sheep) on North American rangelands find the wild horses quite competitive, while sheep growers in Australia have spent a century battling rabbits and kangaroos, both nonruminant herbivores.

Cecal Fermentors Small herbivores like the rabbit have a digestive strategy that emphasizes fermentation of the nonfiber constituents of forages. Even though the **rabbit** is herbivorous, the digestibility of fiber in this species is very low. It has evolved a digestive strategy that allows it to selectively separate and excrete indigestible fiber, while retaining the more digestible nonfiber components for fermentation in the cecum. The cecum in cecal fermentors is a very large, muscular, blind pouch. Digesta moves from the ileum to the colon. Fiber particles, because of their size and low density, tend to congregate in the lumen of the colon. More dense nonfiber material and fluid tend to concentrate at the circumference of the colon. Peristaltic contractions of the colon move fiber through rapidly for excretion in the feces, while "antiperistaltic" contractions of the haustrae, or muscular bands that segment the colon (Fig. 1.7), move nonfiber components (e.g., starch, protein) and fluids into the cecum, where they undergo fermentation. Small herbivores such as the rabbit produce two types of feces: hard and soft. The hard feces, or fecal pellets, consist largely of indigestible fiber. The soft feces **(cecotropes)** are actually cecal contents. These are consumed by the animal directly from the anus, providing it with bacterial protein and vitamins synthesized in the cecum. Thus this digestive strategy allows small herbivores to consume a low-energy herbivorous diet without the disadvantage of having to transport a large quantity of indigestible fiber in the gut. They eliminate fiber rapidly and concentrate their digestive activity on the more nutritious nonfiber components.

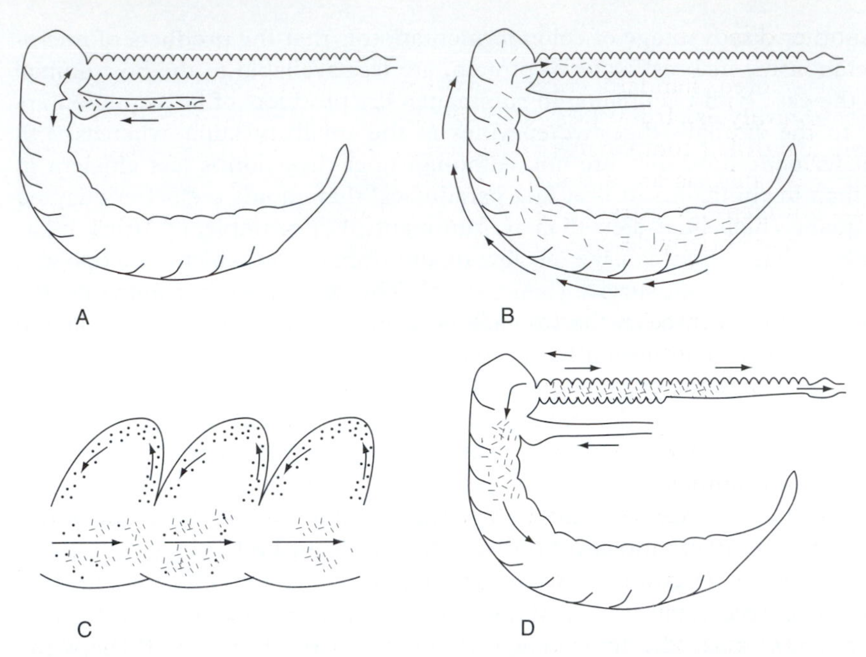

FIGURE 1.7 Mechanisms for the selective excretion of fiber and retention of small particles and solubles for fermentation in the cecum of the rabbit. (A) Intestinal contents enter the hindgut at the ileocecal-colonic junction and uniformly disperse in the cecum and colon. Dashes represent large fiber particles, and dots represent nonfiber particles. (B) Contraction of the cecum moves material into the proximal colon. (C) Peristaltic action moves large fiber particles (dashes) down the colon for excretion as hard feces. Contractions of the haustrae of the colon move small particles (dots) and fluids backwards into the cecum. (D) Small particles and fluids are thus separated from fiber. (Courtesy of Mark A. Grobner.)

TECHNIQUES USED IN EVALUATING FEEDS

Efficient utilization of feedstuffs in animal feeding necessitates knowledge of their nutrient composition, the digestibility of nutrients they contain (bioavailability), their ability to provide energy, the presence of inhibitors or toxins, and the need for processing to improve palatability or nutrient availability. In this section, some of the more common techniques used to acquire this information are briefly described.

Feed Microscopy

Feed microscopy is the technique of evaluating feed samples with a microscope for the purpose of identifying the components. This information can be used to confirm or deny the presence of all of the listed ingredients, noting the presence of extra ingredients, identifying unwanted materials, poisonous weed seeds, and so on. Even if adulterants are finely ground to avoid detection, a feed microscopist may be able to identify them by histological structure using a high-power compound microscope. Feed microscopy is used mainly for regulatory purposes to assess the legitimacy of commercial mixed feeds.

Feed Analysis

The time-honored standard chemical method of feed analysis is called the **proximate analysis.** It involves some simple chemical techniques designed to differentiate nutrient components from nonnutritive material. The components of the proximate analysis are dry matter, crude protein, ether extract, ash, crude fiber, and nitrogen-free extract (NFE). Each of these categories and the information provided are briefly described below.

Dry Matter Dry-matter determination involves drying the feed sample in a drying oven to constant weight. Dry-matter determination is important for two reasons. First, some materials, such as green feeds, silage, food-processing wastes, and so on, have a very high moisture content. Their quality and value depend upon how much of their weight is actually made up of water. This information is important, for example, in balancing rations containing silage and in purchasing feedstuffs. The other reason that dry-matter determinations are important is that feed analyses are performed on the dried sample. Most analyses involve small sample sizes of a few grams or less; micromethods use milligram quantities. To avoid spurious weight changes of the sample due to loss or gain of water from the laboratory environment, it is necessary to conduct the analyses on a dry-weight basis. However, in most cases the livestock producer or nutritionist needs to know the composition on an as-is basis or as-fed basis. Thus it is important to be able to interconvert feed analysis data from a dry-matter to an as-is basis.

1. To convert values from a dry-matter basis to a wet-weight or as-is basis, multiply by the percent dry matter divided by 100.
   Example: On a dry-matter basis, an alfalfa sample contains 16.5% crude protein. The dry-matter content is 92%. The protein content on an as-is basis is $16.5 \times (92/100) = 15.2\%$.
2. To convert from a wet-weight to a dry-weight basis, multiply by 100 divided by the percent dry matter.
   Example: A barley sample contains 11.2% crude protein on a wet-weight basis and contains 96% dry matter. The protein content on a dry-matter basis is $11.2 \times (100/96) = 11.7\%$. Values are *always* higher on a dry-matter basis.

Crude Protein (Kjeldahl Analysis) Crude protein (CP) is defined as the nitrogen content of the feed multiplied by a factor of 6.25. This factor is derived from the generalization that most proteins contain 16 percent nitrogen. Nitrogen is measured by the **Kjeldahl** procedure, which is named after the Danish chemist who developed it. The feed sample is boiled in concentrated sulfuric acid, resulting in the complete oxidation of all organic material. The proteins and amino acids are completely degraded; their nitrogen is released as ammonium ion (NH_4^+). The solution is then made alkaline, converting NH_4^+ to ammonia (NH_3). Steam is passed through the solution (steam distillation), driving off the NH_3, which is trapped in a boric acid solution. The concentration of NH_3 is measured by titration. It is important to recognize that the crude-protein procedure measures nitrogen. Thus, it does not distinguish between high-quality and poor-quality protein, or protein and nonprotein nitrogen.

Ether Extract The fat content of feeds is determined by extracting a feed sample with ether. The change in weight of the sample is due to the loss of ether extract or ether-soluble lipids.

Ash Total mineral content is measured by burning the feed sample in a muffle furnace at 600°C. This combusts all organic matter, leaving a residue of ash or inorganic mineral salts. The ash content is useful as a measure of total mineral content and also is indicative of soil contamination of feedstuffs. A high ash content of a hay sample, for instance, may indicate excessive contamination with soil and dirt.

Crude Fiber The crude-fiber content is measured by boiling an ether-extracted feed sample in dilute acid, then in dilute alkali, drying, and burning in a muffle furnace. The difference in weight before and after burning is the crude-fiber fraction. In the modified crude-fiber determination, the correction for ash content is not done. This procedure was developed in the early days of nutrition research in an attempt to estimate the indigestible portion of food in the human digestive tract by stimulating the acid condition of the stomach and alkaline condition of the intestine. In spite of well-recognized inadequacies, crude-fiber values are still widely used.

New procedures to estimate the fiber content of forages have been developed. These methods, such as acid detergent fiber and neutral detergent fiber, are described in more detail in Chapter 5.

Nitrogen-Free Extract Nitrogen-free extract (NFE) is derived by subtracting the sum of the other proximate components from 100:

$$\%\text{NFE} = 100 = \%\ (\text{water} + \text{crude protein} + \text{ether extract} + \text{ash} + \text{crude fiber})$$

The NFE theoretically represents mainly starch, sugars, and other readily available carbohydrates. In practice, the inaccuracies associated with crude fiber are such that neither crude fiber nor NFE are nutritionally very meaningful.

Determination of Specific Nutrients Although proximate analysis gives a general indication of the value of a feed, it does not really deal with specific nutrients. These can be determined by specific techniques. Amino acids are measured by column chromatography, usually using an automated system known as an **amino acid autoanalyzer.** Individual minerals, such as calcium, phosphorus, copper, and zinc, can be measured by spectrophotometric procedures, or by the now-common technique of atomic absorption spectrometry. Specific vitamins are measured using the chromatographic procedures, such as high-performance liquid chromatography (HPLC). These nutrients are not usually measured on a routine basis because the determinations are expensive and may be time consuming.

Near-infrared reflectance spectroscopy is a widely used technique for feed and forage analysis. Feed samples are exposed to infrared radiation, and the reflectance of the radiation by the sample is measured. The instrument is calibrated by mathematically equilibrating reflectance values with known chemical values for a series of samples of a particular ingredient. Thus the technique is of most value when a large number of similar samples are to be analyzed. Its use in forage testing was reviewed by Brown et al. (1990).

Estimation of Feed Energy

In the United States, feed energy is expressed in calories. In most other countries, the joule is used as the measure of energy. Appropriate definitions and conversions are as follows.

1 calorie = the amount of heat required to raise the temperature of 1 gram of water by 1 degree centigrade, measured from 14.5 to 15.5 degrees centigrade

1 kilocalorie (kcal) = 1,000 calories

1 megacalorie (Mcal) = 1,000 kilocalories

1 calorie = 4.184 joules (J)

1 kilocalorie = 4.184 kilojoules (kJ)

1 kilojoules = 0.239 kcal

The caloric content of biological materials is determined in a **bomb calorimeter.** In brief, the sample is burned in a combustion chamber (bomb) inserted in a vessel containing a known weight of water. As the sample burns, it releases heat, which is taken up by the water. From the weight of the sample, weight of the water, and rise in temperature of the water, the number of calories of heat energy released can be calculated. When a feed sample is burned in a bomb calorimeter, its **gross energy** is determined. To determine the fraction of the gross energy that the animal can actually utilize, a metabolism trial must be conducted to account for various losses. This gives values for digestible, metabolizable, and net energy:

Digestible energy (DE) = Gross energy (GE) − fecal energy

Metabolizable energy (ME) = DE − (urinary energy + rumen gas losses)

Net energy (NE) = ME − heat loss

Thus the NE represents the fraction of the GE actually utilized for productive purposes.

In the preceding determinations, feces and urine arising from test feed are collected and their energy content determined by bomb calorimetry. Since the urinary energy is usually quite low, DE values are satisfactory for expression of energy requirements of most species. In poultry, feces and urine are voided together, so ME values are commonly used. For determination of NE, a respiration chamber allowing measurement of heat loss is needed. Alternatively, the NE can be estimated by measuring energy retention by the animal, using a specific-gravity determination on the carcass.

Total Digestible Nutrients An older method for estimating digestible energy is the **total digestible nutrients** (TDN) system. This is the summation of the digestible energy-yielding constituents. The ether extract is multiplied by 2.25, to account for the higher energy content of fat compared to carbohydrate and protein. (Fats yield about 9 kcal/g DE, whereas protein and carbohydrate yield about 4 kcal/g.) The digestibility values for each fraction are determined experimentally (see the next section), or book values are used.

%TDN = % digestible CP + 2.25 (% digestible ether extract)
+ % digestible crude fiber + % digestible NFE

Determination of Digestibility

An essential feature of feedstuff evaluation is nutrient **digestibility,** which measures the amount of a nutrient in a feed that is digested and absorbed and thus available for metabolism. For example, raw feathermeal contains 90 percent crude protein, but the digestibility of the protein is only 10 percent. Therefore, the potentially usable protein content of raw feathermeal is only 9 percent. (Steam cooking feathers to produce hydrolyzed feathermeal markedly improves the value of the material.) This is an extreme situation but, it is hoped, makes the point.

There are several procedures used to assess digestibility. Direct measurement (the **total collection method**) involves keeping an animal in a metabolism cage or crate and measuring the feed intake and the feces excreted. The feed and feces are analyzed for nutrients of interest. The difference between the amount of nutrient consumed and excreted in the feces is the amount digested and absorbed:

$$\% \text{ digestibility} = \frac{\text{amount of nutrient consumed} - \text{amount of nutrient excreted in feces}}{\text{amount of nutrient consumed}}$$

The following is a simple illustration of a digestibility calculation:

Example: A pig is housed in a metabolism crate. It consumes 20 lb of the test diet, and excretes 10 lb of feces. The percentage crude protein of the feed is 15 and of the feces is 7. Therefore,

1. protein intake = 20 lb × 0.15 = 3 lb
2. fecal protein = 10 lb × 0.07 = 0.7 lb
3. protein digested = 3 lb - 0.7 lb = 2.3 lb
4. % protein digestibility = (2.3/3.0) × 100 = 76.6%

Another method of fecal collection is to equip the animal with a harness and collection bags so that feces and urine (if necessary) can be collected (Fig. 1.8). Animals can be trained to accept this procedure. Digestibility can also be estimated by procedures that do not require a total fecal collection. An indigestible substance (referred to as a **marker** or indicator) can be added to the feed. By measuring the concentration of the marker in the feed and feces, one can calculate the extent of disappearance from the gut of digestible compounds to the feed. This procedure can use "grab samples" of fecal material rather than a total collection. **Chromic oxide,** an indigestible bright green dye, is often used as a marker.

For example, suppose a test feed contains 1 percent chromic oxide. This feed is fed to a pig, and samples of the feces are collected and analyzed for concentration of chromic oxide. If the feces contain 2 percent chromic oxide, the dry-matter digestibility of the test diet is 50 percent. (Because the concentration of the nonabsorbable chromic oxide has doubled during passage through the gut, 50 percent of the dry matter must have disappeared from the gut by being absorbed.)

To calculate the digestibility of individual nutrient categories by the indicator method, the following equation is used:

$$\frac{\% \text{ nutrient}}{\text{digestibility}} = 100 - \left(100 \times \frac{\% \text{ indicator in feed}}{\% \text{ indicator in feces}} \times \frac{\% \text{ nutrient in feces}}{\% \text{ nutrient in feed}}\right)$$

FIGURE 1.8 A horse equipped with a harness and bags for collection of feces and urine. This procedure is used to collect the excreta for conducting nutrient balance and digestibility trials.

When protein digestibility is determined as described above, it is referred to as apparent protein digestibility. **True protein digestibility** includes a correction for **endogenous protein** (or nitrogen), also called *metabolic fecal nitrogen.* Endogenous protein is associated with protein that did not come directly from the diet but is secreted into the gut as digestive enzymes and sloughed-off cells of the gut mucosa. The lining of the gut undergoes continual replacement with new cells and the sloughing off of old cells, which contain protein that is excreted in the feces. For most purposes, apparent protein digestibility values are adequate.

It is becoming increasingly common to measure the digestibility of proteins and bioavailability of amino acids in the stomach and small intestine **(ileal digestion),** especially in pigs (Fuller et al., 1994). This is because dietary proteins are subjected to degradation in the digestive tract by the animal's digestive enzymes in the foregut and by microflora (bacteria) in the hindgut. Proteins and amino acids that enter the colon are of little nutritional benefit to the animal. They are degraded by microbes and either converted to ammonia, which is absorbed and excreted in the urine, or converted to microbial protein, which is excreted in the feces. To more accurately measure protein digestibility, the method of choice is to collect the digesta as it leaves the small intestine (the ileum). This is done by equipping pigs with ileal cannulas, which allow collection of digesta as it leaves the ileum. Ileal re-entrant cannulas are used so that the digesta enters the hindgut normally when ileal collections are not being made. Correction for endogenous nitrogen to measure true protein digestibility and true bioavailability of amino acids is important when ileal digestibilities are being determined (Nyachoti et al., 1997).

Indigestible markers can also be used to **estimate the feed intake** of ruminants. Markers such as chromic oxide are introduced into the rumen, either by infusion through a rumen fistula or by capsule. Total fecal collection is made using fecal collection bags. Using marker concentrations in rumen and fecal contents, and digestibility values obtained using the nylon bag technique (see Chapter 5), it is possible to calculate an estimated feed intake using methods reviewed by Pienaar and van Ryssen (1997). This technique is useful with grazing animals when it is difficult to measure feed intake directly.

CLASSIFICATION OF FEEDSTUFFS

Feeds can be classified according to some of their general properties. The classification used here is typical of that used in the feed industry.

Feedstuffs can be classed as either concentrates or roughages. Concentrates have a low fiber content and a high content of either energy or protein or both. Cereal grains, for example, are considered primarily as energy sources, but also contribute a significant amount of protein. Protein supplements generally are products with more than 20 percent crude protein. Many are by-products of oil seed crops (e.g., soybean meal).

Roughages are bulky materials with a high fiber content and a low nutrient density. Hay, pasture, silage, wheat straw, and cottonseed hulls are examples of roughages. They are used primarily in feeds for ruminants or nonruminant herbivores.

Classification of Feeds

I. Concentrates (energy sources)
- Cereal grains, e.g., corn, sorghum, barley, rye, triticale
- Other grains, e.g., buckwheat
- Grain milling by-products, e.g., wheat bran, corn gluten meal
- Roots and tubers, e.g., cassava, potatoes
- Food processing by-products, e.g., molasses, bakery waste, citrus pulp, distillers and brewers by-products
- Industrial by-products, e.g., wood molasses

II. Concentrates (protein supplements)
- Oilseed meals, e.g., soybean, cottonseed, rapeseed, canola, linseed, peanut, safflower, sunflower meals
- Grain legumes, e.g., beans, peas, lupins
- Animal proteins, e.g., meat meal, tankage, fishmeal, whey, feather meal
- Nitrogen sources for ruminants
  - Nonprotein nitrogen sources, e.g., urea, biuret, dried poultry waste
  - Bypass proteins, e.g., corn gluten meal

III. Roughages
- Pasture
  - Grasses
  - Legumes
- Green chop
- Silage

- Dry forages
  - Hay
  - Straw, stover, chaff
- Agricultural by-products, e.g., corncobs, hulls, bagasse

IV. Feed additives

- Mineral supplements, e.g., salt, limestone
- Vitamin supplements
- Synthetic amino acids
- Drugs, e.g., antibiotics, ionophores
- Preservatives, e.g., antioxidants, mold inhibitors
- Buffers
- Enzymes
- Hormones
- Flavors
- Probiotics
- Pellet binders

SOURCES OF CURRENT INFORMATION ON FEEDS

Sources of current information on feedstuffs and their utilization by animals include popular press publications and scientific journals. The most comprehensive popular press publication is *Feedstuffs.* It is published weekly and is essentially a newspaper of the feed industry (the *Wall Street Journal* of the feed industry). It contains articles on new research findings, current prices, trends of feed commodities, reports of trade meetings, changes in feed industry regulations, and so on. *Feedstuffs* is available from The Miller Publishing Company, 12400 Whitewater Drive, Minnetonka, Minnesota, 55343. Another popular press publication is *Feed International,* published by the Watt Publishing Company, Sandstone Building, Mt. Morris, Illinois, 61054. It deals with an international perspective and emphasizes the technology of feed manufacturing. *Feed Management,* also published by Watt, emphasizes feed manufacturing.

Research on feeds and their utilization is reported in scientific journals. In the United States, most agricultural research is conducted at state experiment stations associated with land-grant universities. Scientists conducting research write up their results, which are then submitted to scientific journals for publication. In most cases, the manuscripts are reviewed by other scientists before publication (peer review) to ensure that the procedures followed and conclusions reached are valid. Some of the major journals to consult for current research findings are the following.

Animal Feed Science and Technology
Australian Journal of Agricultural Research
Canadian Journal of Animal Science
Journal of Animal Science
Journal of Dairy Science
Journal of Nutrition
Nutrition Research
Poultry Science

Information on specific nutrient requirements can be obtained from publications in the series *Nutrient Requirements of Domestic Animals* published by the National Research Council (NRC), a subdivision of the U.S. National Academy of Sciences. The NRC publications are prepared by a subcommittee composed of several internationally recognized authorities on the nutrition of the species involved. The committee members meet several times to prepare their report, assembling data from the world's scientific literature and using their best judgment to establish nutrient requirements of the species. The NRC nutrient requirements are the authoritative figures used in the United States and in much of the rest of the world. They are intended to be true requirements and do not have a "margin of safety." In Britain, the Agricultural Research Council (ARC) publishes recommendations of nutrient requirements for livestock. Several other countries also have their own recommendations. The NRC and ARC values are the most widely used.

QUESTIONS AND STUDY GUIDE

1. The dietary requirements for B-complex vitamins are higher for cats and chickens than for swine and rabbits. Explain why, taking into account differences in digestive tract physiology.
2. The oral administration of magnets to cattle that might be exposed to sharp objects like wire (e.g., dairy cattle fed wire-tied alfalfa hay bales) is helpful in preventing "hardware disease." Explain, and describe what the consequences of hardware disease might be.
3. Some people have claimed that the ruminant animal is a protein factory in reverse, wasting protein that could be more efficiently used if consumed directly by people. Discuss.
4. It has been suggested that the digestive tract of equines is a superior adaptation to high-fiber diets compared with the ruminant stomach when availability of low-quality forage is not limiting. Explain.
5. Some animals engage in coprophagy (consumption of feces) or cecotrophy (consumption of cecal contents). What are the nutritional benefits of these behaviors?
6. Compare and explain the relative efficiencies of fiber digestion in colon versus cecal fermentors.
7. You have submitted some samples of dairy feed to a feed analysis laboratory. You receive the following report.
   a. Hay sample

   % crude protein = 17% (dry-matter basis)
   % crude fiber = 24% (dry-matter basis)
   % moisture = 12%

   b. Silage sample

   % crude protein = 8% (dry-matter basis)
   % crude fiber = 26% (dry-matter basis)
   % moisture = 75%

   Calculate the percent crude-protein and crude-fiber contents of these samples on an as-fed basis.
8. A silage sample contains 40% dry matter. A 15-g sample of the wet silage contains 144 mg N. What is the percent crude-protein content of the silage dry matter?

9. The crude protein of feed grade urea is 282% and of biuret is 230%. Explain how an ingredient can be more than 100% crude protein.

10. A feed sample contains 450 mg N per 25 g of sample (D.M. basis). Calculate the percent crude-protein content of this material.

11. Not all proteins contain 16% N. Thus, under some conditions it may be appropriate to use a factor other than 6.25. For example, for some types of feeds, a factor of 5.55 is used. What percent N would the proteins of these feeds contain?

12. A new crop, favabeans, is being evaluated as a potential protein supplement for swine. The beans contain 40% crude protein on a wet-wt basis (crude protein = 6.25 N). A pig is kept in a metabolism crate and fed a diet in which favabeans provide all the protein. The following data are collected.

%crude protein of diet = 20% (wet-wt. basis)
feed intake = 9.24 lb (wet-wt. basis)
wet wt. of feces excreted = 2800g
a 200 g sample of feces contains 120 g water
10 g fecal dry matter contain 342 mg N
800 ml urine were excreted
20 ml urine contained 200 mg N

Calculate the percent digestibility of the favabean protein and the % digestible protein (swine) content of favabeans. What is the percent of favabean content of the diet?

13. Why is it easier to measure digestible energy content of feeds for swine than for chickens?

14. Suppose you are conducting an experiment to determine the energy value of alfalfa hay for horses. You conduct a metabolism trial, using a harness and collection bag for collecting fecal excretion. The following data are obtained.

alfalfa intake(wet-wt. basis) = 40 lb
% dry matter of alfalfa = 90%
wet wt. of feces = 12 kg
% water in feces = 40%
proximate analysis data of the alfalfa (dry-wt. basis):
ether extract = 3%
crude protein = 16%
crude fiber = 38%
NFE = 32%

For the preceding values the percent digestibility values are 90, 70, 25, and 95% for ether extract, protein, fiber, and NFE, respectively. Bomb calorimetry data was

2 g feed dry-matter yield 8 kcal
3 g fecal dry-matter yield 9.33 kcal

Calculate (1) the gross energy and digestible energy in kcal/kg (dry-weight basis), (2) the lb TDN per 100-lb alfalfa dry matter, (3) the percent ash of the alfalfa dry matter, and (4) the caloric equivalent of TDN (i.e., 1 kg of TDN is equivalent to how many kcal of DE?).

15. A digestibility trial is conducted in water buffaloes. A diet containing chromic oxide is fed. The following data are collected.

    % chromic oxide in feed = 1.2%
    % chromic oxide in feces = 1.7%
    % N in feed = 2.56%
    % crude protein in feces = 8%

    Calculate the % apparent crude protein digestibility.

16. A disreputable salesperson has concocted a mixture of sawdust, cinders, ground rubber, crankcase oil, and leather trimmings, had the mixture pelleted, and is selling it as a "bargain" feed for pigs. Shortly after the salesperson has left town, local farmers begin submitting complaints to their extension service about the feed. It seems to be worthless. However, when a proximate analysis is performed, the results meet the NRC suggested requirements for pigs. A bomb calorimetry test shows it to have a high gross energy content, exceeding 5,000 kcal/kg. Explain.

REFERENCES

Brown, W.F., J. E. Moore, W. E. Kunkle, C. G. Chambliss, and K. M. Portier. 1990. Forage testing using near infrared reflectance spectroscopy. *J. Anim. Sci.* 68:1416–1427.

Fuller, M.F., B. Darcy-Vrillon, J. P. Laplace, M. Picard, A. Cadenhead, J. Jung, D. Brown, and M. F. Franklin. 1994. The measurement of dietary amino acid digestibility in pigs, rats, and chickens: A comparison of methodologies. *Anim. Feed Sci. Tech.* 48:305–324.

Janis, C. 1976. The evolutionary strategy of the Equidae and the origins of rumen and cecal digestion. *Evolution* 30:757–774.

Nyachoti, C. M., C. F. M. de Lange, B. W. McBride, and H. Schulze. 1997. Significance of endogenous gut nitrogen losses in the nutrition of growing pigs: A review. *Can. J. Anim. Sci.* 77:149–163.

Pienaar, J. P. and J. B. J. van Ryssen. 1997. The accuracy of indirect methods for predicting feed intake of sheep at steady-state intake and following abrupt changes in intake. *Livestock Prod. Sci.* 48:117–127.

CHAPTER 2

Concentrates: Energy Sources

Objectives

1. To discuss grains and other concentrates in terms of
   - agronomic features
   - nutritive value
   - special properties
   - processing methods
   - deleterious factors
2. To discuss disorders associated with grain overload, including
   - lactic acidosis
   - laminitis
   - polioencephalomalacia
   - enterotoxemia

CEREAL GRAINS

Cereals are members of the grass family (*Gramineae*). Grains are edible seeds; thus **cereal grains** are the seeds of cultivated grasses. They are the primary energy sources for humans and nonruminant animals. Swine and poultry production is based on cereal grain diets. Approximately 90 percent of the human plant food supply is derived from the following crops, listed in order of importance: wheat, rice, corn, potato, barley, sweet potato, cassava, soybean, oat, sorghum, millet, rye, peanut, field bean, pea, banana, and coconut. Almost half of these crops are the eight major cereal grains: wheat, rice, corn, barley, oats, rye, sorghum, and millet. Wheat, rice, and corn make up about 75 percent of the total world grain production. The properties of cereal grains as feedstuffs will be discussed; the grains are considered in order of their general importance as feed.

General Structure of Grains

The general features of a typical cereal grain seed are shown in Fig. 2.1. The seed consists of the plant embryo (germ), the endosperm, and the outside protective layers. The outermost layer is the **hull,** which is high in fiber and protects the seed from mechanical damage and invasion by pathogens. Beneath the hull is the **aleurone layer (bran),** which contains fiber and protein. It is often pigmented. Inside the aleurone layer is the **endosperm,** consisting mainly of starch. Imbedded in the endosperm is the germ, or seed embryo. The **germ** is high in oil, protein, and other nutrients, which the sprouting seed will use for growth. The endosperm is primarily a reserve of energy for the developing seed. The main difference between a grass seed and a cereal grain seed is that through plant breeding, seeds with a high starch (endosperm) content have been developed.

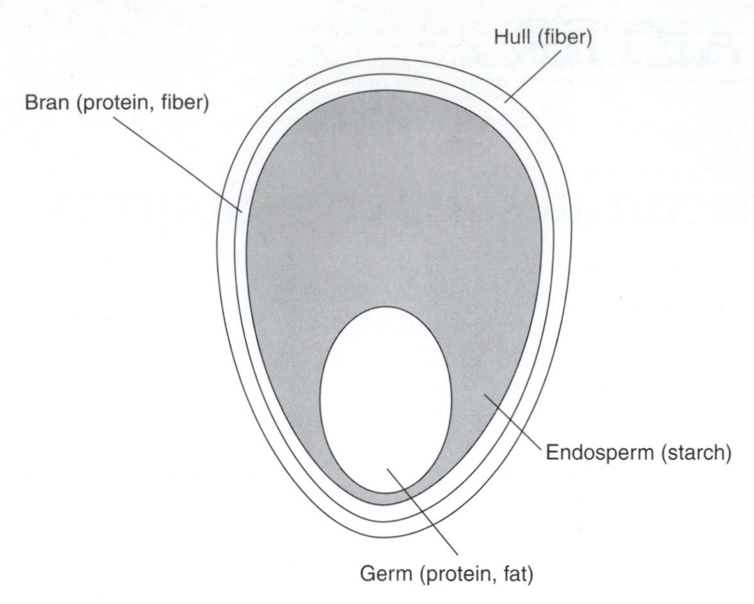

FIGURE 2.1 General features of a cereal grain, showing locations of major nutrient fractions.

Corn (Maize)

Corn (*Zea mays*) is a native of the Americas, domesticated and developed by the American Indians. In many English-speaking countries, corn is called *maize,* and *corn* is used as a general word for grain. In the United States, it was first called Indian corn, meaning Indian grain, before common usage led to acceptance of *corn* as the U.S. term for *Zea mays.*

The origin of corn and the process of domestication remain a mystery. One theory is that corn is derived from **teosinte,** a wild grass that is native to Mexico. It is a close relative of corn, and teosinte and corn cross readily. There is a perennial species of teosinte (*Zea perennis*) that crosses readily with corn. Efforts to develop **perennial corn** by crossing with teosinte have shown some success (Wagoner, 1990).

Corn is the world's most important feed grain for a number of good reasons relating to both its agronomic features and nutritional value. It is adapted over a wide range of climates and environments and can be rapidly modified by plant breeding to produce cultivars (varieties) that are productive in new areas. For example, a few years ago existing cultivars were unsuitable for production in the Pacific Northwest states, because the grain did not mature in the short growing season. New, earlier-maturing cultivars are now available and productive in this region.

Corn can produce more energy per acre than any other cereal grain. The basic reason for this is related to the plant's biochemistry. Corn has the C_4 photosynthetic pathway. These **C_4 plants** have a tropical origin and are inherently more productive than plants with the C_3 pathway. They utilize solar energy with a higher efficiency than **C_3 plants,** so in a given environment they can synthesize more carbohydrate (photosynthate). The term *C_4* means that the first products

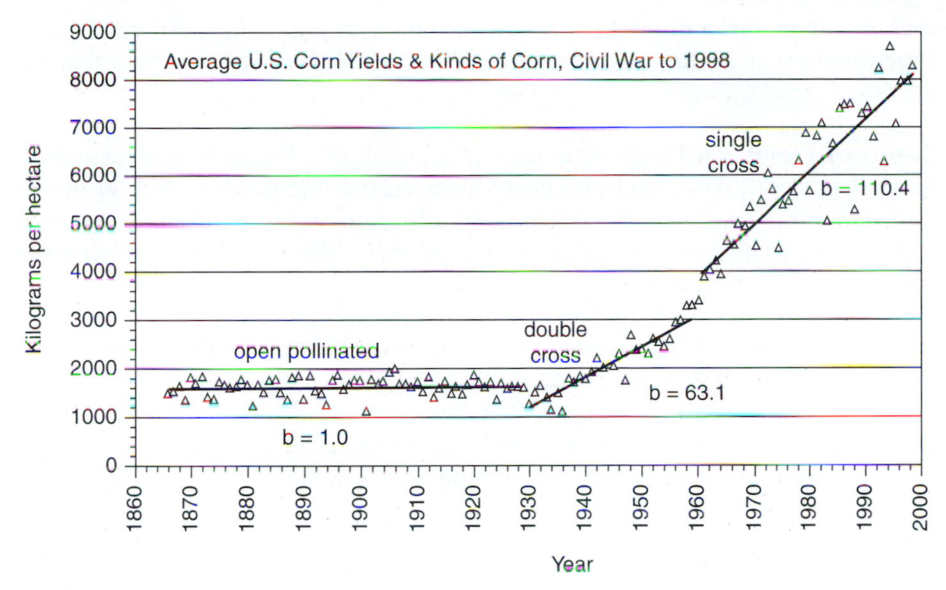

FIGURE 2.2 Average U. S. corn yields since the Civil War. Note that yields have steadily and dramatically increased, largely because of the development of hybrid corn, a trend likely to continue with application of biotechnology to plant breeding.(From A. F. Troyer, 1999.)

of the photosynthetic reactions are compounds with four carbon atoms (oxaloacetic, malic, and aspartic acids), whereas in C_3 plants the first product has three carbons (phosphoglyceric acid). Other advantages of C_4 plants are that they use carbon dioxide and water more efficiently, and use nitrogen for dry-matter accumulation with at least twice the efficiency of C_3 plants at all levels of fertilizer application. The C_4 plants are more tolerant of high temperatures. Thus, for biochemical reasons, corn is more productive than any other grain. This factor alone explains its dominance as a feed grain.

Another factor relating to its high yield is that hybrids can be readily produced. Corn is unique among grains in having the male and female flower organs separately on the same plant; the ear contains the ovary and the tassel at the top of the plant produces pollen. This has allowed the development of high-yielding **hybrid corn** varieties, which now make up the bulk of corn production in the United States. The tremendous increase in corn yields since the introduction of hybrid corn in 1930 is illustrated in Fig. 2.2. Applications of biotechnology to plant breeding are likely to result in continuous yield increases. Further developments in plant breeding include incorporation of genes for herbicide resistance to control weeds without damage to the crop.

In addition to its high-energy yield per acre, corn is the cereal grain with the highest digestible energy content for animals. It is very palatable, and contains no intrinsic toxic or deleterious factors. Thus it can be regarded as the best feed grain, and all others can be described in terms of their value as compared to corn. In addition to its high yield of grain, corn produces large amounts of leaves and stalks. According to Van Soest (1994), if all the crop land in the United States were converted to cereal production, sufficient straw and stover would be produced to support about three times the present ruminant population. Corn produces a larger quantity of higher quality crop residue suitable for ruminant feed than do the other cereal grains.

Types of Corn There are at least seven types of corn, including dent, flint, flour, pop, sweet, waxy, and pod corn. Dent corn is the primary type grown as feed, and sweet corn is the major type grown for human consumption. Dent corn is characterized by a dent or indentation at the top of the kernel, formed when the starch shrinks during drying.

The major parts of the kernel include the germ, which contains the rudiments of the embryo from which a seedling can develop, and the endosperm, a storage reserve of starch that serves as the energy source for the developing seed. The outer layer of the endosperm, called the *aleurone layer,* contains protein. Most feed corn is yellow in color, due to carotenoid pigments. White corn lacks these pigments. Red, purple, or black corn has pigments in the aleurone layer.

Nutrient Content

1. Energy *Carbohydrates* Corn is a high-energy grain because it is high in starch and oil, and low in fiber. The major energy-yielding fraction is the starchy endosperm. Starch consists of two main types: **amylose** and **amylopectin.** These are both composed of glucose but differ in the manner in which glucose molecules are joined together. Amylose is a straight chain of glucoses, whereas amylopectin is a branched structure. Regular hybrid corn contains starch composed of about 25 percent amylose and 75 percent amylopectin. Waxy corn has endosperm composed almost entirely of amylopectin. There appears to be little difference in the feeding value of normal and waxy corn.

Digestion and Metabolism of Corn Carbohydrate In nonruminants, the starch in corn is of high digestibility. It is digested primarily in the small intestine by pancreatic enzymes, such as pancreatic amylase. The net result is that starch is degraded to the glucose molecules of which it is composed. Glucose is the end product of starch digestion in nonruminants and is absorbed from the small intestine into the blood. The absorbed glucose is the primary energy source for cellular metabolism:

$$C_6H_{12}O_6 + 6O_2 + ADP = \xrightarrow[\text{metabolism}]{\text{cellular}} ATP + 6CO_2 + 6H_2O + \text{heat}$$

Cellular metabolism is basically the reverse of photosynthesis, so that the solar energy trapped by the plant in synthesizing carbohydrate is released in animal metabolism in a form (ATP) that can be utilized in biochemical reactions.

In ruminants, corn starch is subject to rumen fermentation. Microbes ferment the starch anaerobically, producing organic acids, the **volatile fatty acids** (VFA), as end products. The principal VFA produced from fermentation of corn is propionic acid. It is absorbed directly from the rumen and utilized in carbohydrate metabolism. Thus, the main absorbed energy source in ruminants is not glucose, but VFA.

Whole grain corn is poorly digested in cattle. At one time, in the Corn Belt, it was common to run pigs in cattle lots to consume the whole corn excreted in cattle feces. It is necessary to break down the waxy external shell of the kernel to permit its degradation in the rumen. Mastication of whole grains during the rumination process can break down corn kernels to some extent, but is ineffective with small-kerneled grains such as wheat and barley (Beauchemin et al., 1994). Some of the common processing procedures include steam rolling, dry rolling, grinding, extruding, popping, and flaking. Besides breaking down the structure of the grain, processing improves corn starch digestibility in ruminants. Starch in grains occurs as starch granules, which are crystal-like structures held together by hydrogen bonding between the starch molecules. The membrane of the granule is resistant to water entry and enzyme attack. Heat treatment of grains causes starch **gelatinization,** in which the hydrogen bonds holding the starch molecules together are denatured. This allows the granules to absorb water and swell. During processing, the swollen starch granules are torn apart by the rolling or flaking, increasing the surface area of starch exposed to digestive enzymes. A key factor in bacterial digestion of grains is attachment of the bacteria to feed particles (McAllister et al., 1994). Whole grains with an intact pericarp are resistant to **bacterial attachment.** Processing treatments increase grain digestibility by providing an opportunity for bacterial attachment to starch granules (Huntington, 1997).

Starch digestion is inherently less efficient in the rumen than in the small intestine of nonruminants. This is because some of the energy is lost via rumen gases, such as methane and carbon dioxide. Rumen carbon dioxide production from microbial metabolism of glucose deprives the animal of the opportunity to oxidize that carbon itself. Thus, grains such as corn are utilized less efficiently by ruminants than by nonruminants. Starch digested in the small intestine has 42 percent more energetic value to the animal than starch digested in the rumen (Owens et al., 1986). Thus, if starch could escape rumen digestion and be digested in the small intestine (postruminal starch digestion), efficiency by ruminants could be improved. However, the efficiency of **postruminal starch digestion** is limited by amylase activity in the intestine. It has been hypothesized that, with high grain diets, production of large quantities of VFA and a rapid passage rate through the rumen would cause the buffering capacity of the small intestine to be exceeded, lowering the intestinal pH below the optima for amylase activity. Thus postruminal starch digestibility would be reduced. Various buffers, such as limestone and cement kiln dust, have been studied as additives to improve intestinal starch digestion. Owens et al. (1986) concluded that energetic efficiency of growing ruminants is greater if starch is digested in the small intestine rather than in the rumen and that digestibility of starch in the intestine is generally at least as high as in the rumen. For best utilization of processed corn diets by cattle, the ruminal outflow of digesta with potential for intestinal digestion should be increased, but without sacrificing total digestion. This can be accomplished with ground corn by feeding long roughage, which optimizes grain utilization by increasing the ruminal washout of small particles (Goetsch et al., 1987). Huntington (1997) stressed the importance of ruminal fermentation of starch as a means of maximizing bacterial protein synthesis. He suggests that because most grain-fed ruminants absorb sufficient glucose from the intestine, it is more important to maximize rumen fermentation of starch for its beneficial effects on protein utilization than to maximize ruminal by-pass of starch.

Corn has a low fiber content, which accounts in part for its high content of digestible energy for nonruminant animals. The fiber in corn occurs mainly in the hull and germ fraction. Removal of the hull and germ by dry milling enhances the nutritional value of corn for swine and poultry (Moeser et al., 2002).

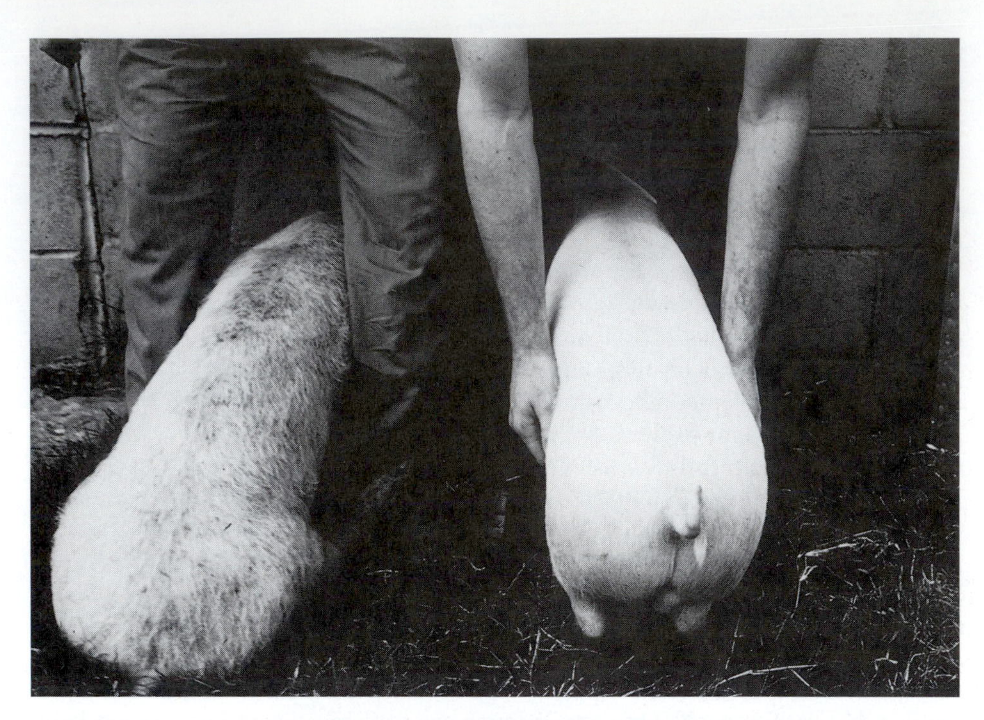

FIGURE 2.3 The corn-fed pig on the right shows the favorable effects of the corn oil on the skin. The pig on the left was fed a diet deficient in essential fatty acids.

Lipids A factor contributing to the high-energy value of corn is its oil content, which averages about 4 percent. With the exception of oats, other cereal grains have a lower lipid content. Corn oil is high in unsaturated fatty acids (**essential fatty acids**). These are important in animal metabolism and are exuded in the hair follicles. This gives corn-fed animals a sleek, shiny appearance. For example, pigs fed corn have a sleeker, glossier hair coat than pigs fed other grains. A lack of essential fatty acids causes dermatitis and a dandruff-like scaly skin. Pigs fed a wheat-based diet will grow as well as those fed corn, but have a dull hair coat and scaly skin (Fig. 2.3). The "bloom" of corn-fed animals is usually considered advantageous by farmers and, therefore, is a factor worthy of consideration in feed manufacture.

There is currently much interest in **high-oil corn (HOC).** The HOC is the culmination of about 90 generations of selection for high oil content. The HOC contains a larger germ than normal corn. The germ is richer than the rest of the kernel, including fat, protein, lysine, and oleic acid contents. Because HOC has a larger germ than regular corn, it has a higher content of these nutrients. The oil content of HOC ranges up to 10 percent of the grain. Advantages of the HOC include its higher energy content, better feed-conversion efficiency in swine and poultry, less dust in feed mills, and improved feed-pellet quality. The net result is improved efficiency in animal production when normal corn is replaced with HOC (Adams et al., 1994; Adeola and Bajjalieh, 1997).

2. Protein It is often difficult for students and livestock producers to appreciate the importance of the energy content of feedstuffs; they tend to think that the quality of feeds is determined by their protein content. The fallacy of this thinking is exemplified by corn. It is the feed grain of choice. If you had a choice of buying any cereal grain at the same cost per ton, you should choose corn, in spite of the fact that it has a lower content and quality of protein than most other grains. This is because the grain portion of the diet serves mainly as a source of energy. Corn grain has a low protein content, ranging from approximately 8 to 10 percent, but is an excellent energy source. Nevertheless, the protein contribution of grains, including corn, is very important in animal feeding.

The quality of a protein refers to its content of essential amino acids. A good-quality protein has an amino acid profile similar to the essential amino acid requirements of animals. Corn-grain protein is recognized as poor quality, being deficient in lysine and tryptophan, and low in methionine. For nonruminants, corn must be supplemented with a protein supplement to supply the deficient amino acids. The poor protein quality of corn is due to the fact that it has a high content of zein. Zein (pronounced zane) is a type of protein called a **prolamine.** Prolamines have a very low lysine content and are almost devoid of tryptophan. The percentage of the total protein in corn of the prolamine fraction is higher than for most other grains (Table 2–1), accounting for its poor protein quality.

Opaque-2 corn is a mutant with a much higher lysine content than regular corn. It is called "opaque" because the endosperm does not transmit light and so does not have the translucence of kernels of normal corn, giving the kernels a faded, dull color. Whereas normal corn protein contains approximately 41 to 52 percent zein and 17 to 28 percent glutelin, opaque-2 corn protein contains only 16 percent zein and 42 percent glutelin. Thus, it has a lower quantity of the low-lysine prolamine. It is well established that because of its superior amino acid balance, opaque-2 corn is a more valuable grain than regular corn for swine and poultry because less protein supplement is needed. However, it has not become widely used because it is lower yielding than normal corn. Economically, it is better to maximize the yield of energy (starch) from corn and obtain the protein supplement needed from another source than to grow a lower yielding, but more nutritionally complete, grain.

The opaque-2 gene is probably of greatest potential significance in human nutrition. In countries where much of the dietary protein of people is derived from corn, as in Mexico for example, incorporation of the high-lysine gene into local corn varieties could have major effects on improving human protein status. However, the lower yield of high-lysine corn limits its application as a human food

TABLE 2–1. Prolamine Content of Cereal Grains

Grain	*Prolamine*	*% of Seed Protein*
Common wheat	Gliadin	45
Durum wheat	Gliadin	60
Rye	Secalin	40
Barley	Hordein	40
Oats	Avenin	12
Rice	—	8
Corn (maize)	Zein	50
Sorghum	Kafirin	60
Common millet	Panicin	60

Adapted from Mosse (1966).

in the same manner as it does with livestock. Graham et al. (1980) demonstrated that although opaque-2 increased the nutritive value of corn for malnourished children, its use " . . . strictly on the basis of the economic consideration of a 10 percent lesser yield, seems to be a large step backward in the effort to improve the lot of millions of children." Thus, as with livestock, it seems more feasible to maximize energy yield with corn and supplement it with a protein supplement than to improve the protein quality of corn at the cost of reducing energy yield per acre. However, a new class of opaque-2 corn called **Quality Protein Maize (QPM)** has been developed (NRC, 1988; Sullivan et al., 1989) that retains the high-protein quality of opaque-2 but has improved agronomic traits, such as higher yields and a hard endosperm that is less susceptible to fungal and insect damage (NRC, 1988). Sullivan et al. (1989) found that QPM had a higher protein quality for pigs than normal feed or food corn. The development of QPM suggests commercial possibilities for high-lysine corn.

Another corn mutant with improved protein quality is floury-2 corn, which is comparable to opaque-2 corn in nutritional value. Floury corn has kernels with soft, fine-textured starch.

3. Vitamins Yellow corn is the only cereal grain to have significant **vitamin A** activity. Its yellow color is due to carotenoid pigments, some of which are precursors of vitamin A, with the main one being β-carotene. Plants do not contain vitamin A as such, but rather contain vitamin A precursors such as β-carotene. **Beta-carotene** is two molecules of vitamin A joined together. An enzyme in animal tissues can split these two molecules apart to liberate vitamin A. Normally this is accomplished in the intestinal mucosa, but in some species the conversion to vitamin A occurs in the liver. The efficiency of conversion of carotene to vitamin A shows species and breed differences. Jersey and Guernsey cattle, for instance, exhibit low conversion efficiency, so their body and milk fats are often yellow because of their carotene content. Cattle are much less efficient than sheep and goats in the conversion of carotene pigments to vitamin A (Yang et al., 1992). Sheep fat is virtually always white, while in grass- or corn-fed cattle, the fat can range from creamy-white to bright orange-yellow from the accumulation of carotenoids in the body fat.

Although corn is the only grain containing β-carotene, its content is low compared to various other sources (Table 2–2). The richest sources are marine animal and fish liver oils. This is because vitamin A accumulates in the liver, and the marine food chain begins with plankton, which, like all green plants, are rich

TABLE 2–2. Vitamin A Content of Some Natural Materials

Item	*Retinol (IU/g)*†	*Item*	*Retinol (IU/g)*
Whale liver oil	400,000	Sun-cured alfalfa meal	150*
Herring liver oil	211,000	Carrots	120*
Shark liver oil	150,000	Corn gluten meal	28*
Cod liver oil	4,000	Yellow corn	8*
Dehydrated alfalfa meal	530*		

Adapted from Scott et al. (1982).
*As provitamin A (beta-carotene).
†IU = International Units (see Chapter 7 for definition).

sources of carotene. Carotene pigments function with chlorophyll in photosynthesis. Since vitamin A is poorly excreted, it builds up in the marine food chain; the carnivorous species such as polar bears at the top of the food chain accumulate very high vitamin A levels in their livers (see Chapter 22).

Corn also contains carotenoids called **xanthophylls.** Xanthophyll pigments do not have vitamin A activity but are important in poultry production for providing the yellow coloration of egg yolks and of the skin of broilers. Poultry products from birds that are fed corn are pigmented, whereas those from birds fed other grains are pale unless other sources of xanthophylls (e.g., alfalfa meal, marigold meal, synthetic xanthophylls) are fed.

Another vitamin that must be considered is niacin (nicotinamide, nicotinic acid). Corn has a low quantity of available niacin. **Niacin** can be synthesized by most animals from the amino acid tryptophan, in which corn is also deficient. An exception is the cat, which is unable to convert tryptophan to niacin. Niacin deficiency in humans was at one time a major public health problem in the southern United States, causing pellagra. **Pellagra** is characterized by the 3D's: dermatitis, diarrhea, and dementia. Before niacin was discovered, thousands of pellagra sufferers in the southern United States were institutionalized for insanity. Their real problem, however, was niacin deficiency induced by a poor diet based on corn and pork fat. Niacin was discovered using dogs as the experimental animal; niacin deficiency in dogs causes a disease called **black tongue.** Corn was not successfully used as a major feed grain for intensive swine and poultry production until the discovery of the need for niacin supplementation. The niacin in corn can be made more available by treatment with alkali. In Latin American countries, where corn is a dietary staple, people have avoided pellagra problems because the traditional way of preparing corn for cooking involves adding lime to ground corn and soaking. This procedure releases bound niacin. Similarly, the traditional diet of the Hopi Indians of Arizona was based on corn, but pellagra was not a problem because of the type of processing used. Their traditional method was to harvest it at the immature, milky stage, bake it in an oven or pit, and then sun-dry it. When processed in this way, corn is rich in available niacin (Carpenter et al., 1988).

With respect to other vitamins, corn is a fair source of vitamin E and low in vitamin D and the B-complex vitamins. In common with all plants, corn is devoid of **vitamin B_{12}** activity.

4. Minerals As is true for all cereal grains, corn contains very little calcium. A calcium supplement is always needed for grain-based diets. Corn is moderately high in phosphorus, with a content of about 0.3 percent. However, for nonruminants, much of this is "organic phosphorus," which is largely unavailable. It is "tied up" in the form of an organic molecule called **phytate.** Grains contain a substance called phytic acid, which binds phosphorus. Each phytic acid molecule can bind six phosphate groups, forming an unabsorbable complex. Phytate phosphorus is of particularly low availability to young animals. In diets for chicks, for example, the phosphorus in corn and other grains is usually ignored in ration formulation. Phytate is digested by microbial phytase in the rumen, so organic phosphorus is bioavailable to ruminants.

New varieties of corn have been developed that have a reduced phytate content, thus improving the bioavailability of phosphorus. The improved varieties are referred to as **high available phosphorus (HAP) corn.**

There are many possibilities for additional genetic engineering of corn. One is the introduction of wheat gluten genes to produce corn with gluten so it could be used in making bread. Herbicide-resistant corn has also been developed.

Deleterious Factors in Corn Unlike most feedstuffs, corn does not intrinsically contain significant toxic or deleterious compounds. However, it is often contaminated with **mycotoxins,** which are of concern in animal feeding. Mycotoxins are toxins elaborated by molds (fungi) growing in or on feedstuffs. They may be produced while the crop is growing or during storage. In general, toxigenic fungi grow best under warm, humid, aerobic conditions. The major mycotoxin of concern with corn is **aflatoxin,** produced by the mold *Aspergillus flavus.* This mold infects corn both in the field and in storage. Drought and insect damage promote infection of the developing grain. Warm, humid storage conditions can promote explosive formation of aflatoxin. Infected corn is not necessarily visibly moldy. Aflatoxin fluoresces under ultraviolet light, so its presence can be detected by examining grain under "black light." Aflatoxin causes reduced feed intake, poor growth, and diarrhea; chronic exposure causes liver damage. Aflatoxin is the most potent carcinogen known; at dietary levels as low as 1 part per billion (ppb), it causes **liver cancer** in rainbow trout, a widely used experimental animal in mycotoxin studies. The maximum level of aflatoxin permitted in feed ingredients by the **Food and Drug Administration (FDA)** in the United States is 20 ppb in corn and other grains intended for human food use, for immature animals, and for dairy cattle. The FDA action level is 100 ppb for corn fed to breeding beef cattle, poultry, and swine; 200 ppb for finishing pigs; and 300 ppb for finishing beef cattle. Contaminated corn may be blended with clean corn for livestock use but not for human use. In some years, because of environmental conditions in the Corn Belt, the average aflatoxin level in U.S. corn exceeds the permitted level. In such cases, the FDA allowable level has been waived because of no reasonable alternative.

Other mycotoxins sometimes associated with corn and other grains are fumonisins, zearalenone (F-2 toxin), ochratoxin, T-2 toxin, vomitoxin, and citrinin. **Fumonisins** cause a variety of toxicity signs in livestock. In horses, a neurological disease called **equine leucoencephalomalacia (ELM)** that causes fatal degeneration of the cerebellum occurs with exposure to fumonisins. Pigs develop pulmonary edema. Poultry and ruminants are relatively resistant to the effects of fumonisins. These mycotoxins are suspected of causing esophageal cancer in humans. Corn is the major grain that may be contaminated with fumonisins. Corn screenings may be particularly toxic because the toxin occurs in the outer layers of the grain, which tend to slough off and be segregated as screenings (fines). **Zearalenone** causes estrogenic effects. Swine are particularly sensitive to it; affected gilts and sows have swelling of the external sex organs and mammary glands and may prolapse. Zearalenone is also produced commercially and is marketed as an estrogenic implant for cattle under the trade name Ralgro. **Vomitoxin** and **T-2 toxin** cause vomiting and feed refusal in swine and esophageal lesions in poultry. **Citrinin** and **ochratoxin** cause kidney lesions. The first signs of mycotoxicosis are poor growth and reduced performance. Mycotoxins are of major concern in animal feeding, causing loss of production efficiency and outbreaks of acute toxicity.

Corn (and other grains, to a lesser extent) may be harvested at a moisture content that is too high for storage as dry grain. Stored grains must contain less than 15 percent moisture, or else molding and/or rotting will occur. Corn may be dried with application of heat, but this is an expensive process. Alternatively, it may be ensiled (see Chapter 6) or preserved in a moist form by addition of preservatives and mold inhibitors (see Chapter 8). Propionic acid, urea, anhydrous ammonia, and sulfur dioxide are substances that have been employed for preservation of moist grains. More information on the use of these substances is given in Chapter 8.

Use of Corn in Animal Diets Corn-soybean meal diets are standard for swine and poultry production in the United States and in much of the rest of the world. Except for very young pigs, the nutritional needs of all classes of swine can be met with a simple mixture of corn and soybean meal, supplemented with salt, minerals, and vitamins. Examples of some typical swine diets (Pond and Maner, 1984) are shown in Table 2–3. Typical poultry diets containing corn are given in Chapter 14.

Corn as a Nutrient and Pharmaceutical Factory These are exciting times for corn breeding. Great strides have been made in genetic modification of corn to improve its nutrient contents, agronomic factors, and industrial uses. For animal feeding, development of high-oil corn, with subsequent incorporation of other traits (stacking traits), such as genes for high lysine, high methionine, low phytate, and so on, have dramatically changed the feed and livestock industries. These **"value-added" grains** have a retained identity so that they can be channeled into specific uses and markets. Corn is also being tested for the production of human antibodies. Human genes for antibody production can be inserted into the corn plant genome, along with genes that stimulate protein production.

Much of the corn now produced in the United States has been genetically modified through the use of gene-transfer techniques—the so-called **GMO corn** (GMO = genetically modified organism). Examples include herbicide-resistant corn and **Bt corn.** *Bacillus thuringiensis* (Bt) is a soil bacterium that produces proteins (e.g., Cry 9C, Cry 1Ab) that are toxic to caterpillars such as the European corn borer. Thus Bt corn produces its own insecticide. The grain and crop residues of Bt corn are safe for animal feeding (Folmer et al., 2002).

TABLE 2–3. Examples of Typical Corn-Based Swine Diets

Ingredient (%)	*Grower Diet*	*Finisher Diet*	*Sow Diet*
Yellow corn	77.5	87.5	54.5
Oats	—	—	25.0
Soybean meal	18.0	8.0	7.0
Meat meal	—	—	7.0
Dicalcium phosphate	2.8	0.5	—
Ground limestone	1.2	0.5	—
Steamed bone meal	—	3.0	1.0
Salt (with trace minerals)	0.5	0.5	0.5
Vitamin supplement	*	*	*
Alfalfa meal	—	—	5.0

Adapted from Pond and Maner (1984).
*To be provided at levels recommended by supplier.

Grain Sorghum and Millet

Sorghum and millet are the major food grains in the semiarid tropics, an ecological zone encircling the globe and including China, India, most of Africa, Australia, Argentina, and parts of the southern United States. In the developed countries, approximately 96 percent of the total sorghum and millet grown is used for animal feed, whereas in the developing countries, only 8 percent of these crops is used for livestock, with the rest used directly as human food. In the Sahelian zone of Africa, approximately 90 percent of the rural population depends on these crops as their major source of food. Sorghum and millet are not closely related botanically, but they are similar in many agronomic respects and grow under similar conditions. Therefore, they will be discussed together.

Sorghum (*Sorghum bicolor*) is a hardy, drought-resistant crop adapted to environmental conditions too harsh for the production of corn (Fig. 2.4). It requires less water than corn and can survive dry conditions and then resume growth when moisture becomes available. In the United States, sorghum is grown as a dryland crop in the Southwest in areas where corn requires irrigation. With the depletion of aquifers in the Southwest, irrigation costs are increasing and a switch from corn to sorghum production in this area is likely. The foliage of sorghum is similar to that of corn in appearance, but the grain is considerably different. It is produced as small, round seeds in an open panicle (Fig. 2.4). There are various subgroups of grain sorghum, such as milo and kafir, but the distinction between them has largely been lost through crosses used in developing new cultivars. Grain sorghum in the United States is commonly referred to as *milo.*

FIGURE 2.4 Sorghum is a grain adapted to semiarid conditions. The grain is produced in an open panicle and thus is susceptible to predation by wild birds. Bird-resistant sorghum has been developed to overcome this problem.

Like corn, sorghum is a C_4 plant, accounting for its high productivity and tolerance to high temperatures.

Millets are relatively minor crops except in localized areas of Asia, Africa, and the former Soviet Union. **Millet** is a collective term used for seeds of several crops that are all members of different grass genera. In order of importance, they include pearl millet (*Pennisetum glaucum*), foxtail millet (*Setaria italica*), proso millet (*Panicum miliaceum*), and several other species. Forage specialists may recognize these genera (*Pennisetum, Setaria, Panicum*) as major tropical forage grasses. In the United States, where millet is grown primarily for use as bird feed, approximately half the millet acreage is pearl millet, a quarter is proso, and the remainder is other species. Millet has traditionally been viewed as a poor person's crop, lacking prestige and appeal, but it holds great potential as a food crop adapted to marginal, drought-stricken areas. Millet grain typically is produced in a spikelike panicle with small, round seeds less than half the diameter of sorghum seeds.

Nutrient Content

1. Energy Grain sorghum is quite similar to corn in its composition. The endosperm is composed primarily of starch. Nonwaxy or normal sorghum contains approximately 27 percent amylose and 73 percent amylopectin; waxy sorghum contains only amylopectin. The waxy type consistently shows a higher feeding value than normal sorghum. Both the starch granules and the protein matrix surrounding them are more digestible in waxy grain.

Sorghum is equal to or slightly lower in value than corn as an energy source. For ruminants, it requires more vigorous **processing,** such as steam-flaking, to achieve optimal digestibility (Reinhardt et al., 1997). However, there is considerable variability in feeding value among sorghum cultivars and types, mainly because of variations in tannin content and seed coat color. Brown, high-tannin, bird-resistant sorghums give poorer animal performance and lower digestibility than low-tannin types. The peripheral endosperm of normal sorghum is extremely dense, hard, and resistant to water penetration and digestion. The effectiveness of processing methods such as steam-flaking in improving utilization of sorghum is due to physical disruption of the peripheral endosperm, not to gelatinization. Traditional methods of preparing sorghum for food in Africa involve malting, fermentation, and cooking with acid or alkali; these treatments help to disrupt the peripheral endosperm.

2. Protein Like corn, sorghum is a fairly poor source of protein. The protein content of sorghum grown in the United States is typically in the range 7 to 10 percent. The quality of the protein is poor; as in corn, protein in sorghum has a high prolamine content (Table 2–1). For swine, lysine is the first limiting amino acid and threonine is the second limiting, followed by tryptophan, methionine, and isoleucine. Numerous studies with swine and poultry indicate that the availability of amino acids in sorghum is low. **Tannins** (phenolics) in sorghum reduce protein availability. Tannins inhibit digestive enzyme activity and form complexes with protein that resist digestion. In ruminants, tannins in bird-resistant sorghum reduce protein availability in both the rumen and small intestine (Streeter et al., 1990a, b), although the reduction in protein digestibility is fairly modest (Streeter et al., 1993). Elkin et al. (1996) reported wide variations in crude protein digestibilities in chickens fed sorghums with similar tannin contents. They suggested that variations in the content of **kafirins** (see Table 2–1), the storage proteins of sorghum, might explain the inconsistent relationships between tannin content and digestibility.

3. Other Nutrients Sorghum has no unusual vitamin or mineral characteristics. Grains typically have very low calcium, high phytate phosphorus, no vitamin B_{12}, and little significant vitamin A activity. Yellow endosperm sorghum contains carotenoid pigments; thus, it presumably does have some vitamin A activity.

Deleterious Factors The principal deleterious factors in grain sorghums are polyphenolic compounds of a type known as **condensed tannins.** These compounds are quite significant nutritionally. Most of the biological effects of tannins are associated with their ability to react with proteins. This effect begins in the mouth, where tannins react with taste receptors to produce the sensation of astringency. The tannin content of unripe fruit is high, causing the astringent effects characteristic of green apples and bananas. During ripening, the tannins are reduced and condensed, which reduces the astringency. Thus, one effect of tannins is to reduce palatability of feedstuffs. In the digestive tract, tannins may react with digestive enzymes, reducing nutrient digestibility. They also react with dietary proteins, forming indigestible complexes (Lizardo et al., 1995).

Sorghum tannins occur primarily in the testa and pericarp, the outermost layers of the seed kernel. The presence of tannins confers a brown pigmentation to the seed, although the seed color is not a reliable indicator of tannin content. Sorghum varieties differ widely in tannin content. **Bird-resistant sorghums,** selected for resistance to wild bird predation, are high in tannins. The first bird-resistant varieties that were developed proved to be unsatisfactory; their tannin content was so high that they gave very poor results when fed to livestock and poultry. Subsequent varieties have been selected so as to have enough tannin to provide adequate bird resistance without markedly reducing their nutritional value. In areas where bird damage is not a problem, low-tannin varieties should be grown. The exposed seeds in the open panicle are very vulnerable to wild bird predation. In Africa, the weaver bird causes great destruction of crops. It is the most numerous bird pest known, with the numbers estimated at 100 billion. The potential crop loss is staggering in its magnitude. In other areas, blackbirds, starlings, sparrows, and crows cause extensive sorghum losses, so the need for bird-resistant sorghum is apparent.

Various **processing methods** can be used to at least partially overcome the effects of sorghum tannins. Treatment with an alkali, such as sodium or ammonium hydroxide, or anhydrous ammonia, is effective. Polyethylene glycol, which forms complexes with tannins, has also been shown to be an effective additive to improve the feeding value of high-tannin sorghum.

Pearl millet has a higher energy and protein content than sorghum and is similar to corn in feeding value for pigs (Lawrence et al., 1995) and cattle (Hill et al., 1996).

Wheat

Wheat is the world's most important crop. It is grown primarily for human consumption and is the preferred grain of most societies. Wheat is the preferred grain for baking, both of bread and other products using flour, such as cookies, cakes, and pasta products (macaroni, spaghetti, noodles, etc.).

Wheat belongs to the genus *Triticum*. There are numerous species; however, only three are of commercial significance. These are *T. durum* (durum), *T. compactum* (club wheat), and *T. aestivum* (common or bread wheat). Minor species include *T. spelta* (spelt) and *T. dicoccum* (emmer), which were formerly grown to a limited extent in the United States but now are virtually unknown. They were grown primarily for feeding purposes.

Wheat can be categorized according to the following characteristics:

1. Growth habit: spring or winter
2. Kernel color: red, white, or amber
3. Kernel texture: hard or soft

Spring wheats are seeded in the spring and harvested in late summer or early autumn. **Winter wheats** are sown in the autumn and harvested the following summer. Winter wheat is higher yielding than spring wheat. The main classes of wheat grown commercially in the United States are hard red winter, hard red spring, durum, soft white winter, and soft white spring. **Hard wheats** are preferred for bread flour because they have a high gluten content. Gluten is a protein that forms a sticky mass of dough, trapping carbon dioxide produced by the yeast, causing bread to rise (leaven). Hard wheats tend to have protein contents of 11 to 14 percent. **Soft wheats** have a lower protein content of 8 to 11 percent. These are suitable for cake, cookie, and pastry flour. **Durum wheat** is used to prepare pasta products. Hard wheats are grown in areas with a fairly dry growing season, such as the Great Plains states and the prairie provinces of Canada. Soft wheats are produced in humid or irrigated areas.

Wheat is generally too expensive to be competitive as a livestock feed. Selling wheat for human consumption is more lucrative for producers than selling it at a price competitive with other feed grains. Wheat is used as a feed mainly when the price is low because of surpluses or when its quality is inadequate for milling purposes, as when damaged by frost, drought, rain, or disease. Drought stress during grain ripening may result in **shriveled wheat,** due to inadequate formation of the starch. The total yield is reduced, but the feeding value of the grain is not usually adversely affected. There is some interest in the development of feed wheats, which would be selected primarily for high yield. They could have a purple gene incorporated into their genotype so they could be identifiable by color and would not enter commercial channels for milling wheat (Myer and Barnett, 1987). As the yield of wheat (and other cereal grains) increases, the percent of protein decreases (Simmonds, 1995). This is because the higher yield is due to greater starch formation, which dilutes the protein present in the seed kernel.

All currently grown wheat varieties are annual grains. Wheat is closely related to the perennial wheatgrasses (*Agropyron* spp.) of North America and can be crossed with several of them. Efforts are underway to develop **perennial wheat** (Wagoner, 1990), using intermediate wheatgrass (*Agropyron intermedium*). Nutritionally, the seeds of wheatgrass are similar to those of wheat (Becker et al., 1986). A perennial grain crop could be helpful in reducing soil erosion, would improve soil texture, and could be grown with a legume that could provide nitrogen. The aftermath could be used for grazing purposes. Of course, development of perennial wheat will be a long-term project, although biotechnology techniques might speed up the process.

Nutrient Content

1. Energy Wheat is virtually identical to corn in digestible energy content, and is equal to corn as an energy source for all livestock. It is more likely than corn to cause digestive disturbances of ruminants because of the rapid digestion rate of the starch (see pp. 59–60).

2. Protein Wheat is superior to corn in terms of its protein content and quality. The soft white wheats of the western United States contain 10 to 11 percent crude protein, whereas the hard spring wheats usually contain 12 to 14 percent protein. Lysine is the most limiting amino acid for swine and poultry, followed by threonine and methionine.

Deleterious Factors As is true for corn, there are no significant toxins in wheat. It is not normally infected in the field with mycotoxin-producing fungi, so mycotoxins are not of concern unless the grain is improperly stored. Occasionally it may be infected with ergot.

Use of Wheat in Animal Diets Wheat is considered the equal of corn in animal feeding. It has a similar energy content and the protein content is higher. Wheat is lower in essential fatty acids than corn, so wheat-fed pigs may not look as sleek as corn-fed animals. Wheat-fed pigs may have rough, scaly skin but their growth performance is not affected. Seerley et al. (1988) noted that the particle size of wheat may affect pig performance, with grain ground to a fine particle size best for young pigs and coarsely ground wheat superior for finishing pigs. Wheat can be substituted for corn in poultry diets with no adverse effects. Where convenient, as with the small-farm poultry flock, whole wheat can be fed to chickens. For ruminants, **lactic acidosis** is more common with wheat-based diets than with other grains because of the rapid fermentation rate in the rumen. This is particularly true of newly harvested or "new" wheat. Dry rolling and steam rolling wheat improves its value for feedlot cattle. Advantages of steam rolling include increased cattle gain and feed efficiency, increased net energy value, and increased ruminal escape and postruminal absorption of the wheat protein (Zinn, 1994).

Wheat contains some **nonstarch polysaccharides** (NSP) such as pentosans, which are poorly digested in nonruminants, and especially in poultry (Choct and Annison, 1992). Supplementation of wheat-based diets with fungal pentosanases can improve the utilization of the NSP fraction by poultry (Campbell and Bedford, 1992; Crouch et al., 1997). The **pentosans** are composed of sugars with five carbons (pentoses); the major pentosans in wheat are xylans, composed of the pentose sugar xylose. Wheat also contains **arabinoxylans,** containing the pentoses arabinose and xylose.

The **gluten proteins** in wheat become sticky when moist. For this reason, wheat and wheat by-products improve **pellet quality;** the glutens cause the ingredients to stick together in the steam-pelleting process.

Wheat gluten proteins cause an inflammatory disease of the intestinal villi, called celiac sprue, in susceptible people. **Celiac sprue** is an autoimmune disease triggered by exposure to a peptide fraction in wheat gluten. Susceptible individuals have unique leukocyte antigens that react with gluten peptides. These individuals must avoid consumption of wheat products. There is also a condition called **tropical sprue,** which is caused by deficiencies of folic acid and vitamin B_{12}.

Barley

Barley (*Hordeum vulgare*) ranks fourth among world crops. It is grown mainly in northerly, cool climates such as the northern United States, Canada, the former Soviet Union, China, and western Europe. In areas with very short growing seasons, either because of latitude or elevation, barley is and will continue to be an important crop. In areas where corn is productive, there is little reason to produce barley. Barley is grown primarily for two purposes: for malting and for feed. **Malt** is produced by soaking barley to initiate germination; after the grain sprouts, germination is stopped by kiln-drying. The product, dried malt, is then mixed with unmalted grain, cooked, drained, and the liquid cooked with hops. It is then fermented to produce beer. The spent grains are used as wet or dried brewers grains for feed.

There are numerous types of barley, including spring and winter varieties. Most barleys are either two-row or six-row, meaning that they have either two or six rows of kernels in the spike. There are five kernel colors: white, black, red, purple, and blue. There are rough-awn and smooth-awn varieties. There are also hull-less (naked) varieties. Kernels from two-row barleys are generally larger and have a higher protein content than those from the six-row types.

Nutrient Content

1. Energy Corn, wheat, triticale, and sorghum are recognized as high-energy grains, whereas barley and oats (**coarse grains**) are lower in energy content, particularly for nonruminants. The lower energy value of barley is due to a lower starch content, a higher content of poorly digested glucans, and a high fiber content. The fiber is associated with the hull and attached awn (Fig. 2.5). Particularly with rough-awn varieties, the hull-awn portion of the seed is very rough and

FIGURE 2.5 Heads of barley with the fibrous and abrasive awn or beard that reduces the palatability of barley grain.

abrasive, making the grain distinctly unpalatable, unless incorporated into a pelleted feed. When fed a nonpelleted, barley-containing diet, pigs may physically separate the hull fraction to avoid having to consume it.

Barley contains water-soluble carbohydrates, called **β-glucans,** which are poorly digested and thus lower the digestible energy content. This is particularly true in poultry and to a lesser extent in swine. The glucans and other gums are found in the hull, in the aleurone, and in the endosperm cell walls. Beta-glucans are composed of glucose molecules linked together by chemical bonds that are different from those found in starch. Animals do not secrete glucanases to break these bonds, so they cannot digest β-glucans. The glucans are viscous, hydroscopic, gummy materials. The viscosity reduces intestinal flow rate and thus feed intake. The viscous glucans also inhibit the formation of lipid micelles and reduce fat absorption. In poultry, the hydroscopic nature of glucans makes the excreta wet and sticky, causing management problems. The manure in poultry houses is normally quite dry and can build up beneath the cages in well-shaped cones, or build up in a deep-litter floor-pen system. When barley is fed, the droppings may become "soupy," increasing environmental humidity and ammonia, and causing problems in manure removal. With young chicks, the glucans in barley may cause a pasty vent condition, in which the sticky excreta dries on the vent, blocking the gut. This material must be removed or the affected bird will die. This problem makes it impractical to use barley in diets for young chicks. The β-glucan content of barley is influenced by genetics and environment. The waxy gene seems to be associated with high glucan content (Newman and Newman, 1987). Hot, dry growing conditions increase β-glucan levels. The effect of β-glucans can be largely overcome by incorporating a source of β-glucanase in the feed. Various bacterial and fungal enzyme sources are useful for this purpose (see Chapter 8). **Soaking** barley in water activates β-glucanases in the seed, reducing the glucan content. These treatments, although effective, may increase diet costs. Soaking or steeping the grain may not be cost-effective unless it can be fed in the wet form.

The glucans are digested by microbial action in the rumen. Engstrom et al. (1989) found that the digestibility in cattle of β-glucans in four types of barley with 3.5 to 4.8 percent glucans was 98 to 99 percent, indicating virtually complete digestion.

The fibrous hull of barley can be removed mechanically to produce **pearled barley.** Although this improves the energy content, it may not be cost-effective. **Hull-less barleys** have a lower fiber content than hulled barley and a higher feeding value. The hull is loosely attached to the kernel and becomes detached during harvesting. They are not extensively grown because of lower yields (actual or perceived; because of lack of a hull, there is less total volume of grain, although the actual yield of hullfree grain may be equivalent), and because barley with hulls is preferred for malting purposes. Classen et al. (1988) concluded that hull-less barley was at least equivalent to wheat as a cereal grain for laying hens.

2. Protein Barley is superior to corn in its protein content and quality. Barley normally ranges from 11 to 13 percent crude protein. As is true with most grains, lysine is the first limiting amino acid and threonine is the second limiting for swine.

Genes in barley have been identified; one high-lysine line of barley is Hiproly. It has 20 to 30 percent more lysine than regular barley but needs agronomic improvement before it can be grown commercially. As with high-lysine corn, **high-lysine barley** has a lower prolamine (hordein) content than regular barley. Newman et al. (1990) reported on a high-lysine barley cultivar that had a grain yield of 91 percent of commercial varieties, and a 27 percent higher lysine yield. High-lysine varieties support improved growth of pigs, as compared with regular barley (Gabert et al., 1996).

3. Minerals As with other grains, much of the phosphorus in barley is bound as phytate. Low-phytic acid barley has been developed, which improves calcium and phosphorus utilization and growth performance in nonruminants (Veum et al., 2002).

Use of Barley in Animal Diets Barley can be usefully employed in swine and beef cattle diets. The problems with β-glucans and the low energy content of the grain generally preclude its use in the high-energy diets typical of modern poultry production. Growing and finishing swine generally perform satisfactorily when barley is the major grain, but growth and feed efficiency are not equal to results with corn-based diets. With weanling pigs, the addition of 20 percent dried whey to barley-based diets stimulates feed intake and growth rate (Goodband and Hines, 1988). These authors also found that fine grinding of barley improved pig performance. Developments in the commercial availability of enzyme preparations containing β-glucanase and other complex-carbohydrate digesting enzymes (e.g., pentosanases, β-glucan solubilase) may result in increased utilization of barley in swine and poultry feeding. Commercial production of high-lysine cultivars should also increase the utilization of barley for feeding purposes.

There is a lot of variability in barley concerning bulk density (bushel weight) and starch content, largely due to variable amounts of fiber. Processing barley for cattle gives variable results. **Dry rolling** can produce excessive amounts of fine particles, which may be fermented too rapidly, causing **grain overload disorders** such as bloat, liver abscesses, and lactic acidosis (Hatfield et al., 1997; Zinn, 1993). **Steam rolling** often gives improved animal performance. **Tempering** of barley is a process in which the grain is soaked for 12 hours and then rolled to produce flakes. Because of its high water content, tempered barley cannot be stored for more than a few days. The advantage of cold tempering is that it does not require specialized and expensive equipment, whereas steam rolling does (Hinman and Sorensen, 1994).

Oats

Oat (*Avena sativa*) is a minor grain and is likely to become even less important as a feedstuff for animals. The importance of grains as sources of energy rather than protein is obvious when considering oats. Oats have the highest protein content and quality of any cereal grain. They are of low importance simply because the yield of energy per acre is much lower than for other grains. For a grain farmer to find it economical to produce oats, the price per ton must be sufficiently higher than for other grains to compensate for the lower yield; otherwise, there is no economic incentive to grow oats. Because of their lower energy content than other grains, there is no valid reason for livestock producers to pay a premium for oats. Thus, they will inevitably decline in importance as livestock feed but, as discussed later, may become more important in human nutrition. In short-season areas where corn cannot be produced, oats and barley may continue to be important crops. Oat production has some value in **crop rotations** of grains. A fungal disease of wheat, "take all," increases in severity when wheat is grown in continuous monoculture. Oats are the only cereal grain resistant to the disease, and when grown in rotation with wheat help to break the "take all" growth cycle. Oats are the only cereal grain that contain saponins, which are natural antifungal agents, causing oats to be resistant to "take all."

Oat production has been a victim of changing times. Oats had their heyday when horses were the main source of mechanical power. Oat grain is an ideal feed for working horses. It contains sufficient energy to meet the demands of hard work; it is very palatable, and the high fiber content reduces the likelihood of "overeating" problems such as colic and founder. In the days of the binder-reaper, straw was harvested from the field and deposited in large piles when the grain was threshed. All farms had livestock (at least horses for power) so the straw was useful for feed and bedding. **Oat straw** is softer and more palatable than other cereal straws and is more suitable for these purposes. The need for straw and grain for horses created a need for a considerable acreage of oats. Farm mechanization and the development of high-yielding hybrid corn doomed oat production.

Oats will continue to be produced for **specialty purposes.** These include the production of breakfast cereal and other food products, where the high palatability of oats is important. Oats may become more important in human nutrition because of the increasing emphasis on fiber in the Western diet. Oats are a good source of fiber, and oat bran, containing **soluble fiber** (β-glucans), has favorable effects on lowering serum cholesterol. For livestock, there is a place for oats in **specialty feeds,** where economics are not as critical as in production situations. For example, there is a considerable light-horse industry; oats are an ideal grain for horses because of their comparatively low energy content. They are useful in creep and starter diets for young animals because of their high palatability and low energy; they are less likely than other grains to cause diseases of overeating such as enterotoxemia, acidosis, colic, and founder. Oats are also produced for forage, and their use for this purpose will be discussed in Chapter 5.

Wild oats are a major weed in wheat and other cereal grains. Because wild oats have a black hull, they must be removed by seed cleaning before the grain can be milled into flour. Thus, considerable quantities of wild oats are available for feed. With processing to remove the fibrous hull, wild oats can be an acceptable replacement for other grains in swine diets (Thacker and Sosulski, 1994).

Nutrient Content

1. Energy Oats are a low-energy grain because of their low starch and high fiber content. They are unusual in having a low-energy content but the highest oil content of any grain, containing 5 percent oil or more. The soft physical nature of the hull and the oil content contribute to the high palatability of oats.

Oats can be processed in various ways to improve their utilization. They are often rolled or crimped. Steam rolling increases their feeding value for cattle (Zinn, 1993). The hull can be mechanically separated, in which case the remaining grain is referred to as **oat groats.** These are a high-quality feed but generally too expensive for routine use. There are also hull-less (naked) varieties of oats.

Hull-less oats are similar to corn in feeding value (Brand and van der Merwe, 1996). A high-fat (9 to 10 percent oil) cultivar of oats has been developed in Canada (Devlin et al., 1977); selection for high fat content is a means of increasing the energy content of oats and perhaps enhancing their potential as a feed grain.

2. Protein Oats have the highest-quality protein of any cereal grain and often have a high protein content as well. They have the lowest prolamine (avenin) content of any grain (Table 2–1). Lysine is the first limiting amino acid. The high protein quality of oats may be of most significance in human nutrition.

3. Other Nutrients According to Frolich and Nyman (1988), oat grain has a higher mineral content than other grains. The **soluble fiber** fraction (β-glucans) has a high mineral content. The β-glucans in oats are of interest in human nutrition because they are a soluble type of fiber (Shinnick et al., 1988). The soluble fiber from oats (e.g., in oat bran) is effective in reducing serum cholesterol levels by stimulating fecal cholesterol excretion and reducing enterohepatic cholesterol recirculation, whereas sources of insoluble fiber (e.g., wheat bran) do not have this effect. On the other hand, insoluble fiber appears to reduce colorectal cancer incidence in humans, whereas soluble fiber is ineffective (Burkitt, 1988).

Rye

Rye (*Secale cereale*) has been an important world crop. It is the only grain besides wheat and triticale that can be used to make leavened bread. Rye bread is dark and heavy and has fallen into disfavor with consumers. World rye production has fallen dramatically in the past few decades (by 24 percent in the 1970s alone).

Rye has a number of positive attributes. It is the grain most tolerant of adverse growing conditions; for example, it is the most cold-tolerant and winter-hardy grain and will grow on infertile acidic soils. At high altitudes and in northern latitudes, where the growing season is short, rye is often the grain of choice. On the other hand, it has a number of negative characteristics. Rye has never given as favorable results in animal feeding as its nutrient composition suggests that it should. Numerous deleterious factors have been suggested as being responsible for this, including alkyl resorcinols, ergot, pectins, pentosans, and water-soluble glucanlike gums (Honeyfield et al., 1983). None of these factors has conclusively been shown to be the responsible factor; rather, they all may be, under various circumstances.

Alkyl resorcinols are phenolic compounds found in most cereal grains, with the highest concentrations in rye and triticale (Sedlet et al., 1984). They are concentrated in the pericarp, so they tend to be in highest concentration in bran and other milling by-products. Sedlet et al. (1984) found that alkyl resorcinols depress feed intake, by mechanisms other than a palatability effect.

Rye is well known for its susceptibility to **ergot** infection. Ergot alkaloids are produced by *Claviceps* fungi, which infect grains and grasses. The fungal spores infect the developing flower and form a large black-purple body in place of the grain kernel (Fig. 2.6). Rye and triticale are more susceptible to ergot infection than other grains because they are cross-pollinated, requiring a longer pollination period than self-pollinated grains; during that time the flowers are susceptible to ergot infection. **Ergotism** was at one time a major public health problem in

Europe because of ergot-contaminated rye bread. The problem declined when wheat replaced rye as the major grain crop. Ergot causes a number of problems in livestock and poultry, including gangrene of the extremities (hooves, legs, tail, ear tips, comb, etc.), reduced weight gains and feed efficiency, abortion and agalactia (lack of milk production) in swine, and convulsions and other neurological effects. Levels as low as 0.1 percent ergot in feeds may adversely affect livestock performance.

Pectins are a group of polysaccharides that function as intercellular cement in plant tissues. They are viscous, gummy compounds. Rye contains a higher pectin content (up to 8 percent) than other grains; some evidence suggests that the poor growth and sticky droppings of chicks fed rye are due to the pectins. By increasing the viscosity of intestinal contents, pectins and glucans may impair nutrient absorption, promoting digestive disturbances, abnormal microbial growth, and diarrhea. Rye also contains **pentosans** that contribute to these problems. Commercial **feed enzymes** (see Chapter 8) with pentosanase activity improve the utilization of rye-based diets by poultry (Campbell and Bedford, 1992; Boros et al., 1995).

Pigs are less sensitive than chickens to the growth-depressing properties of rye (Honeyfield et al., 1983). Subtherapeutic levels of antibiotics in the diet prevent the growth-depressing effects of rye in poultry but not in swine. Proudfoot and Hulan (1986) evaluated the effects of feeding rye to laying hens. Egg production was depressed with all rye levels above 10 percent of the diet. In contrast, Campbell and Campbell (1989) observed reduced egg production only at rye levels above 40 percent. Reduced feed intake may account for the adverse effects at higher levels.

The protein content of rye is similar to that of barley. Rye protein has a higher lysine content than wheat and barley, giving it a higher biological value.

Triticale

Triticale is a hybrid, or synthetic, grain produced by crossing durum wheat and rye. Its name is derived from parts of *Triticum* (wheat) and *Secale* (rye). Triticale has been of interest as a feed grain because of reports of a high protein content and improved amino acid balance as compared to other grains. These virtues have not always been apparent. The objectives of triticale development have been to combine the winter hardiness of rye with the high yields of wheat and the high protein content of wheat with the high lysine of rye.

As with rye, triticale has never quite lived up to expectations. In spite of favorable protein and energy contents, triticale has usually given poorer results than other grains in feeding trials. Factors suggested as being responsible include trypsin and chymotrypsin inhibitors (Erickson et al., 1979), low palatability (Shimada et al., 1974), alkyl resorcinols (Radcliffe et al., 1981), water-soluble pentosans (Rundgren, 1988), and ergot. Triticale genotypes differ in the proportions of wheat and rye genes. Selections with low proportions of the rye genome have superior feeding value (Boros, 1999).

Trypsin and chymotrypsin are important pancreatic enzymes functioning in protein digestion. **Trypsin** (protease) **inhibitors** complex with these enzymes, eliminating their ability to digest protein. Triticale has trypsin inhibitor activity; thus its protein digestibility is reduced. Plant-breeding programs to reduce trypsin inhibitor activity of triticale are in progress. The occurrence of trypsin in-

FIGURE 2.6 Infection of grain with ergot. The ergot bodies contain alkaloids that are toxic to humans and livestock.

hibitors in grains has been reviewed by Sosulski et al. (1988). Trypsin inhibitors are concentrated in cereal embryos, but also constitute 5 to 10 percent of the water-soluble proteins in rye, barley, and wheat. They are stable to heat, but those in grains have only minor effects on protein digestibility, with the exception of those in triticale.

The poor **palatability** of triticale (Brand et al., 1995) may be partially caused by ergot, but even ergot-free triticale gives evidence of reduced palatability, suggesting the involvement of other factors. Coffey and Gerrits (1988) reported research with swine showing that a low trypsin inhibitor, ergot-free cultivar of triticale (B858) gave growth performance equivalent to a corn-soy diet when used in place of corn in swine starter, grower, and finisher diets. The triticale contained 15.9 percent crude protein and 0.46 percent methionine. There was no response to lysine and methionine supplementation, but the amount of supplementary protein needed was reduced.

Rice

Rice (*Oryza sativa*) is a major food grain for millions of people in the tropics, particularly in Asia. It is grown almost exclusively as a human food. Occasionally, small amounts of the unprocessed grain (rough or paddy rice) are available for feeding purposes. Rough rice contains about 8 percent crude protein, 9 percent crude fiber, 1.9 percent ether extract, and 6.5 percent ash. The unprocessed grain contains about 25 percent of its weight as hulls. **Rice hulls** are high in silica and are abrasive to both feed mill equipment and the digestive tract. The hulls are almost totally indigestible. Allowing for the reduction in energy content because of the hulls, the grain is similar to corn in nutritional value. Processing rice for human consumption produces large quantities of rice bran, which is used as a feedstuff for livestock.

Wild rice (*Zizania aquatica*) is an aquatic annual grass native to North America. It is not a member of the rice family but is so named because it grows in water. Traditionally it has been harvested from shallow lakes by Native Americans using canoes. It has now been commercialized and is grown in flooded paddies that are drained for mechanical harvest with combines. Wild rice is an expensive, gourmet food item produced in very limited quantities.

OTHER GRAINS

Teff

Teff (*Eragrostis abyssinica*) is an important cereal grain in Ethiopia. Like other cereal grains, it was developed from a wild grass. Teff is adapted to dry areas with a short rainy season. The straw is used as a livestock feed.

Buckwheat

Buckwheat (*Fagopyrum esculentum*) is not a cereal grain because it is not a member of the grass family. It is a broad-leaved plant. Buckwheat is used primarily in North America for pancake flour; it is also grown to some extent for export to Japan. Cull buckwheat is used as livestock feed. The production and utilization of buckwheat was reviewed by Pomeranz (1983).

Buckwheat is lower yielding than other grains, but it does have some useful attributes. It reaches maturity very rapidly, and so can be sown late in the season. It is used as a **"catch crop"** that is planted to salvage a crop if an earlier sown crop has failed. It is more productive than other grains on infertile soils. It does not respond with a yield increase to fertilization, so its production costs are lower. It is often grown as a cover crop, or green manure crop for plowing under. Because of its rapid growth, it can be used as a smother crop to control weeds. Buckwheat blooms within 4 to 6 weeks after planting. The seeds are dark brown or black and pyramidal in shape. The grain contains approximately 11 to 13 percent crude protein; the protein is of good biological value. Buckwheat has a higher lysine content than other grains, averaging approximately 0.6 percent, as compared to 0.3 to 0.45 percent for wheat and barley and 0.18 percent for corn. The grain has a fairly high fiber content, which reduces its digestible energy value. Thacker et al. (1983b) reported a high nutritive value of **common buckwheat.** When various cereal grains were fed as the sole diet (with a mineral and vitamin supplement) to rats or in combination with buckwheat, growth was superior in each case with the buckwheat-grain mixture (Thacker et al., 1983a), suggesting that the use of buckwheat in cereal, grain-based diets could reduce the amount of protein supplement needed. Anderson and Bowland (1984) evaluated buckwheat in diets for growing pigs, either as the sole grain, or with 25, 50, 75, or 100 percent replacement of the buckwheat by either wheat or barley. There were no differences in performance, indicating that buckwheat can be used as a total replacement for wheat or barley in diets for growing pigs.

Tartary buckwheat (*F. tataricum*) differs from common buckwheat in its growth habit and has small round seeds. Tartary buckwheat is especially tolerant of poor acid soils, is more frost resistant, and yields more under adverse conditions than does common buckwheat (Nicholson et al., 1976). Tartary buckwheat has a lower protein content, less favorable amino acid balance, higher fiber content, and lower feeding value than common buckwheat (Thacker et al., 1983b). Nicholson et al. (1976) found that tartary buckwheat was inferior to barley in digestibility and in supporting growth of steers; it had approximately 85 percent of the digestible energy content of barley. Pelleted whole buckwheat was more palatable than the dry rolled form.

Buckwheat contains a compound, **fagopyrin,** which causes **photosensitization** of light-skinned animals. Photosensitizing agents such as fagopyrin are absorbed and react with ultraviolet (uv) light (sunlight) to produce severe skin lesions (Fig. 2.7). Dark melanin pigments block the uv light, so it is only areas with light skin and hair that are affected. Animals fed buckwheat may develop photosensitization and should be housed without exposure to sunlight if possible. Buckwheat is also used as a forage crop; dermatitis of light-skinned animals grazing buckwheat pasture may occur (Mulholland and Coombe, 1979).

Wild buckwheat (*F. convolvulus*) is a common weed of grain fields and the seeds may be a contaminant of grains. Harrold et al. (1980) determined the feeding

FIGURE 2.7 A sheep with photosensitization showing lesions around the eyes and nostrils. This sheep has facial eczema (see Chapter 5). Some plants, such as buckwheat, contain compounds which, when absorbed, react with sunlight at the surface of light-colored skin and cause severe dermatitis. (Courtesy of H. H. Meyer.)

value of wild buckwheat seeds for rats. Growth rate of rats fed wild buckwheat was approximately 66 percent of that of controls; digestibility of dry matter and kilocalories was 49 percent. Thus, wild buckwheat seeds appear to be of quite low feeding value.

Amaranth

Amaranthus spp. have been cultivated as grains since antiquity in Latin America. The use of the grain in religious ceremonies by the Incas of South America apparently led to partially successful efforts by the Spanish conquistadors to eliminate amaranth production. It has not been an important crop for hundreds of years. **Amaranth** is a vigorous, broad-leaved plant. A wild species, *A. retroflexus,* is commonly known as **red-root pigweed.** Renewed interest in amaranth production has occurred with the recognition that the seed has a very high lysine content and a high protein content. The calcium content is also much higher than for cereal grains. The amaranth seeds are very small, oval, and have a nutty taste. In spite of its excellent amino acid composition, feeding trials with laboratory animals, swine, and poultry have usually given poor results. The raw seeds are toxic, causing feed refusal, liver damage, neurological effects (ataxia) in poultry, and heart failure in swine. Heat treatment of the seed partially overcomes the toxic effects (Tillman and Waldroup, 1987). Boiling the seeds totally inactivates toxins (Fadel et al., 1996). Laovoravit et al. (1986) noted that chicks fed amaranth-containing diets showed convulsive symptoms typical of **thiamin deficiency.** The birds responded dramatically to injected thiamin. The seed contained only about 1 milligram thiamin per kilogram. Tillman and Waldroup (1988) and Ravindran et al. (1996) found that up to 40 percent heat-treated (extruded) amaranth seed could be fed to broilers with satisfactory results.

Because of its very favorable protein and amino acid composition, amaranth may have potential as a feed grain. Plant breeding programs, such as at the Rodale Research Center in Pennsylvania, show progress in improving agronomic features of the crop. Identification of the toxic principles and selection against them is needed if amaranth is to become a useful feed grain. It also has potential as a forage crop (Pond and Lehmann, 1989).

Chenopodium quinoa (quinua, quinoa) is a "pseudocereal" grain somewhat similar to amaranth in several respects. It also has a long history of food use in South America and has a common weed as a relative (lamb's quarters, *C. album*). As with amaranth, there are efforts to promote **quinoa** as a crop in North America. It contains a number of deleterious factors, including saponins and oxalates. Dehulling quinoa seed removes most of the saponins (Gee et al., 1993). Levels up to 15 percent dehulled quinoa seed can be used satisfactorily in broiler diets (Jacobsen et al., 1997). "Sweet" varieties of quinoa with low saponin content have been developed (Gee et al., 1993). **Saponins** are bitter compounds that irritate mucous membranes of the mouth and digestive tract.

Sprouted Grains

Sprouted grains are of interest for two reasons. Inclement harvesting conditions often lead to preharvest germination of grains (sprouting), rendering them unfit for commercial purposes. Interest then arises in use of these grains as feed. For example, more than one million tons of sprouted wheat were available in a wet season in Washington and Idaho (Rule et al., 1986). Secondly, there is periodic interest in and promotion of hydroponic units for sprouting grain for animal feed. Claims are often made that the process can produce huge amounts of feed in a small area at low cost.

During sprouting, the germinating seed metabolizes starch, which produces heat and CO_2 as waste products, thus reducing the amount of energy in the grain. Sprouting uses up the starch. Therefore the protein, fiber, and other remaining nutrients make up a greater percentage of the total. The percentage of protein in sprouted grain is therefore higher than for the unsprouted material. Thus, sprouting reduces the amount of total grain available and increases the concentration (but not the amount) of other nutrients. The vitamin content, especially the carotene, increases with sprouting. However, vitamins can be purchased at a lower cost than the value of the weight of the grain lost by sprouting. Thus the **hydroponic sprouting** of grains appears not to be a useful process in terms of nutritional value; the germinating seed wastes energy that the animals being fed the grain would otherwise use. Mycotoxins may be produced by molds growing on sprouting grains (Capper, 1988).

CEREAL MILLING BY-PRODUCTS

For human consumption, cereal grains are processed in various ways that result in the production of by-products that are usually disposed of by being used as livestock feeds. The initial step in grain processing is cleaning the seed. This is done by passing the grain over shaking screens to sieve out dirt, dust, weed seeds, and small or broken grains. An air stream may be used to blow off chaff and other low-density material. The **screenings** consist of a mixture of these materials. Their feeding value will depend on the relative proportions of dust or other inert material compared to grain fragments and weed seeds. Some weed seeds can be good sources of nutrients, whereas others may be toxic or unpalatable. At one time, contamination of grains with toxic weed seeds was a significant problem. Because contaminants are concentrated in the screenings, they are more toxic than the uncleaned grain. Most grain now is grown with the use of herbicides to control broad-leaved weeds, so toxicities associated with poisonous weed seed in the grain or screenings are infrequent.

In commercial channels in the United States, the content of noxious or toxic **weed seeds** in screenings is regulated. Screenings must not contain any seeds or other material in amounts either injurious to livestock or imparting any objectionable odor or flavor to their milk or meat. The screenings must not contain more than 4 whole prohibited noxious weed seeds per pound and must not contain more than 100 whole restricted noxious weed seeds per pound. The prohibited and restricted weed seeds are those identified by the state or province in which the screenings are sold or used. The screenings must be identified as either grain screenings, mixed screenings, or chaff and/or dust. Grain screenings must consist of 70 percent or more grain, including light and broken grains, wild buckwheat, and wild oats, and not more than 6.5 percent ash (an indicator of the presence of dust and dirt). Mixed screenings must contain not more than 27 percent crude fiber and not more than 15 percent ash. Screenings with a high content of weed seeds can be ground prior to use as a feed. Because of their variable fiber content, screenings are more useful in ruminant diets than as swine or poultry feedstuffs.

A number of **weed seeds** (e.g., wild oats) are reasonably good feedstuffs, whereas others (e.g., redroot pigweed and wild buckwheat) are unpalatable and of low value. Quite extensive data on the composition of grain screenings and weed seeds and their feeding value for livestock have been reported by Beames et al. (1986), Harrold et al. (1980), and Tait et al. (1986).

Grain dust accumulates in elevators and other grain handling facilities; it consists of very fine particulate matter of grain and soil contamination. It may be used as feed, preferably after pelleting to reduce the dustiness. Australian workers (Hogan et al., 1980) found that alkali treatment (3 percent sodium hydroxide) of wheat dust increased its digestibility and also degraded organophosphate insecticides present in the product.

Milling of grains involves processing them to produce products suitable for human consumption or industrial purposes. **Dry milling** is used with fibrous grains (barley, rice, oats) to remove the hull, and with wheat to produce flour. In flour milling, the outer layer of the seed, which contains fiber, is removed to produce a flour composed mainly of the starchy endosperm. **Wet milling,** that is, mixing the ground grain with water, is used primarily with corn to produce corn starch, sugar, syrup, and oil.

Wheat Milling By-Products

In wheat milling, the endosperm is separated from other fractions of the seed. The endosperm consists mainly of starch and gluten. The outermost layer of the seed is highest in fiber and, when removed, constitutes most of the bran fraction. The aleurone layer is just beneath the bran layer in the seed and consists of fiber and protein. It is a layer of thick-walled cells, containing the enzymes that digest starch in the sprouting seed. The wheat germ is at the base of the seed and is high in fat, vitamins, and protein. The germ contains the embryonic wheat plant, whereas the endosperm is a starch reserve to provide energy for the germinating seed.

Wheat milling by-products are classified according to their protein and fiber contents into the following categories.

By-Product	*Minimum Crude Protein (%)*	*Maximum Crude Fiber (%)*
Bran	13.5–15	12.0
Middlings	10–14	9.5
Mill run	14–16	9.5
Shorts	14–16	7
Red dog	13.5–15	4

Wheat bran consists of the outermost layer of the seed along with some flour. It is a flaky, reddish-brown material. Wheat bran is quite palatable and is well known for its ability to prevent constipation because of its swelling and water-holding capacities. This is important in certain situations, such as with sows confined in farrowing crates. Bran has a high capacity to absorb water and swell because of its fiber and nonstarch carbohydrates (e.g., glucans), and so it has a bulk effect in the colon, giving it laxative properties. Bran has an amino acid balance superior to that of whole wheat. A notable characteristic of bran and other milling by-products is that the phosphorus content is very high and the calcium low. This results in an unfavorable calcium:phosphorus ratio that, if uncorrected with proper calcium supplementation, can cause adverse effects on animals. A low-calcium:high-phosphorus ratio in the diet can provoke a condition called **nutritional secondary hyperparathyroidism.** The high blood phosphorus levels stimulate the parathyroid gland to increase its secretion of **parathyroid hormone (PTH).** Parathyroid hormone stimulates the excretion of the excess phosphorus by the kidney. However, it also stimulates the mobilization of calcium from bone, since the usual demand for PTH is to mobilize bone calcium when serum calcium is low. The excessive mobilization of bone calcium results in the bones becoming demineralized and fibrotic. In horses, the frontal bones of the face are particularly affected. The demineralized, fibrotic bone enlarges and is referred to as **"big head"** in horses. In the past, it was often called **"miller's disease."** When farmers took their grain to the flour mill, they would haul it in horse-drawn wagons. As the grain was being unloaded, the horses were fed bran or other by-products at the mill. If they were used on a regular basis for this purpose, they developed nutritional secondary hyperparathyroidism, or miller's disease. (*Hyper* means excessive, and *secondary* refers to the condition being induced by the diet rather than being a primary malfunction of the gland.)

Because of their bulky nature and high fiber content, bran and other millfeeds are not usually fed to swine and poultry, except in gestation-lactation diets for sows to prevent constipation. They are most suitable for ruminants and nonruminant herbivores like horses and rabbits. For these species, the low energy content helps to prevent carbohydrate overload conditions such as founder and colic in horses and enterotoxemia in rabbits.

Bran and other millfeeds are good sources of the water-soluble vitamins except niacin, which is in an unavailable form. Much of the phosphorus exists as **phytate,** which is of low bioavailability to swine and poultry. **Wheat middlings** (midds) are similar to bran except that they have lower fiber and higher flour contents, so they are higher in digestible energy than bran. For cattle, wheat middlings can be regarded more as a roughage source than a grain. It can be substituted for alfalfa in feedlot cattle diets without affecting performance, but it is not suitable as a substitute for corn (Dalke et al., 1997). In England and Australia, the midds are called *pollard.* **Wheat mill run** is a blend of bran and midds. Wheat shorts are lower in fiber and contain a higher flour content. **Red dog,** of which only small quantities are available, contains more flour than other millfeeds.

Processing procedures can increase the digestibility of mill feeds. Pelleting wheat bran ruptures the aleurone-layer cells and improves digestibility. Treatment with cellulolytic enzymes also improves bran utilization. There is a high degree of variability in millfeeds. They are usually marketed as bran and wheat mill run (or middlings) and contain all of the other fractions. In other words, most mills do not produce red dog or shorts separately but combine these fractions and market either midds or mill run. The bran is usually marketed as a separate entity.

Corn Mill Feeds

When corn is dry milled, the major products are cornmeal, hominy, grits, and flour. These are all used for human consumption. The major corn dry milling by-product is **hominy feed,** consisting of the bran, germ, and some of the flour. It is higher in protein and fiber than corn grain. Leeson et al. (1988) evaluated hominy and grits as poultry feed. Regular hominy is a by-product of the dry milling process that includes fat extraction. Leeson et al. (1988) found that regular hominy was inferior to corn in supporting growth and feed of broilers but was adequate for turkeys. A high-fat hominy from which the oil had not been extracted was equivalent to corn in feeding value.

Corn gluten feed and **corn gluten meal** are produced as by-products of the wet milling of corn for starch or syrup production. Corn gluten meal has approximately twice the protein content of gluten feed. These products will be further discussed in Chapter 4.

Rice Bran

In areas where rice is produced, rice bran is a major by-product. It consists of the fibrous outer layer of the grain, some hull, chipped grain, and calcium carbonate, which is added during the milling process. The feeding value of rice bran has been reviewed by Farrell (1994) and Warren and Farrell (1990). Rice bran is a major feedstuff in tropical countries, where, in many cases, cereal grains per se are too expensive to be employed in animal feeding. In such countries (e.g., Indonesia and Thailand), rice bran is a major ingredient of poultry and swine diets. Rice bran has several notable features as a feed ingredient. It has a high oil content (13 percent) that increases its energy value to the level of that of the grain itself. It is a good source of B vitamins, protein, and amino acids. However, it also contains factors that promote **rancidity,** especially under the warm, humid climatic conditions that favor autooxidation. These include lipoxidases, which are enzymes that promote oxidation of unsaturated fatty acids. Rancid feeds are unpalatable and potentially toxic. Rice bran has a high phytate content, and zinc deficiency signs (**parakeratosis**) may occur in swine fed rice-bran-containing diets unless a higher than normal level of supplementary zinc is used. Rice bran contains variable quantities of hulls. **Rice hulls** are high in silica, which makes them very abrasive. Because of their high silica content, they have extremely low digestibility and are virtually inert. They increase the wear on feed mill equipment and may cause damage to the gut lining. Inflammation of the mucosa and bleeding may occur, with consequent adverse effects on animal performance. Rancidity in conjunction with hulls may intensify gastrointestinal disturbances. Rice hulls find limited application as a source of indigestible fiber, for example in rabbit diets. Heat treatment of rice bran may improve its utilization by swine and poultry by inactivating lipoxidases and trypsin inhibitors.

Rice bran has a high fiber content composed of hemicelluloses containing highly branched arabinoxylans (pentosans) (Annison et al., 1995). Addition of pentosanases to diets containing rice bran improves its utilization by poultry (Wang et al., 1997).

Other Grain Milling By-Products

The milling of other grains such as barley, sorghum, rye, and oats produces similar types of by-products. With barley and oats, the hulls are removed when the grains are used for human consumption, producing pearled barley and oat groats. Oat and barley hulls may be used as feed; oat hulls are more palatable and find application in situations where a fiber source is needed.

In general, grain by-products are most useful for ruminants and nonruminant herbivores. Although they can give good results with swine and poultry, the cost per kilocalorie of digestible energy in the developed countries is usually cheaper from feed grains (e.g., corn). In developing countries, where grains are used to a greater extent for direct human consumption, grain by-products are major energy sources for swine and poultry production.

GRAIN OVERLOAD OF RUMINANTS

Ruminants fed diets high in grain are subject to a number of nutritional or metabolic problems, including acidosis, rumenitis, laminitis, displaced abomasum, liver abscesses, polioencephalomalacia, and enterotoxemia. Many of these conditions are secondary to acute acidosis.

Acidosis

Lactic acidosis occurs when there is an abrupt increase in the intake of readily fermentable carbohydrates. A typical example would occur when range- or pasture-fed cattle, which may never have been fed grain in their lives, are brought into a feedlot and fed grain. If the amount of grain fed exceeds the ability of the rumen microbes to make an orderly change from a primarily cellulolytic microflora to a starch-digesting (amylolytic) microflora, a disturbance in the rumen flora develops. When ruminants are switched abruptly from a cellulose-based to a starch-based diet, there is an initial shortage of amylolytic organisms. This void is quickly occupied by fast-growing amylolytic bacteria, particularly *Streptococcus bovis*, which has a generation interval of only a few minutes. *S. bovis* proliferates and becomes the dominant organism. It produces lactic acid as a fermentation end product, particularly D-lactate, which is poorly absorbed and metabolized. Lactate-producing *Lactobacillus* spp. also proliferate. *S. bovis* increases the proportion of lactic acid in its fermentation products as rumen pH drops, causing a spiraling effect of ever-increasing lactate and lower pH (Russell and Hino, 1985). The accumulation of lactic acid in the rumen and blood causes rumen acidosis and metabolic acidosis. **Lactic acid** is a stronger acid than the principal VFA (pKa of 3.0 vs. 4.8 for acetic acid) and causes a severe drop in rumen pH to values as low as 4.0. This has a corrosive effect on the rumen wall, causing the papillae to peel off **(parakeratosis in ruminants).** Absorption is impaired, and bacteria invade the rumen wall and gain systemic entrance. A high incidence of liver abscesses may result. The low rumen pH inhibits the growth of all organisms, including *S. bovis*, leaving the rumen as a stagnant vat of strong acid. Recovery at this stage is unlikely. Chronic **liver abscess,** associated with feeding high-energy diets, reduces growth rate, feed efficiency, and carcass dressing percentage. The incidence of liver abscess in American feedlot cattle is typically 12–32 percent (Nagaraja and Chengappa, 1998). It has a major economic impact because of carcass condemnation and reduced animal performance. Ruminal lesions from lactic acidosis are the main predisposing factor leading to invasion by bacteria. *Fusobacterium necrophorum* is the principal bacteria involved (Nagaraja and Chengappa, 1998).

The accumulation of lactic acid in the rumen increases the osmolality of the rumen contents, drawing water from the blood and causing dehydration and hemoconcentration. Absorbed acid may cause systemic acidosis with a lowered blood pH, electrolyte imbalance, and kidney failure. Hemoconcentration may cause arterioles in the extremities to rupture, causing **laminitis,** or **founder.** This condition is characterized by sore, tender feet, unnatural stance, and abnormal overgrowth of hoof tissue. Dehydration and impaired kidney function may cause secondary neurological syndromes such as lack of coordination, depression, and coma. Surviving animals may suffer a long period of unthriftiness. Rumen stasis and altered microbial populations may result in polioencephalomalacia, or induced thiamin deficiency. Organisms such as *Clostridium sporogenes*, which produces thiaminase, may proliferate.

Acute lactic acidosis causes dramatic increases in rumen acidity and osmolality, causing severe rumenitis, decreased blood pH, and fatal dehydration (Owens et al., 1998). **Chronic acidosis** reduces feed intake and animal performance in feedlot cattle (Chapter 15) and dairy cattle (Chapter 17).

Acidosis is not readily treated but is best prevented by making the change gradually from a roughage-based to a concentrate-based diet. Feedlots use **"warm-up rations,"** so that incoming cattle receive gradual increases in the proportion of grain in the diet. Grains highest in starch, such as corn and wheat, are most likely to cause problems. Because of the rapid fermentation of its starch, wheat is probably the grain most likely to cause acidosis. **Processing** methods such as steam-flaking that increase starch fermentation rates also increase the potential for acidosis. Feeding **buffers** may help to buffer the lactic acid, but at the same time may prolong the period of proliferation of *S. bovis*. The **ionophores** rumensin and lasalocid, when administered to cattle for 7 days before switching to high-grain diets, aid in preventing acidosis by inhibiting *S. bovis* and *Lactobacillus* spp. (Nagaraja et al., 1981). Some rumen bacteria such as *Megasphaera elsedenii* use lactate as a substrate. It may be possible to administer preparations of lactate-utilizing bacteria as a direct-fed microbial product to alleviate lactic acidosis and to speed up the adaptation to high-concentrate diets (Kung and Hession, 1995). **Rumen protozoa** seem to have protective effects against acidosis by engulfing starch granules and reducing the rate of starch fermentation, particularly during the "warm-up" period (Goad et al., 1998; Owens et al., 1998).

The rumen can be "primed" for high concentrate diets by exposure to lactate to promote populations of **lactate-metabolizing microbes** (Huntington and Britton, 1979). Condensed distillers by-products contain organic acids (fumaric, lactic) that can promote growth of lactate-utilizing rumen bacteria when distillers by-products are included in feedlot warm-up diets (Fron et al., 1996).

Lactic acidosis also occurs in **horses** fed high-grain diets (Garner et al., 1977). These investigators developed a protocol for inducing laminitis in horses by dosing with corn starch to produce carbohydrate overload of the hindgut. Signs of acidosis are similar to those of ruminants, with increased blood lactate, hemoconcentration and dehydration, circulatory acidosis, and laminitis. Laminitis in horses may also occur with the ingestion of lush grass or clover containing a high content of readily fermentable carbohydrates (see Chapter 18).

Feeding the antibiotic **virginiamycin** helps to reduce or prevent the development of grain overload acidosis in ruminants (Godfrey et al., 1995) and lactic acidosis and laminitis in horses (Rowe et al., 1994). The mode of action is the suppression of *S. bovis* and *Lactobacillus* spp., the major lactate-producing microbes in both the rumen and the hindgut. Virginiamycin inhibits the growth of these Gram-positive bacteria by penetrating the bacterial cell wall and blocking protein synthesis (Rogers et al., 1995).

Other Deleterious Factors

Enterotoxemia in sheep and goats (overeating disease, pulpy kidney) is another disorder associated with grain overload. An abundance of fermentable carbohydrate may promote proliferation of *Clostridium perfringens* in the intestine. This organism produces extremely potent neurotoxins that are absorbed and kill the affected animal. Vaccines containing an antitoxin against enterotoxemia are available for vaccination. Overeating of grain may induce *C. perfringens* enterotoxemia in horses in conjunction with enteritis (Carroll et al., 1987). Enterotoxemia is caused by clostridial toxins, produced when *Clostridia* sp. proliferate in the hindgut when a rich substrate of readily fermentable carbohydrate is present. Enterotoxemia is particularly a problem with lambs.

Feedlot bloat is associated with the feeding of high-grain, low-roughage diets to cattle. Feedlot bloat may be of the free-gas type rather than the frothy bloat associated with lush pasture legumes. Free-gas bloat may be a result of excessive consumption of dense feed, which depresses the cardia below the level of the rumen fluid level, thereby causing the accumulation of gas. However, about 90 percent of feedlot bloat cases are of the frothy type (Cheng et al., 1998). Prevention of feedlot bloat is best accomplished by provision of a small amount of coarse roughage to cattle on high-energy diets and by avoiding excessively fine grinding of grains, which causes an excessive rate of fermentation. Lactic acid-producing bacteria such as *S. bovis* have been implicated in the etiology of frothy feedlot bloat; monensin and poloxalene are effective as feed additives in preventing this type of feedlot bloat (Bartley et al., 1983).

An induced thiamin (vitamin B_1) deficiency, called **polioencephalomalacia** (PEM) in the United States and *cerebrocortical necrosis* in Europe, can occur in cattle fed high-concentrate diets. The condition is apparently caused by the proliferation of microbes (e.g., *Clostridium spirogenes*) that produce thiaminase, a thiamin-degrading enzyme. Avoiding sudden exposure to high-grain diets and administration of thiamin to affected animals are recommended control measures. High sulfur intakes from feed or water can also induce PEM (Gould, 1998).

QUESTIONS AND STUDY GUIDE

1. Why is corn (*Zea mays*) more productive than other cereal grain crops?
2. Why is corn considered the best feed grain for livestock production?
3. How does the prolamine content of corn influence its feeding value?
4. Using typical yields of corn and soybeans in your area, calculate the potential production of digestible energy and crude protein for swine feeding in the case of a farmer with 1,000 acres, who has the choice of growing (a) 800 acres of hybrid corn and 200 acres of soybeans; (b) 1,000 acres of high-lysine (opaque-2) corn; or (3) 900 acres opaque-2 corn and 100 acres of soybeans. Assume that opaque-2 corn will have 10 percent lower yield than hybrid corn.
5. Why is pellagra not a public health problem in Mexico?
6. Approximately 1 ppb of aflatoxin in the feed will cause liver cancer in rainbow trout. How many grams of aflatoxin would have to be in 100 tons of corn to yield a level of 1 ppb?
7. How would you expect the protein content of drought-stressed, shriveled wheat to compare to that of normal, plump kernels?

8. Oat grain has the highest protein content of any cereal grain, yet oats are a very minor feedstuff. Explain. What developments in human nutrition might affect the importance of oats and barley as crops?
9. Oats are a useful grain for horses. Why? Traditionally, horses have been fed oats and timothy hay. Some horse raisers insist that they must have these feeds, regardless of cost. Can you raise horses successfully without using these feedstuffs?
10. Why is buckwheat considered a grain but not a cereal grain?
11. What is a green manure crop?
12. What is the difference between sunburn and photosensitization?
13. What is the commercial potential for hydroponically sprouted grains as animal feed?
14. What is the cause of parakeratosis in ruminants?
15. What are the consequences, if any, of shifting cattle abruptly from a high-grain diet to a grain-free, high-roughage diet (e.g., pasture)?
16. Why is bran a laxative? For what type of livestock are bran and other milling by-products most suitable? How can these products be processed to improve their feeding value?
17. How does rice bran compare to wheat bran in composition and feeding value?
18. If a horse breaks into a feedroom and consumes a large amount of grain, it may in a few hours show signs of abdominal and hoof pain. Its hooves may be very hot and tender. Explain.

REFERENCES

General

National Research Council. 1988. *Nutrient Requirements of Swine*. Washington, DC: National Academy Press.

Scott, M. L., M. C. Nesheim, and R. J. Young. 1982. *Nutrition of the Chicken.* Ithaca, NY: M. L. Scott and Associates.

Stoskopf, N. C. 1985. *Cereal Grain Crops.* Reston, VA: Reston Publishing Co.

Corn

Adams, M. H., S. E. Watkins, A. L. Waldroup, P. W. Waldroup, and D. L. Fletcher. 1994. Utilization of high-oil corn in diets for broiler chickens. *J. Appl. Poult. Res.* 3:146–156.

Adeola, O. and N. L. Bajjalieh. 1997. Energy concentration of high-oil corn varieties for pigs. *J. Anim. Sci.* 75:430–436.

Beauchemin, K. A., T. A. McAllister, Y. Dong, B. I. Farr, and K.-J. Cheng. 1994. Effect of mastication on digestion of whole cereal grains by cattle. *J. Anim. Sci.* 72:236–246.

Carpenter, K. J., M. Schelstraete, V. C. Vilicich, and J. S. Wall. 1988. Immature corn as a source of niacin for rats. *J. Nutr.* 118:165–169.

Folmer, J. D., R. J. Grant, C. T. Milton, and J. Beck. 2002. Utilization of Bt corn residues by grazing beef steers and Bt corn silage and grain by growing beef cattle and lactating dairy cows. *J. Anim. Sci.* 80:1352–1361.

Goetsch, A. L., F. N. Owens, M. A. Funk, and B. E. Doran. 1987. Effects of whole or ground corn with different forms of hay in 85% concentrate diets on digestion and passage rate in beef heifers. *Anim. Feed Sci. Tech.* 18:151–164.

Graham, G. G., D. V. Glover, G. Lopez de Romana, E. Morales, and W. C. MacLean, Jr. 1980. Nutritional value of normal, opaque-2, and sugary-2 opaque-2 maize hybrids for infants and children. I. Digestibility and utilization. *J. Nutr.* 110:1061–1069.

Huntington, G. B. 1997. Starch utilization by ruminants: From basics to the bunk. *J. Anim. Sci.* 75:852–867.

McAllister, T. A., H. D. Bae, G. A. Jones, and K.-J. Cheng. 1994. Microbial attachment and feed digestion in the rumen. *J. Anim. Sci.* 72:3004–3018.

Moeser, A. J., I. B. Kim, E. van Heugten, and T. A. T. G. Kempen. 2002. The nutritional value of degermed, dehulled corn for pigs and its impact on the gastrointestinal tract and nutrient excretion. *J. Anim. Sci.* 80:2629–2638.

Mosse, J. 1966. Alcohol-soluble proteins of cereal grains. *Fed. Proc.* 25:1663–1669.

National Research Council. 1988. *Quality Protein Maize*. Washington, DC: National Academy Press.

Orskov, E. R. 1986. Starch digestion and utilization in ruminants. *J. Anim. Sci.* 63:1624–1633.

Owens, F. N., R. A. Zinn, and Y. K. Kim. 1986. Limits to starch digestion in the ruminant small intestine. *J. Anim. Sci.* 63:1634–1648.

Pond, W. G. and J. H. Maner. 1984. *Swine Production and Nutrition*. Westport, CT: AVI Publishing.

Sullivan, J. S., D. A. Knabe, A. J. Bockholt, and E. J. Gregg. 1989. Nutritional value of quality protein maize and food corn for starter and grower pigs. *J. Anim. Sci.* 67:1285–1292.

Theurer, C. B. 1986. Grain processing effects on starch utilization by ruminants. *J. Anim. Sci.* 63:1649–1662.

Troyer, A. F. 1999. Background of U. S. hybrid corn. *Crop Sci.* 39:601–626.

Van Soest, P. J. 1994. *Nutritional Ecology of the Ruminant*. Ithaca, NY: Cornell University Press.

Wagoner, P. 1990. Perennial grain development: Past efforts and potential for the future. *Crit. Rev. Plant Sci.* 9:381–408.

Yang, A., T. W. Larsen, and R. K. Tume. 1992. Carotenoid and retinol concentrations in serum, adipose tissue and liver and carotenoid transport in sheep, goats and cattle. *Aust. J. Agr. Res.* 43:1809–1817.

Sorghum and Millet

Elkin, R. G., M. B. Freed, B. R. Hamaker, Y. Zhang, and C. M. Parsons. 1996. Condensed tannins are only partially responsible for variations in nutrient digestibilities of sorghum grain cultivars. *J. Agric. Food Chem.* 44:848–853.

Hill, G. M., G. L. Newton, M. N. Streeter, W. W. Hanna, P. R. Utley, and M. J. Mathias. 1996. Digestibility and utilization of pearl millet diets fed to finishing beef cattle. *J. Anim. Sci.* 74:1728–1735.

Lawrence, B. V., O. Adeola, and J. C. Rogler. 1995. Nutrient digestibility and growth performance of pigs fed pearl millet as a replacement for corn. *J. Anim. Sci.* 73:2026–2032.

Lizardo, R., J. Peiniau, and A. Aumaitre. 1995. Effect of sorghum on performance, digestibility of dietary components and activities of pancreatic and intestinal enzymes in the weaned piglet. *Anim. Feed Sci. Tech.* 56:67–82.

Reinhardt, C. D., R. T. Brandt, Jr., K. C. Behnke, A. S. Freeman, and T. P. Eck. 1997. Effect of steam-flaked sorghum grain density on performance, mill production rate, and subacute acidosis in feedlot steers. *J. Anim. Sci.* 75:2852–2857.

Rooney, L. W. and R. L. Pflugfelder. 1986. Factors affecting starch digestibility with special emphasis on sorghum and corn. *J. Anim. Sci.* 63:1607–1623.

Streeter, M. N., G. M. Hill, D. G. Wagner, F. N. Owens, and C. A. Hibberd. 1993. Effect of bird-resistant and non-bird-resistant sorghum grain on amino acid digestion by beef heifers. *J. Anim. Sci.* 71:1648–1656.

Streeter, M. N., D. G. Wagner, C. A. Hibberd, and F. N. Owens. 1990a. The effect of sorghum grain variety on site and extent of digestion in beef heifers. *J. Anim. Sci.* 68:1121–1132.

Streeter, M. N., D. G. Wagner, E. D. Mitchel, Jr., and J. W. Oltjen. 1990b. Effect of variety of sorghum grain on digestion and availability of dry matter and starch in vitro. *Anim. Feed Sci. Tech.* 29:279–287.

Wheat

Bell, J. M. and D. M. Anderson. 1984. Comparisons of wheat cultivars as energy and protein sources in diets for growing and finishing pigs. *Can. J. Anim. Sci.* 64:957–970.

Becker, R., G. D. Hanners, D. W. Irving, and R. M. Saunders. 1986. Chemical composition and nutritional qualities of five potential perennial grains. *Food Sci. Tech.* 19:312–315.

Campbell, G. L. and M. R. Bedford. 1992. Enzyme applications for monogastric feeds: A review. *Can. J. Anim. Sci.* 72:449–466.

Choct, M. and G. Annison. 1992. The inhibition of nutrient digestion by wheat pentosans. *Brit. J. Nutr.* 67:123–132.

Crouch, A. N., J. L. Grimes, P. R. Ferket, L. N. Thomas, and A. E. Sefton. 1997. Enzyme supplementation to enhance wheat utilization in starter diets for broilers and turkeys. *J. Appl. Poult. Res.* 6:147–154.

Myer, R. O. and R. D. Barnett. 1987. Evaluation of purple-seeded, high-protein, soft wheats in diets for young, growing swine. *Nutr. Rep. Int.* 35:819–824.

Seerley, R. W., W. L. Vandergrift, and O. M. Hale. 1988. Effect of particle size of wheat on performance of nursery, growing and finishing pigs. *J. Anim. Sci.* 66:2484–2489.

Simmonds, N. W. 1995. The relation between yield and protein in cereal grain. *J. Sci. Food Agric.* 67:309–315.

Wagoner, P. 1990. Perennial grain development: Past efforts and potential for the future. *Crit. Rev. Plant Sci.* 9:381–408.

Zinn, R. A. 1994. Influence of flake thickness on the feeding value of steam-rolled wheat for feedlot cattle. *J. Anim. Sci.* 72:21–28.

Barley

Classen, H. L., G. L. Campbell, B. G. Rossnagel, and R. S. Bhatly. 1988. Evaluation of hull-less barley as replacement for wheat or conventional barley in laying hen diets. *Can. J. Anim. Sci.* 68:1261–1266.

Engstrom, D. F., G. W. Mathison, and L. A. Goonewardene. 1989. Effect of Beta-glucans, processing method and other factors on barley utilization by cattle. *Can. J. Anim. Sci.* 69:293 (Abst.).

Fengler, A. I., F. X. Aherne, and A. R. Robblee. 1990. Influence of germination of cereals on viscosity of their aqueous extracts and nutritive value. *Anim. Feed Sci. Tech.* 28:243–253.

Gabert, V. M., H. Jorgensen, G. Brunsgaard, B. O. Eggum, and J. Jensen. 1996. The nutritional value of new high-lysine barley varieties determined with rats and young pigs. *Can. J. Anim. Sci.* 76:443–450.

Goodband, R. D. and R. Hines. 1988. An evaluation of barley in starter diets for swine. *J. Anim. Sci.* 66:3086–3093.

Hatfield, P. G., J. A. Hopkins, G. T. Pritchard, and C. W. Hunt. 1997. The effects of amount of whole barley, barley bulk density, and form of roughage on feedlot lamb performance, carcass characteristics, and digesta kinetics. *J. Anim. Sci.* 75:3353–3366.

Hesselman, K., K. Elwinger, and S. Thomke. 1982. Influence of increasing levels of β-glucanase on the productive value of barley diets for broiler chickens. *Anim. Feed Sci. Tech.* 7:351–358.

Hinman, D. D. and S. J. Sorensen. 1994. Comparison of different tempering times for barley based feedlot diets and their influence on animal performance, carcass characteristics, diet digestibility and *in vitro* barley digestibility. *Proc. West. Sect. Am. Soc. Anim. Sci.* 45:309–312.

Newman, C. W., M. Overland, R. K. Newman, K. Bang-Olsen, and B. Pedersen. 1990. Protein quality of a new high-lysine barley derived from RISO 1508. *Can. J. Anim. Sci.* 70:279–285.

Newman, R. K. and C. W. Newman. 1987. Beta-glucanase effect on the performance of broiler chicks fed covered and hulless barley isotypes having normal and waxy starch. *Nutr. Rep. Int.* 36:693–699.

Truscott, D. R., C. W. Newman, and N. J. Roth. 1988. Effects of hull and awn type on the feed value of Betzes barley. 3. Growth performance, carcass characteristics and nitrogen and energy digestibility by pigs. *Nutr. Rep. Int.* 38:221–230.

Veum, T. L., D. R. Ledoux, D. W. Bollinger, V. Raboy, and A. Cook. 2002. Low-phytic acid barley improves calcium and phosphorus utilization and growth performance in growing pigs. *J. Anim. Sci.* 80:2663–2670.

Zinn, R. A. 1993. Influence of processing on the comparative feeding value of barley for feedlot cattle. *J. Anim. Sci.* 71:3–10.

Oats

Brand, T. S. and J. P. van der Merwe. 1996. Naked oats (*Avena nuda*) as a substitute for maize in diets for weanling and grower-finisher pigs. *Anim. Feed Sci. Tech.* 57:139–147.

Burkitt, D. P. 1988. Dietary fiber and cancer. *J. Nutr.* 118:531–533.

Campbell, G. L., F. W. Sosulski, H. L. Classen, and G. M. Ballance. 1987. Nutritive value of irradiated and β-glucanase-treated wild oat groats (*Avena fatuva L.*) for broiler chickens. *Anim. Feed Sci. Tech.* 16:243–252.

Cave, N. A. and V. D. Burrows. 1993. Evaluation of naked oat (*Avena nuda*) in the broiler chicken diet. *Can. J. Anim. Sci.* 73:393–399.

Christison, G. I. and J. M. Bell. 1980. Evaluation of Terra, a new cultivar of naked oats (*Avena nuda*) when fed to young pigs and chicks. *Can. J. Anim. Sci.* 60:465–471.

Devlin, T. J., J. R. Ingalls, and H. R. Sharma. 1977. Evaluation of high fat oats in rations of growing and finishing ruminants. *Can. J. Anim. Sci.* 57:735–743.

Frolich, W. and M. Nyman. 1988. Minerals, phytate and dietary fibre in different fractions of oatgrain. *J. Cereal Sci.* 7:73–82.

Morris, J. R. and V. D. Burrows. 1986. Naked oats in grower-finisher pig diets. *Can. J. Anim. Sci.* 66:833–836.

Myer, R. O., R. D. Barnett, and W. R. Walker. 1985. Evaluation of hull-less oats (*Avena nuda L.*) in diets for young swine. *Nutr. Rep. Int.* 32:1273–1277.

Pettersson, D., H. Graham, and P. Aman. 1987. The productive value of whole and dehulled oats in broiler chicken diets, and influence of β-glucanase supplementation. *Nutr. Rep. Int.* 36:743–750.

Shinnick, F. L., M. J. Longacre, S. L. Ink, and J. A. Marlett. 1988. Oat fiber: Composition versus physiological function in rats. *J. Nutr.* 118:144–151.

Thacker, P. A. and F. W. Sosulski. 1994. Use of wild oat groats in starter diets for swine. *Anim. Feed Sci. Tech.* 46:229–237.

Zinn, R. A. 1993. Influence of processing on the feeding value of oats for feedlot cattle. *J. Anim. Sci.* 71:2303–2309.

Rye

Boros, D., R. R. Marquardt, and W. Guenter. 1995. Rye as an alternative grain in commercial broiler feeding. *J. Appl. Poult. Res.* 4:341–351.

Campbell, G. L. and M. R. Bedford. 1992. Enzyme applications for monogastric feeds: A review. *Can. J. Anim. Sci.* 72:449–466.

Campbell, G. L. and L. D. Campbell. 1989. Rye as a replacement for wheat in laying hen diets. *Can. J. Anim. Sci.* 69:1041–1047.

Fengler, A. I. and R. R. Marquardt. 1986. An antinutritional factor in rye: Isolation and activity of the water-soluble pentosans. *Can. J. Anim. Sci.* 66:330 (Abst.).

Honeyfield, D. C., J. A. Froseth, and J. McGinnis. 1983. Comparative feeding value of rye for poultry and swine. *Nutr. Rep. Int.* 28:1253–1260.

Pawlik, J. R., A. I. Fengler, and R. R. Marquardt. 1986. Improvement of the nutritional value of rye: Destruction or reduction of the antinutritive activity by various treatments. *Can. J. Anim. Sci.* 66:331 (Abst.).

Proudfoot, F. G. and H. W. Hulan. 1986. The nutritive value of ground rye as a feed ingredient for adult Leghorn hens. *Can. J. Anim. Sci.* 66:311–315.

Sedlet, J., M. Mathias, and K. Lorenz. 1984. Growth-depressing effects of 5-n-pentadecylresorcinol: A model for cereal allkylresorcinols. *Cereal Chem.* 61:239–241.

Triticale

Batterham, E. S., H. S. Saine, and L. M. Andersen. 1989. The effect of mild heat on the nutritional value of triticale for growing pigs. *Anim. Feed Sci. Tech.* 26:191–205.

Boros, D. 1999. Influence of R genome on the nutritional value of triticale for broiler chicks. *Anim. Feed Sci. Tech.* 76:219–226.

Brand, T. S., R. C. Olckers, and J. P. van der Merwe. 1995. Triticale (*Tritico secale*) as substitute for maize in pig diets. *Anim. Feed Sci. Tech.* 53:345–352.

Coffey, M. T. and W. J. Gerrits. 1988. Digestibility and feeding value of B858 triticale for swine. *J. Anim. Sci.* 66:2728–2735.

Erickson, J. P., E. R. Miller, F. C. Elliott, P. K. Ku, and D. E. Ullrey. 1979. Nutritional evaluation of triticale in swine starter and grower diets. *J. Anim. Sci.* 48:547–553.

Farrell, D. J., C. Chan, and F. McCrae. 1983. A nutritional evaluation of triticale with pigs. *Anim. Feed Sci. Tech.* 9:49–62.

Hale, O. M., D. D. Morey, and R. O. Myer. 1985. Nutritive value of Beagle 82 triticale for swine. *J. Anim. Sci.* 60:503–510.

Radcliffe, B. C., C. F. Driscoll, and A. R. Egan. 1981. Content of 5-alkyl resorcinols in selection lines of triticale grown in South Australia. *Aust. J. Exp. Agric. Anim. Husb.* 21:71–74.

Radcliffe, B. C., A. R. Egan, and C. J. Driscoll. 1983. Nutritional evaluation of triticale grain as an animal feedstuff. *Aust. J. Exp. Agric. Anim. Husb.* 23:419–425.

Rundgren, M. 1988. Evaluation of triticale given to pigs, poultry and rats. *Anim. Feed Sci. Tech.* 19:359–375.

Shimada, A., T. R. Cline, and J. C. Rogler. 1974. Nutritional evaluation of triticale for the nonruminant. *J. Anim. Sci.* 38:935–940.

Sosulski, F. W., L. A. Minja, and D. A. Christensen. 1988. Trypsin inhibitors and nutritive value in cereals. *Plant Foods Human Nutr.* 38:23–34.

Buckwheat

Anderson, D. M. and J. P. Bowland. 1984. Evaluation of buckwheat (*Fagopyrum esculentum*) in diets of growing pigs. *Can. J. Anim. Sci.* 64:985–995.

Farrell, D. J. 1987. A nutritional evaluation of buckwheat (*Fagopyrum esculentum*). *Anim. Feed Sci. Tech.* 3:95–108.

Harrold, R. L., D. L. Craig, J. D. Nalewaja, and B. B. North. 1980. Nutritive value of green or yellow foxtail, wild oats, wild buckwheat or redroot pigweed seed as determined with the rat. *J. Anim. Sci.* 51:127–131.

Mulholland, J. G. and J. B. Coombe. 1979. A comparison of the forage value for sheep of buckwheat and sorghum stubbles grown on the Southern Tablelands of New South Wales. *Aust. J. Exp. Agric. Anim. Husb.* 19:297–302.

Nicholson, J. W., R. McQueen, E. A. Grant, and P. L. Burgess. 1976. The feeding value of tartary buckwheat for ruminants. *Can. J. Anim. Sci.* 56:803–808.

Pomeranz, Y. 1983. Buckwheat: Structure, composition, and utilization. *CRC Crit. Rev. Food Sci. Nutr.* 19:213–258.

Thacker, P. A., D. M. Anderson, and J. P. Bowland. 1983a. Nutritive value of common buckwheat as a supplement to cereal grains when fed to laboratory rats. *Can. J. Anim. Sci.* 63:213–219.

Thacker, P. A., D. M. Anderson, and J. P. Bowland. 1983b. Chemical composition and nutritive value of buckwheat cultivars for laboratory rats. *Can. J. Anim. Sci.* 63:949–956.

Amaranth and Quinoa

Acar, N. and P. Vohra. 1988. Nutritional value of grain amaranth for growing chickens. *Poult. Sci.* 67:1166–1173.

Fadel, J. G., W. G. Pond, R. L. Harrold, C. C. Calvert, and B. A. Lewis. 1996. Nutritive value of three amaranth grains fed either processed or raw to growing rats. *Can. J. Anim. Sci.* 76:253–257.

Gee, J. M., K. R. Price, C. L. Ridout, G. M. Wortley, R. F. Hurrell, and I. T. Johnson. 1993. Saponins of quinoa (*Chenopodium quinoa*): Effects of processing on their abundance in quinoa products and their biological effects on intestinal mucosal tissue. *J. Sci. Food Agric.* 63:201–209.

Jacobsen, E. E., B. Skadhauge, and S. E. Jacobsen. 1997. Effect of dietary inclusion of quinoa on broiler growth performance. *Anim. Feed Sci. Tech.* 65:5–14.

Laovoravit, N., F. H. Kratzer, and R. Becker. 1986. The nutritional value of amaranth for feeding chickens. *Poult. Sci.* 65:1365–1370.

National Academy of Sciences. 1984. *Amaranth. Modern prospects for an ancient crop.* Washington, DC: National Academy Press.

Pond, W. G. and J. W. Lehmann. 1989. Nutritive value of a vegetable amaranth cultivar for growing lambs. *J. Anim. Sci.* 67:3036–3039.

Ravindran, V., R. L. Hood, R. J. Gill, C. R. Kneale, and W. L. Bryden. 1996. Nutritional evaluation of grain amaranth (*Amaranthus hypochondriacus*) in broiler diets. *Anim. Feed Sci. Tech.* 63:323–331.

Tillman, P. B. and P. W. Waldroup. 1987. Effects of feeding extruded grain amaranth to laying hens. *Poult. Sci.* 66:1697–1701.

Tillman, P. B. and P. W. Waldroup. 1988. Performance and yields of broilers fed extruded grain amaranth and grown to market weight. *Poult. Sci.* 67:743–749.

Sprouted Grains

Capper, A. L. 1988. Fungal contamination of hydroponic forage. *Anim. Feed Sci. Tech.* 20:163–169.

Peer, D. J. and S. Leeson. 1985a. Feeding value of hydroponically sprouted barley for poultry and pigs. *Anim. Feed Sci. Tech.* 13:183–190.

_______ 1985b. Nutrient content of hydroponically sprouted barley. *Anim. Feed Sci. Tech.* 191–202.

_______ 1985c. Nutrient and trypsin inhibitor content of hydroponically sprouted soybeans. *Anim. Feed Sci. Tech.* 13:203–214.

Rule, D. C., R. L. Preston, R. M. Koes, and W. E. McReynolds. 1986. Feeding value of sprouted wheat (*Triticum aestivum*) for beef cattle finishing diets. *Anim. Feed Sci. Tech.* 15:113–121.

Cereal Milling By-Products

Annison, G., P. J. Moughan, and D. V. Thomas. 1995. Nutritive activity of soluble rice bran arabinoxylans in broiler diets. *Brit. Poult. Sci.* 36:479–488.

Beames, R. M., R. M. Tait, and D. J. Litsky, 1986. Grain screenings as a dietary component for pigs and sheep. I. Botanical and chemical composition. *Can. J. Anim. Sci.* 66:473–481.

Bersch, S., F. H. Kratzer, and P. Vohra. 1989. Necessity of heat processing for improving nutritional value of rice bran for chickens. *Nutr. Rep. Int.* 40:827–830.

Castell, A. G. 1984. Response of growing-finishing pigs to dietary inclusion of green foxtail (*Setaria viridis*). *Can. J. Anim. Sci.* 64:1063–1066.

Dalke, B. S., R. N. Sonon, Jr., M. A. Young, G. L. Huck, K. K. Kreikemeier, and K. K. Bolsen. 1997. Wheat middlings in high-concentrate diets: Feedlot performance, carcass characteristics, nutrient digestibilities, passage rates, and ruminal metabolism in finishing steers. *J. Anim. Sci.* 75:2561–2566.

Erickson, J. P., E. R. Miller, P. K. Ku, G. F. Collings, and J. R. Black. 1985. Wheat middlings as a source of energy, amino acids, phosphorus and pellet binding quality for swine diets. *J. Anim. Sci.* 60:1012–1020.

Farrell, D. J. 1994. Utilization of rice bran in diets for domestic fowl and ducklings. *World's Poult. Sci. J.* 50:115–131.

Goodband, R. D. and R. H. Hines. 1988. An evaluation of barley in starter diets for swine. *J. Anim. Sci.* 66:3086–3093.

Green, D. A., R. A. Stock, F. K. Goedeken, and T. J. Klopfenstein. 1987. Energy value of corn wet milling by-product feeds for finishing ruminants. *J. Anim. Sci.* 65:1655–1666.

Harrold, R. L., D. L. Craig, J. D. Nalewaja, and B. B. North. 1980. Nutritive value of green or yellow foxtail, wild oats, wild buckwheat or redroot pigweed seed as determined with the rat. *J. Anim. Sci.* 51:127–131.

Hogan, J. P., P. Davis, and J. R. Lindsay. 1980. Nutritional evaluation of alkali-treated wheat dust. *J. Aust. Inst. Agr. Sci.* 46:249–251.

Leeson, S., N. Hussar, and J. D. Summers. 1988. Feeding and nutritive value of hominy and corn grits for poultry. *Anim. Feed Sci. Tech.* 19:313–325.

O'Hearn, V. L. and R. A. Easter. 1983. Evaluation of wheat middlings for swine diets. *Nutr. Rep. Int.* 28:403–411.

Patterson, P. H., M. L. Sunde, E. M. Schieber, and W. B. White. 1988. Wheat middlings as an alternate feedstuff for laying hens. *Poult. Sci.* 67:1329–1339.

Proudfoot, F. G. and H. W. Hulan. 1988. Nutritive value of wheat screening as a feed ingredient for broiler chickens. *Poult. Sci.* 67:615–618.

Robles, A. and R. C. Ewan. 1982. Utilization of energy of rice and rice bran by young pigs. *J. Anim. Sci.* 55:572–577.

Tait, R. M., R. M. Beames, and J. Litsky. 1986. Grain screenings as a dietary component for pigs and sheep. II. Animal utilization. *Can. J. Anim. Sci.* 66:483–494.

Wang, G. J., R. R. Marquardt, W. Guenter, Z. Zhang, and Z. Han. 1997. Effects of enzyme supplementation and irradiation of rice bran on the performance of growing Leghorn and broiler chicks. *Anim. Feed Sci. Tech.* 66:47–61.

Warren, B. E. and D. J. Farrell. 1990. The nutritive value of full-fat and defatted Australian rice bran. I. Chemical composition. II. Growth studies with chickens, rats and pigs. III. The apparent digestible energy content of defatted rice bran in rats and pigs and the metabolisability of energy and nutrients in defatted and full-fat bran in chickens and adult cockerels. IV. Egg production of hens on diets with defatted rice bran. *Anim. Feed Sci. Tech.* 27:219–228; 229–246; 247–257; 259–268.

Grain Overload of Ruminants

Bartley, E. E., T. G. Nagaraja, E. S. Pressman, A. D. Dayton, M. P. Katz, and L. R. Fina. 1983. Effects of lasalocid or monensin on legume or grain (feedlot) bloat. *J. Anim. Sci.* 56:1400–1406.

Carroll, C. L., G. Hazard, P. J. Coloe, and P. T. Hooper. 1987. Laminitis and possible enterotoxaemia associated with carbohydrate overload in mares. *Equine Vet. J.* 19:344–346.

Cheng, K. -J., T. A. McAllister, J. D. Popp, A. N. Hristov, Z. Mir, and H. T. Shin. 1998. A review of bloat in feedlot cattle. *J. Anim. Sci.* 76:299–308.

Fron, M., H. Madeira, C. Richards, and M. Morrison. 1996. The impact of feeding condensed distillers by-products on rumen microbiology and metabolism. *Anim. Feed Sci. Tech.* 61:235–245.

Garner, H. E., D. P. Hutcheson, J. R. Coffman, A. W. Hahn, and C. Salem. 1977. Lactic acidosis: A factor associated with equine laminitis. *J. Anim. Sci.* 45:1037–1041.

Goad, D. W., C. L. Goad, and T. G. Nagaraja. 1998. Ruminal microbial and fermentative changes associated with experimentally induced subacute acidosis in steers. *J. Anim. Sci.* 76:234–241.

Godfrey, S. I., J. B. Rowe, G. R. Thorniley, M. D. Boyce, and E. J. Speijers. 1995. Virginiamycin to protect sheep fed wheat, barley or oats from grain poisoning under simulated drought feeding conditions. *Aust. J. Agr. Res.* 46:393–401.

Gould, D. H. 1998. Polioencephalomalacia. *J. Anim. Sci.* 76:309–314.

Huntington, G. B. and R. A. Britton. 1979. Effect of dietary lactic acid on rumen lactate metabolism and blood acid base status of lambs switched from low to high concentrate diets. *J. Anim. Sci.* 49:1569–1576.

Kung, Jr., L. and A. O. Hession. 1995. Preventing in vitro lactate accumulation in ruminal fermentations by inoculating with *Megasphaera elsedenii. J. Anim. Sci.* 73:250–256.

Nagaraja, T. G. and M. M. Chengappa, 1998. Liver abscesses in feedlot cattle: A review. *J. Anim. Sci.* 76:287–298.

Nagaraja, T. G., T. B. Avery, E. E. Bartley, S. J. Galitzer, and A. D. Dayton. 1981. Prevention of lactic acidosis in cattle by lasalocid or momensin. *J. Anim. Sci.* 53:206–216.

Owens, F. N., D. S. Secrist, W. J. Hill, and D. R. Gill. 1998. Acidosis in cattle: A review. *J. Anim. Sci.* 76:275–286.

Rogers, J. A., M. E. Branine, C. R. Miller, M. I. Wray, S. J. Bartle, R. L. Preston, D. R. Gill, R. H. Pritchard, R. P. Stilborn, and D. T. Bechtol. 1995. Effects of dietary virginiamycin on performance and liver abscess incidence in feedlot cattle. *J. Anim. Sci.* 73:9–20.

Rowe, J. B., M. J. Lees, and D. W. Pethick. 1994. Prevention of acidosis and laminitis associated with grain feeding in horses. *J. Nutr.* 124:2742S–2744S.

Russell, J. B. and T. Hino. 1985. Regulation of lactate production in *Streptococcus bovis:* A spiraling effect that contributes to rumen acidosis. *J. Dairy Sci.* 68:1712–1721.

CHAPTER 3

Other Concentrate Energy Feeds

Objectives

1. To discuss feedstuffs other than grains that are used as energy sources in animal feeding, including
   - roots and tubers
   - food processing by-products
   - fats and oils
   - milk by-products
2. To highlight areas that are currently of widespread interest, including
   - molasses-urea blocks for developing countries
   - high-fat gestation diets to improve baby pig survival
   - rumen-protected fats for dairy cattle
   - nutritional roles of fish oils (omega-3 fatty acids)

ROOTS AND TUBERS

Cassava

Cassava (*Manihot esculenta*), also known as manioc and tapioca, is a tropical plant grown in the lowland tropics for human food and to a lesser extent for animal feed. It has coarse, woody stems and large pinnate leaves that can be used as forage. The enlarged roots (Fig. 3.1) are rich in starch, and are harvested for food and feed. The roots contain approximately 83 percent nitrogen-free extract, which is mostly starch; approximately 3 to 4.5 percent crude protein; and 3.7 percent crude fiber (dry-matter basis). The fresh roots contain an average of 65 percent water. Cassava is primarily an energy source; the protein content is low and severely deficient in sulfur amino acids.

Cassava can be fed fresh to animals or chopped and sun dried to produce cassava meal. Cassava meal is produced in Southeast Asian countries, such as Thailand, and exported, particularly to western European countries, where it is widely used in swine and poultry feeding. Its importation is based more on trade agreements than on a necessity of the European Union (EU) countries to import feedstuffs. In fact, the EU has a surplus of feed grains.

Cassava leaves and roots contain toxic cyanogenic glycosides. **Glycosides** contain substances linked by an ether bond (-0-) with a carbohydrate fraction. The noncarbohydrate portion is called an *aglycone*. Glycosides are metabolized by enzymatic action to release the aglycone from the carbohydrate. The aglycone may then be metabolized further. **Cyanogenic glycosides** are those that contain cyanide as part of the aglycone. When they are metabolized, free cyanide, a deadly poison, is released. **Cyanide** is extremely toxic to cellular metabolism. Plants containing glycosides also contain enzymes to hydrolyze them, but in different

FIGURE 3.1 Tubers of cassava, also known as manioc and tapioca. Cassava is an important energy source for both humans and livestock in many tropical countries.

compartments of the plant cell. Thus, cyanide is not released until the plant cells are disrupted in some way.

Cassava is processed to reduce its cyanide content. When the roots are chopped, pulped, or mashed, the cell structure is macerated and the glycoside and enzymes come together, releasing cyanide:

$$\text{Cyanogenic glycosides} \xrightarrow[\text{enzymes}]{\text{plant}} \text{aglycone} + \text{carbohydrate}$$

$$\text{Aglycone} \xrightarrow[\text{enzymes}]{\text{plant}} \text{HCN} + \text{nontoxic compounds}$$

The HCN is hydrocyanic or **prussic acid;** the ionic form is cyanide (CN^-). Cyanide is volatile, so when the cassava roots are pulped and sun dried, free cyanide is formed, much of which is then volatilized into the air. Extraction of the pulped root with water will also wash out much of the cyanide.

Adequate processing of cassava will prevent acute cyanide poisoning. However, small amounts of residual cyanide remain. Methionine and vitamin B_{12} function in the detoxification and excretion of cyanide, so the requirements for these nutrients are elevated with cassava-based diets.

Cassava is often regarded as a "poor person's food" and may be relegated to infertile soils where other crops are not productive. Cassava is tolerant of acid, infertile soils and has a great deal of potential as a tropical feedstuff. It is an excellent energy source for swine and poultry, if properly supplemented with protein, sulfur amino acids, and vitamins (Lekule et al., 1988; Walker, 1985). The peels are also acceptable feed for poultry (Osei and Twumasi, 1989).

Dried **cassava meal** can be used to replace grain in high-concentrate diets for cattle (Zinn and DePeters, 1991). Because cassava has a low protein content, more protein supplementation is needed than with grain-based diets. Dried poultry waste (see Chapter 4) has a high nitrogen content but is low in energy; cassava is high in energy but low in nitrogen, so a combination of broiler litter and cassava can be used effectively by ruminants (Holzer et al., 1997).

Potatoes

The potato (*Solanum tuberosum*) is a native of the Andes highlands of South America, but it is now grown extensively in all of the temperate regions of the world. Russia, Poland, Germany, the United States, France, and China are leading producers. The edible parts of the potato plant are the tubers beneath the surface of the ground. Tubers on the surface exposed to sunlight become green, bitter, and toxic due to alkaloid formation. There are many varieties of potatoes, with different shapes and colors. Only white-fleshed potatoes are grown commercially in the United States, but yellow-fleshed varieties are popular in Europe. There are even blue and purple-fleshed varieties. Skin colors include white, red, and blue. In the United States, potatoes are grown specifically for human consumption. Cull, surplus, and small potatoes, and potato processing wastes are fed to livestock. In Europe, some high-yielding varieties are grown specifically for animal feed. Per unit of land area, potatoes can produce 1.7 times the dry matter, 2 times the energy, and 1.3 times the protein of cereal grains (Whittemore, 1977).

On a dry-matter basis, whole potatoes contain 4 to 5 percent ash, 10 to 11 percent crude protein, 4 percent crude fiber, and 82 to 83 percent nitrogen-free extract, and they compare favorably to grains in protein content. The lysine content is twice that of corn. Approximately 70 percent of the potato dry matter is starch, containing about 75 percent amylopectin and 25 percent amylose. **Raw potato starch** is poorly digested, so potatoes should be cooked prior to their use as feed. Cooking breaks down the structure of the starch granules and improves digestibility and palatability. Potatoes contain trypsin and chymotrypsin inhibitors throughout the tuber and in the peel, with some cultivar differences in their concentration (Nicholson and Allen, 1989). Cooking destroys these protease inhibitors. Potatoes contain an alkaloid called solanine. **Alkaloids** are toxins containing nitrogen in a carbon ring. **Solanine** is bitter, is an irritant to the mucous membranes of the gut, and inhibits nerve function (it is a cholinesterase inhibitor). Normally the solanine content is not a concern, but in tubers exposed to sunlight, either pre- or postharvest, solanine is synthesized and may be present at toxic levels. Solanine and chlorophyll are synthesized under similar conditions; thus green potatoes are green because of their chlorophyll content, but they are also high in solanine (Dao and Friedman, 1994).

A variety of potato-processing by-products are produced and used as feed. French fries and other processed products that do not meet quality-control specifications are used as feed. In Oregon, Washington, and Idaho, a major potato-producing area, **potato wastes** are an important component of diets for feedlot cattle. Batches of french fries, for instance, may be packaged and boxed, but if the test kitchen results do not pass quality standards, the entire batch, including packages and boxes, is sent off to the feedlots where it is fed, cardboard and all! **French fries** are partially cooked in vegetable oil during their production, so they have a high energy content and are well utilized by ruminants (Rooke et al., 1997). **Potato chip waste** can be substituted for corn at up to 20 percent of the diet for pigs without reducing performance (Rahnema and Borton, 2000).

Large amounts of water are used in potato-processing plants. Strong solutions of alkali (sodium hydroxide) are used to remove peels. It is a common practice to locate feedlots near the processing plants and store the wet potato by-products in lagoons or pits. The material is frequently quite alkaline and ferments vigorously because of the high starch and glucose content. The resulting product is often offensive in appearance, consistency, and smell, but when mixed with grain and other ingredients in a feed wagon, it seems to be highly palatable to cattle.

Potato steam peel is another by-product of french fry production. It is produced by subjecting the potatoes to pressurized steam at about 200°C for a few seconds (Van Lunen et al., 1989). No hydroxide is used; the peel is removed by a scrubber and is a slurry containing approximately 15 percent dry matter, approximately 15 percent crude protein, 6 percent ash (dry-matter basis), and a pH of 5.8 (Nicholson et al., 1988). It can be used at levels up to 30 percent of the dry-matter intake of pigs with no adverse effects on performance (Van Lunen et al., 1989; Nicholson et al., 1988).

Other Roots and Succulents

Root crops were at one time grown quite extensively in North America and Europe as sources of winter feed for livestock. Because of high labor costs for producing them and lower yields of nutrients compared to hybrid corn and other crops, there is very little production of root crops in the United States, although they are still important in northern Europe. Some of the more important root crops include mangels (*Beta vulgaris*), turnips (*Brassica compestris* var. *rapa*), rutabagas or swedes (*Brassica rapus* var. *napobrassica*), and field carrots. **Mangels** are similar to sugar beets except that they are longer, higher yielding, and have a lower sugar content. Approximately 40 percent of the total length of the root extends above the soil surface. After harvest, root crops are stored in cellars or outdoor pits, and protected from freezing. They can also be used as green feed and harvested directly by grazing stock. Because most of the carbohydrate in fodder beets exists as sugars rather than as starch, the beets are very rapidly fermented in the rumen (Sabri et al., 1988a), and there is a potential for acidosis to occur. Their digestibility in ruminants is very high, with the digestibilities of dry matter and gross energy near 96 percent in sheep (Sabri et al., 1988b).

Sweet Potatoes and Yams Sweet potatoes (*Ipomoea batatas*) are grown primarily for human consumption. A member of the morning-glory family, the sweet potato grows in warm climates with long growing seasons, primarily in tropical and subtropical areas. Cull sweet potatoes are used as animal feed. The tubers contain trypsin inhibitors and so must be cooked before using as feed (Bouwkamp, 1985). They are similar to potatoes in feeding value. Sweet potato chips have about 79 percent of the metabolizable energy content of corn for swine (Wu, 1980) and can replace corn in swine finisher diets (Tor-Agbidye et al., 1990). Dried **sweet potato meal** can replace up to 40 percent corn in broiler diets with no adverse effects on performance (Ravindran and Sivakanesan, 1996). In Asia, it is common to feed the sweet potato vines, which are very palatable, as well as roots, to pigs. The vines are also used as forage for ruminants. If they become moldy, both the roots and vines can be toxic, producing **lung disease**

(pulmonary emphysema). This is caused by 3-substituted furans (e.g., ipomeanol) produced as stress metabolites in response to fungal infection.

Yams (*Dioscorea* spp.) are similar to sweet potatoes in appearance and nutrient content. They are used primarily for human consumption. Grown mainly in the tropics, yams can be used as feed for pigs or other livestock if they are available in surplus quantities. They contain trypsin inhibitors and must be cooked (Panigrahi and Francis, 1982). **Arrowroot** (*Maranta arundinacea*) is another tropical tuber with potential as a feedstuff (Erdman and Erdman, 1984).

Aroids are root crops that grow in swampy areas of the humid tropics. Cocoyams or **taro** (*Colocasia esculenta*) is grown quite extensively as a food crop and may have potential as an animal feed. Raw taro contains crystals of **calcium oxalate**, which have a strong irritant effect on mucous membranes of the mouth and digestive tract. Boiling the raw taro lowers the oxalate content and eliminates the growth-inhibitory effect of raw taro when included in poultry diets (Ravindran et al., 1996).

Jerusalem Artichoke There has been sporadic interest in the Jerusalem artichoke (*Helianthus tuberosus*) as a food and feed crop and, most recently, for fermentation to produce ethanol for fuel consumption. The plant is neither an artichoke nor from Jerusalem. It is a member of the sunflower family and is native to North America. The taste is said to resemble that of the true artichoke, and Jerusalem is a corruption of *girasol,* the Italian name for sunflower. The plant grows to a height of approximately 6 feet, and the flowers are similar to those of sunflowers, but smaller. It produces irregularly shaped knobby tubers that are rich in **inulin**, a polysaccharide composed of **fructose**. Although animals do not produce an enzyme to digest inulin, it is hydrolyzed by acid in the stomach (Graham and Aman, 1986). Both the tubers and the vegetative matter can be used as feed, but the plant appears to have little to offer to justify its use in place of traditional crops.

Farnworth et al. (1995) found that dietary Jerusalem artichoke reduced the fecal odor of pigs. Polymers of fructose (fructans or fructooligosaccharides) are growth factors for bifidobacteria, whereas they cannot be fermented by many common gut microbes. Changes in gut microbes can alter fecal odor, which is caused largely by the compounds produced by intestinal bacteria.

Vegetables and Fruit A number of other succulent materials that serve primarily as energy sources include cull fruits and vegetables, such as onions, pumpkins, apples, and pears. In the Pacific Northwest of the United States, **cull onions** are used fairly extensively as a winter supplement for sheep and cattle. They contain n-propyl disulfide, which can induce hemolytic anemia in animals, and so should be fed at levels not exceeding 15 percent of the total diet. Another hazard associated with onion feeding (and with other material of a similar size) is the possibility of choking or blockage of the esophagus.

At one time, pumpkins and squash were grown as animal feeds and were often interplanted with corn. This practice is no longer economically viable (or practical with the use of mechanical corn harvesters).

Large quantities of reject **bananas** (*Musa* spp.) are available for feeding purposes in many tropical countries. Green bananas are somewhat unpalatable because of their tannin content (Dhua and Sen, 1989). Better results with nonruminants are usually obtained with ripe fruit. Like cassava, bananas are low in crude protein and high in starch and so serve primarily as an energy source. Pond and Maner (1984) summarized a number of studies indicating that, with proper protein supplementation, banana fruit can replace most or all of the grain in swine diets without adverse effects on performance. Preston and Leng (1987) emphasize that it is better to feed ruminants the green fruit, while the carbohydrate is still in the form of starch rather than sugars. The skin (peel) is rich in tannins, which might have some value in ruminants by decreasing rumen degradability of dietary protein. Bananas are deficient in fermentable nitrogen, which can be readily supplied with a urea supplement. By-pass protein can be provided with protein-rich legume foliage (e.g., leucaena, gliricidia) or with cottonseed meal. Plantains are similar to bananas in appearance and feeding value, although they are somewhat higher in dry matter and starch content.

The foliage of bananas can also be used as feed. Bananas have a pseudostem that is quite digestible in ruminants. The leaves are tough and fibrous but are consumed by livestock and can be used as a roughage source. The stalk, which is very high in cellulose and low in crude protein (7 percent) has limited value as a roughage source in ruminants, being equivalent to tropical grass hay in feeding value (Viswanathan et al., 1989). **Enset** (*Ensete vertricosum*) is a large banana-like plant grown in Ethiopia as a food crop. The leaves and pseudostem are similar to those of the banana in terms of feeding value for ruminants (Fekadu and Ledin, 1997).

Breadfruit meal, prepared by drying the fruit of the breadfruit tree (*Artocarpus communis*), is an excellent source of starch and can replace corn in poultry diets (Ravindran and Sivakanesan, 1996).

FOOD PROCESSING AND INDUSTRIAL BY-PRODUCTS

Molasses

Molasses is a product of the sugar-refining industry. The principal types are cane and beet molasses refined from sugarcane and sugar beets, respectively. They are similar in composition and feeding value.

Sugarcane (*Saccarum officinarum*) is produced in tropical and subtropical regions, particularly in the Caribbean countries, Latin America, Southeast Asia, and the southern United States. India, Brazil, Cuba, Mexico, China, and the United States are the leading sugar-producing countries. Sugarcane is a perennial grass, with thick-sugar-rich stems and abundant leaves. The cane is harvested when the sugar content is at a maximum and transported by truck, railroad, or oxcart to a refining plant. The stems are pressed to squeeze out the juice, containing the sugar. The fibrous residue of the stalks is called **bagasse** and is burned or used as a low-quality roughage for feed. The juice is concentrated by boiling to a solution of approximately 50 percent sucrose, after previous clarification with lime to remove impurities. The sugar crystalizes out of the concentrated juice and is collected as raw sugar. The juice residue is the molasses. From each ton of sugarcane, approximately 100 kg of refined sugar and 25 to 50 kg of molasses are produced. The use of whole sugarcane as a feedstuff will be discussed in Chapter 5.

Liquid molasses contains 15 to 25 percent water. It is a black, syrupy sweet solution, containing at least 46 percent sugars. It can be dried to produce dried molasses, but the added cost usually does not warrant drying. It is very low in protein content. Molasses functions primarily as an energy source and can be fed at levels up to 30 percent of the diet. At higher levels, it has laxative properties because of its high mineral content. It is particularly high in potassium. Bayley et al. (1983) fed cane molasses at 68.5 percent of the diet of pigs, as the sole source of dietary carbohydrate; the feces were black and liquid, but there were no other adverse effects.

Molasses is often included in manufactured feeds at levels of 2 to 5 percent to increase palatability. It reduces dustiness and "fines" and acts as a pellet binder to improve pellet quality. At levels above 5 to 10 percent, it may cause milling problems because of its stickiness and may form large clumps in the mixer or stick to the equipment in the feed mill. Molasses is suitable for use in liquid feeding systems for ruminants. Urea or other sources of nitrogen can be added and the liquid feed offered free-choice to roughage-fed animals to provide supplementary protein and energy. Pate et al. (1990) found that natural protein (cottonseed meal) was superior to urea as a component of a molasses supplement for beef cows fed low-quality forage.

In many developing countries, abundant supplies of molasses are available but often are not used efficiently in animal feeding. It is often considered a nuisance and is spread on fields or even on roads just to get rid of it (Sansoucy, 1986). It is difficult in these countries to transport, store, handle, and feed liquid molasses. A solution to this problem is to prepare **molasses-urea blocks**, which are much easier to transport and feed. These can be manufactured by very simple means using modest equipment. A suitable recipe is as follows: 50 percent molasses, 10 percent urea, 10 percent quicklime, 5 percent salt, 25 percent wheat bran. Another suitable recipe is as the preceding, but with 5 percent cement in place of 5 percent quicklime; still another modification is 45 percent molasses and 15 percent cement. Substitutions of other ingredients, such as peanut hulls for wheat bran, can be made. The ingredients are mixed by hand and placed in molds, which can simply be plastic pails or wooden boxes. The blocks harden in about 15 hours. Blocks prepared in this manner can be very effective as sources of supplementary energy and nitrogen for ruminant animals (Fig. 3.2).

There is much interest in the tropics in developing beef production systems based on molasses feeding. In sugar-producing countries, there are large quantities of molasses available. This high-energy feedstuff is not utilized directly by humans in large amounts. In Cuba, for instance, intensive beef production has been developed on a molasses-urea based diet. Molasses provides the major source of fermentable carbohydrate, and urea provides fermentable nitrogen. A small amount of fishmeal or other bypass protein is used to provide supplementary amino acids. A roughage source, such as sugarcane tops, bagasse, or elephant grass, is used to provide sufficient fiber to maintain normal rumen function. Protein-rich legume forages, such as leucaena, may be used as sources of both roughage and the bypass protein.

In molasses-based feeding systems, a problem referred to as **"molasses toxicity"** may develop, characterized by neurological defects such as incoordination and blindness. The clinical syndrome is identical to **polioencephalomalacia** or cerebrocortical necrosis associated with induced thiamin deficiency

FIGURE 3.2 Use of molasses-urea blocks for supplementing the feed of sheep and goats in Africa. These blocks can be easily prepared by farmers using readily available ingredients. (Courtesy of R.A. Leng.)

in ruminants (see Chapter 2). Preston and Leng (1987) have reviewed the literature on the condition. Molasses toxicity has a complicated etiology and involves an inadequate supply of glucose for the brain, induced thiamin deficiency, and rumen stasis. Inadequate glucose status occurs because molasses fermentation produces a high ratio of butyrate to propionate as end-products. Butyrate is ketogenic and propionate is glucogenic. An excess of butyrate relative to propionate results in inadequate glucose synthesis and a shortage of glucose for brain metabolism. Molasses toxicity occurs when the roughage component of the diet is insufficient. Low fiber intake results in rumen stasis and the proliferation of slow-growing microbes that produce thiaminase, destroying thiamin. The combined thiamin-glucose deficiency results in brain damage. Provision of adequate roughage is effective in preventing molasses toxicity.

In some tropical areas, sugarcane may grow productively, but there may not be a local sugar industry. In this case, the energy of sugarcane for animal feeding can be made available by crushing the cane in small presses to squeeze out the juice. The juice can be used in place of molasses in mixed feeds. The fibrous residue is useful as a feed for ruminants. McMeniman et al. (1990) developed a process for **fractionating sugarcane** to make more efficient use of it. In this process, the mature stalks are shredded to separate the sugar-rich cells from the vascular bundles and rind fibers. The shredded material is dehydrated and then screened into three fractions. The first fraction contains sugar-rich cells, the second contains less sugar and more fiber bundles, and the third fraction is predominantly fibrous rind. The sugar-rich fraction can be used as an energy source similar to grain for high-producing animals, and the other fractions can be used in maintenance diets. This or other fractionation techniques, if commercially adopted, should increase the efficiency of tropical livestock production.

Other types of molasses include citrus and wood molasses. **Citrus molasses** is a by-product of citrus processing. It has a bitter taste because of its organic acid content and may be unpalatable. **Wood molasses** is a mixture of hemicelluloses and other water-soluble carbohydrates produced in the manufacture of particle board from wood. It is used primarily as a pellet binder (hemicellulose extract). It contains pentose and hexose sugars, including xylose, which is toxic to nonruminants, causing poor growth, diarrhea, and eye cataracts in pigs. In ruminants, xylose is fermented in the rumen and is not toxic. Chalupa and Montgomery (1979) found that Masonex (a hemicellulose extract of hardboard manufacturing containing carbohydrate and polyphenolics) was highly fermentable in ruminants, producing an equivalent amount of energy as cane molasses. Zinn (1993) reported that a wood sugar concentrate by-product of paper manufacturing had about 70 percent of the net energy value of sugar cane molasses.

Table Sugar (Sucrose) and Other Simple Carbohydrates

Table sugar (sucrose) is composed of two sugars, glucose and fructose, joined together. Table sugar is not commonly fed to livestock because of its cost, but it is an excellent source of energy. Although disparaged as a source of "empty calories" in human nutrition, sucrose is an excellent source of energy when used as a component of a balanced diet (Beech et al., 1991). Animals almost universally (except for carnivores) have a "sweet tooth," and, for this reason, small amounts of sugar may be used to increase the palatability of feeds. Because of its lack of other nutrients, refined sugar is often used in purified diets, in which nutrient content is carefully controlled. With pelleted diets, levels of sucrose above approximately 10 percent of the diet result in **caramelization** of the sugar and conversion of the feed to a glasslike consistency, plugging the pellet die and thus restricting its use in pelleted diets. Caramel is burnt sugar produced after sucrose is heated in the presence of alkaline substances.

Newborn animals do not secrete **sucrase**, the enzyme that digests sucrose into glucose and fructose. Osmotic diarrhea can occur if large amounts of sucrose are included in milk replacers for pigs and calves. However, sucrase activity increases rapidly in the first few weeks of life.

Fructose has been utilized experimentally as a feed ingredient, usually in the form of high-fructose corn syrup. Coffey et al. (1987) compared high-fructose corn syrup and added fat as additives to sow diets in late gestation and lactation. Fructose has a diabetic-like effect in causing elevated plasma glucose and lowered insulin concentrations. These changes could influence energy transfer to offspring via the placenta and mammary glands. Pigs from fructose-fed sows had higher plasma-free fatty acids, which could be important to survival.

Butanediol (1,3-butanediol) is a chemically synthesized carbohydrate-like compound that has been tested as a potential feedstuff. Stahley et al. (1985) and Spence et al. (1985) independently evaluated supplementation of sow diets with butanediol as a potential means of increasing fetal energy stores and baby pig survival. In both studies, beneficial responses to butanediol were noted.

Dried By-Product Feeds

A variety of dried by-products from food processing are used for feed. **Dried bakery product** consists of reclaimed bakery products, such as bread and cookies, that have exceeded the allowable shelf life. These products are high in energy because they are often high in sugar and fat and are very palatable. Dried bakery product contains approximately 9 percent protein and 12 percent fat. Some swine enterprises near large cities utilize stale bread, pastries, pies, cookies, and other baked products in a slop feeding system, combining the bakery products with brewers grains, liquid whey, or other liquid by-products to produce a balanced diet. At least 25 percent of the grain can be replaced by dried bakery product in broiler diets with no adverse effects on bird performance (Saleh et al., 1996).

Beet pulp is the residue remaining after sugar beets have been pressed to remove the sugar-containing juice. Although it has a fairly high fiber content (approximately 16 percent crude fiber), the fiber is easily digested, even by nonruminants. It is a palatable feed and may be reconstituted with water to produce a highly palatable moist feed. Often molasses is added back to it, which further increases its energy content and palatability.

Dried citrus pulp is the residue remaining after manufacture of orange and grapefruit juice. It consists of the peel, pulp, and seeds. During the processing of citrus fruits, lime is added to inactivate the pectins in the peel, so the dried pulp has a high ash and calcium content. Because of the organic acids, citrus pulp is not very palatable and is primarily useful in ruminant feeds. It can be used at levels up to 10 percent in swine feeds; higher levels reduce feed intake and weight gain. It is utilized in dairy cattle feeding in areas where citrus is produced, such as California, Arizona, and Florida. The pectins in dried citrus pulp are fermented even more rapidly than starch. Dried citrus pulp does not have the fiber digestion-depressing effect noted with starch sources, but it creates favorable conditions for cellulolysis in the rumen (Ben-Ghedalia et al., 1989).

In many areas, by-products of other types of fruit processing are available, including **cull fruits** and **fruit pomace** (the residue remaining after preparation of juice and/or wine from apples, pears, grapes, and other fruits). Pomace is generally quite high in fiber, particularly when it consists mainly of the seeds, such as **raspberry pomace** (McDougall and Beames, 1994). Such products have very low nutritive value for livestock, however. **Grape pomace** is high in fiber because of the seeds, skins, and stems, and also has a high tannin content. The tannins reduce palatability and impair protein digestibility. **Olive by-products** such as pits and pulp have a very low energy content and are not generally worth using as feed (Al Jassim et al., 1997). **Almond hulls** (the dried, fleshy pericarp of the almond fruit) have a high content of water-soluble carbohydrates and are well utilized by cattle (Aguilar et al., 1984) and goats (Reed and Brown, 1988). Almond hulls can be used in place of alfalfa, with appropriate protein supplementation. **Cull nuts** (e.g., walnuts, hazelnuts, pistachios) are excellent feedstuffs for ruminants. They are high-energy feeds because of their high oil content. **Ground mango kernels** have potential as a feedstuff in tropical countries (Teguia, 1995).

Apple pomace has been evaluated quite extensively as an energy source for ruminants (Givens and Barber, 1987). The crude protein in apple pomace is almost completely indigestible because of the effects of tannins. Supplementation of apple pomace, with urea as the only source of added nitrogen, gives poor results in cows (Rumsey, 1979) and ewes (Rumsey and Lindahl, 1982), apparently because of a lack of nondegradable (bypass) protein. Calves and lambs were lighter and less vigorous from dams fed apple pomace-urea diets than from those receiving diets with adequate protein. Skeletal deformities in calves from cows fed apple pomace-urea diets were noted (Bovard et al., 1977), but the cause has not been identified.

Vegetable canning by-products, either in the wet or dehydrated form, are available in large quantities in some areas, such as the vegetable production areas of California. These products are normally disposed of by being sold as animal feedstuffs. Gasa et al. (1989) reported on the nutritive value of artichoke by-product, pepper and cauliflower residues, dried tomato pulp, and pea vines. Organic matter digestibilities were approximately 70 to 80 percent except for tomato pulp, which had 47 percent digestibility. These products should be fed at levels below 40 percent of the diet so as to minimize digestive disturbances and feed refusal. They are most suitable for ruminants and most commonly are fed to dairy cattle. Wet tomato pomace is produced as a by-product of tomato juice manufacture and in many areas is available at the same time as corn is ready to harvest for silage. **Wet tomato pomace** can be mixed with corn plant material to be ensiled, producing a product equivalent in nutritional value to corn silage (Weiss et al., 1997).

Mint by-product is the fibrous residue remaining after the steam distillation of mint oil from peppermint and spearmint. Because of the heat from the steaming process, mint by-product has much of its protein bound to fiber (acid detergent insoluble nitrogen). Mint by-product can be ensiled, but the silage has a very low feeding value for ruminants (Mustafa et al., 2001).

Other Industrial Processing Wastes

In particular localities, industrial by-products may be available for potential application as feedstuffs. For example, large amounts of **dried coffee residue** are a by-product of the manufacture of instant coffee. It contains about 20 percent crude protein and 18 percent oil and has some value as a ruminant feed (Ali et al., 1977), although animal acceptability of it is not good. Cacao bean processing wastes from chocolate and cocoa manufacture are another potential by-product feed. Cacao and coffee beans contain theobromine and caffeine and can be toxic. Calves fed 5 to 10 percent **chocolate waste** exhibited hyperexcitability, exaggerated gaits, excessively alert appearance, and mortality (Curtis and Griffiths, 1972). McNaughton et al. (1997) found that a chocolate confectionery product not meeting standards for human consumption could be fed to pigs at inclusion levels up to 30 percent of the diet with no adverse effects on growth or carcass quality. Yang et al. (1997) used a **milk chocolate product** containing about one third each whole milk, cocoa, and sucrose. They found it could be used as a replacement for dried whey in pig starter diets, up to 5 percent of the diet. Higher levels reduced pig performance. Perhaps not unexpectedly, pigs strongly preferred diets with milk chocolate over dried whey.

Fats and Oils

Fats and oils, which contain about 2.25 times as much digestible energy as carbohydrates, are concentrated sources of energy for animal feeds. High-energy diets must contain added fat to achieve a high-energy concentration. Besides contributing energy, fats increase palatability of feeds to most animals. Fat acts as a lubricant in feed manufacturing and improves pelleting efficiency. It reduces dustiness of feeds and "fines" in pelleted diets.

The distinction between **fats** and **oils** is based on their physical properties at room temperature: fats are solid and oils are liquid. Two main chemical factors are involved in determining if a lipid is a fat or an oil. The more unsaturated a lipid, the lower its melting point. Thus a lipid containing mainly saturated fatty acids will be a solid at room temperature, a lipid with a moderate degree of unsaturation may turn solid in a refrigerator, whereas a highly unsaturated lipid may remain a liquid in a freezer. **Unsaturation** refers to the presence of double bonds in a fatty acid. The other factor affecting the physical state of a lipid is the number of carbon atoms in the fatty acids. The shorter the "carbon chain" of the fatty acid, the greater the tendency of a lipid containing it to be a liquid at room temperature. Thus coconut and palm oils are liquid, even though they are almost completely saturated, because these **tropical oils** contain fatty acids with 12 or 14 carbons. In contrast, most fatty acids in fats and oils have 16, 18, or 20 carbons.

Most fats are of animal origin. **Animal fats** include tallow (from beef and mutton/lamb) and lard (from pork). At one time, "lard-type" pigs were produced because lard was a valuable commodity. It has largely been replaced by lipids of plant origin. Lean meat is now desired by consumers. Immense quantities of waste fat are trimmed off carcasses and recycled (inefficiently) as feed ingredients.

Animal fats used in feeding are mainly beef tallow and mixed animal fat and grease. **Grease** is animal fat that solidifies at a lower temperature than tallow. Sources of the fat include slaughterhouse waste, trimmings, and fat from rendering plants. Animals processed at rendering plants include those dying on farms from various causes. In most areas, laws require that they be removed and disposed of in an acceptable manner, which generally means being rendered or buried. Products from rendering plants include hides, meat, and bone meal (to be discussed in Chapter 4) and animal fat. **Restaurant fat** and grease are also disposed of by rendering. The feed industry provides the main outlet for recycled restaurant fat and grease. A new calorie-free fat substitute (trade name Olestra) has been developed; it is produced by a chemical reaction between sucrose and soybean oil. Olestra would reduce the fat and calorie consumption of people when used as a substitute for fat. However, its widespread use in the restaurant business could cause considerable problems in the use of recycled fat by the feed industry because an indigestible product would be of no value as an energy source for animals.

Because saturated fatty acids are absorbed less efficiently, fats high in saturated fatty acids usually have a somewhat lower digestible energy value than oils. Addition of some unsaturated lipid improves the digestibility of fats. **Hydrolyzd fats** are by-products of soap manufacture and consist primarily of free fatty acids (by law, not less than 85 percent total fatty acids).

Oils are of vegetable (e.g., corn oil) or marine animal (e.g., cod liver oil) origin. They have a high content of unsaturated fatty acids (except for coconut and palm oils). Many plants are grown to produce edible and/or industrial oils. Some common **edible oils** include corn, cottonseed, soybean, olive, safflower, sunflower, rapeseed (canola), and peanut oil. **Industrial oils** include castorbean, rapeseed, jojoba, linseed (flax), and tung oil.

Most edible oils are used for human consumption as cooking and salad oil and as margarine. They are usually too expensive for use in animal feeding. Great expansion of **palm oil production** in tropical countries, coupled with efforts of the U.S. oil crop producers to discourage use of palm oils in edible applications, may result in a surplus of palm and coconut oils, which will become economically competitive as feed ingredients. Because palm and coconut oils are saturated, they do not have the serum-cholesterol-lowering properties of the unsaturated vegetable oils.

Problems Associated with Use of Fats and Oils Lipids containing unsaturated fatty acids are susceptible to the development of **rancidity.** This is a process by which oxygen reacts with double bonds and produces various peroxides and free radicals that are chemically very reactive. The products of rancidity have an objectionable odor and adversely affect palatability of a feed. They may also be toxic. Rancidity causes destruction of a number of vitamins, particularly the fat-soluble ones. Because rancidity involves oxidation of double bonds, vegetable oils are more susceptible to this problem than tallow and other animal fats. Rancidity can be prevented or slowed by the addition of **antioxidants** to fats and feeds. Vitamin E is the major natural antioxidant, although a number of synthetic antioxidants such as ethoxyquin (santoquin), butylated hydroxyanisole (BHA), and butylated hydroxytoluene (BHT) are also used.

In whole grains, the oil is protected against rancidity by being compartmentalized in the cell structure. Grinding of grains with a high oil content, such as corn, results in their becoming susceptible to rancidity. Grains and seeds like soybeans contain the enzyme lipoxidase (lipoxygenase), which stimulates rancidity. When the structure of the seed is broken by feed processing, lipoxidase, oxygen, and oil are mixed together and rancidity occurs rapidly. Heat processing of full fat soybeans reduces their susceptibility to rancidity by inactivating lipoxidase.

Use of Fats in Animal Diets Added fat is normally used at levels of 3 to 5 percent of the diet. At higher levels, there are problems in maintaining pellet quality in pelleted diets, with a tendency for the pellets to crumble and break apart. In nonpelleted diets, higher fat levels cause the feed to "bridge" in storage bins and feeders, preventing free flow of the feed, particularly in cold weather. In hot weather, it causes the feed to be greasy, which may present handling and aesthetic problems (causing feed bags to be greasy, etc.).

Nonruminant Diets So long as a diet is kept balanced with respect to other nutrients, fat can be fed at high levels to nonruminants. Diets with 20 to 30 percent fat are palatable and readily consumed by swine and poultry. The upper limit of fat inclusion for monogastrics is determined by feed manufacturing and handling problems, not by animal acceptance. The use of high-fat diets necessitates recognition of the relationship between dietary energy level and feed intake. Nonruminant animals regulate their voluntary feed intake to consume sufficient feed to meet their energy requirements. If they are fed a low-energy diet, they increase feed intake to compensate. If a high-energy diet is used, feed intake is depressed. (These relationships are discussed in Chapter 9.) Thus, when high-fat diets are used, the concentration of protein and other nutrients must be increased because the amount required is consumed in a smaller amount of feed.

A special application of dietary fat is the feeding of **high-fat diets** to sows near the end of gestation. Baby pigs are born in an immature state and have a limited amount of stored energy (as liver glycogen) at birth. High mortality occurs in the neonatal period. Feeding fat to sows in late gestation can increase the survival rate of baby pigs by causing them to be born with a higher level of stored energy. A minimum of 7.5 percent added fat is needed, and levels higher than 15 percent provide no additional advantage. High-fat diets should be fed for at least 7 days prior to farrowing. Although this procedure has a favorable effect on baby pig survival, it is not always economical, depending on current prices of fat sources. An analysis of prevailing conditions needs to be made to determine if added fat is cost-effective. Fat feeding in early lactation increases milk production, which contributes to improved baby pig survival (Coffey et al., 1987).

Medium-chain triglycerides (MCT), such as those in coconut oil, are especially effective in increasing baby pig survival when included in sow diets (Azain, 1993).

A product known as **"dried fat"** is available commercially. Dried-fat products are composed of carrier substances that absorb and retain the fat, such as whey, wheat bran, and vermiculite. Dried fat can be bagged and handled like conventional feed ingredients. It is more expensive than other fat sources, so it is usually used only in high-cost diets, such as pig starter diets.

The nature of the dietary fat influences the properties of the body fat of swine and poultry. The **fatty acid composition of body fat** resembles that of dietary fat. Thus, if a pig is fed corn oil, its body fat will resemble corn oil. Diets containing high levels of unsaturated fats should not be fed to market animals for several weeks before slaughter. Animals synthesize fat from carbohydrate; the synthesized fat is primarily saturated, which results in a hard or firm body fat. The "soft pork" condition is produced when pigs are fed sources of unsaturated fats, such as "hogging off" peanuts, soybeans, and other oil-rich seeds (Fig. 3.3). This procedure is not commonly carried out now. In times past, hogs were often turned out into fields to harvest such crops as peanuts, which is a less efficient method than mechanical harvesting of crops and feeding the animals in confinement. Pigs may be used to glean peanuts remaining in the field after harvest (West and Myer, 1987; Myer et al., 1985); from 500 to 1000 kg/ha (hectare) of peanuts commonly remain after harvesting.

Another way of increasing dietary energy and fat contents is through the use of **full-fat oilseeds,** such as soybeans, rapeseed (canola), and sunflower seed. Oilseeds can be processed with an extruder, and the extruded seeds can be used as a protein-energy source. Extruded soybeans, also known as full-fat soybeans, are a common feed ingredient. It is necessary to crush or grind the seeds, particularly seeds with a hard seed coat. If the seed coat is not broken, the seed may pass through the gut undigested. Use of full-fat oilseeds as feedstuffs is discussed in Chapter 4.

Ruminant Diets Ruminants are less tolerant of dietary fat than are nonruminants. Fat levels exceeding 5 to 7 percent of the diet may result in digestive disturbances, diarrhea, and reduced feed intake. Fat digestion occurs primarily in the small intestine, where the emulsifying agent, bile, is secreted. Bile solubilizes fat in an aqueous medium by emulsifying it and is, therefore, necessary for fat digestion. In the rumen, there is a lack of **emulsifying agents** and fat-digesting enzymes (lipases); therefore, high-fat intake causes rumen contents to be coated

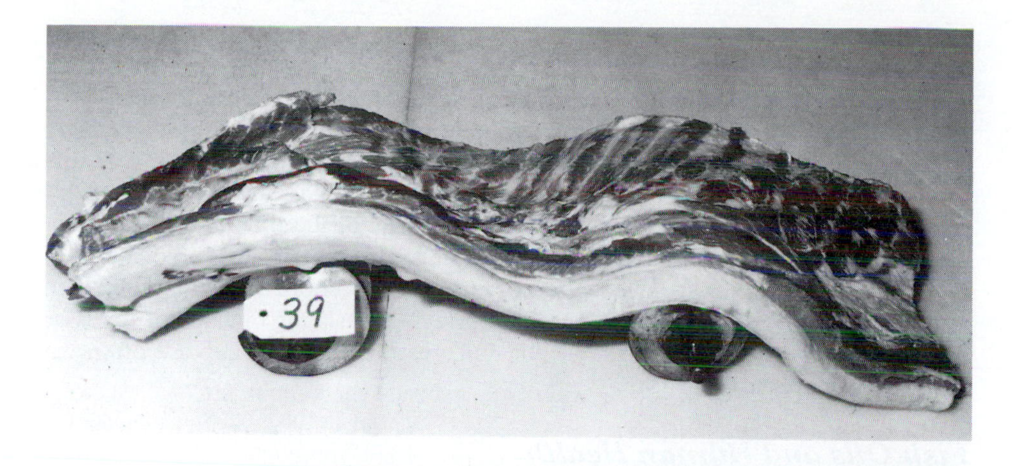

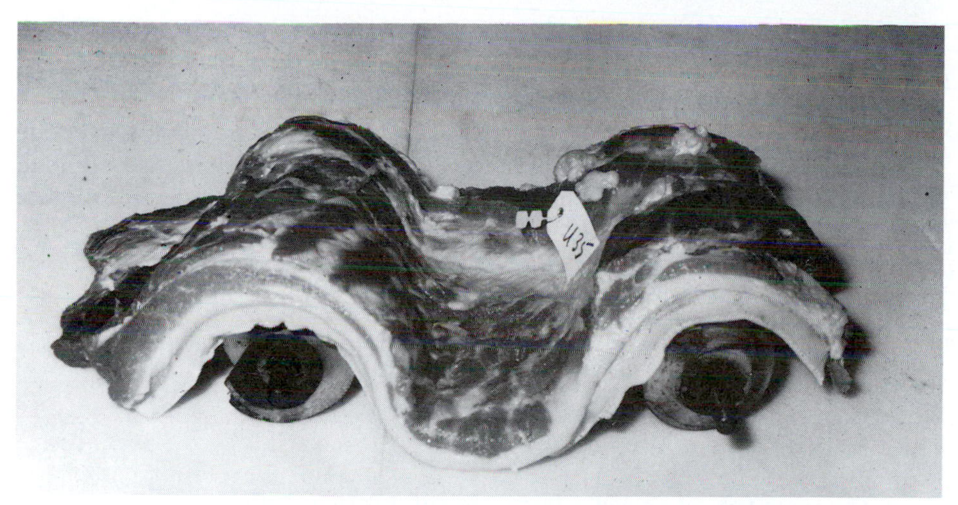

FIGURE 3.3 Examples of bacon strips from a pig fed a corn-soy diet (top), and from a pig used to "hog off" peanuts (bottom) that shows the soft pork condition caused by the unsaturated fatty acids in the peanut oil.(Courtesy of R.O. Myer.)

with indigestible fat, which interferes with normal digestive processes, particularly of fiber particles. Fats also have an inhibitory effect on rumen microorganisms. In the rumen, unsaturated fatty acids are converted to saturated fats by a process called **hydrogenation** (the addition of hydrogen to double bonds). For this reason, the body fat of ruminants tends to be fairly saturated, regardless of the type of diet fed. The hydrogenation of linoleic acid in the rumen results in the formation of some isomers of linoleic acid that are collectively referred to as **conjugated linoleic acid (CLA).** The CLA have numerous health benefits in humans, such as reduction of cardiovascular disease. CLA has been shown in laboratory animals to have potent anticancer activity (Belury, 2002). Food products of ruminant origin such as dairy products and beef are excellent dietary sources of CLA. CLA can also be fed to swine and poultry to increase the CLA content of eggs and meat.

The negative effects of fat on rumen function can be partially overcome by adding a source of cations, such as limestone. Calcium forms insoluble **calcium soaps** with fatty acids and prevents them from inhibiting rumen microbes. Calcium soaps are a good source of fat in cattle rations because they do not adversely affect rumen fermentation (Schneider et al., 1988). Calcium soaps are well utilized in ruminants; the pH of the ingesta exiting the abomasum is very low, and it remains low in the proximal small intestine because of the limited buffering activity of pancreatic secretions in ruminants. The low pH causes the dissociation of calcium soaps, allowing the absorption of fatty acids. In nonruminants, soaps are poorly digested. A variety of commercial products, consisting of calcium soaps or prilled free fatty acids, are available for dairy cattle feeding. They are often referred to as **rumen inert fats.**

Fish Oils and Human Health

There has been interest in fish oils **(omega-3 fatty acids)** for their apparent beneficial effects on human health. Several investigators have noted that groups of people, such as the Greenland Eskimos, who have a high intake of fish, have a low incidence of coronary heart disease, low serum cholesterol, and low triglyceride levels. These effects have been associated with the consumption of fish oils high in omega-3 fatty acids. The common unsaturated fatty acids in vegetable oils are omega-6. The omega designation refers to the location of the double bond closest to the terminal methyl (CH_3) group of the fatty acid. Thus, in an omega-3 fatty acid, the first double bond is on the third carbon from the terminal end:

$$CH_3{-}CH_2{-}CH{=}CH{-}(CH_2)_n{-}COOH$$

omega-3 fatty acid

$$CH_3{-}CH_2{-}CH_2{-}CH_2{-}CH_2{-}CH{=}CH{-}(CH_2)_n{-}COOH$$

omega-6 fatty acid

The main fish oil omega-3 fatty acids of nutritional interest are eicosapentaenoic acid (EPA; 20:5, n-3) and docosahexanoic acid (DHA; 22:6, *n*-3). The beneficial effects of fish oils on cardiovascular disease seem to be associated mainly with the EPA. The omega-3 fatty acids lower total serum lipid and cholesterol levels and also modulate the conversion of arachidonic acid to prostaglandins, prostacyclins, thromboxanes, and leukotrienes. These substances have hormonal or hormonelike roles in vasoconstriction and blood platelet aggregation.

The enthusiasm for the healthful effects of fish oils should be tempered by the fact that adverse effects are also noted. These include prolonged bleeding times, reduced platelet counts, and reduced platelet adhesiveness, associated with a change in the platelet fatty acid composition from the omega-6 to omega-3 fatty acids. The effects on platelets seem to result in a low rate of thrombic disorders in Greenlanders and Japanese with a high fish intake, but the tendency to bleed readily cannot be considered desirable.

The omega-3 fatty acid content of animal products can be increased by the use of fishmeal as a feed ingredient. Hulan et al. (1988, 1989) demonstrated that feeding redfish (*Sebastes* spp.) meal to broilers enhanced the omega-3 fatty acid content of their tissues to levels comparable to levels in fish, without producing fishy flavors or any other undesirable effects on the meat. Omega-3 fatty-acid-enriched eggs (so-called **designer eggs**) can be produced by feeding laying hens full-fat fish meal (Nash et al., 1995) or full-fat flaxseed (Leskanich and Noble, 1997; Scheidler and Froning, 1996). **Flaxseed** is one of the richest plant sources of omega-3 fatty acids, such as linolenic acid (Van Elswyk, 1997). These "designer eggs" are acceptable in taste to consumers (Scheidler et al., 1997) and may have value in healthful diets for avoiding coronary heart disease.

The omega-3 fatty acids are dietary requirements of fish, and a source of linolenic acid (*w*-3) needs to be included in fish diets (see Chapter 23). Signs of deficiency include poor growth, liver and heart lesions, and a shock syndrome. The omega-3 fatty acids originate in plankton and accumulate in the food chain of marine fish.

Omega-3 and omega-6 fatty acids are especially important as essential fatty acids for **dogs.** Sources of these fatty acids (e.g., evening primrose oil) are used in dog feeding to prevent and treat dermatitis.

Milk By-Products

The principal milk by-product that is used as an energy source is **whey,** the liquid residue remaining after protein and fat are removed as curd in the manufacture of cheese. Most of the lactose, minerals, and water-soluble protein are left in the whey fraction. **Liquid whey** is 93 to 96 percent water and is a serious pollution problem for cheese factories. The high water content is an impediment to its use as a feed, both because of the low nutrient density and the expense of transporting water from the cheese factory to the farm. When used as a sole source of drinking water, whey can be consumed at sufficient levels for it to contribute significantly as a nutrient source (Maswaure and Mandisodza, 1995).

Condensed whey is prepared by removing part of the water and usually has a dry-matter content of 40 to 60 percent. Dried whey is a dry white powder, containing 70 to 75 percent lactose and 12 to 13 percent crude protein. Dried whey is used in milk replacers and starter diets for baby pigs. Whey is digested very efficiently by young animals but may cause digestive disturbances due to lactose intolerance in older animals. The physiological basis of this is related to the activity of lactase enzyme in the intestine. Because lactose (milk sugar) is the primary energy source for suckling animals, it is essential that they be capable of digesting lactose. In weaned animals (except humans), milk is not a component of the diet, so lactase activity in the gut is unnecessary and as a result is low or absent. In weaned animals, **lactose intolerance** is caused by the lack of lactase secretion. As a result, lactose cannot be digested in the small intestine and reaches the large intestine. Here it is a substrate for bacterial fermentation, causing production of gas and irritant organic acids, leading to digestive disturbances, flatulence, and diarrhea. Additionally, the soluble lactose in the contents of the large intestine draws water from the blood through osmotic action, further contributing to diarrhea and dehydration. For growing-finishing pigs, levels of 20 to 30 percent dried whey in the diet are acceptable; higher levels may cause digestive disturbances and reduce performance.

Liquid whey can be stored in tanks or pits. If it is allowed to ferment, organic acids are produced that act as preservatives, but the palatability of this "sour whey" is lower than for fresh or "sweet whey." Formaldehyde (0.1 percent) is an effective preservative, but it is not approved for this use in the United States.

Liquid whey can be processed by ultrafiltration to increase the proportion of solids, producing **whey protein concentrate (WPC)** (Cinq-Mars, 1986a,b). These authors found WPC to be a useful protein supplement for baby pigs. Fermentation with **Lactobacillus** spp. to reduce the lactose level did not alter its feeding value.

Modern milk processors are producing a variety of new products for the feed industry, including crystalline whey, deproteinized whey, and whey permeate. These products have numerous applications in the feeding of baby pigs (Nessmith et al., 1997).

QUESTIONS AND STUDY GUIDE

1. Why is it advisable to process cassava and potatoes before using them as feeds? How are these products usually processed?
2. High levels of molasses in the diet may cause diarrhea in weaned pigs. Why? To what extent, if any, would the sucrose content contribute to this problem?
3. Why is coconut oil a liquid even though it is almost entirely saturated? Why is there concern about the use of coconut oil in the human diet?
4. Which is more likely to go rancid: beef tallow or soybean oil? Why?
5. Explain the effects of grinding and extrusion on the susceptibility of corn, wheat, and soybeans to rancidity.
6. Discuss the importance of omega-3 fatty acids in human nutrition and their potential implications in animal nutrition.
7. Suppose you are a swine producer. A cheese factory 10 miles from your farm has offered you liquid whey; the whey is free, but you have to sign a contract agreeing to take it all. You have enough pigs that you could use the entire quantity of whey produced. Is this a deal you cannot refuse? Explain.

REFERENCES

Cassava, Potatoes, Sweet Potatoes, Other Roots and Tubers, and Other Fruits and Vegetables

Bouwkamp, J. C. ed. 1985. *Sweet Potato Products: A Natural Resource for the Tropics*. Boca Raton, FL: CRC Press.

Clark, P., D. I. Givens, and J. M. Brunnen. 1987. The chemical composition, digestibility and energy value of fodder-beet roots. *Anim. Feed Sci. Tech.* 18:225–231.

Dao, L. and M. Friedman. 1994. Chlorophyll, chlorogenic acid, glycoalkaloids and protease inhibitor content of fresh and green potatoes. *J. Agric. Food Chem.* 42:633–639.

Dhua, R. S. and S. K. Sen. 1989. Seasonal changes in active tannin content in pulp and peel of the Giant Governor banana during fruit growth and maturity. *Trop. Agric.* 66: 284–285.

Edwards, S. A., C. B. Fairbairn, and A. L. Capper. 1986. Liquid potato feed for finishing pigs: Feeding value, inclusion rate and storage properties. *Anim. Feed Sci. Tech.* 15:129–139.

Erdman, M. D. and B. A. Erdman. 1984. Arrowroot (*Maranta arundinacea*), food, feed, fuel and fiber resource. *Econ. Bot.* 38:332–341.

Farnworth, E. R., H. W. Modler, and D. A. Mackie. 1995. Adding Jerusalem artichoke (*Helianthus tuberosa* L.) to weanling pig diets and the effect on manure composition and characteristics. *Anim. Feed Sci. Tech.* 55:153–160.

Fekadu, D. and I. Ledin. 1997. Weight and chemical composition of the plant parts of enset (*Ensete ventricosum*) and the intake and degradability of enset by cattle. *Livestock Prod. Sci.* 49:249–257.

Graham, H. and P. Aman. 1986. Composition and digestion in the pig gastrointestinal tract of Jerusalem artichoke tubers. *Food Chem.* 22:67–76.

Holzer, Z., Y. Aharoni, V. Lubimov, and A. Brosh. 1997. The feasibility of replacement of grain by tapioca in diets for growing-fattening cattle. *Anim. Feed Sci. Tech.* 64:133–141.

Hulan, H. W., F. G. Proudfoot, and C. G. Zarkadas. 1982a. Potato waste meal. I. Compositional analyses. *Can. J. Anim. Sci.* 62:1161–1169. 1982b. Potato waste meal. II. The nutritive value and quality for broiler chicken. *Can. J. Anim. Sci.* 62:1171–1180.

Job, T. A., J. A. Oluyemi, and S. Entonu. 1979. Replacing maize with sweet potato in diets for chicks. *Brit. Poult. Sci.* 20:515–519.

Lekule, F. P., L. A. Mtenga, and A. Just. 1988. Total replacement of cereals by cassava and rice polishings in diets of growing-finishing pigs. *Trop. Agric.* 65:321–324.

Nicholson, J. W. G. and J. G. Allen. 1989. The distribution of trypsin and chymotrypsin inhibitors in potato tubers. *Can. J. Anim. Sci.* 69:513–515.

Nicholson, J. W. G., P. M. Snoddon, and P. R. Dean. 1988. Digestibility and acceptability of potato steam peel by pigs. *Can. J. Anim. Sci.* 68:233–239.

Oke, O. L. 1984. The use of cassava as a pig feed. *Nutr. Abst. Rev.* 54:301–314.

Osei, S. A. and I. K. Twumasi. 1989. Effects of oven-dried cassava peel meal on the performance and carcass characteristics of broiler chickens. *Anim. Feed Sci. Tech.* 24:247–252.

Panigrahi, S. and B. Francis. 1982. Digestibility and possible toxicity of the yam *Dioscorea alata. Nutr. Rep. Int.* 26:1007–1013.

Pond, W. G. and J. H. Maner. 1984. *Swine Production and Nutrition.* Westport, CT: AVI Publishing Company.

Preston, T. R. and R. A. Leng. 1987. *Matching Ruminant Production Systems with Available Resources in the Tropics and Sub-Tropics.* Armidale, Australia: Penambul Books.

Rahnema, S. and R. Borton. 2000. Effect of consumption of potato chip scraps on the performance of pigs. *J. Anim. Sci.* 78:2021–2025.

Ravindran, V. and R. Sivakanesan. 1996. Replacement of maize with sweet potato (*Ipomea batatas* L.) tuber meal in broiler diets. *Brit. Poult. Sci.* 37:95–103.

Ravindran, V., R. Sivakanesan, and H. W. Cyril. 1996. Nutritive value of raw and processed colocasia (*Colocasia esculenta*) corn meal for poultry. *Anim. Feed Sci. Tech.* 57: 335–345.

Rooke, J. A., A. R. Moss, A. I. Mathers, and R. Crawshaw. 1997. Assessment using sheep of the nutritive value of liquid potato feed and partially fried potato chips (french fries). *Anim. Feed Sci. Tech.* 64:243–256.

Sabri, M. S., N. W. Offer, and D. J. Roberts. 1988a. Effects of fodder beet on rumen metabolism. *Anim. Prod.* 47:429–434. 1988b. A note on the apparent digestibility of fodder beet roots in sheep. *Anim. Prod.* 47:509–511.

Taylor, J. A. and I. G. Partridge. 1987. A note on the performance of growing pigs given diets containing manioc. *Anim. Prod.* 44:457–459.

Tor-Agbidye, Y., S. Gelaye, S. L. Louis, and G. E. Cooper. 1990. Performance and carcass traits of growing-finishing swine fed diets containing sweet potato meal or corn. *J. Anim. Sci.* 68:1323–1328.

Van Lunen, T. A., D. M. Anderson, A. M. St. Laurent, J. W. G. Nicholson, and P. R. Dean. 1989. The feeding value of potato steam peel for growing-finishing pigs. *Can. J. Anim. Sci.* 69:225–234.

Viswanathan, K., R. Kadirvel, and D. Chandrasekaran. 1989. Nutritive value of banana stalk (*Musa cavendishi*) as a feed for sheep. *Anim. Feed Sci. Tech.* 22:327–332.

Walker, N. 1985. Cassava and tallow in diets for growing pigs. *Anim. Prod.* 40:345–350.

Whittemore, C. T. 1977. The potato (*Solanum tuberosum*) as a source of nutrients for pigs, calves and fowl—A review. *Anim. Feed Sci. Tech.* 2:171–190.

Wu, J. F. 1980. Energy value of sweet potato chips for young swine. *J. Anim. Sci.* 51:1261–1265.

Zinn, R. A. and E. J. DePeters. 1991. Comparative feeding value of tapioca pellets for feedlot cattle. *J. Anim. Sci.* 69:4726–4733.

Molasses, Sucrose, and Other Simple Carbohydrates

Baker, D. H. 1977. Xylose and xylan utilization by the chick. *Poult. Sci.* 56:2105–2107.

Bayley, H. S., V. Figueroa, J. Ly, A. Maylin, and A. Perez. 1983. Utilization of sugar cane final molasses by the pig: Energy metabolism. *Can. J. Anim. Sci.* 63:455–462.

Beech, S. A., R. Elliot, and E. S. Batterham. 1991. Sucrose as an energy source for growing pigs:

A comparison of the effects of sucrose, starch and glucose on energy and protein retention. *Anim. Prod.* 53:383–393.

Chalupa, W. and A. Montgomery. 1979. Fermentability of Masonex and cane molasses. *J. Anim. Sci.* 48:393–400.

Coffey, M. T., J. A. Yates, and G. E. Combs. 1987. Effects of feeding sows fat or fructose during late gestation and lactation. *J. Anim. Sci.* 65:1249–1256.

Chen, M. C., C. B. Ammerman, P. R. Henry, A. Z. Palmer, and S. K. Long. 1981. Citrus condensed molasses solubles as an energy source for ruminants. *J. Anim. Sci.* 53:253–259.

McMeniman, N. P., R. Elliott, and M. O'sullivan. 1990. The use of dried sugar cane fractions as the principal energy source in sheep rations. *Anim. Feed Sci. Tech.* 28:155–168.

Pate, F. M., D. W. Sanson, and R. V. Machen. 1990. Value of a molasses mixture containing natural protein as a supplement to brood cows offered low quality forages. *J. Anim. Sci.* 68:618–623.

Sanchez, M. and T. R. Preston. 1980. Sugar cane juice as cattle feed: Comparisons with molasses in the absence or presence of protein supplement. *Trop. Anim. Prod.* 5:117–124.

Sansoucy, R. 1986. Manufacture of molasses-urea blocks. *World Anim. Rev.* 57:40–48.

Spence, C. A., R. D. Boyd, C. D. Wray, and D. M. Whitehead. 1985. Effect of 1,3-butanediol and short chain acids in sow gestation diets on maternal plasma metabolites and fetal energy storage. *J. Anim. Sci.* 60:1280–1287.

Stahley, T. S., G. L. Cromwell, and H. J. Monegue. 1985. Effects of prepartum administration 1,3-butanediol to sows on growth and survival of neonatal pigs. *J. Anim. Sci.* 61:1485–1491.

Zinn, R. A. 1993. Comparative feeding value of wood sugar concentrate and cane molasses for feedlot cattle. *J. Anim. Sci.* 71:2297–2302.

By-Product Feeds

Aguilar, A. A., N. E. Smith, and R. L. Baldwin. 1984. Nutritional value of almond hulls for dairy cows. *J. Dairy Sci.* 67:97–103.

Ali, M. M., J. H. Topps, and T. B. Miller. 1977. Evaluation of dried coffee residues as a component of diet for ruminants. *Proc. Nutr. Soc.* 36:67A.

Al Jassim, R. A. M., F. T. Awadeh, and A. Abodabos. 1997. Supplementary feeding value of urea-treated olive cake when fed to growing Awassi lambs. *Anim. Feed Sci. Tech.* 64:287–292.

Ben-Ghedalia, D., E. Yosef, J. Miron, and Y. Est. 1989. The effects of starch- and pectin-rich diets on quantitative aspects of digestion in sheep. *Anim. Feed Sci. Tech.* 24:289–298.

Bhattacharya, A. N., T. M. Khan, and M. Uwayjan. 1977. Dried beet pulp as a sole source of energy in beef and sheep rations. *J. Anim. Sci.* 41:616–621.

Bovard, K. P., T. S. Rumsey, P.R. Oltjen, J. P. Fontenot, and B.M. Priode. 1977. Supplementation of apple pomace with nonprotein nitrogen for gestating beef cows. II. Skeletal abnormalities of calves. *J. Anim. Sci.* 46:523–531.

Champe, K. A. and D. C. Church. 1980. Digestibility of dried bakery product by sheep. *J. Anim. Sci.* 51:25–27.

Curtis, P. E. and J. E. Griffiths. 1972. Suspected chocolate poisoning of calves. *Vet. Rec.* 90:313.

Gasa, J., C. Castrillo, M. D. Baucells, and J. A. Guada. 1989. By-products from the canning industry as feedstuffs for ruminants: Digestibility and its prediction from chemical composition and laboratory assays. *Anim. Feed Sci. Tech.* 25:67–77.

Givens, D. I. and W. P. Barber. 1987. Nutritive value of apple pomace for ruminants. *Anim. Feed Sci. Tech.* 16:311–315.

Kelly, P. 1983. Sugar beet pulp—A review. *Anim. Feed Sci. Tech.* 8:1–18.

McDougall, N. R. and R. M. Beames. 1994. Composition of raspberry pomace and its nutritive value for monogastric animals. *Anim. Feed Sci. Tech.* 45:139–148.

McNaughton, E. P., R. O. Ball, and R. M. Friendship. 1997. The effects of feeding a chocolate product on growth performance and meat quality of finishing swine. *Can. J. Anim. Sci.* 77:1–8.

Mustafa, A. F., J. J. McKinnon, and D. A. Christensen. 2001. Effects of feeding ensiled spearmint (*Mentha spicata*) by-product on nutrient utilization and ruminal fermentation of steers. *Anim. Feed Sci. Tech.* 92:33–43.

National Research Council (NRC). 1983. *Underutilized Resources as Animal Feedstuffs*. Washington, DC: National Academy Press.

Ravindran, V. and R. Sivakanesan. 1995. Breadfruit (*Artocarpus communis*) meal: Nutrient composition and feeding value for broilers. *J. Sci. Food Agric*. 69:379–383.

Reed, B. A. and D. L. Brown. 1988. Almond hulls in diets for lactating goats: Effects on yield and composition of milk, feed intake, and digestibility. *J. Dairy Sci*. 71:530–533.

Rumsey, T. S. 1979. Addition of trace minerals, starch, and straw to apple pomace-urea diets of gestating beef cows. *J. Anim. Sci*. 48:495–499.

Rumsey, T. S. and I. L. Lindahl. 1982. Apple pomace and urea for gestating ewes. *J. Anim. Sci*. 54:221–234.

Saleh, E. A., S. E. Watkins, and P. W. Waldroup. 1996. High-level usage of dried bakery product in broiler diets. *J. Appl. Poult. Res*. 5:33–38.

Teguia, A. 1995. Substituting ground mango kernels (*Mangifera indica* L.) for maize in broiler starter diets. *Anim. Feed. Sci. Tech*. 56:155–158.

Weiss, W. P., D. L. Frobose, and M. E. Koch. 1997. Wet tomato pomace ensiled with corn plants for dairy cows. *J. Dairy Sci*. 80:2896–2900.

Yang, H., J. A. Kerber, J. E. Pettigrew, L. J. Johnston, and R. D. Walker. 1997. Evaluation of milk chocolate product as a substitute for whey in pig starter diets. *J. Anim. Sci*. 75:423–429.

Fats

Azain, M.J. 1993. Effects of adding medium-chain triglycerides to sow diets during late gestation and early lactation on litter performance. *J. Anim. Sci*. 71:3011–3019.

Belury, M.A. 2002. Inhibition of carcinogenesis by conjugated linoleic acids: Potential mechanisms of action. *J. Nutr*. 132:2995–2998.

Coffey, M. T., J. A. Yates, and G. E. Combs. 1987. Effects of feeding sows fat or fructose during late gestation and lactation. *J. Anim. Sci*. 65:1249–1256.

Fraley, J. R., D. A. Cook, C. L. Johnson, and A. H. Jensen. 1988. An evaluation of a dry-fat product as a source of supplemental energy in pig diets. *J. Anim. Sci*. 66:1697–1702.

Hulan, H. W., R. G. Ackman, W. M. N. Ratnayake, and F. G. Proudfoot. 1988. Omega-3 fatty acid levels and performance of broiler chickens fed redfish meal or redfish oil. *Can. J. Anim. Sci*. 68:533–547. 1989. Omega-3 fatty acid levels and general performance of commercial broilers fed practical levels of redfish meal. *Poult. Sci*. 68:153–162.

Leskanich, C. O. and R. C. Noble. 1997. Manipulation of the n-3 polyunsaturated fatty acid composition of avian eggs and meat. *World's Poult. Sci. J*. 53:155–183.

Myer, R. O., R. L. West, D. W. Gorbet, and C. L. Brasher, 1985. Performance and carcass characteristics of swine as affected by the consumption of peanuts remaining in the field after harvest. *J. Anim. Sci*. 61:1378–1386.

Nash, D. M., R. M. G. Hamilton, and H. W. Hulan. 1995. The effect of dietary herring meal on the omega-3 fatty acid content of plasma and egg yolk lipids of laying hens. *Can. J. Anim. Sci*. 75:247–253.

Palmquist, D. L. and T. C. Jenkins. 1980. Fat in lactation rations: A review. *J. Dairy Sci*. 63:1–14.

Pettigrew, J. E. 1981. Supplemental dietary fat for peripartal sows: A review. *J. Anim. Sci*. 53:107–117.

Scheidler, S. E. and G. W. Froning. 1996. The combined influence of dietary flaxseed variety, level, form, and storage conditions on egg production and composition among vitamin E-supplemented hens. *Poult. Sci*. 75:1221–1226.

Scheidler, S. E., G. Froning, and S. Cuppett. 1997. Studies of consumer acceptance of high omega-3 fatty acid-enriched eggs. *Poult. Sci*. 6:137–146.

Sharma, H. R., B. White, and J. R. Ingalls. 1986. Utilization of whole rape (canola) seed and sunflower seeds as sources of energy and protein in calf starter diets. *Anim. Feed Sci. Tech*. 15:101–112.

Schneider, P., D. Sklan, W. Chalupa, and D. S. Kronfeld. 1988. Feeding calcium salts of fatty acids to lactating cows. *J. Dairy Sci*. 71:2143–2150.

Van Elswyk, M. E. 1997. Nutritional and physiological effects of flax seed in diets for laying fowl. *World's Poult. Sci. J*. 53:253–264.

West, R. L. and R. O. Myer. 1987. Carcass and meat quality characteristics and backfat fatty acid composition of swine as affected by the consumption of peanuts remaining in the field after harvest. *J. Anim. Sci.* 65:475–480.

Milk By-Products

Cinq-Mars, D., B. Belanger, B. LaChance, and G. J. Brisson, 1986a. Fermented whey protein concentrate fed to weaned piglets. *Can. J. Anim. Sci.* 66:1117–1123.

Cinq-Mars, D., B. Belanger, B. LaChance, and G. J. Brisson. 1986b. Performance of early weaned piglets fed diets containing various amounts of whey protein concentrate. *J. Anim. Sci.* 63:145–150.

Maswaure, S. M. and K. T. Mandisodza. 1995. An evaluation of the performance of weaner pigs fed diets incorporating fresh sweet liquid whey. *Anim. Feed Sci. Tech.* 54:193–201.

Nessmith, W. B., Jr., J. L. Nelssen, M. D. Tokach, R. D. Goodband, and J.R. Bergstrom. 1997. Effects of substituting deproteinized whey and (or) crystalline lactose for dried whey on weanling pig performance. *J. Anim. Sci.* 75:3222–3228.

CHAPTER 4

Protein Sources

Objectives

1. To describe important features of common protein supplements:
   - oilmeals that are by-products of vegetable oil production
   - grain legumes
   - meat meal, fish meal, and other animal protein sources
   - new and unconventional protein sources
2. To highlight areas of current widespread interest in livestock production and research:
   - ecological roles of plant toxins
   - processing methods to inactivate deleterious factors
   - role of plant breeding to improve nutritional value and reduce toxicity problems
   - nonprotein nitrogen in ruminant nutrition
   - rumen degradable versus rumen undegradable proteins
   - improvement of protein status through applications of biotechnology

Protein supplements are integral components of animal diets. Swine and poultry are fed diets based on cereal grains, which provide needed energy and also some protein. Protein supplements are needed to increase the total quantity of protein as well as to satisfy essential amino acid requirements. For ruminants, amino acid needs are largely met by microbial synthesis in the rumen, so dietary protein quality is of less concern. However, there is an increasing trend to partition the protein requirements of ruminants into two categories: fermentable nitrogen for rumen microbes, and dietary protein which escapes rumen fermentation (nondegradable, escape, or bypass protein) and serves as a direct source of absorbed amino acids.

Protein supplements are arbitrarily defined as having at least 20 percent crude protein. Most protein sources contribute other nutrients as well and many, particularly plant proteins, contain deleterious or toxic factors. These properties will be considered as each protein source is discussed.

OILSEED MEALS

Oilseed meals are by-products of vegetable-oil production for edible and industrial purposes. Vegetable oils have been used since antiquity for food, soapmaking, illumination, and lubrication. **Edible oils** are used as salad oils, in cooking, baking, in many other types of food, and in margarine. Many plant oils have fatty acid structures that are uniquely suited to certain **industrial applications**. Rapeseed oil is used in lubricants for marine and high-speed jet engines because the oil clings to metal surfaces, even when washed by steam. Linseed oil and

TABLE 4–1. Physical and Chemical Properties of Vegetable Oils Derived from Crops Yielding Oilmeal Protein Concentrates

			*Fatty Acids, % of Total**					
Fat or Oil	*Iodine Number*	*Melting Point, °C*	*16:0*	*18:0*	*18:1*	*18:2*	*18:3*	*20:4*
Canola	114	—	4.3	1.7	**59.1**	22.8	8.2	0.5
Coconut[†]	8–10	20–35	8.0	2.8	5.6	1.6	—	—
Corn	115–127	−10	12.0	2.7	30.1	**54.7**	1.4	0.2
Cottonseed	97–115	10–16	20.9	1.9	16.0	**59.6**	0.1	—
Linseed	—	—	6.4	3.3	17.0	15.6	**57.7**	—
Olive	79–90	0	14.0	2.6	**74.0**	8.1	1.0	0.4
Palm	48–56	27–50	**42.0**	5.4	39.1	10.6	—	0.2
Peanut	84–100	−2	11.1	3.0	**52.1**	27.8	0.5	0.7
Rapeseed	81	−9	1.7	0.1	14.3	13.4	8.9	0.9
Safflower	145	−17	12.3	1.8	11.2	**74.3**	—	0.5
Soybean	130–138	−21	11.5	4.3	27.3	**49.7**	6.9	0.2
Sunflower	125–136	−17	6.8	3.9	15.7	**73.5**	—	—

*The major fatty acids in each oil are boldfaced.
[†]The major fatty acids in coconut oil have fewer than 16 carbons.

other drying oils are used in paints and varnishes. Vegetable oils are used in manufacturing plastics and nylon. Oils are used in the manufacture of soap, and specialty oils, such as meadowfoam and jojoba oil, are used in **cosmetics**. There are many other industrial uses. On a worldwide basis, there is a tremendous industry in industrial and edible vegetable oils. Composition and properties of some natural fats and oils are given in Table 4–1.

Oils are extracted by either of two main processes: mechanical expression **(expeller process)** and **solvent extraction**. Most commercial oil is now obtained by solvent extraction or by a prepress solvent extraction method, in which an expeller process is used to remove part of the oil, followed by solvent extraction. **Hexane** is the main solvent used. Other solvents, such as ethanol and isopropyl alcohol, are being investigated as alternatives to hexane; hexane is extremely flammable, is nonbiorenewable, poses health risks, and is an air pollutant (O'Quinn et al., 1997). Oilseed crops are grown primarily for their oil, but the protein-rich by-product meals remaining after oil extraction are also valuable commodities. They are the main plant protein supplements now used in animal feeding. In some countries, they are referred to as oilmeals (e.g., soybean oilmeal), whereas in others they are known as cake (e.g., cottonseed cake, palm kernel cake).

Soybean Meal

Soybean meal is the most important protein supplement for livestock feeding in the United States and in many other countries. Soybeans (*Glycine max*) are an ancient crop and have been grown in China for thousands of years. Soybeans were introduced into the United States about 1800, but little attention was paid to them for many years. Until 1920, they were grown mainly for hay. Since 1950, U.S. soybean production has expanded dramatically in the Corn Belt and southern states. Other major producers of soybeans are Brazil, Argentina, and China. Soybeans are annual legumes and are adapted to conditions similar to those required for corn. Through plant breeding and the use of germ plasm from different climatic areas of China, cultivars have been developed that are adapted to most environments in the United States. Soybeans are harvested by combine after the pods are fully mature and the leaves have yellowed and dropped.

Deleterious Factors Raw soybeans are toxic to most animals. They contain a variety of toxins, including protease inhibitors, lectins, phytoestrogens, saponins, goitrogens, and several others. Fortunately, the nutritionally significant toxins are readily destroyed by heat treatment. Feeding raw soybeans to most species, particularly young animals, results in poor growth, rough hair or feathers, and pancreatic enlargement because of the effects of protease inhibitors. **Protease inhibitors**, also known as *trypsin inhibitors,* inhibit the pancreatic enzymes trypsin and chymotrypsin. These proteolytic enzymes are very important in the digestion of proteins in the small intestine. The protease inhibitors are themselves proteins. In simple terms, when trypsin or chymotrypsin attempts to digest the protease-inhibitor protein, the inhibitor binds to the enzyme irreversibly. Both the enzyme and the inhibitor are excreted, causing a reduction in protein digestibility and an increased loss of enzyme protein. The pancreas gland attempts to compensate for the reduction in protein digestion by enlarging (**hypertrophy**) to produce more enzymes. The increased production of pancreatic enzymes is a factor in the growth depression because it represents an increased loss of essential amino acids that are excreted as endogenous protein. **Endogenous protein** is that protein in the feces that originates from the animal's internal secretions rather than directly from the diet. Soybeans can be fed to livestock after appropriate heat treatment to destroy trypsin-inhibitor activity. A common method of accomplishing this is by the use of an **extruder**, a machine in which the beans are forced through a die. The friction involved produces sufficient heat to destroy the inhibitors. Extruded soybeans, often referred to as *full-fat soybeans,* contain approximately 18 percent oil and 38 percent crude protein. Economic conditions dictate whether it is cost-effective to feed full-fat beans or if it is more profitable to extract the oil and use only the meal for feed. The heat treatment of soybeans or soybean meal necessitates striking a balance between adequate heating to inactivate toxins and not overheating to cause damage to the protein. Excessive heating causes lysine tie-up by **Maillard (browning) reactions**, in which free amino groups react with sugars to produce indigestible brown polymers. Several tests are used to assess the adequacy of heat treatment. A simple one, widely used in soybean meal processing plants, is the urease index or test. Soybeans contain **urease**, the enzyme that converts urea to ammonia. Although urease in soybeans is probably not nutritionally significant, it is inactivated at a temperature similar to that required for destruction of trypsin inhibitors so that loss of urease activity is indicative of inactivation of trypsin inhibitors as well. In the **urease test**, urea and phenol red are incubated with soybean product. If active urease is present, the urea is converted to ammonia, raising the pH and changing phenol red from colorless to bright red. A simple spot test is often used. A trypsin-inhibitor value can also be determined by measuring the inhibitory activity of the sample against the action of trypsin *in vitro*. A color score can also be used. These tests are discussed in more detail by McNaughton et al. (1981).

Soybeans low in trypsin-inhibitor activity have been developed (Cook et al., 1988; Herkelman et al., 1992) and can be fed raw without adverse effects. However, trypsin inhibitors protect the seeds from insect damage by inhibiting digestive enzymes that insects secrete. Production of **low trypsin-inhibitor soybeans** might require increased use of pesticides.

Soybeans contain storage proteins (**glycinin** and **conglycinin***) that may cause allergenic reactions in animals, particularly in preruminant calves (Pedersen and Sissons, 1984; Kilshaw and Sissons, 1979) and baby pigs, resulting in atrophy of intestinal villi and impaired nutrient absorption. Gastrointestinal hypersensitive reactions arising from absorption of **soybean antigens** are well-known. Soybean products used in milk replacers should be heated sufficiently to denature antigenic proteins. Baby pigs fed creep diets containing soy products may become sensitized to the soy proteins and exhibit gastrointestinal distress when exposed to the same ingredients after weaning (see Chapter 13).

Soybeans contain **lectins**, which are glycoproteins that bind to carbohydrates in the intestinal mucosa (brush border), causing digestive disturbances. Soybean meal may contain sufficient lectins to have detrimental effects on animal health and productivity (Maenz et al., 1999).

Isoflavones are estrogenic compounds found in soy products. Soy isoflavones include genistein, daidzein, and glycitein. Although of interest as nutraceuticals for human use, soy isoflavones appear to have little effect on animal performance (Payne et al., 2001).

Nutrient Content Soybean meal is the major protein supplement used for nonruminant animals in most areas of the world with intensive animal and poultry production. Swine and poultry are fed corn-soy diets, with corn as the energy source and soybean meal as the protein supplement. In the same way that corn is the standard to which other feed grains are compared, soybean meal is the standard to which other protein supplements are compared. Soybean meal is highly palatable and has a high digestibility, a high protein content of 44 to 50 percent, a good amino acid balance, a low fiber content, and a high digestible energy content. A corn-soy diet will meet the protein and energy requirements of virtually all classes of swine, except perhaps for the very young early-weaned pig. Some examples of simple **corn-soy swine diets** are shown in Table 4–2. These examples illustrate how simple in composition an adequate diet can be. Until the discovery of vitamin B_{12}, the use of all-plant diets was not successful for swine and poultry production. However, as long as vitamin B_{12} is added, a properly balanced corn-soy diet is ideal for poultry and swine production.

Soybean meal is totally satisfactory as a protein supplement for ruminants. It is usually economically more efficient to use poorer quality proteins such as cot-

TABLE 4–2. Examples of Swine Diets Based on Corn and Soybean Meal

Ingredient	*Starter Diet, % (18% CP)*	*Grower Diet, % (16% CP)*	*Finisher Diet, % (14% CP)*
Salt, minerals, vitamins	5	5	3
Corn	69	74	82
Soybean meal	26	21	15

**Con* is a prefix meaning "with"; thus conglycinin is found with glycinin.

tonseed meal. However, soybean meal may have application in rations for highly productive ruminants because of its rumen bypass potential (see the section Bypass, Escape, Nondegradable Proteins, and Amino Acids later in this chapter).

Methionine is the first-limiting amino acid for nonruminants. Although corn-soy diets are adequate for swine, for poultry they usually require methionine supplementation with synthetic methionine or with other protein supplements such as fish meal. Soybean meal has a fairly high phytic acid content, which reduces the availability of phosphorus and zinc. This factor should be taken into consideration in diet formulation. Soybean meal is marketed at standard protein contents, either 44 or 49 percent for dehulled soybean meal. An additional advantage of soybean meal as a protein supplement is its consistency. With some ingredients, variability in nutrient content is a problem in ration formulation.

Soybean meal contains approximately 5 to 6 percent **oligosaccharides**, mainly raffinose and stachyose, which are short-chain polysaccharides containing a number of different sugars. Nonruminants, and especially poultry, are unable to digest the oligosaccharide fraction efficiently. As a result, this carbohydrate material is fermented in the hindgut, and, because it is soluble, it has an osmotic effect in increasing fecal moisture content, causing diarrhea and wet droppings. This is particularly important in the nutrition of young turkey poults. Because of the high-protein requirement of poults, the soybean meal content of the diet may be as high as 50 percent, and the indigestible oligosaccharides cause significant wet-litter problems. In addition, the metabolizable energy content of the soybean meal is lower than if the oligosaccharides were a digestible component. **Ethanol extraction** of soybean meal removes the oligosaccharides and increases the energy and protein content of the meal (Leske et al., 1988). Commercialization of this process would produce a superior quality soybean meal, especially for turkeys and fish, which are fed diets high in protein. Soy oligosaccharides have few if any detrimental effects in pigs (Smiricky et al., 2002), although Zhang et al (2003) reported that stachyose contributes to diarrhea in baby pigs fed soybean meal.

Soybeans contain hemicelluloses such as galactomannan in the cell wall structure. Galactomannan is a polymer of D-mannose with sidechains of D-galactose units. Addition of the enzyme β-mannanase to corn-soy diets improves growth performance of pigs and poultry (Pettey et al., 2002).

Cottonseed Meal

Cottonseed meal is the second most important plant protein supplement used in the United States. It is a by-product of the second order: cotton is grown primarily for its fiber; the oil extracted from the seed is the major by-product, and the cottonseed meal remaining after oil extraction is a secondary by-product. It is also the one that receives the least concern in quality control. Cotton (*Gossypium hirsutum*) has been grown for several thousand years. It is a semitropical plant grown as an annual. Major cotton-producing areas are the United States, former Soviet Union, China, India, Brazil, Mexico, Egypt, and most other tropical countries. "King Cotton" was an economic mainstay of the southern United States before the Civil War; major cotton producing states now are Texas, Mississippi, California, and Arkansas.

Deleterious Factors: Gossypol A major constraint to the use of cottonseed and cottonseed meal as feedstuffs is the presence of a toxic constituent called *gossypol.* **Gossypol** is a phenolic compound, which means that it contains aromatic (benzene) rings with hydroxyl (–OH) groups attached. Its name is derived from *Gossypium* phenol. It is a yellow pigment that occurs in pigment glands scattered throughout the seed (Fig. 4.1). Although there are glandless cotton varieties that lack gossypol, they are not often produced commercially. Gossypol, like most toxins in plants, is a natural pesticide and helps protect the plant against insects and other pests. **Glandless cotton** requires excessive treatment with pesticides, which generally makes it uneconomical to produce. Cottonseed meal prepared from glandless cotton gives better results than regular cottonseed meal for nonruminants (LaRue et al., 1985), but cannot completely replace soybean meal without depressing performance. The nutritive value of glandless cottonseed meal is significantly greater than for normal cottonseed meal, with higher digestibility of protein and amino acids. An interesting possibility for producing gossypol-free cottonseed meal is described by Dilday (1986), who found a single cotton plant that had gossypol in all parts but the seeds. A plant breeding program has been initiated to transfer the gene(s) for glandless seeds into commercial cotton varieties (Dilday, 1986). This plan would utilize the insecticidal properties of gossypol in protecting the rest of the plant; at the same time, it would produce gossypol-free meal.

Gossypol has a number of adverse or toxic effects in mammals and poultry. One unique effect is **olive-green yolks** in stored eggs (Fig. 4.2) caused by a chemical reaction between gossypol and iron in the egg yolk. Gossypol also causes reduced growth and feed intake. Over a prolonged period, it causes damage to the heart, liver, and lungs, resulting in cardiac irregularity, cardiac failure (heart attack), pulmonary edema, and labored breathing. Because gossypol reacts with iron, it causes anemia by tying up iron in an unavailable form.

An interesting application of gossypol is in the development of a **male birth control pill**. A Chinese scientist reported that in a 10-year period, not a single child had been born in Wang village in Jiangsu (Hron et al., 1987). It was discovered that, during this period, crude cottonseed oil containing gossypol had been used for cooking. Subsequent studies revealed that gossypol is a potent contraceptive agent in the human male and, at levels lower than those implicated in cardiac and lung lesions, blocks sperm production. Research is in progress in China on the development of a male birth control pill from gossypol. There do not seem to be any fertility problems in livestock that are fed cottonseed meal, although damage to germinal epithelium of rams and bulls has been observed (Randel et al., 1992).

In the commercial extraction of cottonseed oil, the seed is processed so as to leave the gossypol in the cottonseed meal. The cottonseed is subjected to heat treatment before extraction; in the presence of heat, gossypol reacts with proteins to produce "**bound gossypol**." The reaction, resulting in a lowered lysine availability, is mainly with the amino acid lysine. The toxicity of cottonseed meal is associated with nonbound or "**free gossypol**." For nonruminants, the level of free gossypol in the diet should not exceed 0.01 percent, or approximately 9 percent dietary cottonseed meal. The meal is monitored for its free gossypol content. The toxic effects of gossypol can be reduced by the addition of iron salts to the diet. Free gossypol reacts with iron and thus becomes "bound" or physiologically inactive. Ruminants are more tolerant of gossypol, but even in ruminants prolonged feeding of whole cottonseed can cause heart and liver damage. Gossypol damage

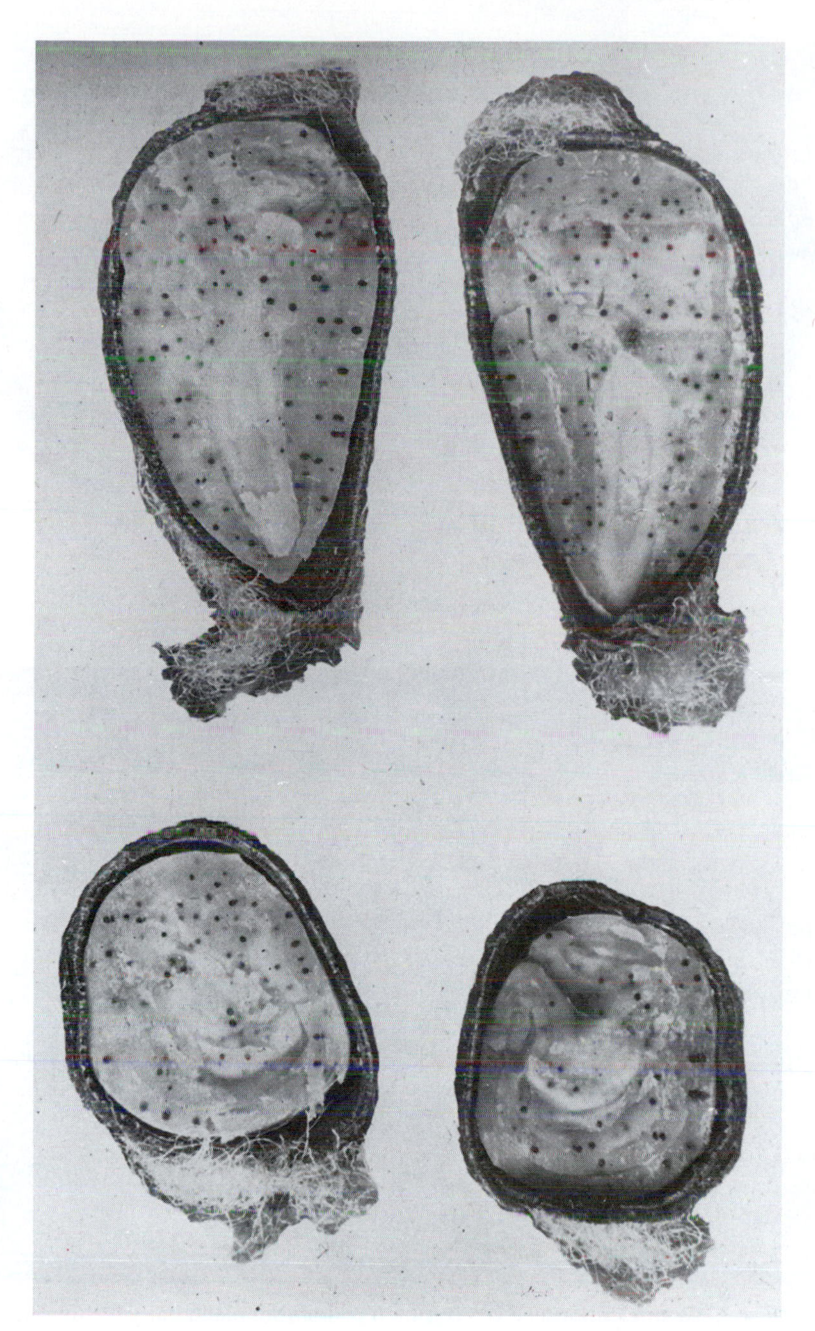

FIGURE 4.1 Pigment glands in cottonseed. These glands, distributed throughout the seed and containing gossypol, cause very adverse effects in livestock fed cottonseed products. (Courtesy of R. J. Hron, Sr.)

is cumulative, and adverse effects may not be observed until after many weeks or months of feeding livestock cottonseed. Calves are more sensitive than mature cows to gossypol toxicity (Holmberg et al., 1988; Risco et al., 1992). Gossypol occurs in cottonseed as a mixture of **(+) and (−) isomers** (Percy et al., 1996). The (−) isomer is the more toxic, as reviewed by Cheeke (1998).

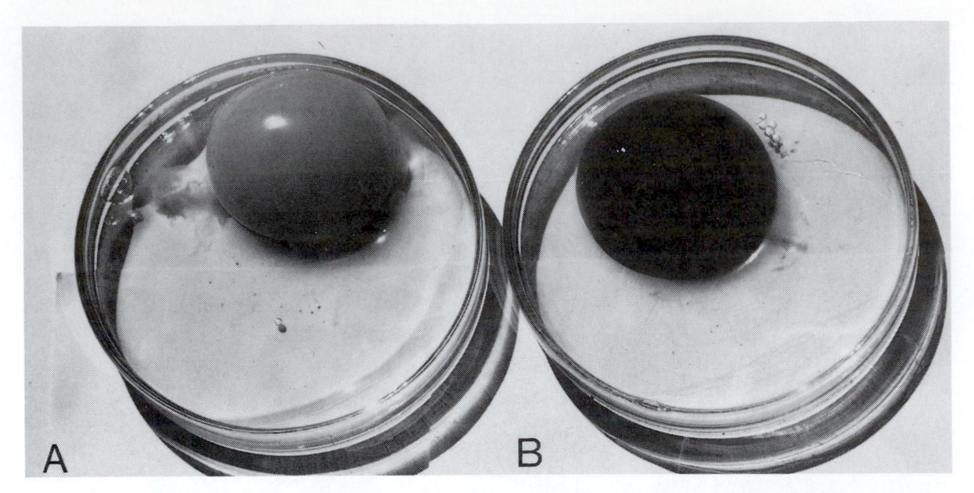

FIGURE 4.2 The effect of gossypol on egg yolk pigmentation. (A) Egg from control bird. (B) Egg from a bird fed gossypol.

FIGURE 4.3 Whole cottonseed is a common protein source in dairy feeding. It must be used with caution because of the potential for gossypol toxicity.

Nutrient Content **Whole cottonseed (WCS)** is a widely used protein supplement in the dairy industry (Fig. 4.3). It has an excellent nutrient balance for dairy cattle, containing 23 percent crude protein, 23 percent fat, and 17 percent fiber. Thus it is a good source of protein, energy, and fiber. Because of the possibility of gossypol toxicity, whole cottonseed and cottonseed meal should be fed to dairy cows at levels no higher than 4 kg per head per day. Replacement heifers can be safely fed diets with 15 percent WCS (Arieli, 1998). The use of cottonseed products in dairy feeding has been reviewed by Coppock et al. (1987) and Arieli (1998). Holmberg et al. (1988) described in detail an outbreak of gossypol toxicity in a dairy-calf-rearing facility; more than 500 calves were lost. The diet contained 27 percent cottonseed meal. Major lesions included ascites, edema, liver damage (acute centrilobular necrosis), cardiovascular lesions, and kidney damage. Losses of sheep to gossypol activity have also been reported (Morgan et al., 1988).

The protein quality and content of **cottonseed meal** are lower than for soybean meal. It contains approximately 41 percent crude protein and 12 percent crude fiber. It is lower in lysine and sulfur amino acids than soybean meal. It is of lower palatability than soybean meal to poultry and swine but is well accepted by ruminants. Considering all of these factors (gossypol, palatability, and protein quality), cottonseed is more suitable for ruminants than for nonruminants. In addition, it is a good source of bypass protein.

Cottonseed oil contains **cyclopropene fatty acids** (sterculic and malvalic acids) that have some unique nutritional properties, including a synergistic effect on the carcinogenicity of aflatoxin. They inhibit desaturase enzymes, which are involved in increasing the unsaturation of fatty acids. Thus, cottonseed meal prepared by the expeller process, containing residual oil, tends to cause a hard body fat in pigs and other nonruminants. A unique effect of cyclopropenoid fatty acids in poultry is the production of pink egg albumen (egg white).

Rapeseed and Canola Meal

Rape is a member of the *Brassica* genus of the Cruciferae (Brassicaea) or cabbage family. The terms **crucifers** and **brassicas** are used interchangeably to refer to these plants. Common cruciferous vegetables include cabbage, cauliflower, brussels sprouts, and broccoli. *Brassica* crop spp. include rape, kale, turnips, and mustard. There are both forage and oilseed rapes. Rapeseed meal is produced from the oilseed varieties, *B. napus* and *B. campestris*. Rapeseed is an important crop in China, India, Western Europe, and Canada.

Much of the work on the development of rapeseed as a major crop has been conducted in Canada (Bell, 1993). Canadian plant breeders have been very successful in modifying rapeseed to give it desirable agronomic, industrial, and nutritional properties. **Rapeseed oil** contains a high content of a fatty acid called **erucic acid,** which is useful in a number of industrial applications. The term *rapeseed* is now used primarily for the high erucic acid industrial oil material. Erucic acid is toxic; therefore, a low erucic acid rape has been developed for edible oil purposes. The low erucic acid oil is used for margarine and cooking/salad oil applications.

Rapeseed contains another group of toxic compounds called *glucosinolates.* **Glucosinolates** inhibit the metabolism of the thyroid gland and may induce **goiter** (enlargement of the thyroid gland). Hence they are called *goitrogens* or *goitrogenic agents.* Glucosinolates have a "hot" taste and are the active ingredients in condiments such as mustard and horseradish. They are an undesirable component of rapeseed meal, reducing its palatability and having goitrogenic activity. Canadian researchers have developed rapeseed varieties that are low in both erucic acid and glucosinolates. These "double zero" varieties have been named **canola** to distinguish them from toxic rapeseed. (The term *canola* was derived from "Canadian low acid," referring to low erucic acid.) Canola meal is a protein supplement derived from low glucosinolate rapeseed. It is the principal rapeseed product used as feed; thus the remainder of the discussion will focus on canola meal.

Canola meal contains 38 to 46 percent crude protein and 10 to 13 percent crude fiber. Its amino acid balance is less favorable than that of soybean meal. For swine and poultry, a mixture of canola and soybean meal usually gives better performance than that obtained with canola meal alone, although Leeson et al. (1987) found that canola meal could be used as a complete replacement for soybean meal in poultry diets with no adverse effects. Extensive studies on the feeding value of canola meal for livestock have been conducted in Canada, and nearly every issue of the *Canadian Journal of Animal Science* since 1970 contains articles on rapeseed/canola.

Because of the unfavorable results originally obtained with rapeseed meal as a feed ingredient, there was resistance to the use of the nontoxic canola meal, through guilt by association. Rapeseed meal is dark in color; feed manufacturers associated the dark color with a poor-quality ingredient and preferred the yellow color of soybean meal. Canola breeders responded by developing yellow-seeded types of canola, which have gained greater acceptance in the feed industry. Yellow-seeded rapeseed has a lower weight of hull than brown-seeded cultivars, the hull has less fiber, and the digestible energy and protein contents are higher (Bell and Shires, 1982). The negative effects of the brown-seeded varieties are attributable to tannins in the dark seed coats.

Canola has a high content of a methylated substance called **sinapine**. In the cecum of the chicken, sinapine is converted to trimethylamine, which is absorbed. Most breeds of chickens convert the trimethylamine to trimethylamine oxide and excrete it. Birds with the Rhode Island Red genes for brown eggshells lack the enzyme that converts the amine to the oxide. As a result, trimethylamine builds up in the blood and is transferred to the egg. Trimethylamine, like amines in general, has a fishy odor. Canola meal should not be fed to layers of the Rhode Island Red breed to avoid producing **fishy-flavored eggs**. Sinapine is an astringent phenolic compound and may have adverse effects on palatability of canola meal (Ismail et al., 1981).

Part of the growth-depressing effects of canola meal relative to soybean meal is the high sulfur content. (*Brassica* spp. typically have a high content of sulfur-containing compounds, such as glucosinolates, and nonprotein sulfur amino acids, such as S-methyl cysteine sulfoxide.) The **high sulfur content** alters anion-cation balance (see Chapter 7) by increasing the anion fraction. Supplementation with sodium, potassium, or calcium adds cations and improves growth of canola-fed broilers (Summers and Bedford, 1994).

The production of rapeseed/canola in Canada has increased dramatically; it is now the second most important Canadian crop after wheat. Rapeseed production is expanding in many other parts of the world also, so the meal will become an increasingly important commodity. Canola production in the United States is increasing rapidly. The oil is in high demand, as it is considered to be the most desirable vegetable oil in terms of human health (heart disease). Also, canola lends itself well to double-cropping; in the southern United States, fall-seeded canola can be double-cropped with soybeans. Continual improvements in the crop are made through plant breeding, such as the development of herbicide resistant cultivars (Blair et al., 1986; Bell and Keith, 1987).

Full-fat canola seed is a high-energy, high-protein feedstuff. It must be processed by extrusion, grinding, jet-sploding, or other process to break the hard seed coat; otherwise the whole seeds will be poorly digested. Shaw et al. (1990) found that up to 15 percent raw canola seed could be fed to pigs with no adverse effects on performance.

Linseed Meal

Linseed meal is the residue remaining after extraction of linseed oil from **flax** (*Linum usitatissum*) seed. Linseed oil is a drying oil used for paint manufacture. With the development of petroleum-based paints, the demand for linseed oil has decreased and acreage has declined accordingly. Flax grown for oil production is of a short-straw type; long-straw flax is grown for the production of fiber from which linen is manufactured. Canada, the United States, the former Soviet Union, India, and Argentina are the major producers of flax seed; the former Soviet Union, Poland, and several western European countries are the main producers of fiber flax. Flaxseeds are small, oblong, and shiny in appearance. The flowers are an attractive blue color.

Linseed meal has a fairly low protein content of approximately 35 percent and is severely deficient in lysine. For this reason, it is more suitable for ruminants and horses than for swine and poultry. Linseed meal, particularly that prepared by the expeller process, is reputed to have favorable effects on the hair coat of animals, producing a good "bloom." For this reason, it is often used as a supplement for show animals. These effects may be due to the residual linseed oil and also have been attributed to mucilaginous gums present in the meal. These gums also contribute to a laxative effect of linseed meal.

Flaxseed has the highest **omega-3 fatty acid** content of any oil seed (Cunnane et al., 1990). These authors found that pigs fed a diet containing 5 percent flaxseed had elevated omega-3 fatty acid levels in their tissues. These fatty acids are characteristic of fish oils and have been linked to beneficial effects on cardiovascular health of humans (see Chapter 3).

Linseed meal contains two types of toxic factors. One is a dipeptide called **linatine**, composed of glutamic acid and 1-amino-D-proline. The latter amino acid is an antagonist of pyridoxine (vitamin B_6). Thus, in swine and poultry linseed meal may induce a **pyridoxine deficiency**. Growth of swine and poultry that are fed linseed-meal-containing diets is improved with supplementary pyridoxine (Bishara and Walker, 1977; Klosterman, 1974). The other type of toxin occurring in linseed meal is **cyanogenic glycosides**, similar to those found in cassava. The level is not sufficiently high to cause cyanide poisoning (Oomah et al., 1992), but in the detoxification of cyanide, thiocyanate is formed. Thiocyanate is a goitrogenic substance; feeding linseed meal to pregnant ewes has resulted in goiter in their fetuses.

The cyanogenic glycosides in linseed meal have protective effects against **selenium toxicity**. Selenium is an essential trace element (see Chapter 7). However, in parts of the United States, such as the Northern Great Plains (North and South Dakota, Wyoming), the soils have a high selenium content, and crops and forages may contain toxic levels of selenium. Signs of selenium toxicity include hoof deformities, hair loss, and gastrointestinal tract irritation. It has been known for many years in these areas that feeding linseed meal has protective effects against selenium toxicity. The cyanogenic glycosides increase the excretion of selenium in the urine, thus decreasing its toxicity (Palmer et al., 1980). Dietary cyanide also increases the severity of selenium deficiency by promoting selenium excretion (Gutzwiller, 1993).

"Linola" is the trade name of a flaxseed developed for human consumption, with reduced levels of linolenic acid. **Linola meal** has similar feeding value to linseed meal for pigs; both are quite inferior to soybean meal as protein supplements for nonruminants (Batterham et al., 1991, 1994). Supplementation of linola meal with pyridoxine did not overcome the growth inhibition in pigs (Batterham et al., 1994). These authors suggested that **mucilage**, a water-soluble gum in flaxseed, may limit feed intake because of its water-absorbing and swelling capacity in the gut.

Peanut Meal

Peanuts (*Arachis hypogoae*) are grown for human consumption and for extraction to produce peanut oil. Peanut meal is the residue remaining after oil extraction. In many countries, peanuts are called *groundnuts.* Principal producers are India, China, Nigeria, the United States, and Brazil. Peanuts are annual legumes, produced mainly in tropical and semitropical environments. The peanut plant has abundant cloverlike leaves, and the peanuts are produced in underground pods (Fig. 4.4).

FIGURE 4.4 Peanuts (groundnuts) are produced in pods beneath the surface of the soil.

Peanut meal contains approximately 45 to 50 percent crude protein, and is quite deficient in lysine. Orok et al. (1975) noted that peanut meal was markedly inferior to soybean meal as a protein source for swine, even when supplemented with lysine. Peanut meal is not extensively used in the United States but has been an important protein supplement in Europe, imported from Africa and Brazil. In this regard, groundnut meal played an important role in the identification of aflatoxins. In 1961, extensive losses of turkey poults occurred in Britain. It was found to be associated with the use of peanut meal imported from Brazil. Further research led to the identification of toxins in the meal, produced by the fungus *Aspergillus flavus*. The toxin was named **aflatoxin** after the fungus that produces it. Aflatoxin has subsequently been found to be a major problem in many other feedstuffs, especially corn. The fungus grows under warm, humid conditions, typical of the areas where peanuts are produced and stored.

At one time it was common to "hog off" peanuts by turning pigs into the fields to root up the crop. This produces soft pork (Fig. 3.3) because of the high intake of unsaturated oil. Pigs should not be fed full-fat peanuts for several weeks prior to slaughter to avoid the soft pork condition (West and Myer, 1987).

Peanuts are well known for their **allergenic effects** in humans (Hopkins, 1995) producing skin rash, respiratory distress, nausea and vomiting, with anaphylactic shock and collapse in severe cases.

Sunflower Meal

The sunflower (*Helianthus annus*) is a native of the Americas. It is a stout, erect annual plant, 5 feet or more in height. The large yellow flowers produce seed heads from several inches up to 2 feet in diameter. The seeds are enclosed in a fibrous hull. The sunflower heads characteristically face toward the sun during the day, due to a bending of the stem producing heliotropic movement (from which the genus name is derived).

Sunflowers are grown mainly in northerly climates; the former Soviet Union, eastern and western European countries, Canada, Argentina, and North Dakota and Minnesota in the United States are major areas of production. They are grown mainly for production of sunflower oil, widely used for edible purposes. The meal has quite a high fiber content (11 to 13 percent crude fiber) that, along with a very low lysine content, limits its use for nonruminants. The combination of high protein (40 to 45 percent) and high fiber makes it quite suitable for ruminants, particularly for dairy cattle and goats, which have a high protein requirement. Heat treatment of sunflower meal for ruminants is desirable to reduce degradability of the crude protein in the rumen, thus increasing its bypass potential (Schroeder et al., 1996).

Sunflower meal contains phenolic compounds, which have an adverse effect on palatability and may reduce protein digestibility. The leaves and stalks also contain phenolics, accounting for their low palatability. Sunflowers have been grown for silage (Thomas et al., 1982) but have largely been replaced by other crops (e.g., corn) that are more palatable and higher yielding. They are grown to a limited extent for silage in northerly areas where corn does not thrive.

Decorticated sunflower meal is prepared by removing as much of the hull as possible using screens. The decorticated meal is higher in protein (45 to 47 percent) and lower in fiber (10 to 12 percent) than regular sunflower meal.

Safflower Meal

Safflower (*Carthamus tinctorius*) is an annual plant resembling a thistle in appearance. The leaves have short, prickly spines. The flowers are usually yellow or red and have been used for production of a red dye (carthamin). Safflower is grown as an oilseed crop in the Middle East, India, and the United States, principally in California, Nebraska, Montana, and North Dakota. Safflower oil is used for edible purposes.

The seeds have a high fiber content due to the prominent hull. **Safflower meal** contains approximately 40 percent crude fiber and only 18 to 22 percent crude protein. The protein is deficient in sulfur amino acids and lysine. Thus safflower meal is useful primarily for ruminants. Decorticated safflower meal, with a lower fiber content, can be used at modest (10 to 15 percent) levels of the diet for swine and poultry. Safflower meal contains two phenolic glycosides that are bitter and cathartic (laxative properties), which limits its use for nonruminants (Lyon et al., 1979).

Sesame Meal

Sesame (*Sesamum indicum*) has been cultivated in Asia and Egypt for thousands of years. It is now produced mainly in India, China, northern Africa, and Mexico. It is grown for oil production, and the decorticated seeds are sprinkled on the surface of certain types of bread, buns, and rolls. The seeds are produced in capsules or pods that split open readily at maturity, so most of the harvesting is done by hand. It is not produced in the United States to any extent because of its unsuitability for mechanical harvesting. Sesame meal is notable for having a very low lysine content.

Coconut Meal and Palm Kernel Meal

Coconut meal or **copra** is the meal remaining after extraction of oil from the dried endosperm of the coconut (*Cocos nucifera*). Coconut meal is an important protein supplement for livestock in many tropical countries. It contains 20 to 26 percent crude protein and approximately 10 percent crude fiber. The lysine content is lower than for soybean meal, but the methionine content is higher. If properly supplemented with lysine, copra is a useful protein supplement for swine (Thorne et al., 1988, 1989).

Palm kernel meal is the residue remaining after extraction of the oil from the seeds of the oil palm (*Elaecis guineensis*). The oil palm is a short tree extensively grown throughout the tropics. Major producers of palm oil and palm kernel meal include Nigeria, Malaysia, Indonesia, and a number of west African countries besides Nigeria. The fruit of the oil palm grows in large bunches that contain from several hundred to 2,000 individual fruits (Fig. 4.5). The fruit consists of a fleshy, oily outer pulp, the shell, and the kernel. The kernels are separated out from the shell and pulp, and the oil is extracted from them. The shells are used for fuel. The kernels contain approximately 45 to 50 percent oil. Palm kernel meal contains approximately 18 to 19 percent crude protein and 13 percent crude fiber, so it is slightly lower in quality than copra. It is used primarily in feeds for ruminants. Rhule (1996) obtained satisfactory growth of pigs with up to 20 percent dietary palm kernel meal. It can also be used in broiler diets with good results (Panigrahi and Powell, 1991).

FIGURE 4.5 Bunches of oil palm fruits (left). An individual fruit (right) showing the fleshy, oily outer pulp and the kernel.

Other Oilseed Meals

A few other minor oilseed meals are available in small quantities for animal feeding. **Mustard meal** is a by-product of the production of brown or Oriental mustard (*Brassica juncea*) for oil extraction. It has a higher crude-protein content than canola meal and a relatively high level of glucosinolates, which account for the "hot" taste of mustard. It is inferior to canola meal as a protein supplement for pigs (Bell et al., 1984) and chickens (Blair, 1984). Ammoniation of mustard meal and other brassica meals reduces the glucosinolate content (Bell et al., 1984). **Crambe** (*Crambe abyssinica*) has been grown to a limited extent as an oilseed crop in the northern United States and Canada. It is a crucifer and contains toxic glucosinolates, as do most other cruciferae. Crambe oil is high in erucic acid, which is used in rubber additives, plastics, coatings, and lubricants. The use of crambe meal as a protein supplement has been reviewed by Yong-Gang et al. (1993). The high level of glucosinolates severely limits its value for nonruminants. Heat-treated crambe meal is a satisfactory protein source for young ruminants (Caton et al., 1994) as well as for gestating and lactating beef cows (Anderson et al., 2000). Current FDA regulations limit crambe meal to 4.2 percent of the diet dry matter (Anderson et al., 2000). **Castor bean meal** is available to a limited extent but is quite toxic. The castor bean (*Ricinus communis*) contains a very toxic lectin called *ricin.* Castor oil is used for industrial uses, paints, and medicinals. The meal is generally used as a fertilizer because of its toxicity to animals.

Meadowfoam (*Limnanthes alba*) is a winter annual plant native to the states of the Pacific Northwest and is being developed in Oregon as an oilseed crop. The meal is low in protein (21 to 23 percent) and high in fiber (26 to 28 percent), so it is mainly suitable for ruminants. It also contains glucosinolates. Meadowfoam oil and **jojoba** (*Simmondsia californica*) oil are of interest because they can replace sperm whale oil in cosmetics. Jojoba is a desert shrub that grows wild in the southwest United States; it is now grown to a limited extent on plantations. Jojoba meal is toxic because it contains the cyanogenic glycoside simmondsin, which causes unpalatability and poor animal performance. Fermentation of jojoba meal with

Lactobacillus acidophilus and *L. bulgaricus* has been shown to improve its feeding value (Verbiscar et al., 1981). However, the detoxified meal still has a fairly poor feeding value because it has a low digestible energy content and a low content and availability of lysine (Ngoupayou et al., 1982).

Guar (*Cyamopsis tetragonoloba*) is an annual legume, grown for its seed, from which a galactomannan gum is obtained. Guar gum is used in a variety of industrial applications, including oil drilling, refining of mineral oils, paper manufacture, and emulsion stabilization in foods. The guar meal remaining after gum extraction contains about 35 percent crude protein and has been used as a feedstuff. It gives better results with ruminants than with swine and poultry. Growth inhibition in nonruminants may be due to the effects of residual gum in the meal. Treatment of guar meal with cellulolytic enzymes eliminates its growth inhibitory properties in poultry.

Rubber seed meal has potential as a protein source in developing countries. The seeds are rich in oil; the meal after oil extraction contains about 28 percent crude protein. Although the protein is low in lysine, it can be used at levels up to 10 percent in swine diets without adversely affecting performance (Babatunde et al., 1990). Other tropical seed meals have been evaluated for poultry. Both **mango seed kernel meal** and **jackseed meal** contain heat-labile toxic components (Ravindran and Sivakanesan, 1996; Ravindran et al., 1996).

GRAIN LEGUMES

Grain legumes, often referred to as **pulse crops**, are members of the *Leguminosae* family. They produce seeds that can be harvested for feed and food. The major grain legumes produced in the industrialized temperate countries are soybeans, field beans, lupins, peanuts, lentils, and peas. In tropical areas, there are a number of other species that are grown for food. Most grain legumes are produced specifically for human food; livestock are fed cull seeds and sometimes the forage (leaves and stems). In general, legume seeds are first-limiting in methionine. Most grain legume seeds contain antinutritive factors (ANF's). Ruminants are usually less sensitive to these deleterious factors than are swine and poultry (Cheeke, 1998).

Soybeans

As discussed previously, soybean products such as soybean meal are extensively used as protein supplements. **Full-fat soybeans** are also used in feeding. Since they contain oil, they are high in energy. Full-fat soybeans are usually prepared for feeding by the extrusion process. Sufficient heat is generated to inactivate trypsin inhibitors and other toxins in extruded soybeans. Sufficient heating of full-fat soybeans is necessary to prevent rancidity. Soybeans contain an enzyme, **lipoxygenase** (lipoxidase), that stimulates reaction of unsaturated fatty acids with oxygen (autooxidation or rancidity). This enzyme must be denatured by heat, or full-fat soybeans will quickly become rancid. If properly processed, they should be stable for at least 2 months.

There is some interest in feeding **raw soybeans**. Use of the raw beans would eliminate the need for processing equipment and the energy expenditure to operate it. Raw soybeans contain a number of toxic factors, which are most significant for young animals. Under some conditions, it may be feasible to feed raw soybeans to adult swine or poultry. Crenshaw and Danielson (1985a) replaced soybean meal with raw soybeans in diets for gestating sows and found no adverse effects on reproductive performance. However, Crenshaw and Danielson (1985b) and Pontif et al. (1987) found that raw soybeans were unsatisfactory for growing-finishing pigs, causing reduced growth, feed efficiency, and carcass quality as compared to soybean-meal-fed controls. Soybeans low in trypsin-inhibitor activity have been developed (Cook et al., 1988). Although, when fed raw, they support better growth of chicks (Anderson-Hafermann et al., 1992) and pigs (Herkelman et al., 1992) than raw conventional soybeans, they still require heat treatment to maximize their feeding value (Zhang et al., 1993). Further development of low-trypsin-inhibitor soybeans could enhance the potential for feeding raw soybeans.

Dry Beans

Many of the common beans used as food, such as the common snap, kidney, pinto, and navy bean, are varieties of *Phaseolus vulgaris*. Other *Phaseolus* spp. include *P. aureus* (mung bean), *P. lunatus* (lima bean) and *P. calcaratus* (rice bean). The common bean is a major vegetable crop, usually harvested in the immature state (e.g., green beans). Beans are not normally grown as animal feeds. Dry beans, such as many varieties of kidney, pinto, and navy beans, are often available as feed, including cull beans and crop surpluses. In general, dry beans contain 23 to 25 percent crude protein.

Since early studies on the nutritional value of proteins, it has been known that **dry beans** contain heat-labile toxic factors with protease (trypsin) inhibitors and lectins of most significance. **Lectins** are particularly important in kidney beans (Van der Poel, 1990). They cause damage to the brush border of the intestine (Fig. 4.6) and thus impair nutrient absorption. Poor growth, signs of protein deficiency (rough hair coat, poor feathering), and diarrhea result. Heat treatment inactivates the inhibitory factors. Heat-treated cull beans, such as extruded beans, can be usefully employed in swine and poultry feeds (Myer and Froseth, 1983a,b). Raw cull beans, even at dietary levels as low as 5 percent, impair growth

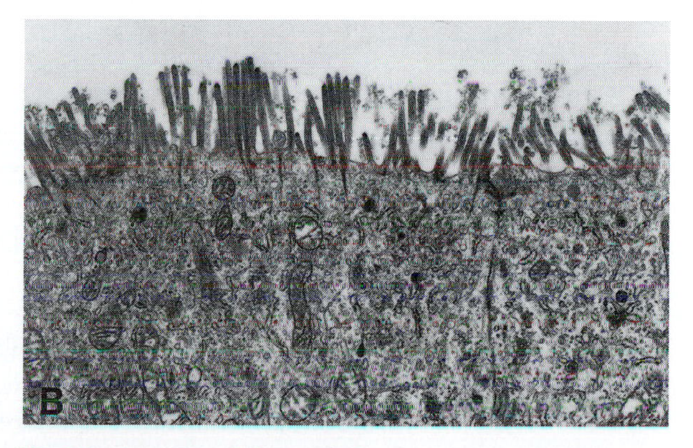

FIGURE 4.6 Effect of kidney bean lectins on the microvilli of the intestine. Microvilli from intestinal cell of a control pig (A) and from a pig fed raw kidney beans (B). The severe disruption of the microvilli causes impaired nutrient absorption and digestive disturbances such as diarrhea. (Courtesy of R. Begbie and T. P. King.)

performance. **Raw kidney beans** are toxic to ruminants as well. Williams et al. (1984) fed diets containing 15.5 and 25 percent raw kidney beans to yearling beef cattle. Reduced growth and feed intake were noted, as well as diarrhea. The animals had a high titer of circulating antibodies to the bean lectins, indicating damage to the intestinal mucosa and absorption of lectin protein. The absorbed lectins could cause systemic effects, including immunosuppression.

The **mung bean** is a large-seeded bean that is important as a protein source in the human diet in China, India, and Africa and is the source of bean sprouts used in Chinese cuisine (e.g., in chop suey). In the United States, production is mainly in Oklahoma. Surplus, undersized, and split beans are often available for feeding purposes. Luce et al. (1989) fed raw mung beans to bred gilts and noted some decrease in reproductive performance as compared to those fed a soybean meal control diet. Mung beans contain trypsin inhibitors and lectins and generally should be heat-treated prior to feeding, although Wiryawan et al. (1997) fed up to 30 percent raw mung beans to finishing pigs with no adverse effects.

The carbohydrate fraction of beans is of lower digestibility than the energy fraction of grains. Beans contain **oligosaccharides** (carbohydrates composed of 3 to 10 simple sugars) such as stachyose and raffinose that are poorly digested. Animals do not have enzymes to digest these carbohydrates. Therefore, they pass undigested through the small intestine and are fermented in the hindgut, causing irritation of the mucosa, abnormal microbial populations, and gas production. These substances are the well-known **flatulence factors** of beans.

Fava Beans

Also known as *faba beans, horse beans,* and *broad beans,* fava beans (*Vicia faba*) are grown extensively in Europe and the Middle East as a human food. In North America, there is considerable interest in producing fava beans for livestock feeding, particularly in Canada. In humans, the consumption of fava beans may cause a disease called **favism**, in which hemolysis of the red blood cells and anemia occur. Favism occurs only in people with a genetically caused low activity of the enzyme glucose-6-phosphate dehydrogenase in their red blood cells. This enzyme is involved in the detoxification of the toxic glycosides vicine and convicine, which fava beans contain. Favism has been an important public health problem in Mediterranean countries. Fava beans contain 24 to 30 percent crude protein and can be effectively utilized in livestock diets. In laying hens, egg size is depressed in birds fed fava beans. This appears to be caused by the vicine and convicine glycosides. Best results are generally obtained by using fava beans to replace only part of the total supplementary protein.

Lupins

There are several common types of lupins (*Lupinus* spp.), including those grown as garden ornamentals, grain lupins, and range lupins. Many species of lupins contain **alkaloids**. Some of the range lupins in North America have teratogenic effects when consumed by cattle. (**Teratogens** are substances that cause birth defects.) Cattle consuming these toxic lupins between the days 40 and 70 of pregnancy may give birth to calves with severe skeletal deformities **(crooked calf disease)**. Grain lupins are grown for their seed, which is used for human and animal consumption. They are referred to as "sweet lupins," having been selected for low alkaloid content. **Sweet lupins** have a tendency to mutate back to the high alkaloid content; continual monitoring of alkaloid content by seed companies is necessary to maintain the low alkaloid status. Sweet lupins are grown quite extensively in Australia and are used in large quantities in the feed industry there. They contain 25 to 35 percent crude protein. Swine are unusually sensitive to lupin alkaloids and may show depressed growth, feed aversion, and vomiting when fed lupin-containing diets. Godfrey et al. (1985) showed that the growth rate of pigs fed lupin-containing diets was inversely proportional to the alkaloid content. Poultry are much more tolerant of lupin in their diet. In addition to alkaloids, lupin seed contains poorly digested carbohydrates, such as galactans (Aguilera et al., 1985). These may accumulate in the hindgut, stimulating fermentation and enlargement of the cecum in swine (Batterham et al., 1986a). Heat treatment of lupin seed does not improve its feeding value for pigs (Batterham et al., 1986a,b). These workers reported evidence of a low availability of lysine in lupin seed. Halvorson et al. (1983) observed that when white lupin seed (*L. albus*) was used in turkey diets, growth rate was somewhat decreased with levels above 15 percent lupin seed, to 94, 89, and 85 percent of control values, with 30, 45, and 60 percent dietary lupin, respectively. Up to 30 percent lupin meal can be used in diets for layers with no adverse effect on production (Watkins and Mirosh, 1987). Lupin seed contains poorly digested **oligosaccharides**. Supplementation of a lupin-seed-containing diet with an oligosaccharidase enzyme preparation increased feed intake and weight gain of chicks (Brenes et al., 1993) and pigs (Gdala et al., 1997). Lupins contain **saponins**, but the levels are not high enough to have antinutritional effects (Muzquiz et al., 1993).

Lupins grown on acid soils may accumulate high levels of **manganese**, which may be toxic to animals. Excess manganese can be counteracted by increasing the iron content of the diet. Australian researchers have shown that the broad-leaved lupin *L. albus* is the main species to accumulate manganese to levels as high as 4,500 parts per million in the seed (Hung et al., 1987). The narrow-leaved lupin *L. augustifolius*, when grown under similar conditions, had low manganese content (60 parts per million).

There are three main species of lupins grown for food and feed: *Lupinus albus, L. angustifolius*, and *L. luteus* (Gladstones et al, 1998). Flower color is white, blue, and yellow, respectively. *L. luteus* is commonly grown as a food in the Andean regions of South America. In a study of the three species, Gdala et al. (1996) found that *L. albus* was less satisfactory as a protein source for pigs than the other species. *L. albus* has generally given somewhat poor results with swine, which may relate to its alkaloid profile and prolonged gut retention time (Dunshea et al, 2001).

Grain lupins are well utilized by ruminants. They are commonly used as a protein supplement for cattle and sheep in Australia. Heat treatment (roasting) improves their nutritional value for high producing dairy cows (Singh et al., 1995), probably by reducing the microbial degradation of protein in the rumen.

Lupins have some desirable agronomic characteristics. They thrive under more adverse conditions than do soybeans. In Australia, they are grown on light, sandy soils in rotation with wheat. Because of their nitrogen-fixing capacity, they improve soil fertility and reduce fertilizer needs for the subsequent wheat crop. In the United States, there is interest in lupins in northern areas and the Pacific Northwest states where soybeans are unproductive. At one time, bitter (high alkaloid) lupins were grown in the southern United States as a green manure crop to plow under to improve soil fertility. This is much less common now than in the past, although with current interest in sustainable agriculture, green manuring may increase. **Bitter lupins** have been grown for centuries in the Andean regions of South America. They are consumed as food, after a prolonged period of soaking and washing with water to remove the bitter alkaloids.

In Australia, where lupins are extensively grown, the stubble is used for grazing by cattle and sheep. There have been toxicity problems with lupin stubble, producing a disease called **lupinosis**. It is caused by a mycotoxin, phomopsin, produced by the fungus *Phomopsis leptostromiformis*. The fungus infects the green plant and persists on the stubble. It causes severe liver damage and fatty infiltration of the liver in animals grazing the stubble. The liver becomes greatly enlarged, greasy, and bright yellow or orange in color. Lupinosis has been a problem not only in Australia but also in Germany, Poland, South Africa, and other areas where lupins are grown extensively. The development of *Phomopsis*-resistant varieties of lupin appears to be the most promising way of reducing losses from the disease.

Field Peas and Lentils

Field peas (*Pisum sativum*) are grown to produce dry, edible peas for human consumption, and some varieties are grown for forage (see Chapter 5.) Peas are best adapted to moist, cool climates. Major areas of field pea production for dry edible peas include the U.S. states of Washington, Oregon, Idaho, Minnesota, and North Dakota, and the northern European countries. Considerable quantities of **cull peas** are available for livestock feeding in these areas. Peas are fairly low in protein (20 to 29 percent); they might be viewed as high protein grains to be used as a combined protein-energy source. They are palatable and do not contain significant amounts of toxic factors that require heat treatment for inactivation, although there is a small amount of trypsin-inhibitor activity (Johns, 1987). As is true in general for legume seeds, they are fairly low in **methionine** content. Peas can be used to provide all of the supplemental protein in cereal-based swine diets, except for the early-weaned pig, with no adverse effects on performance. For young pigs, steam pelleting and methionine addition are necessary for best results. In the Pacific Northwest, where soils are low in **selenium**, selenium supplementation of cull pea diets for swine has been shown to be beneficial. Peas are probably of most value in swine feeding; for poultry production, higher energy and protein requirements and industry demand for high-energy diets may make it impractical to use peas in many cases. Johns (1987) found that growth of chicks was lower with peas than with soybean meal; there was a substantial response to methionine supplementation, and the presence of trypsin inhibitors had some negative influence. Hlodversson (1987) in Sweden reported that white-flowered peas had a higher nutritive value for growing-finishing pigs than dark-flowered varieties. This was associated with the presence of tannins in the testa (hull) of dark-flowered peas. Protein digestibility was lower (70.8 percent) for dark than for white (83.2 and 89.9 percent for two varieties) flowered peas, suggesting a tie-up of protein by tannins. Energy digestibility was also reduced. Total polyphenol content was 2.7 percent in the dark and 1.3 to 1.4 percent in the white flowered varieties.

Peas contain a substantial quantity of **non-starch polysaccharides** (Igbasan and Guenter, 1996; Igbasan et al., 1997). The addition of feed enzymes such as pectinases can increase digestibility. Toasting or other heat treatment can reduce digestibility of the oligosaccharides in peas (Canibe and Knudsen, 1997).

Lentils (*Lens culinaris*) are similar to peas in production and food value. In the United States, they are grown primarily in the Pacific Northwest states and are used mainly in soups. Cull lentils may be fed to livestock and are similar to field peas in feeding value. Bell and Keith (1986) suggested that cull lentils were worth from 10 to 50 percent more than good quality grain, depending on relative costs of grain and protein supplement. The 10 percent inclusion level was optimal (Castell and Cliplef, 1988).

Other Grain Legumes

A number of other grain legumes are grown to a limited extent for feed, particularly in tropical countries. These include cowpeas, pigeonpeas, winged bean, chickpeas, and jackbeans.

The **cowpea** (*Vigna sinensis* or *unguiculata*) is an annual vine or semiviny plant grown mainly in tropical areas. Cowpeas are the "black eyed pea" of Southern cuisine in the United States. More than 90 percent of the world cowpea production is in Africa. The chemical composition is similar to that of other legume seeds. They contain protease inhibitors; feeding value is improved by heat treatment. With heat treatment and methionine addition, they can be used in place of soybean meal for swine and poultry.

In the 1980s, there was a surge of interest in the **winged bean** (*Psophocarpus tetragonolobus*) as a potential answer to protein deficiency in the tropics. All parts of the plant can be consumed. It produces a tuber that can be eaten like potatoes; the leaves, flowers, and tender pods can be used as vegetables, and the seeds can be eaten as a bean. In the enthusiasm of the times to solve the "world food problem," the winged bean was extensively promoted. However, as Henry et al. (1985) pointed out in an article titled "The Winged Bean: Will the Wonder Crop Be Another Flop?" there are no simple solutions to world food production needs, and the introduction of a new crop into an existing agricultural scene usually has unanticipated consequences. In the case of the winged bean, one of these was that the bean requires a prolonged cooking time because of its fibrous husk, placing a major burden on fuel supplies, which are in critically short supply in many tropical areas. Additionally, the extra fuel, cooking time, and preparation needed places a major labor burden on women, making the crop unacceptable in many areas. Jaffe and Korte (1976) reported that winged bean seeds contain trypsin inhibitors and lectins. Raw winged beans were toxic as the sole source of protein in rat diets; cooked seeds supported growth and the addition of methionine improved performance. Fernando and Bean (1986) reported that behenic acid (22:0 fatty acid) is an antinutritional factor found in all parts of the winged bean plant, with the highest concentration in the mature seed.

The **chickpea** (*Cicer arietinum*), commonly known as *garbanzo* in the United States and *gram* or *bengal gram* in India, is grown as a human food. Like most pulse crops, chickpeas are a good source of lysine, limiting in methionine, and require heat treatment to inactivate protease inhibitors. The protein quality is similar to that of soybean meal (Newman et al., 1987). Chickpeas are a satisfactory replacement for soybean meal in the concentrate mix for dairy cattle (Hadsell and Sommerfeldt, 1988).

Another legume, the **pigeon pea** (*Cajanus cajan*), is grown quite extensively in the tropics. The pigeonpea is a shrubby perennial. The seeds require cooking to inactivate protease inhibitors. The grain is deficient in tryptophan as well as sulfur amino acids.

Canavalia ensiformis (**jackbean**) is a tropical viny legume grown to some extent as a food, feed, and fodder plant. Jackbeans contain a number of toxins, including lectins and a toxic amino acid, canavanine. They are also a rich source of the urea-hydrolyzing enzyme urease.

Lathyrus spp. (*L. cicera* and *L. sativus*), also known as **sweet peas**, are grain legumes that have been grown for food since the Neolithic period (Hanbury et al., 2000). They contain toxic amino acids that cause a neurodegenerative disease called **lathyrism** (reviewed by Cheeke, 1998). Cultivars with low lathrogen levels have been developed and can be safely incorporated into animal diets (Hanbury et al., 2000). Because of the irreversible, severe paralysis typical of lathyrism, caution should be exercised to ensure that low lathyrogen varieties are used.

For those with a particular interest in tropical legumes, the National Academy of Sciences (1979) publication *Tropical Legumes: Resources for the Future* should be consulted.

MILLING BY-PRODUCTS

The main milling by-product used as a protein supplement is **corn gluten meal**, produced in the wet milling of corn. In this process, corn is made into a slurry with water and processed with heat and chemical treatments to produce corn starch, fructose, corn syrup, and corn oil. Two types of residues are produced: corn gluten meal (CGM) and corn gluten feed (CGF). The CGM contains 40 to 60 percent crude protein, whereas CGF has 20 to 25 percent crude protein. The lower protein content of CGF is due to its higher bran content. The CGM is a good source of methionine. CGF is especially useful in dairy cattle diets as a good source of energy, fiber, and bypass protein. Firkins et al. (1985) and Ohajuruka and Palmquist (1989) reported excellent results with corn gluten feeds for ruminants when used to provide 15 to 20 percent of feed dry matter. Corn gluten feed can be used at levels up to 25 percent in diets for laying hens without adversely affecting egg production (Castanon et al., 1990).

DISTILLERY AND BREWERY BY-PRODUCTS

Distillery by-products are residues of grains that have been fermented to produce whiskey and other liquor or ethanol for fuel (gasohol). Most of the starch and sugars are fermented to produce alcohol; the residue consists largely of unfermented protein and fiber. The residue, whole spent stillage, is separated into distillers' grains and clarified stillage. The latter fraction can be dried and marketed as distillers' dried solubles, or added back to distillers' grains (distillers' grains with solubles). Small distilleries often dispose of the distillers' solubles by making them available as livestock feed. Condensed distillers' solubles can be used to provide up to 20 percent of the dry-matter intake of feedlot cattle when offered free choice in place of water (Rust et al., 1990). **Distillers' dried grains** contain 9 to 13 percent crude fiber and 26 to 35 percent crude protein. The protein is derived from the grain and the yeast. Because of the yeast, the B-vitamin content is high. **Brewers' grains**, derived from barley fermented to produce beer, has higher crude-fiber (19 percent) and similar crude-protein (26 to 29 percent) contents as compared to distillers' grains. Because of their bulky, fibrous nature and low energy content, distiller and brewer by-products are more suitable for ruminants than nonruminants. Sorghum distillers' by-products (from fuel ethanol production) are similar in feeding value to corn distillers' by-products (Lodge et al., 1997).

ANIMAL PROTEIN SOURCES

Prior to 1948, the inclusion of animal protein sources in poultry and swine diets was essential to provide the **"animal protein factor."** This factor was identified in 1948 and named vitamin B_{12}, the last vitamin to be discovered. Plant materials do not contain vitamin B_{12}; use of all-plant diets for nonruminants (including humans) has become feasible only since the commercial availability of vitamin B_{12}. With the use of the synthetic vitamin, corn-soy diets have become the norm for swine and poultry production, and animal protein sources are much less important than formerly.

Many of the animal protein sources are of high quality, containing a good balance of amino acids for protein synthesis in animals. This is not surprising, when one considers the origin of the animal protein sources. (You are what you eat, in this case!) Animal protein sources are generally by-products and may present waste disposal problems, so their use as feedstuffs has additional value in terms of waste disposal costs. In other words, directing these products toward the feed industry by establishing competitive prices for them is more economically attractive to industries than paying the costs for disposal of them as liquid or solid wastes.

Food safety concerns regarding the feeding of animal protein sources (e.g., meat meal) have resulted in severe restrictions on the use of these products. Several outbreaks of *E. coli*-induced food poisoning associated with the consumption of beef, and the determination that "mad cow disease" has probably been introduced into humans via consumption of beef, has raised widespread public opposition to the recycling of animal proteins in animal feeding.

Meat Meal and Meat and Bone Meal

These products, derived from slaughterhouse wastes and products of rendering plants, contain carcass trimmings, condemned carcasses and livers, inedible offal (lungs, gastrointestinal tract), tendons, ligaments, hides, horns, hair, wool, and blood. If bone is included, it is known as **"meat and bone meal."** The quality of these products is variable, depending on the proportions of particular by-products. One of the problems of using meat and bone meal in ration formulation is its inherent variability, although efforts are made to standardize products to a consistent nutrient composition. Much of the protein in meat and bone meal is derived from collagen, a protein that is a major constituent of connective tissue, including bone. **Collagen** has an unusual amino acid composition, characterized by a high content of hydroxyproline and a complete absence of cysteine, cystine, and tryptophan. Hair, wool, horns, and hooves are composed of **keratin**, a poorly digested protein with a very high content of unavailable cystine. Thus the relative amounts of collagens and keratins in meat meal can have a pronounced effect on its protein quality. In general, meat and bone meal is less palatable and has a lower protein quality than soybean meal.

Meat and bone meal is an excellent source of calcium (7 to 10 percent) and phosphorus (3.8 to 5 percent). Meat is very low in calcium and high in phosphorus, and the reverse is true for bone. Thus the percentage of these minerals is variable, depending on the proportions of meat and bone in the product. Variability in calcium and phosphorus content is another problem in using meat and bone meal, as improper calcium-to-phosphorus ratios can be achieved.

Tankage is similar to meat and bone meal and is an older term used for the product produced when bones and meat scraps are wet-rendered by cooking in tanks. Most meat and bone meal is now dry-rendered, that is, cooked in a steam-jacketed vessel until the moisture has evaporated.

Tibbetts et al. (1987) prepared silage from poultry offal (heads, feet, and viscera) using *Lactobacillus acidophilus* as a fermentation aid. The ensiled material gave satisfactory results when used at levels up to 20 percent in swine diets; higher levels reduced growth. However, Urlings et al. (1993) questioned the safety of fermentation of raw slaughter by-products. These by-products (offal) are highly contaminated with bacteria and viruses; fermentation alone cannot be relied upon for sterilization of offal material.

It is very important that meat and bone meal and other animal by-products be properly heat-sterilized during processing. Contamination of the product with **Salmonella** species is a potential hazard. In 1988–1989, major problems occurred in the United Kingdom from improperly processed renderers' products. Contamination of meat meal with *Salmonella*-infected poultry by-products resulted in widespread *Salmonella* contamination of eggs and broiler meat when the meat meal was used in poultry feeds. An even more serious problem also occurred. The disease agent that causes scrapie in sheep was transferred to cattle through the feeding of meat meal prepared from condemned scrapie-infected sheep. This has resulted in a new disease of cattle, bovine spongiform encephalopathy (BSE) or **"mad cow disease."** Affected animals develop various neurological symptoms, including exaggerated limb movements, muscular jerking, anxiety, and frenzied movements culminating in death (Fig. 4.7). The brain shows extensive damage, with neuron degeneration and a spongelike appearance (spongiosis) (Fig. 4.8). The disease is caused by an unusual agent called a **prion**. A prion contains no DNA; it is an unusually folded protein that causes other proteins in nerve cells to unravel, causing a chain reaction of prion formation in the brain (Fig. 4.9). This "circle of death" spreads throughout the brain, causing the small holes that create the spongiform appearance.

BSE is one of numerous transmissible spongiform encephalopathy (TSE) conditions caused by prions, including the **Creutzfeldt-Jakob disease (CJD)** in humans. CJD occurs at a low rate in the human population worldwide. The concern with "mad cow disease" is that there is definite evidence that BSE has been transferred to humans via the consumption of beef, producing a new variant form of the disease, VCJD (Hill et al., 1997; Bruce et al., 1997).

The development of BSE in British cattle was associated with the feeding of meat meal produced by a new extraction process (Taylor et al., 1995). This rendering process, as well as the feeding of ruminant-derived meat meal back to ru-

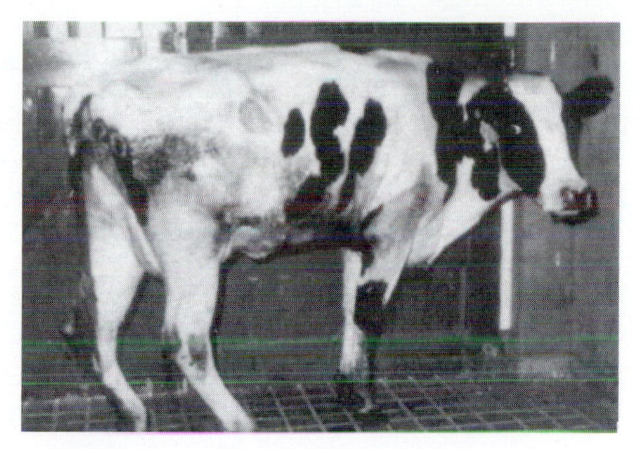

FIGURE 4.7 Holstein cows exhibiting signs of BSE (mad cow disease). Typically, animals show bulging eyes (left) and emaciation, weakness and staggering (right). (Courtesy of S. Franklin.)

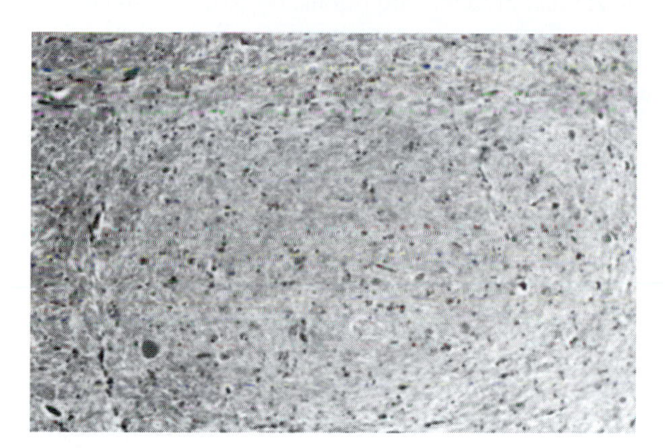

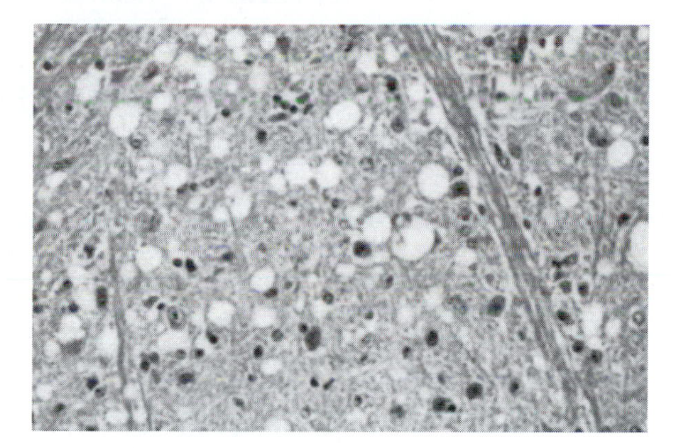

FIGURE 4.8 Normal bovine brain tissue (left) and vacuolated spongiosis of brain tissue from a cow affected with BSE (right). (Courtesy of S. Franklin.)

minants, has been banned in most countries, hopefully bringing the BSE crisis to a conclusion. The occurrence of BSE and the new human disease, VCJD, have been confined almost entirely to Great Britain. The British beef industry has been extremely hard hit by the BSE outbreak and may never fully recover. In 2003, the first cases of BSE in North America occurred in Canada and the United States.

Blood Meal

Dried blood meal contains approximately 80 percent crude protein, is high in lysine, and is severely deficient in isoleucine. The digestibility of blood meal is often low because of heat damage occurring in the drying process. Processing methods using low drying temperatures have improved the feeding value of blood meal. However, it should be used at levels not exceeding 6 to 8 percent of the diet because of amino acid imbalances induced by higher levels (Wahlstrom and Libal, 1977). Spray-dried blood has become very important in diets for very young baby pigs, raised under a **Segregated Early Weaning** program (see Chapter 13).

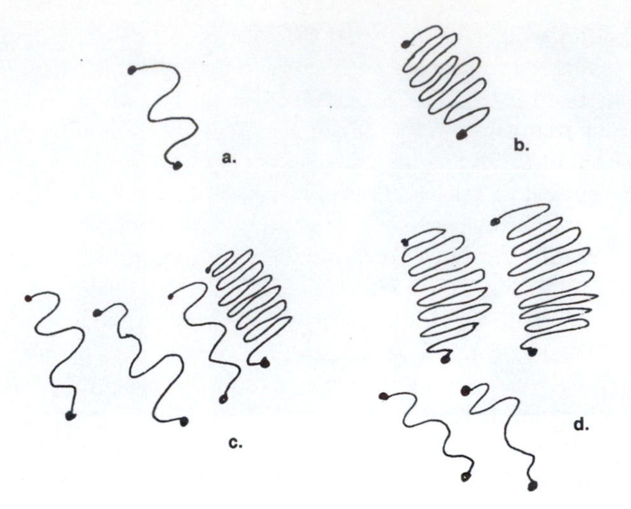

FIGURE 4.9 Conversion of normal brain proteins into prions. (a) Normal protein in a brain cell. (b) The same protein, twisted into a stable configuration called a prion. (c) A prion touches another protein and converts it to a prion. (d) Normal protein that has been "prionized" by contact with a prion. (From Cheeke, 2004.)

Fish Meal

Good quality fish meal is an exceptionally good source of well-balanced protein, and the price usually is reflective of this. Fish meals are often very expensive and are employed mainly in diets for poultry and young animals where high quantity and quality of protein are needed. Fish meal is prepared either from whole fish specifically caught for this purpose or from by-products from the use of fish prepared for human consumption. Some of the common types of fish meal include herring meal from Canada, the United States, Iceland, and Norway, menhaden meal from the Gulf and Atlantic coasts of the United States, anchovy meal from Peru, Chile, and Ecuador (Peruvian fish meal), and pilchard meal from South Africa. Peruvian fish meal was a major factor in the world feed industry in the 1960s, but overfishing and changes in the Humboldt Current have led to a serious decline in the Peruvian industry.

In the preparation of fish meal, the oil is usually extracted because it is a valuable commodity. Since fish oils are highly unsaturated, they are very susceptible to development of **rancidity**. High-fat fish meals may heat in storage and spontaneous combustion may even occur. Well-prepared fish meals are among the highest quality protein sources available. In addition to a high content and quality of protein, fish meal provides abundant amounts of minerals (calcium, phosphorous, and trace elements) and vitamins. It is highly palatable to swine and poultry and is generally used in starter diets for swine, chickens, and turkeys, for which amino acid and protein requirements are high.

Fish tissue contains abundant amounts of amines, which have a "fishy" smell. Fish meal should not be fed to swine and poultry in the immediate period before slaughter because the flesh may have a **fishy flavor**. The same is true for laying hens; if they are fed fish meal, the eggs may have an undesirable fish flavor. Fishy flavor is not a problem if the fish meal has been properly and promptly processed, avoiding putrefaction. Fishy flavored eggs can also occur when canola or rapeseed meal is fed to layers (see Rapeseed and Canola Meal, this chapter).

There has been interest in the use of fish meal in diets for ruminants as a source of **bypass protein** (see pp. 126–129). Because of its high protein content, excellent amino acid balance, and low degradation rate in the rumen, fish meal is a very effective source of bypass protein (Gibb and Baker, 1987; Newbold et al., 1987). Preston and Leng (1987) noted that fish meal supplementation of a molasses-urea diet for cattle in Cuba markedly increased cattle growth because of its bypass protein contribution.

Fish is, of course, a good source of "fish oils" or **omega-3 fatty acids**, which are of interest in human nutrition. Hulan et al. (1988, 1989) have shown that use of fish meal in poultry diets can markedly increase the omega-3 fatty acid content of the chicken meat.

Use of Fish in Diets for Fur Animals Mink and foxes, being carnivores, require high protein diets generally based on animal products such as meat and fish. "Trash fish" are often sold to fur ranchers; these are fish that are not marketable for human consumption for some reason. Some are very bony; others are unattractive in taste, flavor, or physical texture. Fish of the carp family, among others, contain an enzyme (thiaminase) that splits apart and destroys vitamin B_1 (thiamin). Feeding raw fish to mink or foxes may induce a thiamin deficiency, known as **Chastek's paralysis**, named after the Minnesota fox farmer who first observed it. Thiamin functions as a constituent of several enzymes necessary for energy metabolism. Thiamin deficiency results in signs such as convulsions, head retraction (star-gazing posture), reduced body temperature, paralysis, and death. Affected animals respond very rapidly to thiamin administration. It is interesting that the thiamin-destroying enzyme occurs not only in fish, but also in some plants such as bracken fern (*Pteridium* spp.) and horsetails (*Equisetum* spp.), and in rumen microbes (see Grain Overload of Ruminants, Chapter 2).

Mink fed certain marine fish may become anemic and have unpigmented hair, producing a condition called *cotton fur syndrome* (see Chapter 21). Fish such as the Pacific hake contain an amine, trimethylamine, and its oxide, trimethylamine oxide, that, during cold storage of the fish, are converted to formaldehyde. Formaldehyde and trimethylamine oxide impair iron absorption and cause **iron deficiency anemia**. Because iron is a cofactor of enzymes that convert the amino acid tyrosine to the hair pigment melanin, iron deficiency results in achromatrichia (lack of hair pigmentation).

Fish are sometimes preserved at sea by the use of nitrates. Nitrates react with amines in fish to produce **nitrosamines**, which are carcinogens. Outbreaks of liver cancer in mink and foxes in the United States, Canada, and Norway have been traced to nitrosamines in nitrate-preserved fish.

Other Fish and Marine Products

Liquified fish is prepared by allowing fish or fish by-products to liquify as a result of hydrolytic enzymes in the fish. The fish is ground and mixed with acid (e.g., formic acid at 3.5 percent of total mix) to lower the pH to a level (pH 2 to 4) that prevents bacterial spoilage. Liquified fish can be used as a protein supplement for wet (slop) diets for pigs and can be used in liquid feeding systems. **Fish silage** is made by the addition of lactic acid bacteria such as *Lactobacillus acidophilus* to ground fish, resulting in fermentation and stabilization by the organic acids so produced (Tibbetts et al., 1981). The preparation and properties of fish silage have been reviewed by Raa and Gildberg (1982). Liquified fish and fish silage are stable and may be stored for extensive periods. Because of the extensive degradation of the fish proteins to short peptides and amino acids in these products, the bioavailability of the protein is very high.

Other fishery by-products used as feedstuffs include shrimp and crab meal. **Shrimp meal** is dried waste of the shrimp industry, consisting of heads, shells, and cull whole shrimp. Shrimp, crabs, other crustacea, and many insects have an exoskeleton composed of **chitin**, a cellulose-like polysaccharide that is of low digestibility in animals. The value of shrimp and crab meal depends on the amount of shell, and thus chitin, present. Chitin contains nitrogen, which is nutritionally unavailable. The protein content should be corrected to account for the chitin content. For example, the protein content of shrimp hulls is 45 to 46 percent, but the nonchitin protein content is only 23 percent. A crude-fiber determination on these products provides a reasonable estimate of the chitin content. Chitin contains 7 percent nitrogen (43.75 percent crude-protein equivalent). **Chitosan** is a derivative of chitin, prepared chemically by removing acetyl groups from the acetylglucosamine groups in chitin. Chitin and chitosan have a high viscosity, increasing the water content of the gut contents of poultry that are fed these products and reducing nutrient digestibility by impeding mucosal nutrient uptake (Razdan and Pettersson, 1994).

Crab meal consists of the shells, viscera, and unextracted meat of crabs. The feeding value depends largely on the chitin content. Crab meal has a very high calcium content that must be balanced with phosphorus sources to maintain a desirable calcium-to-phosphorus ratio. Crab meal can provide approximately half the supplementary protein in swine rations without depressing performance (Husby, 1980). There is some evidence that feeding a source of lactose (e.g., whey) in diets containing chitin will increase microbial digestion of chitin in the gut in poultry (Austin et al., 1981). Chitin can be used fairly well by ruminants because of the chitinase activity of rumen microbes. Laflamme (1988) recommended that crab meal constitute not more than 10 percent of the total ration of beef cattle or 15 percent of the concentrate mix. There seems to be a period of adaptation necessary to obtain adequate levels of chitin-utilizing microbes in the rumen (Bunting et al., 1994; Nicholson et al., 1996a,b).

Myer et al. (1988) ensiled **scallop processing waste**, mainly viscera, by mixing the macerated viscera with formic acid (3.5 percent in total mix) and storing under anaerobic conditions. The silage was a satisfactory feedstuff for pigs, and using it in swine diets could offer a solution to a waste disposal problem of the fishing industry.

Crab meal protein is highly resistant to degradation in the rumen but is digestible in the small intestine (Nicholson et al., 1996a). The main problem in the use of **crustacean processing wastes** (crab, shrimp, crawfish, lobster, etc.) for ruminants is their low palatability. Abazinge et al. (1994) found that ensiling crab waste with straw increased its palatability. One of the problems with ensiling crustacean by-products is that they have a very high buffering capacity. Therefore, it is necessary to add considerable fermentable carbohydrate in order for there to be sufficient lactic acid produced to stabilize the silage (Evers and Carroll, 1996). Pelleting a crab-meal-containing diet increases its acceptability to cattle (Nicholson et al., 1996b).

By grinding and sieving, crab waste can be separated into coarse (high in chitin and ash) and fine (high in protein) fractions. The fine particles could be used in diets for swine, whereas the high-chitin material could be used for ruminants or as a fertilizer (Nicholson et al., 1996a).

Krill (Fig. 4.10) are shrimplike crustacea found in immense quantity in the waters off Antarctica. They are a major food source for a number of whale species. With the cessation of the whaling industry, interest has developed in the direct harvest of krill as a food or feedstuff. A potential catch of 100 to 400 million tons per year is estimated (Martin, 1979). Krill meal compares favorably with fish meal in composition. Krill is rich in carotenoid pigments (e.g., astaxanthin); krill meal is used quite often in fish diets to enhance red and pink pigmentation of the skin and flesh.

Meat meal prepared from marine mammals is sometimes used as a protein source. At one time, **whale meal** was an available feedstuff. In Eastern Canada, seal carcasses are a potentially abundant protein by-product of the Newfoundland sealing industry (Anderson et al., 1997; Charmley and O'Reilly, 1997; Robinson, 1996).

In other areas, such as Norway, seals that are inadvertently caught in fishing nets are used in animal feeding. **Seal meal** contains about 65 percent crude protein and 14 percent fat. The use of these kinds of products as animal feed is not viewed favorably by many people, in view of environmental and animal rights considerations with respect to conservation of marine mammals such as seals and whales.

FIGURE 4.10 Krill are shrimplike crustacea that are utilized in fish diets to enhance red and pink pigmentation of the skin and flesh (see Chapters 8 and 23).

Hydrolyzed Feathermeal and Leathermeal

Feathers are almost pure protein, with a crude protein content of 85 to 90 percent. However, this protein is primarily keratin, which has a very low digestibility. **Raw feathers** are almost completely indigestible. If they are cooked with steam, internal bonds in the keratin molecule are broken and the product, **hydrolyzed feathermeal**, is of high digestibility. Well-controlled processing is necessary; excessive steam hydrolysis converts cystine to **lanthionine** and lysine to **lysinoalanine**. Neither of these amino acids is used effectively in metabolism (Latshaw et al., 1994). Feathermeal is deficient in both lysine and methionine but has a high content of cystine. Hydrolyzed feathermeal is somewhat unpalatable and, because of this and the poor amino acid balance, should not be fed at more than 5 to 7 percent of the diet for swine and poultry (Chiba et al., 1996). **Hydrolyzed hair** from hogs and other animals is similar in feeding value to hydrolyzed feathermeal. Feathermeal and hairmeal compare favorably with cottonseed meal as nitrogen sources for ruminants (Aderibigbe and Church, 1983). Feathermeal can be used effectively in liquid supplements for ruminants because it is high in crude protein, is stable in molasses without suspending agents, and has 70 percent rumen escape protein (Pate et al., 1995).

Hydrolyzed leathermeal is a by-product of the tanning industry. It is a poor-quality protein consisting largely of collagen. Chromium compounds are used in the tanning process; **chromium poisoning** is a concern when tannery by-products are used. The chromium level should not exceed 2.75 percent of the by-product. Hydrolyzed leathermeal contains approximately 75 percent crude protein and can be used successfully at levels up to 6 percent in diets for ruminants (Knowlton et al., 1976).

SYNTHETIC AMINO ACIDS

The main reason that protein supplements are used in diets for nonruminants is to supply the essential amino acids in which the grain portion of the diet is deficient. Instead of using intact proteins, synthetic amino acids can be used. Sources of feed-grade lysine and methionine have been available for many years, and threonine and tryptophan have more recently become available. These amino acids are produced by bacterial fermentation; new techniques in biotechnology are likely to increase the number of amino acids available and to reduce their cost.

Amino acids exist as optical isomers, as either D- or L-amino acids. Only the L-form is used in protein synthesis. In some cases, the D-isomer can be converted to the L- (e.g., methionine), whereas in other cases (e.g., lysine) it is not utilized. Thus the common forms available in the feed industry are **DL-methionine** and **-L-lysine**. Methionine is also available in a form called **methionine hydroxy analog (MHA)**, which has a hydroxy group instead of the amine group. MHA is converted to methionine in the liver. MHA is available in a liquid form, which may have advantages in some cases and be inconvenient in others. On a methionine equivalency basis, allowing for the difference in molecular weights, there is no appreciable difference between MHA and DL-methionine in their nutritional value.

Another amino-acid-like compound available commercially is **taurine**. Taurine is not a dietary essential for livestock, but it is for cats. Taurine deficiency in cats results in degeneration of the retina of the eye and impairment of vision. Taurine is a sulfur-containing compound (an amino sulfonic acid) that is a derivative of sulfur amino acids. Taurine is added to pet foods.

An interesting concept for meeting amino acid needs is the use of **genetically modified bacteria** that excrete large amounts of a specific amino acid. Genetically altered *Lactobacilli* strains that secrete lysine have been developed (Newman and Sands, 1984; Newman et al., 1988). These organisms can be used to produce fermented grains with superior lysine contents. It should be possible, using biotechnology techniques, to introduce genes such as these into normal intestinal flora so that amino acid supplementation could be provided by the gut microflora.

When synthetic amino acids are used, consideration should be given to the adequacy of the diet for meeting metabolic requirements of the nonessential amino acids. **Nonessential amino acids** are not required per se in the diet, but they are required metabolically for protein synthesis. They are formed in the liver by interconversions of amino acids, so that if there is an excess of one and a deficiency of another, the deficient amino acid can be formed from the ones in excess. But for this to occur, there must be an amino acid pool. If the essential amino acids are provided with synthetic amino acids instead of with intact protein, there may be a shortage of amino acids for interconversions to form the nonessentials. This is referred to as a requirement for **nonspecific nitrogen**.

The amount and source of nonessential nitrogen may influence the growth performance of chickens (Bedford and Summers, 1986). Under certain conditions, such as heat stress, it may be desirable to minimize the level of excess nonessential amino acids, as energy is required to deaminate them and shunt them into pathways of carbohydrate or fat metabolism.

NITROGEN SOURCES FOR RUMINANTS

Because of microbial fermentation in the rumen, ruminant animals can utilize **nonprotein nitrogen** (NPN) sources to satisfy part of their protein requirement. In the rumen, microbial enzymes degrade dietary protein to amino acids and then ferment the amino acids as energy sources, excreting the nitrogen as ammonia. For their protein needs, microbes synthesize the amino acids they require, beginning from inorganic nitrogen (ammonia) and carbon fragments from dietary amino acids and products of carbohydrate digestion. Microbial protein becomes available to the host animal following passage of the microbes from the rumen. In the acid environment of the abomasum, the microbes are killed and then are digested by the animal's proteolytic enzymes in the small intestine. Thus many of the amino acids absorbed in ruminants originate from microbial protein and not directly from dietary protein.

Nonprotein Nitrogen Sources

Nonprotein nitrogen sources can be utilized by ruminants if they can be converted to ammonia in the rumen. Ammonia nitrogen is used by the microbes for amino acid synthesis, in the same manner as ammonia from degraded amino acids is used.

Urea The most common NPN source is urea. Urea is a normal product of metabolism. It is produced in the liver from ammonia arising from amino acid metabolism in animals. Urea is secreted from the liver into the blood, filtered out in the kidney, and excreted in the urine. Urea used in animal feeding (and fertilizers) is produced chemically, using nitrogen and carbon dioxide in the air and the chemical energy of natural gas. Thus urea manufacture is allied to the petroleum industry. Feed-grade urea contains 45 percent nitrogen, which is equivalent to 281 percent protein (N × 6.25).

Urea is utilized by being converted to ammonia in the rumen by the action of microbial urease. Conversion of urea to ammonia is rapid, so there is a rapid increase in the rumen ammonia levels after consumption of urea-containing feed.

Factors Influencing Urea Utilization A major factor influencing the efficiency of urea utilization is the amount of readily available carbohydrate in the diet. Carbohydrates that are readily fermented are starches and sugars in cereal grains. Because urea is rapidly converted to ammonia, it is important that the microbial population be able to utilize it rapidly. Otherwise, rumen ammonia levels increase, ammonia is absorbed, and ammonia may be excreted in the urine. This represents a loss of dietary nitrogen. The optimal concentration of rumen ammonia for maximizing efficiency of microbial fermentation is approximately 20 mg ammonia nitrogen per 100 ml rumen fluid (National Research Council, 1985). A vigorous microbial population will utilize ammonia rapidly and will also produce large quantities of organic acids. Absorption of ammonia is reduced to a low level when the rumen pH is 6.0 or lower.

Ammonia released from urea breakdown increases the pH of the rumen. When urea is fed in diets high in grain, this effect is useful in helping to prevent rumen acidosis caused by the large quantities of organic acids produced during fermentation.

Urea is not usually considered a satisfactory nitrogen supplement for **low-quality roughages**. These highly fibrous feedstuffs do not support vigorous microbial growth, so the amount of ammonia utilized for microbial protein synthesis is low. Digestion of poor-quality roughages takes a fairly prolonged period; urea does not provide a continuous level of nitrogen in the rumen to sustain microbial growth. There is a peak of ammonia production after urea is consumed; it is either utilized immediately in microbial metabolism or is absorbed. Contrary to this view, Preston and Leng (1987) propose that urea is one of the best nitrogen supplements to use for low-quality forages and fibrous by-products in tropical developing countries, provided that a delivery system is used that ensures a steady intake of urea. One method of doing this is to provide liquid supplements of urea-molasses mixtures, to provide molasses blocks containing urea (see Chapter 3), or to add urea to the drinking water (Stephenson et al., 1981). With low-quality roughages, fermentable nitrogen is usually the first-limiting factor in rumen fermentation. The level of rumen ammonia necessary to support optimal fermentation is approximately 20 mg ammonia nitrogen per 100 ml of rumen fluid; after a single administration of urea, the rumen ammonia level peaks soon after feeding but is below the optimal concentration for the rest of the day. Increased frequency of urea intake can maintain the rumen ammonia at adequate levels for rumen fermentation. Fiber digestion reaches a peak approximately 5 hours after ingestion of the feed; the urea delivery method must assure supplementary nitrogen during that time. It is metabolically more efficient to use urea to provide fermentable nitrogen than to use intact protein, particularly in areas where protein sources are expensive or not readily available. Under these conditions, the objective in diet formulation should be to maximize the rumen bypass potential of dietary protein and minimize the conversion of intact protein to ammonia in the rumen.

Although it is conventional wisdom and intuitively logical that a continuous supply of urea should be superior to intermittently available urea in the rumen, it is interesting that Mizwicki et al. (1980) found no difference in response to urea in steers fed low-quality range grass when the urea was infused continuously into the rumen or when given at intervals, including once per day (85 g urea in 1 hour). Dry-matter digestibility and nitrogen retention were increased by approximately 5 percent when urea was supplemented, regardless of the rate of administration.

According to Mathison et al. (1994), the ideal **slow-release NPN source** should provide branched-chain fatty acids for amino acid synthesis, in addition to a slow release of ammonia. These investigators found that isobutyraldehyde monourea had some effectiveness in this way as a slow-release NPN source. In the 1980s, there was considerable interest in and commercial development of isoacids as feed additives to maximize microbial protein synthesis (see Chapter 8), but their use has been discontinued.

When NPN sources are used to provide much of the **fermentable nitrogen** in the rumen, it may be necessary to add an **inorganic sulfur** source to provide for sulfur amino acid synthesis. In the industrialized countries, acid rain usually provides sufficient sulfur in the soil, but in tropical countries with heavily leached soils or soils of volcanic origin, supplementary sulfur may be needed because of low sulfur content of forages.

Urea Usefulness The circumstances under which dietary urea (or NPN in general) is not useful is when the rumen contains an adequate quantity of fermentable nitrogen for maximum microbial growth. Providing further fermentable nitrogen will be of no benefit. The dietary protein level above which urea is useless is not a constant but varies with dietary energy level, feed intake, and the degradability of dietary protein. As the energy level or intake increases, microbial growth is stimulated and more fermentable nitrogen can be incorporated into microbial protein. Various systems have been developed to estimate the productive value of urea. Further information on this subject is provided by the NRC (1985).

Urea Toxicity Urea is nontoxic but can cause **ammonia toxicity**. When the level of urea in the diet is excessive, the amount of ammonia absorbed can overwhelm the liver's capacity to detoxify it. Ammonia may then enter the general circulation. The brain is very sensitive to ammonia, and neurological disturbances result when blood ammonia levels are elevated. Signs of toxicity include labored breathing, incoordination, staggering, profuse slobbering, and bloating. Usually mortality occurs. Urea toxicity can be treated by the administration of an acid (e.g., vinegar) in cold water; the acid reduces ammonia absorption and the cold water dilutes the rumen ammonia and retards conversion of urea to ammonia. Unless only a few animals are involved, treatment is impractical. To prevent urea toxicity, urea should not be used at levels greater than 2 percent of the diet. Usually it should not be used to provide more than one-third of the total dietary nitrogen. Glucose metabolism is impaired in animals affected by subclinical ammonia toxicity (Fernandez et al., 1988); on a chronic basis, this could adversely affect the performance of NPN-fed animals.

Urea can be chemically modified to reduce the rate at which it is hydrolyzed in the rumen. A product known as *starea* is formed from heating corn and urea. Urea can be complexed with formaldehyde to reduce its rate of hydrolysis (Makkar et al., 1988). These slow-release forms of NPN do not improve the efficiency of NPN use, but they can be useful in reducing urea toxicity. They are not used commercially to much extent.

Bacteria associated with the rumen epithelium may get to preferentially utilize urea secreted from the blood to the rumen (Bunting et al., 1989). Hence, when rumen ammonia concentrations are limiting, bacteria associated with the rumen wall may rapidly utilize recycled urea and limit its diffusion into the bulk of the rumen contents, thus limiting microbial fermentation.

Blood and Milk Urea Concentrations Urea is the major form of excretory nitrogen in mammals. The level of **blood urea nitrogen (BUN)** in ruminants reflects the efficiency of utilization of dietary crude protein. Urea rapidly equilibrates throughout body fluids, including the milk. Thus, **milk urea nitrogen (MUN)** also may serve as an index of efficiency of protein utilization, particularly in the lactating dairy cow (Broderick and Clayton, 1997). For field testing purposes, milk samples are easily obtained. Elevated levels of BUN are associated with decreased conception rates of dairy cattle. BUN or MUN levels exceeding 20 mg/dL are indicative of potentially reduced reproductive rates in dairy cattle (Butler et al., 1996).

Other Nonprotein Nitrogen Sources

1. Biuret Biuret is produced by heating urea to a high temperature, resulting in the condensation of two urea molecules. Biuret is hydrolyzed to ammonia in the rumen more slowly than is urea. This should result in greater efficiency of nitrogen utilization because the ammonia is released at about the same rate as the microbes use it. This is particularly important in roughage-based diets; biuret is considered to be a much better nitrogen supplement for roughages than is urea, although data do not always support this supposition (Oltjen et al., 1969). Thus, biuret is a **slow-release form of nitrogen.** Biuret is more palatable than urea; urea is bitter, whereas biuret is tasteless.

Biuret is converted to ammonia in the rumen by the microbially produced enzyme biuretase. The organisms produce this enzyme only after a fairly long and continuous period of exposure to biuret. It may require an adaptation period of up to 6 weeks for biuret to be used efficiently. If it is withdrawn from the diet for a few days, a new adaptation period may be needed. Biuret that is not metabolized in the rumen is absorbed and excreted as such in the urine.

2. Dried Poultry Waste Another source of NPN is dried poultry waste (DPW). This is the manure and litter from poultry houses. Birds excrete a concentrated source of nitrogen, uric acid, in their urine (which is excreted with the feces). The uric acid in poultry excreta can be used as a nitrogen source by rumen microbes. Uric acid is degraded to ammonia more slowly in the rumen than is urea. At the levels used as a protein supplement, DPW is not unpalatable. The main concern is with the possible presence of drugs in the manure. Copper sulfate, for example, is sometimes used as a feed additive for poultry. Sheep, in particular, are very susceptible to **copper toxicity** and should not be fed DPW if it is from birds fed high copper levels. This aspect is considered in more detail in Chapter 8. DPW is a suitable supplement for beef cows wintered on low-quality roughage such as native range or cornstalks (Jordon et al., 2002). Poultry litter has a high calcium content and should not be fed to cows in late gestation because of the role of high calcium prepartum in development of milk fever (see Chapter 17). The utilization of animal wastes as feedstuffs has been covered in symposium proceedings published in the *Journal of Animal Science* (48:111–233, 1979).

A potential hazard with utilization of DPW as a feedstuff is the possible presence of pathogenic organisms such as *Salmonella.* Deep stacking of broiler litter to raise the internal temperature to 60°C or ensiling are both effective means of eliminating all pathogens (Chaudhry et al., 1996). Addition of low-quality roughages (e.g., straw) to broiler litter before it is deep-stacked can improve the nutritive value of roughages for ruminants (Park et al., 1995). Deep stacked poultry litter should be covered with plastic to reduce exposure to air and excessive heating (Rankins et al., 1993).

Ammoniation of low-quality roughages improves their digestibility (see Chapter 6). Treatment of straw with DPW has been found to be an effective means of providing ammonia, particularly in developing countries. Ensiling rice straw or other fibrous materials with DPW results in ammonia being released by bacterial action. Urea can also be used or sources of urea such as human or animal urine. Other NPN sources occasionally used, mainly for experimental purposes, are diammonium citrate and diammonium phosphate.

Nonprotein Nitrogen Utilization by Nonruminant Herbivores

Microbes in the cecum and colon of nonruminant herbivores can utilize NPN as a nitrogen source. This process is less efficient as a means of using NPN in place of dietary protein than it is in ruminants, primarily because the site of fermentation is at the end rather than at the beginning of the digestive tract. This presents difficulties in getting the NPN source to the hindgut, and in the animal's utilization of microbial protein. Dietary urea is used inefficiently by horses and rabbits because it is hydrolyzed to ammonia by bacterial urease in the small intestine. The ammonia is absorbed and reconverted to urea in the liver. Some urea does reach the hindgut, but by secretion from the blood into the cecum rather than by transit through the small intestine. Much of the absorbed ammonia is excreted in the urine and wasted.

Nonruminant herbivores are less susceptible to urea toxicity than cattle, presumably because some of the urea is absorbed directly and excreted in the urine, rather than all of it being converted to ammonia. **Ammonia toxicity** in horses causes neurological symptoms, such as head pressing, with the head pressed against a solid object. Biochemically, the condition is caused by the metabolism of ammonia in the brain. The ammonia reacts with α-keto glutaric acid to form glutamine; the depletion of α-keto glutaric acid, an intermediate of the citric acid cycle, impairs ATP production by brain tissue, leading to coma and death. Similar signs occur with liver dysfunction, impairing the liver's ability to convert ammonia to urea. This occurs with poisoning of horses by toxic plants containing hepatotoxic alkaloids, such as *Senecio* species.

Biuret is more suitable than urea as an NPN source for horses and rabbits because it is not hydrolyzed in the small intestine and so can reach the hindgut intact. In rabbits, microbial protein synthesized in the cecum can be made available to the animal via cecotrophy (coprophagy), the process by which cecal contents are consumed. It is less certain how horses can derive benefit from NPN sources because the microbial protein synthesized in the hindgut is excreted in the feces. Horses fed low-protein diets do practice coprophagy, which may be a mechanism to conserve nitrogen when poor-quality diets are consumed.

Bypass, Escape, Nondegradable Proteins, and Amino Acids

Microbial protein has a less favorable amino acid balance than many dietary proteins. Thus, in many cases, protein would be used more efficiently if it "bypassed" rumen fermentation and went directly to the small intestine for digestion by the animal itself. The terms *bypass, escape,* and *nondegradable* protein are used synonymously. There has been much interest in attempting to determine the bypass potential of protein supplements; however, total success has not been achieved to date. Estimates of ruminal escape or bypass values are given in Table 4–3. It should be recognized that there is a great deal of variability in experimental values both within and among feedstuffs. Among grains, corn has the greatest bypass protein potential. However, the quality (amino acid profile) of the corn protein is low. Corn protein is deficient in two of the amino acids, lysine and methionine, that are most important in a bypass protein. High-lysine corn is a good source of bypass lysine and methionine; feedlot cattle fed high-lysine corn show improved performance over those fed normal corn (Ladely et al., 1995). They attributed this result to improved energy utilization and not to the additional lysine. Some

TABLE 4–3. Estimates of Bypass Protein Content of Common Feedstuffs*

Feedstuff	*% Nondegradability**	*Feedstuff*	*% Nondegradability**
Grains		*By-Product Feeds (cont.)*	
Barley	21	Corn gluten meal	55
Corn	65	Distillers' dried grains	62
Sorghum	52	Fish meal	80
Lupin	35	Meat meal	76
		Meat and bone meal	60
Oil Meals			
Cottonseed (Solvent)	41	*Forages*	
Soybean	28	Alfalfa hay	28
Linseed	44	Dehydrated alfalfa	62
Peanut	30	Bromegrass hay	32
Rapeseed	23	Corn silage	27
Sunflower	24	Timothy hay	42
		Subterranean clover (immature)	27
By-Product Feeds			
Blood meal	82		
Brewers dried grains	53		

Adapted from National Research Council (1985).
*Percent of total crude protein that escapes rumen fermentation

of the animal protein sources have the greatest bypass potential; blood meal, meat meal, and fish meal are among the highest in nondegradable protein content as well as being of high protein quality. Drying of forages (e.g., dehy) increases their bypass potential because of the denaturation of soluble cytoplasmic leaf proteins. Methods of estimating rumen-undegraded protein have been described by NRC (1985).

Treatment of proteins with **formaldehyde** or **tannins** renders them resistant to digestion in the rumen without impairing intestinal digestion. **Heat treatment** of protein sources increases their bypass potential. Soybean meal (Coenen and Trenkle, 1989) and cottonseed meal (Goetsch and Owens, 1985) prepared by the mechanical expeller or screw-press process, in which considerable heat is generated, have greater bypass activity than those prepared by solvent extraction. Controlled **nonenzymatic browning,** by reacting soybean meal with xylose in the presence of heat, increased the flow of undegraded protein to the small intestine (Cleale et al., 1987) and may offer potential as a processing method to improve bypass activity of proteins.

Mir et al. (1984b) compared several methods of treating soybean and canola proteins to reduce their degradation in the rumen. Treatment with formaldehyde can cause overprotection of the protein, reducing its digestibility in the small intestine. Sodium hydroxide (2 g/100 g dry matter) was effective in reducing ruminal degradation of protein. Sodium hydroxide treatment of proteins may result in crosslinkages between amino acids in proteins, thus protecting them from microbial attack (Mir et al., 1984b). These workers also found that treatment of proteins with whole blood protected them from rumen degradation, possibly by coating the protein sources and providing a physical barrier to microbial action (blood protein is not easily degraded in the rumen). A similar but lesser effect was achieved by using fish hydrolysate, which also has a high resistance to rumen degradation. Feeding trials with dairy cattle confirmed that treatment of dietary proteins with sodium hydroxide and fresh blood improved protein utilization and increased milk production (Mir et al., 1984a). Another means of protecting proteins from rumen digestion is to coat them with calcium soaps of long-chain fatty acids (rumen inert fats, see Chapter 3); this procedure supplies both bypass protein and energy (Sklan, 1989).

Lysine and methionine are the two major amino acids that are often limiting in the amino acids absorbed from the intestine. Besides supplementing high-performing ruminants with bypass proteins having a high concentration of lysine and methionine, it is also possible to use rumen bypass sources of these amino acids. If lysine and methionine are added directly to the diet, they will be fermented and degraded in the rumen. **Rumen protected amino acids (RPAA)** are commercially available. Polymers that are pH sensitive have been used to encapsulate methionine and lysine (Rulquin and Delaby, 1997; Merchen and Titgemeyer, 1992). The polymers are stable in the rumen but break down in the highly acidic conditions of the abomasum. Commercial sources of RPAA include Smartamine M (70 percent methionine) and Smartamine ML (15 percent methionine, 50 percent lysine), marketed by Rhone-Poulenc; and Mepron M85, marketed by Degussa Corporation. Methionine hydroxy analog (MHA) is a source of RPAA because, compared to methionine, it has a very slow rate of degradation in the rumen. Organic **metal chelates** of amino acids, such as zinc methionine, are sources of RPAA.

Another approach to increasing the rumen bypass of amino acids is the feeding of inhibitors of amino acid deamination (Ramirez Pineres et al., 1997). Diarylliodonium compounds such as diphenyliodonium chloride are effective inhibitors of microbial deamination of amino acids. These compounds are not yet used commercially in ruminant feeding.

Although young succulent grass has a high protein content, protein supplementation sometimes improves cattle performance. For example, even though wheat pasture may contain 20 to 30 percent crude protein, steers grazing wheat pastures may show positive growth responses to supplementation with bypass protein (Andersen et al., 1987). Because the protein in the forage has a high solubility, it is highly degraded in the rumen to ammonia (Vogel et al., 1987). As much as 30 percent of the nitrogen of fresh forages may disappear between the mouth and duodenum, presumably as absorbed ammonia (Beever and Siddons, 1986). This inefficiency of **forage protein utilization** reflects an imbalance between rumen ammonia and the energy supply for optimal microbial utilization of the degraded nitrogen. A protein supplement with high bypass potential will provide supplementary protein in the small intestine. The increased growth rate is probably due to the positive effects of bypass protein on feed intake. The mechanism by which **bypass protein** stimulates feed intake is not known; the effect is most pronounced when low-quality roughages are fed (see Chapter 9). It is possible that growth rate of ruminants on pasture may sometimes be limited by lack of bypass protein. Anderson et al. (1988) supplemented yearling steers grazing spring and fall smooth brome pasture with various levels of escape protein (blood meal and corn gluten meal). In both of these periods of active grass growth, gains of the cattle were increased by provision of bypass protein. Thus, ruminants that are grazing actively growing, protein-rich pasture may in fact be deficient in nonfermentable protein and will respond to supplementation with an escape or bypass protein source. Abdalla et al. (1988) examined the seasonal variation in protein and fiber fractions of pasture grazed to keep it in a vegetative stage of growth. Protein solubility (hence rumen degradability) was highest in the first and last months of grazing, whereas the highest concentrations of nondegradable protein occurred during midsummer.

Worrell et al. (1990) found that the growth response of cattle grazing cereal forage to supplementation with protein (cottonseed meal) was enhanced and prolonged for the entire grazing season by also supplementing with an ionophore (lasalocid).

Stobbs et al. (1977) and Flores et al. (1979) obtained positive lactational responses when cows grazing high-protein tropical grass pasture were supplemented with a source of bypass protein, either as formaldehyde-treated casein or as the high-tannin content tropical legume forage *Leucaena leucocephala* (see Chapter 5). The response to bypass protein was primarily due to a higher daily intake of herbage. Responses to supplements of bypass protein have also been observed in animals fed conserved forage. Veira et al. (1988) noted that supplementation of grass silage with fish meal improved growth rate and feed efficiency of beef calves. Laflamme (1988) noted similar results with calves fed hay.

The suckling stimulus causes closure of the esophageal groove and the passage of milk directly from the esophagus to the omasum. Milk, therefore, is an excellent source of bypass protein. Suckling calves and lambs on pasture benefit from this effect. Similarly, in tropical countries or other areas where ruminants are fed low-quality roughages, the bypass potential of milk protein and the ability of the dam to lactate on a poor-quality diet are significant in improving the utilization efficiency of low-quality feeds.

The concept of **bypass energy** has not been investigated with the same vigor as has bypass protein. As discussed earlier (Chapter 2), postruminal digestion of starch results in up to 30 percent more energy made available to the animal than when it is fermented in the rumen. Methods of protecting cereal grain starch from rumen fermentation should therefore increase digestible energy. Fluharty and Loerch (1989) evaluated the treatment of corn with formaldehyde and other additives (glyoxal, propionaldehyde, tannic acid, and masonex—wood sugars and soluble phenolics) on starch digestion. Ruminal starch digestion decreased by 30 and 41.5 percent with treatments of 1 and 2 percent formaldehyde, respectively, compared to untreated corn, while whole tract starch digestion was not affected. Formaldehyde forms crosslinkages with hydroxyl groups in starch, thus reinforcing hydrogen bonds holding the starch granule together (Fluharty and Loerch, 1989). This inhibits swelling and disruption of the starch granule; however, acid treatment reverses the aldehyde effects by rendering the treated starch digestible in the small intestine after passing through the highly acidic abomasum. Rumen inert fats, such as calcium soaps (see Chapter 3), are also sources of bypass energy.

NEW AND UNCONVENTIONAL SOURCES OF PROTEIN

Leaf Protein Concentrates

Green leafy plants such as alfalfa contain large amounts of protein. In tropical areas, where there are often protein shortages, tropical forages (particularly legumes) may contain abundant amounts of protein. The protein in forages has a good amino acid balance for nonruminant animals and humans. However, it is in a package unsuitable for direct use except by ruminants and other herbivores. Forages are unpalatable to swine, poultry, and humans; their fibrous nature makes them of low digestibility, and the digestive tract of simple nonruminants cannot accommodate a large volume of low-energy forages. The challenge of maximizing utilization of forage protein can be met by fractionation of forages to separate the protein from the fiber. **Fractionation of forages** involves pressing or squeezing the juice out of green plants, followed by coagulation of the protein in the juice by heat to produce a protein curd. The curd can be separated from the solubles by settling, filtration, or centrifugation. The protein fraction is referred to as **leaf protein concentrate (LPC).** It can be fed in the wet form, especially to pigs, or can be dried. The dried LPC has a crude protein content of 50 to 60 percent. Although LPC can be equivalent to soybean meal in feeding value, it has not become a widely used product because the economics of production are not favorable. However, if the need for protein increases greatly, which seems inevitable as world population is still increasing at a rapid rate, LPC will undoubtedly make a contribution. Leaf protein concentrate is a bright green color, but, by appropriate processing, can be bleached to produce a white protein concentrate suitable for direct incorporation into human foods.

Alfalfa juice, containing soluble proteins without the fiber, has been evaluated as a protein source for swine (Barber et al., 1981). Use of alfalfa juice as the sole protein supplement gave very poor results, with poor growth and signs of salt toxicity. Alfalfa juice has a very high potassium concentration, which along with its high content of other cations such as sodium, calcium and magnesium, causes toxicity due to electrolyte overload. Alfalfa juice is thus unsatisfactory as a protein source for nonruminants, except at low levels. The deproteinized juice remaining after removal of the protein to produce LPC can be used as a feed for ruminants. Referred to as **alfalfa molasses,** it has a feeding value of about 78 percent of that of cane molasses for feedlot steers (Zinn, 1988).

A few animal species have evolved a feeding strategy that accomplishes green-leaf fractionation. Lowry (1989) reported that some **folivorous herbivores** such as the fruit bat chew green leaves to a bolus, swallow the liquid portion (the leaf protein concentrate), and expel the pellet of extracted fiber. This is a unique adaptation for utilization of fibrous feeds.

Single Cell Protein

Single cell protein (SCP) refers to bacteria, yeast, algae, and fungi. These organisms have a high protein content, and the potential yield per unit area is very high. Besides their high protein productivity, single-cell organisms are of interest because they can be used to purify sewage and animal and industrial wastes. Because of their rapid rate of cell division, they have a high DNA content and a high nucleic acid level. **Nucleic acids** contain nitrogen and are digested in the rumen and small intestine by nucleases. In ruminants, nucleic acids can be used as sources of NPN, whereas in simple nonruminants they are not useful nutritionally. Nucleic acids are excreted by being metabolized to uric acid. In most animals, uric acid is further metabolized to allantoin, a more water-soluble compound, which is excreted in the urine. In some humans, the conversion of uric acid to allantoin is inefficient. Uric acid may precipitate out of the blood and cause accumulation of uric acid crystals in the joints. This condition, known as **gout,** is aggravated by consumption of foods with a high nucleic acid content.

Algae Algae are grouped into two categories: macroalgae (seaweeds) and microalgae (e.g., *Chlorella* and *Spirulina* spp.). Both classes are used for human food, animal feed, fertilizers, biochemicals, and pharmaceuticals. Use of seaweeds as feedstuffs is discussed in Chapter 6. The green algae *Chlorella* has been evaluated as a feedstuff and as an organism for purifying wastes. *Chlorella* can be grown in lagoons or vats fertilized with sewage, swine manure, or other animal wastes. *Chlorella* has even been of interest in the space program; algae production might be a feasible way of recycling human wastes in spacecrafts on extended missions or on permanent space colonies. Algae have a high concentration of protein and are excellent sources of xanthophyll and carotene pigments.

Another type of algae with potential as a feedstuff is the blue-green algae *Spirulina maxima*. *Spirulina* has been used as food by people in Chad in Africa and in Mexico by Aztecs living around Lake Texcoco. *Spirulina* is sold in "health food" stores as a food supplement for humans. It contains 65 to 70 percent crude protein and can be grown in lakes and waters too alkaline to be used for irrigation. The high pH ensures an abundant supply of carbon dioxide for photosynthesis, making *S. maxima* very productive. According to Contreras et al. (1979), *Spirulina* can produce 125 times as much protein per acre as corn, 70 times as much as fish farming, and 600 times as much as cattle production. Of course, the inputs are high also, but if maximum production of food is needed, algae production has great potential. The major expense in the use of algae as an SCP source is the harvesting of the small cells by centrifugation, precipitation, or flocculation. Another problem is that cultures of algae tend to "crash" with infection by wild species; thus a continuous culture of beneficial algae requires constant management skill.

Other SCP Sources Besides algae, other SCP sources with potential as feedstuffs include bacteria, fungi, and yeasts. In Britain, bacteria have been grown commercially on by-products of petroleum refining, such as paraffins, to produce a high-protein animal feed. **Yeasts** have also been produced on substrates of petroleum hydrocarbons and wastes of sulfite-process paper manufacture. Mixed populations of microbes are produced in wastewater treatment processes, yielding activated sludges that might be employed as feeds. These include activated sewage sludge and sludges produced in the cleanup of industrial and food-processing liquid wastes. So long as toxic heavy metals or other toxic wastes are not introduced into the system, activated sewage sludge can be a potentially useful feedstuff.

Aquaculture-Produced Products

Aquaculture may be used in the purification of wastewater. In the Orient, **carp** have been used for thousands of years in an integrated system in which human and animal organic wastes are used to fertilize fish ponds, thereby providing an effective waste treatment system and fish for food or feed. A **polyculture** of carp species is used, so all ecological niches of the pond are utilized. For example, grass carp consume filamentous algae and aquatic plants, silver carp utilize phytoplankton, bighead carp feed on zooplankton, and the omnivorous common carp feed on molluscs, bottom organisms, and fecal matter of other fishes (Buck et al., 1978). This kind of system is high yielding without the need for inputs of expensive feedstuffs. *Tilapia* spp., fast-growing warm-water fish, are also useful in polyculture systems.

In developing countries, such an aquaculture system could be used in **recycling animal wastes** to produce high-protein fish that could be fed to swine or poultry. The enriched water is useful in crop irrigation. Moav et al. (1977) used a polyculture of common, silver, and grass carp and tilapia in ponds fertilized with liquid cow manure to produce a high yield of fish. Though this system is not presently economically attractive in the United States, it might be in countries with greater need for maximizing efficiency of food production.

Insect Meals

Insects have been used as food sources by humans in some parts of the world and are readily consumed by free-ranging chickens, turkeys, and ducks. **Fly meal** (dried housefly pupae) has been tested as a feedstuff for poultry. The main interest in fly pupae has been as a means of disposal of poultry manure, with the production of fly meal as a secondary consideration. Fly meal can substitute for soybean meal in poultry diets with no adverse effects on performance (Calvert et al., 1969; Teotia and Miller, 1974). Dehydrated fly larvae compare favorably to soybean meal as a protein supplement for poultry (Zuidhof et al., 2003). In Asia, silkworm pupae are a by-product of the silk industry and have been utilized in poultry feeding. Because insects are part of the natural diet of poultry, they might utilize insects more efficiently than other animals (Finke et al., 1987), although evidence for this is lacking. **Chitinase,** the enzyme capable of hydrolyzing the chitinous exoskeleton of invertebrates, does occur in the stomach of some insect-eating birds (Robbins, 1993).

Soldier fly (*Hermetia illucens*) **larvae** grown on cattle manure were an acceptable protein source for swine (Newton et al., 1977). The larvae contained 42 percent crude protein and 35 percent ether extract. **Dried bee meal** containing 62 percent crude protein has also been tested (Salmon and Szabo, 1981).

QUESTIONS AND STUDY GUIDE

1. What factors make soybean meal the most important protein supplement for swine and poultry diets?
2. What is the purpose of the urease test?
3. Glandless cottonseed meal supports better growth of pigs than regular cottonseed meal. Why is it not the major cottonseed meal used as a feed?
4. Heat treatment of cottonseed meal has both positive and negative effects. Discuss.
5. Consult a current issue of the *Canadian Journal of Animal Science*, and locate an article dealing with rapeseed or canola meal. Provide a brief synopsis of the article.
6. What is the significance of sinapine in poultry nutrition?
7. Why may feeding linseed meal to pregnant sheep cause goiter in their lambs?
8. Arachidonic acid is an unsaturated fatty acid. From what oilseed crop do you think the name *arachidonic* is derived?
9. Visit a grocery store and make a list of all of the types of vegetable oil available. How many of the oils listed in Table 4–1 can you find?
10. What is the main factor limiting the extent to which sunflower meal and safflower meal can be used in swine and poultry diets?
11. Is copra an important protein source in the continental United States? Why or why not?
12. What relevance do the terms *wonder crop* and *flop* in the title "The Winged Bean: Will the Wonder Crop Be Another Flop?" have in the context of foreign aid or development programs (e.g., U.S. Agency for International Development programs, Peace Corps, etc.)?
13. During the "energy crunch" of the 1970s, numerous entrepreneurs constructed distillery units for sale to farmers so they could ferment grain to produce fuel (alcohol) for their tractors. Farmers were often told that they could use some of their corn to produce fuel. After fermentation, the distillers' grains remaining would have a much higher protein content than the original grain. Thus, the farmers not only would get fuel for their tractor but would improve the feeding value of their grain in the process. What is the fallacy in this proposal?
14. What might be some potentially damaging ecological effects of large-scale harvesting of krill?
15. Suppose you have a swine farm, and on the farm there is a shallow lake with a large population of common carp. You contemplate harvesting the

carp as a protein source for your pigs. Is this a practical idea? What are some of the factors to consider when using carp as a feed?

16. Hydrolyzed feathermeal is probably a better protein source for ruminants than for nonruminants. Why?

17. A graduate student is conducting nutritional studies with pigs. She formulates a purified diet by using crystalline essential amino acids at levels to provide exactly the essential amino acid requirements. Growth of the pigs is quite poor on this diet. When it is supplemented with either urea or glutamic acid (a nonessential amino acid), growth is improved. Explain.

18. Why is urea more suitable as a nitrogen supplement for cattle than for horses? Would biuret be better than urea as a supplement for horses? Why?

19. A trial was conducted with lambs, with the following results. Wool growth was measured by tattooing a patch on each animal and weighing the weight of wool clipped off the patch every 30 days. Explain the results.

Diet	*Avg. Daily Gain, g*	*Wool Growth g per patch*
1. Basal (meets NRC requirements)	110	50
2. Basal + 0.2% methionine	112	51
3. Basal + 0.2% MHA	135	67
4. Basal + 5% soybean meal	120	58
5. Basal + 3% feathermeal	125	65

20. What is alfalfa molasses?

21. Humans consuming large amounts of *Chlorella* may develop gout, whereas pigs fed the same diet do not. Explain.

22. Comfrey (see Chapter 5) is promoted in "health food" stores with the implication that it has medicinal properties because it contains allantoin. Assuming that comfrey does contain allantoin, is this likely to be of much significance? Why?

23. The nutritive value of proteins is often assessed by conducting a nitrogen balance trial in which intake and excretion (fecal and urinary) of nitrogen are measured. Because the nitrogen in feeds is associated mainly with amino acids, measuring nitrogen balance gives a good indication of the metabolic fate of amino acids. The term *biological value* (BV) refers to the percentage of absorbed nitrogen retained in the body:

$$\text{BV} = \frac{\text{N retained}}{\text{N absorbed}} \times 100$$

The *net protein utilization* (NPU) refers to the percentage of ingested nitrogen that is retained:

$$\text{NPU} = \frac{\text{N retained}}{\text{N absorbed}} \times 100$$

a. Explain in words the biological significance of the "N retained" fraction. What specific nutrients does it correspond to? How do you measure the N retained fraction?

b. Two potential protein sources for swine are being evaluated in a nitrogen balance trial. A corn-soy control diet is used for comparative purposes. The following data, the average values for six pigs per treatment, were collected. The animals were kept in metabolism crates for feces and urine collection.

	Treatment		
Diet	*Control*	*Protein A*	*Protein B*
% protein in diet (DM basis)	15.1	15.3	15.2
Dry-matter intake (kg)	16.0	15.8	12.6
Fecal dry matter excreted (g)	3200	3000	5600
% N in fecal DM	1.8125	2.447	3.830
Total urine excreted (1)	10.0	10.2	8.4
Urinary N (mg/ml)	3.2	13.8	0.548

For each diet, calculate the following:

a. % digestibility of crude protein
b. Biological value
c. Net protein utilization

Describe the value of proteins A and B as protein sources for swine. What are the positive and negative aspects of each protein source? How might the nutritive value of each of the test proteins be improved?

24. A poultry scientist is measuring the digestibility of amino acids and crude protein in sorghum grain. He is comparing the digestibility in surgically altered birds in which the urinary excretion is separated from the fecal excretion, with normal birds in which the feces and urine are excreted together. The digestibilities of sorghum amino acids are very similar between the control and surgically altered birds. However, there is a large difference in apparent digestibility of crude protein. Explain. Would you expect the protein digestibility to be higher or lower in the surgically altered birds?

25. What are some nutritional effects of oligosaccharides in feedstuffs? What common feeds contain significant oligosaccharide levels?

REFERENCES

Soybean Meal

Cook, D. A., A. H. Jensen, J. R. Fraley, and T. Hymowitz. 1988. Utilization by growing and finishing pigs of raw soybeans of low Kunitz trypsin inhibitor content. *J. Anim. Sci.* 66:1686–1691.

Herkelman, K. L., G. L. Cromwell, T. S. Shahley, T. W. Pfeiffer, and D. A. Knabe. 1992. Apparent digestibility of amino acids in raw and heated conventional and low trypsin-inhibitor soybeans for pigs. *J. Anim. Sci.* 70:818–826.

Kilshaw, P. J. and J. W. Sissons. 1979. Gastrointestinal allergy to soyabean protein in preruminant calves. Allergic constituents of soybean products. *Res. Vet. Sci.* 27:306–371.

Leske, K., C. Coon, O. Akavanichan, and T. Cheng. 1988. Nutritional effects of oligosaccharide-free soybean meal. *Poult. Sci.* 67:109 (Abst).

Maenz, D. D., G. G. Irish, and H. L. Classen. 1999. Carbohydrate-binding and agglutinating lectins in raw and processed soybean meals. *Anim. Feed Sci. Tech.* 76:335–343.

McNaughton, J. L., F. N. Reece, and J. W. Deaton. 1981. Relationships between color, trypsin inhibitor contents, and urease index of soybean meal and effects on broiler performance. *Poult. Sci.* 60:393–400.

O'Quinn, P. R., D. N. Knabe, E. J. Gregg, and E. W. Lusas. 1997. Nutritional value for swine of soybean meal produced by isopropyl alcohol extraction. *J. Anim. Sci.* 75:714–719.

Payne, R. L., T. D. Bidner, L. L. Southern, and J. P. Geaghan. 2001. Effects of dietary soy isoflavones on growth, carcass traits, and meat quality in growing-finishing pigs. *J. Anim. Sci.* 79:1230–1239.

Pedersen, H. E. and J. W. Sissons. 1984. Effect of antigenic soyabean protein on the physiology and morphology of the gut in the preruminant calf. *Can. J. Anim. Sci.* 64 (Suppl. 1):183–184.

Pettey, L. A., S. D. Carter, B. W. Senne, and J. A. Shriver. 2002. Effects of β-mannanase addition to corn-soybean meal diets on growth performance, carcass traits, and nutrient digestibility of weanling and growing-finishing pigs. *J. Anim. Sci.* 80:1012–1019.

Sidhu, G. S. and D. G. Oakenfull. 1986. A mechanism for the hypocholesterolaemic activity of saponins. *Brit. J. Nutr.* 55:643–649.

Smiricky, M. R., C. M. Grieshop, D. M. Albin, J. E. Wubben, V. M. Gabert, and G. C. Fahey, Jr. 2002. The influence of soy oligosaccharides on apparent and true ileal amino acid digestibilities and fecal consistency in growing pigs. *J. Anim. Sci.* 80:2433–2441.

Wormsley, K. G. 1988. Trypsin inhibitors: Potential concern for humans. *J. Nutr.* 118:134–136.

Zhang, L. Y., D. F. Li, S. Qiao, E. W. Johnson, L. I. Baoyu, P. A. Thacker and I. K. Han, 2003. Effects of stachyose on performance, diarrhoea incidence and intestinal bacteria in weanling pigs. *Arch. Anim. Nutr.* 57:1–10.

Cottonseed Meal

Arieli, A. 1998. Whole cottonseed in dairy cattle feeding: A review. *Anim. Feed Sci. Tech.* 72:97–110.

Cheeke, P. R. 1998. *Natural Toxicants in Feeds, Forages, and Poisonous Plants.* Upper Saddle River, NJ: Prentice Hall, Inc.

Coppock, C. E., J. K. Lanham, and J. L. Horner. 1987. A review of the nutritive value and utilization of whole cottonseed, cottonseed meal and associated by-products for dairy cattle. *Anim. Feed Sci. Tech.* 18:89–129.

Dilday, R. H. 1986. Development of a cotton plant with glandless seeds, and glanded foliage and fruiting forms. *Crop Sci.* 26:639–641.

Holmberg, C. A., L. D. Weaver, W. M. Guterbock, J. Genes, and P. Montgomery. 1988. Pathological and toxicological studies of calves fed a high concentration cotton seed meal diet. *Vet. Pathol.* 25:147–153.

Hron, R. J., S. P. Koltun, J. Pominski, and G. Abraham. 1987. The potential commercial aspects of gossypol. *J. Am. Oil Chem. Soc.* 64:1315–1319.

LaRue, P. C., D. A. Knabe, and T. D. Tanksley, Jr. 1985. Commercially processed glandless cottonseed meal for starter, grower and finisher pigs. *J. Anim. Sci.* 60:495–502.

Morgan, S., E. L. Stair, T. Martin, W. C. Edwards, and G. L. Morgan. 1988. Clinical, clinicopathologic, pathologic, and toxicologic alterations associated with gossypol toxicosis in feeder lambs. *Am. J. Vet. Res.* 49:493–499.

Morgan, S. E. 1989. Gossypol as a toxicant in livestock. *Vet. Clin. North Am.: Food Anim. Pract.* 5:251–262.

Percy, R. G., M. C. Calhoun, and H. L. Kim. 1996. Seed gossypol variation within *Gossypium barbadense* L. cotton. *Crop Sci.* 36:193–197.

Randel, R. D., C. C. Chase, Jr., and S. J. Wyse. 1992. Effects of gossypol and cottonseed products on reproduction of mammals. *J. Anim. Sci.* 70:1628–1638.

Risco, C. A., C. A. Holmberg, and A. Kutches. 1992. Effect of graded concentrations of gossypol on calf performance: Toxicological and pathological considerations. *J. Dairy Sci.* 75:2787–2798.

Canola (Rapeseed) Meal

Bell, J. M. 1993. Factors affecting the nutritional value of canola meal: A review. *Can. J. Anim. Sci.* 73:679–697.

Bell, J. M. and M. O. Keith. 1987. Feeding value for pigs of canola meal derived from Westar and Triazine-tolerant cultivars. *Can. J. Anim. Sci.* 67:811–819.

Bell, J. M. and A. Shires. 1982. Composition and digestibility by pigs of hull fractions from rapeseed cultivars with yellow or brown seed coats. *Can. J. Anim. Sci.* 62:557–565.

Blair, R., R. Misir, J. M. Bell, and D. R. Clandinin. 1986. The chemical composition and nutritional value for chickens of meal from recent cultivars of canola. *Can. J. Anim. Sci.* 66:821–825.

Ismail, F., M. Vaisey-Genser, and B. Fyfe. 1981. Bitterness and astringency of sinapine and its components. *J. Food Sci.* 46:1241–1244.

Leeson, S., J. O. Atteh, and J. D. Summers. 1987. The replacement value of canola meal for soybean meal in poultry diets. *Can. J. Anim. Sci.* 67:151–158.

Shaw, J., D. K. Baidoo, and F. X. Aherne. 1990. Nutritive value of canola seed for young pigs. *Anim. Feed Sci. Tech.* 28:325–331.

Summers, J. D. and M. Bedford. 1994. Canola meal and diet acid-base balance for broilers. *Can. J. Anim. Sci.* 74:335–339.

Linseed Meal

Batterham, E. S., L. M. Andersen, D. R. Baigent and A. G. Green. 1991. Evaluation of meals from Linola™ low-linolenic acid linseed and conventional linseed as protein sources for growing pigs. *Anim. Feed. Sci. Tech.* 35:181–190.

Batterham, E. S., L. M. Andersen, and A. G. Green. 1994. Pyridoxine supplementation of Linola™ meal for growing pigs. *Anim. Feed Sci. Tech.* 50:167–174.

Bishara, H. N. and H. F. Walker. 1977. The vitamin B_6 status of pigs given a diet containing linseed meal. *Brit. J. Nutr.* 37:321–331.

Cunnane, S. C., P. A. Stitt, J. Ganguli, and J. K. Armstrong. 1990. Raised omega-3 fatty acid levels in pigs fed flax. *Can. J. Anim. Sci.* 70:251–254.

Gutzwiller, A. 1993. The effect of a diet containing cyanogenic glycosides on the selenium status and the thyroid function of sheep. *Anim. Prod.* 57:415–419.

Klosterman, H. J. 1974. Vitamin B_6 antagonists of natural origin. *J. Agr. Food Chem.* 22:13–19.

Oomah, B. D., G. Mazza, and E. O. Kenaschuk. 1992. Cyanogenic compounds in flaxseed. *J. Agric. Food Chem.* 40:1346–1348.

Palmer, I. S., O. E. Olson, A. W. Halverson, R. Miller, and C. Smith. 1980. Isolation of factors in linseed oil meal protective against chronic selenosis in rats. *J. Nutr.* 110:145–150.

Peanut Meal

Hopkins, J. 1995. The very intolerant peanut. *Food Chem. Toxic.* 33:81–86.

Orok, E. J., J. P. Bowland, and C. W. Briggs. 1975. Rapeseed, peanut and soybean meals as protein supplements with or without added lysine: Biological performance and carcass characteristics of pigs and rats. *Can. J. Anim. Sci.* 55:135–146.

West, R. L. and R. O. Myer. 1987. Carcass and meat quality characteristics and backfat fatty acid composition of swine as affected by the consumption of peanuts remaining in the field after harvest. *J. Anim. Sci.* 65:475–480.

Sunflower Meal

Kepler, M., G. W. Libal, and R. C. Wahlstrom. 1982. Sunflower seeds as a fat source in sow gestation and lactation diets. *J. Anim. Sci.* 55:1082–1086.

Marchello, M. J., N. K. Cook, V. K. Johnson, W. D. Slauger, D. K. Cook, and W. E. Dinusson. 1984. Carcass quality, digestibility and feedlot performance of swine fed various levels of sunflower seed. *J. Anim. Sci.* 58:1205–1210.

Richardson, C. R., R. N. Beville, R. K. Ratcliff, and R. C. Albin. 1981. Sunflower meal as a protein supplement for growing ruminants. *J. Anim. Sci.* 53:557–563.

Schroeder, G. E., L. J. Erasmus, and H. H. Meissner. 1996. Chemical and protein quality parameters of heat processed sunflower oilcake for dairy cattle. *Anim. Feed Sci. Tech.* 58:249–265.

Seerley, R. W., D. Burdick, W. C. Russom, R. S. Lowrey, H. C. McCampbell, and H. E. Amos. 1974. Sunflower meal as a replacement for soybean meal in growing swine and rat diets. *J. Anim. Sci.* 38:947–953.

Thomas, V. M., D. N. Sneddon, R. E. Roffler, and G. A. Murray. 1982. Digestibility and feeding value of sunflower silage for beef steers. *J. Anim. Sci.* 54:933–937.

Safflower Meal

Lyon, C. K., M. R. Gumbmann, A. A. Betschart, D. J. Robbins, and R. M. Saunders. 1979. Removal of deleterious glucosides from safflower meal. *J. Am. Oil Chem. Soc.* 56:560–564.

Coconut Meal and Palm Kernel Meal

Onwudike, O. C. 1986. Palm kernel meal as a feed for poultry. 1. Composition of palm kernel meal and availability of its amino acids to chicks. *Anim. Feed Sci. Tech.* 16:179–186.

Osei, S. A. and J. Amo. 1987. Palm kernel cake as a broiler feed ingredient. *Poult. Sci.* 66:1870–1873.

Panigrahi, S. and C. J. Powell. 1991. Effects of high rates of inclusion of palm kernel meal in broiler chick diets. *Anim. Feed Sci. Tech.* 34:37–47.

Rhule, S. W. A. 1996. Growth rate and carcass characteristics of pigs fed diets containing palm kernel cake. *Anim. Feed Sci. Tech.* 61:167–172.

Thorne, P. J., J. Wiseman, D. J. A. Cole, and D. H. Machin. 1988. Use of diets containing high levels of copra meal for growing/finishing pigs and their supplementation to improve animal performance. *Trop. Agric.* 65:197–201.

______ . 1989. The digestible and metabolizable energy value of copra meals and their prediction from chemical composition. *Anim. Prod.* 49:459–466.

Miscellaneous Oilseed Meals

Anderson, V. L., J. S. Caton, J. D. Kirsch, and D. A. Redmer. 2000. Effect of crambe meal on performance, reproduction, and thyroid hormone levels in gestating and lactating beef cows. *J. Anim. Sci.* 78:2269–2274.

Babatunde, G. M., W. G. Pond, and E. R. Peo, Jr. 1990. Nutritive value of rubber seed (*Hevea brasiliensis*) meal: Utilization by growing pigs of semipurified diets in which rubber seed partially replaced soybean meal. *J. Anim. Sci.* 68:392–397.

Bell, J. M., M. O. Keith, J. A. Blake, and D. I. McGregor. 1984. Nutritional evaluation of ammoniated mustard meal for use in swine feeds. *Can. J. Anim. Sci.* 64:1023–1033.

Blair, R. 1984. Nutritional evaluation of ammoniated mustard meal for chicks. *Poult. Sci.* 63:754–759.

Caton, J. S., V. I. Burke, V. L. Anderson, L. A. Burgwald, P. L. Norton, and K. C. Olson. 1994. Influence of crambe meal as a protein source on intake, site of digestion, ruminal fermentation, and the microbial efficiency in beef steers fed grass hay. *J. Anim. Sci.* 72:3238–3245.

Lambert, J. L., D. C. Clanton, I. A. Wolff, and G. C. Mustakas. 1970. Crambe meal protein and hulls in beef cattle rations. *J. Anim. Sci.* 31:601–607.

Miller, R. and P. R. Cheeke. 1986. Evaluation of meadowfoam (*Limnanthes alba*) meal as a feedstuff for beef cattle. *Can. J. Anim. Sci.* 66:567–568.

Ngoupayou, J. D. N., P. M. Maiorino, and B. L. Reid. 1982. Jojoba meal in poultry diets. *Poult. Sci.* 61:1692–1696.

Ravindran, V. and R. Sivakanesan. 1996. The nutritive value of mango seed kernels for starting chicks. *J. Sci. Food Agric.* 71:245–250.

Ravindran, V., G. Ravindran, and R. Sivakanesan. 1996. Evaluation of the nutritive value of jackseed (*Artocarpus heterophyllus*) meal for poultry. *J. Agri. Sci.* 127:123–130.

Throckmorton, J. C., P. R. Cheeke, D. C. Church, D. W. Holtan, and G. D. Jolliff. 1982. Evaluation of meadowfoam (*Limnanthes alba*) meal as a feedstuff for sheep. *Can. J. Anim. Sci.* 62:513–520.

Verbiscar, A. J., T. F. Banigan, C. W. Weber, B. L. Reid, R. S. Swingle, J. E. Trei, and E. A. Nelson. 1981. Detoxification of jojoba meal by Lactobacilli. *J. Agric. Food Chem.* 29:296–309.

Yong-Gang, L., A. Steg, and V. A. Hindle. 1993. Crambe meal: a review of nutrition, toxicity and effect of treatments. *Anim. Feed Sci. Tech.* 41:133–147.

Grain Legumes

Cheeke, P. R. 1998. *Natural Toxicants in Feeds, Forages, and Poisonous Plants*. Upper Saddle River, NJ: Prentice Hall, Inc.

Soybeans

Anderson-Hafermann, J. C., Y. Zhang, and C. M. Parsons. 1992. Effects of heating on nutritional quality of conventional and Kunitz trypsin inhibitor-free soybeans. *Poult. Sci.* 71:1700–1709.

Cook, D. A., A. H. Jensen, J. R. Fraley, and T. Hymowitz. 1988. Utilization by growing and finishing pigs of raw soybeans of low Kunitz trypsin inhibitor content. *J. Anim. Sci.* 66:1685–1691.

Crenshaw, M. A. and D. M. Danielson. 1985a. Raw soybeans for gestating swine. *J. Anim. Sci.* 60:163–170.

______ . 1985b. Raw soybeans for growing-finishing pigs. *J. Anim. Sci.* 60:725–730.

Herkelman, K. L., G. L. Cromwell, T. S. Stahly, T. W. Pfeiffer, and D. A. Knabe. 1992. Apparent digestibility of amino acids in raw and heated conventional and low trypsin-inhibitor soybeans for pigs. *J. Anim. Sci.* 70:818–826.

Pontif, J. E., L. L. Southern, D. F. Coombs, K. W. McMillin, T. D. Bidner, and K. L. Watkins. 1987. Gain, feed efficiency and carcass quality of finishing swine fed raw soybeans. *J. Anim. Sci.* 64:177–181.

Zhang, Y., C. M. Parsons, K. E. Weingartner, and W. B. Wijeratne. 1993. Effect of extrusion and expelling on the nutritional quality of conventional and Kunitz trypsin inhibitor-free soybeans. *Poult. Sci.* 72:2299–2308.

Dry Beans

Luce, W. G., C. V. Maxwell, D. S. Buchanan, R. O. Bates, M. D. Woltmann, S. A. Norton, and G. N. Dietz. 1989. Raw mung beans as a protein source for bred gilts. *J. Anim. Sci.* 67:329–333.

Myer, R. O. and J. A. Froseth. 1983a. Heat-processed small red beans (*Phaseolus vulgaris*) in diets for young pigs. *J. Anim. Sci.* 56:1088–1096.

______ . 1983b. Extruded mixtures of beans (*Phaseolus vulgaris*) and soybeans as protein sources in barley-based swine diets. *J. Anim. Sci.* 57:296–306.

Myer, R. O., J. A. Froseth, and C. N. Coon. 1982. Protein utilization and toxic effects of raw beans (*Phaseolus vulgaris*) for young pigs. *J. Anim. Sci.* 55:1087–1098.

Rodriguez, J. P. and H. S. Bayley. 1987. Steam-heated culled beans: Nutritional value and digestibility for swine. *Can. J. Anim. Sci.* 67:803–810.

Van der Poel, A. F. B. 1990. Effect of processing on antinutritional factors and protein nutritional value of dry beans (*Phaseolus vulgaris* L.): A review. *Anim. Feed Sci. Tech.* 29:179–208.

Williams, P. E. V., A. J. Pusztai, A. MacDearmid, and G. M. Innes. 1984. The use of kidney beans (*Phaseolus vulgaris*) as protein supplements in diets for young, rapidly growing beef steers. *Anim. Feed Sci. Tech.* 12:1–10.

Fava Beans

Aherne, F. X., A. J. Lewis, and R. T. Hardin. 1977. An evaluation of faba beans (*Vicia faba*) as a protein supplement for swine. *Can. J. Anim. Sci.* 57:321–328.

Campbell, L. D., G. Olaboro, R. R. Marquardt, and D. Waddell. 1980. Use of fababeans in diets for laying hens. *Can. J. Anim. Sci.* 60:395–405.

Gardiner, E. E., S. Dubetz, and G. A. Kemp. 1980. Growth responses of chicks fed fababean diets. *Can. J. Anim. Sci.* 60:433–439.

Muduuli, D. S., R. R. Marquardt, and W. Guenter. 1981. Effect of dietary vicine on the productive performance of laying chickens. *Can. J. Anim. Sci.* 61:757–764.

Olaboro, G., L. D. Campbell, and R. R. Marquardt. 1981. Influence of fababean fractions on egg weight among laying hens fed test diets for a short time period. *Can. J. Anim. Sci.* 61:751–755.

Lupin Seed

Aguilera, J. F., E. Molina, and C. Prieto. 1985. Digestibility and energy value of sweet lupin seed (*Lupinus albus* var. Multolupa) in pigs. *Anim. Feed Sci. Tech.* 12:171–178.

Batterham, E. S., L. M. Andersen, B. V. Burnham, and G. A. Taylor. 1986a. Effect of heat on the nutritional value of lupin (*Lupinus angustifolius*) seed meal for growing pigs. *Brit. J. Nutr.* 56:169–177.

Batterham, E. S., L. M. Andersen, R. F. Lowe, and R. E. Darnell. 1986b. Nutritional value of lupin (*Lupinus albus*)-seed meal for growing pigs: Availability of lysine, effect of autoclaving and net energy content. *Brit. J. Nutr.* 56:645–659.

Brenes, A., R. R. Marquardt, W. Guenter, and B. A. Rotter. 1993. Effect of enzyme supplementation on the nutritional value of raw, autoclaved, and dehulled lupins (*Lupinus albus*) in chicken diets. *Poult. Sci.* 72:2281–2293.

Dunshea, F. R., N. J. Gannon, R. J. van Barneveld, B. P. Mullen, R. G. Campbell and R. H. King. 2001. Dietary lupins (*Lupinus angusti folius* and *Lupinus alba*) can increase digesta retention in the gastrointestinal tract of pigs. *Aust. J. Agr. Res.* 52:593–602.

Gdala, J., A. J. M. Jansman, P. van Leeuwen, J. Huisman, and M. W. A. Verstegen. 1996. Lupins (*L. luteus, L. albus, L. angustifolius*) as a protein source for young pigs. *Anim. Feed Sci. Tech.* 62:239–249.

Gdala, J., A. J. M. Jansman, L. Buraczewska, J. Huisman, and P. van Leeuwen. 1997. The influence of α-galactosidase supplementation on the ileal digestibility of lupin seed carbohydrates and dietary protein in young pigs. *Anim. Feed Sci. Tech.* 67:115–125.

Gladstones, J. S., C. A. Atkins and J. Hamblin (Eds.) Lupins as crop plants. Biology, Production and Utilization. 1998 Wallingford, U. K. CAB International.

Godfrey, N. W., A. R. Mercy, Y. Emms, and H. G. Payne. 1985. Tolerance of growing pigs to lupin alkaloids. *Aust. J. Exp. Agric.* 25:791–795.

Hale, O. M. and J. D. Miller. 1985. Effects of either sweet or semi-sweet blue lupin on performance of swine. *J. Anim. Sci.* 60:989–997.

Halvorson, J. C., M. A. Shehata, and P. E. Waibel. 1983. White lupins and triticale as feedstuffs in diets for turkeys. *Poult. Sci.* 62:1038–1044.

Hung, T. V., P. D. Handson, V. C. Amenla, W. S. A. Kyle, and R. S. T. Yu. 1987. Content and distribution of manganese in lupin seed grown in Victoria and in lupin flour, spray-dried powder and protein isolate prepared from the seeds. *J. Sci. Food Agric.* 41:131–139.

Musquiz, M., C. L. Ridout, K. R. Price, and G. R. Fenwick. 1993. The saponin content and composition of sweet and bitter lupin seed. *J. Sci. Food Agric.* 63:47–52.

Singh, C. K., P. H. Robinson, and M. A. McNiven. 1995. Evaluation of raw and roasted lupin seeds as protein supplements for lactating cows. *Anim. Feed Sci. Tech.* 52:63–76.

Watkins, B. A. and L. W. Mirosh. 1987. White lupin as a protein source of layers. *Poult. Sci.* 66:1798–1806.

Peas and Lentils

Bell, J. M. and M. O. Keith. 1986. Nutritional and monetary evaluation of damaged lentils for growing pigs and effects of antibiotic supplements. *Can. J. Anim. Sci.* 66:529–536.

Canibe, N. and K. E. Bach Knudsen. 1997. Digestibility of dried and toasted peas in pigs. 1. Ileal and total tract digestibilities of carbohydrates. *Anim. Feed Sci. Tech.* 64:293–310. 2. Ileal and total tract digestibilities of amino acids, protein and other nutrients. *Anim. Feed Sci. Tech.* 64:311–325.

Castell, A. G. and R. L. Cliplef. 1988. Live performance, carcass and meat quality characteristics of market pigs self-fed diets containing cull-grade lentils. *Can. J. Anim. Sci.* 68:265–273.

Hlodversson, R. 1987. The nutritive value of white- and dark-flowered cultivars of pea for growing-finishing pigs. *Anim. Feed Sci. Tech.* 17:245–255.

Igbasan, F. A. and W. Guenter. 1996. The evaluation and enhancement of the nutritive value of yellow-, green-, and brown-seeded pea cultivars for unpelleted diets given to broiler chickens. *Anim. Feed Sci. Tech.* 63:9–24.

Igbasan, F. A., W. Guenter, and B. A. Slominski. 1997. Field peas: Chemical composition and energy and amino acid availabilities for poultry. *Can. J. Anim. Sci.* 77:293–300.

Johns, D. C. 1987. Influence of trypsin inhibitors in four varieties of peas (*Pisum sativum*) on the growth of chickens. *New Z. J. Agric. Res.* 30:169–176.

Pearson, G. and W. C. Smith. 1989. Effect of inclusion rate of peas (*Pisum sativum* var. Pania) in the diet on the performance of growing pigs. *New Z. J. Agric. Res.* 32:117–120.

Savage, G. P. 1988. The composition and nutritive value of lentils (*Lens culinaris*). *Nutr. Abst. & Rev.* 58 (Series A):319–343.

Other Grain Legumes

Fernando, T. and G. Bean. 1986. The reduction of antinutritional behenic acid in winged bean (*Psophocarpus tetragonolobus* L. D C) seeds. *Qual. Plant Foods Hum. Nutr.* 36:93–96.

Hadsell, D. L. and J. L. Sommerfeldt. 1988. Chickpeas as a protein and energy supplement for high producing dairy cows. *J. Dairy Sci.* 71:762–772.

Hanbury, C. D., C. L. White, B. P. Mullin, and K. H. M. Siddique. 2000. A review of the potential of *Lathyrus sativus* L. and *L. cicera* L. grain for use as animal feed. *Anim. Feed Sci. Tech.* 87:1–27.

Henry, C. J. K., P. A. Donachie, and J. P. W. Rivers. 1985. The winged bean: Will the wonder crop be another flop? *Ecol. Food Nutr.* 16:331–338.

Jaffe, W. G. and R. Korte. 1976. Nutritional characteristics of winged beans in rats. *Nutr. Rep. Int.* 14:439–445.

National Academy of Sciences. 1984. *The Winged Bean: A High Protein Crop for the Tropics.* Washington, DC: National Academy Press.

______. 1979. *Tropical Legumes: Resources for the Future.* Washington, DC: National Academy of Sciences.

Newman, C. W., N. J. Roth, R. K. Newman, and R. H. Lockerman. 1987. Protein quality of chickpea (*Cicer arietinum* L.). *Nutr. Rep. Int.* 36:1–8.

Nwokolo, E. 1987. Nutritional evaluation of pigeon pea meal. *Plant Foods Human Nutr.* 37:238–290.

Visitpanich, T., E. S. Batterham, and B. W. Norton. 1985a. Nutritional value of chickpea (*Cicer arietinum*) and pigeon pea (*Cajanus cajan*) meals for growing pigs and rats. I. Energy content and protein quality. *Aust. J. Ag. Res.* 36:327–335.

______ . 1985b. Nutritional value of chickpea (*Cicer arietinum*) and pigeonpea (*Cajanus cajan*) meals for growing pigs and rats. II. Effect of autoclaving and alkali treatment. *Aust. J. Ag. Res.* 36:337–445.

Wiryawan, K. G., H. M. Miller, and J. H. G. Holmes. 1997. Mung beans (*Phaseolus aureus*) for finishing pigs. *Anim. Feed Sci. Tech.* 66:297–303.

Milling and Fermentation By-Products

Castanon, F., R. W. Leeper, and C. M. Parsons. 1990. Evaluation of corn gluten feed in the diets of laying hens. *Poult. Sci.* 69:90–97.

Firkins, J. L., L. L. Berger, and G. C. Fahey, Jr. 1985. Evaluation of wet and dry distillers' grains and wet and dry corn gluten feeds for ruminants. *J. Anim. Sci.* 60:847–860.

Lodge, S. L., R. A. Stock, T. J. Klopfenstein, D. H. Shain, and D. W. Herold. 1997. Evaluation of corn and sorghum distillers by-products. *J. Anim. Sci.* 75:37–43.

Ohajuruka, O. A. and D. L. Palmquist. 1989. Response of high-producing dairy cows to high levels of dried corn gluten feed. *Anim. Feed Sci. Tech.* 24:191–200.

Rust, S. R., J. R. Newbold, and K. W. Metz. 1990. Evaluation of condensed distillers' solubles as an energy source for finishing cattle. *J. Anim. Sci.* 68:186–192.

Animal Protein Sources

Bruce, M. E., R. G. Will, J. W. Ironside, I. McConnell, D. Drummond, A. Suttie, L. McCardle, A. Chree, J. Hope, C. Birkett, S. Cousens, H. Fraser, and C. J. Bostock. 1997. Transmissions to mice indicate that 'new variant' CJD is caused by the BSE agent. *Nature* 389:498–501.

Cheeke, P. R. 2004. *Contemporary Issues in Animal Agriculture.* Upper Saddle River, N.J.: Prentice Hall, Inc.

Hill, A. F., M. Desbruslais, S. Joiner, K. C. L. Sidle, I. Gowland, J. Collinge, L. J. Doey, and P. Lantos. 1997. The same prion strain causes VCJD and BSE. *Nature* 389:448–450.

Taylor, D. M., S. L. Woodgate, and M. J. Atkinson. 1995. Inactivation of the bovine spongiform encephalopathy agent by rendering procedures. *Vet. Rec.* 137:605–610.

Tibbetts, G. W., R. W. Seerley, and H. C. McCampbell. 1987. Poultry offal ensiled with *Lactobacillus acidophilus* for growing and finishing swine diets. *J. Anim. Sci.* 64:182–190.

Wahlstrom, R. C. and G. W. Libal. 1977. Dried blood meal as a protein source in diets for growing-finishing swine. *J. Anim. Sci.* 44:778–783.

Fish and Other Marine Products

Abazinge, M. D. A., J. P. Fontenot, and V. G. Allen. 1994. Digestibility, nitrogen utilization, and voluntary intake of ensiled crab waste-wheat straw silage mixtures fed to sheep. *J. Anim. Sci.* 72:565–571.

Anderson, D. M., L. White, and J. MacLean. 1997. Determination of the true metabolizable energy (TMEN) by roosters of feedstuffs made from seal by-products. *Can. J. Anim. Sci.* 77:165–167.

Austin, P. R., C. J. Brine, J. E. Castle, and J. P. Zikakis. 1981. Chitin: New facets of research. *Science* 212:749–753.

Bunting, L. D., L. S. Sticker, L. C. Kappel, and Y. Zhang. 1994. Growth responses and ruminal adaptation of lambs fed crustacean processing wastes. *Anim. Feed Sci. Tech.* 45:229–241.

Charmley, E. and E. O'Reilly. 1997. Evaluation of seal meal as a protein supplement for growing steers. *Can. J. Anim. Sci.* 77:529–531.

Evers, D. J. and D. J. Carroll. 1996. Preservation of crab or shrimp waste as silage for cattle. *Anim. Feed Sci. Tech.* 59:233–244.

Gibb, M. J. and R. D. Baker. 1987. Performance of young steers offered silage or thermo-ammoniated hay with or without a fishmeal supplement. *Anim. Prod.* 45:371–381.

Hulan, H. W., R. G. Ackman, W. M. N. Ratnayake, and F. G. Proudfoot. 1988. Omega-3 fatty acid levels and performance of broiler chickens fed redfish meal or redfish oil. *Can. J. Anim. Sci.* 68:533–547.

______ . 1989. Omega-3 fatty acid levels and general performance of commercial broilers fed practical levels of redfish meal. *Poult. Sci.* 68:153–162.

Husby, F. M. 1980. King crab meal: A protein supplement for swine. *Agroborealis (Alaska)* 12:4–8.

Laflamme, L. F. 1988. Utilization of crab meal fed to young beef cattle. *Can. J. Anim. Sci.* 68:1237–1244.

Martin, R. E. 1979. Krill. *Food Tech.* 33:46–51.

Myer, R. O., D. D. Johnson, W. S. Otwell, W. R. Walker, and G. E. Combs. 1988. Potential utilization of scallop viscera silage for solid waste management and as a feedstuff for swine. *Nutr. Rep. Int.* 37:499–514.

Newbold, J. R., P. C. Garnsworthy, P. J. Buttery, D. J. A. Cole, and W. Haresign. 1987. Protein nutrition of growing cattle: Food intake and growth responses to rumen degradable protein and undegradable protein. *Anim. Prod.* 45:383–394.

Nicholson, J. W. G., R. E. McQueen, J. G. Allen, and R. S. Bush. 1996a. Composition, digestibility and rumen degradability of crab meal. *Can. J. Anim. Sci.* 76:89–94.

______ . 1996b. Effect of mash or pelleted supplements containing crab meal on intake and weight gains of beef cattle. *Can. J. Anim. Sci.* 76:95–103.

Offer, N. W. and R. A. K. Husain. 1987. Fish silage as a protein supplement for early-weaned calves. *Anim. Feed Sci. Tech.* 17:165–177.

Preston, T. R. and R. A. Leng. 1987. *Matching Ruminant Production Systems with Available Resources in the Tropics and Subtropics.* Armidale, Australia: Penambul Books.

Raa, J. and A. Gildberg. 1982. Fish silage: A review. *CRC Critical Rev. Food Sci. Nutr.* 16:383–419.

Razdan, A. and D. Pettersson. 1994. Effect of chitin and chitosan on nutrient digestibility and plasma lipid concentrations in broiler chickens. *Brit. J. Nutr.* 72:277–288.

Robinson, P. H. 1996. Evaluation of a seal by-product meal as a feedstuff for dairy cows. *Anim. Feed Sci. Tech.* 63:51–62.

Shqueir, A. A., R. O. Kellems, and D. C. Church. 1984a. Effects of liquified fish, cottonseed meal and feather meal on *in vivo* and *in vitro* rumen characteristics of sheep. *Can. J. Anim. Sci.* 64:881–888.

Shqueir, A. A., D. C. Church, and R. O. Kellems. 1984b. Evaluation of liquified fish in digestibility and feedlot performance trials with sheep. *Can. J. Anim. Sci.* 64:889–898.

Stout, F. M., J. E. Oldfield, and J. Adair. 1960. Nature and cause of the cotton fur abnormality in mink. *J. Nutr.* 70:421–426.

Tibbetts, G. W., R. W. Seerley, H. C. McCampbell, and S. A. Vezey. 1981. An evaluation of an ensiled waste fish product in swine diets. *J. Anim. Sci.* 52:93–100.

Feather, Hair, and Leathermeal

Aderibigbe, A. O. and D. C. Church. 1983. Feather and hair meals for ruminants. I. Effect of degree of processing on utilization of feather meal. *J. Anim. Sci.* 56:1198–1207.

______ . 1983. Feather and hair meals for ruminants. II. Comparative evaluation of feather and hair meals as protein supplements. *J. Anim. Sci.* 57:473–482.

Chiba, L. I., H. W. Ivey, K. A. Cummins, and B. E. Gamble. 1996. Hydrolyzed feather meal as a source of amino acids for finisher pigs. *Anim. Feed Sci. Tech.* 57:15–24.

Knowlton, P. H., W. H. Hoover, C. J. Sniffen, C. S. Thompson, and P. C. Belyea. 1976. Hydrolyzed leather scrap as a protein source for ruminants. *J. Anim. Sci.* 43:1095–1103.

Latshaw, J. D., N. Musharof, and R. Retrum. 1994. Processing of feather meal to maximize its nutritional value for poultry. *Anim. Feed Sci. Tech.* 47:179–188.

Papadopoulos, M. C., A. R. El Bousky, A. E. Roodbeen, and E. H. Ketelaars. 1986. Effects of processing time and moisture content on amino acid composition and nitrogen characteristics of feather meal. *Anim. Feed Sci. Tech.* 14:279–290.

Pate, F. M., W. F. Brown, and A. C. Hammond. 1995. Value of feather meal in a molasses-based liquid supplement fed to yearling cattle consuming a forage diet. *J. Anim. Sci.* 73:2865–2872.

Urlings, H. A. P., P. G. H. Bijker, and J. G. van Logtestijn. 1993. Fermentation of raw poultry byproducts for animal nutrition. *J. Anim. Sci.* 71:2420–2426.

Amino Acids

Bedford, M. R. and J. D. Summers. 1986. Effect of altering the source of nonessential nitrogen on performance and carcass composition of the broiler chick. *Can. J. Anim. Sci.* 66:1097–1105.

Newman, C. W., D. C. Sands, M. E. Megeed, and R. K. Newman. 1988. Replacement of soybean meal in swine diets with L-lysine and *Lactobacillus fermentum. Nutr. Rep. Int.* 37:347–355.

Newman, R. K. and D. C. Sands. 1984. Nutritive value of corn fermented with lysine excreting lactobacilli. *Nutr. Rep. Int.* 30:1287–1293.

Ruminant Nitrogen Sources and Bypass Proteins

Abdalla, H. O., D. G. Fox, and R. R. Seaney. 1988. Variation in protein and fiber fractions in pasture during the grazing season. *J. Anim. Sci.* 66:2663–2667.

Andersen, M. A., G. J. Vogel, and G. W. Horn. 1987. Effect of meat meal supplementation of steers grazing wheat pasture on forage intake and post-ruminal flow of non-ammonia nitrogen. *J. Anim. Sci.* 65:341–342 (Abst.).

Anderson, S. J., T. J. Klopfenstein, and V. A. Wilkerson. 1988. Escape protein supplementation of yearling steers grazing smooth brome grass pastures. *J. Anim. Sci.* 66:237–242.

Beever, D. E. and R. C. Siddons. 1986. Digestion and metabolism in the grazing ruminant. In *Control of Digestion and Metabolism in Ruminants*, L. P. Milligan, W. L. Grovum, and A. Dobson, eds., pp. 479–497. Englewood Cliffs, NJ: Prentice Hall.

Bhattacharya, A. N. and J. C. Taylor. 1975. Recycling animal waste as a feedstuff: A review. *J. Anim. Sci.* 41:1438–1457.

Broderick, G. A. and M. K. Clayton. 1997. A statistical evaluation of animal and nutritional factors influencing concentrations of milk urea nitrogen. *J. Dairy Sci.* 80:2964–2971.

Bunting, L. D., J. A. Boling, C. T. MacKown, and G. M. Davenport. 1989. Effect of dietary protein level on nitrogen metabolism in the growing bovine. II. Diffusion into and utilization of endogenous urea nitrogen in the rumen. *J. Anim. Sci.* 67:820–826.

Butler, W. R., J. J. Calaman, and S. W. Beam. 1996. Plasma and milk urea nitrogen in relation to pregnancy rate in lactating dairy cattle. *J. Anim. Sci.* 74:858–865.

Calvert, C. C. 1979. Use of animal excreta for microbial and insect protein synthesis. *J. Anim. Sci.* 48:178–192.

Chaudhry, S. M., J. P. Fontenot, Z. Naseer, and C. S. Ali. 1996. Nutritive value of deep stacked and ensiled broiler litter for sheep. *Anim. Feed Sci. Tech.* 57:165–173.

Cleale, R. M., R. A. Britton, T. J. Klopfenstein, M. L. Bauer, D. L. Harmon, and L. D. Satterlee. 1987. Induced non-enzymatic browning of soybean meal. II. Ruminal escape and net portal absorption of soybean protein treated with xylose. *J. Anim. Sci.* 65:1319–1326.

Coenen, D. J. and A. Trenkle. 1989. Comparisons of expeller-processed and solvent-extracted soybean meals as protein supplements for cattle. *J. Anim. Sci.* 67:565–573.

Fernandez, J. M., W. J. Croom, Jr., A. D. Johnson, R. D. Riquette, and F. W. Edens. 1988. Subclinical ammonia toxicity in steers: Effects on blood metabolite and regulatory hormone concentrations. *J. Anim. Sci.* 66:3259–3266.

Flores, J. F., T. H. Stobbs, and D. J. Minson. 1979. The influence of the legume *Leucaena leucocephala* and formal-casein on the production and composition of milk from grazing cows. *J. Agric. Sci.* 92:351–357.

Fluharty, F. L. and S. C. Loerch. 1989. Chemical treatment of ground corn to limit ruminal starch digestion. *Can. J. Anim. Sci.* 69:173–180.

Goetsch, A. L. and F. N. Owens. 1985. The effects of commercial processing method of cottonseed meal on site and extent of digestion in cattle. *J. Anim. Sci.* 60:803–813.

Jordon, D. J., T. J. Klopfenstein, and D. C. Adams. 2002. Dried poultry waste for cows grazing low-quality winter forage. *J. Anim. Sci.* 80:818–824.

Ladely, S. R., R. A. Stock, T. J. Klopfenstein, and M. H. Sindt. 1995. High-lysine corn as a source of protein and energy for finishing calves. *J. Anim. Sci.* 73:228–235.

Laflamme, L. F. 1988. Fishmeal as a supplement to hay-based cattle diets. *Can. J. Anim. Sci.* 68:1323 (Abst.).

Makkar, H. P. S., D. Lall, and S. S. Negi. 1988. Complexes of urea and formaldehyde as non-protein nitrogen compounds in ruminant rations: A review. *Anim. Feed Sci. Tech.* 20:1–12.

Mathison, G. W. R. Soofi-Siawash, and M. Worsley. 1994. The potential of isobutyraldehyde monourea (propanal, 2-methyl-monourea) as a nonprotein nitrogen source for ruminant animals. *Can. J. Anim. Sci.* 74:665–674.

McCaskey, T. A. and W. B. Anthony. 1979. Human and animal health aspects of feeding livestock excreta. *J. Anim. Sci.* 48:163–177.

Merchen, N. R. and E. C. Titgemeyer. 1992. Manipulation of amino acid supply to the growing ruminant. *J. Anim. Sci.* 70:3238–3247.

Mir, Z., G. K. MacLeod, J. G. Buchanan-Smith, D. G. Grieve, and W. L. Grovum. 1984a. Effect of feeding soybean meal protected with sodium hydroxide, fresh blood, or fish hydrolysate to growing calves and lactating dairy cows. *Can. J. Anim. Sci.* 64:845–852.

———. 1984b. Methods for protecting soybean and canola proteins from degradation in the rumen. *Can. J. Anim. Sci.* 64:853–865.

Mizwicki, K. L., F. N. Owens, K. Poling, and G. Burnett. 1980. Timed ammonia release for steers. *J. Anim. Sci.* 51:698–703.

National Research Council. 1985. *Ruminant Nitrogen Usage.* Washington, DC: National Academy Press.

Newbold, J. R., P. C. Garnsworthy, P. J. Buttery, D. J. A. Cole, and W. Haresign. 1987. Protein nutrition of growing cattle: Food intake and growth responses to rumen degradable protein and undegradable protein. *Anim. Prod.* 45:383–394.

Oltjen, R. R., E. E. Williams, Jr., L. L. Slyter, and G. V. Richardson. 1969. Urea versus biuret in a roughage diet for steers. *J. Anim. Sci.* 29:816–822.

Park, K. K., A. L. Goetsch, A. R. Patil, B. Kouakou, and Z. B. Johnson. 1995. Composition and in vitro digestibility of fibrous substrates placed in deep-stacked broiler litter. *Anim. Feed Sci. Tech.* 54:159–174.

Preston, T. R. and R. A. Leng. 1987. *Matching Ruminant Production Systems with Available Resources in the Tropics and Subtropics.* Armidale, Australia: Penambul Books.

Ramirez Pineres, M. A., W. C. Ellis, G. Wu, and S. C. Ricke. 1997. Effects of diphenyliodium chloride on proteolysis and leucine metabolism by rumen microorganisms. *Anim. Feed Sci. Tech.* 65:139–149.

Rankins, D. L. Jr., J. T. Eason, T. A. McCaskey, A. H. Stephenson, and J. G. Floyd, Jr. 1993. Nutritional and toxicological evaluation of three deep-stacking methods for processing of broiler litter as a foodstuff for beef cattle. *Anim. Prod.* 56:321–326.

Rulquin, H. and L. Delaby. 1997. Effects of the energy balance of dairy cows on lactational responses to rumen-protected methionine. *J. Dairy Sci.* 80:2513–2522.

Sklan, D. 1989. *In vitro* and *in vivo* rumen protection of proteins coated with calcium soaps of long-chain fatty acids. *J. Agric. Sci. Camb.* 112:79–83.

Smith, L. W. and W. E. Wheeler. 1979. Nutritional and economic value of animal excreta. *J. Anim. Sci.* 48:144–156.

Stephenson, R. G. A., J. C. Edwards, and P. S. Hopkins. 1981. The use of urea to improve milk yield and lamb survival of merinos in a dry tropical environment. *Aust. J. Agr. Res.* 32:497–509.

Stobbs, T. H., D. J. Minson, and M. N. McLeod. 1977. The response of dairy cows grazing a nitrogen fertilized grass pasture to a supplement of protected casein. *J. Agric. Sci.* 89:137–141.

Veira, D. M., J. G. Proulx, G. Butler, and A. Fortin. 1988. Utilization of grass silage by cattle: Further observations on the effect of fishmeal. *Can. J. Anim. Sci.* 68:1225–1235.

Vogel, G. J., M. A. Andersen, and G. W. Horn. 1987. Kinetics of ruminal nitrogen disappearance from wheat forage in situ. *J. Anim. Sci.* 65:341 (Abst.).

Worrell, M. A., D. Undersander, C. E. Thompson, and W. C. Bridges, Jr. 1990. Effects of time of season and cottonseed meal and lasalocid supplementation of steers grazing rye pastures. *J. Anim. Sci.* 68:1151–1157.

Leaf Protein Concentrates

Ameenuddin, S., H. R. Bird, D. J. Pringle, and M. L. Sunde. 1983. Studies on the utilization of leaf protein concentrates as a protein source in poultry nutrition. *Poult. Sci.* 62:505–511.

Barber, R. S., R. Braude, K. G. Mitchell, and I. G. Partridge. 1981. Lucerne juice as a protein supplement for growing pigs: Effects of mineral content of the diet and of the water supply. *Anim. Feed Sci. Tech.* 6:35–41.

Lowry, J. B. 1989. Green-leaf fractionation by fruit bats: Is this feeding behavior a unique nutritional strategy for herbivores? *Aust. Wildl. Res.* 16:203–206.

Myer, R. O., P. R. Cheeke, and W. H. Kennick. 1975. Utilization of alfalfa protein concentrate by swine. *J. Anim. Sci.* 40:885–891.

Telek, L. and H. D. Graham, eds. 1983. *Leaf Protein Concentrates.* Westport, CT: AVI Publishing Co.

Zinn, R. A. 1988. Comparative feeding value of alfalfa presscake and alfalfa molasses in finishing diets for feedlot cattle. *J. Anim. Sci.* 66:151–158.

Single Cell Proteins

Cheeke, P. R., E. Gaspar, L. Boersma, and J. E. Oldfield. 1977. Nutritional evaluation with rats of algae (chlorella) grown on swine manure. *Nutr. Rep. Int.* 16:579–585.

Chung, P., W. G. Pond, J. M. Kingsbury, E. F. Walker, and L. Krook. 1978. Production and nutritive value of *Arthrospira plantesis,* a spiral blue-green algae grown on swine wastes. *J. Anim. Sci.* 47:319–330.

Contreras, A., D. C. Herbert, B. G. Grubbs, and I. L. Cameron. 1979. Blue-green alga, *Spirulina,* as the sole dietary source of protein in sexually maturing rats. *Nutr. Rep. Int.* 19:749–763.

Kellems, R. O., M. S. Aseltine, and D. C. Church. 1981. Evaluation of single cell protein from pulp mills: Laboratory analyses and *in vivo* digestibility. *J. Anim. Sci.* 53:1601–1608.

Lipstein, B. S. and S. Hurwitz. 1980. The nutritional value of algae for poultry: Dried Chlorella in broiler diets. *Brit. Poult. Sci.* 21:9–21.

Lipstein, B., S. Hurwitz, and S. Bornstein. 1980. The nutritional value of algae for poultry: Dried Chlorella in layer diets. *Brit. Poult. Sci.* 21:23–27.

Aquaculture-Produced Proteins

Buck, D. H., R. J. Baur, and C. R. Rose. 1978. Utilization of swine manure in a polyculture of Asian and North American fishes. *Trans. Am. Fish. Soc.* 107:216–222.

Moav, R., G. Wohlfarth, G. L. Schroeder, G. Hulata, and H. Barash. 1977. Intensive polyculture of fish in freshwater ponds. I. Substitution of expensive feeds by liquid cow manure. *Aquaculture* 10:25–43.

Insect Meals

Calvert, C. C., R. D. Martin, and N. O. Morgan. 1969. House fly pupae as food for poultry. *J. Econ. Entomol.* 62:938–939.

Finke, M. D., G. R. DeFoliart, and N. J. Benevenga. 1987. Use of a four-parameter logistic model to evaluate the protein quality of mixtures of Mormon cricket meal and corn gluten meal in rats. *J. Nutr.* 117:1740–1750.

Finke, M. D., M. L. Sunde, and G. R. DeFoliart. 1985. An evaluation of the protein quality of Mormon crickets (*Anabrus simplex* Haldeman) when used as a high protein feedstuff for poultry. *Poult. Sci.* 64:708–712.

Newton, G. L., C. V. Booram, R. W. Barker, and O. M. Hale. 1977. Dried *Hermetia illucens* larvae meal as a supplement for swine. *J. Anim. Sci.* 44:395–400.

Robbins, C. T. 1993. *Wildlife Feeding and Nutrition.* San Diego: Academic Press.

Salmon, R. E. and T. I. Szabo. 1981. Dried bee meal as a feedstuff for growing turkeys. *Can. J. Anim. Sci.* 61:965–968.

Teotia, J. S. and B. F. Miller. 1974. Nutritive content of house fly pupae and manure residue. *Brit. Poult. Sci.* 52:1830–1835.

Zuidhof, M. J., C. L. Molnar, F. M. Morley, T. L. Wray, F. E. Robinson, B. A. Khan, L. Al-Ani, and L. A. Goonewardene. 2003. Nutritive value of house fly (*Musca domestica*) larvae as a feed supplement for turkey poults. *Anim. Feed Sci. Tech.* 105:225–230.

CHAPTER 5

Roughages

Objectives

1. To describe the nutritive value and other properties of various classes of forages:
   - grasses
   - legumes
   - tropical forages
2. To discuss factors influencing forage quality:
   - structure of plant tissue
   - environmental, maturity, and seasonal effects
3. To highlight areas of current interest:
   - nutritive value of C_3 vs C_4 forages
   - detergent methods for fiber analysis
   - associative effects of forages and concentrates
   - forage-related animal disorders
     - grass tetany
     - tall fescue toxicosis
     - forage mycotoxins
4. To identify problems with and methods to improve utilization of tropical forages

Roughages are bulky feeds, high in fiber and low in energy. They are used mainly by herbivorous animals, both the ruminant and nonruminant herbivores. The nutrients in roughages are made available primarily by microbial digestion, either in the rumen or in the hindgut. Types of roughages used as feedstuffs include pasture and other grazed forages, hay and dehydrated forage, silage, and crop residues and by-products (straw, stover, hulls, stubble, and the like).

FORAGE PLANTS

Forages are plants grown or used specifically for feeding to animals. Cultivated or tame forages are (mainly) grasses and legumes planted for pasture, hay, or silage, whereas millions of acres of native rangelands contain wild grasses, legumes, and browse species that are utilized by grazing animals. The major component of forages and other fibrous feeds is cellulose, the most abundant organic compound on earth. Approximately 100,000,000,000 tons of cellulose are synthesized by plants every year (Goodwin and Mercer, 1983). Cellulose is composed entirely of glucose. The only practical way of utilizing this glucose as human food is through the production of meat and milk using ruminants and other herbivorous animals.

CHEMICAL AND STRUCTURAL CHARACTERISTICS OF FORAGES

The predominant chemical characteristic of forages is their high content of cell wall material. The structure of plant cells is shown in Fig. 5.1. They consist of a fibrous cell wall, with the cell contents inside. The cell wall is composed of metabolically inert, highly fibrous material, whereas the cell contents contain the metabolically active components. Dominant **cell contents** components include the chloroplasts, in which photosynthesis occurs, and the mitochondria and cellular protoplasm containing the enzymes involved in carbohydrate and protein synthesis. The major components of the **cell wall** include cellulose, lignin, hemicellulose, and silica. Cellulose and lignin provide structural strength to the cell wall. The polysaccharides (complex carbohydrates) of the cell wall occur either in crystalline (microfibrils) or noncrystalline (matrix) form.

The microfibrils consist of cellulose molecules ordered together in rodlike chains to form bundles of fibers. They form a three-dimensional crystal lattice. The matrix is made up of hemicelluloses and pectins. **Hemicelluloses** are complex carbohydrates containing mixtures of sugars; the main types are xylans, mannans, and galactans, containing the sugars xylose, mannose, and galactose,

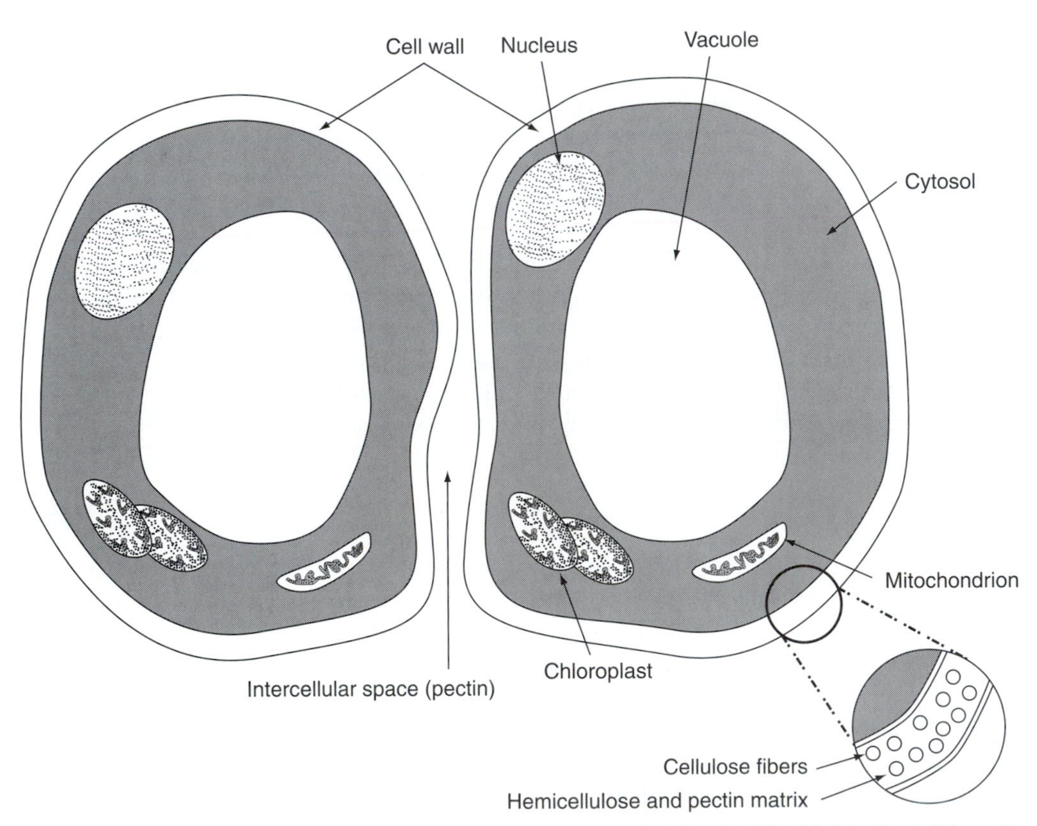

FIGURE 5.1 Simplified diagram of forage plant tissue cells. The highly digestible cell contents are encased in the less digestible cell wall. The nutritive value of a forage depends on the accessibility of the cell contents to microbial enzymes, which in turn depends on the structure of the cell wall.

respectively. The **cell wall** can be compared to reinforced concrete: the matrix polysaccharides are the cement and the cellulose fibers are the reinforcing rod. The matrix contains a considerable amount of water, forming a viscous gel. As the growth of the cell ceases, the space occupied by water becomes progressively filled with lignin so that the matrix becomes permeated with this compound. This process is called **lignification** and it occurs after the deposition of the polysaccharide components. In some plants, especially some grasses and *Equisetum* (horsetail) species, the matrix may contain considerable quantities of silica. The space between cells (intercellular space) is called the middle lamella. This is composed largely of pectin. In woody tissues, the pectin is permeated with lignin.

The **cell contents** are the metabolically active part of the plant cell and consist of the cytosol and vacuole(s) (Fig. 5.1). The cytosol contains the nucleus, mitochondria, and chloroplasts and is the part of the cell where the photosynthetic work is taking place. The interior vacuole, which may be one large vacuole or numerous small ones, is an aqueous phase that helps keep the cell in turgid condition. It is also a storage area for cell wastes, food reserves, and toxic substances, such as alkaloids, tannins, and glycosides. It has lysosomal activity for intracellular digestion. Thus, the vacuoles are a water solution of free sugars, organic acids, minerals, and secondary compounds (secondary compounds are alkaloids, glycosides, and tannins, which are chemical defenses from the plant perspective and toxins from the animal perspective).

The nutritional value of forages depends largely on the relative proportions of cell contents and cell wall constituents, and on the degree of lignification of the cell walls.

NUTRIENT COMPONENTS OF FORAGES

Protein and Nitrogenous Compounds

The nitrogen in forages consists of proteins and various nonprotein nitrogen compounds, such as amino acids, amides, nitrates, and ammonia. Several hundred different **amino acids** have been isolated from plants. Some of these are toxic to animals. The nonprotein nitrogen components may comprise as much as 25 percent of the total nitrogen. The major protein in plant cells is the chloroplast protein. The second major protein category is the cytosolic fraction or soluble proteins found in the cytosol, consisting mainly of enzymes. The amino acid composition of proteins is very similar among plant species. The reason for this is that one enzyme, D-ribulose 1,5-diphosphate carboxylase (RUDP carboxylase), makes up more than 50 percent of the cell protein. This is a key enzyme in photosynthesis. **Leaf proteins** have a high quality (good amino acid balance) with high lysine levels. Methionine is the first limiting amino acid. Amino acid balance of dietary protein is not important in ruminants directly, but since leaf protein is of higher quality than microbial protein, it would be desirable, for maximum utilization of forage protein, for it to have a high rumen bypass potential. Forages that contain tannins, such as many legumes, have a high level of protected protein and may be superior protein sources to those with a high content of readily degradable protein.

Carbohydrates in Forages

The **nonstructural carbohydrates** are the starches (polymers of glucose), fructosans (polymers of fructose), and simple sugars (glucose, fructose, sucrose). These are all readily fermented in the rumen. Tropical grasses (C_4 plants) and legumes accumulate starch and sucrose in their stems (e.g., sugarcane) and leaves. Temperate grasses (C_3 plants) accumulate sucrose and fructosans, mainly in the stems.

The **structural carbohydrates** are found in the cell wall. Tropical forages are higher in cell wall components (fiber) than are temperate forages. The high cell wall content of tropical grasses is related to their C_4 type of photosynthesis and the high proportion of vascular tissue.

Minerals in Forages

The mineral content of forages depends on both plant species and soil mineral content. The concentration of most of the trace elements is largely a function of the mineral status of the soil in which the forage is growing. Legumes have a higher calcium content than grasses. Tropical forages contain less calcium than temperate species, and, in addition, many tropical grasses contain **oxalates** that reduce calcium availability to animals. Insoluble calcium oxalate is formed when calcium reacts with oxalic acid in tropical grasses. Legumes have a higher magnesium content than grasses, and both legumes and grasses in the tropics have higher magnesium contents than their temperate counterparts. Magnesium deficiency is of most importance in animals grazing temperate grasses, causing the metabolic disorder of grass tetany (hypomagnesemia), which is associated with very low blood magnesium levels. Tropical legumes and grasses are generally lower in phosphorus than temperate species. Pastures often have inadequate phosphorus levels for grazing animals. Beef cows on low-phosphorus pasture may not have normal estrus-cycle activity, resulting in poor fertility.

Vitamins in Forages

Herbivorous animals obtain adequate quantities of the B vitamins as a result of microbial activity in the gut, so the vitamin B content of forages, although generally high, is not of concern. Vitamin K is synthesized in the gut also; vitamin D is obtained through exposure to sunlight by grazing animals. Only vitamins A and E need to be provided in the diet; green forages are an excellent source of vitamins E and β-carotene, the precursor of vitamin A. With animals on green pasture, no vitamin supplementation is needed. When a prolonged period of grazing mature, dry, bleached forage is anticipated, supplementary vitamins A and E should be provided.

Deleterious Factors

Many forages contain deleterious factors, such as alkaloids, glycosides, toxic amino acids, and mycotoxins. Examples of toxins include alkaloids in reed canary grass, cyanogenic glycosides in white clover, estrogens in clover and alfalfa, tannins in trefoil, and the toxic amino acid mimosine in leucaena. Examples of problems caused by mycotoxins in forage include ryegrass staggers, facial eczema, sweet clover poisoning, paspalum staggers, red clover slobbers, and moldy hay toxicosis. These **natural toxicants** are discussed in detail by Cheeke (1998).

DETERGENT SYSTEM FOR FORAGE ANALYSIS

The proximate analysis system is inaccurate when applied to roughages because it does not accurately separate the carbohydrate fraction into the true fibrous fractions and the nonfiber components. Van Soest (1994) of Cornell University has developed a better system based on the use of detergents. Refluxing (boiling) a forage sample in a detergent solution solubilizes proteins, sugars, minerals, starch, and pectins. These compounds are highly digestible in the rumen. The detergent-insoluble fraction corresponds to the fibrous cell wall material.

In the Van Soest procedure, a dried, ground forage sample is refluxed first in a neutral-detergent solution of sodium lauryl sulfate and ethylenediaminetetraacetate (EDTA) at pH 7.0. The detergent solubilizes the proteins and dissolves minerals, sugars, starch, and pectins. The dissolved material is called "cell contents" and is completely digestible in ruminants. The insoluble residue is called **neutral detergent fiber** (NDF). The NDF (or a new sample of the forage) is then boiled in an acid detergent solution (cetyl trimethylammonium bromide in 1N H_2SO_4). In addition to solubilizing the cell contents, the acid detergent dissolves hemicellulose. The residue is called **acid detergent fiber** (ADF). Thus hemicellulose = NDF − ADF. The ADF fraction consists largely of cellulose, lignin, silica, and cutin (cutin is the waxy material on the surface of leaves). The ADF fraction can be further categorized by boiling in concentrated (72 percent) H_2SO_4. This dissolves cellulose and leaves a residue of lignin, silica, and cutin. This can be treated with a permanganate solution that oxidizes and destroys lignin. The **silica** content can be measured by ashing a sample of ADF; the residue is silica. Thus, by these methods, the plant cell wall material can be completely classified.

FACTORS AFFECTING FORAGE QUALITY

Plant Anatomy and Morphology

Leafiness in pasture plants is positively associated with forage quality. In general, leaves are higher in protein and energy, lower in fiber, and of higher digestibility than stems. The quality of forages decreases with increasing maturity because of an increasing proportion of stem and increasing lignification of the cell walls. The effect of lignification of the cell wall on forage digestibility can be visualized by imagining a small inflated balloon inside a larger one. The inner balloon cannot be popped without breaking down the outer one. Similarly, a lignified cell wall prevents microbial access to the nutritious cell contents within. In a few cases, where the stem serves as a storage organ for energy, it may have higher nutritive value than the leaves. Examples of this situation are the grasses timothy and sugarcane.

Temperate versus Tropical Forages

It is well known that tropical grasses are of lower nutritional value than temperate forages. They can produce tremendous amounts of low-quality forage. In spite of major research efforts, the productivity of livestock in tropical countries has not been much improved. As quoted in Mahadevan (1982), "Half a century of animal disease prevention and eradication measures, several decades of grassland and pasture investigations and of research into most aspects of animal nutrition and genetics have scarcely affected animal production (in the tropics)." A major challenge is to find ways of exploiting the high yield potential of tropical forages to improve efficiency of animal production.

Tropical grasses have the C_4 type of photosynthesis, producing compounds with four carbon atoms as the first end products of photosynthesis. The C_4 pathway is an adaptation that permits tolerance to high temperatures and efficient nitrogen utilization. The anatomy of the plant tissue has modifications attributable to the C_4 pathway. Tropical grasses have a high content of vascular tissue to transport photosynthate; this vascular tissue is encased in parenchymal bundle sheaths. They have a low content of **mesophyll cells**, which are the cells where most of the photosynthesis occurs. In temperate grasses the mesophyll cells are in a loose, irregular arrangement with air spaces within the tissue, whereas in tropical grasses the mesophyll cells are densely packed around the vascular tissue (Fig. 5.2). This impedes their digestion. Thus the tropical grasses have a high content of the slowly digested vascular and parenchyma sheath tissue and a low content of readily digested mesophyll cells. Electron microscope studies (Akin, 1982) have shown that the anatomy of tropical grasses is a major impediment to the rumen microbes; the rate of digestion of tropical grasses in the rumen is much slower than for temperate species. Additionally, most tropical grasses exhibit con-

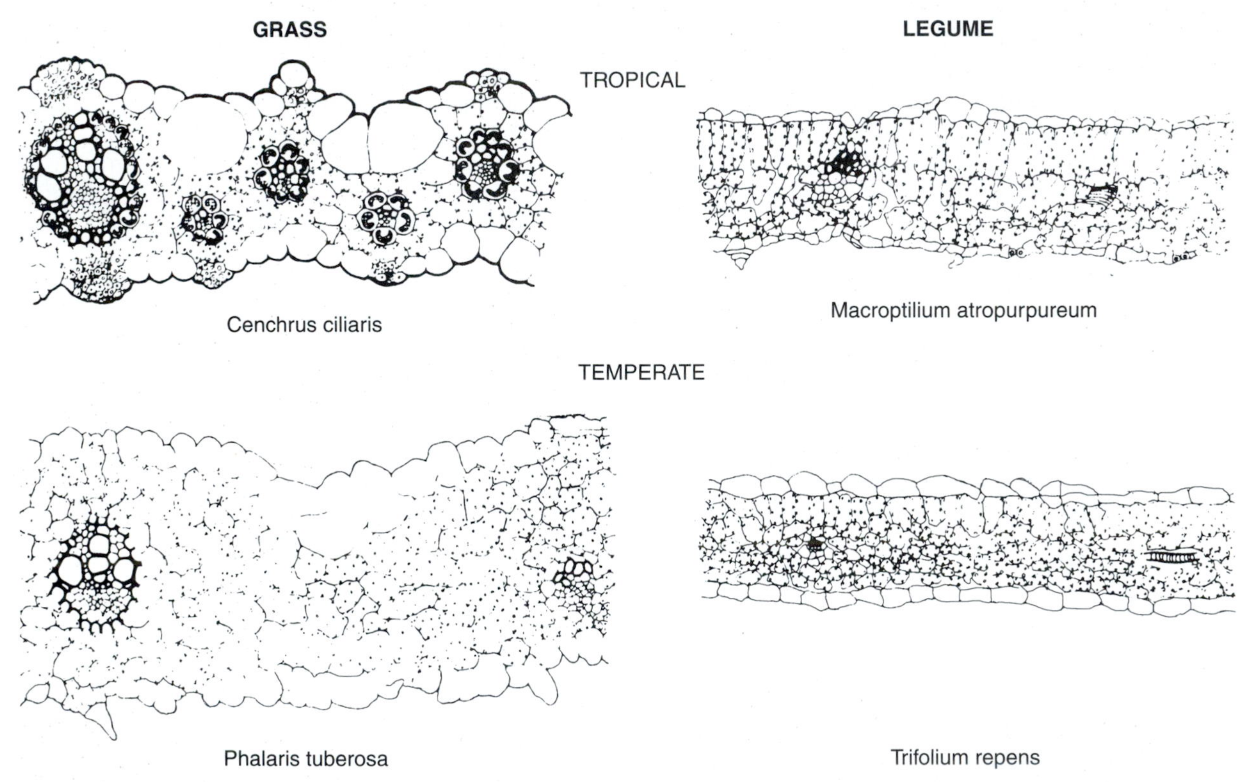

FIGURE 5.2 Cross sections of the leaves of tropical and temperate grasses and legumes. Tropical grasses have a high content of vascular bundles and tightly packed mesophyll cells. These materials present a high resistance to mechanical and microbial breakdown in the rumen, which accounts for their low digestibility. (Courtesy of D. J. Minson and J. R. Wilson in *J. Australian Institute of Agricultural Science* 46:247–249, 1980.)

tinuous stem elongation and flowering throughout the growing season; this results in a low leaf-to-stem ratio. The stemminess reduces palatability and digestibility. Stems are lower in both nutrient content and digestibility than leaves.

Grasses with C_4 photosynthesis are less preferred by herbivores than C_3 grasses. The C_4 plants have a competitive advantage over C_3 plants under low atmospheric CO_2 conditions (Ehleringer et al., 1997). Therefore, they have tended to expand in number during glacial periods when CO_2 levels are low. Ehleringer et al. (1997) and Cerling et al. (1997) speculate that changes in the ratio of C_3/C_4 plants within ecosystems during periods of climatic change could impact both the evolution and composition of mammalian grazing systems. For example, the loss of many **megaherbivores** from North America during the last glaciation period coincided with the expansion of low nutritive value C_4 grasses. The C_3 and C_4 plants incorporate different ratios of $^{13}CO_2$ and $^{14}CO_2$ in their tissues. The $^{13}C/^{14}C$ ratio in teeth of fossilized animals can be used to determine the nature of their diet (MacFadden and Cerling, 1994). These authors correlated this ratio in the teeth of extinct ancestral horses with changes in the dominant vegetation pattern (change from forests to grasslands). The present period of rising atmospheric CO_2 favors the expansion of C_3 plants. Thus, **global climate changes** can exert diverse ecological effects.

Environmental and Seasonal Factors

High environmental temperatures promote lignification and reduce forage digestibility. For this reason, first cutting hay is usually of higher nutritive value than second cutting material. High temperatures promote **lignification** of the cell walls of both leaves and stems and, by increasing metabolic rate of the tissue, reduce the pool size of metabolites in the cell contents, thus decreasing protein and soluble carbohydrate and increasing structural carbohydrate and cell wall. Day length has a similar effect; short days and long nights reduce the total photosynthesis and increase the amount of photosynthate used in cellular respiration. Spring grass generally has a higher nutritive value than autumn grass from the same field (MacRae et al., 1985), probably because of higher soluble carbohydrate and amino acid content. Climatic factors also contribute to the lower nutritive value of tropical grasses. High environmental temperatures promote development of stems and increased lignification of plant tissues. For every degree increase in temperature, the digestibility of tropical grasses decreases by 0.6 to 1.0 percent (Mannetje, 1984).

Nitrogen fertilization increases plant protein levels as well as yields of grasses. Because tropical grasses use nitrogen more efficiently than temperate grasses, they have less total nitrogen in their tissues; therefore they are lower in crude protein content.

Genetic Factors and Forage Quality

Species and cultivar differences in forage quality of grasses are well known (Collins and Casler, 1990). Grass species differ in the rate at which quality declines with maturation. For example, Collins and Casler (1990) observed that tall fescue and reed canarygrass declined more rapidly in quality than timothy, smooth bromegrass, and orchardgrass. Timothy maintained its quality with maturity better than the other grasses.

Forage quality can be influenced by plant selection. Anderson et al. (1988) developed a line of switchgrass (*Panicum virgatum*) with increased *in vitro* dry-matter digestibility. Higher animal gains were observed with the improved cultivar (Trailblazer) than with the unselected line. Hall et al. (1994) modified the rate of *in vitro* digestion of alfalfa to lower its bloat-producing potential. Numerous other examples of the use of plant breeding to improve forage quality exist, including low-alkaloid reed canarygrass, low-endophyte tall fescue, and low-estrogen subterranean clover.

ASSESSMENT OF FORAGE QUALITY

Forage quality refers to the totality of factors that influence the nutritive value of forages for herbivorous animals (primarily ruminants). The major factors relating to forage quality are digestibility, feed consumption, and the provision of nutrients.

Digestibility and Forage Quality

The digestibility of forages is an important factor in determining their nutritive value for two reasons. (1) The higher the digestibility, the more nutrients are liberated for use by the animal. (2) As digestibility increases, feed intake can increase because turnover rate in the rumen increases. Ingesta does not leave the rumen until it has been digested to a small particle size. If this process is rapid, the digested feed can be replaced by further feed intake. If digestion is prolonged, rumen fill limits further intake. Determination of digestibility *in vivo* (in the living animal) was described in Chapter 1. Either metabolism crates or fecal collection bags are needed to accurately collect excreta.

For the determination of forage digestibility, *in vitro* (literally, "in glass"; or, in other words, in a test tube) techniques are extensively used. In simplest terms, ***in vitro* digestibility** determinations involve test tubes containing a buffer solution, the test forage, and rumen microbes incubated at body temperature under anaerobic conditions (Fig. 5.3). The buffer solution represents artificial saliva and buffers the acids produced during fermentation. The rumen microbes are obtained by collecting rumen fluid from a fistulated animal and straining the fluid to remove particles. The test forages are added individually in weighed amounts to tubes, which are gassed with CO_2 to create an anaerobic condition. The tubes are placed in a water bath and incubated at 37°C, usually for 24 to 48 hours. At the end of incubation, the tubes are filtered and the indigestible residue measured. The difference between the starting and ending weight represents the amount digested. By measuring the amount of dry-matter, fiber, and other nutrient fractions in the forage before and after incubation, the digestibility of each fraction can be determined. The values correlate well with data obtained in *in vivo* determinations (Judkins et al., 1990). The advantage of the *in vitro* technique is that hundreds of samples can be run easily using only a few grams of each feedstuff, whereas metabolism trials are lengthy, expensive, and involve large amounts of feed.

The **nylon bag technique** for determination of digestibility involves inserting the test forage sample in a nylon bag, which is then suspended in the rumen via a rumen fistula. After 24 or 48 hours the bag is withdrawn and, by

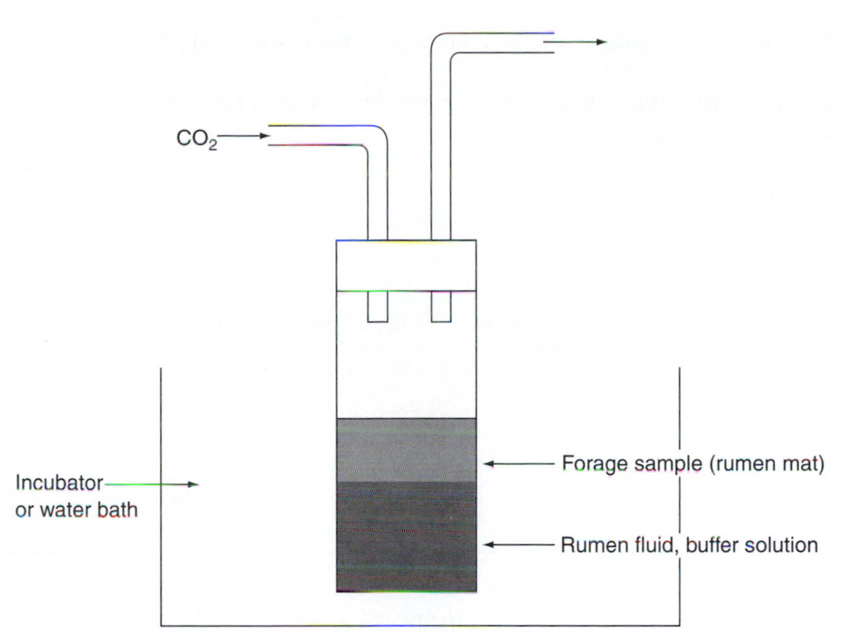

FIGURE 5.3 A simplified diagram of an *in vitro* rumen fermentation flask for measuring digestibility of feedstuffs for ruminants. The feed or forage sample is incubated with rumen fluid and a buffer solution (artificial saliva) under anaerobic conditions, at body temperature (37°C). Such a system simulates the digestive activity of the rumen and is much simpler and less expensive to perform than *in vivo* digestibility trials with animals.

analysis of the sample residue, digestibility can be determined. This is also known as an ***in sacco*** or ***in situ*** procedure.

Another technique that can be used to assess digestibility of feedstuffs in ruminants is the measurement of ***in vitro* gas production**. When feeds are fermented by rumen microbes, gases are liberated, with a high correlation between gas production and digestibility. Measurement of either gas pressure or gas volume can be accomplished with simple equipment (Theodorou et al., 1994). Detailed discussion of the kinetics of rumen gas production is provided by Cone et al. (1997) and Schofield et al. (1994). Stern et al. (1997) have comprehensively reviewed methods for determining digestibility in ruminants.

Factors Affecting Digestibility of Roughages The population of rumen microbes reflects the nature of the diet consumed. Roughage diets are high in cellulose, low in starch, and intermediate in soluble sugars, thus supporting a microbial flora of mainly **cellulolytic** and **saccharolytic** (sugar-digesting) **bacteria**. These organisms produce acetate as their main fermentation end product. With high starch diets, **amylolytic bacteria** predominate; they ferment starch, sugars, and hemicellulose to propionate. The rate of digestion of cellulose is much slower than for soluble carbohydrates. The rumen pH is lower when high-concentrate diets are fed than when high-roughage diets are fed (Fig. 5.4). There are several reasons for this. Decreased rumination time and reduced salivation associated with concentrate feeding lower the rumen buffering capacity. When

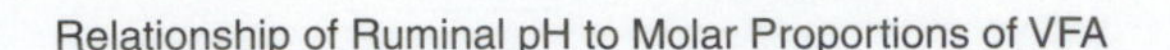

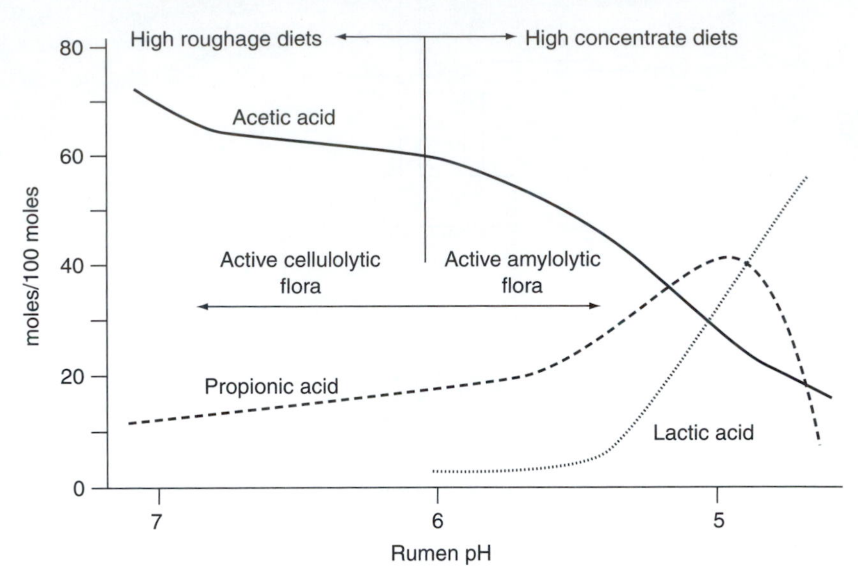

FIGURE 5.4 With high-fiber diets, cellulolytic microbes produce mainly acetic acid, which maintains the rumen pH above 6.0. With high-concentrate diets, the starch digesting, or amylolytic, microbes produce propionic and lactic acids, which are stronger acids than acetic acid, and which lower the rumen pH below the optimal for cellulose digestion. This result accounts for the negative associative effects of concentrates on roughage utilization.

coupled with the high rate of propionate production from rapid fermentation of starch, this causes a depression in rumen pH. Because lactic acid is of greater acidity than VFA (pk 3.8 vs 4.7), lactate production causes a marked drop in pH (lactic acidosis).

When feed is consumed, most feed particles are buoyant and become part of the mat or **raft** floating on top of the rumen fluid (Fig. 5.5). As particles are disintegrated by microbial action, they sink and exit the rumen. Entrapped gases and CO_2 produced by microbial fermentation keep plant material buoyant until it has been digested. Most of the digestion of fibrous feeds is accomplished by microbes that adhere to the feed particles. Cellulolytic organisms have structures, such as the glycocalyx (filamentous glycoprotein web), that adhere to cell wall material. A **consortium** of different microbes is involved in digestion of different plant tissues, so there is a sequential progression of microbes accomplishing fiber digestion. Bacteria initially invade the plant tissues through stomata and physical fractures caused by processing or mastication. The bacteria colonize the inside of the plant cells, where they are protected from protozoal predation. Various types of **rumen fungi** (Fig. 5.6) are intimately involved in digestion of highly lignified cell wall material (Grenet and Barry, 1988; Akin, 1987). The action of fungi may be to degrade lignin, thus increasing accessibility of cell wall material to bacterial digestion. Rumen fungi have a good amino acid balance and high digestibility (Gulati et al., 1989), so they should be a valuable source of digestible amino acids in ruminants fed low-quality roughages.

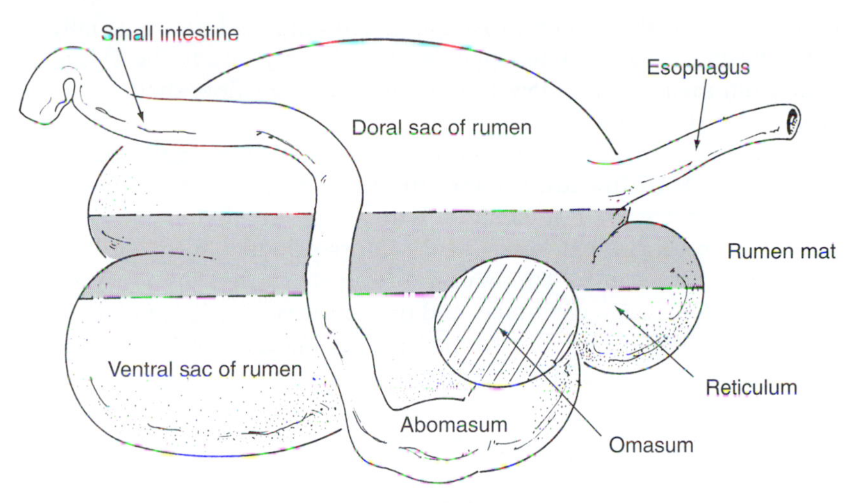

FIGURE 5.5 The rumen contents are stratified with a mat or raft of fibrous material floating on liquid. During rumination, a bolus is formed at the base of the esophagus and regurgitated. As forage is digested, the particle size is reduced, and the small particles drop to the bottom of the rumen and flow through the omasum.

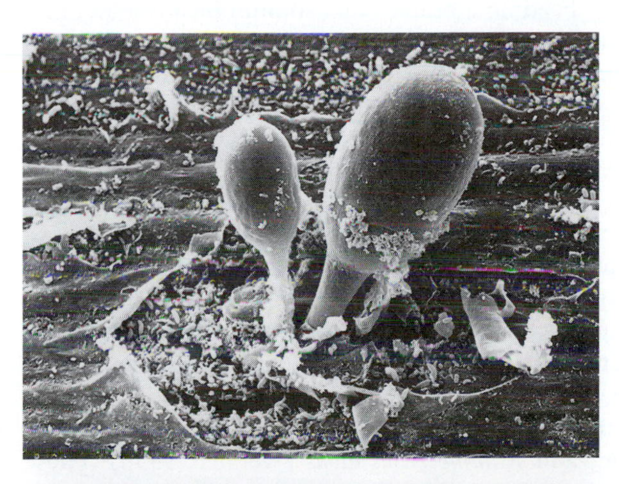

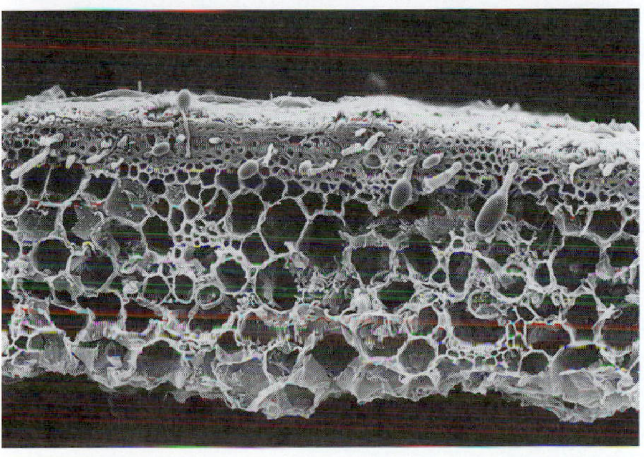

FIGURE 5.6 Rumen fungi have an important role in digestion of low-quality roughages. (Top) Sporangia or fruiting body of a rumen fungus. The sporangia produce zoospores that infect other feed particles in the rumen. (Courtesy of K. Joblin.) (Bottom) Fungi growing on the cell walls of wheat straw in the rumen. Breakdown of the lignified cell wall by the rumen fungi renders the remainder of the plant cell more accessible to bacteria. (Courtesy of E. Grenet, P. Barry, and INRA [France].)

Protozoa thrive in the rumen of roughage-fed animals but are usually absent on high-concentrate diets. This appears to be largely due to "washing out" of protozoa from the rumen on low-fiber diets; with high-fiber diets, protozoa are retained within the floating mat.

A major animal factor influencing efficiency of roughage utilization is the rumination rate. **Rumination** is important in helping to degrade fibrous material into smaller particle sizes, facilitating their digestion by microbes. The maximum time spent in rumination is 8 to 9 hours per day (Welch, 1982), so the more roughage that can be ruminated during that time, the greater the efficiency of digestion, roughage intake, and productivity. Cattle are markedly more efficient in rumination (amount ruminated per kilogram $BW^{.75}$ per minute)* than sheep, whereas goats are intermediate between sheep and cattle (Welch, 1982). Calves ruminate less efficiently than mature bovines. Thus, younger cattle should be fed better-quality hay than cows, and cattle can utilize low-quality roughage better than sheep. This is a reflection of the digestive capabilities of grazers compared to concentrate selectors or intermediate feeders (Chapters 1 and 9).

Supplementation of roughage diets with readily fermentable carbohydrate (e.g., grains) usually depresses fiber digestibility. As discussed previously, fermentation of concentrate results in a drop in rumen pH. When the pH drops below 6.1, the cellulolytic organisms are severely impaired (Mould et al., 1983a), thus reducing fiber digestibility. Feeding a buffer (e.g., sodium bicarbonate) and feeding the roughage in a long form (e.g., hay), which stimulates salivation and buffer secretion, are effective in preventing the negative effects of concentrate on roughage digestion (Mould et al., 1983b). When starch is present, mixed rumen microbes digest starch in preference to fiber (Van der Linden et al., 1984). For example, Bradford (1989) points out that in the Mediterranean and Middle East countries it is a widespread practice to supplement straw-based diets with government-subsidized barley or other cereal grain. This results in inefficient use of both the grain and the straw. Use of a protein supplement to replace most or all of the grain would increase the efficiency of resource utilization and animal production. Thus, there is a negative associative effect of concentrate on forage utilization. **Associative effects** occur when the addition of one ingredient modifies the utilization of another.

The predominant cellulolytic bacteria in the rumen are *Fibrobacter succinogenes, Ruminococcus flavefaciens,* and *Ruminococcus albus* (Weimer, 1996). These species have extreme nutritional specialization and use cellulose and its hydrolytic products as their sole energy source. They attach to plant fibers by a glycoprotein glycocalyx. This close association with fiber protects the cells from protozoa and prevents the cellulolytic enzymes from ruminal proteases. Cellulolytic bacteria have not evolved mechanisms to grow under strongly acidic conditions (Russell and Wilson, 1996). Acid-resistant amylolytic bacteria have evolved the capacity to let their intracellular pH decrease, maintain a small pH gradient across the cell membrane, and prevent an intracellular accumulation of VFA anions (Russell and Wilson, 1996).

*$BW^{.75}$ is metabolic body size (see Chapter 9).

Feed Consumption

A major factor determining forage quality is the intake. Animals cannot consume sufficient amounts of any forage to satisfy their energy requirements for maximum growth or lactational output, but the higher the consumption of a nutritious forage, the closer the nutrient intake will come to satisfying the requirements. Intake of a forage by ruminants is closely related to digestibility. The rate at which material leaves the rumen by digestion determines the rate at which new feed can be added. Besides the effect of digestion rate, other factors influence forage intake. The voluntary intake also depends on palatability, which in turn is influenced by taste (pleasant vs bitter, etc.) and physical properties (prickly, coarse, spines, etc.). Factors influencing feed intake and palatability will be discussed in more detail in Chapter 9.

Provision of Nutrients

Assessment of forage quality implicitly requires consideration of nutrient content. The nutrient contribution of forages to the diet can be assessed in several ways. These include the proximate analysis, modified to utilize the NDF and ADF determinations. In some situations, it may be desirable to include analysis of cellulose and lignin. Where forage is a major component of the diet, mineral analysis, especially for calcium and phosphorus, should be conducted. Vitamin analysis is generally not performed, for reasons of expense and because vitamin needs of herbivores are generally met by nondietary means.

In many cases, determination of ADF and crude protein is sufficient to give an adequate assessment of forage quality. If specific nutritional problems are encountered (e.g., outbreak of grass tetany, appearance of signs of copper deficiency, etc.), then more specific analyses are needed.

FORAGES FOR PASTURE, SILAGE, AND HAY

Forages grown for feeding purposes are usually grasses or legumes, although a few other types, including broad-leaved nonleguminous plants, such as buckwheat, *Brassica* spp. (kale, rape, turnips, etc.), and comfrey are grown to some extent. Forages may be annuals (e.g., annual ryegrass, forage rape, subterranean clover), biennials (e.g., sweetclover), or perennials (most grasses and legumes). Annuals live 1 year or less, biennials 2 years, and perennials 3 years or more.

The agronomic features, nutritional value, and toxins or deleterious factors associated with some of the more important forages will be discussed.

Grasses as Forage

Grasslands have long been major ecological systems. The prairies of North America, plains of Africa, steppes of Russia, pampas of Argentina, and outback of Australia are examples. Many species of grazing herbivores, including cattle and the equines, have evolved on grasslands. Many of the grassland ecosystems have been subjected to the plow, sometimes with disastrous results as in the infamous 1930s dust bowl of the Great Plains in the United States. The high yields of modern American agriculture have come at the expense of the native grasslands and have involved paying a price of massive soil erosion, loss of soil texture, and the need for heavy inputs of agricultural chemicals and fertilizers. Some agricultural specialists are concerned that the soils of the Midwest are being mined by a type of short-term agriculture that cannot be sustained for generations. **Sustainable agriculture**, on the other hand, involves crop rotations incorporating forages. Grasses are particularly effective in contributing organic matter, improving soil texture, increasing water-holding capacity, and reducing wind and water erosion. The major economic use of the grass, of course, is for livestock production. In some countries, notably New Zealand, the agricultural economy is based on grass and the production of meat, milk, and wool.

General Nutritional and Toxic Properties of Grasses In comparison to legume forages, grasses tend to be lower in crude protein, digestible energy, calcium, and magnesium. Their nutritional value tends to decrease markedly with maturity, due to the effects of lignification. Palatability follows a similar trend.

Two nutritional problems that are mainly associated with grasses are nitrate poisoning and grass tetany. **Nitrate poisoning** can occur when grasses are heavily fertilized with nitrogen. Nitrogen in excess of the plants' requirement for protein synthesis is accumulated as nitrate ion. In the rumen, nitrate (NO_3^-) is reduced to nitrite (NO_2^-).

Nitrite is the actual toxic entity; it is absorbed and reacts with hemoglobin to form a compound called methemoglobin (pronounced met-hemoglobin). Because methemoglobin cannot carry oxygen, nitrate poisoning causes death due to oxygen starvation of the tissues. Signs of toxicity include labored breathing, bluish mucous membranes, convulsions, and blood that is dark brown or chocolate colored (the color of methemoglobin). Levels of 0.5 percent or more nitrate in forage are potentially dangerous.

Grass tetany is a metabolic disorder mainly associated with consumption of grasses by ruminants. The disease is caused by very low blood levels of magnesium (**hypomagnesemia**). Signs of grass tetany include incoordination and convulsions, especially when animals are disturbed, and high mortality of affected individuals. Grasses in the spring tend to have a low magnesium content and a high content of organic acids that tie up magnesium in the rumen in an unavailable form. High nitrogen fertilization and cold, wet weather intensify the problem. Grass tetany can be prevented by providing mineral mixes with added magnesium or by dusting the grass with magnesium oxide. High forage potassium concentrations can impair magnesium absorption (Dalley et al., 1997a,b). **High-potassium forage** is an increasing problem in the dairy industry. The disposal of dairy manure on cropland is in many cases increasing forage potassium to levels greater than 3 percent, with consequent adverse effects on magnesium metabolism (Fisher et al., 1994). The major sites of **magnesium absorption** in cattle are the rumen and the hindgut. High dietary potassium tends to increase rumen pH, which reduces ruminal magnesium absorption and may shift it to the hindgut (Dalley et al., 1997a,b). High forage potassium and low solubility of magnesium in the rumen could be of importance in triggering hypomagnesemia (Dalley et al., 1997b). High forage fermentable nitrogen would also tend to increase rumen pH, further reducing magnesium solubility. A comprehensive review of plant and animal factors affecting magnesium metabolism and grass tetany is provided in a series of symposium papers (*J. Anim. Sci.* 67:3437–3494, 1989).

Forage grasses can be classified as temperate (cool-season), warm-season, and tropical. **Cool-season grasses**, such as orchardgrass, ryegrass, and tall fescue, are species that make their most rapid growth during cool, moist seasons of the year and usually go dormant during hot, dry periods. **Warm-season grasses** (e.g., bermudagrass) make their major growth during the warm or hot periods and are dormant during cool periods. **Tropical grasses**, such as napiergrass, guineagrass, and pangolagrass, are adapted to tropical regions only; they are not frost hardy. In addition, there are some **summer annual grasses** and **cereals** grown for forage. These include the forage sorghums (e.g., sudangrass) and corn, oats, barley, wheat, and rye, which are sometimes grown for forage purposes, especially oats for hay and corn for silage.

Properties of Temperate (Cool-Season) Grasses **Timothy** (*Phleum pratense*) was formerly grown extensively in the United States for hay used in feeding the millions of draft horses that once powered American agriculture and transportation. To many people, timothy is still synonymous with horse hay. It is less important than it once was, but timothy is still a major hay and pasture species in the cool, humid regions of the United States. It is generally grown as a component of a mixture of grasses and legumes. Timothy is a bunch grass and has a characteristic cylindrical seed head. It is intolerant of heavy grazing pressure. Timothy has low concentrations of magnesium at all stages of growth (Reid et al., 1984), so grass tetany is a concern where this grass is used. Supplemental magnesium can be provided in various ways (see Chapter 7). Winter and Gupta (1983) reported that timothy grown in the Canadian province of Prince Edward Island was deficient in most mineral elements, including magnesium. Selection of timothy for improved nutritional value should include estimates of nutrient digestibility and indigestibility, as well as voluntary intake (Mason and Flipot, 1988).

Orchardgrass (*Dactylis glomerata*) is quite tolerant of shade, which probably accounts for its common name. It grows in clumps, producing an open sod. Like timothy, it is a common component of hay and pasture mixtures in the Midwest, Northeast, and Pacific Coast states of the United States. Known as *cocksfoot* in Europe, orchardgrass is productive but loses palatability rapidly as it matures. In areas where rainy conditions are prevalent in spring, late maturing varieties should be used for hay production.

Tall fescue (*Festuca arundinacea*) is an important pasture and hay species of the southeastern United States and the Pacific Northwest states. It is a deep-rooted, drought-resistant perennial bunchgrass. It produces a dense sod, resistant to trampling, and continues to grow and produce pasture when most grasses are dormant during the summer. Tall fescue is somewhat unpalatable but, in contrast to most grasses, retains its palatability when mature. Because of these factors, it is one of the most valuable grass species in areas where it is adapted.

Although it has many desirable properties, tall fescue can also have deleterious effects on livestock production. In many areas of the southeast states, problems in tall fescue pastures have included fescue foot, summer fescue toxicosis, fat necrosis, and reproductive problems. **Fescue foot** is very similar to ergotism and in fact may be ergotism. It is characterized by loss of blood supply to the extremities, with necrosis of the hooves, lower limbs, tail, and tips of the ears. Cattle are the major species affected. Early signs of fescue foot are lameness and reluctance to walk, which may be followed by loss of the hooves and gangrene of the lower legs and tail. Although the condition is virtually identical to effects of ergot, no evidence of the typical ergot fungi infection of the seed is evident. In tall fescue, the ergot alkaloids are produced by **endophytes**, which are fungi that live within plant tissue. The history of the discovery of the roles of endophytes in tall fescue toxicoses has been reviewed by Bacon (1995). The endophyte in tall fescue is called *Neotyphodium coenophialum* (formerly *Acremonium coenophialum*).

Summer fescue toxicosis of cattle is characterized by poor growth or milk production, a rough hair coat, unthrifty appearance, and an obvious sensitivity to heat stress. In hot weather, affected animals show excessive respiration rate and elevated body temperature, and seek relief from heat stress by standing in water or mud and staying in shade as much as possible. The causative factors of summer fescue toxicosis are the **ergot alkaloids** produced when the plants are infected with an endophyte fungus. In areas where the condition is a problem, tall fescue pastures should be replaced with endophyte-free grass. The fungus spreads only through infected seed; therefore, if a new pasture is established with endophyte-free fescue seed, it will be nontoxic. Although endophyte-free fescue is nontoxic, the grass is less vigorous and productive than infected fescue. Some producers prefer to use endophyte-infected grass and minimize toxicity problems by careful animal management.

Reproductive problems are noted with livestock on endophyte-infected tall fescue pastures. **Horses** are particularly susceptible. Mares consuming tall fescue may experience prolonged gestation, agalactia (lack of lactation), and thickened edematous placenta, and have large, weak foals with elongated hooves. Survival rate of the foals is very low. Abortion may also occur. These effects in horses have been reviewed by Cross et al. (1995).

Fat necrosis is another condition sometimes seen in cattle on tall fescue pastures. It may be a chronic effect of summer toxicosis. Clumps of hard fat form around internal organs, such as the intestine and reproductive tract, as well as subcutaneously. Constriction of the intestine by the fatty deposits may eventually cause mortality.

Most, if not all, of these various conditions caused by endophyte-infected tall fescue can be explained by the effects of **ergot alkaloids** (e.g., ergovaline, the main alkaloid in tall fescue), on prolactin secretion by the brain. Animals consuming toxic tall fescue have extremely low serum prolactin levels, as low as 1 percent of normal (Porter et al., 1990). Ergot alkaloids inhibit **prolactin secretion** by the pituitary gland by binding to dopamine receptors. Dopamine is a brain neurotransmitter that regulates prolactin secretion. By mimicking dopamine, ergot alkaloids suppress prolactin secretion. Prolactin has a number of physiological roles, including effects on milk secretion, reproduction, gut motility, and the appetite center. These effects are reviewed by Cheeke (1998).

Fescue foot, summer toxicosis, and fat necrosis are all related to infection of tall fescue by endophytic fungi and can be eliminated by replacing toxic pastures with new varieties of endophyte-free fescue. In many areas, this represents a serious challenge because of the steep topography of many of the fescue pastures in the southeastern United States. Fescue toxicity has not often been observed in the Pacific Northwest states, where fescue is used extensively as a forage, because of a fortuitous lack of presence of the endophytes.

Smooth bromegrass (*Bromus inermis*) is grown extensively in the northern United States and in Canada. It is a leafy, sod-forming perennial grass spreading by underground rhizomes. It is similar to other temperate species in nutritive value and has no major toxic or deleterious factors.

Perennial ryegrass (*Lolium perenne*) and annual ryegrass (*Lolium multiflorum*) are important forages in cool temperate climate areas, including much of the United States, Australia, New Zealand, Great Britain, and western Europe. They are less winter hardy than many of the other temperate grasses and are intolerant of drought but can survive flooding and are productive on poorly drained soils. **Annual ryegrass** is often used for overseeding bermudagrass or other warm-season grasses to provide winter pasture in the southern United States. In the Pacific Northwest, where ryegrass is grown extensively for seed, ryegrass fields are used for winter pasture for sheep.

Much of the livestock industry of New Zealand is based on perennial ryegrass-white clover pastures. Several toxicity problems with ryegrass are encountered in New Zealand, including facial eczema, ryegrass staggers, and annual ryegrass toxicity. **Facial eczema** is caused by fungal spores produced by fungi inhabiting dead plant material at the base of the pasture. The spores contain a toxin called sporidesmin. When consumed by sheep and cattle, sporidesmin causes photosensitization or skin lesions occurring with exposure to sunlight (see Fig. 2.7). **Ryegrass staggers**, as the name implies, is characterized by incoordination and neural dysfunction in affected animals. It is caused by mycotoxins called *tremorgens* (e.g., ***lolitrem B***) that are produced by an endophyte (*Neotyphodium lolii*). The neural effects are temporary, and affected animals usually recover. **Annual ryegrass toxicity**, involving permanent nerve damage and mortality, is a problem mainly in Australia and South Africa. It is caused by a toxin produced by a bacterium that lives in nematode galls on the ryegrass seed heads. Thus the ryegrass must be infected with both the nematodes and bacteria for the toxins to be produced. Endophyte-related problems with tall fescue and ryegrasses have been reviewed by Cheeke (1998).

Reed canarygrass (*Phalaris arundinacea*) is very tolerant of flooding and poorly drained soils and is mainly grown in wet areas where other forages cannot be produced. It is a vigorous perennial grass spreading by underground rhizomes, forming a heavy sod that helps to reduce trampling damage when cattle are grazed on wet pastures. It begins growth early in the spring, often growing up through standing water. It is quite a coarse grass and should be cut for hay before seed head emergence if possible. Reed canarygrass is widely viewed as being unpalatable and of low nutritional value. It contains a number of alkaloids that appear to be responsible for the low palatability and also have neurological effects and induce diarrhea. **Low alkaloid cultivars** have been developed that are palatable (Fig. 5.7) and give satisfactory animal performance (Duynisveld and Wittenberg, 1993; Wittenberg et al., 1992). The availability of these cultivars should increase the use of reed canarygrass in areas where it is well adapted. Environmental stresses such as drought and continuous grazing increase the alkaloid content of reed canarygrass (Duynisveld and Wittenberg, 1993). The alkaloid content is lower in hay than in the fresh plant, but is not reduced by ensiling (Tosi and Wittenberg, 1993). Even the low-alkaloid cultivars of reed canarygrass are less palatable and have lower digestibility than other cool-season grasses such as timothy and tall fescue (Narasimhalu et al., 1995). ***Phalaris* poisoning** of livestock caused by the alkaloids is quite common in Australia, New Zealand, and South Africa. Cases of **canarygrass poisoning** have also occurred in the United States

FIGURE 5.7 Reed canarygrass contains alkaloids that adversely affect its palatability. In this plant breeding program, sheep are rejecting high-alkaloid plants to graze more palatable low-alkaloid selections.(Courtesy of G. C. Marten, USDA-ARS, University of Minnesota, St. Paul.)

(East and Higgins, 1988; Nicholson et al., 1989). Pronounced neurological signs such as convulsions, tremors, and incoordination occur in *Phalaris* poisoning.

Other Temperate Grasses Other temperate grasses of lesser importance include **Kentucky bluegrass** (*Poa pratensis*), **bentgrasses** (*Agrostis* spp.), and **redtop** (*Agrostis alba*). These are sod-forming grasses and are relatively unproductive. Their use for forage production has declined markedly in recent years. **Meadow foxtail** (*Alopecurus pratensis*) is a perennial bunchgrass adapted to cooler regions of North America. Its main attribute is that it begins growth very early in the spring and, therefore, has good potential for early spring grazing. Canadian studies (Rode, 1986; Rode and Pringle, 1986) have indicated that performance of cattle grazing meadow foxtail is lower than would be expected from its nutrient content, suggesting the possible presence of deleterious factors. **Quackgrass** (*Agropyron repens*) is an aggressive plant, spreading by underground rhizomes. It is usually viewed as a weed, but where it occurs naturally in pastures, it may be useful. It forms a dense sod and has good forage qualities (Stoszek et al., 1979).

Many other grasses are important in the dryland and semiarid range areas of the western United States, including many native species. The principal cultivated grass used in reseeding rangelands is **crested wheatgrass** (*Agropyron desertorum*), a hardy, drought-resistant bunchgrass native to Central Asia. The native grasslands of the North American prairies were of two major ecological types: the tallgrass prairie and the mixed prairie (short, mid, and tallgrass). Pioneers used big bluestem (*Andropogon gerardi*) of the tallgrass prairie as an indicator of land suitable for corn, whereas to the west, western wheatgrass (*Agropyron smithii*) was an indicator of good wheat land. Much of the native prairie grasslands has been lost to the plow; perhaps if efforts to produce perennial wheat are successful (Chapter 2), the grasslands will return.

Small Grains as Forage All of the cereal grains (Chapter 2) can be utilized as forage, by harvesting for hay, silage, or use as pasture. Those most commonly used are corn, wheat and oats, and forage sorghums.

Millions of weaned calves (stockers) are grazed each year on winter **wheat pastures** of the southern Great Plains, in Oklahoma, Kansas, Nebraska, and Texas (Mader et al., 1983). The usual grazing period is from November to mid-March, at which time the cattle are removed and the wheat produces a grain crop. Grazing wheat often increases the grain yield by reducing lodging (plants falling or blown over by wind), especially with tall-stalk cultivars (Redmon et al., 1995). Grazing may reduce the yield of semidwarf wheat cultivars by reducing leaf area. Typically, wheat is heavily fertilized with nitrogen, producing a lush, succulent pasture high in soluble protein and carbohydrates. These pastures can produce excellent cattle gains of 0.65 to 0.9 kilograms a day. Three major problems encountered with cattle on wheat pasture are bloat, grass tetany, and nitrate poisoning.

Frothy bloat is caused by foam-producing soluble proteins released into the rumen from succulent forages high in cell content material. Normally bloat is of concern only with legume pastures. The heavy nitrogen fertilization of wheat produces a high content of cytoplasmic proteins that, along with the succulent nature of the material, results in rapid release of a high concentration of cytoplasmic proteins in the rumen. Wheat pasture bloat is one of the few examples of bloat problems associated with nonlegumes. Wheat pasture bloat can be prevented by the use of **antibloat agents** such as poloxalene (bloat guard), generally provided in blocks distributed in the pasture. Feeding some coarse roughage may be helpful in reducing bloat incidence. Feeding the ionophore **monensin** (see Chapter 8) reduces wheat pasture frothy bloat, probably by modifying rumen microbial populations to maintain the ruminal pH above that required for maximum foam stability and strength (Branine and Galyean, 1990).

Even though wheat pasture often has a very high crude protein content, approximately 20 to 30 percent (Bohman et al., 1983a), improved cattle performance with the provision of protein supplements has been noted. This is an effect of bypass protein. The protein in wheat forage is very soluble in the rumen, and there is a high content of NPN. Thus, the wheat crude protein is highly degradable in the rumen, with a high ammonia production. Supplementation with bypass protein may be useful under such circumstances (Phillips et al., 1995).

Wheat pasture poisoning occurs mainly in mature beef cows grazing winter wheat. The symptoms include excitement, incoordination, and staggering, with increased signs of nervousness as the disease progresses, including muscle twitching, teeth grinding, and profuse salivation. A comatose condition follows, with death six to ten hours after symptoms first develop (Bohman et al., 1983b). The condition appears to involve both hypocalcemia and hypomagnesemia (Bohman et al., 1983a,b). Bohman et al. (1983a) characterized the composition of wheat forage throughout the grazing season. Prior to tetany episodes, there was an elevation in *trans*-aconitic acid in the forage. This is an organic acid in grasses that impairs magnesium absorption and metabolism. The etiology seemed to involve several other plant factors. Cows with wheat poisoning had severe hypocalcemia (Bohman et al., 1983b). Affected animals should receive therapy with both calcium and magnesium. Provision of mineral mixes with calcium and magnesium would presumably have preventative value.

The performance of sheep and cattle during the first few weeks on wheat pasture is often poor (Gallavan et al., 1989). This could be due to the need to adapt to a high-moisture diet or one high in NPN. In the case of sheep, Gallavan et al. (1989) found that rumen capacity limits the dry-matter intake of animals fed wheat forage. A dry-matter content of 17 percent was necessary for maintenance. In the early grazing period, the dry-matter content of wheat pasture is in the area of 18 to 20 percent and increases to 25 to 33 percent later in the grazing season (Stewart et al., 1981; Gallavan et al., 1989). The increasing dry-matter content could account for the improved animal performance as grazing period increases. The soluble nitrogen content can be as high as 70 to 90 percent of total nitrogen, with NPN constituting up to 75 percent of total nitrogen. Bloat occurs most often with wheat pasture when the soluble nitrogen and NPN contents exceed 60 and 40 percent, respectively (Stewart et al., 1981).

Oats have been used extensively for forage production. At one time, a mixture of oats, peas, and vetch was a standard hay crop. The oats provide grain and relatively palatable and digestible straw, whereas the peas and vetch, being legumes, provide a good source of protein. **Peas and vetch** (*Vicia sativa* and *Vicia villosa*) are viny annual legumes. Vetch seeds contain toxic lathyrogenic amino acids, and vetch forage has caused "vetch poisoning" in Oklahoma and South Africa (Cheeke, 1998). Clinical signs of vetch poisoning include dermatitis, eye edema, and diarrhea. Mortality may occur. Oat forage or oats, peas, and vetch are much less commonly used today because of the availability of higher yielding crops such as corn silage.

A number of varieties of **forage sorghums**, such as sudangrass, are used for hay, silage, and pasture. These are warm-season grasses and produce high yields of succulent forage. Forage sorghums contain cynanogenic glycosides; these are of concern primarily when the forage has been damaged by frost. The breakdown of the cell structure mixes the glycosides with cyanide-liberating enzymes, with the result that the frosted forage can have a high content of free **cyanide**. Because cyanide is volatile, the forage is most toxic the day after frost damage has occurred.

Corn is a widely grown silage crop; corn silage will be discussed later (Chapter 6). Stubble and stover, the field residues remaining after harvesting small grains and corn, respectively, may be grazed by livestock. The use of these crop residues in beef cattle production is described further in Chapter 15. Cattle make more effective use of these low-quality roughages than do sheep (Coombe and Mulholland, 1988). This is largely because of differences in grazing behavior and feed selection; cattle more readily consume the coarse dead plant material (Mulholland et al., 1977).

Warm-Season Grasses In the southern United States, a variety of warm-season grasses are grown for hay and pasture. **Bermudagrass** (*Cynodon dactylon*) is one of the most important. It grows most rapidly when the mean daily temperature is above 24°C (75°F). It is a sod-forming grass, spreading by stolons (aboveground stems) and rhizomes (belowground stems). Numerous improved cultivars of bermudagrass are available; coastal is the most important. It is much more productive than common bermudagrass. Bermudagrass is most commonly established by planting sprigs (stolons and rhizomes) with special planters, or by discing them in.

Bahiagrass (*Paspalum notatum*) is native to South America (Bahia is a state in Brazil). It is widely grown in the southern United States. It has an extensive rhizomatous root system, forming a dense sod. It tolerates a wider range of soil types and conditions than bermudagrass and produces moderately well, even on extremely infertile soils. Its dense sod is resistant to weed encroachment and withstands heavy grazing pressure. The Pensacola variety is the most widely grown. **Carpetgrass** (*Axonopus affinis*) is a sod-forming, low-growing grass common throughout the Gulf Coast and southern Atlantic coast states. It is relatively unproductive and is not generally planted intentionally. It has become established by natural seeding throughout the southern states.

Other warm-season grasses grown in the southern states include johnsongrass (*Sorghum halepense*) and dallisgrass (*Paspalum didatalum*). **Johnsongrass**, a forage sorghum, contains cyanogenic glycosides that may occasionally cause toxicity, particularly following a frost. It is quite productive and is used for pasture and production of hay and silage. However, because it is very aggressive and spread rapidly by rhizomes, it is often regarded as a noxious weed. **Dallisgrass** is most suitable for pasture use, having most of its leaves close to the ground. It is susceptible to infection with ergot; dallisgrass is a major cause of ergot problems in the United States.

Tropical Grasses There is no clear distinction between warm-season and tropical grasses. The terminology used here is to consider the tropical grasses to be those that grow only in tropical regions.

As mentioned previously, tropical grasses have a leaf anatomy characterized by a high content of poorly digested vascular tissue and a low content of the more readily digested mesophyll cells. Thus, tropical grasses have approximately 12 to 15 percent lower dry-matter digestibility in ruminants than the temperate grasses. Some of the more common tropical grasses are the following.

Botanical Name	***Common Name***
Brachiaria decumbens	Signalgrass
Cenchrus ciliaris	Buffelgrass
Chloris gayana	Rhodesgrass
Digitaria decumbens	Panogolagrass
Melinis minutiflora	Molassesgrass
Panicum maximum	Guineagrass, green panicgrass
Paspalum dilatatum	Paspalum, dallisgrass
Pennisetum clandestinum	Kikuyugrass
Pennisetum purpureum	Elephantgrass, napiergrass
Setaria sphacelata	Setaria

A number of tropical grasses such as buffelgrass, pangola, setaria, and kikuyugrass contain sufficient quantities of oxalates to impair calcium metabolism. In tropical regions of Australia, "big head" of horses is a common form of hyperparathyroidism induced by high levels of oxalate in tropical grass pastures.

Sugarcane is a tropical grass with tremendous potential as a feedstuff for ruminant production in the tropics and might be considered the tropical equivalent of feed grain. Some of the advantages of sugarcane are that it has a very high dry-matter yield, its digestibility increases with plant maturity, and it retains its nutritive value for many months after maturity (Pate, 1981). Whole chopped sugarcane can be used, supplemented with urea as a source of fermentable nitrogen and a source of bypass protein, such as rice polishings or legume forage (e.g., leucaena). The use of **sugarcane-urea diets** has been reviewed by Preston and Leng (1987). Pate et al. (1985) reported that urea was inferior to cottonseed meal as a nitrogen source for cattle that were fed sugarcane-based diets, probably because of a deficit of nondegradable protein. The efficiency of utilization of sugarcane can be improved by fractionating it to separate the juice from the fiber. Sugarcane fiber (**bagasse**) has a very low digestibility and reduces voluntary intake because of rumen fill and prolonged turnover time. Inexpensive cane crushers are available in many areas for removal of the juice. Animal productivity on sugarcane juice diets is higher than on molasses-based diets, because all of the sucrose is still present, whereas molasses is the residue remaining after removal of much of the sucrose. Sugarcane juice can be effectively utilized in diets for both ruminants and monogastrics.

Legumes as Forage

Legumes hold an esteemed position as forages, for both agronomic and nutritional considerations. Legumes are plants that, via a symbiotic relationship with bacteria in root nodules, are able to utilize atmospheric nitrogen. The bacteria involved are of the *Rhizobium* genus. The rhizobia enter the root hairs of compatible legume species, initiating nodule formation by plant defense mechanisms. The nodules contain a red pigment, leghemoglobin, that participates in nitrogen fixation. In this process, atmospheric nitrogen is reduced, with ammonia the first stable intermediate formed. Ammonia reacts with organic acids to produce amino acids, which can be used by the plant for protein synthesis. The plant provides the carbon portion (photosynthate) and the rhizobia add the nitrogen, producing amino acids to serve the needs of both the plant and the microbes.

The ability of legumes to fix atmospheric nitrogen is of tremendous importance. The nitrogen the plant needs is provided "free," and legumes can increase soil nitrogen to fertilize other plants. When the plant is stressed, by cutting for hay for example, the root nodules are shed, thus increasing the nitrogen content of the surrounding soil. Legumes are highly valued for their soil-enriching properties. Before chemical fertilizers were available, legumes were a major source of nitrogen. Crop rotations were practiced so that corn or wheat might be grown following alfalfa or clover, to benefit from the nitrogen fixation. Legume crops were often plowed under as "green manure." A partial return to these practices may be desirable in terms of sustainable agriculture.

Nutritionally, legumes are generally superior to grasses as sources of protein, energy, calcium, and magnesium. The inclusion of legume species in pasture seedings usually results in greater productivity or performance of grazing animals. Cattle and sheep consuming legumes grow more rapidly than animals consuming grasses of equal energy digestibility (Glenn et al., 1989). Among the reasons are lower ruminal acetate-to-propionate ratios, higher microbial efficiency, and a greater flow rate of dry matter and nitrogen to the duodenum.

Temperate Legumes As corn is the standard to which other feed grains are compared, **alfalfa** is the standard for forages. Known as **lucerne** in many countries, alfalfa (*Medicago sativa*) was first domesticated in Persia. It is now grown worldwide, thriving in all climates except the tropics. It is best adapted to dry climates, provided that ample water is supplied. Alfalfa has an extensive root system with a taproot extending deep into the subsoil. It will not tolerate poorly drained conditions. Where it is well adapted, it can be extremely productive, producing several times as much protein per acre as corn or soybeans.

Sometimes called the "Queen of the Forages," alfalfa is grown for hay, pasture, silage, and dehydrated meal. It can be used in the diets of all types of domestic animals. Alfalfa is often used as pasture for dairy and beef cattle and sheep. It is desirable to strip graze alfalfa (see Chapter 6) to harvest the plant efficiently and provide adequate recovery from grazing. Alfalfa is quickly eliminated by continuous grazing, through depletion of root reserves of carbohydrate. Late fall grazing also contributes to premature loss of alfalfa stands.

Although alfalfa pastures can be highly productive, there is a severe bloat hazard. **Bloat** occurs when gases produced in rumen fermentation become trapped in the rumen in the form of a stable foam. The presence of foam at the base of the esophagus inhibits the eructation mechanism. As the gas pressure in the rumen builds up, pressure on the diaphragm prevents breathing, and death occurs. The basic causes of bloat include the rapid release of cytoplasmic proteins from lush plant tissues, combined with an abundant supply of fermentable carbohydrate, which promotes gas production. Legumes like alfalfa are bloat-producing, because they have a high concentration of soluble protein, and the cell contents are released rapidly in the rumen. Canadian plant breeders have shown that selection of alfalfa for a slow initial rate of breakdown in the rumen can reduce its bloat-producing potential and may lead to development of "bloat-safe" alfalfa cultivars (Hall et al., 1994).

Bloat can be prevented with antifoaming agents. The major commercial product is **poloxalene** blocks, marketed as "Bloat Guard," that are distributed in areas where livestock congregate and throughout the pasture. Animals will lick the blocks regularly (see Fig. 8.2), maintaining a sufficient concentration of poloxalene in the rumen to prevent froathy bloat. When used according to the manufacturer's directions, Bloat Guard blocks are very successful in preventing bloat. Administration of ionophores is also partially effective in reducing bloat. Bagley and Feazel (1989) reported that giving monensin as a bolus in a slow-release delivery device significantly reduced the incidence of bloat in stocker cattle. Majak and Hall (1990) noted that occurrence of bloat was associated with low rumen sodium and high potassium levels. Potassium stabilizes foam in colloidal suspensions while sodium disperses colloids. Supplementation of feedlot diets with 4 percent salt reduces bloat (Majak and Hall, 1990). These authors suggest that manipulation of sodium and potassium by use of dietary supplements may aid in controlling bloat. The effectiveness of bloat-preventing blocks containing antifoaming agents and salt might in part be due to their sodium contribution.

There are a number of **nonbloating legumes** such as birdsfoot trefoil, cicer milkvetch, arrowleaf clover, sainfoin, and crown vetch. **Condensed tannins** in trefoil and sainfoin protect against bloat, both by precipitating soluble proteins and by inhibiting microbial enzymes, thus slowing the rate of release of cell contents. Although arrowleaf clover and cicer milkvetch do not contain tannins

(Howarth et al., 1986), their rate of breakdown in the rumen is slower than for bloat-producing legumes like alfalfa, red clover, and white clover. The cell walls of bloat-causing legumes are digested more quickly than those of bloat-safe species. Some animals are more susceptible to bloat than others. This may be due in part to differences in secretion of salivary mucins, which act as antifoaming agents. Despite the fact that sheep tend to be more selective than cattle in grazing and select a diet higher in leaves and lower in stems, they are more resistant than cattle to bloat (Colvin and Backus, 1988) for reasons that are not entirely clear but may involve differences in rumen motility.

Alfalfa hay is very important to the dairy industry as a source of protein and fiber. It is also extensively used for beef cattle, sheep, and horses. Because of its high calcium content, it is an excellent feedstuff for rapidly growing or lactating animals. Although alfalfa has a high calcium content, approximately 20 to 33 percent of it is present as insoluble calcium oxalate (Ward et al., 1979, 1984). Calcium oxalate has low bioavailability as a source of calcium. However, the nonoxalate calcium in alfalfa is highly available. Therefore, alfalfa calcium in total has approximately 80 to 95 percent the bioavailability of calcium carbonate (Ward et al., 1984).

Biotechnology techniques are being used to increase the nutritive value of alfalfa. These include the introduction of genes for tannin production to produce nonbloating alfalfa. Another example is the introduction of genes from sunflowers for a rumen-stable sulfur amino acid-rich protein, to serve as a source of high-quality rumen bypass protein (Tabe et al., 1995).

Clovers **Red clover** (*Trifolium pratense*) is the most widely grown clover in North America. It is similar to alfalfa in palatability and nutritional value but is not as productive. Red clover can be grown in cool, humid, temperate regions, including the midwestern and northeastern states and the Pacific Northwest of the United States. It is more tolerant of poorly drained soils than alfalfa and is suitable for both pasture and hay production. As with alfalfa, continuous grazing will rapidly eliminate red clover from a pasture. Most varieties are biennials or short-lived perennials. In some areas, red clover may cause profuse salivation (slobbers) in livestock, caused by the mycotoxin slaframine (see Chapter 8).

Red clover has a tendency to turn dark brown or black after being mowed for hay or silage. This is caused by a high content of **polyphenol oxidase**, an enzyme that causes browning. The brown color does not reduce the nutritional value and may even enhance it by inhibiting proteases that degrade proteins.

White clover (*T. repens*) is mainly a pasture species, having a prostrate growth habit, spreading by stolons (Fig. 5.8). It is widely used throughout the temperate countries of the world and is the major pasture legume in North America, Europe, New Zealand, and parts of Australia and South America. It is quite resistant to grazing pressure and can often be continuously grazed without harm. In a grass-white clover pasture, the clover provides nitrogen to stimulate grass growth and complements the grass nutritionally. Careful livestock management is needed to maintain an optimal balance of clover and grass. Undergrazing will lead to dominance by the grass through shading of the clover, whereas overgrazing may lead to the elimination of the grass. As with alfalfa, bloat is a hazard with white clover. Bloat Guard blocks should be used in pastures with a significant percentage of white clover. White clover also contains cyanogenic glycosides, but these do

FIGURE 5.8 White clover, a low-growing pasture legume tolerant of high grazing pressure.

not usually constitute much of a hazard for cyanide (prussic acid) poisoning (Vickery et al., 1987).

Alsike clover (*T. hybridum*) is intermediate in general properties between red and white clovers. It is an important forage legume in the clover-timothy area of the United States and Canada, and especially in the northern regions of Canada. It is especially well adapted to cool climates and wet, poorly drained soils. In Canada, a condition known as **alsike clover poisoning** of horses has been described (Nation, 1989). Signs of poisoning include photosensitization, liver failure, and neurological impairment. The liver is greatly enlarged, and gray-brown or yellow-green in color. Although commonly referred to as alsike clover poisoning, the evidence linking it to consumption of the plant is not totally conclusive (Nation, 1989).

Sweetclover (*Melilotus officinalis*) (yellow-flowered) and *Melilotus alba* (white-flowered) is a biennial thriving under a wide range of soil and climatic conditions. It is commonly grown in the Great Plains states, the Midwest, and the prairie provinces of Canada. In the past, it was extensively grown as a soil-improvement crop. It has a thick, extensive taproot that opens up the subsoil and provides a large amount of nitrogen when plowed under as a green manure crop.

Sweetclover is quite coarse and stemmy and does not produce hay of the quality obtainable with alfalfa. Although having an attractive, vanillalike fragrance, sweetclover is bitter and somewhat unpalatable. The odor is due to coumarin present in the plant. Coumarin is the plant factor responsible for **sweetclover disease** or "bleeding disease." Moldy sweetclover hay fed to cattle may cause hemorrhage and uncontrollable bleeding. Animals dehorned or castrated may bleed to death or spontaneous internal hemorrhaging may occur. This is caused by a substance called **dicumarol**, produced from coumarin by the metabolism of fungi (molds). Dicumarol is an inhibitor of vitamin K and leads to impaired formation of prothrombin, one of the essential factors for blood clotting. Thus sweetclover disease is basically an induced **vitamin K deficiency**. Cultivars of sweetclover with a low coumarin content have been developed. However, whenever sweetclover is used as a hay crop, there should be the awareness of potential problems if molding occurs.

Subterranean clover (*T. subterraneum*) is a winter annual of Mediterranean origin. In the United States, it is adapted mainly to the western valleys of California and Oregon, where it germinates in the fall and produces the bulk of its forage output in late winter and spring. It is extensively grown in Western Australia, often in rotation with wheat. Sub clover, as it is commonly known, produces a bristly seed bur, which is pushed into the ground at maturity, accounting for its name. Thus it reseeds itself readily, and once it is established, further reseeding is unnecessary. Sub clover is well known for its **phytoestrogen** activity. After it was introduced into Australia, reproductive problems in sheep, called *clover disease,* were observed. These problems, manifested by infertility and genital tract abnormalities, were found to be caused by a number of estrogenic compounds present in sub clover. Low estrogen cultivars that do not cause reproductive problems have been developed. Other annual clovers include crimson (*T. incarnatum*), arrowleaf (*T. vesiculosum*), and berseem (*T. alexandrinum*) clovers. These are winter annuals, which in the United States are grown primarily in the southern states. The flowers of **crimson clover** are bright red (crimson) and quite fibrous or hairy. When used for hay, the flower heads may form impactions (phytobezoars) in the stomach of horses. Crimson and berseem clovers grow upright, whereas arrowleaf has long viny stems. These annual clovers are productive and are used for pasture and hay. They are also excellent soil-building legumes.

Other Temperate Legumes Trefoils, the major one being **birdsfoot trefoil** (*Lotus corniculatus*), are perennial forage legumes grown primarily in areas that are too poorly drained, acidic, or infertile for the production of alfalfa. In the United States, trefoil is grown mainly in the Northeast, Midwest, and Pacific Coast states. Trefoil is grown quite extensively in New Zealand on infertile acidic soils. Birdsfoot trefoil derives its common name from the appearance of the seed pods. Trefoil provides satisfactory pasture and hay, but it is less productive than alfalfa and less tolerant of grazing than white clover. In New Zealand, trefoil growing on infertile acid soils has a high tannin content, which reduces the performance of sheep. Soil fertilization with phosphorus and sulfur results in a sharp reduction in the tannin content and an improvement in growth of lambs. At moderate levels, the **tannins** in trefoil are advantageous. They reduce the fermentation of protein in the rumen, giving high bypass protein content. By complexing with soluble proteins in the rumen and inhibiting microbial digestion, tannins prevent bloat. Trefoils are nonbloating legumes for this reason. The ideal tannin content of trefoil is 0.2 to 0.4 percent of dry matter (Barry, 1985); on low fertility soils the levels can reach 9 percent or more of dry matter. Chiquette et al. (1988) found that the *in vitro* digestibility of a high tannin (5 percent) line of birdsfoot trefoil was lower than for a low tannin (1 percent) line. According to these authors, in the presence of high tannin in forage, rumen bacteria may form glycocalyx-enclosed microcolonies that provide protection against tannins, but that also limit the secretion of bacterial exoenzymes and limit the penetration and thus digestion of the plant tissue. Tannins in trefoil are located in vesicles just under the epidermal layer, often adjacent to the stomata, that aid in the chemical defense of the plant against microorganisms.

Condensed tannins in trefoil and some other forage legumes are effective in suppressing **gastrointestinal parasites** (roundworms) in sheep and other grazing animals (Butter et al., 2000). Condensed tannins are nematocidal; many internal parasites such as roundworms are nematodes.

Lespedezas are important legumes grown for pasture and hay in the southeastern states. These are warm-season legumes, native to Asia. The major lespedezas grown in the United States are Korean (*Lespedeza stipulacea*), striate (*L. striata*), and sericea (*L. cuneata*). Sericea lespedeza is a perennial and the other two are annuals. Lespedeza is quite stemmy, being almost shrubby in appearance. The annual species, which can be maintained in pastures through natural reseeding, are superior in nutritive value to the perennial sericea. Sericea lespedeza contains tannins that reduce palatability and digestibility. Newer cultivars have been developed, with lower tannin levels, finer stems, and higher feeding value.

Other temperate legumes are of minor importance in the United States. **Crown vetch** (*Coronilla varia*) is grown quite extensively in the Midwest and Northeast as a ground cover and beautification plant along highways, reclaimed stripmine spoils, and other disturbed areas. It contains glycosides that reduce its palatability and the performance of animals consuming it. **Cicer milkvetch** (*Astragalus cicer*) has been grown to a limited extent in Montana and Wyoming; it is rhizomatous with coarse, succulent, hollow stems. It is a nontoxic species of the *Astragalus* genus, which is better known for its toxic species. These include locoweeds and *Astragalus* species that accumulate toxic levels of selenium. Marten et al. (1987) reported that cicer milkvetch was much less palatable than other common legumes and caused photosensitive reactions in Holstein and Shorthorn heifers being used in grazing trials. The average daily gains of heifers were 670, 810, 800, and 420 grams for alfalfa, birdsfoot trefoil, sainfoin (*Onobrychis viciifolia*), and cicer milkvetch, respectively. The superior gains with trefoil and sainfoin are probably because they contain tannins, which lead to a greater bypass protein content. The poor growth with cicer milkvetch was a consequence of its low palatability, causing low feed intake. The factors in milkvetch responsible for photosensitization and low palatability were not identified. **Sainfoin** is a perennial forage legume of Russian origin. It is deep-rooted with stout, erect stems arising from a crown. It is similar to alfalfa in culture and feeding value but is seldom grown in the United States except in Montana and adjacent states, where it is produced on irrigated land.

Kudzu (*Pueraria lobata*) is a rapidly growing vine that was introduced into the southern United States and now is considered a pest in most areas because of its aggressive growth. It is a nutritious forage (Corley et al., 1997), but it is difficult to manage for this purpose because of its viny growth pattern and intolerance of heavy grazing.

Perennial peanut (*Arachis glabrata*) is a fine-stemmed, leafy, warm-season legume adapted to Florida and other southern states. It is a suitable substitute for alfalfa in these regions; alfalfa does not thrive under wet subtropical conditions. Perennial peanut (cv. Florigraze) is similar to alfalfa in nutritive value for ruminants (Romero et al., 1987). Perennial peanut is established by planting rhizomes rather than seed. The rhizomes (underground stems) spread rapidly, establishing a thick sod.

Lathyrus spp. (perennial and annual sweet peas) are best known for their toxic properties. Lathyrus seeds are used as a human foodstuff in India; under some conditions, they may result in neural and skeletal abnormalities referred to as lathyrism (see Chapter 4). The toxic agents are a number of lathyrogenic amino acids. Some *Lathyrus* spp. are suitable for use as forage. The flatpea (*L. sylvestris*) has some potential as a forage for ruminants. Foster (1990) found that flatpea hay was an acceptable forage for sheep, although it has a somewhat lower nutrient digestibility than alfalfa. *Lathyrus* spp. are vigorous, viny plants adapted to poor soils and may have limited application in areas where other forages are not productive (Foster, 1990).

Tropical Legumes Tropical forage legumes are often quite different in appearance and growth habit from the temperate legumes, such as clover and alfalfa. Many are vines, shrubs, or trees. Tree legumes, if cut regularly, can be coppisced to produce lush young growth, which is harvested repeatedly. In the tropics, it is common to keep animals in confinement and bring the forage to them (the **cut and carry system**). This is practiced on small farms, where only a few goats, sheep, rabbits, or a cow may be raised. Forage, consisting of roadside grasses, weeds, surplus vegetables, and tree leaves, is collected daily.

Tree legumes are very useful in the tropics. They capture a large amount of solar energy, produce large yields of biomass harvestable in both the wet and dry seasons, improve soil fertility, and exploit moisture and mineral reserves deep in the soil profile. In addition to forage, they produce timber, fuel, and shade for crops when used in alley farming systems. One of the most promising tropical tree legumes is *Leucaena leucocephala*. It is a rapidly growing tree, producing leaflets with 25 percent or more crude protein. It can be harvested manually or, as it is grown in Australia, grazed by cattle to keep it in a shrubby form. It is especially useful as a supplement to tropical grasses during the dry season. **Leucaena** has sometimes been promoted as a "miracle tree" for the tropics. However, agricultural specialists are increasingly aware that there are no miracle crops. Much of the "miracle" status of leucaena was lost in the 1980s when a worldwide epidemic of psyllids (aphidlike insects) decimated many stands of leucaena.

Leucaena contains mimosine, a toxic amino acid. **Mimosine toxicity** (Fig. 5.9) causes alopecia (loss of hair), dermatitis, esophageal lesions, and poor growth. In ruminants, mimosine is metabolized in the rumen to a metabolite (dihydropyridone, DHP), which is goitrogenic, causing animals grazing leucaena to develop enlarged thyroid glands. A major advance in the utilization of leucaena has been the discovery that ruminants in some geographical areas have rumen microbes that degrade mimosine and DHP, and detoxify them. Jones et al. (1986) demonstrated that cattle and goats in Hawaii and Indonesia do not develop toxicity symptoms when fed leucaena, whereas cattle in Australia developed ill-thrift, dermatitis, and goiter. Administration of rumen contents from Hawaiian and Indonesian ruminants into Australian cattle eliminated their susceptibility to leucaena toxicity (Quirk et al., 1988). Hammond et al. (1989) introduced a mimosine-metabolizing rumen microbe into cattle in Florida and showed that it successfully colonized the animals and persisted over the winter when there was no leucaena in the diet. Leucaena is native to the southern United States and has

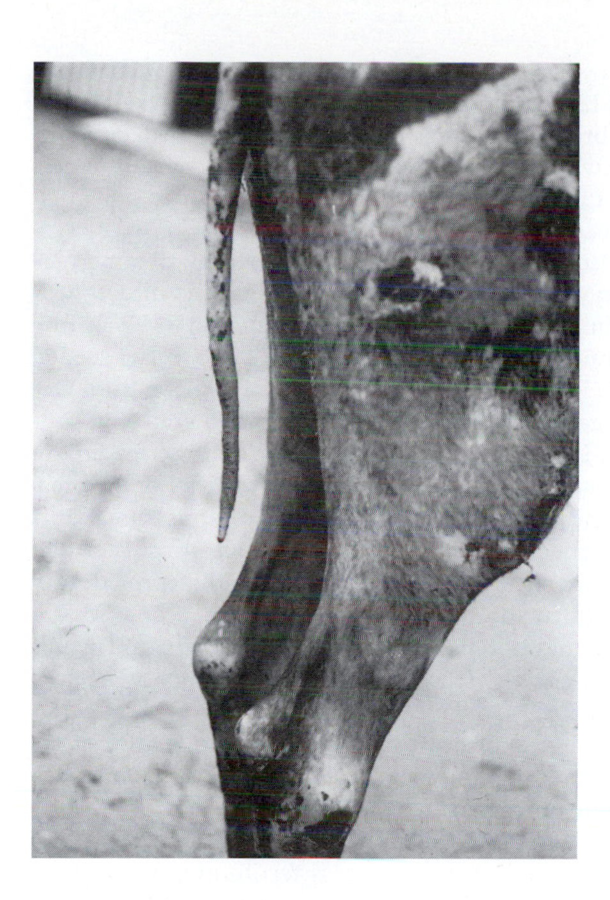

FIGURE 5.9 Mimosine toxicity in a bovine. The result is loss of hair and dermatitis. (Courtesy of R. J. Jones.)

potential as a browse species. Once the detoxifying microbes have been introduced into a few animals in a herd, they rapidly inoculate the rest of the cattle, via contamination of feed and water or by wind-spread spores (Quirk et al., 1988). This procedure of introducing **toxin-degrading rumen microbes** into animals in areas where the organisms are not indigenous offers great potential as a means of improving the utilization of toxin-containing forages. Leucaena toxicosis has been reviewed by Hammond (1995).

A number of other tropical tree legumes, such as *Gliricida sepium*, are useful tropical legume forages. A wide variety of nontree tropical legumes are also grown for forage. Some of the more important include centro (*Centrosema pubescens*), desmodium (*Desmodium* spp.), siratro (*Macroptilium atropurpureum*), and stylo (*Stylosanthes* spp.). These forages are quite comparable to temperate clovers and alfalfa in feeding value.

Other Forages

Most cultivated forages are grasses or legumes. There are a few other broad-leaved forage plants of limited importance. The *Brassica* spp. are probably the most significant. These include kale, rape, turnips, and Chinese radish (Tyfon). **Forage brassicas** are of importance in Great Britain and New Zealand as winter pasture; they are planted in the autumn and provide pasture, mainly for sheep, during the winter months. In North America, they are grown mainly as temporary pastures to fill in during limited periods when other forage is not available. For example, **kale** and **rape** may be sown in late summer for fall pasture. In the dryland areas of the U.S. Pacific Northwest, brassicas such as **turnips** are often seeded by airplane on wheat fields; after the wheat is harvested, the stubble is irrigated to germinate the turnips. These provide forage for cattle and stabilize sandy soils that are very susceptible to wind erosion.

Forage brassicas can be used to extend the pasture season by treating permanent pastures in midsummer with herbicide to suppress grass and direct seeding of the brassica crop into the sod (Jung et al., 1986; Reid et al., 1994). In a study in New Hampshire, **tyfon** (*Brassica campestris sensulato* x *B. rapa*) seeded into a paraquat-sprayed pasture in mid-July provided grazing for sheep from October to December, with lamb gains exceeding those of a hay and grain-fed control group (Koch et al., 1987). Lambert et al. (1987) found that the feeding of grass hay to lambs on a forage rape pasture improved performance. Forage brassicas have a low fiber content and the fiber has a fast rate of disappearance during *in vitro* fermentation. Lambert et al. (1987) suggested that an all-brassica diet may have insufficient fiber for normal rumen function, and that providing a source of roughage improves performance. Hart and Horn (1987) proposed that ensiling turnips with wheat straw may be a useful way of conserving them. Turnips can be seeded in warm-season grass pastures during cool seasons when the grasses are dormant or unproductive.

A hazard associated with the use of brassica pastures is that *Brassica* spp. contains a toxic amino acid (*S*-methyl cysteine sulfoxide: SMCO), commonly referred to as the **brassica anemia factor**. This amino acid causes hemolysis of the red blood cells, producing anemia, red urine (hemoglobinuria or "red water disease"), and poor animal performance. Excess fertilization with nitrogen and sulfur promotes high levels of SMCO in brassica plants. Other potential hazards with forage brassicas include pulmonary emphysema, bloat, polioencephalomalacia, and hypothyroidism. Pulmonary emphysema, caused by 3-methyl indole derived from rumen metabolism of tryptophan, is probably the most serious of the brassica forage disorders. The succulent, high-protein forage that is rapidly digested in the rumen provokes not only emphysema but also bloat. Both problems can be reduced by allowing a period of adaptation to a brassica diet, by limiting intake, and by providing hay or other dry forage as a supplement. This will also help to reduce PEM, which is also induced by a sudden high intake of readily fermentable carbohydrate. Hypothyroidism, caused by glucosinolates, may occur with long-term consumption of brassica forage, with reduced growth and goiter being the principal signs. Ensiling brassica forage reduces the concentrations of SMCO and glucosinolates, thus reducing potential toxicoses (Fales et al., 1987).

Buckwheat, previously discussed as a grain crop, is also grown to a limited extent for pasture. Its main virtue is that it has a very rapid growth rate, producing usable forage within a few weeks of seeding. Light-skinned animals may develop photosensitization when grazing buckwheat or buckwheat stubble.

Chicory (Cichorium *intybus*) is a perennial forb grown quite extensively as a forage crop in New Zealand. It supports better animal performance than ryegrass-white clover pastures (Kusmartono and Barry, 1997), because of a rapid degradation rate in the rumen. This allows faster digesta clearance from the rumen and, thus, a higher feed intake (McCoy et al., 1997). Chicory persists well in grass-legume pastures and increases the seasonal distribution of forage during periods when grass and legume growth is declining (Kunelius and McRae, 1999).

Comfrey (*Symphytum officinale*) is a broad-leaved herb sometimes promoted as a forage plant. It is propagated by root cuttings. Although it produces a high yield of green forage, it is very high in moisture, making it difficult to cure for hay. It also contains mucilaginous gums, which hinder chopping and drying of the crop. It is unsuitable for ensiling, because of its high moisture content, low concentration of fermentable carbohydrate and high buffering capacity (Wilkinson, 2003). Comfrey has been extensively promoted on the basis of alleged medicinal properties and is a popular item in "health food" stores, where it is marketed as a tea and a component of various creams, ointments, and pills. It is reputed to contain allantoin, a compound with very mild pharmacological properties. Allantoin is a normal excretory product in the urine of humans, being produced from metabolism of nucleic acids. Comfrey contains a number of toxic alkaloids (Culvenor et al., 1980) that are carcinogenic (Hirono et al., 1978). Therefore, it should be used with caution as an animal feed and as an herb.

Bamboo (*Phyllostachys* spp.) is, its appearance notwithstanding, a member of the grass family. There has been some interest in the use of bamboo foliage as a forage. Although it can have a high crude-protein content, bamboo is unpalatable to livestock and often not well accepted. The *in vitro* rumen digestibility of bamboo forage dry matter is low, about 40 percent (Greenway et al., 1999). Even in the giant panda, a vegetarian carnivore that feeds extensively on bamboo, the digestibility is very low, with dry-matter digestibility being less than 20 percent (Dierenfeld et al., 1982).

The **prickly pear cactus** (*Opuntia* spp.) is used as a forage plant in Mexico, Brazil, and the southwestern United States. Although there are spineless varieties, in most cases spined varieties are used; the spines are burned off with a propane torch before use as feed (Fig. 5.10). Prickly pear is especially useful under drought conditions when other forage may not be available. Consumption of the plant can be regulated by spine burning. The plants accumulate nutrients during favorable years and can be stored on the range until needed. Prickly pear is high in readily fermentable carbohydrate but low in crude protein; with adequate protein supplementation, it can support excellent performance of cattle (Shoop et al., 1977). Provision of **spineless cactus** as a supplement to sheep fed wheat straw in North Africa eliminated the need for drinking water and improved the intake of the low-quality roughage (Ben Salem et al., 1996). Russell and Felker (1987) reviewed the characteristics of prickly pear and suggested it should be given serious consideration as a crop in arid areas. *Opuntia* utilize water for dry matter production more efficiently than most other plants. Cacti and many other plants adapted to arid environments have a type of photosynthesis known as the crassulacean acid metabolism (CAM). Plants with the CAM pathway fix carbon dioxide at night, so their stomata are open at night and closed during the day. This reduces water loss, accounting for the efficient utilization of water for dry-matter production. Prickly pear shows dramatic yield responses to nitrogen and phosphorus fertilization (Nobel et al., 1987). Russell and Felker (1987) have proposed that a combination of prickly pear cacti with shrubby woody legumes (e.g., mesquite—*Prosopis* spp.—*Acacia* spp.) may be a productive system for arid environments, with the cacti providing energy and the legumes protein for grazing livestock. Such nontraditional approaches are difficult for

FIGURE 5.10 Prickly pear cactus can be used as forage for cattle. Burning the spines off with a propane torch (top) allows cattle to consume the plant (bottom). (Courtesy of C. E. Russell.)

many animal and range scientists to accept but may offer much potential in the race against desertification in Africa, South America, and Mexico.

Kochia (*Kochia scoparia*) is a drought-resistant plant widely distributed in the arid and semiarid southwest United States. It thrives on arid, infertile, high-sodium soils where few other plants will grow. There has been interest in commercial development of kochia as a forage plant. A number of toxicity syndromes have resulted from its use. These include signs of thiamin deficiency suggesting the presence of a thiaminase (Dickie and James, 1983), and photosensitization suggestive of liver damage (Dickie and Berryman, 1979). Identification of toxic principles should aid in plant-breeding programs to produce a nontoxic strain of kochia, which could be a valuable forage plant in many areas. Canadian researchers (Iwaasa et al., 1989) reported that kochia hay could be fed to cattle at up to 40 percent of the diet without adverse effects on dry-matter intake or nitrogen retention. However, at higher levels, performance and nitrogen retention of beef cattle are reduced (Cohen et al., 1989). Performance was not improved by feeding calcium carbonate to counteract the soluble oxalates in kochia. Kochia seed contains saponins and is toxic to poultry (Coxworth and Salmon, 1972). Treatment with sodium hydroxide partially overcomes the toxicity. **Halophytes** (plants tolerant to saline soils) often accumulate high sodium concentrations (Grimson et al., 1989), which may adversely affect their palatability. *Salicornia bigelovii* is a halophytic salt-marsh plant with oil-rich seeds. It can be irrigated with seawater and has potential as an oilseed crop in saline, desert areas (Glenn et al., 1991). The meal has some growth-depressing properties attributed to saponins. When chicks were fed *Salicornia* meal-containing diets, with supplementary cholesterol to bind saponins, growth was equal to that obtained with the control (soybean meal) diet (Attia et al., 1997).

Saltbush (*Atriplex* spp.) is a halophyte that is used in arid areas as a browse species for ruminants. It can be irrigated with seawater.

Pineapple postharvest forage, consisting of the leaves, stems (ratoons), stump, and roots, has been evaluated as a potential feedstuff for beef cattle (Kellems et al., 1979). The forage was reasonably high in energy (58 percent TDN) and low in protein (4 to 7 percent, with a digestibility of only 35 percent). When supplemented with protein and minerals, pineapple silage and green chop was a satisfactory feedstuff for cattle. Pineapple bran, the residue remaining after pineapple juice extraction, is also an acceptable feedstuff for ruminants. Banana by-products (fruit, peel, leaf, pseudostem) are widely used as feedstuffs in the tropics (see Chapter 3).

QUESTIONS AND STUDY GUIDE

1. What is lignification?
2. In what part of the plant tissue cells of cassava, white clover, and sudangrass are the cyanogenic glyocosides found? Why does a frost increase the toxicity of sudangrass?
3. Why is the amino acid composition of different forages quite similar?
4. What factors influence the value of forages as sources of calcium, phosphorus, magnesium, and trace elements?
5. Tropical grasses have a lower protein content than temperate grasses. Why? Explain why tropical grasses have a lower digestibility in ruminants than temperate grasses have.
6. What effects do environmental temperature and day length have on forage quality?
7. Is the material in the rumen a homogeneous mixture? Explain.
8. What is the role of fungi in rumen fermentation?

9. What effect does feeding rolled corn as a supplement have on the digestibility of a grass-hay diet by cattle? Why?
10. Heavy nitrogen fertilization of grass pastures, especially in the spring, can result in some metabolic problems in livestock. Discuss.
11. Name three forage grasses commonly grown in your area.
12. Many horse raisers have said that horses should be fed a ration of oats and timothy hay. What are the advantages of these feedstuffs in horse feeding? Can horses be successfully raised without using oats and timothy?
13. What are some of the nutritional problems associated with tropical grasses?
14. Name three forage legumes commonly grown in your area. How can you tell if a plant is a legume?
15. A report in the journal *Animal Science* (a British publication) describes research using lucerne-cocksfoot hay. What would this be called in the United States?
16. What causes bloat? How can it be prevented?
17. What two forage legumes can have a high tannin content?
18. What is the cut and carry system?
19. Comfrey is an herb that contains carcinogenic alkaloids. Visit a "health food" store and determine if any comfrey-containing products are for sale. If so, what is their intended purpose? Do you think that the sale of products containing carcinogens is justified, or should it be banned by regulatory agencies?
20. It has been observed that Brahman-type cattle are less susceptible to summer fescue toxicosis than are European breeds. What might explain this difference?
21. Alfalfa leaves normally consist of three leaflets. New varieties of multileaflet alfalfa have been produced, with up to nine leaflets per leaf. What nutritional benefits, if any, would multileaflet alfalfa have over normal trifoliate alfalfa?

REFERENCES

General

Barnes, R. F., D. A. Miller, and C. J. Nelson, eds. 1995. *Forages*. Vol. I. *An Introduction to Grassland Agriculture*. Vol. II. *The Science of Grassland Agriculture*. Ames: Iowa State University Press.

Cheeke, P. R. 1998. *Natural Toxicants in Feeds, Forages, and Poisonous Plants*. Upper Saddle River, NJ: Prentice Hall, Inc.

Church, D. C., ed. 1988. *The Ruminant Animal. Digestive Physiology and Nutrition*. Englewood Cliffs, NJ: Prentice Hall.

Gilchrist, F. M. C. and R. I. Mackie, eds. 1984. *Herbivore Nutrition in the Subtropics and Tropics*. Craighall, South Africa: The Science Press.

Hacker, J. B. and J. H. Ternouth, eds. 1987. *The Nutrition of Herbivores*. Marrickville, Australia: Academic Press.

Van Soest, P. J. 1994. *Nutritional Ecology of the Ruminant*. Ithaca, NY: Cornell University Press.

Forage Composition and Quality

Akin, D. E. 1982. Microbial breakdown of feed in the digestive tract. In *Nutritional Limits to Animal Production from Pastures*, J. B. Hacker, ed. pp. 201–223. Farnham Royal, England: Commonwealth Agricultural Bureaux.

______ . 1987. Association of rumen fungi with various forage grasses. *Anim. Feed Sci. Tech.* 16:273–285.

______ . 1989. Histological and physical factors affecting digestibility of forages. *Agron. J.* 81:17–25.

Anderson, B., J. K. Ward, K. P. Vogel, M. G. Ward, H. J. Gorz, and F. A. Haskins. 1988. Forage quality and performance of yearlings grazing switchgrass strains selected for differing digestibility. *J. Anim. Sci.* 66:2239–2244.

Bradford, E. 1989. Animal agriculture and development: Challenges and opportunities. *Can. J. Anim. Sci.* 69:847–856.

Cerling, T. E., J. R. Ehleringer, and J. M. Harris. 1997. Carbon dioxide starvation, the development of C_4 ecosystems, and mammalian evolution. *Phil. Trans. Roy. Soc. Lond.* B353:159–171.

Collins, M. and M. D. Casler. 1990. Forage quality of five cool-season grasses. I. Cultivar effects. II. Species effects. *Anim. Feed Sci. Tech.* 27:197–207; 209–218.

Cone, J. W., A. H. van Gelder, and F. Driehuis. 1997. Description of gas production profiles with a three-phasic model. *Anim. Feed Sci. Tech.* 66:31–45.

Ehleringer, J. R., T. E. Cerling, and B. R. Helliker. 1997. C_4 photosynthesis, atmospheric CO_2, and climate. *Oecologia* 112:285–299.

Goodwin, T. W. and E. I. Mercer. 1983. *Introduction to Plant Biochemistry*. London: Pergamon Press.

Grenet, E. and P. Barry. 1988. Colonization of thick-walled plant tissues by anaerobic fungi. *Anim. Feed Sci. Tech.* 19:25–31.

Gulati, S. K., J. R. Ashes, G. L. R. Gordon, P. J. Connell, and P. L. Rogers. 1989. Nutritional availability of amino acids from the rumen anaerobic fungus *Neocallimastix* sp. LMI in sheep. *J. Agr. Sci.* 13:383–387.

Hall, J. W., W. Majak, D. G. Stout, K.-J. Cheng, B. P. Goplen, and R. E. Howarth. 1994. Bloat in cattle fed alfalfa selected for a low initial rate of digestion. *Can. J. Anim. Sci.* 74:451–456.

Judkins, M. B., L. J. Krysl, and R. K. Barton. 1990. Estimating diet digestibility: A comparison of eleven techniques across six different diets fed to rams. *J. Anim. Sci.* 68:1405–1415.

MacFadden, B. J. and T. E. Cerling. 1994. Fossil horses, carbon isotopes and global change. *Trends in Ecol.* 9:481–486.

MacRae, J. C., J. S. Smith, P. J. S. Dewey, A. C. Brewer, D. S. Brown, and A. Walker. 1985. The efficiency of utilization of metabolizable energy and apparent absorption of amino acids in sheep given spring- and autumn-harvested dried grass. *Brit. J. Nutr.* 54:197–209.

Mahadevan, P. 1982. Pastures and animal production. In *Nutritional Limits to Animal Production from Pastures,* J. B. Hacker, ed. pp. 1–17. Farnham Royal, England: Commonwealth Agricultural Bureaux.

Mannetje, L. 1984. Nutritive value of tropical and subtropical pastures, with special reference to protein and energy deficiency in relation to animal production. In *Herbivore Nutrition in the Subtropics and Tropics,* F. M. C. Gilchrist and R. I. Mackie, eds. pp. 51–66. Craighall, South Africa: The Science Press.

Mould, F. L., E. R. Orskov, and S. O. Mann. 1983a. Associative effects of mixed feeds. I. Effects of type and level of supplementation and the influence of the rumen fluid pH on cellulolysis *in vivo* and dry matter digestion of various roughages. *Anim. Feed Sci. Tech.* 10:15–30.

Mould, F. L., E. R. Orskov, and S. A. Gauld. 1983b. Associative effects of mixed feeds. II. The effect of dietary addition of bicarbonate salts on the voluntary intake and digestibility of diets containing various proportions of hay and barley. *Anim. Feed Sci. Tech.* 10:31–47.

Russell, J. B. and D. B. Wilson. 1996. Why are ruminal cellulolytic bacteria unable to digest cellulose at low pH? *J. Dairy Sci.* 79:1503–1509.

Schofield, P., R. E. Pitt, and A. N. Pell. 1994. Kinetics of fiber digestion from *in vitro* gas production. *J. Anim. Sci.* 72:2980–2991.

Stern, M. D., A. Bach, and S. Calsamiglia. 1997. Alternative techniques for measuring nutrient digestion in ruminants. *J. Anim. Sci.* 75:2256–2276.

Theodorou, M. K., B. A. Williams, M. S. Dhanoa, A. B. McAllan, and J. France. 1994. A simple gas production method using a pressure transducer to determine the fermentation kinetics of ruminant feeds. *Anim. Feed Sci. Tech.* 48:185–197.

Van der Linden, Y., N. O. van Gylswk, and H. Schwarz. 1984. Influence of supplementation of corn stover with corn grain on the fibrolytic bacteria in the rumen of sheep and their relation to the intake and digestion of fiber. *J. Anim. Sci.* 59:772–783.

Van Soest, P. J. 1994. *Nutritional Ecology of the Ruminant*. Ithaca, NY: Cornell University Press.

Weimer, P. J. 1996. Why don't ruminal bacteria digest cellulose faster? *J. Dairy Sci.* 79:1496–1502.

Welch, J. G. 1982. Rumination, particle size and passage from the rumen. *J. Anim. Sci.* 54:885–894.

Cool-Season Grasses and Cereal Grains

Bacon, C. W. 1995. Toxic endophyte-infected tall fescue and range grasses: Historic perspectives. *J. Anim. Sci.* 73:861–870.

Bohman, V. R., F. P. Horn, B. A. Stewart, A. C. Mathers, and D. L. Grunes. 1983a. Wheat pasture poisoning. I. An evaluation of cereal pastures as related to tetany in beef cows. *J. Anim. Sci.* 57:1352–1363.

Bohman, V. R., F. P. Horn, E. T. Littledike, J. G. Hurst, and D. Griffin. 1983b. Wheat pasture poisoning. II. Tissue composition of cattle grazing cereal forages and related to tetany. *J. Anim. Sci.* 57:1364–1373.

Branine, M. E. and M. L. Galyean. 1990. Influence of grain and monensin supplementation on ruminal fermentation, intake, digesta kinetics and incidence and severity of frothy bloat in steers grazing winter wheat pasture. *J. Anim. Sci.* 68:1139–1150.

Coombe, J. B. and J. G. Mulholland. 1988. Food intake and levels of rumen metabolites in cattle grazing wheat or oat stubble. *Aust. J. Agr. Res.* 39:629–638.

Cross, D. L., L. M. Redmond, and J. R. Strickland. 1995. Equine fescue toxicosis: Signs and solutions. *J. Anim. Sci.* 73:899–908.

Dalley, D. E., P. Isherwood, A. R. Sykes, and A. B. Robson. 1997a. Effect of intraruminal infusion of potassium on the site of magnesium absorption within the digestive tract of sheep. *J. Agric. Sci.* 129:99–105.

_______ . 1997b. Effect of *in vitro* manipulation of pH on magnesium solubility in ruminal and caecal digesta in sheep. *J. Agric. Sci.* 129:107–111.

Duynisveld, G. W. and K. M. Wittenberg. 1993. Evaluation of Rival, Venture and Frontier reed canarygrass as pasture forage. *Can. J. Anim. Sci.* 73:89–100.

East, N. E. and R. J. Higgins. 1988. Canarygrass (*Phalaris* sp.) toxicosis in sheep in California. *J. Am. Vet. Med. Assoc.* 192:667–669.

Fisher, L. J., N. Dinn, R. M. Tait, and J. A. Shelford. 1994. Effect of level of dietary potassium on the absorption and excretion of calcium and magnesium by lactating cows. *Can. J. Anim. Sci.* 74:503–509.

Gallavan, R. H., Jr., W. A. Phillips, and D. L. Van Tungeln. 1989. Forage intake and performance of yearling lambs fed harvested wheat forage. *Nutr. Rep. Int.* 39:643–648.

Hemken, R. W., J. A. Jackson, Jr., and J. A. Boling. 1984. Toxic factors in tall fescue. *J. Anim. Sci.* 48:1011–1016.

Mader, T. L., G. W. Horn, W. A. Phillips, and R. W. McNew. 1983. Low quality roughages for steers grazing wheat pasture. I. Effect on weight gains and bloat. *J. Anim. Sci.* 46:1021–1028.

Mason, W. N. and P. M. Flipot. 1988. Evaluation of timothy cultivars for voluntary intake and nutrient components. *Can. J. Anim. Sci.* 68:1121–1129.

Mulholland, J. G., J. B. Coombe, and W. R. McManus. 1977. Diet selection and intake by sheep and cattle grazing together on stubbles of wheat, oats or barley. *Aust. J. Exp. Agric. Anim. Husb.* 17:224–229.

Narasimhalu, P., K. B. McRae, and H. T. Kunelius. 1995. Hay composition, and intake and digestibility in sheep of newly introduced cultivars of timothy, tall fescue, and reed canarygrass. *Anim. Feed Sci. Tech.* 55:77–85.

Nicholson, S. S., B. M. Olcott, E. A. Usenik, H. W. Casey, C. C. Brown, L. E. Urbatsch, S. E. Turnquist, and S. C. Moore. 1989. Delayed phalaris grass toxicosis in sheep and cattle. *J. Am. Vet. Med. Assoc.* 195:345–346.

Phillips, W. A. 1986. Adaptation of stocker calves to non-protein nitrogen diets as a result of grazing wheat forage. *J. Anim. Sci.* 62:464–472.

Phillips, W. A., G. W. Horn, and M. E. Smith. 1995. Effect of protein supplementation on forage intake and nitrogen balance of lambs fed freshly harvested wheat forage. *J. Anim. Sci.* 73:2687–2693.

Porter, J. K., J. A. Stuedemann, F. N. Thompson, Jr., and L. B. Lipham. 1990. Neuroendocrine measurements in steers grazed on endophyte-infected tall fescue. *J. Anim. Sci.* 68:3285–3292.

Redmon, L. A., G. W. Horn, E. G. Krenzer, Jr., and D. J. Bernardo. 1995. A review of livestock grazing and wheat grain yield: Boom or bust? *Agron. J.* 87:137–147.

Reid, R. L., B. S. Baker, and L. C. Vona. 1984. Effects of magnesium sulfate supplementation and fertilization on quality and mineral utilization of timothy hays by sheep. *J. Anim. Sci.* 59:1403–1410.

Rode, L. M. 1986. Inhibitory effect of meadow foxtail (*Alopecurus pratensis*) on the growth of steers. *Can. J. Anim. Sci.* 66:303–305.

Rode, L. M. and W. L. Pringle. 1986. Growth, digestibility and voluntary intake by yearling steers grazing timothy (*Phleum pratense*) or meadow foxtail (*Alopecurus pratensis*) pastures. *Can. J. Anim. Sci.* 66:463–472.

Stewart, B. A., D. L. Grunes, A. C. Mathers, and F. P. Horn. 1981. Chemical composition of winter wheat forage grown where grass tetany and bloat occur. *Agron. J.* 73:337–347.

Stoszek, M. J., J. E. Oldfield, G. E. Carter, and P. H. Weswig. 1979. Effect of tall fescue and quackgrass on copper metabolism and weight gains of beef cattle. *J. Anim. Sci.* 48:893–899.

Tosi, H. R. and K. M. Wittenberg. 1993. Harvest alternatives to reduce the alkaloid content of reed canarygrass forage. *Can. J. Anim. Sci.* 73:373–380.

Winter, K. A. and U. C. Gupta. 1983. The mineral content of timothy grown in Prince Edward Island. *Can. J. Anim. Sci.* 63:133–139.

Wittenberg, K. M., G. W. Duynisveld, and H. R. Tosi. 1992. Comparison of alkaloid content and nutritive value for tryptamine- and β-carboline-free cultivars of reed canarygrass (*Phalaris arundinacea* L.). *Can. J. Anim. Sci.* 72:903–909.

Warm-Season and Tropical Grasses

Akin, D. E. 1979. Microscopic evaluation of forage digestion by rumen microorganisms—A review. *J. Anim. Sci.* 48:701–710.

______ . 1982. Microbial breakdown of feed in the digestive tract. In *Nutritional Limits to Animal Production from Pastures,* J. B. Hacker, ed. pp. 201–223. Farnham Royal, England: Commonwealth Agricultural Bureaux.

Humphreys, L. R. 1987. *Tropical Pastures and Fodder Crops.* Burnt Mill, England: Longman Scientific and Technical.

Pate, F. M. 1981. Fresh chopped sugarcane in growing-finishing steer diets. *J. Anim. Sci.* 53:881–888.

Pate, F. M., P. M. Fairhurst, and J. T. K. Munthali. 1985. Urea levels and supplemental energy sources in sugarcane diets. *J. Anim. Sci.* 61:252–259.

Preston, T. R. 1982. Nutritional limitations associated with the feeding of tropical forages. *J. Anim. Sci.* 54:877–884.

Preston, T. R. and R. A. Leng. 1987. *Matching Ruminant Production Systems with Available Resources in the Tropics and Subtropics.* Armidale, Australia: Penambul Books.

Temperate Legumes

Bagley, C. P. and J. I. Feazel. 1989. Influence of a monensin ruminal bolus on the performance and bloat prevention of grazing steers. *Nutr. Rep. Int.* 40:707–716.

Barry, T. N. 1985. The role of condensed tannins in the nutritional value of *Lotus pedunculatus* for sheep. 3. Rates of body and wool growth. *Brit. J. Nutr.* 54:211–217.

Butter, N. L., J. M. Dawson, D. Wakelin, and P. J. Buttery. 2000. Effect of dietary tannin and protein concentration on nematode infection (*Tricostronglylus colubriformis*). *J. Agric. Sci.* 134:89–99.

Chiquette, J., K.-J. Cheng, J. W. Costerton, and L. P. Milligan. 1988. Effect of tannins on the digestibility of two isosynthetic strains of birdsfoot trefoil (*Lotus corniculatus L.*) using *in vitro* and *in sacco* techniques. *Can. J. Anim. Sci.* 68:751–760.

Colvin, H. W. and R. C. Backus. 1988. Bloat in sheep (*Ovis aries*). *Comp. Biochem. Physiol.* 91:635–644.

Corley, R. N., A. Woldeghebriel, and M. R. Murphy. 1997. Evaluation of the nutritive value of kudzu (*Puerasia lobata*) as a feed for ruminants. *Anim. Feed Sci. Tech.* 68:183–188.

Foster, J. G. 1990. Flatpea (*Lathyrus sylvestris* L.): A new forage species? A comprehensive review. *Adv. in Agron.* 43:241–313.

Glenn, B. P., G. A. Varga, G. B. Huntington, and D. R. Waldo. 1989. Duodenal nutrient flow and digestibility in Holstein steers fed formaldehyde- and formic acid-treated alfalfa or orchardgrass silage at two intakes. *J. Anim. Sci.* 67:513–528.

Howarth, R. E., K.-J. Cheng, W. Majak, and J. W. Costerton. 1986. Ruminant bloat. In *Control of Digestion and Metabolism in Ruminants,* L. P. Milligan, W. L. Grovum, and A. Dobson, eds. pp. 516–527. Englewood Cliffs, NJ: Prentice Hall.

Kudo, H., K.-J. Cheng, M. R. Hanna, R. E. Howarth, B. P. Goplen, and J. W. Costerton. 1985. Ruminal digestion of alfalfa strains selected for slow and fast initial rates of digestion. *Can. J. Anim. Sci.* 65:157–161.

Majak, W. and J. W. Hall. 1990. Sodium and potassium concentrations in ruminal contents after feeding bloat-inducing alfalfa to cattle. *Can. J. Anim. Sci.* 70:235–241.

Marten, G. C., F. R. Ehle, and E. A. Ristau. 1987. Performance and photosensitization of cattle related to forage quality of four legumes. *Crop Sci.* 27:138–145.

Nation, P. N. 1989. Alsike clover poisoning: A review. *Can. Vet. J.* 30:410–415.

Romero, F., H. H. Van Horn, G. M. Prine, and E. C. French. 1987. Effect of cutting interval upon yield, composition and digestibility of Florida 77 alfalfa and Florigraze rhizoma peanut. *J. Anim. Sci.* 65:786–796.

Tabe, L. M., T. Wardley-Richardson, A. Ceriotti, A. Aryan, W. McNabb, A. Moore, and T. J. V. Higgins. 1995. A biotechnological approach to improving the nutritive value of alfalfa. *J. Anim. Sci.* 73:2752–2759.

Vickery, P. J., J. L. Wheeler, and C. Mulcahy. 1987. Factors affecting the hydrogen cyanide potential of white clover (*Trifolium repens* L.). *Aust. J. Agr. Res.* 38:1053–1059.

Ward, G., L. H. Harbers, and J. J. Blaha. 1979. Calcium-containing crystals in alfalfa: Their fate in cattle. *J. Dairy Sci.* 62:715–722.

Ward, G., L. Harbers, A. Kahrs, and A. Dayton. 1984. Availability of calcium from alfalfa for chicks. *Poult. Sci.* 63:82–88.

Tropical Legumes

Jones, R. J. and R. G. Megarrity. 1986. Successful transfer of DHP-degrading bacteria from Hawaiian goats to Australian ruminants to overcome the toxicity of leucaena. *Aust. Vet. J.* 63:259–262.

Hammond, A. C. 1995. Leucaena toxicosis and its control in ruminants. *J. Anim. Sci.* 73:1487–1492.

Hammond, A. C., M. J. Allison, M. J. Williams, G. M. Prine, and D. B. Bates. 1989. Prevention of leucaena toxicosis of cattle in Florida by ruminal inoculation with 3-hydroxy-4-(IH)-pyridone-degrading bacteria. *Am. J. Vet. Res.* 50:2176–2180.

National Academy of Sciences. 1984. *Leucaena: Promising Forage and Tree Crop for the Tropics.* Washington, DC: National Academy Press.

Quirk, M. F., J. J. Bushell, R. J. Jones, R. G. Megarrity, and K. L. Butler. 1988. Live-weight gains on leucaena and native grass after dosing cattle with rumen bacteria capable of degrading DHP, a ruminal metabolite from leucaena. *J. Agri. Sci.* 111:165–170.

Other Forages

Attia, F. M., A. A. Alsobayel, M. S. Kriadees, M. Y. Al-Saiady, and M. S. Bayoumi. 1997. Nutrient composition and feeding value of *Salicornia bigelovii* Torr meal in broiler diets. *Anim. Feed Sci. Tech.* 65:257–263.

Ben Salem, H., A. Nefzaoui, H. Abdouli, and E. R. Orskov. 1996. Effect of increasing level of spineless cactus (*Opuntia ficus indicus* var. *inermis*) on intake and digestion by sheep given straw-based diets. *Anim. Sci.* 62:293–299.

Cohen, R. D. H., A. D. Iwaasa, M. E. Mann, E. Coxworth, and J. A. Kernan. 1989. Studies on the feeding value of *Kochia scoparia* (L) Schrad. hay for beef cattle. *Can. J. Anim. Sci.* 69:735–743.

Coxworth, E. C. M. and R. E. Salmon. 1972. Kochia seed as a component of the diet of turkey poults: Effects of different methods of saponin removal or inactivation. *Can. J. Anim. Sci.* 52:721–729.

Culvenor, C. C. J., M. Clarke, J. A. Edgar, J. L. Frahn, M. V. Jago, J. E. Peterson, and L. W. Smith. 1980. Structure and toxicity of the alkaloids of Russian comfrey (*Symphytum* × *uplandicum Nyman*), a medicinal herb and item of the human diet. *Experentia* 36:377–379.

Dickie, C. W. and J. R. Berryman. 1979. Polioencephalomalacia and photosensitization associated with *Kochia scoparia* consumption in range cattle. *J. Am. Vet. Med. Assoc.* 175:463–465.

Dickie, C. W. and L. F. James. 1983. *Kochia scoparia* poisoning in cattle. *J. Am. Vet. Med. Assoc.* 183:765–768.

Dierenfeld, E. S., H. F. Hintz, J. B. Robertson, P. J. Van Soest, and O. T. Oftedal. 1982. Utilization of bamboo by the giant panda. *J. Nutr.* 112:636–641.

Fales, S. L., D. L. Gustine, S. C. Bosworth, and R. J. Hoover. 1987. Concentrations of glucosinolates and *S*-methylcysteine sulfoxide in ensiled rape (*Brassica napus* L.). *J. Dairy Sci.* 70:2402–2405.

Glenn, E. P., W. O. Fames, M. C. Watson, T. L. Thompson, and R. O. Kuehl. 1991. *Salicornia bigelovii* Torr: An oilseed halophyte for seawater irrigation. *Science* 251:1065–1067.

Greenway, S. L., M. Keller, G. Nelson, P. R. Cheeke, and D. J. Carroll. 1999. Evaluation of temperate bamboo forage as an animal feedstuff: Seasonal changes in composition and digestibility in ponies. *Proc. West. Sec. Am. Soc. Anim. Sci.* 50:193–195.

Grimson, R. E., G. E. Riemer, R. P. Stilborn, R. J. Volek, and P. K. Gummeson. 1989. Agronomic and chemical characteristics of *Kochia scoparia* (L.) *Schrad.* and its value as a silage crop for growing beef cattle. *Can. J. Anim. Sci.* 69:383–391.

Hart, S. P. and F. P. Horn. 1987. Ensiling characteristics and digestibility of combinations of turnips and wheat straw. *J. Anim. Sci.* 64:1790–1800.

Hirono, I., H. Mori, and M. Haga. 1978. Carcinogenic activity of *Symphytum officinale*. *J. Nat. Cancer Inst.* 61:865–868.

Iwaasa, A. D., M. E. Manns, R. Cohen, E. Coxworth, and J. Kernan. 1989. Studies on the voluntary intake and digestibility of *Kochia scoparia* fed at various levels. *Can. J. Anim. Sci.* 69:277–278 (Abst.).

Jung, G. A., R. A. Byers, M. T. Panciera, and J. A. Shaffer. 1986. Forage dry matter accumulation and quality of turnip, swede, rape, Chinese cabbage hybrids, and kale in the eastern USA. *Agron. J.* 78:245–253.

Kellems, R. O., O. Wyman, A. H. Nguyen, J. C. Nolan, C. M. Campbell, J. R. Carpenter, and E. B. Ho-a. 1979. Post-harvest pineapple plant forage as a potential feedstuff for beef cattle: Evaluated by laboratory analyses, *in vitro* and *in vivo* digestibility and feedlot trials. *J. Anim. Sci.* 48:1040–1048.

Knipfel, J. E., J. A. Kernan, E. C. Coxworth, and R. D. H. Cohen. 1989. The effect of stage of maturity on the nutritive value of kochia. *Can. J. Anim. Sci.* 69:1111–1114.

Koch, D. W., F. C. Ernst, N. R. Leonard, R. R. Hedberg, T. J. Blenk, and J. R. Mitchell. 1987. Lamb performance on extended-season grazing on Tyfon. *J. Anim. Sci.* 64:1275–1279.

Kunelius, H. T. and K. B. McRae. 1999. Forage chicory persists in combination with cool season grasses and legumes. *Can J. Plant Sci.* 79:197–200.

Kusmartono, A. S. and T. N. Barry. 1997. Rumen digestion and rumen outflow rate in deer fed fresh chicory (*Cichorium intybus*) or perennial ryegrass (*Lolium perenne*). *J. Agric. Sci.* 128:87–94.

Lambert, M. G., S. M. Abrams, H. W. Harpster, and G. A. Jung. 1987. Effect of hay substitution on intake and digestibility of forage rape (*Brassica napus*) fed to lambs. *J. Anim. Sci.* 65:1639–1646.

MacDearmid, A., G. M. Innes, P. E. V. Williams, and M. Kay. 1982. Kale for beef production. *Anim. Prod.* 34:191–196.

McCoy, J. E., M. Collins, and C. T. Dougherty. 1997. Amount and quality of chicory herbage ingested by grazing cattle. *Crop Sci.* 37:239–242.

Nobel, P. S., C. E. Russell, P. Felker, J. G. Medina, and E. A. Cuna. 1987. Nutrient relations and productivity of prickly pear cacti. *Agron. J.* 79:550–555.

Reid, R. L., J. R. Puoli, G. A. Jung, J. M. Cox-Ganser, and A. McCoy. 1994. Evaluation of brassicas in grazing systems for sheep. I. Quality of forage and animal performance. *J. Anim. Sci.* 72:1823–1831.

Russell, C. E. and P. Felker. 1987. The prickly-pears (*Opuntia* spp., Cactaceae): A source of human and animal food in semiarid regions. *Econ. Bot.* 41:433–445.

Shoop, M. C., E. J. Alford, and H. F. Mayland. 1977. Plains prickly pear is a good forage for cattle. *J. Range Manage.* 30:12–17.

Wilkinson, J. M. 2003. A laboratory evaluation of comfrey (*Symphytum officinale* L.) as a forage crop for ensilage. *Anim. Feed. Sci. Tech.* 104:227–233.

CHAPTER 6

Forage Harvesting and Utilization

Objectives

1. To discuss forage harvesting systems:
   - grazing management
   - agroforestry
   - hay and silage
   - low-quality crop residues
   - unconventional feeds: wood products, aquatic weeds
2. To highlight areas of current interest:
   - intensive grazing management systems
   - agroforestry in temperate and tropical regions
   - hay dessicants and preservatives
   - chemical treatment (ammoniation) of low-quality roughages
   - the hazards of ammoniated hay toxicosis
   - use of wood by-products, fast-growing trees, and aquatic weeds in livestock production

Forages are harvested either directly by grazing animals or mechanically. Direct harvest by animals is the simplest and requires the lowest capital investment. However, it is not necessarily the most efficient, either in resource utilization or in economic efficiency.

GRAZING SYSTEMS

Grazing consists of allowing livestock to harvest forage. **Extensive grazing systems,** such as those on the rangelands of North America, Australia, and Latin America, may involve little in animal or plant management other than periodically rounding up animals for market. This is particularly true in Australia and some South American countries, such as Brazil. In North America, management of livestock and plant resources on rangelands may be relatively sophisticated, involving rotational grazing, fertilization, and range reseeding. Typically, extensive grazing occurs on arid and semiarid rangelands.

Intensive grazing involves utilization of productive pastures with a fairly high density of animals per unit of land area. Pastures are fenced, and management procedures such as fertilization and clipping are often utilized. Intensive pastures are classified either as permanent or temporary. **Permanent pastures** are composed of forage grass and legume species established by seeding them or native pastures consisting of wild or indigenous species or domesticated species that are sufficiently aggressive to invade without human assistance. Species in permanent pastures are perennial and/or self-reseeding annuals, such as subterranean clover. **Temporary pastures** consist of high-yielding annual species, such as sudangrass, rape, kale, and turnips, planted to provide pasture for short periods of time when other pastures are unproductive. These periods vary, depending upon geographical area, but include examples such as sudangrass for midsummer pasture when cool-season grasses are dormant, rape for late summer and fall pasture in temperate areas, and kale for winter grazing in Great Britain and New Zealand.

Grazing Management

Efficient forage utilization requires good animal and pasture management. Maximum production per animal and per unit of land cannot be obtained concurrently; the most efficient method of production usually involves a compromise between these two ideals. A major challenge in grazing management is that the seasonal production of forage does not coincide with animal requirements. In both tropical and temperate regions, there are seasonal peaks of forage production and seasons when little growth occurs. In temperate areas, there is a spring flush of forage, during which excess quantities of high-quality material are produced, followed by summer growth that is inadequate in quantity and quality, and little or no forage production during winter. The aim of grazing management is to maximize the use of forage while maintaining the botanical integrity of the pasture so it will be productive in future years.

The quality of herbage declines with maturity; effective management, therefore, involves striving for maximum yield of digestible nutrients rather than maximum dry matter. The limiting nutritional component of pastures is usually energy. Vigorous well-managed pastures usually produce adequate protein but inadequate energy for maximum animal performance. It is impossible to achieve maximum growth rates or lactation with pasture alone. Supplementation with an energy source, such as corn, will increase productivity but is not always economical and decreases efficiency of utilization of the forage.

Systems of grazing management include continuous, rotational, and strip grazing. **Continuous grazing** involves stocking a pasture with animals continually. It is the least costly system in terms of capital investment for fencing and stock watering and is the easiest to perform. Disadvantages of continuous grazing are that some parts of the pasture are grazed excessively and other parts are underutilized, resulting in patches of overgrazed pasture and patches of mature, unpalatable, poorly digestible forage. In each case, adverse effects on botanical composition may follow. Many grasses and legumes will not tolerate continual defoliation and will die out in overgrazed pastures. In the underutilized areas, valuable low-growing species, such as white clover, may be crowded out. Thus the long-term effect of continuous grazing may be to eliminate the most nutritious, palatable, and productive pasture species. Species with a prostrate or semiprostrate growth pattern, such as white clover, Kentucky bluegrass, and bentgrasses, are tolerant of continuous grazing. With some species, such as tall fescue, excess spring growth may be **"stockpiled"** by allowing it to accumulate; it will be consumed later in the season, even though its nutritional value will be reduced. Other species, such as orchardgrass, cannot be handled in this way, because the mature forage is extremely unpalatable and indigestible.

Although tall fescue that is "stockpiled" will be consumed, its nutritive value is low, with a low content of total sugars and a high cellulose content. An interesting approach to improving the quality of fescue is to spray it with a plant growth regulator to retard seedhead production and prolong the period of vegetative growth. Such a chemical is mefluidide, which inhibits reproduction and enhances total sugar and crude-protein content of the forage (Glenn et al., 1980; Robb et al., 1982).

Rotational grazing avoids many of the disadvantages of continuous grazing. In rotational grazing a pasture is divided into a number of smaller paddocks, and high stocking density rapidly grazes the forage in the paddock off to a uniform height. The animals are then moved on to another paddock, allowing the grazed pasture a period for regrowth. Selective grazing does not occur, so all species and all plant parts are harvested. A disadvantage of rotational grazing is that, unless the management is good, animal performance may suffer. Over a period of several days, the quantity and quality of the forage will change. When animals are first turned into a paddock, they will have maximum intake of high-quality forage. As the leafy material is eaten, the remaining forage will be of lower digestibility, and a lower quantity will be available. Thus, unless animals are carefully managed, there may be a "boom and bust" cycle. During periods of very rapid spring growth, it is often best to allow access to the entire pasture to get maximum animal performance, and then use rotational grazing later in the spring.

The most intensive rotational grazing systems involve dividing a pasture into many paddocks (grazing cells) so that animals are moved daily. This may give the highest yield of animal products per unit of land area but requires extra capital for fencing and watering. High-voltage "New Zealand" electric fencing is useful in permitting paddock construction inexpensively. Often it is feasible to use a wheel system, where water and salt are provided in the center, and the paddocks are formed on a circular basis with electric fences as the wheel spokes.

The **"Savory system"** has been developed by Allan Savory from southern Africa (Savory, 1988). This grazing management system is based on the grazing behavior of wild African herbivores (and the North American bison prior to extirpation). In the wild, large concentrations of animals intensively graze a small area for a short time and then move on. The plants are defoliated, both by grazing and trampling action. The old dead plant material is removed, the top layer of soil is stirred up by hoof action, and the grasses uniformly regenerate. They are not grazed again until they have recovered from defoliation. Overgrazing involves repeated defoliation without sufficient time for plant recovery. The grazing action of migrating wild herbivores results in efficient use of grassland resources, regeneration, and revigorization of the range, with an ample recovery time. In the Savory system, the range or pasture area is divided into grazing cells that are grazed intensively for a short period and then permitted adequate recovery time. It is basically an intensive rotational grazing system. A disadvantage is a higher incidence of disease (e.g., pneumonia) in animals that are continuously kept together in a high-density situation.

A modified form of rotational grazing is **strip grazing.** Animals are given a new strip of pasture each day or several times a day by moving an electric fence across the pasture and back-fencing them out of the previously grazed portion. This system has been used extensively with dairy cattle in New Zealand. It is labor intensive but gives maximum forage utilization. Bloat can be controlled by spraying the daily strip of forage with an antifoaming agent.

Mixed-Species Grazing

Herbivores differ in their grazing and browsing habits, so different animal species tend to select different plants and plant parts. Horses and cattle are grazers, preferring grasses, whereas sheep and goats are browsers, preferring more tender and succulent plant parts than do cattle. Thus in a pasture, different animal species may not be directly competitive. **Mixed grazing** of sheep and cattle, particularly if browse plants or shrubs are present, may produce higher yields of animal products per unit of land area than single-species grazing (Nolan and Connolly, 1977, 1989). Furthermore, mixed grazing of sheep and cattle has beneficial effects on maintaining desirable forage species in a pasture and improves the soil environment for root growth and soil organisms (Abaye et al., 1997). Goats are particularly useful in areas where there is a lot of brush. Mixed grazing is also useful in some areas in controlling poisonous plants. Sheep and goats are resistant to toxic plants, such as *Senecio* spp. and larkspurs. Combined grazing of sheep or goats with cattle may control cattle-poisoning plants. Further information on the effects of poisonous plants is available in Cheeke (1998).

Complex mixed-grazing systems have evolved in natural ecosystems, exemplified by the Serengeti plains of Africa. In this ecosystem, many species of herbivores, including zebra, elephants, rhinos, giraffes, and numerous antelope species, share the same habitat. Each species occupies a specialized niche, consuming forage unavailable or unutilized by other species. The total meat yield in such a system exceeds that which can be achieved with livestock, suggesting that this resource could be most efficiently utilized on a sustainable basis by game ranching (McNaughton, 1979).

Grazing Disease: Acute Bovine Pulmonary Emphysema

Acute bovine pulmonary emphysema (ABPE) is a respiratory disease of cattle associated with grazing management. It occurs when there is an abrupt change of diet from a sparse feed to a lush succulent pasture. Typical examples would include range cattle that are moved from dry range to irrigated pastures or hay meadows. In Britain, it occurs frequently when livestock are turned into lush fields of kale, rape, or other succulent feeds. Animals affected by ABPE develop severe respiratory problems, with labored breathing, an audible expiratory grunt, and frothing at the mouth. The lungs are grossly enlarged and edematous. ABPE is caused by 3-methyl indole (3-MI) produced from ruminal metabolism of tryptophan. The 3-MI is absorbed and metabolized to an active toxic metabolite in the lungs. The sudden change in diet causes an increase in tryptophan intake and altered rumen microflora. The feeding of ionophores, such as rumensin and lasalocid, has protective activity by preventing proliferation of 3-MI-producing microbes. ABPE can be best prevented by avoiding a sudden dietary change from a sparse to a lush pasture and by providing supplementary feed before the animals are moved. Moldy sweet potato vines and tubers contain a mycotoxin that causes ABPE in cattle. A common weed in much of the United States, purple mint (*Perilla frutescens*) also contains compounds that cause ABPE.

AGROFORESTRY TECHNIQUES

Trees and livestock can be produced concurrently under many situations, a system known as **agroforestry.** In temperate areas, agroforestry usually involves dual use of logged-off land for forage production and regeneration of timber trees. Thus, while conifers are being reestablished, the land can be grazed by cattle, sheep, or goats. This may benefit tree production by removal of competing grasses and broad-leaved weeds and brush and by recycling of nutrients via manure. On the negative side, animal management must be good to avoid damage to the trees from animal browsing, debarking, and trampling. The reestablishment of conifer forests after logging is sometimes difficult because of competition between brush, grass, and other vegetation and the young trees. Suppression of competing vegetation with herbicides, a standard practice, has come under increasing restriction and prohibition. Livestock grazing has been found to be an effective tool for brush control. Leininger and Sharrow (1987) found that sheep grazing caused little damage to new plantings of Douglas fir; the trees were generally avoided and never comprised more than 3 percent of sheep diets. Tree height and diameter growth rate were significantly increased in grazed as compared to ungrazed areas (Sharrow et al., 1989), and total productivity (pasture plus tree biomass) was 30 percent greater with agroforestry than with either pasture or forest alone (Sharrow et al., 1996). Agroforestry systems involving livestock and conifers (e.g., pines) are widely used in Australia and New Zealand (Anderson et al., 1988). (see Fig. 6.1.)

In tropical areas, agroforestry is of great interest (Zimmerman, 1986). Deforestation is occurring rapidly, and there is an urgent need to produce wood for fuel and timber. In addition, tropical soils are readily leached and highly erodable. There are numerous tropical tree species, most of them legumes, that can be

FIGURE 6.1 Agroforestry in Australia involving production of sheep and Radiata pine. (Courtesy of J.D. Kellas.)

grown for fodder as well as for fuel and timber. Leucaena is one of the most promising species. These tree legumes can be grown for multiple purposes, including forage, fuel, timber, fences, and soil improvement, with regular crops produced beneath them. One system used is **alley cropping,** where the trees are planted in rows with vegetables or other crops grown between the rows. The trees can be coppiced for forage production, to be used in a cut and carry system. In many tropical countries, it is a common practice to plant **living fences,** using leguminous trees, such as *Gliricidia sepium.* Besides providing a fence, the trees can be lopped periodically to provide forage. In areas with a Mediterranean climate (humid winter, dry summer), a leguminous tree called **tagasaste** or tree lucerne (*Chamaeacytisus palmensis*) is a useful forage source during the dry season (Borens and Poppi, 1990; Lefroy et al., 2001). It contains alkaloids such as sparteine, which because of their bitter taste may reduce intake and result in low animal performance (Ventura et al., 2000). Tagasaste is quite extensively used in alley cropping in Western Australia (Lefroy et al., 2001).

Legume trees, such as leucaena and other species, can be grown in tropical pastures as **protein banks.** A protein bank is a small area of land relative to the total pasture that is fertilized and managed to maintain productive, high-quality legumes to provide supplementary protein to animals grazing the larger area of unimproved tropical grass pasture. The protein bank can either be harvested manually or grazed judiciously with good management to prevent overgrazing the legume species.

HAY AND HAYMAKING

Hay is produced by dehydrating green forage to a moisture content of 15 percent or less. Preserving forage as hay is a means of distributing forage throughout the year. Forage is usually in excess in spring and early summer and in deficit for the rest of the year. In temperate regions, hay is needed for winter feeding, whereas in the tropics, forage conservation is desirable to provide feed during the dry season.

The curing of hay is usually accomplished by mowing forage and allowing wind and sun action to dehydrate it. Depending upon the crop and weather conditions, this may take from 2 days to a week or more. Mowing is usually done with a mower-conditioner that has rollers through which the cut plants pass. This causes crimping or crushing of the stems, speeding up the drying process. Usually the hay is turned at least once with a rake, to expose the bottom of the windrow to drying action. At one time, most hay was "put up" loose, but now it is almost universally baled. The size of the bales varies from a standard two-twine bale weighing 50 to 75 lb, to three-wire bales of 120 to 150 lb, to large round bales that may contain a ton or more of hay (Fig. 6.2).

Either baling twine or wire may be used. Wire allows a tighter bale, which is desirable when hay is to be transported considerable distances by truck. A disadvantage of wire is that pieces of wire inevitably get into the hay. This may cause **"hardware disease";** cows should be given magnets if wire-tie bales are used (see Chapter 1). A disadvantage of sisal twine is that it may rot in the stack so that the bales break when handled. Plastic twine is durable to a fault, but it is nonbiodegradable and may cause environmental contamination problems.

FIGURE 6.2 A round bale being discharged from a baler. Round bales can be produced rapidly and are suitable for automated handling and feeding. (Courtesy of Case International.)

Plastic twine is particularly undesirable for sheep hay; bits of plastic in the wool markedly reduce its value. In areas where the climate is not too wet, large round bales can be left in the field to be utilized by livestock during the winter. There is considerable waste when this system is used. It is better to haul the bales to a central feeding area and feed using a round bale feeder. Round bales may be covered with plastic to reduce rain damage. Brasche and Russell (1988) in Iowa found that covering round bales did not improve performance of cattle fed the hay but increased dry-matter recovery by approximately 4.5 percent compared to unprotected bales.

The **optimum plant maturity time to cut hay** is when the yield of digestible dry matter is highest. This does not correspond to maximum dry-matter yield. Usually the period of early bloom is the optimal stage. In practice, the weather has a major influence; cutting often must be delayed because of rain. In wet climates, short-range weather forecasts are useful. If hay is cut just after a storm has ended and several days of clear weather are predicted, there is usually time to cure the material.

The **optimal daily maturity time to cut hay** is in the afternoon when the concentration of sugars (photosynthate) is highest. During the night, plant tissues "burn up" sugars to support their metabolism. During daylight hours, photosynthesis proceeds, leading to an accumulation of nonstructural carbohydrates. Animals show a preference for hay cut in the afternoon rather than in the morning because of its higher concentration of sugars (Fisher et al., 1999).

There are a number of additives available that can reduce the time taken to cure hay. **Dessicants** such as alkali, when sprayed on the hay at cutting, help to break down the outer cutin layer and promote drying. Norton and Gondipon (1984) found that various alkalis (potassium carbonate, sodium carbonate, and sodium hydroxide) reduced the drying time of several tropical grasses and legumes, with sodium hydroxide and potassium carbonate being the most effective. Alkali solutions dissolve the waxy cuticle layer on the leaf surface, allowing greater water transpiration from the leaves. Wieghart et al. (1983) found that spraying on a mixture of potassium carbonate and methyl esters of long-chain fatty acids hastened the drying of alfalfa hay. A concentration of approximately 0.2 to 0.4 *M* sodium or potassium carbonate is effective. Mullahey et al. (1988) found that potassium carbonate treatment was effective in reducing the drying time of alfalfa hay but that frequent monitoring of the dry-matter content was necessary to detect when the hay reached 80 percent dry matter and was ready for baling. Other chemicals such as sodium azide increase drying rate by inhibiting the closure of stomata in the leaves (Norton and Gondipon, 1984). By speeding up drying time, dessicants also reduce nutrient losses from cellular respiration. **Hay preservatives,** such as organic acids, especially propionic acid, inhibit mold growth and allow baling at a higher moisture content.

A new development in haymaking is a machine that shreds forages, presses them into a mat, and lays the mat on the field to dry. The shredded mat is dry enough to bale in four to six hours and has higher nutritional value than hay prepared in the traditional manner (Chiquette et al., 1994).

Nutrient Losses in Haymaking

Losses of nutrients occur when forage is conserved as hay (or in any other form, for that matter). When growing plants are cut, they continue to be metabolically active until the cell sap has been dessicated. Thus the cellular respiration uses up stored carbohydrates in the plant tissues. From 4 to 15 percent of the initial dry matter may be lost through cell respiration during wilting. Another major loss of nutrients, particularly with alfalfa, is leaf shattering. When the hay is handled during raking and baling, the leaves may shatter off the stems and be lost. **Rain damage** is a major factor reducing the quality of hay, particularly if the hay is quite dry when the rain occurs. Rain leaches out soluble nutrients including protein, soluble carbohydrates, and minerals. After hay is cut, proteases in plants break down proteins to amino acids, which are water soluble. Rained-on hay may become moldy, reducing its palatability and possibly forming mycotoxins. Rain-damaged hay usually needs to be raked at least one additional time, resulting in greater shattering losses.

Vitamin losses may occur in hay curing and storage. The vitamin A activity, associated with β-carotene and other carotenoid pigments, may decline markedly during curing of hay, particularly when it is bleached by sunlight. The loss of carotene roughly parallels the loss of green pigmentation; bright green hay is an excellent source of vitamin A activity, whereas yellow weathered hay may be devoid of carotene. In contrast, green forage has no vitamin D content; vitamin D is formed by activation of plant sterols by sunlight after plants are cut.

If hay is not sufficiently dry when put into storage, it may heat. The heating is due to microbial respiration (heat of fermentation). Hay can heat to a moderate degree without damage. Excessive heat results in nonenzymatic browning, which is a heat-induced reaction between protein and carbohydrate. This browning, known as the Maillard or **browning reaction,** reduces the availability of protein. The brown products have a pronounced tobaccolike odor, which at moderate levels may actually increase palatability. The degree of **heat damage** can be assessed by measuring the nitrogen content of ADF. The browning reaction products are polymers that are associated with the ADF fraction. As the nitrogen content of ADF increases, the apparent protein digestibility of the forage decreases. When the amount of ADF-bound nitrogen exceeds approximately 30 percent of total nitrogen, the hay has received a detrimental degree of heat damage, with lowered digestibility and feeding value.

A major concern with storage of high-moisture hay is the build-up of sufficient heat to cause **spontaneous combustion** and a barn fire. Moist hay undergoes two periods of heating. The first appears a few days after storage, and is due to plant cell respiration. Moisture is driven off by the heat; farmers often refer to this as the hay going through a sweat. The maximum temperature reached is about 35°C (95°F). The second temperature rise is due to microbial activity, due to the growth of fungi. High temperatures may be reached. When the temperature inside the stack reaches 60°C (140°F), the hay should be checked frequently. If it reaches 160°F, it should be checked hourly. If it reaches 180°F, it is likely to ignite and should be removed from the barn with fire-fighting equipment standing by. When hot hay is exposed to oxygen, it may spontaneously ignite.

When hay is rain-damaged in the field, there may be growth of saprophytic "field fungi" such as *Fusarium, Alternaria,* and *Cladosporium* spp. (Kaspersson et al., 1984). Their growth ceases under storage conditions, and "storage fungi" such as *Aspergillus, Penicillium,* and *Trichothecium* spp. become dominant. These organisms can produce mycotoxins such as aflatoxin and trichothecenes (Clevstrom et al., 1981). Moldy hay may be toxic to both livestock and humans. "Farmer's lung" disease is caused by inhaling dust from very moldy hay (Lacey, 1968). Large spore concentrations are found in the air when such hay is moved or fed. Cattle may develop respiratory disease when fed the material.

Hay Preservatives

In areas with humid climates, there has been much interest in the use of hay preservatives to reduce spoilage of moist hay and to allow putting up moist hay to shorten the field curing time when there is danger of inclement weather. A variety of chemicals have been used with varying degrees of success. The oldest method, which has been used for many years, is the sprinkling of common salt (sodium chloride) on hay as it is stored. **Salt** has a dessicating effect and draws moisture out of the hay, helping to reduce spoilage. Salted hay is also of increased palatability to livestock. Salt does not have antimicrobial activity. **Formic acid** and **propionic acid** have antifungal activity and may have favorable effects in preventing spoilage. Propionic acid is quite effective in small-scale laboratory studies, but under practical conditions a very high application rate is necessary to prevent deterioration (Lacey et al., 1981). Formic acid may inhibit many fungi but permits the growth of *Aspergillus flavus,* leading to aflatoxin production (Clevstrom et al., 1981). The most effective preservative is **anhydrous ammonia.** Its application during hay storage is difficult and hazardous. Urea can be used as a source of ammonia, which is formed as a result of urease activity in the hay. Urea or ammonia treatment not only acts as a preservative but improves fiber digestibility and nitrogen content of the hay (Belanger et al., 1987; Thorlacius and Robertson, 1984). In general, 3 percent anhydrous ammonia is optimal; urea addition is not as effective as direct ammoniation (Craig et al., 1988). **Biological preparations,** such as *Lactobacillus acidophilus* products, do not seem to be effective hay preservatives (Baron and Greer, 1988).

Horse owners have questioned the safety, acceptability, and feed value of preservative-treated hays for horses (Battle et al., 1988). Lawrence et al. (1987)

found that horses preferred untreated hay over hay preserved with propionic acid, but they readily consumed acid-treated hay at the same rate as untreated hay when a choice was not available. Battle et al. (1988) also found that hay treated with preservatives was readily consumed by horses with no ill effects.

Several commercial hay preservative products containing the bacterium *Pediococcus pentosaceus* are available (Wittenberg, 1994, 1995). When applied to moist hay, this organism grows and produces metabolites such as hydrogen peroxide, organic acids, and antibacterial bacterocins that inhibit fungal and bacterial growth. Under practical conditions with hay exposed to rain, these inoculants were not effective as preservatives (Wittenberg, 1995).

Another approach for protecting hay bales from rain damage is the application of a layer of feed-grade fat on the outside of the stack to cause rain to run off rather than soak in. White (1997) investigated this method and found that it was not effective, and even promoted hay spoilage by retarding migration of internal moisture in the stack to the air.

DEHYDRATED FORAGES

Artificial dehydration of forage is accomplished by drying the material in rotating drums heated by natural gas. Although a high-quality product can be obtained, the process is relatively expensive. In the United States, alfalfa is the only major crop that is dehydrated in this manner. Dehydrated alfalfa is commonly referred to as **dehy.** The material is green-chopped and transported directly to the dryer. Because it is dried very quickly after harvest, losses of nutrients are minimized. Dehy is usually bright green in color and has a high vitamin-A activity. Because it is an expensive product, dehy is mostly used in specialty feeds, such as horse and rabbit diets. It is used at low (1 to 3 percent) levels in poultry diets as a source of xanthophyll pigments. Increasingly, alfalfa (Fig. 6.3) is field-wilted before artificial drying to reduce the energy cost.

FIGURE 6.3 Alfalfa is a high-yielding and nutritious forage. In this picture, alfalfa is being swathed, and after field wilting it will be picked up and taken to a dehydration plant to produce dehy (dehydrated alfalfa). (Courtesy of American Alfalfa Processors Association.)

THE USE OF SILAGE

Silage is forage conserved by anaerobic fermentation. Plant material is chopped and packed into a silo to exclude oxygen. The respiring plant cells use up the existing oxygen and then die. This is followed by the proliferation of lactic-acid-producing anaerobic bacteria; fermentation continues until the accumulation of lactic acid depresses the pH to approximately 4, at which point further bacterial growth is inhibited. The silage is thus stabilized.

For production of good silage, the forage must contain sufficient fermentable carbohydrate to allow enough lactic acid to be produced to lower the pH to 4. With some materials, such as grass, it may be necessary to add grain or molasses as a source of substrate for fermentation. The moisture content is important; with low-moisture material, exclusion of oxygen is difficult and molding and heating may occur. A dry-matter content of 25 to 35 percent is considered optimal for silage making. With high-moisture material, there may be putrefaction, with *Clostridia* fermentation producing large amounts of butyric acid and offensive-smelling amines, such as tryptamine and histamine. Legumes have a higher buffering capacity than grasses and require a higher level of soluble carbohydrate to support sufficient fermentation to lower the pH adequately.

Various types of silos can be used. **Tower** or **upright silos** have been widely employed. Packing to exclude oxygen is readily accomplished with an upright silo. Losses from seepage of liquids are a greater problem with upright silos than with horizontal silos. These include **trench** or **bunker silos** and "ag bags," or large **plastic bags** into which chopped forage is blown (Fig. 6.4). Robinson et al. (1988) evaluated several commercial brands of silage bags for their efficacy in ensiling alfalfa. After 8 months of storage, significant degradation of some of the bags occurred, with consequent loss of silage quality. The polyethylene film used in silage bags should have ultraviolet stabilizers, be impermeable to oxygen, and resist elongation in filling. Large round bales can be wrapped in plastic or covered with a plastic bag for ensiling (balage). Trench or bunker silos are packed by using a crawler tractor or similar heavy equipment. They are covered with poly-

FIGURE 6.4 Large plastic bags for storage of moist hay (balage) or silage.

ethylene, which is often held down with old tires. Ricketts et al. (1984) in England reported that "big-bale silage" was implicated in outbreaks of botulism in horses. *Clostridium botulinum* is a soilborne microbe that is a frequent contaminant of forage. Ricketts et al. (1984) suggest than when silage is prepared in plastic bags, fermentation may be insufficient to prevent the growth of *C. botulinum.* Horses are particularly sensitive to the botulism toxin. These authors suggest that horse owners be particularly aware of the importance of avoiding spoilage in silage, and that they realize that big-bale silage inherently carries an element of risk. *Listeria monocytogenes* is a pathogenic bacteria that can grow in silage (Caro et al., 1990). **Listeriosis** causes abortion and neurological problems in livestock. It is an aerobic organism that may grow when silage has been inadequately packed.

The main crops used for silage are grass and legume forages and field corn. An extensive review of the production and utilization of **grass silage** is provided by Harrison et al. (1994). **Corn silage,** a high-quality feed containing abundant energy from the corn grain, is widely used as a roughage and energy source for dairy cattle. Because of its high fermentable-carbohydrate content, corn is readily ensiled. Advantages of corn silage include the high yield per acre (through ensiling the whole plant instead of harvesting only the grain, which increases the energy yield by 40 to 50 percent) and the high-energy content of corn silage. The time of harvest is very important. To maximize nutrient yield, the corn should be harvested when it reaches physiological maturity. When corn is mature, the kernels will be dented. A test often used is the **black layer test** (Fig. 6.5). When the grain reaches physiological maturity, several layers of cells near the tip of the kernel turn black. When kernels in the middle of the ear of corn have a black layer, the corn is ready for harvest. Corn stover is often ensiled and is referred to as *stalklage.* Addition of molasses and ammonia during the ensiling process improves utilization of corn stalklage by dairy cattle (Hargreaves et al., 1984).

Brown midrib mutants (bmr) of corn and sorghum are available (Cherney et al., 1991). These mutants have reduced lignin (11–17 percent reduction) compared to the normal phenotype and also have less cross-linking of phenolic acids, such as ferulic acid, with hemicellulose (Goto et al., 1994). Consequently, they have higher forage quality than regular corn, with increased digestibility and cell wall degradation in the rumen. The bmr corn silage improves voluntary feed intake of dairy cattle because of the higher digestibility of the NDF and a more rapid turnover in the rumen. Another mutant, high-oil corn, when ensiled does not seem to give improved performance by dairy cattle (La Count et al., 1995).

Sunflowers are grown to some extent as a silage crop (Thomas et al., 1982), especially in areas where corn is not productive because of a short growing season. In northern Europe, **Jerusalem artichoke** may have some potential

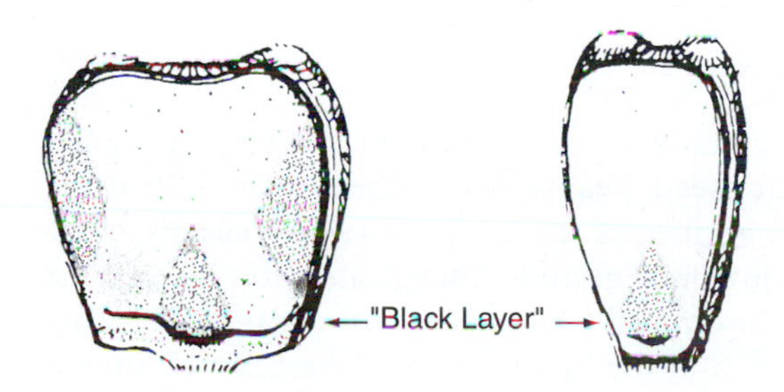

FIGURE 6.5 Silage corn that is ready for harvest with the maximum yield of digestible nutrients has a black layer visible at the base of the kernel.

as a perennial silage crop (Hay and Offer, 1992); the tubers sprout every year. In areas with wet spring weather, this would be an advantage over annual crops that would require soil preparation under wet conditions.

A variety of **silage additives** is available (Henderson, 1993). Acids can be used to lower the pH to inhibit fermentation. Since fermentation results in a loss of energy content when carbohydrate is being used as a substrate to support bacterial metabolism, addition of acids preserves carbohydrates that would otherwise be fermented to lactic acid. In Europe, it has been a common practice to use a mixture of **inorganic acids** (sulfuric, hydrochloric, and phosphoric acids) to lower the pH to 4. In the United States, it has been more common to use **organic acids,** such as formic, acetic, and propionic acids. These are metabolizable in animals, so they contribute to the energy content of the silage directly while reducing fermentation losses. With corn silage, it is often necessary to supply nitrogen to get adequate fermentation. Urea can be used for this purpose. A number of bacterial cultures are available as silage additives. In general, they do not seem to have an appreciable effect; abundant anaerobic microbes develop in fermenting silage without the need for an exogenous source.

Lactic acid bacteria are unable to digest starch; their main substrates are simple sugars (Leahy et al., 1990). Silage additives containing amylase to convert starch to glucose can improve fermentation of corn silage. Leahy et al. (1990) found that the growth rates and feed conversion ratios of beef heifers were improved when whole corn plant silage was prepared with 0.05 percent alpha-amylase as an additive, compared to similar silage without amylase. Commercial silage additives with amylase activity are available. Leahy et al. (1990) also found that 0.1 percent sorbic acid was an effective silage additive in preventing surface spoilage; sorbic acid is a selective inhibitor of yeasts, molds, and heterolactic bacteria.

Forage conservation is a problem in tropical countries. Typically, there is a wet and a dry season. Animal performance in the dry season is poor and would be enhanced if the higher-quality forage produced in the wet season could be preserved. For climatic and cultural reasons, haymaking is impractical in the humid tropics. Silage would appear to have more potential. Problems with **ensiling tropical forages** include their low content of fermentable carbohydrate, their coarse structure that impedes packing and oxygen exclusion, and their high water content. High environmental temperatures promote a high rate of bacterial fermentation; the heat of fermentation may be high enough to cause substantial heat damage to the protein and even to ignite the material. Sugarcane, which is high in fermentable carbohydrate, can be added to tropical grasses to improve their silage-making potential. Cassava meal and coconut meal are also suitable additives for improving silage production with tropical grasses (Panditharatne et al., 1986).

USE OF LOW-QUALITY CROP RESIDUES

Low-quality fibrous crop residues include straw, chaff, corn stover (corn stalks and leaves), sugarcane bagasse, cottonseed, peanut and soybean hulls, cotton gin trash, corn cobs, and so on. Vast quantities of these types of materials are available. Unfortunately, they are very low in digestible energy and protein and are highly lignified. They are digested slowly in the rumen because of their low content of readily available carbohydrate and the lignification of the cell wall material. The slow turnover in the rumen results in rumen fill becoming a limiting factor. Crop residues contain insufficient nitrogen to support adequate microbial growth in the rumen. This can be rectified by adding a protein supplement; however, the low fermentable energy is less easily overcome. Another factor that must be considered is the possible presence of toxic residues of fungicides, pesticides, and herbicides. In grass seed production, for example, use of these agricultural chemicals may cause the straw to be unfit for feeding purposes.

Treatments to Improve Utilization of Crop Residues

Chemical Treatment The main factor responsible for the very low digestibility of fibrous crop residues is **lignification** of the cell wall material. Fibrous materials like straw and hulls are primarily cell wall; because these tissues are no longer metabolically active, there is very little cell contents fraction. Lignin is very indigestible, and, by encrusting cellulose fibers and other cell wall constituents, impedes their digestion by microbial enzymes. However, lignin is very susceptible to degradation by oxidation and is destroyed by oxidizing agents, such as hypochlorite (bleach) and sulfites. These products are used in the treatment of wood pulp to isolate cellulose for the purposes of paper manufacture.

It has been known for many years that treatment of fibrous materials with alkali improves their digestibility. Sodium hydroxide (lye) is one of the most commonly used alkaline substances for this purpose. **Alkali** dissolves lignin and renders the cell wall constituents susceptible to microbial digestion. Alkali treatment of straw is used extensively in Europe. In the United States, it is generally not economically viable; the cost of treating straw or other crop residues with alkali exceeds the benefits gained. Sodium hydroxide treatment of roughages, however, has several disadvantages in addition to cost: the need for supplemental nitrogen, the increased sodium load on the animal, sodium contamination of soil, and the hazardous nature of the chemical.

An alternative to sodium hydroxide treatment is **ammoniation.** Ammonia treatment, either as ammonium hydroxide or gaseous ammonia, is effective in dissolving lignin. Ammonia also solubilizes hemicellulose and causes swelling of the cellulose, improving its accessibility to microbial enzymes and fiber digestibility. Ammoniation provides a source of supplementary nitrogen that can be used by rumen microbes. Other advantages are that ammonia is an effective preservative, there is no excretory load of sodium, ammoniation can be used on long unchopped material, and ammonia is safer and easier to use than caustic alkali. The procedure generally used is to stack the straw or other low-quality roughage, cover the stack with plastic tucked in well around the edges, and release gaseous ammonia into the stack. Over a period of several weeks, the ammonia permeates the stack and dissolves lignin.

During ammoniation, there is an obvious change in color of treated straw to a more intense yellow-brown. The physical texture is visibly altered, which is particularly evident when the straw is chopped or ground. Ammoniated straw is much more readily reduced to a small particle size than the untreated material, facilitating its digestion in the rumen. During ammoniation, some of the nitrogen becomes bound to the fiber. Ammonia reacts with hemicelluloses released after cross-links between hemicellulose chains are hydrolyzed. This bound nitrogen is largely unavailable to the rumen microbes.

Urea can be used as a source of ammonia; most roughages have sufficient bacterial urease present to convert urea to ammonia. Urea may be slightly less effective than anhydrous ammonia because of the formation of ammonium carbonate, which decreases the pH of the straw, thus reducing the alkalinity effect on conformational changes in the fiber (Preston and Leng, 1987). Sanderson et al. (1985) found that ammoniation of sweet clover hay suppressed the formation of dicumarol, thus aiding in prevention of sweetclover poisoning. Untreated and propionic-acid-treated hay had dicumarol levels of 52 and 38 milligrams per kilogram, levels that are potentially toxic. Ammonia-treated hay had a much lower dicumarol level of 15 mg/kg. Ammoniation is effective in reducing the effects of a number of other toxins such as aflatoxin. Bell et al. (1984) showed that ammoniation of mustard meal reduced the glucosinolate content by more than 80 percent. The ergot alkaloids in endophyte-infected tall fescue are detoxified by ammoniation of tall fescue hay (Chestnut et al., 1991). Examples of the types of responses obtained with ammoniation of low-quality forage are shown in Table 6–1. Note the enhancement in fiber digestibility with ammoniation and the improvement in growth rate of animals fed the treated material.

Silanikove et al. (1988) utilized urea treatment to conserve moist green grass for periods of at least a month for feeding to dairy cattle. Grass was field wilted to 60 percent dry matter and then chopped with urea added (35 g urea N per kg DM). The grass was stored in large open stacks. Advantages of this system are the preservation of grass at a growth stage of high nutrient density and digestibility, lower nutrient losses than in field drying or silage making, addition of rumen-fermentable nitrogen, and improved fiber digestibility. An entire field can be harvested at one time, which is much simpler than a green-chop system in which a strip of forage is mechanically harvested every day.

TABLE 6–1. Effect of Ammoniation on Digestibility in Cattle of Low-Quality Forages

	Treatment		
Item and Author	***Nontreated + Urea***	***Nontreated + Urea + Molasses***	***Ammoniated***
Brown et al., 1987			
Organic matter dig., %	48.9	47.9	57.5
NDF dig., %	57.0	51.0	68.7
ADF dig., %	46.8	38.8	56.6
Avg. daily gain, g	270	390	540
Avg. daily feed, kg	4.30	5.19	5.20
Item and Author	***Untreated***	***Ammoniated***	
Brown, 1988			
Organic matter dig., %	48.5	62.1	
NDF dig., %	57.0	71.4	
ADF dig., %	51.2	66.7	
Avg. daily gain, g	100	340	

Toxicity of Ammoniated Roughages Ammoniation of feeds may result in production of toxins and toxicity symptoms in animals consuming the ammoniated feed. This toxicity has been observed in numerous countries, and has been variously termed *bovine hysteria, bovine bonkers,* ***ammoniated hay toxicosis,*** *hyperexcitability,* and *crazy cow syndrome.* Neurological signs such as hyperactivity, incoordination, tremors, and convulsions occur in affected animals. Cattle may become extremely nervous and difficult to handle. Affected animals may suddenly gallop in circles and run into fences, gates, and other objects, often causing themselves injury (Kerr et al., 1987; Perdok and Leng, 1987; Weiss et al., 1986). The causative factor is 4-methyl-imidazole, formed by a reaction between reducing sugars (e.g., glucose) and ammonia in the presence of heat. Toxicity does not occur if the temperature of ammoniated roughage remains below 70°C during ammoniation (Perdok and Leng, 1987). Although some of the symptoms resemble those of hypomagnesemia (grass tetany) where serum magnesium is depressed, administration of magnesium does not alter the symptomology (Perdok and Leng, 1987). The symptoms are also similar to those of simple ammonia toxicity; a high level of ammonia in the material could presumably aggravate the symptoms. The toxins are transferred in the milk, suggesting that caution should be used when feeding ammoniated roughages to dairy cattle (Perdok and Leng, 1987). Ammoniation of molasses may also produce bovine hysteria; a commercial ammoniated molasses product was introduced in the United States in the 1950s but withdrawn because of a high incidence of the hysteria disorder. Because of the involvement of reducing sugars, ammoniation of good-quality roughages with abundant reserves of carbohydrate should be avoided.

Other Chemical Treatments Treatment of low-quality agricultural byproducts with an **alkaline hydrogen peroxide** solution increases digestibility (Kerley et al., 1985, 1987). Natural degradation of lignin by fungi is believed to involve a role of natural oxidants, such as hydrogen peroxide. In the alkaline peroxide treatment, the lignin-carbohydrate bonds are disrupted and the ordered structure of the cellulose fibers is altered. These effects increase the accessibility of the cellulose to cellulolytic enzymes, increasing cellulose digestion. Digestibility of the fiber fractions and performance of sheep were dramatically improved when the wheat straw was treated with alkaline peroxide (Table 6–2). Similar results were obtained in a subsequent study (Kerley et al., 1987). The alkaline peroxide treatment also appears to cleave linkages between phenolic acids and structural polysaccharides and lignin (Kerley et al., 1988). Digestibility of structural polysaccharides is inhibited when they are bound to phenolic acids such as coumaric and ferulic acids. Kerley et al. (1988) suggest that bonding between phenolic compounds and lignin may be an important factor limiting feed intake, digestibility, and performance of animals fed low-quality roughages. Treatment of roughages with **ozone** has also been shown to degrade lignin and improve fiber digestibility (Miron and Yokoyama, 1990).

Physical Treatment The digestibility of low-quality roughages can be improved by **grinding or chopping** the material to reduce particle size. This exposes more terminal ends of cellulose fibers to cellulase. However, the benefits gained in improved digestibility are usually offset by a more rapid passage of small

TABLE 6–2. Effect of Alkaline Hydrogen Peroxide Treatment of Wheat Straw on Its Nutritive Value to Sheep

	Treatment and % of Straw in Diet			
	Alkaline Peroxide		*Control*	
Item	*36*	*72*	*36*	*72*
Feed intake, g/day	985	769	863	434
Apparent digestibility, %				
Dry matter	72.2	67.4	53.4	48.5
NDF	63.4	73.6	28.0	43.0
ADF	60.2	72.6	30.0	44.3
Cellulose	64.8	76.6	26.8	43.4
kcal	72.1	66.5	52.8	48.8
Weight gain, g/day	241	235	202	−106

Adapted from Kerley et al. (1985).

particles from the rumen, which thus escape microbial digestion. The major benefits of chopping or grinding roughage are reduced feed wastage.

Biological Treatment Lignin can be degraded by various aerobic fungi and bacteria. A typical study is that of Ward and Perry (1982), who observed that culture of the fungus *Trichoderma viride* on corn cobs increased their nutritional value for sheep. White-rot fungi, which typically grow on wood, specifically degrade lignin and have shown effectiveness in improving the digestibility of straw (Jalc et al., 1994). Bacterial preparations digest primarily cellulose and hemicellulose in straw, whereas fungi digest lignin as well.

Improvement of the Rumen Environment Ndlovu and Buchanan-Smith (1985) suggested that although traditional approaches to improving utilization of poor-quality roughages have mainly involved physical or chemical treatments, an alternative approach would be proper supplementation to optimize the rumen environment for fiber digestion. Low-quality roughages lack adequate fermentable nitrogen, true protein, readily available carbohydrates, and minerals for optimal microbial activity. Slow rates of fiber digestion and outflow from the rumen are major constraints to increasing voluntary intake of poor-quality roughages. Ndlovu and Buchanan-Smith (1985) observed that supplementation of low-quality roughages (barley straw, bromegrass hay, and corncobs) with ground alfalfa hay improved fiber digestibility and rumen-environment parameters, such as concentration of branched-chain fatty acids and ruminal ammonia-nitrogen. Some cellulolytic microbes require isobutyric and valeric acids and preformed peptides and amino acids, and ruminal degradation of alfalfa could have provided these. Readily fermentable cell wall constituents of alfalfa might stimulate cellulolytic microbes with increased colonization of the poor-quality roughage. Subsequent work (Ndlovu and Buchanan-Smith, 1987a,b) indicated that the primary effect of alfalfa supplementation was intraruminal, and it may have been largely due to increasing the rumen level of fermentable nitrogen. Hunt et al. (1988) similarly noted a positive associative effect of alfalfa on wheat straw intake. The presence of alfalfa or clover in the diet increases sporangial numbers of rumen fungi, even though the fungi do not colonize alfalfa leaflets (Akin, 1987). This effect could contribute to the positive associative effect noted when alfalfa is used to supplement straw.

Ndlovu and Buchanan-Smith (1987a) observed that postruminal infusion of protein in the abomasum increased the intake of a corncob-based diet. The positive effects of bypass protein on intake of low-quality roughages are well-known and discussed further in Chapter 9.

For maximum utilization of low-quality roughages, it is desirable to integrate the various methods discussed. Physical and chemical treatments (e.g., chopping and ammoniation) improve accessibility of the fiber to cellulolytic microbes, whereas optimizing dietary supplementation with rumen microbe growth factors may enhance growth of the cellulolytic organisms. Supplementation with bypass protein stimulates intake of low-quality roughage, perhaps by affecting the "set point" at which rumen distention causes cessation of feed intake (Chapter 9).

Economic Feasibility of Crop Residue Treatment Millions of tons of straw and other crop residues are potentially available as feedstuffs. Without treatment, their digestibility is so low that they have little value as feed. Alkali treatment and ammoniation of straw have been fairly extensively employed, particularly in Europe. **Alkali-treated straw** is a useful and acceptable roughage source. In areas where feedstuffs are expensive or are in limited supply, chemical treatment of straw is economically viable. In the United States, the availability of surplus grains generally makes treated straw economically unattractive. In developing countries, chemical treatment of low-quality grasses and crop residues may be feasible. **Wood ash** can be used as a source of alkali and cattle urine as a source of ammonia. On a small scale, simple treatment of straw with these materials may be useful in improving efficiency of livestock production in developing countries. Nolte et al. (1987) found wood ashes to be as effective as sodium hydroxide in improving fiber utilization in ruminants.

OTHER POTENTIAL ROUGHAGE SOURCES

Wood and Wood Products

Wood and wood products such as newspaper, pulp, and cardboard are cellulosic substances that can be digested by microbial cellulases and can be used as feedstuffs for ruminants. Although there has been some interest in the use of newspaper and cardboard as feed ingredients, they are generally more expensive sources of calories than other roughages. **Cardboard** and corregated paper have a fairly low energy value for ruminants and are most useful in maintenance diets for mature animals (Aderibigbe and Church, 1987; Belyea et al., 1987).

Certain trees have potential as feedstuffs. **Aspen** (*Populus tremuloides*) grows in extensive tracts in the northern United States and in Canada. Aspen and other poplars have a rapid growth rate and the wood is relatively free of toxic substances. Aspens can be harvested, ground up, treated with sodium hydroxide or ammonia, and ensiled or fed directly. Feeding trials with aspen show promising results (Singh and Kamstra, 1981). Millett et al. (1970) determined the *in vitro* digestibility of wood, bark, and alkali-treated wood of a number of tree species. In general, the digestibility of untreated wood was very low, with aspen and maple showing the most promise. In most cases bark had a higher digestibility than wood. Alkali-treatment of the wood substantially improved its digestibility. **Aspen leaves** have a high content of tannins, which markedly impair their digestibility (Bas et al., 1985). The digestibility of hybrid poplar foliage is lower than for alfalfa, due to the effects of tannins (Ayers et al., 1996).

Forest tree foliage has been evaluated as a feedstuff. Much of the work has been done in the former Soviet Union, so the Russian word *muka* is used to describe this product. Muka is not suitable for use with nonruminants (Hunt and Barton, 1978) but may have some application in ruminant diets (Tait et al., 1982). Steam cooking the product improves its utilization. Tait et al. (1982) concluded that a level of 20 percent lodgepole pine (*Pinus contorta*) muka could be utilized in sheep diets without adverse effects. Inclusion of muka depressed digestibility of protein and fiber. The pine muka contained about a 5 percent crude protein and 46 percent ADF. Ponderosa pine (*Pinus ponderosa*) needles should not be used because they contain factors that cause abortion (Stegelmeier et al., 1996).

In general, wood and wood by-products have not been economically competitive as feedstuffs, but they have been used in times of emergency. For example, during World War II, 1.5 million tons of wood pulp were fed to cattle and horses in the Scandinavian countries because of a lack of other feedstuffs (NRC, 1983).

Kenaf (*Hibiscus cannabinus*) is a giant grass species, grown to produce fiber for the paper industry. The leaves and plant tips (apical fraction) are by-products potentially available for livestock feed. Kenaf silage is of low digestibility and palatability (Xiccato et al., 1998).

Water Weeds

In some areas, particularly in the tropics and subtropics, aquatic vegetation is a major weed problem. One of the most notable is **water hyacinth** (*Eichhornia crassipes*). This rapidly growing plant thrives in polluted waterways, choking them with dense vegetation that impedes navigation and plugs hydroelectric systems. The decaying material depletes dissolved oxygen, resulting in fish kills. Millions of dollars are spent in removing water hyacinth from canals, lakes, and rivers. The harvested material must be disposed of in some way, and its use as a feed is one possibility. The main problem in utilizing it is its high water content. Processing methods could no doubt be devised to circumvent this problem. Water hyacinth has been investigated as a means of treating sewage and animal wastes in lagoons by removing soluble nutrients and purifying the water. The harvested material could be utilized as a feed.

Duckweeds, which are small floating aquatic plants, have been examined as a means of recycling nutrients in animal wastes (Rusoff et al., 1980) and sewage (Haustein et al., 1990, 1994). Dried duckweed contains up to 40 percent crude protein. Especially in tropical or semitropical areas where growth rate of aquatic plants is high, plants such as water hyacinth or duckweed may be useful in removing pollutants in wastewater and providing forage to be used in animal feeding. Harvesting water weeds with herbivorous fish such as grass carp (Reichert and Trede, 1977), and preparing fish meal for feeding purposes, may be ways of integrating livestock production and aquaculture in intensive systems, particularly in tropical countries.

Linn et al. (1975a, b) and Heffron et al. (1977) evaluated a number of temperate aquatic weeds as feedstuffs. The plants in general were similar to alfalfa in chemical composition but inferior in feeding value; drying or ensiling did not improve their acceptability or usefulness. Aquatic plants generally have a high ash content due predominantly to contamination with sand or silt (Lizama et al., 1988). Lizama et al. (1988) found that inclusion of elodea in chick diets did not adversely affect growth. Cai and Curtis (1989) found that elodea gave excellent results when fed as the sole diet to young grass carp, surpassing the growth rate of carp fed a commercial fish pellet. These reports suggest that elodea may be one of the best aquatic plants in terms of feeding value. Ali and Leeson (1994) provided a good review of the nutritional value of aquatic weeds and their potential utilization as feedstuffs.

Seaweed

Seaweeds, which are giant algae (macroalgae), are plentiful in many coastal areas of the world. They might be potential feedstuffs. Seaweed has a fairly low crude-protein content and is characterized by a high ash content. The carbohydrate fraction consists of poorly digested glycoprotein polymers and polysaccharides containing unusual sugars. Most of the protein is bound in an unavailable form to polysaccharide polymers. Seaweeds are a poor source of digestible protein and energy (Ventura et al., 1994). They are primarily of value as sources of minerals, having a high content of sodium, potassium, iodine, and chlorine. These elements are readily obtainable from other sources, so seaweeds would appear to have little potential as feedstuffs for commercial application. Seaweed (e.g., Pacific kelp meal) is often sold as a sequestering agent (see Chapter 8) to improve mineral availability. However, copper chelated with Pacific Coast kelp was not any more efficacious than inorganic copper sulfate in meeting the copper requirements of baby pigs (Stansbury et al., 1990). Seaweed has laxative and diuretic effects, and it should be limited to about 10 percent of the diet (Beames, 1990). According to Beames (1990), dietary seaweed may reduce the incidence of internal parasites in pigs.

An interesting use of seaweed as feed occurs on North Ronaldsay Island off the coast of Scotland. The North Ronaldsay breed of sheep feeds almost exclusively on seaweed because the animals are restricted to living on the beach areas by a stone wall surrounding the island. This breed has adapted to the very low copper content of seaweed (about 2 ppm) but, as a consequence, is very susceptible to copper toxicity. Efforts to preserve flocks of the North Ronaldsay breed in other locations have been impeded by the virtual inability of the sheep to consume normal vegetation without developing copper toxicity (MacLachlan and Johnston, 1982). The North Ronaldsay sheep consume mainly brown kelps, which have a dry matter digestibility of about 70 percent (Hansen et al., 2003). The sheep are able to meet their energy needs from a 100 percent seaweed diet.

QUESTIONS AND STUDY GUIDE

1. What are the merits of continuous versus rotational grazing?
2. A herd of cows is moved from a dry nonirrigated grass pasture into a lush, heavily fertilized grass-clover pasture. After approximately 1 week, many of the cows are coughing and showing signs of acute respiratory distress. What is the problem?
3. What is a protein bank and how is it useful?
4. Discuss factors that cause nutrient losses in hay and methods to reduce these losses.
5. Damp hay may heat and even spontaneously ignite. Why? How can heating of moist hay be reduced?
6. Cattle fed moldy hay may develop respiratory disease. Why?
7. What are the advantages and disadvantages of silage compared to other forms of conserved forage?
8. Why is straw a poor-quality feed? What can be done to improve its feeding value?
9. What are the advantages of ammoniation of low-quality roughage as compared to other treatments? What are some disadvantages?
10. Which would you expect to have a higher feeding value for cattle: newspaper or ground wood (both prepared from the same tree species)? Why?

REFERENCES

Grazing and Agroforestry

Abaye, A.O., V.G. Allen, and J.P. Fontenot. 1997. Grazing sheep and cattle together or separately: Effect on soils and plants. *Agron. J.* 89:380–386.

Anderson, G.W., R.W. Moore, and P.J. Jenkins. 1988. The integration of pasture, livestock and widely spaced pine in South West Western Australia. *Agroforestry Syst.* 6:195–211.

Borens, F.M.P. and D.P. Poppi. 1990. The nutritive value for ruminants of tagasaste (*Chamaecytisus palmensis*), a leguminous tree. *Anim. Feed Sci. Tech.* 28:275–292.

Cheeke, P.R. 1998. *Natural Toxicants in Feeds, Forages, and Poisonous Plants.* Upper Saddle River, NJ: Prentice Hall, Inc.

Glenn, S., C.E. Rieck, D.G. Ely, and L.P. Bush. 1980. Quality of tall fescue forage affected by mefluidide. *Agric. Food Chem.* 28:391–393.

Hodgson, J. and A.W. Illius., eds. 1996. *The Ecology and Management of Grazing Systems.* Wallingford, UK: CAB International.

Lefroy, E.C., R.J. Stirzaker, and J.S. Pate. 2001. The influence of tagasaste (*Chamaecytisus proliferus*) trees on the water balance of an alley cropping system on deep sand in southwestern Australia. *Aust. J. Agric. Res.* 52:235–246.

Leininger, W.C. and S.H. Sharrow. 1987. Seasonal diets of herded sheep grazing Douglas-fir plantations. *J. Range Manage.* 40:551–555.

McNaughton, S.J. 1979. Grazing as an optimization process: Grass-ungulate relationships in the Serengeti. *Amer. Nat.* 113:691–703.

Nolan, T. and J. Connolly. 1977. Mixed stocking by sheep and steers—A review. *Herbage Abst.* 47:367–374.

______ . 1989. Mixed *vs* monograzing by steers and sheep. *Anim. Prod.* 48:519–533.

Robb, T.W., D.G. Ely, C.E. Rieck, R.J. Thomas, B.P. Glenn, and S. Glenn. 1982. Mefluidide treatment of fall fescue: Effect on nutrient utilization. *J. Anim. Sci.* 54:155–163.

Savory, A. 1988. *Holistic Resource Management.* Washington, DC: Island Press.

Sharrow, S.H., D.H. Carlson, W.H. Emmingham, and D. Lavender. 1996. Productivity of two Douglas fir/subclover/sheep agroforests compared to pasture and forest monocultures. *Agroforestry Syst.* 34:305–313.

Sharrow, S.H., W.C. Leininger, and B. Rhodes. 1989. Sheep grazing as a silvicultural tool to suppress brush. *J. Range Manage.* 53:2–4.

Valentine, J.F. 1990. *Grazing Management.* San Diego: Academic Press.

Ventura, M.R., J.I.R. Castanon, M. Muzquiz, P. Mendez, and M.P. Flores. 2000. Influence of alkaloid content on intake of subspecies of *Chamaecytisus proliferus. Anim. Feed Sci. Tech.* 85:279–282.

Zimmerman, T. 1986. Agroforestry—A last hope for conservation in Haiti? *Agroforestry Syst.* 4:255–268.

Hay and Hay Preservatives

Baron, V.S. and G.G. Greer. 1988. Comparison of six commercial hay preservatives under simulated storage conditions. *Can. J. Anim. Sci.* 68:1195–1207.

Battle, G.H., S.G. Jackson, and J.P. Baker. 1988. Acceptability and digestibility of preservative-treated hay by horses. *Nutr. Rep. Int.* 37:83–89.

Belanger, G., A.M. St. Laurent, C.A. Esau, J.W.G. Nicholson, and R.E. McQueen. 1987. Urea for the preservation of moist hay in big round bales. *Can. J. Anim. Sci.* 67:1043–1053.

Brasche, M.R. and J.R. Russell. 1988. Influence of storage methods on the utilization of large round hay bales by beef cows. *J. Anim. Sci.* 66:3218–3226.

Chiquette, J., P. Savoie, and A. Lirette. 1994. Effects of maceration at mowing on digestibility and ruminal fermentation of timothy hay in steers. *Can. J. Anim. Sci.* 74:235–242.

Clevstrom, G., B. Goransson, R. Hlodversson, and H. Pettersson. 1981. Aflatoxin formation in hay treated with formic acid and in isolated strains of *Aspergillus flavus. J. Stored Prod. Res.* 17:151–161.

Coombs, D.F., W.M. Craig, A.F. Loyacano, M.A. Jordan, and L.R. Verma. 1989. In vitro and in vivo digestibility of urea and ammonia treated high-moisture ryegrass hay. *Nutr. Rep. Int.* 39:51–60.

Craig, W.M., J.A. Ulloa, K.L. Walkins, and B.D. Nelson. 1988. Effects of level, form and time of urea application on the nutritive value of Alicia bermudagrass. *J. Anim. Sci.* 66:185–193.

Fisher, D.S., H.S. Mayland, and J.C. Burns. 1999. Variation in ruminants preference for tall fescue hays cut either at sundown or at sunup. *J. Anim. Sci.* 77:762–768.

Hlodversson, R. and A. Kaspersson. 1987. Nutrient losses during deterioration of hay in relation to changes in biochemical composition and microbial growth. *Anim. Feed Sci. Tech.* 15:149–165.

Kaspersson, A., R. Hlodversson, U. Palmgren, and S. Lindgren. 1984. Microbial and biochemical changes occurring during deterioration of hay and preservative effect of urea. *Swed. J. Agric. Res.* 14:127–133.

Lacey, J. 1968. The microflora of fodders associated with bovine respiratory disease. *J. Gen. Microbiol.* 51:173–177.

Lacey, J., K.A. Lord, and G.R. Cayley. 1981. Chemicals for preventing moulding in damp hay. *Anim. Feed Sci. Tech.* 6:323–336.

Lawrence, L.M., K.J. Moore, H.F. Hintz, E.H. Jaster, and L. Wischover. 1987. Acceptability of alfalfa hay treated with an organic acid preservative for horses. *Can. J. Anim. Sci.* 67:217–220.

Mullahey, J.J., J.P. Mueller, and J.T. Green. 1988. Enhancement of field hay curing with potassium carbonate. *Agron. J.* 80:186–192.

Norton, B.W. and R. Gondipon. 1984. Effects of alkali treatment on the drying rate and nutritive value of some tropical grasses and legumes. *J. Aust. Inst. Agri. Sci.* 49:55–58.

Thorlacius, S.O. and J.A. Robertson. 1984. Effectiveness of anhydrous ammonia as a preservative for high-moisture hay. *Can. J. Anim. Sci.* 64:867–880.

White, J.S. 1997. Nutrient conservation of baled hay by sprayer application of feed grade fat with or without barn storage. *Anim. Feed Sci. Tech.* 65:1–4.

Wieghart, M., J.W. Thomas, M.B. Tesar, and C.M. Hansen. 1983. Acceleration of alfalfa drying in the field by chemical application at cutting. *Crop Sci.* 23:225–229.

Wittenberg, K.M. 1994. Nutritive value of high moisture alfalfa hay preserved with *Pediococcus pentosaceus. Can. J. Anim. Sci.* 74:229–234.

______. 1995. Efficacy of *Pediococcus pentosaceus* for alfalfa forage exposed to precipitation during field wilting. *Can. J. Anim. Sci.* 75:303–308.

Silage

Caro, M.R., E. Zamora, L. Leon, F. Cuello, J. Salinas, D. Megias, M.J. Cubero, and A. Contreras. 1990. Isolation and identification of *Listeria monocytogenes* in vegetable by-product silages containing preservative additives and destined for animal feeding. *Anim. Feed Sci. Tech.* 31:285–291.

Cherney, J.H., D.J.R. Cherney, D.E. Akin, and J.D. Axtell. 1991. Potential of brown-midrib, low-lignin mutants for improving forage quality. *Adv. Agron.* 46:157–198.

Fisher, L.J., G.C.L. Pennells, and J.A. Shelford. 1984. The effect of the additive "silogen" on the intake and digestibility of grass silage. *Can. J. Anim. Sci.* 64:709–715.

Goto, M., J. Matsuoka, T. Sato, H. Ehara, and O. Morita. 1994. Brown midrib mutant maize with reduced levels of phenolic acids ether-linked to the cell walls. *Anim. Feed Sci. Tech.* 48:27–38.

Hargreaves, A., J.T. Huber, J. Arroyoluna, and L. Kung. 1984. Influence of adding ammonia to corn stalkage on feeding value for dairy cows and on fermentation changes. *J. Anim. Sci.* 59:567–575.

Harrison, J.H., R. Blauwiekel, and M.R. Stokes. 1994. Fermentation and utilization of grass silage. *J. Dairy Sci.* 77:3209–3235.

Hay, R.K.M. and N.W. Offer. 1992. *Helianthus tuberosus* as an alternative forage crop for cool maritime regions: A preliminary study of the yield and nutritional quality of shoot tissues from perennial stands. *J. Sci. Food Agric.* 60:213–221.

Henderson, N. 1993. Silage additives. *Anim. Feed Sci. Tech.* 45:35–56.

La Count, D.W., J.K. Drackley, T.M. Cicela, and J.H. Clark. 1995. High-oil corn as silage or grain for dairy cows during an entire lactation. *J. Dairy Sci.* 78:1745–1754.

Leahy, K.T., K.M. Barth, P.P. Hunter, and S.A. Nicklas-Bray. 1990. Effects of treating corn silage with alpha-amylase and (or) sorbic acid on beef cattle growth and carcass characteristics. *J. Anim. Sci.* 68:490–497.

McDonald, P. 1981. *The Biochemistry of Silage.* New York: John Wiley.

Panditharatne, S., V.G. Allen, J.P. Fontenot, and M.C.N. Jayasuriya. 1986. Ensiling characteristics of tropical grasses as influenced by stage of growth, additives and chopping length. *J. Anim. Sci.* 63:197–207.

Ricketts, S.W., T.R.C. Greet, P.J. Glyn, C.D.R. Ginnett, E.P. McAllister, J. McCaig, P.H. Skinner, P.M. Webbon, D.L. Frape, G.R. Smith, and L.G. Murray. 1984. Thirteen cases of botulism in horses fed big bale silage. *Equine Vet. J.* 16:515–518.

Robinson, P.H., J.J. Kennelly, and G.W. Mathison. 1988. Influence of type of silage bag on chemical composition of alfalfa silage. *Can. J. Anim. Sci.* 68:831–837.

Thomas, V.R., D.N. Sneddon, R.E. Roffler, and G.A. Murray. 1982. Digestibility and feeding value of sunflower silage for beef steers. *J. Anim. Sci.* 54:933–937.

Woolford, M.K. 1984. *The Silage Fermentation.* New York: Marcel Dekker.

Crop Residues

Akin, D.E. 1987. Association of rumen fungi with various forage grasses. *Anim. Feed Sci. Tech.* 16:273–285.

Bell, J.M., M.O. Keith, J.A. Blake, and D.I. McGregor. 1984. Nutritional evaluation of ammoniated mustard meal for use in swine feeds. *Can. J. Anim. Sci.* 64:1023–1033.

Brown, W.F. 1988. Maturity and ammoniation effects on the feeding value of tropical grass hay. *J. Anim. Sci.* 66:2224–2232.

Brown, W.F., J.D. Phillips, and D.B. Jones. 1987. Ammoniation or cane molasses supplementation of low quality forages. *J. Anim. Sci.* 64:1205–1214.

Chestnut, A.B., H.A. Fribourg, K.D. Gwinn, P.D. Anderson, and M.A. Cochran. 1991. Effects of ammoniation on toxicity of *Acremonium coenophialum* infested tall fescue. *Anim. Feed Sci. Tech.* 35:227–236.

Hunt, C.W., T.J. Klopfenstein, and R.A. Britton. 1988. Effect of alfalfa addition to wheat straw diets on intake and digestion in beef cattle. *Nutr. Rep. Int.* 38:1249–1257.

Jalc, D., R. Zitnan, and F. Nerud. 1994. Effect of fungus-treated straw on ruminal fermentation in vitro. *Anim. Feed Sci. Tech.* 46:131–141.

Kerley, M.S., G.C. Fahey, Jr., L.L. Berger, J.M. Gould, and F.L. Baker. 1985. Alkaline hydrogen peroxide treatment unlocks energy in agricultural by-products. *Science* 23:820–822.

Kerley, M.S., G.C. Fahey, Jr., L.L. Berger, N.R. Merchen, and J.M. Gould. 1987. Effects of alkaline hydrogen peroxide treatment of wheat straw on site and extent of digestion in sheep. *J. Anim. Sci.* 63:868–878.

Kerley, M.S., K.A. Garleb, G.C. Fahey, Jr., L.L. Berger, K.J. Moore, G.N. Phillips, and J.M. Gould. 1988. Effects of alkaline hydrogen peroxide treatment of cotton and wheat straw on cellulose crystallinity and on composition and site and extent of disappearance of wheat straw cell wall phenolics and monosaccharides by sheep. *J. Anim. Sci.* 66:3235–3244.

Kerr, L.A., W. Groce, and K.W. Kersting. 1987. Ammoniated forage toxicosis in calves. *J. Am. Vet. Med. Assoc.* 191:551–552.

Miron, J. and M.T. Yokoyama. 1990. Ozone-treated lucerne hay as a model to study lucerne degradation and utilization by rumen bacteria. *Anim. Feed Sci. Tech.* 27:269–280.

Ndlovu, L.R. and J.G. Buchanan-Smith. 1985. Utilization of poor quality roughages by sheep: Effects of alfalfa supplementation on ruminal parameters, fiber digestion and rate of passage from the rumen. *Can. J. Anim. Sci.* 65:693–703.

______ . 1987a. Alfalfa supplementation of corn cob diets for sheep: Effect of ruminal or postruminal supply of protein on intake, digestibility, digesta, passage and liveweight changes. *Can. J. Anim. Sci.* 67:1075–1082.

______ . 1987b. Alfalfa supplementation of corn cob diets for sheep: Comparison with higher volatile fatty acids, soybean protein and alfalfa fiber on intake, environment and digestion in the rumen and digesta passage. *Can. J. Anim. Sci.* 67:1083–1091.

Nolte, M.E., J.H. Cline, B.A. Dehority, S.C. Loerch, and C.F. Parker. 1987. Treatment of wheat straw with alkaline solutions prepared from wood ashes to improve fiber utilization by ruminants. *J. Anim. Sci.* 64:669–677.

Perdok, H.B. and R.A. Leng. 1987. Hyperexcitability in cattle fed ammoniated roughages. *Anim. Feed Sci. Tech.* 17:121–143.

Preston, T.R. and R.A. Leng. 1987. *Matching Ruminant Production Systems with Available Resources in the Tropics and Sub-Tropics.* Armidale, N.S.W., Australia: Penambul Books.

Sanderson, M.A., D.W. Meyer, and H. Casper. 1985. Dicumarol concentrations and forage quality of sweet clover forage treated with propionic acid or anhydrous ammonia. *J. Anim. Sci.* 61:1243–1252.

Shand, W.J., E.R. Orskov, and L.A.F. Morrice. 1988. Rumen degradation of straw. 5. Botanical fractions and degradability of different varieties of oat and wheat straws. *Anim. Prod.* 47:387–392.

Silankove, N., O. Cohen, D. Levanon, T. Kipnis, and Y. Kugenheim. 1988. Preservation and storage of green panic (*Panicum maximum*) as moist hay with urea. *Anim. Feed Sci. Tech.* 20:87–96.

Sundstol, F. and E. Owen, eds. 1984. *Straw and Other Fibrous By-Products as Feed.* Amsterdam: Elsevier.

Ward, J.W. and T.W. Perry. 1982. Enzymatic conversion of corn cobs to glucose with *Trichoderma viride* fungus and the effect on nutritional value of the corn cobs. *J. Anim. Sci.* 54:609–617.

Weiss, W.P., H.R. Conrad, C.M. Martin, R.F. Cross, and W.L. Shockey. 1986. Etiology of ammoniated hay toxicosis. *J. Anim. Sci.* 63:525–532.

Wood and Wood Products

Aderibigbe, A.O. and D.C. Church. 1987. Evaluation of cardboard and dried poultry waste as feed ingredients for ruminants. *Anim. Feed Sci. Tech.* 18:209–224.

Ayers, A.C., R.P. Barrett, and P.R. Cheeke. 1996. Feeding value of tree leaves (hybrid poplar and black locust) evaluated with sheep, goats and rabbits. *Anim. Feed Sci. Tech.* 57:51–62.

Bas, F.D., F.R. Ehle, and R.D. Goodrich. 1985. Evaluation of pelleted aspen foliage as a ruminant feedstuff. *J. Anim. Sci.* 61:1030–1036.

Baertsche, S.R., M.T. Yokoyama, and J.W. Hanover. 1986. Short rotation, hardwood tree biomass as potential ruminant feed—Chemical composition, nylon bag ruminal degradation and ensilement of selected species. *J. Anim. Sci.* 63:2028–2043.

Belyea, R.L., F.A. Martz, M. Madhisetty, and T.P. Davis. 1987. Intake, digestibility and energy utilization of corrugated paper diets. *Anim. Feed Sci. Tech.* 17:57–64.

Croy, D.S. and L.M. Rode. 1988. Nutritive value of pulp mill fiber waste for ruminants. *Can. J. Anim. Sci.* 68:461–470.

Drevjany, L.A., G.S. Hooper, and W.V. Candler. 1984. Processed poplar wood pellets as a source of energy for Holstein bull calves. *Can. J. Anim. Sci.* 64:1035–1044.

Hunt, J.R. and G.M. Barton. 1978. Nutritive value of spruce muka (foliage) for the growing chick. *Anim. Feed Sci. Tech.* 3:63–72.

Millett, M.A., A.J. Baker, W.C. Feist, R.W. Mellenberger, and L.D. Satter. 1970.

Modifying wood to increase its *in vitro* digestibility. *J. Anim. Sci.* 31:781–788.

National Research Council. 1983. *Underutilized Resources as Animal Feedstuffs.* Washington, DC: National Academy Press.

Singh, M. and L.D. Kamstra. 1981. Utilization of whole aspen tree material as a roughage component in growing cattle diets. *J. Anim. Sci.* 53:551–556.

Stegelmeier, B.L., D.R. Gardner, L.F. James, K.E. Panter, and R.J. Molyneux. 1996. The toxic and abortifacient effects of ponderosa pine. *Vet. Pathol.* 33:22–28.

Tait, R.M., J.R. Hunt, C. Gaston, and G.M. Barton. 1982. Utilization of lodgepole pine (*Pinus contorta*) muka (foliage) by sheep. *Can. J. Anim. Sci.* 62:467–471.

Xiccato, G., A. Trocino, and A. Carazzolo. 1998. Ensiling and nutritive value of kenaf (*Hibiscus cannabinus*). *Anim. Feed Sci. Tech.* 71:229–240.

Aquatic Plants and Seaweed

Ali, M.A. and S. Leeson. 1994. Nutritional value and utilization of aquatic weeds in the diet of poultry. *World's Poult. Sci. J.* 50:237–251.

Beames, R.M. 1990. Seaweed. In *Nontraditional Feed Sources for Use in Swine Production,* P.A. Thacker and R.N. Kirkwood, eds. pp. 386–396. London: Butterworth.

Beames, R.M., R.M. Tait, J.N.C. Whyte, and J.R. Englar. 1977. Nutrient utilization experiments with growing pigs fed diets containing from 0 to 20% kelp (*Nereocystis leutkeana*) meal. *Can. J. Anim. Sci.* 57:121–129.

Brewer, D., J. McLachlan, A.C. Neish, A. Taylor, and T.M. MacIntyre. 1979. Effects of feeding *Chondrus crispus* (Irish moss) during the grazing season on the condition and fertility of Shropshire ewes. *Can. J. Anim. Sci.* 59:95–100.

Cai, Z. and L.R. Curtis. 1989. Effects of diet on consumption, growth and fatty acid composition in young grass carp. *Aquaculture* 81:47–60.

Hansen, H.R., B.L. Hector, and J. Feldmann. 2003. A qualitative and quantitative evaluation of the seaweed diet of North Ronaldsay sheep. *Anim. Feed Sci. Tech.* 105:21–28.

Haustein, A.T., R.H. Gilman, P.W. Skillicorn, H. Hannan, F. Diaz, V. Guevara, V. Vergara, A. Gastanaduy, and J.B. Gilman. 1994. Performance of broiler chicks fed diets containing duckweed. *J. Agric. Sci.* 122:285–289.

Haustein, A.T., R.H. Gilman, P.W. Skillicorn, V. Vergara, V. Guevara, and A. Gastanaduy. 1990. Duckweed, a useful strategy for feeding chickens: Performance of layers fed with sewage-grown Lemnacea species. *Poult. Sci.* 69:1835–1844.

Heffron, C.L., J.T. Reid, W.M. Haschek, A.K. Furr, T.F. Parkinson, C.A. Bache, W.H. Gutenmann, E.E. St. John, and D.J. Lisk. 1977. Chemical composition and acceptability of aquatic plants in diets of sheep and pregnant goats. *J. Anim. Sci.* 45:1166–1172.

Herbert, J.G. and G. Wiener. 1978. The effect of breed and of dried seaweed meal in the diet on the levels of copper in liver, kidney and plasma of sheep fed on a high copper diet. *Anim. Prod.* 26:193–201.

Linn, J.G., E.J. Staba, R.D. Goodrich, J.C. Meiske, and D.E. Otterby. 1975a. Nutritive value of dried or ensiled aquatic plants. I. Chemical composition. *J. Anim. Sci.* 41:601–609.

Linn, J.G., R.D. Goodrich, D.E. Otterby, J.C. Meiske, and E.J. Staba. 1975b. Nutritive value of dried or ensiled aquatic plants. II. Digestibility by sheep. *J. Anim. Sci.* 41:610–615.

Lizama, L.C., J.E. Marion, and L.R. McDowell. 1988. Utilization of aquatic plants *Elodea canadensis* and *Hydrilla verticillata* in broiler chick diets. *Anim. Feed Sci. Tech.* 20:155–161.

MacLachlan, G.K. and W.S. Johnston. 1982. Copper poisoning in sheep from North Ronaldsay maintained on a diet of terrestrial herbage. *Vet. Rec.* 111:299–301.

National Academy of Sciences. 1976. *Making Aquatic Weeds Useful: Some Perspectives for Developing Countries.* Washington, DC: National Academy Press.

Riechert, C. and R. Trede. 1977. Preliminary experiments on utilization of water hyacinth by grass carp. *Weed Res.* 17:357–360.

Rusoff, L.L., E.W. Blakeney, and D.D. Culley. 1980. Duckweeds (*Lemnacea* Family), a potential source of protein and amino acids. *J. Agric. Food Chem.* 28:848–850.

Stansbury, W.F., L.F. Tribble, and D.E. Orr, Jr. 1990. Effect of chelated copper sources on performance of nursery and growing pigs. *J. Anim. Sci.* 68:1318–1322.

Ventura, M.R., J.I.R. Castanon, and J.M. McNab. 1994. Nutritional value of seaweed (*Ulva rigida*) for poultry. *Anim. Feed Sci. Tech.* 49:87–92.

Wolverton, B.C. and R.C. McDonald. 1978. Nutritional composition of water hyacinths grown on domestic sewage. *Econ. Bot.* 32:363–370.

CHAPTER 7

Micronutrients: Mineral and Vitamin Supplements

Objectives

1. To outline metabolic roles of minerals and vitamins and describe major deficiency signs
2. To put in perspective the relative practical importance of individual minerals and vitamins
3. To highlight areas of current interest:
   bioavailability of minerals
   dietary electrolyte balance and its role in milk fever, leg disorders, and heat stress
4. To describe mineral deficiencies of tropical grasslands

The feedstuffs previously discussed function primarily in providing protein and energy. In addition, they contain variable quantities of minerals and vitamins. Usually, supplementation with sources of these nutrients is required to produce balanced, nutritionally adequate diets. This chapter will emphasize the sources of minerals and vitamins and their bioavailability. However, practical aspects of metabolic functions, deficiency and toxicity symptoms, presence in feedstuffs, and nutrient interrelationships are also considered.

MINERAL SUPPLEMENTS

Minerals are the inorganic elements required as nutrients. (The macro- and microelements that are nutritionally essential are listed in Chapter 1.)

Calcium and Phosphorus The requirements and metabolism of these nutrients are closely interrelated; therefore, they will be discussed together. Bone mineral consists mainly of tricalcium phosphate and other salts of these two minerals. Approximately 99 percent of the calcium and 80 percent of the phosphorus in the animal body occur in the bones and teeth. **Phosphorus** is very important in cellular metabolism; many of the intermediates of carbohydrate metabolism are phosphorylated (e.g., glucose-6-phosphate, fructose-1,6-diphosphate, phosphoglyceric acid, and so forth). The central compound in energy metabolism, adenosine triphosphate (ATP), is a phosphorylated compound. Thus, reactions involving phosphorus are crucial to animal metabolism. Increasingly, calcium is being found to be involved in processes of energy metabolism as well.

Calcium and/or phosphorus deficiency symptoms reflect the metabolic functions of these nutrients as well as **vitamin D.** Vitamin D functions in calcium and phosphorus absorption and in bone mineralization. In growing animals, the primary sign of deficiency of calcium, phosphorus, and/or vitamin D, is rickets.

Rickets is characterized by spongy, poorly mineralized bones, leading to lameness, fractures, misshapen bones, and death. In adults, bone demineralization occurs with a deficiency of these nutrients. This condition is known as **osteomalacia. Osteoporosis** is a bone disorder in which the bone mass decreases, although its mineral composition is normal. Osteoporosis is common in elderly women.

Acute metabolic calcium deficiency causes an impairment of nerve and muscle function because of the role of calcium in the transmission of nerve impulses to muscle (neuromuscular junction). This situation results when serum calcium levels are depressed **(hypocalcemia).** This does not usually develop from a simple dietary deficiency of calcium, but from impaired calcium mobilization from bone. Hypocalcemia is common in dairy cattle; it is referred to as **milk fever** or **parturient paresis.** It occurs most frequently soon after lactation begins, because of a hormonal insufficiency (see Chapter 17). Normally, bone mineral is mobilized to meet the high calcium requirement for lactation. Mobilization of bone mineral is stimulated by various hormones. It is desirable to "prime" dry cows on a low-calcium diet to stimulate endocrine activity so that when lactation begins a high rate of mobilization of bone calcium can occur. The bone mineral is replenished during the dry (nonlactating) period.

Phosphorus deficiency of livestock is relatively common in various parts of the world where soil levels of the mineral are low. Such conditions are prevalent in parts of South America (e.g., the savannahs of Colombia, Venezuela, and Brazil) where the soils are acid and infertile. Phosphorus-deficient grazing animals often have a depraved appetite **(pica)** and consume wood, bones, and so on. Death from botulism may occur from bone chewing. Deficiencies this severe rarely occur in the United States. Less obvious phosphorus inadequacy may impair reproductive performance; cattle may not have normal estrous cycles and may not experience estrus or heat periods. Thus, conception rates are low and calving intervals are extended. In parts of the tropics, cows may not produce a calf more often than once every 2 or 3 years or may not even come into estrus because of severe phosphorus deficiency. Phosphorus nutrition of grazing cattle has been reviewed by Karn (2001).

The utilization of calcium and phosphorus is influenced by the relative amounts of each in the diet. The optimal **calcium-to-phosphorus ratio** is considered to be in the range 2:1 to 1:2. A high ratio of phosphorus to calcium may cause nutritional secondary hyperparathyroidism.

Sources of Calcium and Phosphorus Some of the common sources of calcium and phosphorus, and their mineral content, are shown in Table 7–1. Most calcium sources have high bioavailability. To supply only calcium, **ground limestone** (calcium carbonate) is usually the best choice. **Dicalcium phosphate** is commonly used to supply both elements. **Bone meal** was a common mineral supplement at one time but is less used now. Crushed **oyster shell** is extensively used in poultry diets to meet the very high calcium requirements of the laying hen for eggshell formation. The availability of calcium from **dolomitic limestones** is lower than for pure calcitic limestone (Ross et al., 1984). A pure calcitic limestone consists of crystals with alternating layers of calcium and carbonate ions. Substitution of magnesium ions for some of the calcium in dolomites results in a denser, less-soluble crystal having lower bioavailability. This also results in a low bioavailability of mag-

TABLE 7–1. Calcium and Phosphorus Sources Used as Feed Supplements or in Mineral Mixes

Source	*% Calcium*	*% Phosphorus*	*% Magnesium*
Calcium carbonate (limestone)	36	—	—
Dolomitic limestone (dolomite)	22	—	10
Oyster shell	35	—	0.3
Calcium sulfate (gypsum)	29	—	—
Bonemeal, steamed	29	14	0.6
Dicalcium phosphate	25–28	18–21	—
Defluorinated rock phosphate	32	18	—
Diammonium phosphate	—	23.5	—
Phosphoric acid	—	31.6	—
Sodium phosphate, monobasic	—	22.4	—
Sodium tripolyphosphate	—	30.85	—

nesium in dolomite. Considerable work has been conducted to determine the **optimal particle size** for calcium carbonate, particularly for dairy cattle and laying hens. Although some studies suggest that larger particle sizes may tend to "meter out" calcium over a prolonged period in hens, tending to improve eggshell quality, in general, the particle size appears not to be of much significance.

Rock phosphate is a commonly used source of phosphorus. Phosphate is mined in various parts of the world, including the United States (e.g., in Montana and Tennessee), the Caribbean region (e.g., Curacao), and the South Pacific islands. Deposits of phosphorus are relatively scarce. Rock phosphates must be processed at high temperature (calcining) to remove **fluorine,** which is often present at toxic levels in phosphate deposits. This treatment also increases the bioavailability of the phosphorus. Some rock phosphate deposits in the United States (e.g., Idaho and Wyoming) contain toxic levels of **vanadium** (1,400 ppm) and have caused vanadium toxicity when used in poultry diets (Ammerman et al., 1977). **Phosphoric acid** is used as a phosphorus source in liquid supplements (see Chapter 15). Phosphorus supplements tend to be somewhat unpalatable. Sources of phosphorus are usually quite expensive, so accurate formulation of diets to avoid using an excess is important in minimizing diet costs. Along with nitrogen, phosphorus is a major water pollutant, and livestock manure is a significant source of water pollution. Because of problems with water quality due to nitrogen and phosphorus contamination, limits on the number of livestock that can be raised in some European countries, such as Holland, have been legislated by the government. Thus, close attention to phosphorus in feed formulation is important. Excretion of phosphorus by swine and poultry can be minimized by the use of **phytase** as a feed additive (see Chapter 8).

Magnesium Magnesium is somewhat similar to calcium in its distribution in the body and in feeds. Approximately 70 percent of the magnesium in the body occurs in bone. Magnesium is a cofactor or activator of numerous enzymes involved in energy metabolism and functions in neuromuscular activity. Magnesium deficiency results in symptoms of hyperirritability, convulsions, muscle twitching, and death. **Hypomagnesemia** (grass tetany) is a common metabolic disease of cattle (see Chapter 5).

The sources of magnesium commonly used in diet supplementation and preparation of mineral mixes are magnesium oxide (MgO) and magnesium sulfate. The bioavailability of the element is high from these sources. For prevention of grass tetany, magnesium can be provided in mineral mixes (e.g., 50:50 MgO:salt), by dusting MgO on pastures, and by administering magnesium bullets consisting of magnesium alloy. The bullets are administered orally; they lodge in the reticulo-rumen and release magnesium continuously as they dissolve (Stuedemann et al., 1984).

Sodium, Potassium, and Chlorine Sodium, potassium, and chlorine are considered together because they are closely interrelated in their metabolism. **Common salt** (NaCl) is routinely added to animal feeds and provided free choice to grazing animals. In North America and Europe, provision of salt to livestock is almost universal. One of the principal reasons for this is that animals, particularly ruminants, have an innate desire to consume salt, which has often been interpreted as reflecting a physiological and nutritional need. However, as Morris (1980) points out, animals (particularly herbivores) have a salt appetite and consume much more salt than required for optimal health. In contrast, the salt appetite in humans is acquired primarily as a result of social and dietary customs. In humans, excessive salt consumption has been linked with hypertension, but in livestock, there is no evidence that "luxury consumption" of salt has harmful effects so long as adequate water is available. If water availability is limited, salt toxicity can occur. Diets of herbivores tend to be high in potassium; the sodium-to-potassium imbalance may be a contributing factor in the salt appetite of these animals.

Ruminants have a pronounced ability to conserve sodium. The rumen acts as a sodium storehouse. When sodium-deficient diets are consumed, rumen sodium is drawn upon to counteract blood sodium depletion, and potassium, which is abundant in grass, replaces sodium in saliva. These mechanisms allow ruminants to adapt to the large sodium-deficient areas of the world, which include most noncoastal land areas. Northern ruminants, such as moose, eat large amounts of sodium-containing aquatic plants during a 3-month season and draw on the rumen sodium pool during the rest of the year (Botkin et al., 1973).

Under range conditions, strategic placement of salt can be used to control the distribution of livestock. Salt is often used as a vehicle for providing trace elements, worming agents, and antibloat compounds. Iodized salt and trace mineralized salt are the forms normally used. For supplementing range cattle with protein, protein supplements can be mixed with salt to limit consumption. A mixture containing 40 percent salt will usually result in adequate intake. Supplements containing salt can also be used to increase water intake of cattle to aid in preventing urinary calculi (see Chapter 15). A supplement containing 15 percent salt is adequate (Bailey, 1981).

For swine and poultry diets, the addition of 0.25 to 0.5 percent salt is a standard practice. Usually the salt is trace mineralized. Many **trace-mineralized salt** preparations are available commercially.

Potassium deficiency is not commonly encountered in livestock production, but it can occur under certain circumstances. Most feedstuffs contain adequate **potassium,** with grains tending to be lower than good-quality roughages. Mature, weathered, poor-quality roughages may lack adequate amounts. Feedlot cattle on high-concentrate diets sometimes respond to dietary potassium supplementation. There has been interest in the effect of stress on potassium. Stress conditions, such as in transporting and marketing feeder cattle and pigs, may stimulate adrenal cortex activity and increase potassium excretion. Dehydration and diarrhea increase loss of the element. Hutcheson and Cole (1986) reported that supplementation of receiving diets for feedlot calves with potassium to provide a total daily intake of 26 g potassium per 100 kg body weight was necessary for optimal weight gains. In contrast, Brumm and Schricker (1989) found no beneficial effects from supplementation of transport-stressed feeder pigs with potassium chloride and concluded that corn-soy diets contain sufficient potassium for pigs experiencing stress in marketing and transport. Further work is necessary to conclusively demonstrate a role for potassium supplementation under practical conditions.

Because alfalfa and grains have a fairly high potassium content, forage crops grown in association with dairy farms may have a high potassium content. As the dairy industry intensifies, there is a trend toward large numbers of cows being kept in confinement, with the manure being disposed of on surrounding crop land, which is often used for forage (e.g., corn silage) production. Forage potassium levels in excess of 5 percent are being encountered (Fisher et al., 1994). When these **high-potassium forages** are fed to dairy cattle, there is an increase in water consumption and urine volume, creating management problems. High potassium intakes interfere with magnesium and calcium absorption and may promote hypomagnesemia and milk fever. The high potassium intake may adversely affect cation-anion balance (see next section).

Dietary Electrolyte Balance Sodium, potassium, and chlorine are interrelated in regulating **electrolyte balance** of animal tissues. **Electrolytes** are electrically charged dissolved substances. The animal body is electrically neutral; acid-base balance is determined as the difference between total cation and anion intake and excretion. **Dietary fixed ions** are bioavailable ions that are not metabolized (e.g., Ca^{2+}, Cl^-), whereas the term **dietary undetermined anion** refers to ions, such as bicarbonate, that balance the dietary fixed cations:

$$\text{Dietary undetermined anion} = (Na^+ + K^+ + Ca^{2+} + Mg^{2+}) - (Cl^- + S^{2-} + P^{2-})$$

Dietary cations are usually present in excess of dietary anions; therefore, to maintain electrical neutrality, they are balanced by the dietary undetermined anions.

The **dietary electrolyte balance** can be calculated as the sum of cations minus the sum of nonmetabolizable anions:

$$\text{Macromineral balance (meq/kg)} = (Na^+ + K^+ + Ca^{2+} + Mg^{2+}) - (Cl^- + P^{2-} + S^{2-})$$

The ions are expressed as milliequivalents because electrolyte balance deals with the equivalence of electrical charges.

The acidogenicity or alkalinogenicity of the diet is largely determined by the relative deficit or excess, respectively, of **metabolizable anions,** such as acetate and citrate (Patience et al., 1987; Patience and Chaplin, 1997). Under most circumstances, the dietary mineral balance is adequately expressed as Na + K - Cl (meq/kg). For poultry, the optimal electrolyte balance is about 250 meq/kg (Hulan et al., 1987; Johnson and Karunajeewa, 1985; Mongin, 1981); electrolyte imbalance results in reduced growth and induces leg abnormalities such as tibial dyschondroplasia (slipped tendon). For swine, the Na + K - Cl (meq/kg) should be in the 100 to 200 range for optimal performance.

The number of milliequivalents per kilogram of diet can be calculated as follows.

Na: meq/kg diet = mg Na/kg ÷ 23.0
K: meq/kg diet = mg K/kg ÷ 39.1
Cl: meq/kg diet = mg Cl/kg ÷ 35.5
Ca: meq/kg diet = mg Ca/kg ÷ 20.0
Mg: meq/kg diet = mg Mg/kg ÷ 12.15
P: meq/kg diet = mg P/kg ÷ 10.3
S: meq/kg diet = mg S/kg ÷ 16.0

The electrolyte balance is influenced by the form in which minerals are provided, that is, inorganic (e.g., NaCl) or organic (e.g., Na citrate) salts. Organic salts contain a metabolizable anion (e.g., citrate is converted to CO_2 and H_2O); thus diets high in metabolizable anions will have an excess of cations relative to anions, giving an alkalinogenic diet.

Amino acid status can influence electrolyte balance, particularly in poultry. In birds, certain amino acids are degraded in the kidney; these include arginine, lysine, histidine, isoleucine, and tyrosine. The enzyme involved is kidney arginase. An excess of dietary lysine causes marked elevation of kidney arginase, resulting in excessive degradation of arginine and arginine deficiency. This lysine-arginine antagonism is influenced by electrolyte status (Austic and Calvert, 1981). The effect of excess lysine is magnified by an excess of chloride. Conversely, responses to dietary supplements of sodium or potassium bicarbonate in swine and poultry fed lysine-deficient diets have been reported (Patience et al., 1987). When the dietary electrolyte balance decreases below about 175 meq/kg, blood pH and bicarbonate drop, indicating a metabolic acidosis. Thus, there is a complex interrelationship among dietary electrolytes, amino acids, and acid-base balance. Typical corn-soy swine diets have an electrolyte balance of approximately 175 meq/kg. The trend to replacement of part of the soybean meal with synthetic lysine (lysine · HCl), thus reducing potassium and increasing chloride, could lead to increased concerns with acid-base balance (Patience et al., 1987).

Electrolyte balance in ruminants is complicated by a number of factors, which are reviewed by Fredeen et al. (1988). Diets containing corn silage and grains have cation deficits and are acidogenic, whereas alfalfa has a cation excess and is alkalinogenic. Ruminant diets are often supplemented with sodium and potassium-containing buffers, calcium carbonate, and magnesium oxide. Prepartum alkalosis may increase the incidence of parturient paresis (milk fever), whereas acidosis may help prevent it (Fredeen et al., 1988; Leclerc and Block, 1989). Diets that are acidic (excess anions) have greater calcium absorption. Those with excess cations may have reduced calcium availability. Thus, in ruminants, the complete macromineral balance equation, which includes calcium, magnesium, phosphorus, and sulfur, probably should be used in calculating electrolyte balance.

There is increasing recognition of the importance of electrolyte balance in the feeding of all types of livestock and poultry. Acid-base balance can influence growth and appetite, leg disorders in poultry, response to thermal stress, incidence of milk fever in dairy cattle, and the metabolism of various amino acids, minerals, and vitamins (Patience, 1990).

Sulfur Sulfur is widely distributed in animal tissues as a component of other nutrients. These include the S-amino acids (methionine, cysteine, cystine, taurine), the vitamins thiamin and biotin, and mucopolysaccharides in connective tissue and mucus secretions. Inorganic sulfate, which is a component of mucopolysaccharides, can be derived from the metabolism of S-amino acids. Thus sulfur per se is not a dietary requirement for nonruminants, although inorganic sulfate in the diet has a sparing effect on the S-amino acid requirement.

In ruminants, sulfur is a dietary requirement under some circumstances. It is required for microbial synthesis of S-amino acids. It is usually derived in adequate quantity from dietary protein, but when NPN constitutes a major part of the dietary nitrogen, supplementation with sulfur may be needed. Elemental sulfur or sulfur salts (sulfates) may be used as sources of the element. A dietary level of 0.2 to 0.3 percent sulfur is adequate for ruminants. Levels above 0.4 percent may be toxic.

Manganese Manganese has a metabolic role in the synthesis of connective tissue; it is a cofactor for an enzyme involved in synthesis of mucopolysaccharides and glycoproteins. **Manganese deficiency symptoms** include defective bone formation, perosis (slipped tendon) in poultry, and reduced growth. Poultry have a high requirement for manganese, probably due in part to high dietary calcium levels, which impair manganese absorption. High dietary iron levels also reduce manganese absorption. The bioavailability of manganese in the diet is quite low; in cattle, only 1 to 4 percent of dietary manganese is absorbed (Hurley and Keen, 1987). Manganese oxide, manganese carbonate, manganese sulfate, and manganese chloride all have similar bioavailability. Manganous sulfate is the commonly used salt in animal diets.

Zinc Zinc is a constituent or cofactor of a variety of enzymes in the body, including a number involved in nucleic acid and protein metabolism. Of key importance are zinc-dependent enzymes functioning in **DNA synthesis.** In zinc deficiency, cell replication is impaired. Therefore, the most rapidly growing tissues are the first to show signs of deficiency. These include the skin, gastrointestinal tract, wound-repair tissue, and the reproductive tract. Growth retardation is the first observable sign (in growing animals) followed by dermatitis.

Zinc is absorbed equally well when provided as the oxide, carbonate, sulfate, or metallic zinc. Zinc oxide and zinc sulfate are the most commonly used forms in the feed industry.

Zinc is one of the trace elements most likely to be deficient in swine and poultry diets because of the effect of dietary phytates on **zinc bioavailability.** Zinc deficiency first became a problem in the United States in the early 1950s, when the use of soybean meal as a protein supplement became widespread.

Soybean meal has a higher content of phytate than many other protein sources, so its use increases the dietary zinc requirement. The signs of zinc deficiency observed under practical conditions when soybean meal was introduced were poor growth, dermatitis (parakeratosis), poor feathering, and general ill-thriftiness. National Research Council zinc requirement figures for swine and poultry have a built-in allowance for zinc tie-up by soybean meal. Zinc deficiency in ruminants is uncommon in the United States but is encountered quite frequently in the tropics (Fig. 7.1).

The availability of zinc and other trace elements is influenced by the process of chelation. A **chelate** is formed by the complexing of a mineral element with an organic compound (the ligand). **Ligands** are referred to as chelating agents or metal-binding agents. The strength with which a ligand binds a mineral is expressed by the **stability constant.** If the stability constant is high, the chelate will be very stable and may bind the mineral in an unavailable form. The stability constants of various elements with particular ligands vary. An element with a high stability constant will displace another element with a lower stability constant. A substance called ethylenediaminetetraacetic acid (EDTA) is an excellent chelating agent; chelates of trace elements with EDTA have high stability constants. Because EDTA is absorbable, addition of EDTA to the diet may increase the absorption of trace elements. This process is known as sequestering minerals, and the ligand is referred to as a **sequestering agent.**

There is some interest in the feed industry in sequestering agents, but in general it is probably cheaper and easier to increase the level of dietary trace elements than to add a sequestering agent if low mineral availability is anticipated.

FIGURE 7.1 Cattle in Brazil suffering from multiple trace element deficiencies, including copper, zinc, and cobalt, as well as phosphorous deficiency. (Courtesy of Lee R. McDowell.)

Seaweed (kelp) is sometimes promoted as being a sequestering agent. Spears (1989) reported that zinc chelated with methionine (zinc methionine) was of similar bioavailability in beef cattle as zinc oxide but was used somewhat more efficiently after absorption. Urinary excretion of zinc was less with zinc methionine, and animal growth rate was slightly higher. A number of organic substances in feedstuffs reduce mineral availability by chelation. These include phytates in cereal grains and plant protein supplements, and oxalic acid in tropical grasses and some poisonous plants (e.g., halogeton).

Iron Iron is a constituent of several metallo-proteins, including the respiratory pigments, such as hemoglobin, cytochrome c, and myoglobin, and enzymes, such as peroxidase and catalase. A principal sign of iron deficiency is **anemia.** Compared to most other minerals, iron is absorbed to a very low extent. The body conserves iron tenaciously and regulates absorption to the amount necessary to replenish losses. In general, ferrous (Fe^{2+}) salts are absorbed more efficiently than ferric (Fe^{3+}) salts. Iron in heme compounds is absorbed efficiently, so meat is a good source of iron in the human diet. Meat in the diet improves the absorption of iron from other sources as well. Organic acids, especially ascorbic acid (vitamin C), improve iron absorption. Ferrous sulfate is the major iron salt used in mineral mixes and diet formulation.

Iron deficiency is not a major problem in livestock. Soil is an excellent source of the element; contamination of feedstuffs with soil often supplies sufficient iron, as does iron consumed by herbivores while grazing. Because **milk** is low in iron, animals such as baby pigs may be susceptible to iron-deficiency anemia if they are not exposed to an environmental source of iron such as soil. Baby pigs are routinely given iron injections at or near birth to prevent anemia.

Copper Copper is a constituent of numerous enzymes, including some of those involved in iron metabolism. Thus **anemia** may be a sign of copper deficiency. Other signs include hair depigmentation (achromotrichia), reduced growth, aortic rupture, abnormal bone formation, ataxia (incoordination), and gastrointestinal disturbances. Most of these can be explained in terms of the metabolic roles of copper. **Hair depigmentation** results because copper is a cofactor of an enzyme for the conversion of tyrosine to the melanin pigment. Usually there is a graying of dark hair in copper-deficient animals. Hereford cattle develop yellow instead of red hair in manifestation of copper deficiency, while the normally black hair of Holsteins may be red. **Aortic rupture** and skeletal deformity are a reflection of the role of copper in the synthesis of collagen and other connective tissues. Copper is an activator of lysyl oxidase, an enzyme involved in the crosslinking of collagen and elastin fibers. **Ataxia** in copper-deficient animals seems to be caused by a deficiency of cytochrome oxidase activity in the motor neurons of the central nervous system. In cattle, **severe diarrhea** may accompany copper deficiency. This is probably a malabsorption syndrome, caused by atrophy of villi and diminished cytochrome oxidase activity in mucosal cells (Fell et al., 1975).

The primary sources of copper used in diet formulation are copper sulfate and copper oxide. Copper sulfide is less bioavailable than other forms. Copper is available to the feed industry in chelated forms, such as copper-lysine. There is little difference in the bioavailability of copper from copper sulfate and copper lysine in pigs (Apgar et al., 1995; Coffey et al., 1994) and ruminants (Luo et al., 1996).

In many parts of the world, **chronic copper deficiency** problems occur with grazing animals. These include Florida in the United States, western Europe, Australia, New Zealand, and South America. Signs include a wasting condition or emaciation, diarrhea, depigmented hair, and anemia. In some cases, it is due to uncomplicated copper deficiency. In others, it is a result of a complex interaction between **copper, molybdenum,** and **inorganic sulfate.** High molybdenum in forage intensifies copper deficiency or may induce a copper deficiency even though forage copper levels may be normal. The mechanism of action appears to be that sulfide produced from sulfates or sulfur amino acids in the rumen reacts with molybdate ions to produce **thiomolybdates.** The thiomolybdates react with copper to form products in which the copper is physiologically unavailable. According to Ward (1978), there are four general situations in which hypocuprosis may be induced by molybdenum: (1) high dietary molybdenum levels in excess of 100 ppm, (2) a low copper-to-molybdenum ratio of 2:1 or less, (3) forage copper of 5 ppm or less, and (4) high protein (20 to 30 percent) in fresh forage. In the latter case, very soluble cytoplasmic proteins of pasture are readily fermented in the rumen, producing sulfide from the sulfur amino acids, followed by formation of unabsorbable copper thiomolybdate. Further discussion of these interrelationships is given by Suttle (1991) and Ward (1978).

Selenium Selenium and vitamin E interact nutritionally by preventing damage to cell membranes from peroxides and other reactive compounds formed during normal metabolic processes. **Vitamin E** acts as an antioxidant within cell membranes and combines with free radicals and other reactive compounds, thus preventing oxidation processes. The process of autooxidation of lipids in animal tissues is similar to the development of rancidity in feeds; both are prevented by vitamin E. **Selenium** helps protect against autooxidation of cell membranes by virtue of being a component of an enzyme, **glutathione peroxidase,** that reduces (provides hydrogen) peroxides, thus converting them to innocuous products. Glutathione peroxidase occurs mainly in the cytosol and reduces peroxides before they can attack cell membranes, whereas vitamin E acts within the membrane itself as a second line of defense. Selenium also functions as a component of a **deiodinase** that converts the thyroid hormone thyroxin into its metabolically active form (triiodothyronine) by the removal of one of the four iodines in thyroxin (Burk and Hill, 1993). Thus selenium has a nutritional interrelationship with iodine. Selenium is also a component of some **selenoproteins** in blood and muscle, for which functions have not yet been identified (Yeh et al., 1997).

There are problems with both deficiency and toxicity of selenium. Selenium-deficient soils are often of volcanic origin; selenium is a volatile element and is volatilized into the atmosphere during volcanic episodes. Areas where it has reentered the soil often have toxic selenium levels. For example, the U.S. Pacific Coast states are selenium deficient, whereas the Great Plains states (North Dakota, South Dakota, Wyoming, and Nebraska) have soils with high selenium levels (Fig. 7.2). Certain plants growing in these areas (e.g., *Astragalus* spp.) may accumulate very high toxic levels of selenium and cause livestock poisoning. **Selenium toxicity** causes alkali disease and blind staggers, which are characterized by abnormal hoof and hair growth, abdominal distress, diarrhea, difficult breathing, prostration, and death. **Selenium deficiency** conditions include

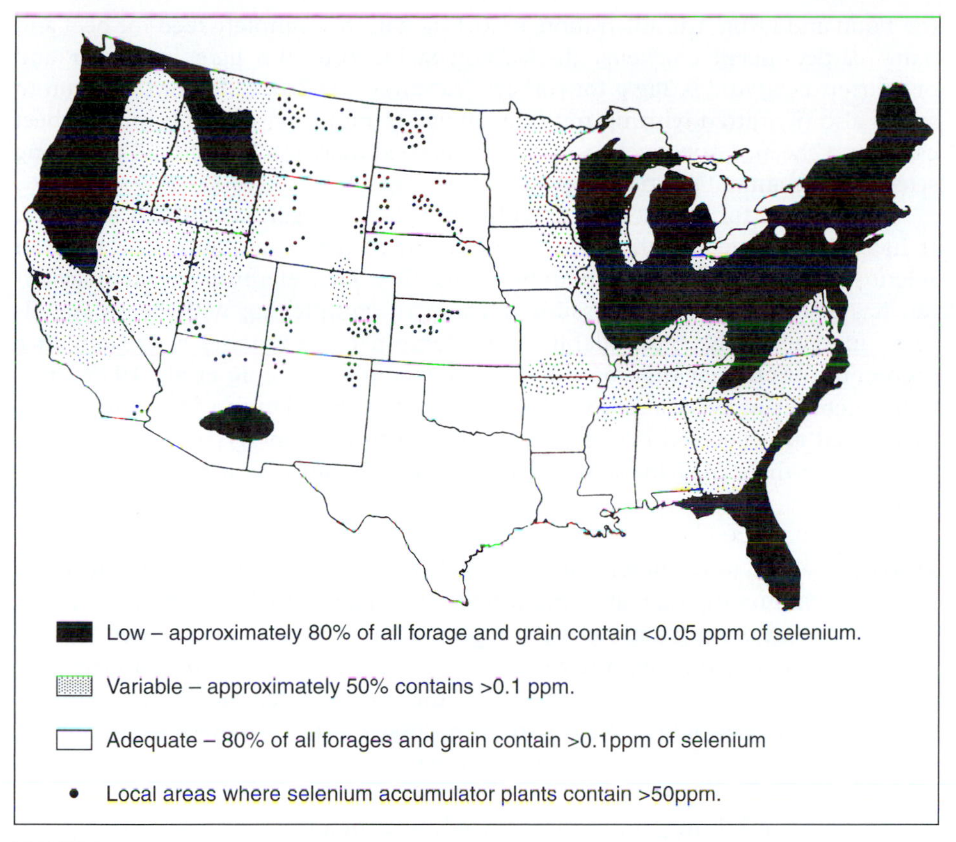

FIGURE 7.2 Selenium status of feedstuffs throughout the United States. Selenium-deficient areas include the Pacific Northwest, Great Lakes region and Northeast, and Florida. Areas with selenium toxicity occur throughout the Great Plains states. (Courtesy of J. Kubota, U.S. Plant, Soil and Nutrition Laboratory, Ithaca, N.Y.)

nutritional muscular dystrophy in all species (white muscle disease, stiff lamb disease), exudative diathesis in chickens (subcutaneous hemorrhages), liver necrosis in swine and rats, unthriftiness (ill-thrift) in ruminants, and reproductive disorders in many species (see Fig. 16.1).

Subclinical selenium deficiency may increase the incidence of **retained placenta** in dairy cattle. Selenium supplementation of animals on diets of marginal selenium status enhances immune status and the ability of the immune system to respond to disease challenges (Knight and Tyznik, 1990).

For most livestock species, a dietary level of 0.1 ppm selenium is adequate to prevent signs of deficiency with an added margin of safety. The true metabolic requirement is closer to 0.05 ppm. Thus, selenium is required at an extremely low level.

Various methods of administration of selenium to livestock include its use in salt mixtures, as a direct additive to the feed, by periodic injections or oral doses, and by administration of selenium "bullets" to ruminants. The "bullets" may be pellets formed from iron or grass boluses, in each case containing a soluble selenium salt. In the United States, selenium levels in the feed permitted by the Food and Drug Administration are 0.1 μg/g in the complete feed for beef and dairy cattle, sheep, chickens, ducks, and swine (but 0.3 μg/g in starter and prestarter diets) and 0.2μg/g for turkeys (Levander, 1986). Selenium addition to salt is also permitted with up to 20 or 30 μg selenium per g of salt mix for beef cattle and sheep, respectively, so as not to exceed daily intakes of 0.23 and 1 mg selenium per animal per day.

The usual form of selenium addition to feeds or salt is **sodium selenite.** It has a high bioavailability. The selenium in plants is largely in the form of seleno-amino acids and has a high bioavailability. The selenium in fish meals often has a low availability, possibly because of complexing with heavy metals (e.g., mercury). The bioavailability of selenium to ruminants is greater on a high-concentrate diet than on a high-roughage diet (Koenig et al., 1997). Rumen microbes convert a large proportion of dietary selenium into unavailable forms that are excreted in the feces (Hartmann and van Ryssen, 1997). High concentrate diets may increase selenium bioavailability as a result of a lower rumen pH.

Iodine The only known metabolic role of iodine is as a component of the thyroid hormones thyroxin and triiodothyronine. The **thyroid hormones** play an integral role in the regulation of the rate of cellular metabolism. When thyroid activity is inadequate (hypothyroidism), the metabolic rate (measured by O_2 consumption or rate of heat production), the growth rate, and productivity are reduced. Hyperthyroidism results in an excessive rate of metabolism, and the body tissues are oxidized to support the excessive metabolism, producing extreme emaciation.

Iodine deficiency results in a lack of the element for thyroid hormone synthesis. In a compensatory response, the body reacts by increasing the amount of thyroid tissue in an effort to produce more hormone. This enlargement of the thyroid gland is called **goiter** and is a classic sign of iodine deficiency. This is most evident in the offspring of iodine-deficient animals. In calves, lambs, and kids born to iodine-deficient dams the enlarged thyroid is very obvious at birth. The young are born weak or dead. In pigs, hairlessness at birth is the prominent sign, whereas in foals the main sign is extreme weakness. Animals born with goiter usually fail to survive.

Many areas of the United States and other countries have iodine-deficient soil and water. Routine use of **iodized salt** for humans and livestock is recommended as an inexpensive, convenient means of preventing iodine deficiency. The direct addition of iodine (e.g., potassium iodide) to the diet should be done cautiously. Iodine toxicity may occur; similar to the deficient state, iodine excess causes goiter in fetuses. Iodine requirements can be increased by the presence of **goitrogens** (e.g., glucosinolates) present in some feeds, particularly *Brassica* spp. Some goitrogenic agents interfere with iodine uptake by the thyroid gland, so their effects can be overcome with supplementary iodine.

Seaweed (kelp) meal is sometimes promoted as being a good source of chelated trace minerals. It is high in iodine, but there is no advantage to its use over the simple employment of iodized salt. Organic iodine compounds (e.g., ethylenediamine dihydroiodide, EDDI) are sometimes used as feed additives for the control of disease such as foot rot and lumpy jaw in cattle.

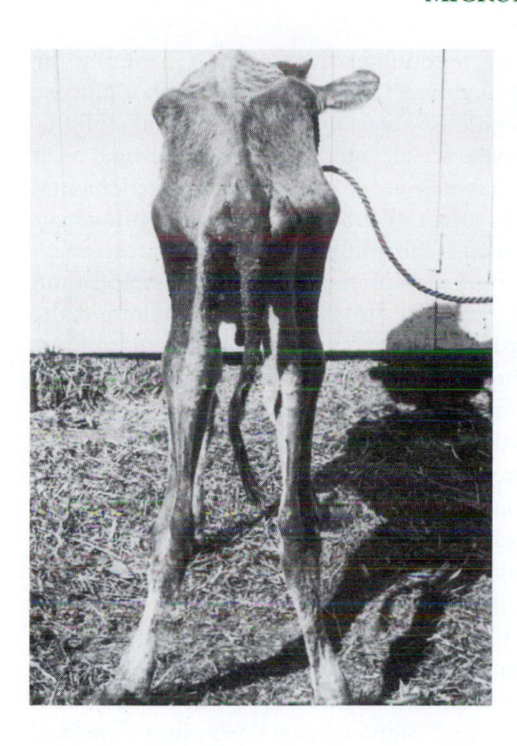

FIGURE 7.3 Cobalt-deficient cow. Cobalt-deficient ruminants are unable to metabolize VFA and become extremely emaciated as a result.

Cobalt A wasting disease of cattle and sheep characterized by severe emaciation is caused by a deficiency of cobalt (Fig. 7.3). Forages in parts of the United States (e.g., Florida, northeastern states), Australia, and South America are often cobalt-deficient because of the low cobalt content of the soil. Cobalt is a dietary essential for ruminants and other herbivores. It is required for the microbial synthesis of **vitamin B_{12}.** This vitamin has a very complex chemical structure and contains cobalt in a chelated form. In ruminants, the vitamin B_{12} requirement is met by microbial synthesis of the vitamin in the rumen, so dietary cobalt is required for this synthesis to occur. Thus **cobalt deficiency symptoms** in ruminants are actually a result of vitamin B_{12} deficiency. Because this vitamin is required for normal metabolism of VFA absorbed from the rumen, cobalt-deficient animals suffer from a cellular lack of energy and become extremely emaciated (Fig. 7.3). They are also anemic, because vitamin B_{12} is required for formation of hemoglobin. Appetite is markedly depressed, probably because of high blood levels of unmetabolized VFA.

In areas where cobalt deficiency occurs, it is easily corrected by the provision of cobaltized salt. A mixture of 150 g of cobalt sulfate per 100 kg of salt can be used. In Australia, cobalt "bullets" have been used quite successfully. Glass boluses containing trace elements have also been developed (Zervas, 1988). These are administered orally and lodge in the reticulorumen, gradually releasing trace elements.

Chromium Chromium has been identified as an essential nutrient for humans and animals (Mertz, 1993). It is a component of the **glucose tolerance factor,** a complex containing trivalent chromium, nicotinic acid (niacin), and amino acids. The glucose tolerance factor potentiates the action of insulin in the uptake of glucose by cells. **"Glucose tolerance"** refers to the ability of glucose to be taken up by cells. With impaired glucose tolerance, cellular uptake of glucose is subnormal, and blood glucose levels are elevated **(hyperglycemia),** as in the disease diabetes. Some cases of diabetes are responsive to supplemental chromium (Mertz, 1993). Because of its effect on insulin, chromium influences carbohydrate, fat, and protein metabolism. Chromium is absorbed most effectively in an organic form, with chromium picolinate being the most common form. Picolinic acid is an organic acid derived from the vitamin niacin and the amino acid tryptophan. **Chromium picolinate** is approved as a feed additive for swine. When fed to growing-finishing pigs, it increases leanness and reduces carcass fat (Kornegay et al., 1997; Mooney and Cromwell, 1997).

In cattle, chromium supplementation has been of interest as a means of increasing the immunocompetence of stressed calves entering the feedlot (Mowat et al., 1993; Chang et al., 1995). Chromium supplementation has beneficial effects in reducing morbidity (illness) from pneumonia, shipping fever, and other forms of respiratory disease. Van Heugten and Spears (1997) did not find an effect of chromium as Cr chloride, Cr picolinate, or Cr-nicotinic acid complex on the immune status of stressed pigs. Data that unequivocally show an immunostimulatory effect of chromium are lacking, but it is documented that under stressful conditions there is increased loss of chromium in the urine (Van Heugten and Spears, 1997). Kegley et al. (1997) found that supplemental chromium improved the growth rate of steers that were stressed by transportation but did not have any effect on immune responses, suggesting other modes of action may be involved.

Bioavailability of Minerals

Bioavailability of mineral sources refers to their ability to supply absorbable mineral elements in a chemical form that can participate in metabolic functions. In most cases, if an element is absorbed, it can function physiologically. Salts of elements differ in their solubility and other properties that affect their bioavailability. Comparative bioavailabilities of sources of minerals commonly available for diet formulation are given in Table 7–2. Chelating agents may improve mineral bioavailability, as discussed for zinc.

TABLE 7–2. Percent of Mineral Element in Some Sources Commonly Used in Mineral Supplements and Relative Bioavailability

Element	*Source Compound*	*% of Element in Compound*	*Bioavailability*
Calcium	Steamed bone meal	29.0(23–37)	High
	Defluorinated rock phosphate	29.2(19.9–35.7)	Intermediate
	Calcium carbonate	40.0	Intermediate
	Soft phosphate	18.0	Low
	Ground limestone	38.5	Intermediate
	Dolomitic limestone	22.3	Intermediate
	Monocalcium phosphate	16.2	High
	Tricalcium phosphate	31.0–34.0	
	Dicalcium phosphate	23.3	High
	Hay sources	variable	Low

Cobalt	Cobalt carbonate	46.0–55.0	Critical tests not
	Cobalt sulfate	21.0	done but com-
	Cobalt chloride	24.7	pound effective
Copper	Cupric sulfate	25.0	High
	Cupric carbonate	53.0	Intermediate
	Cupric chloride	37.2	High
	Cupric oxide	80.0	Low
	Cupric nitrate	33.9	Intermediate
Iodine	Calcium iodate	63.5	Available but
	EDDI	80.0	unstable
	Potassium iodide, stabilized	69.0	High
	Pentacalcium orthoperiodate		
Iron	Iron oxide	46.0–60.0	Unavailable
	Ferrous sulfate	20.0–30.0	High
	Ferrous carbonate	36.0–42.0	Low
Magnesium	Magnesium carbonate	21.0–28.0	High
	Magnesium chloride	12.0	High
	Magnesium oxide	54.0–60.0	High
	Magnesium sulfate	9.8–17.0	
	Potassium and magnesium sulfate	11.0	High
Manganese	Manganous sulfate	27.0	High
	Manganous oxide	52.0–62.0	High
Phosphorus	Defluorinated rock phosphate	13.3(8.7–21.0)	Intermediate
	Calcium phosphate	18.6–21.0	High
	Dicalcium phosphate	18.5	Intermediate
	Tricalcium phosphate	18.0	
	Phosphoric acid	23.0–25.0	High
	Sodium phosphate	21.0–25.0	High
	Potassium phosphate	22.8	
	Soft phosphate	9.0	Low
	Steamed bone meal	12.6(8–18)	High
Potassium	Potassium chloride	50.0	High
	Potassium sulfate	41.0	High
	Potassium and magnesium sulfate	18.0	High
Selenium	Sodium selenate	40.0	High
	Sodium selenite	45.6	High
Sulfur	Calcium sulfate (gypsum)	12.0–20.1	Low
	Potassium sulfate	28.0	High
	Potassium and magnesium sulfate	22.0	High
	Sodium sulfate	10.0	Intermediate
	Anhydrous sodium sulfate	22.0	
	Sulfur, flowers of	96.0	Low
Zinc	Zinc carbonate	52.0	High
	Zinc chloride	48.0	Intermediate
	Zinc sulfate	22.0–36.0	High
	Zinc oxide	46.0–73.0	High

Adapted from McDowell (1985).

Mineral Supplementation of Ruminants in the Tropics

One of the most important limitations to efficient livestock production in tropical areas is mineral deficiencies in the forage. Often simultaneous deficiencies of several minerals are involved, producing symptoms of wasting (emaciation), loss of hair, hair depigmentation, dermatitis, bone abnormalities, abortion, and poor reproductive performance. The high incidence of mineral deficiency reflects geologic and climatic factors. Younger geological formations are higher in minerals than are soils derived from older, more acid, coarse, sandy rock formations. Under the conditions of high rainfall and temperature characteristic of much of the tropics, there is heavy leaching of minerals from soils. In tropical rain forests, almost all of the minerals available are in the plants themselves and are recycled from decaying vegetation. When these lands are cleared for agriculture, the soil is often severely deficient in minerals. Many tropical soils are acid to varying degrees. Uptake of most minerals by plants is impaired under acidic conditions. High crop and forage yields increase the rate at which minerals are removed from the soil and intensify mineral-deficiency problems.

Some of the most widespread mineral deficiencies in the tropics are those of phosphorus, sodium, copper, selenium, cobalt, iodine, and zinc. Phosphorus deficiency is responsible for many fertility problems, with cattle in many areas having a calving interval of several years. Copper deficiency is very widespread, second only to phosphorus deficiency in importance. Salt deficiency is common because tropical forages have a low sodium content, and the high temperatures increase losses in sweat. Goiter due to iodine deficiency is very common. Zinc deficiency in grazing animals is very rare in temperate regions but is quite common in the tropics. Severe dermatitis and reproductive failure occur in zinc deficiency. A complete mineral mixture containing all of the nutritionally essential minerals should be provided to ruminants grazing tropical grasslands. Further information is provided by McDowell (1985).

VITAMIN SUPPLEMENTS

As discussed in Chapter 1, vitamins are classified as fat soluble (A, D, E, K), and water soluble (B-complex, C). There are some general comments on vitamins that are relevant to vitamin requirements and their occurrence in feeds.

Animals with microbial digestion (ruminants and nonruminant herbivores) generally do not require dietary sources of the water-soluble vitamins and vitamin K because they are synthesized by gut microbes. In ruminants, the vitamins synthesized in the rumen are absorbed in the small intestine. In nonruminant herbivores that practice coprophagy or cecotrophy, such as the rabbit, microbially synthesized vitamins are made available to the host when the cecal contents (cecotropes) are consumed. In the case of animals like horses that do not routinely engage in coprophagy, there is probably adequate absorption of water-soluble vitamins in the cecum and colon. Thus for grazing animals, microbial synthesis provides all but vitamins A, D, and E. Vitamin D is synthesized in the skin upon exposure to sunlight, and green forage is an excellent source of vitamins A and E. Thus under many conditions, grazing animals require no vitamin supplementation.

Swine and poultry kept under modern confinement systems of management no longer have access to sunlight and to green plants and insects that at one time served as their vitamin sources. Thus vitamin supplementation for these animals is very important. It is commonly believed that exposure to stress increases vitamin requirements, but data supporting this contention are lacking. Most feed manufacturers use vitamin premixes prepared by companies that specialize in micronutrient premixes. Usually these are formulated to provide a "margin of safety" over NRC requirement figures, and the livestock or poultry producer will rarely if ever encounter vitamin deficiencies when using reputable commercial diets or vitamin supplements.

Each of the vitamins is discussed in this section, including requirements, sources, bioavailability, deficiency signs, and potential toxicity.

Vitamin A Vitamin A has several major metabolic roles. Various forms of vitamin A are essential for vision, as components of visual pigments that trap light in the eye and trigger nerve impulses via the optic nerve to the brain. Severe **vitamin A deficiency** causes blindness because of lack of visual pigments. The second major metabolic role of vitamin A accounts for a variety of deficiency symptoms. Vitamin A has an essential role in the synthesis of glycoproteins and mucopolysaccharides for the addition of glycosyl (carbohydrate-like) groups to proteins to form glycoproteins that are constituents of cell membranes, connective tissue, and mucous secretions. Vitamin A deficiency thus results in formation of underglycosylated glycoproteins. Signs of vitamin A deficiency attributable to **defective glycoprotein synthesis** are abnormal bone growth and degeneration of mucous membranes in the eye, mouth, digestive tract, and reproductive tract. Vitamin A also has a role in cellular proliferation and differentiation. Thus, deficiency results in impaired growth and reproductive defects, including impaired spermatogenesis and failure of fetal growth, causing abortion or fetal resorption. Vitamin-A-deficient animals are susceptible to infections of mucous membranes of the eye (xerophthalmia), gut, and reproductive tract because of impaired mucous secretion and keratinization of the epithelial tissue. Defective immune system activity may also occur.

Chemical Forms of Vitamin A Vitamin A activity in plant tissues is associated with carotene pigments, mainly β-carotene. **Carotenoid pigments** are one of three classes of plant pigments that function in photosynthesis: chlorophylls, carotenoids, and phycobilins. Thus carotenes are found in green plant tissue. The yellow autumn color of leaves is due to the loss of chlorophyll, allowing the carotenes to become visible. There are approximately 600 carotenoids found in nature; 10 percent of them can serve as precursors of vitamin A (Olson, 1989).

The β-carotene molecule consists of two vitamin A molecules joined together. Enzymes in the intestinal mucosa convert β-carotene to vitamin A by splitting it. The efficiency of this conversion varies among species. In humans, cattle, and horses, significant amounts of carotene may be absorbed. This accounts for the yellow coloration of milk and body fat of Jersey and Guernsey cattle and the yellow skin pigmentation of vegetarians consuming large amounts of carrots and other carotene-rich vegetables. The fat of grass-fed cattle is often very yellow, because of the accumulation of carotenoid pigments. Sheep rarely have yellow fat, because of their efficient metabolism of carotenoids (Yang et al., 1992).

Several other chemical forms of vitamin A exist: vitamin A alcohol (retinol), vitamin A aldehyde (retinene, also known as retinal), and vitamin A acid (retinoic acid). **Retinol** is the most potent form. Several isomers (*cis, trans*) of each form exist. **Retinoic acid** serves all metabolic functions of vitamin A except the visual role, whereas the other forms are active in all the functions.

Vitamin A activity is usually expressed in **international units (IU):**

1 IU = 0.3 μg retinol	1 IU = 0.55 μg retinyl palmitate
1 IU = 0.344 μg retinyl acetate	1 IU = 0.6 μg beta-carotene

(Reminder: 1 μg = 1 microgram = 0.001 mg.)

Carotene and vitamin A are susceptible to destruction by oxidation. Large losses of carotene occur during the curing of hay. This is caused by lipoxygenase activity, as well as by photochemical degradation due to sunlight. Bleaching of hay by sunlight completely destroys carotene activity. Good green leafy hay will contain adequate carotene; the degree of greenness of hay is a good indicator of its carotene content. Yellow or brown hay is devoid of vitamin A activity. Destruction of vitamin A occurs in pelleting and during feed storage. Jones (1986) reported that the loss of vitamin A activity during pelleting varies from 6 to 30 percent, depending on whether the free alcohol or esterified form is used, the presence of an antioxidant, the temperature reached during pelleting, and so on. Rancidity of feeds and presence of trace elements, such as copper, accelerate destruction of vitamin A. Vitamin A is degraded by rumen fermentation. Cattle fed high-concentrate diets have higher vitamin A requirements than those fed forage. Rode et al. (1990) demonstrated a high degree of vitamin A destruction in rumen fluid from concentrate-fed cattle. According to these authors, "Conjugated double bonds such as those in vitamin A are excellent electron sinks in the rumen environment. The extensive degradation of vitamin A in concentrate rumen fluid is consistent with the production of more reduced end products in rumen fluid from cattle fed high-concentrate diets."

Sources of Vitamin A Common sources of carotene in livestock diets are green feeds, dehydrated forages, and yellow corn. Supplemental vitamin A is usually provided using synthetic forms. Normally it is provided as esterified vitamin A, with the reactive hydroxyl group on the molecule esterified with an acid. This makes the molecule much less sensitive to oxidation. Retinol acetate, propionate, and palmitate are the ester forms commonly used. Feed grade vitamin A sources are generally in a dry stabilized form, consisting of beadlets with a protective gelatin covering and containing an antioxidant.

There has been considerable interest in the use of β-carotene as a feed additive for dairy cattle based on reports that it improved reproductive performance. The corpus luteum of the cow has a very high β-carotene content. However, most research has not supported the assertion of a function of β-carotene independent of vitamin A (Folman et al., 1987).

Toxicity of Vitamin A Of all vitamins, vitamin A is the most likely to present toxicity problems. Toxic levels are 4 to 10 times the nutritional requirements in non-ruminants and approximately 30 times the nutritional requirements for ruminants (NRC, 1987). Signs of **vitamin A toxicity** are in many respects similar to those of deficiency and may involve similar biochemical defects at the molecular level. Signs include skeletal malformations, reduced growth, conjunctivitis, and reproductive failure. In rabbits, both deficiency and toxicity of vitamin A cause fetal resorption, abortion, hydrocephalus (enlarged fluid-containing head) of fetuses, and small, weak animals at birth. The efficiency of conversion of ß-carotene to vitamin A decreases as dietary carotene level increases so that vitamin A toxicity cannot be induced by high carotene intakes (Olson, 1989).

Vitamin D Vitamin D is aptly known as "the sunshine vitamin." It is formed by the irradiation of sterols in plants and in the skin of animals. Vitamin D occurs in two major forms: vitamin D_2 (ergocalciferol), and vitamin D_3 (cholecalciferol). **Ergocalciferol** is activated plant sterol; **cholecalciferol** is activated animal sterol. They differ slightly in chemical structure. Dietary sources of vitamin D include hay and other sun-cured forage and fish-liver oils. In contrast to their vitamin A activity, plants contain vitamin D activity only after the plant cells have died. Thus, in green pasture, the only dietary source of vitamin D would be dead plant leaves. Green hay, which can be an excellent source of vitamin A activity, may be low in vitamin D.

Animals kept in total confinement, as is the case in modern swine and poultry facilities, are not exposed to sunlight and will require a dietary source of vitamin D. Grazing animals exposed to sunlight normally do not require supplementation with vitamin D. In the winter in northern climates, the amount and intensity of sunlight is often too low for adequate synthesis of vitamin D in the skin, and provision of a supplement is advisable. Smith and Wright (1984) in England demonstrated that the level of metabolically active vitamin D in the blood of sheep decreased drastically in the winter months and that supplementation with the vitamin was advisable. The rapid decline in blood levels in the autumn suggested that vitamin D stores are rapidly depleted, at least in sheep.

In all animals, vitamin D_2 is converted to D_3, with D_3 being the metabolically active form. The efficiency of conversion is very low in poultry. Vitamin D plays an important role in calcium metabolism, being involved in the regulation of calcium absorption and bone mineralization. The **metabolically active form** of the vitamin is 1,25-dihydroxycholecalciferol (1,25-OHD_3). This is formed by the addition of two hydroxyl (-OH) groups to D_3. The first hydroxylation takes place in the liver to produce 25-OHD_3. This compound is then secreted into the blood. In the kidney, it is converted to 1,25-OHD_3 (calcitriol), which in turn is secreted into the blood. It acts like a hormone in regulating calcium absorption and bone mineralization. The rate of formation of 1,25-OHD_3 is regulated by the **parathyroid hormone (PTH)** that, in turn, is regulated by the serum calcium level. Thus, when there is a need for more calcium, calcium absorption is increased by "turning on" 1,25-OHD_3 formation.

The 25-OHD_3 and 1,25-OHD_3 metabolites have useful applications in human and animal medicine. In humans, nonvitamin-D-responsive rickets is caused by a genetic lack of enzymatic formation of 1,25-OHD_3 in the kidney. The direct administration of the metabolite overcomes this barrier. Similarly, in patients with kidney failure, serious bone disease occurs due to lack of 1,25-OHD_3. This can now be administered directly to alleviate the condition. In dairy cattle, administration of vitamin D metabolites prepartum protects against milk fever (Gast et al., 1979). The l-hydroxycholecalciferol form of vitamin D is less costly to synthesize than the 1,25-OHD_3 and is as effective in improving calcium and phosphorus utilization in chicks (Biehl and Baker, 1997) and pigs (Biehl and Baker, 1996). These vitamin D metabolites increase the utilization of phytate phosphorus, apparently by enhancing intestinal phytase activity (Biehl and Baker, 1997).

The presence of mycotoxins such as aflatoxin in the diet increases the vitamin D requirement. This has been of practical importance in poultry. Aflatoxin impairs vitamin D absorption, and its hepatotoxic effects may interfere with the formation of 25-OHD_3 in the liver.

The major physiological role of vitamin D is to regulate the absorption of calcium from the intestine by regulating the mucosal synthesis of **calcium-binding protein** (calbindin). There are some interesting differences among species in vitamin D regulation of calcium absorption, reflecting evolutionary background. For example, **llamas** have a high dietary vitamin D requirement because in their native habitat at high elevations in the Andes, they are exposed to intense solar radiation (Van Saun et al., 1996). Thus, they evolved in a situation where inefficient vitamin D synthesis in the skin was not a problem because of the abundant ultraviolet radiation. At the other extreme, the **African mole-rat** spends its entire life underground and is never exposed to sunlight. In this species, calcium is passively absorbed in an efficient, nonvitamin-D-dependent process, so they have no requirement for vitamin D (Pitcher et al., 1992). **New World primates,** such as marmosets and tamarins, have very high vitamin D requirements, because, like the llama, they have high exposure to solar radiation in their habitat in the tree canopy in equatorial regions. In contrast, **New World rodents,** such as pacas and agoutis, that live on the dark forest floor have very low vitamin D requirements and are very efficient in calcium absorption (Kenny et al., 1993). As discussed further in Chapter 22, these differences are important in **zoo animals.** Kenny et al. (1993) found that vitamin D toxicity occurred in New World rodents that were exhibited with New World primates. They ate the primate food, which had a high vitamin D content required by the primates but that was a toxic level for the rodents.

Vitamin E Vitamin E and selenium function in preventing the breakdown of cell membranes by free radicals and other products of lipid autooxidation. Nutritional muscular dystrophy in most species, liver necrosis in swine and rats, fetal resorption in rats, and encephalomalacia (brain degeneration) in poultry are examples of vitamin E deficiency. In preventing these lesions, vitamin E functions as a biological antioxidant. Vitamin E exists in several chemical forms, known as **tocopherols,** with alpha(α)-tocopherol having the highest potency. One IU of vitamin E is equivalent to 1 mg of *dl*-α-tocopherol acetate; *dl*-α-tocopherol has a potency of 1.1 IU per mg. Activities of *d*-α-tocopherol acetate and *d*-α-tocopherol are 1.36 and 1.49 IU per mg, respectively. Commercially, vitamin E is available as *dl*-α-tocopherol acetate and *d*-α-tocopherol acetate. The esterified acetate group renders the vitamin resistant to oxidation and stable in feeds, but these forms have antioxidant activity only after they are hydrolyzed in the digestive tract, so they do not protect feeds against rancidity. Good sources of vitamin E in animal diets include green forage, good-quality hay, alfalfa meal, and cereal grains.

Currently, there is much interest in the use of high dietary levels of vitamin E for improving meat quality (Liu et al., 1995). Animals fed high vitamin E for several weeks before slaughter store increased quantities of the vitamin in the muscle tissue. Because of its antioxidant activity, vitamin E in the muscle reduces lipid oxidation, delaying the discoloration of fresh and frozen meat, and thus extending **shelf life** (Faustman et al., 1998).

Selenium and vitamin E have been of interest as stimulators of immune responses in animals. According to Finch and Turner (1996), the greatest **immunologic effects** of vitamin E and selenium supplementation are observed in poultry, pigs, and laboratory animals. There is little response or effectiveness in ruminants.

Vitamin K Vitamin K functions in the **blood-clotting** process. It is essential for the activation of prothrombin (a plasma protein) by catalyzing the addition of carbon dioxide to glutamic acid residues in prothrombin, creating calcium-binding sites. During the clotting process, prothrombin is activated to thrombin by addition of calcium; thrombin converts the soluble protein fibrinogen to insoluble fibrin, forming the clot. Vitamin K occurs naturally in plants and bacteria. Vitamin K (phylloquinone) is of plant origin and vitamin K_2 (menaquinone) is from bacterial sources. Vitamin K_3 (menadione) is a synthetic form and is water soluble. In contrast to the other fat-soluble vitamins, vitamin K is synthesized by ruminal and intestinal microbes. Thus, it is rarely necessary to supplement ruminants with vitamin K. An exception is in cattle consuming moldy sweetclover hay containing dicumarol (see Chapter 5). Swine and poultry diets should be supplemented with vitamin K. Aflatoxin and other mycotoxins increase the vitamin K requirement. Coccidiosis infection increases the vitamin K requirement of poultry. The principal commercial forms of vitamin K used in the feed industry are derivatives of **menadione,** including menadione sodium bisulfite and menadione dimethyl-pyrimidinol bisulfite.

B-Complex Vitamins The B vitamins function as cofactors of enzymes involved in energy metabolism; some have additional roles as well. Metabolic functions, deficiency signs, and supplementation are briefly described. Toxicity is rarely of concern because the water-soluble vitamins are readily excreted. The B-complex vitamins are synthesized in adequate amounts in the rumen and, thus, are not normally dietary requirements of ruminant animals. For swine and poultry, a complete vitamin mixture containing all known vitamins is routinely used.

Thiamin (Vitamin B_1) Thiamin functions in enzymes involved in decarboxylation (removal of CO_2) reactions such as the conversion of pyruvate to acetate in carbohydrate metabolism. Thiamin deficiency results in impaired carbohydrate metabolism, with an accumulation of pyruvate in the blood. Elevated pyruvate affects the nervous system, causing loss of appetite (anorexia) and neurological signs such as convulsions, opisthotonus (head retraction, "stargazing posture"), paralysis, reduced heart rate, and subnormal body temperature (Fig. 7.4). Deficiency signs are very quickly reversed when the vitamin is administered.

Most feedstuffs contain adequate levels of thiamin, so uncomplicated thiamin deficiency is rare. Most instances of deficiency are caused by destruction of thiamin in the diet or in the digestive tract. Several materials have thiaminase activity. **Thiaminases** are enzymes that split the thiamin molecule and destroy its activity. Various types of fish (e.g., carp), ferns (e.g., bracken fern), *Equisetum* spp.

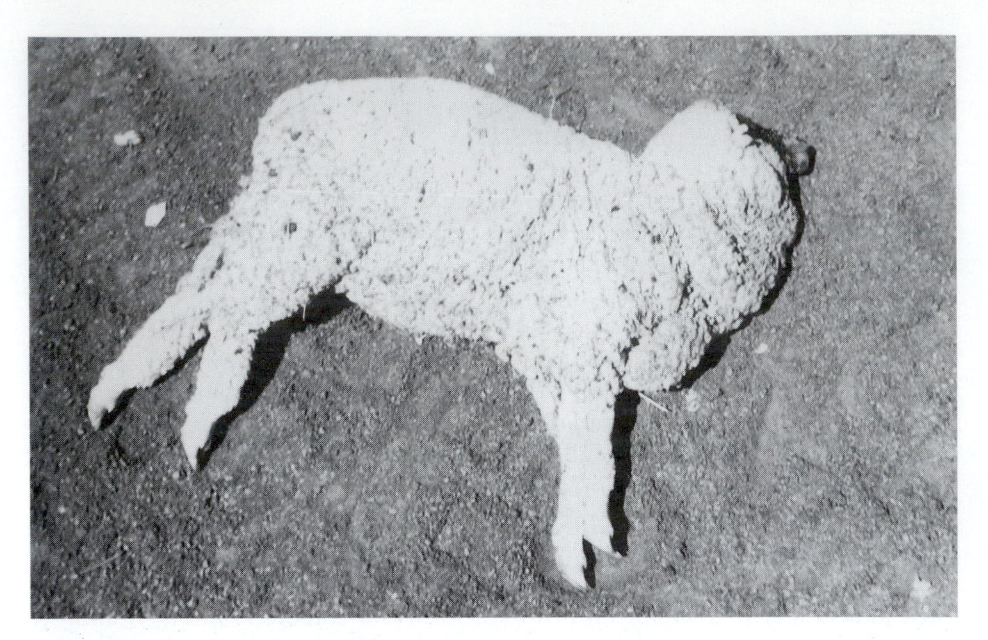

FIGURE 7.4 Polioencephalomalacia (induced thiamin deficiency). This particular case was caused by consumption of ferns high in thiaminase activity. (Courtesy of B. Chick.)

(horsetails), and bacteria produce thiaminases. A condition known as Chastek's paralysis in mink and foxes is an induced thiamin deficiency caused by feeding raw fish. Bracken fern poisoning of horses causes thiamin deficiency.

Thiamin is a B vitamin that is sometimes important in ruminant nutrition. Ruminants fed high-concentrate diets may develop **polioencephalomalacia** (PEM), also called *cerebrocortical necrosis* (see Chapter 2). A number of thiaminase-producing microorganisms, such as *Clostridium sporogenes,* proliferate when high-concentrate diets are fed. They secrete thiaminase that then degrades the vitamin in the rumen. It has been suggested that lactic acidosis is a predisposing factor for the development of PEM (Brent and Bartley, 1984). Under some conditions, feedlot cattle may benefit from the addition of thiamin to the diet. Approximately 1 g thiamin per animal per day is necessary to prevent PEM in the presence of rumen thiaminase. Feeding ionophores (rumensin, lasalocid), which inhibit lactic acid production, may have beneficial effects in prevention of PEM. A PEM-like syndrome affects cattle fed high-molasses diets (Preston and Leng, 1987).

The coccidiostat amprolium is a thiamin antagonist. It is commonly used in poultry diets to treat coccidiosis. Under normal conditions, amprolium does not seem to induce thiamin deficiency when fed to poultry. Thiamin sources include the hydrochloride and mononitrate forms. The mononitrate form is more stable, particularly under humid conditions.

Riboflavin (Vitamin B_2) Riboflavin functions in coenzymes such as flavin adenine dinucleotide (FAD) and flavin mononucleotide (FMN) that are involved in energy metabolism. Riboflavin is one of the vitamins most likely to be deficient in swine and poultry diets because grains and plant protein supplements are poor sources. Alfalfa meal is a good source, but its use in swine and poultry diets is less common than formerly. Signs of riboflavin deficiency include curled toe paralysis in chicks; reduced egg production and hatchability in poultry; and skin lesions, reduced growth, reproductive failure, and high neonatal mortality in pigs. Riboflavin is commercially available as the crystalline vitamin. It is stable to heat but sensitive to degradation by light.

Niacin Like riboflavin, niacin is a constituent of cofactors that function in energy metabolism, such as nicotinamide adenine dinucleotide (NAD) and nicotinamide adenine dinucleotide phosphate (NADP). Niacin in cereal grains (especially corn) is bound to amino acids and has low bioavailability. Corn-based diets in particular require niacin supplementation. Niacin deficiency in humans is responsible for **pellagra,** a condition characterized by a fiery red tongue, dermatitis, dementia, and diarrhea. Deficiency signs in pigs include poor growth, dermatitis, diarrhea, and intestinal lesions. Niacin deficiency in poultry results in leg problems (enlarged hock joint), poor feathering, and dermatitis. Niacin is available commercially in two forms, niacinamide and nicotinic acid. Niacin is stable in the presence of heat, oxygen, moisture, and light.

There is substantial evidence that niacin supplementation of diets for ruminants may be beneficial (Brent and Bartley, 1984). Growth responses of cattle and sheep fed high-concentrate diets have been noted with niacin supplementation. In dairy cattle, niacin responses in the postpartum period occur, with increased milk production. Niacin supplementation also seems to be beneficial in preventing ketosis in dairy cattle. A dose of 12 g niacin per day per cow is typical. Recommended levels of supplementary niacin are 100 ppm for feedlot cattle, 200 ppm for postpartum dairy cows, and 400 ppm for ketosis prevention in high-producing cows (Brent and Bartley, 1984).

Pyridoxine (Vitamin B_6) Pyridoxine has several roles in protein metabolism, being involved in amino acid interconversions (transamination) and decarboxylation. It is found in adequate quantities for swine and poultry in grains and plant protein sources, so it is seldom deficient in diets. Requirements are increased when linseed meal is used. Linseed meal contains linatine, a pyridoxine antagonist. Signs of pyridoxine deficiency include convulsions, dermatitis, and impaired reproduction. Vitamin B_6 is available commercially as crystalline pyridoxine hydrochloride. It is fairly stable in mixed feeds.

Pantothenic Acid Pantothenic acid is widely distributed among feedstuffs. Its bioavailability from barley, wheat, and soybean meal is high, but it is low in corn and sorghum. Pantothenic acid is best known for its metabolic role in coenzyme A. Acetate and other fatty acids occur esterified to coenzyme A (acetyl CoA, fatty acyl CoA) in metabolism. Signs of deficiency include reduced growth, dermatitis, and neurological defects. Pigs exhibit a characteristic goose-stepping gait (leg extended without bending while walking).

Pantothenic acid is available commercially as *d*- and *dl*-calcium pantothenate. Only the *d*- isomer is biologically active. One gram of *d*-calcium pantothenate is equivalent to 0.92 g *d*-pantothenic acid activity, whereas 1 g of *dl*-calcium pantothenate has a potency of 0.46 g *d*-pantothenic acid. Calcium pantothenate has hygroscopic and electrostatic properties that cause it to pick up moisture and stick to metal, causing handling problems. It is usually combined with a carrier that prevents these problems.

Biotin Biotin is important in carbohydrate and fat metabolism, being essential for transfer of carbon dioxide groups in decarboxylations and fatty acid synthesis. It is widely distributed in common feedstuffs but its bioavailability in some feeds (e.g., wheat, barley, sorghum, and oats) is low (Frigg, 1984; Misir and Blair, 1988; Sauer et al., 1988). Biotin deficiency of poultry fed wheat-based diets has been a common problem. Biotin deficiency in non-ruminants causes dermatitis and cracks in the feet (foot pad dermatitis), poor growth, loss of hair or feathers, and poor reproduction. In ruminants (see Chapter 17) and horses (see Chapter 18), biotin supplementation often improves hoof health by reducing hoof cracks and lesions. It also increases fingernail thickness in humans. For experimental purposes, biotin deficiency may be induced by feeding raw egg white, which contains **avidin,** a biotin antagonist. Mink fed raw turkey viscera containing eggs develop "turkey waste greying," characterized by **achromotrichia** (loss of hair pigmentation) and a "spectacle eye" condition caused by loss of hair around the eye (Wehr et al., 1980).

Synthetic biotin is quite expensive and so a decision to use supplementary biotin should be considered carefully. It is available as the synthetic vitamin, either at 100 percent potency or in various dilutions. As it is readily destroyed in feeds by rancidity, it should be protected by the addition of an antioxidant to the feed.

Choline and Betaine Choline is often considered a vitamin, although it is not, strictly speaking, a dietary essential. It can be synthesized in the body using methyl groups from other sources such as methionine. Deficiency symptoms are difficult to produce with practical diets. **Choline** is a constituent of phospholipids in cell membranes and is essential for nerve function as a component of acetylcholine. Choline is added to feeds as choline chloride, either as a liquid or a dry powder. The liquid form is very corrosive and requires special handling equipment. Although choline can be synthesized from methionine, it is less expensive to provide it as supplementary choline rather than by providing extra methionine. Choline supplementation is often used with sows; it seems to improve reproductive performance.

Betaine is structurally similar to choline, and can serve as a source of methyl groups for the synthesis of phospholipids and cell membranes. The use of betaine in poultry diets has been reviewed by Kidd et al. (1997). Betaine is available commercially and may be used to replace choline and methionine when it is economically practical to do so.

Folic Acid (Folacin) Folic acid has an important metabolic role, along with vitamin B_{12}, in the synthesis of hemoglobin. The principal deficiency sign is anemia. Several reports indicate that folic acid supplementation of practical diets enhances reproductive performance of swine with an increase in litter size (see Chapter 13). Folic acid is available commercially as the crystalline vitamin and is often used in the form of the sodium salt. It is sensitive to heat and light but is not affected by oxidation.

Vitamin B_{12} (Cyanocobalamin) Because vitamin B_{12} does not occur in plant materials, it is a dietary essential for swine and poultry. In ruminants and other herbivores, it is synthesized in adequate amounts by gut microflora. **Cobalt** is required in the diet of herbivores to allow for adequate microbial synthesis of the vitamin. Vitamin B_{12} and folic acid have a dual role in formation of erythrocytes. The classical deficiency symptom in humans is pernicious anemia, caused by a genetic lack of the "intrinsic factor" necessary for vitamin B_{12} absorption. Swine and poultry fed diets deficient in the vitamin have reduced growth, anemia, and poor reproduction. Ruminants have a high metabolic requirement for vitamin B_{12}, because it is involved in the metabolism of propionic acid. This requirement is met by ruminal synthesis, but adequate dietary cobalt is required for microbial synthesis of the vitamin. Vitamin B_{12} is available in crystalline form as cyanocobalamin. It has good stability in feeds.

Carnitine Carnitine is a naturally occurring compound with vitamin-like activity. It facilitates the transport of long-chain fatty acids into the mitochondria for ATP production. Carnatine is available as a feed additive. There is some evidence that supplemental carnitine may enhance lipid metabolism in baby pigs (Owen et al., 1996) and reduce carcass fat in growing-finishing pigs (Owen et al., 2001).

Ascorbic Acid Ascorbic acid (vitamin C) has traditionally been regarded as a dietary essential only for primates and guinea pigs. The list of vitamin-C-requiring animals has been expanded to include such exotics as the fruit-eating bat and the red-vented bulbul bird, as well as various fish species. These species lack an enzyme, *L*-gulonolactone oxidase, necessary for synthesis of ascorbic acid from glucose. The species with a dietary vitamin C requirement evolved on diets containing fruit or other good sources of the vitamin and lost the ability to synthesize it.

Although livestock and poultry are considered to be able to synthesize adequate vitamin C, there are numerous reports of positive responses to supplementation with the vitamin. **Stress** has been suggested as a factor increasing requirements. The effects of dietary ascorbic acid on egg shell quality and resistance to heat stress have been reviewed by Pardue and Thaxton (1986) and Gross (1988). However, the conditions under which vitamin C supplementation may be beneficial have not yet been adequately defined, and the vitamin is little used in the feed industry. It is primarily used in fish feeds (see Chapter 23).

QUESTIONS AND STUDY GUIDE

1. If in balancing a ration you find that the phosphorus level is adequate but a calcium supplement is needed, what would be a good source? What are some suitable sources if only phosphorus is needed? If both elements are needed?
2. Of what physiological advantage would it be for a cow on a phosphorus-deficient diet not to come into estrus? Does phosphorus have a specific metabolic role in reproductive processes? (You will need to search other literature sources to answer this question.)
3. Symptoms of grass tetany and milk fever are somewhat similar. Why?
4. Is blood glucose an electrolyte? Serum albumin? Explain. What is the difference among a solute, an electrolyte, and an osmotically active substance?

5. A swine diet contains 0.3 percent sodium, 0.6 percent potassium, and 0.55 percent chlorine. What is the electrolyte balance of this diet? Is it in the optimal range?
6. Use of synthetic lysine in swine and poultry feeds might increase problems with electrolyte balance. Why?
7. Is calcium propionate acidogenic or alkalinogenic? Why?
8. What vitamin might be expected to interact with manganese in its metabolic role? Why?
9. Parakeratosis occurs in swine and cattle. What are the symptoms and causes of the disorder in each species?
10. How can copper deficiency symptoms be explained in terms of the metabolic roles of copper?
11. Selenium is required at a level of 0.1 parts-per-million (ppm) in swine and poultry diets. How many grams (or mg or μg) of selenium should be added to 1 ton of feed to provide 0.1 ppm? How could this amount be accurately weighed out and adequately mixed in a feedmill so that any sample of the 1 ton of feed would contain 0.1 ppm selenium? Suppose you are monitoring the selenium content and are measuring the selenium in a 1-g sample of feed. What is the actual amount of selenium you measure in the sample? Actually, you wouldn't add selenium as the pure element but as the salt sodium selenite (Na_2SeO_3). How much sodium selenite would you add to 1 ton of feed to provide 0.1 ppm Se?
12. Why might stress influence vitamin requirements?
13. The NRC requirement for gestating sows for vitamin A is 4,000 IU/kg diet. Suppose you are formulating a sow diet, and you intend to use synthetic vitamin A to provide the entire requirement. You have a premix available containing 6 million IU retinol/lb of premix. How much of the premix per ton of feed should you use to provide 4,000 IU vitamin A/kg diet?
14. A vitamin A premix contains 20 mg retinol per kg. How many IU per kg is this?
15. How do green leafy forages compare as sources of vitamin D and vitamin A activity? How about dead, brown, dried-up grass leaves?
16. Define a hormone. Describe how the metabolic action of vitamin D may qualify it as a hormone. Consider endocrine glands and feedback mechanisms in your answer.
17. Aflatoxin is a hepatotoxin, meaning that it causes destruction of liver tissue. What effects would you expect aflatoxin to have on the absorption and storage of vitamin A as a consequence of its effects on the liver?
18. Why does feeding sweetclover hay to cattle sometimes increase their vitamin K requirements?
19. Under what circumstances might cattle benefit from supplementation with water-soluble vitamins? Name the specific vitamins that might be of concern.
20. A swine producer obtains broken eggs from an egg processing plant, and feeds the raw eggs to his pigs. What kinds of problems might occur?
21. A farmer has a mixed herd of Holstein and Hereford steers on a pasture. The cattle have scours (diarrhea) and the Herefords have yellow hair; the black hair on the Holsteins has turned red. What is your diagnosis?
22. Why are mineral deficiencies in livestock particularly important in tropical countries?
23. Explain how calcium and vitamin K are interrelated in their metabolism.

REFERENCES

Minerals

Ammerman, C.B., S.M. Miller, K.R. Fick, and S.L. Hansard II. 1977. Contaminating elements in mineral supplements and their potential toxicity: A review. *J. Anim. Sci.* 44:485–508.

Apgar, G.A., E.T. Kornegay, M.D. Lindemann, and D.R. Notter. 1995. Evaluation of copper sulfate and a copper lysine complex as growth promoters for weanling swine. *J. Anim. Sci.* 73:2640–2646.

Austic, R.E. and C.C. Calvert. 1981. Nutritional interrelationships of electrolytes and amino acids. *Fed. Proc.* 40:63–67.

Bailey, C.B. 1981. Silica metabolism and silica urolithiasis: A review. *Can. J. Anim. Sci.* 61:219–235.

Botkin, D.B., P.A. Jordan, A.S. Dominski, H.S. Lowendorf, and G.E. Hutchinson. 1973. Sodium dynamics in a northern ecosystem. *Proc. Natl. Acad. Sci. USA* 70:2745–2748.

Brumm, M.C. and B.R. Schricker. 1989. Effect of dietary potassium chloride on feeder pig performance, market shrink, carcass traits and selected blood parameters. *J. Anim. Sci.* 67:1411–1417.

Burk, R.F. and K.E. Hill. 1993. Regulation of selenoproteins. *Ann. Rev. Nutr.* 13:65–81.

Chang, X., D.N. Mowat, and B.A. Mallard. 1995. Supplemental chromium and niacin for stressed feeder calves. *Can. J. Anim. Sci.* 75:351–358.

Coffey, R.D., G.L. Cromwell, and H.J. Monegue. 1994. Efficacy of a copper-lysine complex as a growth promotant for weanling pigs. *J. Anim. Sci.* 72:2880–2886.

Davis, G.K. and W. Mertz. 1987. Copper. In *Trace Elements in Human and Animal Nutrition,* Vol. 1, pp. 301–364. Orlando, FL: Academic Press.

Fell, B.F., D. Dinsdale, and C.F. Mills. 1975. Changes in enterocyte mitochondria associated with deficiency of copper in cattle. *Res. Vet. Sci.* 18:274–281.

Fisher, L.J., N. Dinn, R.M. Tait, and J.A. Shelford. 1994. Effect of level of dietary potassium on the absorption and excretion of calcium and magnesium by lactating cows. *Can. J. Anim. Sci.* 74:503–509.

Fredeen, A.H., E.J. DePeters, and R.L. Baldwin. 1988. Effects of acid-base disturbances caused by differences in dietary fixed ion balance on kinetics of calcium metabolism in ruminants with high calcium demand. *J. Anim. Sci.* 66:174–184.

Hartmann, F. and J.B.J. van Ryssen. 1997. Metabolism of selenium and copper in sheep with and without sodium bicarbonate supplementation. *J. Agric. Sci.* 128:357–364.

Hulan, H.W., P.C.M. Simons, P.J.W. Van Schagen, K.B. McRae, and F.G. Proudfoot. 1987. Effect of cation-anion imbalance and calcium content on general performance and incidence of leg abnormalities of broiler chickens. *Can. J. Anim. Sci.* 67:165–177.

Hurley, L.S. and C.L. Keen. 1987. Manganese. In *Trace Elements in Human and Animal Nutrition,* Vol. 1, W. Mertz, ed. Orlando, FL: Academic Press.

Hutcheson, D.P. and N.A. Cole. 1986. Management of transit-stress syndrome in cattle: Nutritional and environmental effects. *J. Anim. Sci.* 62:555–560.

Johnson, R.J. and H. Karunajeewa. 1985. The effects of dietary minerals and electrolytes on the growth and physiology of the young chick. *J. Nutr.* 115:1680–1690.

Karn, J.F. 2001. Phosphorus nutrition of grazing cattle: A review. *Anim. Feed Sci. Tech.* 89:133–153.

Kegley, E.B. and J.W. Spears. 1995. Immune response, glucose metabolism, and performance of stressed feeder calves fed inorganic or organic chromium. *J. Anim. Sci.* 73:2721–2726.

Kegley, E.B., J.W. Spears, and T.T. Brown, Jr. 1997. Effect of shipping and chromium supplementation on performance, immune response, and disease resistance of steers. *J. Anim. Sci.* 75:1956–1964.

Knight, D.A. and W.J. Tyznik. 1990. The effect of dietary selenium on humoral immunocompetence of ponies. *J. Anim. Sci.* 68:1311–1317.

Koenig, K.M., L.M. Rode, R.D.H. Cohen, and W.T. Buckley. 1997. Effects of diet and chemical form of selenium on selenium metabolism in sheep. *J. Anim. Sci.* 75:817–827.

Kornegay, E.T., Z. Wang, C.M. Wood, and M.D. Lindemann. 1997. Supplemental chromium picolinate influences nitrogen balance, dry matter digestibility, and carcass traits in growing-finishing pigs. *J. Anim. Sci.* 75:1319–1323.

Leclerc, H. and E. Block. 1989. Effects of reducing dietary cation-anion balance for prepartum dairy cows with specific reference to hypocalcemic parturient paresis. *Can. J. Anim. Sci.* 69:411–423.

Levander, O.A. 1986. Selenium. In *Trace Elements in Human and Animal Nutrition,* 5th ed., W. Mertz, ed. pp. 209–279. Orlando, FL: Academic Press.

Luo, X.G., P.R. Henry, C.B. Ammerman, and J.B. Madison. 1996. Relative bioavailability of copper in a copper-lysine complex or copper sulfate for ruminants as affected by feeding regimen. *Anim. Feed. Sci. Tech.* 57:281–289.

McDowell, L.R. 1985. *Nutrition of Grazing Ruminants in Warm Climates.* Orlando, FL: Academic Press.

Mertz, W., ed. 1988. *Trace Elements in Human and Animal Nutrition,* Vols. I and II. Orlando, FL: Academic Press.

Mertz, W. 1993. Chromium in human nutrition: A review. *J. Nutr.* 123:626–633.

Mooney, K.W. and G.L. Cromwell. 1997. Efficacy of chromium picolinate and chromium chloride as potential carcass modifiers in swine. *J. Anim. Sci.* 75:2661–2671.

Mongin, P.J. 1981. Recent advances in dietary anion-cation balance: Applications in poultry. *Proc. Nutr. Soc.* 40:285–294.

Morris, J.G. 1980. Assessment of sodium requirements of grazing beef cattle: A review. *J. Anim. Sci.* 50:145–152.

Mowat, D.N., X. Chang, and W.Z. Yang. 1993. Chelated chromium for stressed calves. *Can. J. Anim. Sci.* 73:49–55.

Patience, J.F. 1990. A review of the role of acid-base balance in amino acid nutrition. *J. Anim. Sci.* 68:398–408.

Patience, J.F., R.E. Austic, and R.D. Boyd. 1987. Effect of dietary electrolyte balance on growth and acid-base status in swine. *J. Anim. Sci.* 64:457–466.

Patience, J.F. and R.K. Chaplin. 1997. The relationship among dietary undetermined anion, acid-base balance, and nutrient metabolism in swine. *J. Anim. Sci.* 75:2445–2452.

Patience, J.F. and M.S. Wolynetz. 1990. Influence of dietary undetermined anion on acid-base balance and performance in pigs. *J. Nutr.* 120:579–587.

Ross, R.D., G.L. Cromwell, and T.S. Stahly. 1984. Effects of source and particle size on the biological availability of calcium in calcium supplements for growing pigs. *J. Anim. Sci.* 59:125–134.

Spears, J.W. 1989. Zinc methionine for ruminants: Relative bioavailability of zinc in lambs and effects on growth and performance of growing heifers. *J. Anim. Sci.* 67:835–843.

Stuedemann, J.A., S.R. Wilkinson, and R.S. Lowrey. 1984. Efficacy of a large magnesium alloy rumen bolus in the prevention of hypomagnesemic tetany in cows. *Am. J. Vet. Res.* 45:698–702.

Suttle, N.F. 1991. The interactions between copper, molybdenum, and sulphur in ruminant nutrition. *Ann. Rev. Nutr.* 11:121–140.

Van Heugten, E. and J.W. Spears. 1997. Immune response and growth of stressed weanling pigs fed diets supplemented with organic or inorganic forms of chromium. *J. Anim. Sci.* 75:409–416.

Ward, G.M. 1978. Molybdenum toxicity and hypocuprosis in ruminants: A review. *J. Anim. Sci.* 46:1078–1085.

Yeh, J-Y., Q.-P. Gu, M.A. Beilstein, N.E. Forsberg, and P.D. Whanger. 1997. Selenium influences tissue levels of Selenoprotein W in sheep. *J. Nutr.* 127:394–402.

Zervas, G. 1988. Treatment of dairy sheep with soluble glass boluses containing copper, cobalt and selenium. *Anim. Feed Sci. Tech.* 190:79–83.

Vitamins

Biehl, R.R. and D.H. Baker. 1996. Efficacy of supplemental 1α-hydroxycholecalciferol and microbial phytase for young pigs fed phosphorus- or amino acid-deficient corn-soybean meal diets. *J. Anim. Sci.* 74:2960–2966.

______ 1997. Utilization of phytate and nonphytate phosphorus in chicks as affected by source and amount of vitamin D_3. *J. Anim. Sci.* 75:2986–2993.

Brent, B.E. and E.E. Bartley. 1984. Thiamin and niacin in the rumen. *J. Anim. Sci.* 59:813–822.

Faustman, C., W.K.M. Chan, D.M. Schaefer, and A. Havens. 1998. Beef color update: The role for vitamin E. *J. Anim. Sci.* 76:1019–1026.

Finch, J.M. and R.J. Turner. 1996. Effects of selenium and vitamin E on the immune responses of domestic animals. *Res. Vet. Sci.* 60:97–106.

Folman, Y., I. Ascarelli, D. Kraus, and H. Barash. 1987. Adverse effect of β-carotene in diet on fertility of dairy cows. *J. Dairy Sci.* 70:357–366.

Frigg, M. 1984. Available biotin content of various feed ingredients. *Poult. Sci.* 63:750–753.

Gast, D.R., R.L. Horst, N.A. Jorgensen, and H.F. DeLuca. 1979. Potential use of 1,25-dihydroxycholecalciferol for prevention of parturient paresis. *J. Dairy Sci.* 62:1009–1013.

Gross, W.B. 1988. Effect of ascorbic acid on the mortality of Leghorn-type chickens due to overheating. *Avian Dis.* 32:561–562.

Jones, F.T. 1986. Effect of pelleting on vitamin A assay levels of poultry feed. *Poult. Sci.* 65:1421–1422.

Kenny, D., R.C. Cambre, A. Lewandowski, J.A. Pelto, N.A. Irlbeck, H. Wilson, G.W. Mierau, F.G. Sill, and M.V.Z. Alberto Pasas Garcia. 1993. Suspected vitamin D_3 toxicity in pacas (*Cuniculus paca*) and agoutis (*Dasyprocta aguti*). *J. Zoo Wildlife Med.* 24:129–139.

Kidd, M.T., P.R. Ferket and J.D. Garlich. 1997. Nutritional and osmoregulatory functions of betaine. *World's Poult. Sci. J.* 53:125–139.

Liu, Q., M.C. Lanari, and D.M. Schaefer. 1995. A review of dietary vitamin E supplementation for improvement of beef quality. *J. Anim. Sci.* 73:3131–3140.

McDowell, L.R. 1989. *Vitamins in Animal Nutrition.* San Diego: Academic Press.

Misir, R. and R. Blair. 1988. Biotin availability from protein supplements and cereal grains for weanling pigs. *Can. J. Anim. Sci.* 68:523–532.

National Research Council. 1987. *Vitamin Tolerance of Animals.* Washington, DC: National Academy of Sciences.

Olson, J.A. 1989. Provitamin A function of carotenoids: the conversion of β-carotene into vitamin A. *J. Nutr.* 119:105–108.

Owen, K.Q., J.L. Nelssen, R.D. Goodband, T.L. Weeden, and S.A. Blum. 1996. Effect of L-carnitine and soybean oil on growth performance and body composition of early-weaned pigs. *J. Anim. Sci.* 74:1612–1619.

Owen, K.Q., J.L. Nelssen, R.D. Goodband, M.D. Tokach, and K.G. Friesen. 2001. Effect of dietary L-carnitine on growth performance and body composition in nursery and growing-finishing pigs. *J. Anim. Sci.* 79:1509–1515.

Pardue, S.L. and J.P. Thaxton. 1986. Ascorbic acid in poultry: A review. *World Poult. Sci.* 42:107–123.

Pitcher, T., R. Buffenstein, J.D. Keegan, G.P. Moodley, and S. Yahav. 1992. Dietary calcium content, calcium balance and mode of uptake in a subterranean mammal, the Damara mole-rat. *J. Nutr.* 122:108–114.

Preston, T.R. and R.A. Leng. 1987. *Matching Ruminant Production Systems with Available Resources in the Tropics and Sub-Tropics.* Armidale, N.S.W., Australia: Penambul Books.

Rode, L.M., T.A. McAllister, and K.-J. Cheng. 1990. Microbial degradation of vitamin A in rumen fluid from steers fed concentrate, hay or straw diets. *Can. J. Anim. Sci.* 70:227–233.

Sauer, W.C., R. Mosenthin, and L. Ozimek. 1988. The digestibility of biotin in protein supplements and cereal grains for growing pigs. *J. Anim. Sci.* 66:2583–2589.

Smith, B.S.W. and H. Wright. 1984. Relative contributions of diet and sunshine to the overall vitamin D status of the grazing ewe. *Vet. Rec.* 115:537–538.

Van Saun, R.J., B.B. Smith, and B.J. Watrous. 1996. Evaluation of vitamin D status of llamas and alpacas with hypophosphatemic rickets. *J. Am. Vet. Med. Assoc.* 209:1128–1133.

Wang, J.Y., F.G. Owen, and L.L. Larson. 1988. Effect of β-carotene supplementation on reproductive performance of lactating Holstein cows. *J. Dairy Sci.* 71:181–186.

Wehr, N.B., J. Adair, and J.E. Oldfield. 1980. Biotin deficiency in mink fed spray-dried eggs. *J. Anim. Sci.* 50:877–885.

Yang, A., T.W. Larsen, and R.K. Tume. 1992. Carotenoid and retinol concentrations in serum, adipose tissue and liver and carotenoid transport in sheep, goats and cattle. *Aust. J. Agr. Res.* 43:1809–1817.

CHAPTER 8

Feed Additives

Objectives

1. To describe types of feed additives and their uses in livestock production
2. To discuss controversial feed additives:
   antibiotics and growth promotants
   hormones
   probiotics
3. To highlight areas of current concern and activity:
   mode of action of antibiotics
   human health concerns related to use of feed additives
   use of ionophores in ruminant nutrition
   defaunation of ruminants
   repartitioning agents to alter carcass composition
   environmental and ecological impacts of copper sulfate, methane inhibitors, and other additives
   use of chelates and buffers in animal production

Feed additives are nonnutritive substances added to feeds to improve the efficiency of feed utilization and feed acceptance or to be beneficial to the health or metabolism of the animal in some way. In other words, they are substances other than the known nutrients. Some feed additives, such as antibiotics, are controversial, with claims that their use has adverse effects on human health. Regardless of the merits of the arguments, many Americans believe that livestock and poultry are fed diets "laced with chemicals" (Fig. 8.1). This concern is likely to result in increasingly restrictive controls on the use of feed additives for animal production. In the United States, the **Food and Drug Administration (FDA)** must approve a feed additive before it can be used commercially. The FDA approval is given only after extensive testing that establishes the safety of the product and substantiates the claims made for it.

The following classification of feed additives attempts to categorize them by their principal biological or economic effects. It is not an all-inclusive list, but it includes the most prominent feed additives. Each one will be discussed in turn.

Types of Feed Additives

I. Additives that influence feed stability, feed manufacturing, and properties of feeds

  A. Antifungals
  B. Antioxidants
  C. Pellet binders

FIGURE 8.1 Feed additives have become controversial, with fears by the general public that use of additives in animal feeds has adverse effects on human health. Marketing of "natural" products produced without the use of feed additives has become common.

II. Additives that modify animal growth, feed efficiency, metabolism, and performance
- **A.** Feed flavors
- **B.** Digestion modifiers
  1. Enzymes
  2. Buffers
  3. Ion-exchange compounds
  4. Ionophores and methane inhibitors
  5. Isoacids
  6. Probiotics
  7. Acidifiers (organic acids)
  8. Antibloating agents
  9. Salivation inducers
  10. Defaunating agents
- **C.** Metabolism modifiers
  1. Hormones
  2. Beta-adrenergic agents (repartitioning agents)
- **D.** Growth promotants
  1. Antibiotics
  2. Chemotherapeutic agents
  3. Saponins

III. Additives that modify animal health
- **A.** Drugs
- **B.** Environmentally active substances
- **C.** Immunomodulators

IV. Additives that modify consumer acceptance
- **A.** Xanthophylls
- **B.** Saponins

ADDITIVES THAT INFLUENCE FEED STABILITY, FEED MANUFACTURING, AND PROPERTIES OF FEEDS

Antifungals

Antifungal agents are used to prevent fungal (mold) growth in stored feed ingredients and mixed feeds. Molds reduce palatability and may produce mycotoxins. Aflatoxin and *Fusarium* toxins are among those of most concern. Mold growth in stored grains and feeds is prevented by adequate drying (to a moisture content of 12 percent or less), storage under dry conditions, and the use of mold inhibitors (antifungals). The use of a mold inhibitor is strongly recommended when the moisture content of the grain exceeds 13 to 14 percent, the relative humidity is above 80 to 85 percent, the temperature is 55°F or above, or the grain is damaged, broken, or insect-infested. **Propionic acid** or its salts (sodium or calcium propionate) are particularly effective at a level of approximately 1 percent of the grain or diet. Propionates provide protection for at least 90 days. Examples of effective antifungals besides propionic acid are sodium diacetate, sorbic acid, and gentian violet. Efficacy of **gentian violet** was reviewed by Tabib and Hamilton (1988). **Ammonia** treatment of grain inhibits mold growth. In addition, ammonia treatment of grain inactivates aflatoxin. Phosphoric acid is an effective antifungal agent (Lin and Chen, 1995) and has the advantage of also supplying an essential nutrient.

Sodium metabisulfite and sulfur dioxide are useful preservatives for moist grains in small bins when the moisture content is less than 30 percent and the ambient temperatures are cool. Sulfurous acid is formed when **sulfur dioxide** contacts water and inhibits growth of bacteria, yeasts, and molds (Mathison et al., 1988). There are some major problems, however, in the use of sulfur dioxide as a grain preservative, including cost, the poisonous nature of the gas, and the acidic nature of the solution, which can seriously corrode steel grain bins. Palatability of treated grains may also be reduced; however, this can be alleviated by addition of sodium bicarbonate. Gibson et al. (1988) noted that sulfur dioxide was not effective when large grain bins were used because of difficulty in adequately dispersing the material in the grain. Sulfur dioxide treatment degrades or inactivates thiamin in stored grains.

A new approach to grain storage to eliminate both mold growth and insect activity is the use of **low temperature chilling.** Cold air is used to maintain a storage temperature below 50°F (10°C), which inhibits both molds and insects.

Antioxidants

Antioxidants are preservatives that prevent the autooxidation of fats (rancidity). Unsaturated fatty acids may react with oxygen to produce undesirable products with offensive odors and toxic properties, and destruction of nutrients, such as the fat-soluble vitamins. Examples of **natural antioxidants** are vitamin E and vitamin C (ascorbic acid). **Synthetic antioxidants** include ethoxyquin (Santoquin), butylated hydroxytoluene (BHT), and butylated hydroxyanisole (BHA). These synthetic antioxidants are very effective in preventing deterioration of stored feeds and are commonly used for this purpose. Besides protecting against peroxidation in feeds, synthetic antioxidants also prevent these reactions in animal tissues, thus having a **sparing effect** on vitamin E and selenium requirements. (A sparing effect occurs when one nutrient or substance reduces the requirement for another nutrient, usually by replacing it in some aspect of its metabolic function.) In addition to preventing rancidity, synthetic antioxidants help to overcome toxic effects of peroxides. Cabel et al. (1988) fed levels of 50, 100, and 175 meq/kg of diet of peroxide to broilers, with and without 62.5 and 125 ppm of ethoxyquin. Supplementary ethoxyquin prevented the deleterious effects of peroxides on growth performance.

Pellet Binders

Pelleted feeds are widely used. Pelleting increases the density of feed, often resulting in increased feed intake and improved growth and feed efficiency. Pelleting reduces feed wastage and eliminates sorting of ingredients by animals, reduces dust, and increases ease of feed handling. The major disadvantage is the added cost, which may add 10 percent or more to the cost of the feed. Another disadvantage from the livestock producers' point of view is that pelleting may hide a multitude of sins in terms of the quality of the ingredients used.

One of the problems associated with pelleted feeds is the presence of **fines** (small feed particles) and the tendency of pellets to crumble and break apart. Pellet binders are used to give firmer, stronger pellets with a reduced tendency to crumble. **Bentonites** are probably the most widely used binders. They are clay minerals (montmorillonite or hydrated aluminum silicate) with ion-exchange and surface-active properties. Bentonite is usually added at 2 to 3 percent of the diet and is most effective when used in the presence of steam. Bentonite is normally considered to be nutritionally inert. Because of its ion-exchange properties, it may influence metabolism of nitrogen in ruminants by complexing with ammonium ion. Bentonite absorbs ammonia from solution when the ammonium concentration is high and releases it when the ammonia concentration is reduced, which may improve the utilization of urea by ruminants (Britton et al., 1978). Bentonite also absorbs mycotoxins and facilitates their excretion (Carson and Smith, 1983).

Ball clay is a type of clay mined in the southern United States that has been used extensively as an anticaking agent and pellet binder in animal, poultry, and fish feeds. The term *ball clay* originates from the early mining practice of rolling the clay into large balls for transport. In 1997, ball clay from several mines was found to be contaminated with **dioxins**, which are highly toxic chlorine-containing compounds (Fries, 1995). Poultry meat samples from birds fed soybean meal containing ball clay as an anticaking agent had dioxin levels of 3–4 parts per trillion (ppt). The FDA has established that edible meat tissue must contain less than 1 ppt. (1 ppt is equivalent to less than one second in 25,000 years!) A symposium "Is Dioxin Contamination a Problem for Livestock?" was published in *Journal of Animal Science* 76:134–159, 1998, with review papers by Roeder et al. (1998) and Safe (1998).

Hemicellulose extracts and **lignin sulfonate**, by-products of the wood-processing industry, are used as pellet binders. Hemicellulose extracts of wood may have "unidentified growth factor" activity attributed to the biological activity of phenolic compounds (Miller et al., 1979).

ADDITIVES THAT MODIFY ANIMAL GROWTH, FEED EFFICIENCY, METABOLISM, AND PERFORMANCE

Feed Flavors

A number of commercial feed flavors are available. Feed flavors are used to increase the acceptance of diets of low palatability, increase the intake of palatable diets, and increase the intake of diets during periods of stress, such as weaning. Virtually all animals except carnivores have a "sweet tooth" and prefer diets with added sweeteners (sucrose, saccharin, and glucose) compared to the same diet without sweetener. However, addition of a sweetener to the diet does not usually alter feed intake or animal preference. McLaughlin et al. (1983) reviewed flavor preferences of pigs. Campbell (1976) reported that pigs weaned from sows fed a flavored lactation diet had improved feed intake and daily gain when fed a flavored starter diet, suggesting that passage of the flavor through the milk will imprint a pig with a preference for that particular flavor. In general, it appears that feed flavors do not improve performance of animals. Stimulating an animal to consume more of a diet than normal, even if possible, would probably not be advantageous. Feed intake is regulated according to metabolic need; excessive feed intake exceeds protein synthesis capabilities, so the extra nutrients are deposited as fat. Thus the most probable useful role of feed flavors would be to stimulate adequate or normal intake of unpalatable but otherwise nutritious diets. Despite the plausibility of this approach, there is little supportive data.

Provenza et al. (1996) found that when lambs were offered choices of several nutritionally adequate diets with various flavors, they selected a varied diet with a decrease in preference for food just eaten. They suggested that animals in confinement might have increased feed intake if offered a variety of foods, rather than being offered a monotonous, unchanging diet.

Digestion Modifiers

Enzymes Animals produce adequate quantities of digestive enzymes for digestion of the proteins, carbohydrates, and lipids that they are capable of digesting. The main potential of enzyme addition to feed appears to be for digestion of substances that the animal is intrinsically incapable of digesting. For nonruminants, addition of cellulase to feeds would provide a means of digesting cellulose. The technology of production and delivery of cellulase has not reached a practical stage. There are commercial sources of cellulolytic enzymes available, but responses have not been encouraging.

The primary use of enzymes as feed additives has been the provision of sources of **β-glucanase** to swine and poultry to increase the digestion of β-glucans (Campbell and Bedford, 1992). Glucans are an important source of carbohydrate in barley and oats (see Chapter 2). Sources of glucanase, isolated from fermentation products, are available commercially. The addition of β-glucanase to barley-containing poultry diets results in significant improvement in growth rate and feed conversion and reduces or eliminates sticky droppings. Similar improvements, to a lesser degree, have been noted in oat-based diets (Pettersson et al., 1987). Water-soluble pentosans are the major antinutritive factors in rye. Enzyme treatment with fungal enzymes from *Trichoderma viride* reduces the viscosity of rye in the gut and improves its utilization by poultry as a result of hydrolysis of the viscous pentosans. The endosperm cell walls of wheat, triticale, and rye contain pentosans. Feeding a source of pentosanase to broiler chickens improved growth and reduced the severity of wet litter and sticky droppings when rye and triticale (Pettersson and Aman, 1988) and wheat (Choct and Annison, 1992) were fed.

In poultry, β-glucans and pentosans in grain cause an increased viscosity of intestinal contents. These **non-starch polysaccharides (NSP)** form a viscous, gummy layer on the surface of the intestinal mucosa, interfering with the final stages of digestion and absorption. The absorption of large molecules, such as fats, is particularly inhibited, causing a specific **fat malabsorption syndrome**. The adverse nutritional effects of NSP in poultry are explained primarily by their viscosity effects (Campbell and Bedford, 1992). In swine, these NSP have much less effect, because the higher water content of their gut contents dilutes the viscosity effect.

It is generally assumed that feed enzymes would not be effective in ruminants, because rumen microbes secrete a wide variety of enzymes capable of degrading dietary carbohydrates. However, there have been substantial improvements in fiber digestibility in cattle receiving a source of fiber-digesting enzymes as a dietary additive (Beauchemin et al., 1995; Lewis et al., 1996).

New crop (recently harvested) wheat, barley, and oats often support poor growth when fed to poultry and swine. After storage for 4–8 weeks, they give better results. This is probably due to decreases in content and chain length of NSP from activity of carbohydrase enzymes in the grain.

Phytase is the enzyme that digests phytate, an organic complex of inositol and phosphate. Cereal grains and plant proteins such as soybean meal have much of their phosphorus content in the form of phytate (organic phosphate). Phytate phosphorus is significant because it has a low bioavailability to nonruminants and thus is a major contributor to phosphorus pollution from animal wastes. In fact, commercial phytases are used primarily in areas where water pollution from animal waste is important. Bedford (2000) has reviewed the use of feed enzymes such as phytases in animal nutrition.

Buffers A **buffer** is a salt of a weak acid or base that resists a pH change, whereas an alkalizing or **neutralizing agent** neutralizes acid but also increases pH. The use of buffers and neutralizing agents in feeds has been reviewed by Staples and Lough (1989). Sodium bicarbonate, potassium bicarbonate, magnesium bicarbonate, calcium carbonate, and bentonite are true buffers, whereas sodium carbonate, potassium carbonate, magnesium oxide, sodium hydroxide, and calcium hydroxide are alkalizing agents. Buffers are used extensively for ruminants fed high-concentrate diets. The need for buffering capacity is greater with high-concentrate than with high-fiber diets because of the greater rate of fermentation producing more acids, less saliva production (saliva is rich in buffers), and less intrinsic buffering capacity. Forages, especially legumes like alfalfa, have appreciable buffering capacity. **Sodium bicarbonate** is probably the most frequently used buffer; other widely used buffers and neutralizing agents include magnesium oxide, calcium carbonate, cement kiln dust, and tetrasodium pyrophosphate. Added buffers are particularly useful in the adaptation period from high-roughage to high-concentrate diets and aid in the prevention of lactic acidosis. They are also useful in rations for dairy cattle fed high-concentrate diets, particularly when corn silage is the major roughage used. Buffers do not appear to be useful when high-roughage diets are fed (Thomas and Hall, 1984). Use of buffers in diets for dairy cattle has been reviewed by Erdman (1988).

A ruminal pH of about 6.0 is the critical pH below which protein and cellulose digestion rates are reduced (Kovacik et al., 1986). These workers found that various dietary levels (1.5, 3, and 4.5 percent) of sodium bicarbonate greatly increased the amount of time that ruminal pH was above 6.0, but total tract nutrient digestibility was not increased. Boerner et al. (1987a) found that bicarbonate increased dry-matter digestibility in beef cattle.

Sodium sesquicarbonate and trona are two newer buffers available to the feed industry. Sodium sesquicarbonate is produced from trona ore and has a higher degree of purity than trona. Boerner et al. (1987a,b) compared trona to bicarbonate in beef feeding trials. Both buffers shifted starch digestion from the rumen to the small intestine. Both buffers increased dry-matter, crude-protein, and cell-solubles digestibility, with trona more effective than bicarbonate. **Trona** was found to be more effective than bicarbonate in buffering the digestive tract pH throughout diurnal cycles of pH changes because of higher buffering capacity and extended buffer response. It has also been found effective in dairy cattle diets (Coppock et al., 1986).

Limestone exerts its maximum buffering activity at a pH of less than 5.5, so its major effects are postruminal (Keyser et al., 1985). Particle size does not have much influence on its effectiveness as a buffer in the gut. **Magnesium oxide** functions as a rumen buffer; its effectiveness is increased with small particle sizes (Xin et al., 1989). There has been considerable interest in the use of buffers, such as ground limestone and cement kiln dust, to enhance postruminal starch digestion. With high-concentrate diets, the high concentration of VFA produced in the rumen may result in acid overload of the duodenum, lowering the intestinal pH below the optimal level for pancreatic amylase activity. Evidence in support of this theory includes the excretion of a considerable amount of undigested starch in the feces of animals of high-concentrate diets and a lower fecal pH with high-starch diets. Fecal pH can be increased by feeding limestone (Haaland and Tyrrell, 1982), but this does not necessarily reflect increased starch digestibility.

Russell and Chow (1993) offered an alternative explanation for the effect of buffers. They suggested that compared to the effect of bicarbonate transferred into the rumen from the blood, dietary sodium bicarbonate would have little effect on rumen pH. They proposed that the mechanism of action of sodium bicarbonate is to increase water consumption, increasing the dilution of rumen fluid and increasing the rate of escape of starch to the intestine, thus reducing VFA production in the rumen.

Ion-Exchange Compounds The principal ion-exchange compounds used in feeding are **zeolites**, which are clay minerals consisting of hydrated aluminosilicates of various cations. They can gain and lose water reversibly and exchange their constituent cations. Besides natural zeoites, there are also **synthetic zeolites** used as molecular sieves consisting of organic resins or inorganic aluminosilicate gels. One of the major natural zeolites used in feeds is clinoptilolite.

Zeolites are reputed to have beneficial effects on growth, feed efficiency, and incidence of enteric disease. Clinoptilolite has the ability to exchange ammonium ions. Zeolites might improve the utilization of NPN by ruminants by complexing with ammonium ions and releasing them gradually over a period of time, although Pond (1984) found no apparent beneficial effects of inclusion of zeolites in lamb diets. Some aluminosilicates are effective absorbants of aflatoxin, and they have been used in poultry to alleviate the effects of toxic levels of dietary aflatoxin (Phillips et al., 1988). Zeolites bind ammonia and, when used as a feed additive, can reduce ammonia in the air in poultry and swine confinement facilities by keeping ammonia bound in the excreta.

Ionophores and Methane Inhibitors **Ionophores** are a class of antibiotics that are extensively used as feed additives for cattle. Their name is derived from their mode of action in interacting with metal ions and serving as a carrier by which these ions can be transported across membranes. The major ionophores used are produced by various strains of *Streptomyces* fungi and include monensin (rumensin), lasalocid, salinomycin, lysocellin, and narasin. These compounds are also used as coccidiostats in poultry. Monensin is called *monensin* when used in poultry feeding, and *rumensin* when used as a cattle-feed additive.

The feeding of ionophores to cattle consistently improves feed conversion efficiency and often improves daily gain. The improvement in feed efficiency is largely attributed to a change in rumen fermentation resulting in an increased proportion of propionic acid in the end products of fermentation. Propionic acid is used more efficiently in metabolism than acetate and butyrate, and there is less production of carbon dioxide and methane during formation of propionate in the rumen. The result is that from a given quantity of feed, the animal derives more net energy when the proportion of propionate as a percent of total VFA increases.

There is also evidence that ionophores may have a favorable influence on carbohydrate metabolism in cattle by stimulating hepatic gluconeogenesis (Benz et al., 1989). Ionophores inhibit the growth of gram-positive bacteria, so the gram-negative population is enriched when ionophores are fed (Bergen and Bates, 1984). The mechanism of action may be through an alteration of intracellular sodium and potassium concentrations. **Gram-positive organisms** are sensitive because they lack an outer membrane to protect their cell membrane; gram-negative bacteria have a different cell wall structure (Russell and Strobel, 1988). Gram-positive antibiotics disrupt the synthesis of peptidoglycans in the cell wall; in gram-positive bacteria, peptidoglycans can constitute as much as 90 percent of the cell wall, whereas they make up less than 10 percent of the cell wall in gram-negative organisms (Russell and Strobel, 1988). A detailed treatment of the biochemistry involved is given by Russell (1987) and Bergen and Bates (1984).

In addition to positive effects on gain and feed efficiency, ionophores reduce lactic acidosis; aid in the control of coccidiosis, feedlot bloat, and acute bovine pulmonary emphysema; and are toxic to the larvae of face and horn flies in the feces (Goodrich et al., 1984). Ionophores help to reduce lactic acidosis by inhibiting growth of *Streptococcus bovis* and *Lactobacillus* spp. that are the major lactate-producing organisms in grain overload (see Chapter 2). In a similar manner, ionophores reduce feedlot bloat by inhibiting *S. bovis* (Bartley et al., 1983). Acute bovine pulmonary emphysema is associated with abrupt change in pasture (see Chapter 6) and the proliferation of *Lactobacilli,* which produce pneumotoxic compounds; ionophores have protective activity against this condition (Carlson and Breeze, 1984). At the levels approved for use as feed additives, ionophores act as coccidiostats in ruminants and aid in preventing coccidiosis. Monensin is toxic to larvae of face and horn flies and aids in the control of these pests (Herald et al., 1982).

Ionophores are particularly effective in increasing growth rate of pasture-fed cattle, with increased daily gains of 10 to 15 percent (Goodrich et al., 1984). Beef cows may be maintained on approximately 10 percent less feed when they are supplemented with 200 mg monensin daily (Turner et al., 1980). Feeding ionophores has favorable effects on reproduction in cattle, such as reducing time to estrus following parturition (Short and Adams, 1988). These effects appear because of an elevation in blood glucose due to increased rumen production of the gluconeogenic VFA propionate in response to ionophore feeding. Blood glucose concentration appears to be the mediator of the energy status influence on female reproduction (Short and Adams, 1988). Because of their effects on ion transport across membranes, ionophores have an influence on mineral metabolism by animals (Elsasser, 1984). For example, Van Ryssen and Barrowman (1987) noted that liver copper and manganese levels were elevated in sheep fed monensin.

Methane (CH_4) production reduces the efficiency of rumen fermentation. **Methane** (natural gas) is combustible and represents a loss of carbon that could otherwise be metabolized by the host animal. Methane formation represents a "hydrogen sink" by which hydrogen formed in fermentation is removed. The rumen is a reducing environment, having an excess of hydrogen ions, especially with high-fiber diets. Carbon dioxide and methane are produced when cellulose is fermented to acetate and butyrate, whereas, when starch is fermented, all the carbon and hydrogen atoms present in glucose are accounted for in the two propionate molecules produced:

$$\underset{\text{glucose}}{C_6H_{12}O} \xrightarrow[\text{fermentation}]{\text{starch}} \underset{\text{propionic acid}}{2CH_3CH_2COOH} \longrightarrow \text{6 absorbed carbons to be metabolized}$$

$$C_6H_{12}O \xrightarrow[\text{fermentation}]{\text{fiber}} \underset{\text{acetic acid}}{2CH_3COOH} + CO_2 + CH_4 \longrightarrow \text{4 absorbed carbons to be metabolized}$$

Inhibition of methane production would increase the efficiency of ruminant production, as well as reduce methane emissions that have been linked to **global warming**.

Various chemicals inhibit methanogenesis. These include chloroform, iodoform, and other halogenated methane analogs. These have not proven to be practical as feed additives, although combining volatile halogenated methane analogs, such as bromochloromethane, with a cyclodextrin carrier shows promise (McCrabb et al., 1997). The main **methane inhibitors** with practical application are the ionophores. Much of their effect on improving feed efficiency is illustrated by the preceding simplified equations (actual reactions are more involved, but have the same net result). Ionophores shift rumen fermentation to increase propionate production. As illustrated, when glucose is fermented to propionate, all of the carbon and hydrogen of glucose is available for absorption and metabolism by the host animal. In contrast, with acetate production, two of the carbons and four of the hydrogens are lost in carbon dioxide and methane production. Thus, from one mole of glucose, propionate formation yields more net energy so feed efficiency is improved.

Since monensin is toxic to horses, they should not have access to monensin-containing feed. The levels used as feed additives or in blocks for cattle can be toxic to horses.

Isoacids The major volatile fatty acids produced in rumen fermentation are short, straight-chain acids: acetic, propionic, and butyric acids. In addition, small quantities of branched-chain VFA are produced in rumen fermentation from the degradation of branched-chain amino acids (valine, isoleucine, leucine, and proline) during microbial digestion of protein. They are required for the microbial synthesis of branched-chain amino acids. Numerous studies have suggested that under some conditions rumen microbial metabolism may be limited by a deficiency of branched-chain VFA. A feed additive termed **isoacids** has been produced commercially to allow the addition of these compounds to the diet. Isoacids include the following VFAs:

Isobutyric acid	$(CH_3)_2CHCOOH$
2-Methylbutyric acid	$CH_3CH_2CH(CH_3)COOH$
Isovaleric acid	$(CH_3)_2CHCH_2COOH$
Valeric acid	$CH_3CH_2CH_2CH_2COOH$

Andries et al. (1987) have summarized the role of isoacids in ruminant nutrition. In general, supplementation with isoacids does not improve the performance of growing animals. Under some conditions, lactating dairy cattle may show increases in milk production when fed isoacids. Further research is necessary to better determine what these conditions are. Since isoacids are produced mainly from amino acids, their concentration in the rumen might be limiting when diets high in NPN or low in rumen degradable protein are fed. A negative factor associated with use of isoacids is their pronounced odor. This is very unpleasant for feed mill staff who work with the products. Commercial production of isoacids as feed additives has been discontinued.

Probiotics (Direct-Fed Microbials) **Probiotics** is a term coined to describe microbes used as feed additives. The term currently preferred is **direct-fed microbials**. They are defined as live microbial feed supplements that beneficially affect the host animal by improving its gastrointestinal microbial balance (Fuller, 1989). The Russian scientist E. Metchnikoff proposed in 1908 that the gut contains "good" and "bad" microorganisms and that consumption of dairy products such as yogurt promotes the "good" bacteria, prevents aging, and prolongs life. The concept that the gut microflora can be modified by feeding "good" bacteria has attracted a considerable number of adherents, and many companies now produce probiotics as feed additives. One of the major reasons for the interest in probiotics is that they are "natural" alternatives to antibiotics, which, ironically, are also "natural." The majority of probiotic products are based on **Lactobacillus acidophilus**, although other organisms such as *Streptococcus faecium*, *Bacillus subtilis*, and yeasts are also used. Probiotics should be viable and capable of growing in the intestinal tract. Responses to nonviable preparations may occur if their activity is due to enzymes that are released, such as β-galactosidase, that digest lactose. The organisms must be able to survive passage through the highly acid stomach. Most lactobacilli meet this criterion. It is very important that they be resistant to bile if they are to survive in the intestine (Gilliland et al., 1984). Administered organisms should be capable of competing with existing gut microflora and should be host-specific. They should be capable of producing the desired effect in the host animal, such as inhibiting growth of pathogens, providing digestive enzymes, and so on. Despite the intrinsic appeal of direct-fed microbials as replacements for antibiotics as feed additives, the responses to probiotics are often marginal or not observed (Kornegay and Risley, 1996).

Several possible modes of action of probiotics have been suggested by Pollmann et al. (1980a).

1. Change in gut microflora and a reduction in *E. coli*
2. Production of antibiotics
3. Synthesis of lactic acid with consequent reduction in intestinal pH
4. Adhesion to or colonization of the intestinal mucosa
5. Prevention of toxic amine synthesis in the gut
6. Other as yet unidentified modes of action, such as stimulating immune responses in the gut

Attachment of *Lactobacilli* to the intestinal mucosa, which may assist in the exclusion of pathogens, has been demonstrated (Barrow et al., 1980). Most enteric pathogens, such as *E. coli* and *Vibrio cholera*, cannot produce disease without **attachment** to the host intestinal cells, so attachment of probiotic organisms to the intestinal cells could be very important in increasing disease resistance (Watkins and Miller, 1983). Lessard and Brisson (1987) showed that feeding a *Lactobacillus* probiotic to young pigs stimulated growth and increased the serum concentration of IgG, indicating a favorable effect on the immune system.

Yeasts (single-celled fungi) and other fungi are also used as probiotics. The main species used are *Saccharomyces cerevisiae* and *Aspergillus oryzae*. "Yeast culture" refers to a dry product containing yeast and the media on which it was grown, dried to preserve the fermentative capacity of the yeast. Administration of yeast culture improved performance of calves subjected to stress from handling and shipping (Phillips and Von Tungeln, 1985). Increases in feed intake with yeast supplementation have been noted (Phillips and Von Tungeln, 1985). This may be a result of favorable effects on palatability or stimulation of rumen fermentation and improved digestibility. Yeast and fungal probiotics cause changes in rumen fermentation, including stimulation of cellulolytic bacteria and an increase in the rate of cellulose digestion (Frumholtz et al., 1989). Glade and Biesik (1986) found that a yeast product improved nitrogen retention in horses, and Glade and Sist (1988) reported improved fiber digestibility in horses given yeast culture.

There are exciting possibilities for creation of new probiotics using biotechnology. Lysine-secreting *Lactobacilli* have been mentioned earlier (Chapter 4) as one possibility. A new area will be the development of **genetically engineered microbes** to improve efficiency of rumen fermentation (Russell and Wilson, 1988; Wallace, 1994). Examples of some of the possibilities include development of microbes to improve the utilization of high-starch diets (the significant amount of fecal starch in ruminants fed high-grain diets indicates the present inefficiency), microbes with improved cellulolytic capability, high xylanase activity to improve digestion of grain by-products, and increased lactate-utilizing ability to reduce lactic acidosis. Microbes that detoxify plant toxins, such as mimosine in leucaena, exist (see Chapter 5). It should be possible to introduce detoxifying enzymes when appropriate to permit the utilization of forages and feeds that are now toxic. Rumen fungi (see Chapter 4) are important in the digestion of lignified feeds. Perhaps it will be possible to prepare probiotic preparations of rumen fungi that could be administered to animals fed low-quality roughages to improve fiber digestion.

Oligosaccharides Oligosaccharides are short chains of sugars linked together that include **fructooligosaccharides (FOS), mannanoligosaccharides (MOS)**, and **galactooligosaccharides (GOS)**, which are short polymers of fructose, mannose, and galactose, respectively. These oligosaccharides are used as feed additives for the modification of gut microbes and stimulation of the intestinal immune system. The intestinal mucosa contains carbohydrate residues that serve as binding sites for pathogenic bacteria that have receptor sites for oligosaccharides on their cell walls. The proposed mode of action of dietary oligosaccharides is that they bind to the receptor sites on bacteria, thus preventing them from binding to the intestinal mucosa. Pathogenic bacteria must bind to and colonize the gut lining to cause enteric disease. The binding sites on bacteria are called **lectins**. Raw beans also contain lectins (see Chapter 4) that cause damage to the intestinal mucosa by binding to the carbohydrate components of the microvilli (see Fig. 4.6).

Another possible mechanism of action of oligosaccharides is that they may stimulate the growth of beneficial bacteria in the gut, such as *Lactobacillus* and *Bifidobacterium* spp. (Gibson and Roberfroid, 1995; Roberfroid et al., 1998). These bacteria use the oligosaccharides as substrates, so their growth is encouraged by the use of oligosaccharides in the diet. By the process of **competitive exclusion**, stimulation of beneficial bacteria should reduce numbers of detrimental organisms. Beneficial microbes such as *Bifidobacteria* have a greater binding affinity for mucosal attachment than enteric pathogens (Meng et al., 1998). Thus stimulation of *Bifidobacteria* by feeding oligosaccharides may aid in preventing the attachment and colonization of the gut epithelium by pathogenic bacteria (Meng et al., 1998).

Although the proposed mechanisms of action of oligosaccharides are intriguing, these products have not been very effective in improving animal performance (Waldroup et al., 1993; Orban et al., 1997).

Gibson and Roberfroid (1995) have introduced the term **prebiotic** for the oligosaccharides. They define *prebiotic* as nondigestible food ingredients that benefit the host by selectively stimulating the growth and/or activity of bacterial species already in the gut, such as *Lactobacillus* and *Bifiobacteria*.

Acidifiers (Organic Acids) Organic acids, commonly referred to as *acidifiers* or *acidifying agents*, have shown favorable effects when used as additives in diets for weanling pigs. In most animals, the stomach does not become highly acid until after weaning. During the suckling period, fatty acids in milk have antimicrobial activity. After weaning, development of a highly acid stomach aids in killing ingested microbes. During a brief period (the window of enteric vulnerability) between weaning and the development of low stomach acid, young animals are susceptible to encroachment of pathogenic organisms into the gut, leading to the well-known phenomenon of postweaning diarrhea (scours). Feeding organic acids during this period may aid in lowering the stomach pH and preventing digestive upset. **Citric** and **fumaric acids** have been the primary acidifiers tested. Falkowski and Aherne (1984) reported a 4 to 7 percent increase in average daily gain and a 5 to 10 percent improvement in feed efficiency with the feeding of 1 or 2 percent fumaric or citric acid to weanling pigs. Similar results were reported by Edmonds et al. (1985) and Giesting and Easter (1985). Patten and Waldroup (1988) observed a significant increase in growth rate of broiler chicks with dietary levels of 0.5 and 1.0 percent fumaric acid. Radecki et al. (1988) found that fumaric acid increased growth rate and feed efficiency of pigs in the starter period, and the effect was not mediated through changes in nutrient digestibility. Giesting and Easter (1985) suggested that part of the beneficial results of citric and fumaric acids may be attributed to their effects on energy metabolism, since both citric and fumaric acids are intermediates in the tricarboxylic acid cycle of cellular energy metabolism. Edmonds et al. (1985) and Burnell et al. (1988) noted an interaction between copper sulfate and organic acids in growth and feed efficiency responses. In calves, acidification of milk replacers may improve milk clot formation in the abomasum, which reduces the risk of digestive upsets (Fallon and Harte, 1980). Acidified milk replacers help to keep the abomasal pH below 4.2, the minimum pH at which *E. coli* can survive.

While the hypothesis that dietary organic acids lower stomach pH in weanling pigs is reasonable, it has not been confirmed experimentally. Risley et al. (1992) and Gabert and Sauer (1995) found no effect of dietary organic acids on digestive tract pH. The mechanism of the growth-promoting effects of organic acids is not clear.

Antibloating Agents Surface-active agents such as poloxalene, marketed as **Bloat Guard**, are effective in preventing frothy bloat. As described earlier (Chapter 5), frothy bloat is associated with consumption of lush, legume pasture, and the rapid release of fermentable carbohydrate and soluble protein in the rumen. Surface-active agents prevent the formation of a stable foam by preventing development of bubble membranes. Antibloat agents are generally provided as components of blocks (Fig. 8.2).

Salivation Inducers Salivation inducers **(sialagogues)** are substances that increase the production and secretion of saliva. Current interest in these substances centers on **slaframine**. It is a mycotoxin produced by fungi that grow on red clover (black patch disease of clover); consumption of infected clover pasture or hay by livestock causes "slobbers" or profuse salivation (Croom et al., 1990, 1995). Froetschel et al. (1995) have proposed that slaframine might have useful pharmacological properties in ruminants. Inadequate secretion of saliva occurs when low-fiber, high-concentrate diets are fed, leading to metabolic disorders and suboptimal feed utilization. Froetschel et al. (1995) demonstrated that administration of slaframine to ruminants increased the salivation rate, the liquid turnover rate in the rumen, and the rumen pH, and it increases the efficiency of microbial protein synthesis in the rumen (Froetschel et al., 1989). Slaframine and other sialagogues appear to have potential as additives to alter rumen fermentation to favorably influence animal performance.

FIGURE 8.2 Provision of blocks containing antifoaming agents to cattle on lush legume pasture is an effective means of reducing bloat. Here an animal is licking a Bloat Guard block containing poloxalene.

Defaunating Agents **Defaunation** is the process of treating a ruminant to eliminate its rumen protozoa. This can be accomplished with a number of chemicals, including copper sulfate, and nonionic and anionic detergents. Numerous studies have been conducted with defaunated ruminants to determine the roles of protozoa in rumen fermentation and as a possible means of improving animal performance. The roles of protozoa in the rumen have been reviewed by Veira (1986). Protozoa feed on bacteria and small feed particles. They may help to stabilize conditions within the rumen by storing starch particles and keeping the bacterial population from proliferating excessively when high-concentrate diets are fed. They may serve as a reservoir of nitrogen in the rumen, which in wild animals may aid in stabilizing rumen fermentation during periods of inadequate feed (Yokoyama and Johnson, 1988). Protozoa are selectively retained in the rumen, probably by sequestration to large feed particles in the floating raft and in the ruminated bolus. Protozoa feed on rumen bacteria and probably on rumen fungi as well. Defaunation results in an increased population of rumen fungi (Preston and Leng, 1987).

Bird and Leng (1978, 1984) and Bird et al. (1979) presented evidence suggesting that defaunation improves animal performance when high-energy, low-protein diets are fed. This work was stimulated by observations that cattle fed sugarcane and molasses-based diets had high populations of rumen protozoa and low productivity. It was hypothesized that the retention of a large population of protozoa in the rumen could reduce availability of microbial protein to the host animal and limit production. With a high-energy molasses or sugar-based diet, growth rates of cattle and sheep were improved by defaunation; wool growth was enhanced also. On conventional diets, there seems to be little effect of defaunation on animal performance (Veira, 1986).

Practical methods of defaunating ruminants and maintaining the defaunated state have not yet been developed. Under certain conditions, the process could favorably influence animal performance. Keeping defaunated animals from becoming refaunated with protozoa is difficult (Lovelock et al., 1982).

Saponins, which are detergent-like cholesterol-binding compounds in many plants, are natural defaunating agents (Wallace et al., 1994). They react with cholesterol in protozoal cell membranes, causing lysis of the cell. Saponins in **yucca extract** (Wallace et al., 1994) and in tropical forage trees (Newbold et al., 1997) may favorably influence rumen fermentation by suppressing protozoa. A consistent effect of defaunation is a reduction in **rumen ammonia**. A major source of rumen ammonia is nitrogen excreted by protozoa feeding on rumen bacteria (Wallace et al., 1994).

Rumen ciliate protozoa increase sulfide production in the rumen via metabolism of S-amino acids, resulting in copper being bound to the sulfide and becoming unavailable for absorption and utilization (Ivan et al., 1986). Defaunated sheep may be susceptible to copper toxicity (Ivan et al., 1986). Liver copper levels are higher in defaunated than in faunated sheep (Ivan, 1988).

Metabolism Modifiers

Hormones Hormones are neither permitted nor used to any extent as feed additives in the United States. At one time, **diethylstilbestrol (DES)**, a synthetic estrogen, was used primarily for cattle but FDA approval was withdrawn. Estrogens improve growth and feed efficiency in ruminants, particularly in steers. Estrogenic compounds may be administered by implant in the ear. One of the main types used is zearalenone **(Ralgro)**, a mycotoxin with estrogenic activity (see Chapter 2). The only hormone permitted in the United States as a feed additive for beef cattle is **melengestrol acetate (MGA)**, a synthetic progesterone that suppresses estrus and improves weight gain and feed efficiency in feedlot heifers. By suppressing estrus, normal behavior and eating patterns of both steers and heifers are maintained by eliminating the presence of heifers in heat.

Beta-Adrenergic Agents (Repartitioning Agents)

A new class of feed additives, the β-adrenergic agents, β-agonists, or **repartitioning agents,** has been developed. They are norepinephrine (noradrenalin) analogs that stimulate β-adrenergic receptors. They are called repartitioning agents because they result in a repartitioning of nutrients from fat to protein synthesis, causing increased muscle mass and decreased body fat (Fig. 8.3). Clenbuterol, cimaterol, and ractopamine are examples of these compounds.

There is considerable interest in reducing the fat content of meat animals and increasing the yield of lean meat. It is estimated that 5 billion pounds of unwanted fat are trimmed from carcasses in the United States each year to be recycled as rendered fat in animal feeds. This is obviously highly inefficient. Feed additives to reduce fat deposition and increase formation of lean tissue (protein) would be very useful. Repartitioning agents are believed to act by increasing fat mobilization (lipolysis) and stimulating protein synthesis. They may also reduce degradation of muscle protein. The accretion of muscle tissue is the net difference between synthesis and degradation of muscle protein; these processes occur continuously. The mode of action has been reviewed by Ricks et al. (1984).

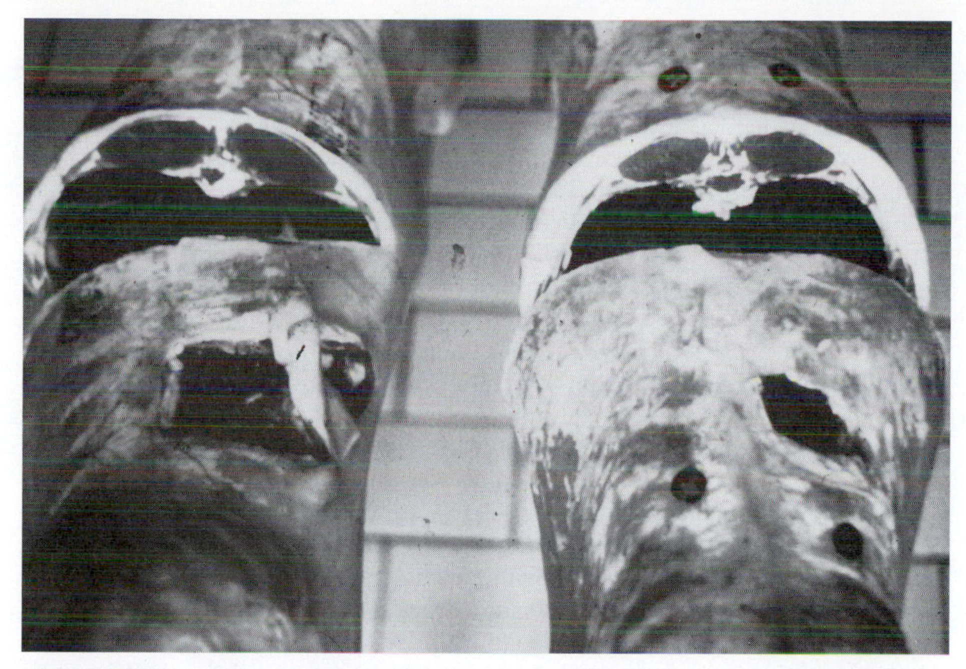

FIGURE 8.3 Effect of a repartitioning agent on protein and fat deposition in lambs. Lamb on left was fed a repartitioning agent (clenbuterol); control lamb on right was not. (Courtesy of American Cyanamid Co.)

A significant problem with repartitioning agents has been their abuse by livestock exhibitors. In numerous instances, grand champion market animals (cattle, sheep, pigs) have been "doped" with clenbuterol to increase muscling. This is a criminal act that has been dealt with by the legal system. Interestingly, **residues of clenbuterol** can be detected many months after its use by the presence of residues in the tissues of the eye and in hair (Dursch et al., 1995; Smith and Paulson, 1997). The clenbuterol is accumulated by melanin, so it occurs in dark hair pigments and the colored retina of the eye.

β-agonists are an example of substances that are potential **health hazards** to humans when present as residues in meat. In Europe, symptoms such as muscle tremors, heart palpitations, headache, dizziness and nausea have occurred in people consuming liver from animals administered clenbuterol. These products, because of their toxicity to humans and potential for abuse, are substances that animal agriculture can do without.

A symposium on the "Pharmacology, Toxicology, and Illegal Use of β-Adrenergic Agonists" was published in *Journal of Animal Science* 76:160–211 (1998), including reviews by Mersmann (1998) and Smith (1998).

Growth Promotants

Antibiotics **Antibiotics** are natural metabolites of fungi that inhibit the growth of bacteria. Their discovery is credited to Alexander Fleming of Great Britain, who noted that bread molds produced a substance that inhibited bacterial growth. This substance was **penicillin**. Since that time, hundreds of antibiotics have been discovered. In the 1940s, commercial production of fermentation products such as penicillin and vitamin B_{12} began, using corn as the substrate for fermentation. Fermentation residues were evaluated as poultry and livestock feeds. Favorable effects on animal performance were noted, with improvement of growth and feed efficiency and a reduction in health problems such as postweaning diarrhea. It was soon discovered that these beneficial effects were associated with antibiotics in the fermentation residues. These findings led to the adoption of the practice of feeding **subtherapeutic levels** of antibiotics to animals, particularly swine and poultry, as growth promotants. Subtherapeutic doses are lower than therapeutic levels used for treatment of disease; prophylactic doses are used for disease prevention. Thus antibiotics used as growth promotants are fed at lower levels than when they are used for the treatment or prevention of disease.

The **mechanism of action** of antibiotics in their growth-promotion effects is not conclusively known but definitely appears to be due to actions on the gut microflora. Some of the most plausible theories identified by Visek (1978) are as follows: (1) Microbes responsible for mild but unrecognized infections are suppressed. (2) Microbial production of growth-depressing toxins is reduced. (3) Antibiotics reduce microbial destruction of essential nutrients in the gut or, conversely, improve the synthesis of vitamins or other growth factors. (4) There is enhanced efficiency of nutrient absorption because the intestinal wall is thinner.

An additional and likely explanation for the antibiotic growth response is that pathogenic gut bacteria produce toxins that, when absorbed, elicit immune responses, increasing the animal's need to synthesize antibodies and other components of the immune response. This diverts nutrients into immune responses instead of growth (see Immunomodulators—this chapter).

Subtherapeutic levels of antibiotics increase an animal's ability to withstand stress and aid in control of postweaning diarrhea. The response to antibiotic feeding is greatest in young, unthrifty, or stressed animals. The response is usually less when animals are housed in a new building; as microbial loads in the environment build up, the response to antibiotics increases. **Broad-spectrum antibiotics** (active against both gram-positive and gram-negative organisms), such as chlortetracycline, are usually more effective than narrower-spectrum agents such as penicillin and bacitracin.

Animals fed antibiotics have a thinner intestinal wall than controls that are not receiving them; this could improve nutrient absorption and also reduce maintenance energy and protein requirements. The intestinal mucosa is the most rapidly regenerating tissue in the body. Enterocytes (see Chapter 1) have a very rapid turnover, and this turnover rate is increased by exposure to bacterial toxins and metabolites, thus substantially increasing endogenous losses of protein and energy. Antibiotics reduce the turnover rate of intestinal mucosal cells. Visek (1978) observed that antibiotics reduce formation of ammonia and amines in the gut; these compounds are irritants that increase the turnover rate of epithelial tissues. Another possibility in explaining the mode of action of antibiotics is that they may reduce the intensity of the immune response (see the section Immunomodulators later in this chapter).

Although antibiotics have been used as feed additives since 1950, their effectiveness has not diminished with time. The magnitude of the growth response is about the same now as when their use first began. This suggests that the development of microbial resistance to antibiotics, which is well documented, does not alter their growth-promoting activity.

Some of the most commonly used growth-promoting antibiotics are bacitracin, bambermycins, chlortetracycline, erythromycin, lincomycin, neomycin, oxytetracycline, oleandomycin, penicillin, streptomycin, tylosin, flavomycin, and virginiamycin.

Antibiotic Feed Additive Use and Human Health The extensive use of subtherapeutic levels of antibiotics as feed additives has sparked controversy as to possible adverse effects of this practice on human health. The major concern is that widespread use of antibiotics may lead to the development of **antibiotic-resistant strains of bacteria**, which could make treatment of human diseases more difficult. The worst-case scenario is that bacteria resistant to all antibiotics could develop and create life-threatening infections in humans (Kunin, 1993). Development of antibiotic-resistant human pathogens, such as penicillin-resistant gonorrhea-causing bacteria, are due primarily to their excessive and/or improper use in medicine, and not to their use as feed additives.

Use of antibiotics in feeds does increase the number of antibiotic-resistant organisms in animals (CAST, 1989; Mathew et al., 1998). The antibiotic resistance can be transferred to other bacteria. The ability of some bacteria to resist antibiotics is carried in nonchromosomal pieces of genetic material called **plasmids**. Transferable plasmids carrying the ability to resist antibiotics are called *R-factors*. The concern is that R-factors could be transferred from bacteria in antibiotic-fed livestock to bacteria in humans. Although this is theoretically possible, unequivocal evidence that it has occurred in the period of antibiotic use has not yet been found. Although the potential problems that could result from antibiotic use should receive continual monitoring, there is no evidence that use of antibiotics as feed additives has harmed human health or that a ban on their use would improve human health.

Continued use of antibiotics as growth promotants is of economic benefit to consumers and to antibiotic manufacturers. The impact of their use on the net income of livestock producers is minimal. Increased costs and lowered efficiency of livestock production from a ban on antibiotic use would be passed on to consumers.

> The principal beneficiary of the present reduced costs of production is the consumer, not the producer. Once the majority of livestock and poultry producers accept the technology of feed additive usage, the savings in cost of production are passed on to the consumer. The end result of unnecessary regulation of feed additive usage of antibiotics or other technology is higher prices of meat, milk and eggs to the consumer.
>
> Hays and Muir (1979)

In a comprehensive study of human health risks associated with subtherapeutic use of penicillin or tetracyclines in animal feed, it was concluded:

> . . . we were unable to find a substantial body of direct evidence that established the existence of a definite human health hazard in the use of subtherapeutic concentrations of penicillin and the tetracyclines in animal feeds. . . . The committee believes that there is indirect evidence implicating subtherapeutic use of antimicrobials in producing resistance in infectious bacteria that causes a potential human health hazard.
>
> National Academy of Science, Institute of Medicine (1989)

One of the ways in which antibiotic resistance could be transferred from microbes in livestock to bacteria infecting humans is via contamination of carcasses in animal and poultry processing plants with *E. coli*, salmonella, and other enteric bacteria. Improvements in processing techniques could reduce or eliminate this potential hazard.

Chemotherapeutic Agents A variety of compounds that have growth-promoting activity are not antibiotics. These include arsenicals, nitrofurans, and sulfonamides. **Arsenicals** are organic compounds containing arsenic, such as arsanilic acid, sodium arsanilate, and roxarsone. Arsenicals have a growth-promoting effect similar to that of antibiotics. An antagonism may exist between arsenicals and copper sulfate; when growth-promoting levels of each are combined, growth depression is often observed (Czarnecki et al., 1984). Roxarsone reduces liver copper in chicks receiving copper supplementation.

Nitrofurans are synthetic compounds with antimicrobial activity. Chemically they are based on a nitrofuran ring (a five-membered ring with an NO_2-group attached). Furazolidone is an example of a commonly used nitrofuran. Nitrofurans inhibit bacterial reproduction. Their primary use as feed additives is in controlling enteric diseases. Furazolidone has growth-promoting activity in poultry and swine.

Sulfonamides are derivatives of sulfanilic acid (a benzene ring with $-NH_2$ and $-SO_3H$ groups attached). Sulfonamides have broad-spectrum bacteriostatic effects (bacteriostats inhibit bacterial reproduction). They are also effective against protozoa that cause coccidiosis, and some viruses. Sulfaquinoxaline has been extensively used in poultry as a coccidiostat. Sulfonamides are competitive inhibitors of vitamin K and may induce a deficiency of the vitamin if low-vitamin-K diets are used.

Carbadox is a synthetic antibacterial compound that is similar in action to antibiotics. It is extensively used in swine diets for its growth-promotant effects.

Copper sulfate has been extensively studied as a feed additive. The dietary requirement for copper for normal health and production is approximately 10 to 20 ppm in most species. At much higher levels (125–250 ppm), copper sulfate is a growth promoter for swine and poultry and is now used extensively for this purpose. This is particularly true in Great Britain, where use of antibiotics as feed additives has been curtailed. Copper sulfate is an effective substitute for antibiotics and probably has a similar mode of action. Copper sulfate produces a 5 to 10 percent improvement in growth and a 4 to 8 percent improvement in feed conversion. There is often a favorable effect in reduction of enteric disease.

Although copper is an effective growth promotant, there are several concerns regarding its use. Copper accumulates in the liver when feed additive levels of the element are fed. There is concern as to whether these high copper levels in the liver represent a potential human health hazard. However, the livers of pigs fed 250 ppm copper are of similar copper content as normal beef liver (200–600 ppm), so the hazard is likely slight. Since copper is primarily effective during the immediate postweaning period, it is probably desirable not to use it in finisher diets, thereby ensuring normal liver copper levels at slaughter. The other major concern with use of copper as a feed additive is the possible environmental effects. Swine and poultry manure are often used to fertilize crop or pastureland, and there is the possibility that crops and pastures could take up excessive levels of copper. Sheep are particularly sensitive to copper toxicity. In addition, broiler litter or dried poultry waste (DPW) is often used as a feedstuff for ruminants (see Chapter 4); manure from birds fed copper sulfate may cause **copper toxicity** in ruminants. Banton et al. (1987) described two outbreaks of copper toxicosis in cows fed chicken litter. One involved a dairy herd fed chicken litter containing 620 ppm copper, with a copper concentration in the complete diet of approximately 300 ppm. Two cows died of copper-induced hemolytic crisis. In the other incident, 15 steers fed a diet containing high copper (700–900 ppm) broiler litter died with hemolytic crisis. It is apparent that high-copper poultry litter is a hazardous feed.

Zinc oxide has similar growth promotant properties as copper sulfate. Levels of 1,500–3,000 ppm zinc oxide are used in nursery pig diets to improve growth and reduce diarrhea and mortality (Hill et al., 2001). As with copper sulfate, there are environmental pollution concerns with the use of zinc.

Saponins Steroid saponins in yucca, a desert plant of the southwestern United States (Fig. 8.4), have been given the trade name of **sarsaponin** and are sold commercially as agents to improve animal performance. Mader and Brumm (1987) found an improvement in gains of steers fed urea-containing diets supplemented with sarsaponin and suggested a role of sarsaponins in ammonia metabolism in the rumen. These authors also noted a slight improvement of feed efficiency in finishing pigs fed sarsaponin. Preston et al. (1987) investigated the effect of 120 ppm dietary sarsaponin on nitrogen metabolism in rats, and they concluded that it has little effect, if any. Nitrogen digestibility and nitrogen balance, growth, and feed efficiency were unaffected by dietary sarsaponin. Yucca saponins reduce rumen ammonia concentrations, primarily by suppressing rumen protozoa (Wallace et al., 1994). Much of the free ammonia in the rumen arises from protozoal digestion of bacterial protein.

FIGURE 8.4 Sarsaponins are obtained from the yucca plant, common in the southwestern United States and in Mexico. Yucca saponins may have beneficial effects as feed additives by reducing ammonia release from the manure in confinement livestock units.

ADDITIVES THAT MODIFY ANIMAL HEALTH

Drugs

A variety of drugs may be added to the feed for the treatment and/or prevention of disease or internal parasites. Some of the drugs commonly used in feeds include worming agents (e.g., phenothiazine, dichlorovos, piperizine) and coccidiostats (e.g., monensin, amprolium, sulfaquinoxalene). A detailed discussion of these is inappropriate here. Regulations involving their use change frequently, and permitted products, dosages, withdrawal times before slaughter and so on vary from country to country. In the United States, complete information on medicated feed additives can be obtained from the *Feed Additive Compendium*, published annually by the Miller Publishing Co. Purchasers of the *Compendium* receive monthly update reports on any additions or changes to the regulations. Each year the magazine *Feedstuffs* (also published by Miller, see Chapter 1) has a special reference issue with current information on feed milling regulations, feed additives, and the like. These publications are indispensable to anyone in the U.S. feed business.

Environmentally Active Substances

A few feed additives are used for their beneficial effects on the environment, particularly the interior environment of poultry, rabbit, and swine confinement facilities. These products are primarily used for the control of ammonia. **Ammonia** is an irritating and toxic gas that is released by bacterial or fungal action on the urea or uric acid excreted by mammals or birds, respectively. Problems linked to atmospheric ammonia have been reviewed by Carlile (1984). In poultry, air ammonia levels above approximately 20 ppm can cause such problems as keratoconjunctivitis (eye damage); damp, matted feathers; and respiratory disease. Ammonia impairs mucus flow and cilia action in the nasal passages and trachea, reducing resistance to respiratory infection. Animal productivity is reduced when detectable levels of ammonia are present in the environment on a chronic basis and the health of workers in the building is compromised.

A number of feed additives aid in controlling ammonia production. **Zeolites** may reduce ammonia to a minor extent when fed to poultry and are more effective when spread directly on the litter or beneath the cages (Nakaue and Koelliker, 1981). **Sarsaponin** in the feed has beneficial effects (Johnston et al., 1981) and is more effective when applied to the excreta directly.

Immunomodulators

Instead of using antibiotics and other feed additives to help control microbes that reduce animal performance, an alternative approach is to use compounds to stimulate an animal's immune system, enhancing resistance to microbial effects. Young animals, such as baby pigs at weaning, have an immature immune system and thus are susceptible to postweaning enteritis and morbidity. Although Thaler et al. (1989) found that an immunomodulator (FK-565) given in the diet did not enhance the performance of baby pigs, the concept has merit and may offer a viable alternative to antibiotics if effective compounds can be developed.

Klasing and coworkers at the University of California have studied the nutritional interrelationships with the immune response (Klasing, 1988; Klasing and Barnes, 1988). Immune responses result in a variety of metabolic adjustments that are mediated by cytokines of leukocytic origin. **Cytokines** are hormonelike molecules involved in cell-to-cell communication. Examples are interleukins and interferons. Substances that elicit immune responses (**immunogens**) result in decreased animal performance. Metabolic adjustments include changes in protein, fat, carbohydrate, and mineral metabolism. These have been reviewed by Klasing (1988). **Stress** may actually reduce some nutrient requirements, rather than increase them, which is the common perception. For example, Klasing and Barnes (1988) showed that immunological stress reduces the amino acid requirements of growing chicks because of reduced growth rate.

Klasing (1988) has proposed a mechanism of action to explain the **role of antibiotics** in improving growth. The antibiotic response is greater in a dirty environment than in a clean one. Antibiotics may act by reducing the microbial burden of the animal and, consequently, the number and intensity of immune responses that result in leukocytic-cytokine release. The synthesis of these substances is metabolically expensive, so the need for lower levels of cytokines provides greater nutrient availability for growth.

Johnson (1997) and Spurlock (1997) have reviewed the effects of immune system stimulation on animal performance. Even subclinical infection has a significant effect in diverting nutrients away from growth and prevents an animal from achieving its genetic potential. In response to antigenic stimuli, a variety of immune system cells, such as activated macrophages, **secrete proinflammatory cytokines** that alter the animal's metabolism. Among the cytokines are **tumor necrosis factor alpha (TNF-α)**, **interleukin-1 (IL-1)**, and **interleukin-6 (IL-6)**. These cytokines have profound behavioral, neuroendocrinological, and metabolic effects. They alter the metabolism of lipids, carbohydrates, and protein. They act in the brain to cause reduced feed intake. They activate other immune modulators such as glucocorticoids, prostaglandins, and catecholamines, which affect cell metabolism and growth. In essence, during an immune challenge, proinflammatory cytokines initiate responses that divert nutrients away from growth in support of immune function. A new system of raising pigs, called **segregated early weaning (SEW)**, is designed to protect young pigs from being immunologically challenged. Their immune systems are not activated, and they use absorbed nutrients with maximum efficiency for growth (see Chapter 13).

Numerous nutrients have immunomodulating effects; they are referred to as **immunonutrients** (Suchner et al., 2000). These include some amino acids (e.g., arginine, cysteine, tryptophan, and taurine), some polyunsaturated fatty acids, vitamins (e.g., A, C, and E), and trace elements (Cr, Zn, Se). The dietary levels of these nutrients needed for optimal immune function exceeds the amounts needed to prevent classical deficiency symptoms.

Some dietary factors have negative effects on the immune system, especially various **mycotoxins**. Examples include aflatoxin, ochratoxin, vomitoxin, and fumonisin. Mycotoxin-induced immunosuppression can result in reduced resistance of the host to infectious disease and also reduction in efficacy of vaccines.

ADDITIVES THAT MODIFY CONSUMER ACCEPTANCE

Xanthophylls

Xanthophylls are red and yellow carotenoid pigments that are important in poultry nutrition because they are deposited in egg yolk and body fat, producing the yellow pigmentation of eggs and of the skin and shanks of broilers. Sources of xanthophylls include green plants such as alfalfa, yellow corn, and a variety of yellow or red plant materials. The richest source is **marigold petal meal**, which is produced commercially for poultry feeding. It contains 6,000 to 10,000 mg xanthophyll per kg; in comparison, dehydrated alfalfa meal contains approximately 200 mg/kg, and corn contains approximately 20 mg/kg. The major xanthophyll in alfalfa is **lutein**, which is yellow, whereas in corn and corn gluten meal the major pigment is **zeaxanthin**, which imparts an orange-red color.

Xanthophylls have no known nutritional value other than as a source of pigment, although Bendich and Shapiro (1986) suggested that they may stimulate the immune system. Consumers prefer pigmented poultry products because poultry that were raised on range before the development of intensive confinement-rearing systems obtained large amounts of xanthophylls from the green vegetation they consumed. Also, most birds were and still are fed corn-based diets, providing a source of pigments. In most parts of the United States, pigmented birds and eggs are preferred; however, in the Pacific Northwest, the preference is for unpigmented products. This probably stems from the tradition of feeding birds wheat in this region because of the lack of locally produced corn.

A number of **synthetic carotenoids** (e.g., canthaxanthin) have been developed and are marketed as pigmenting agents for poultry. Until recently, alfalfa meal was a principal source of xanthophylls in poultry diets. With the increasing emphasis on high-energy diets for poultry production, however, even the use of 1 to 3 percent alfalfa meal is considered undesirable by the industry because it reduces feed conversion efficiency. Synthetic xanthophylls are increasingly used in poultry diets. Xanthophyll pigments are also important in fish nutrition (see Chapter 23).

Dried marine algae are good sources of both xanthophyll pigments and omega-3 fatty acids (Herber-McNeill and Van Elswyk, 1998) and can be used as feed additives to enhance both omega-3 fatty acid and xanthophyll contents of egg yolks. Marine microalgae are the original source of the omega-3 fatty acids in fish and therefore represent a more direct dietary source of these fatty acids than fish meal. Carotenoids (e.g., astaxanthin, canthaxanthin) are also used as pigmenting agents in diets for trout and salmon (see Chapter 23).

Saponins

Saponins complex with cholesterol and have been shown to have blood-cholesterol-lowering properties. Several studies have been performed to determine if feeding saponins will lower the cholesterol content of meat and eggs, but with negative results. Nakaue et al. (1980) used high-saponin and low-saponin alfalfa in layer diets and found no differences in egg cholesterol levels. Sim et al. (1984) found that dietary saponin reduced transfer of dietary cholesterol to egg yolk but failed to lower egg cholesterol content. Apparently, synthesis of cholesterol within the bird obviates any dietary effect.

QUESTIONS AND STUDY GUIDE

1. There are three main groups of people or industries that are affected by the use of feed additives—consumers, manufacturers of additives, and livestock producers. Of these groups, who benefits most from their use? Who benefits the least? Discuss this issue in class.
2. Critique this statement: The elimination of the use of feed additives would markedly increase the cost of meat, milk, and eggs to the consumer; would reduce the profits to the shareholders of multinational drug companies; and would have comparatively little influence on the net income of livestock producers.
3. If the use of feed additives that increase animal performance were banned by regulatory agencies, what effect do you think this would have on the number of livestock-producing farms and the number of animals per farm? What would be the positive and negative effects of these changes in farm size and number of producers?
4. Sodium and potassium propionate are used as mold inhibitors in feeds. Analysis of blood of wild bighorn sheep shows that they have sodium and potassium propionate in their blood. Where did this propionate come from?
5. Why is bentonite used as a feed additive?
6. Suppose a feed flavor were discovered that would cause pigs to eat more of a corn-soy diet than they would without the added flavor. Would this be advantageous or not? Why? What effect might it have on carcass composition?
7. Discuss the effects of buffers and ion-exchange compounds on rumen fermentation.
8. What are ionophores? What effects do they have on animal production? Discuss their mode of action.
9. Discuss the potential of probiotics as replacements for antibiotics in swine and poultry nutrition. Compare their modes of action.
10. How might biotechnology be used to improve the efficacy of probiotics?
11. Why is fumaric acid used as a feed additive?
12. Slaframine is an example of a natural toxin with possible application as a feed additive. What is another example of a natural toxicant that has useful application in animal production?
13. Discuss the use of hormones as feed additives.
14. What are the potential advantages of raising defaunated ruminants?
15. What are the economic implications of the use of repartitioning agents in animal production?
16. What is your opinion of antibiotics as feed additives? Should their use be banned? On what basis should a decision be made on whether or not to ban use of antibiotics? Who should make that decision?
17. Feed-grade copper sulfate has five water of hydration molecules per $CuSO_4$. If a diet has 2 lbs of feed-grade copper sulfate added per ton, what is the amount of added copper in ppm?
18. Which is more or less hazardous as a feed additive: antibiotics, probiotics, or copper sulfate? Explain your choice.
19. What is the source of ammonia in confinement swine and poultry houses? What effect does dietary protein level have on the ammonia level of the environment? Are any feed additives effective in reducing air ammonia levels?

20. What is the difference in nutritional value between eggs with a bright yellow yolk and those with a pale yolk? Between white- and brown-shelled eggs?
21. What is the effect of stress on nutrient requirements? Some research shows that exposure to stress reduces the requirements for some nutrients. How might these observations be explained?

REFERENCES

Antifungals, Antioxidants, and Preservatives

Cabel, M. C., P. W. Waldroup, W. D. Shermer, and D. F. Calabotta. 1988. Effects of ethoxyquin feed preservative and peroxide level on broiler performance. *Poult. Sci.* 67:1725–1730.

Gibson, D. M., J. J. Kennelly, F. X. Aherne, and G. W. Mathison. 1988. Efficacy of sulfur dioxide as a grain preservative. *Anim. Feed Sci. Tech.* 19:203–218.

Lin, C. D. and T. C. Chen. 1995. Relative antifungal efficacies of phosphoric acid and other compounds on fungi isolated from poultry feed. *Anim. Feed Sci. Tech.* 54:217–226.

Mathison, G. W., L. P. Milligan, and R. M. Elofson. 1988. Feeding value and preservation of high moisture grain treated with sulfur dioxide. *Can. J. Anim. Sci.* 68:1209–1223.

Tabib, Z. and P. B. Hamilton. 1988. Factors influencing antifungal activity of gentian violet in poultry feed and ingredients. *Poult. Sci.* 67:58–63.

Pellet Binders

Britton, R. A., D. P. Colling, and T. J. Klopfenstein. 1978. Effect of complexing sodium bentonite with soybean meal or urea on *in vitro* ruminal ammonia release and nitrogen utilization in ruminants. *J. Anim. Sci.* 46:1738–1747.

Carson, M. S. and T. K. Smith. 1983. Role of bentonite in prevention of T-2 toxicosis in rats. *J. Anim. Sci.* 57:1498–1506.

Fries, G. F. 1995. A review of the significance of animal food products as potential pathways of human exposures to dioxins. *J. Anim. Sci.* 73:1639–1650.

Martin, L. C., A. J. Clifford, and A. D. Tillman. 1969. Studies on sodium bentonite in ruminant diets containing urea. *J. Anim. Sci.* 29:777–782.

Miller, B. L., G. C. Fahey, Jr., and S. L. Schussler. 1979. Isolation of biologically active fractions from wood hemicellulose extracts. *J. Anim. Sci.* 48:1129–1134.

Roeder, R. A., M. J. Garber, and G. T. Schelling. 1998. Assessment of dioxins in foods from animal origins. *J. Anim. Sci.* 76:142–151.

Safe, S. H. 1998. Development validation and problems with the toxic equivalency factor approach for risk assessment of dioxins and related compounds. *J. Anim. Sci.* 76:134–141.

Skoch, E. R., S. F. Binder, C. W. Deyoe, G. L. Allee, and K. C. Behnke. 1983. Effects of steam pelleting conditions and extrusion cooking on a swine diet containing wheat middlings. *J. Anim. Sci.* 47:929–935.

Wood, J. F. 1987. The functional properties of feed raw materials and their effect on the production and quality of feed pellets. *Anim. Feed Sci. Tech.* 18:1–17.

Feed Flavors

Campbell, R. G. 1976. A note on the use of a feed flavour to stimulate the feed intake of weaner pigs. *Anim. Prod.* 23:417.

Corkum, M. J., L. A. Bate, A. Lirette, and T. Tennessen. 1994. Effects of flavouring agents on intake of silage by feedlot steers. *Can. J. Anim. Sci.* 74:387–389.

Kornegay, E. T., S. E. Tinsley, and K. L. Bryant. 1979. Evaluation of rearing systems and feed flavors for pigs weaned at two to three weeks of age. *J. Anim. Sci.* 48:999–1006.

McLaughlin, C. L., C. A. Baile, L. L. Buckholtz, and S. K. Freeman. 1983. Preferred flavors and performance of weanling pigs. *J. Anim. Sci.* 56:1287–1293.

Provenza, F. D., C. B. Scott, T. S. Phy, and J. J. Lynch. 1996. Preference of sheep for foods varying in flavors and nutrients. *J. Anim. Sci.* 74:2355–2361.

Enzymes

Beauchemin, K. A., L. M. Rode, and V. J. H. Sewalt. 1995. Fibrolytic enzymes increase fiber digestibility and growth rate of steers fed dry forages. *Can. J. Anim. Sci.* 75:641–644.

Bedford, M. R. 2000. Exogenous enzymes in monogastric nutrition—Their current value and future benefits. *Anim. Feed Sci. Tech.* 86:1–13.

Campbell, G. L. and M. R. Bedford. 1992. Enzyme applications for monogastric feeds: A review. *Can. J. Anim. Sci.* 72:449–466.

Choct, M. and G. Annison. 1992. The inhibition of nutrient digestion by wheat pentosans. *Brit. J. Nutr.* 67:123–132.

Lewis, G. E., C. W. Hunt, W. K. Sanchez, R. Treacher, G. T. Pritchard, and P. Feng. 1996. Effect of direct-fed fibrolytic enzymes on the digestive characteristics of a forage-based diet fed to beef steers. *J. Anim. Sci.* 74:3020–3028.

Pettersson, D. and P. Aman. 1988. Effects of enzyme supplementation of diets based on wheat, rye, or triticale on their productive value for broiler chickens. *Anim. Feed Sci. Tech.* 20:313–324.

Pettersson, D., H. Graham, and P. Aman. 1987. The productive value of whole and dehulled oats in broiler chicken diets, and influence of β-glucanase supplementation. *Nutr. Rep. Int.* 36:743–750.

Buffers and Ion-Exchange Compounds

Boerner, B. J., F. M. Byers, G. T. Schelling, C. E. Coppock, and L. W. Greene. 1987a. Trona and sodium bicarbonate in beef cattle diets: Effects on site and extent of digestion. *J. Anim. Sci.* 65:303–308.

______ . 1987b. Trona and sodium bicarbonate in beef cattle diets: Effects on pH and volatile fatty acid concentrations. *J. Anim. Sci.* 65:309–316.

Coppock, C. E., G. T. Schelling, F. M. Byers, J. West, J. M. Labore, and C. E. Gates. 1986. A naturally occurring mineral as a buffer in the diet of lactating dairy cows. *J. Dairy Sci.* 69:111–123.

Erdman, R. A. 1988. Dietary buffering requirements of the lactating dairy cow: A review. *J. Dairy Sci.* 71:3246–3266.

Haaland, G. L. and H. F. Tyrrell. 1982. Effects of limestone and sodium bicarbonate buffers on rumen measurements and rate of passage in cattle. *J. Anim. Sci.* 55:935–942.

Keyser, R. B., C. H. Noller, L. J. Wheeler, and D. M. Schaefer. 1985. Characterization of limestones and their effects *in vivo* and *in vitro* in dairy cattle. *J. Dairy Sci.* 68:1376–1389.

Kovacik, A. M., S. C. Loerch, and B. A. Dehority. 1986. Effect of supplemental sodium bicarbonate on nutrient digestibilities and ruminal pH measured continuously. *J. Anim. Sci.* 62:226–234.

Nakaue, H. S. and J. K. Koelliker. 1981. Studies with clinoptilolite in poultry. I. Effect of feeding varying levels of clinoptilolite (zeolite) to dwarf single comb White Leghorn pullets and ammonia production. *Poult. Sci.* 60:944–949.

Phillips, T. D., L. F. Kubena, R. B. Harvey, D. R. Taylor, and N. D. Heidelbaugh. 1988. Hydrated sodium calcium aluminosilicate: A high affinity sorbent for aflatoxin. *Poult. Sci.* 67:243–247.

Pond, W. G. 1984. Response of growing lambs to clinoptilolite or zeolite NaA added to corn, corn-fish meal and corn-soybean meal diets. *J. Anim. Sci.* 59:1320–1328.

Russell, J. B. and J. M. Chow. 1993. Another theory for the action of ruminal buffer salts: Decreased starch fermentation and propionate production. *J. Dairy Sci.* 76:826–830.

Staples, C. R. and D. S. Lough. 1989. Efficacy of supplemental dietary neutralizing agents for lactating dairy cows. A review. *Anim. Feed Sci. Tech.* 23:277–303.

Thomas, E. E. and M. W. Hall. 1984. Effect of sodium bicarbonate and tetrasodium pyrophosphate upon utilization of concentrate- and roughage-based cattle diets: Cattle studies. *J. Anim. Sci.* 59:1309–1319.

Xin, Z., W. B. Tucker, and R. W. Hemken. 1989. Effect of reactivity rate and particle size of magnesium oxide on magnesium availability, acid-base balance, mineral metabolism, and milking performance of dairy cows. *J. Dairy Sci.* 72:462–470.

Ionophores and Methane Inhibitors

Bartley, E. E., T. G. Nagaraja, E. S. Pressman, A. D. Dayton, M. P. Katz, and L. R. Fina. 1983. Effects of lasalocid or monensin on legume or grain (feedlot) bloat. *J. Anim. Sci.* 56:1400–1406.

Benz, D. A., F. M. Byers, G. T. Schelling, L. W. Greene, D. K. Lunt, and S. B. Smith. 1989.

Ionophores alter hepatic concentrations of intermediary carbohydrate metabolites in steers. *J. Anim. Sci.* 67:2393–2399.

Bergen, W. G. and D. B. Bates. 1984. Ionophores: Their effect on production efficiency and mode of action. *J. Anim. Sci.* 58:1465–1483.

Carlson, J. R. and R. G. Breeze. 1984. Ruminal metabolism of plant toxins with emphasis on indolic compounds. *J. Anim. Sci.* 58: 1040–1049.

Elsasser, T. H. 1984. Potential interactions of ionophore drugs with divalent cations and their function in the animal body. *J. Anim. Sci.* 59:848–853.

Goodrich, R. D., J. E. Garrett, D. R. Gast, M. A. Kirick, D. A. Larson, and J. C. Meiske. 1984. Influence of monensin on the performance of cattle. *J. Anim. Sci.* 58:1483–1498.

Herald, F., F. W. Knapp, S. Brown, and N. W. Bradley. 1982. Efficacy of monensin as a cattle feed additive against the face fly and horn fly. *J. Anim. Sci.* 54:1128–1131.

McCrabb, G. J., K. T. Berger, T. Magner, C. May, and R. A. Hunter. 1997. Inhibiting methane production in Brahman cattle by dietary supplementation with a novel compound and the effects on growth. *Aust. J. Agr. Res.* 48:323–329.

Russell, J. B. 1987. A proposed mechanism of monensin action in inhibiting ruminal bacterial growth: Effects on ion flux and protonmotive force. *J. Anim. Sci.* 64:1519–1525.

Russell, J. B. and S. A. Martin. 1984. Effects of various methane inhibitors on the fermentation of amino acids by mixed rumen microorganisms in vitro. *J. Anim. Sci.* 59:1329–1338.

Russell, J. B. and H. J. Strobel. 1988. Effects of additives on in vitro ruminal fermentation: A comparison of monensin and bacitracin, another gram-positive antibiotic. *J. Anim. Sci.* 66:552–558.

Short, R. E. and D. C. Adams. 1988. Nutritional and hormonal interrelationships in beef cattle reproduction. *Can. J. Anim. Sci.* 68:29–39.

Turner, H. A., D. C. Young, R. J. Raleigh, and D. Zobell. 1980. Effect of various levels of monensin on efficiency and production of beef cows. *J. Anim. Sci.* 50:385–390.

Van Ryssen, J. B. J. and P. R. Barrowman. 1988. Effect of ionophores on the accumulation of copper in the livers of sheep. *Anim. Prod.* 44:255–261.

Isoacids

Andries, J. I., F. X. Buysse, D. L. De Brabander, and B. G. Cottyn. 1987. Isoacids in ruminant nutrition: Their role in ruminal and intermediary metabolism and possible influences on performances—A review. *Anim. Feed Sci. Tech.* 18:169–180.

McCollum, F. T., Y. K. Kim, and F. N. Owens. 1987. Influence of supplemental four- and five-carbon volatile fatty acids on forage intake and utilization by steers. *J. Anim. Sci.* 65:1674–1679.

Mir, P. S., Z. Mir, and J. A. Robertson. 1986. Effect of branched-chain amino acids or fatty acid supplementation on in vitro digestibility of barley straw or alfalfa hay. *Can. J. Anim. Sci.* 66:151–156.

Probiotics

Barrow, P. A., B. E. Brooker, R. Fuller, and M. J. Newport. 1980. The attachment of bacteria to the gastric epithelium of the pig and its importance in the microecology of the intestine. *J. Appl. Bacteriol.* 48:147–154.

Frumholtz, P. P., C. J. Newbold, and R. J. Wallace. 1989. Influence of *Aspergillus oryzae* fermentation extract on the fermentation of a basal ration in the rumen simulation technique (Rusitec). *J. Agr. Sci.* 113:169–172.

Fuller, R. 1989. Probiotics in man and animals: A review. *J. Appl. Bacteriol.* 66:365–378.

Gibson, G. R., and M. B. Roberfroid. 1995. Dietary modulation of the human colonic microbiota: Introducing the concept of prebiotics. *J. Nutr.* 125:1401–1412.

Gilliland, S. E., T. E. Stahly, and L. J. Bush. 1984. Importance of bile tolerance of *Lactobacillus acidophilus* used as a dietary adjunct. *J. Dairy Sci.* 67:3045–3051.

Glade, M. J. and L. M. Biesik. 1986. Enhanced nitrogen retention in yearling horses supplemented with yeast culture. *J. Anim. Sci.* 62:1635–1640.

Glade, M. J. and M. D. Sist. 1988. Dietary yeast culture supplementation enhances urea recycling in the equine large intestine. *Nutr. Rep. Int.* 37:11–17.

Grieve, D. G. 1979. Feed intake and growth of cattle fed liquid brewers yeast. *Can. J. Anim. Sci.* 59:89–94.

Hale, O. M. and G. L. Newton. 1979. Effects of a nonviable *Lactobacillus* species fermentation product on performance of pigs. *J. Anim. Sci.* 48:770–775.

Kornegay, E. T. and C. R. Risley. 1996. Nutrient digestibilities of a corn-soybean meal diet as influenced by *Bacillus* products fed to finishing swine. *J. Anim. Sci.* 74:799–805.

Lessard, M. and G. J. Brisson. 1987. Effect of a *Lactobacillus* fermentation product on growth, immune response and fecal enzyme activity in weaned pigs. *Can. J. Anim. Sci.* 67:509–516.

Meng, Q., M. S. Kerley, T. J. Russel, and G. L. Allee. 1998. Lectin-like activity of *Escherichia coli* K88, *Salmonella choleraesuis*, and *Bifidobacteria pseudolongum* of porcine gastrointestinal origin. *J. Anim. Sci.* 76:551–556.

Orban, J. I., J. A. Patterson, O. Adeola, A. L. Sutton, and G. N. Richards. 1997. Growth performance and intestinal microbial populations of growing pigs fed diets containing sucrose thermal oligosaccharide caramel. *J. Anim. Sci.* 75:170–175.

Phillips, W. A. and D. L. Von Tungeln. 1985. The effects of yeast culture on the post-stress performance of feeder calves. *Nutr. Rep. Int.* 32:287–294.

Pollmann, D. S., D. M. Danielson, and E. R. Peo, Jr. 1980a. Effects of microbial feed additives on performance of starter and growing-finishing pigs. *J. Anim. Sci.* 51:577–581.

Pollmann, D. S., D. M. Danielson, W. B. Wren, E. R. Peo, Jr., and K. M. Shahani. 1980b. Influence of *Lactobacillus acidophilus* inoculum on gnotobiotic and conventional pigs. *J. Anim. Sci.* 51:629–637.

Pollmann, D. S., D. M. Danielson, E. R. Peo, Jr. 1980c. Effect of *Lactobacillus acidophilus* on starter pigs fed a diet supplemented with lactose. *J. Anim. Sci.* 51:638–644.

Roberfroid, M. B., J. A. E. Van Loo, and G. R. Gibson. 1998. The bifidogenic nature of chicory inulin and its hydrolysis products. *J. Nutr.* 128:11–19.

Russell, J. B. and D. B. Wilson. 1988. Potential opportunities and problems for genetically altered rumen microorganisms. *J. Nutr.* 118:271–279.

Tagg, J. R., A. S. Dajani, and L. W. Wannamaker. 1976. Bacteriocins of gram-positive bacteria. *Bacteriol. Rev.* 40:722–756.

Waldroup, A. L., J. T. Skinner, R. E. Hierholzer, and P. W. Waldroup. 1993. An evaluation of fructooligosaccharides in diets for broiler chickens and effects on *Salmonella* contamination of carcasses. *Poult. Sci.* 72:643–650.

Wallace, R. J. 1994. Ruminal microbiology, biotechnology, and ruminant nutrition: Progress and problems. *J. Anim. Sci.* 72:2992–3003.

Watkins, B. A. and B. F. Miller. 1983. Competitive gut exclusion of avian pathogens by *Lactobacillus acidophilus* in gnotobiotic chicks. *Poult. Sci.* 62:1772–1779.

Acidifying Agents

Burnell, T. W., G. L. Cromwell, and T. S. Stahly. 1988. Effects of dried whey and copper sulfate on the growth responses to organic acid in diets for weanling pigs. *J. Anim. Sci.* 66:1100–1108.

Edmonds, M. S., O. A. Izquierdo, and D. H. Baker. 1985. Feed additive studies with newly weaned pigs: Efficacy of supplemental copper, antibiotics and organic acids. *J. Anim. Sci.* 60:462–469.

Falkowski, J. F. and F. X. Aherne. 1984. Fumaric and citric acid as feed additives in starter pig nutrition. *J. Anim. Sci.* 58:935–938.

Fallon, R. J. and F. J. Harte. 1980. Effect of feeding acidified milk replacer on calf performance. *Anim. Prod.* 30:459 (Abst.).

Gabert, V. M. and W. C. Sauer. 1995. The effect of fumaric acid and sodium fumarate supplementation to diets for weanling pigs on amino acid digestibility and volatile fatty acid concentrations in ileal digesta. *Anim. Feed Sci. Tech.* 53:243–254.

Giesting, D. W. and R. A. Easter. 1985. Response of starter pigs to supplementation of corn-soybean meal diets with organic acids. *J. Amin. Sci.* 60:1288–1294.

Henry, R. W., D. W. Pickard, and P. E. Hughes. 1985. Citric acid and fumaric acid as food additives for early-weaned piglets. *Anim. Prod.* 40:505–509.

Patten, J. D. and P. W. Waldroup. 1988. Use of organic acids in broiler diets. *Poult. Sci.* 67:1178–1182.

Radecki, S. V., M. R. Juhl, and E. R. Miller. 1988. Fumaric and citric acids as feed additives in starter pig diets: Effect on performance and nutrient balance. *J. Anim. Sci.* 66:2598–2605.

Risley, C. R., E. T. Kornegay, M. D. Lindemann, C. M. Wood, and W. N. Eigel. 1992. Effect of feeding organic acids on selected intestinal content measurements at varying times postweaning in pigs. *J. Anim. Sci.* 70:196–206.

Salivation Inducers

Croom, W. J., Jr., M. A. Froetschel, and W. M. Hagler, Jr. 1990. Cholinergic manipulation of digestive function in ruminants and other domestic livestock: A review. *J. Anim. Sci.* 68:3023–3032.

Croom, W. J., Jr., W. M. Hagler, Jr., M. A. Froetschel, and A. D. Johnson. 1995. The involvement of slaframine and swainsonine in slobbers syndrome: A review. *J. Anim. Sci.* 73:1499–1508.

Froetschel, M. A., H. E. Amos, J. J. Evans, W. J. Croom, Jr., and W. M. Hagler, Jr. 1989. Effects of a salivary stimulant, slaframine, on ruminal fermentation, bacterial protein synthesis and digestion in frequently fed steers. *J. Anim. Sci.* 67:827–834.

Froetschel, M. A., M. N. Streeter, H. E. Amos, W. J. Croom, Jr., and W. M. Hagler, Jr. 1995. Effects of abomasol slaframine infusion on ruminal digesta passage and digestion in steers. *Can. J. Anim. Sci.* 75:157–163.

Defaunating Agents

Bird, S. H. and R. A. Leng. 1978. The effects of defaunation of the rumen on the growth of cattle on low-protein high-energy diets. *Brit. J. Nutr.* 40:163–167.

______ . 1984. Further studies on the effects of the presence or absence of protozoa in the rumen on live-weight gain and wool growth of sheep. *Brit. J. Nutr.* 52:607–611.

Bird, S. H., M. K. Hill, and R. A. Leng. 1979. The effects of defaunation of the rumen on the growth of lambs on low-protein-high-energy diets. *Brit. J. Nutr.* 42:81–87.

Ffoulkes, D. and R. A. Leng. 1988. Dynamics of protozoa in the rumen of cattle. *Brit. J. Nutr.* 59:429–436.

Hsu, J. T., G. C. Fahey, N. R. Merchen, L. L. Berger, and J. H. Clark. 1989. The efficacy of alkanate 35L3, calcium peroxide and nystatin as defaunating agents in sheep. *Nutr. Rep. Int.* 39:205–213.

Ivan, M. 1988. Effect of faunation on ruminal solubility and liver content of copper in sheep fed low or high copper diets. *J. Anim. Sci.* 66:1496–1501.

Ivan, M., D. M. Veira, and C. A. Kelleher. 1986. The alleviation of chronic copper toxicity in sheep by ciliate protozoa. *Brit. J. Nutr.* 55:361–367.

Lovelock, L. K. A., J. G. Buchanan-Smith, and C. W. Forsberg. 1982. Difficulties in defaunation of the ovine rumen. *Can. J. Anim. Sci.* 62:299–303.

Newbold, C. J., S. M. El Hassan, J. Wang, M. E. Ortega, and R. J. Wallace. 1997. Influence of foliage from African multipurpose trees on activity of rumen protozoa and bacteria. *Brit. J. Nutr.* 78:237–249.

Preston, T. R. and R. A. Leng. 1987. Matching *Ruminant Production Systems with Available Resources in the Tropics and Subtropics*. Armidale, Australia: Penambul Books.

Veira, D. M. 1986. The role of ciliate protozoa in nutrition of the ruminant. *J. Anim. Sci.* 63:1547–1560.

Wallace, R. J., L. Arthaud, and C. J. Newbold. 1994. Influence of *Yucca shidigera* extract on ruminal ammonia concentrations and ruminal microorganisms. *Appl. Environ. Micro.* 60:1762–1767.

Yokoyama, M. T. and K. A. Johnson. 1988. Microbiology of the rumen and intestine. In *The Ruminant Animal*, D. C. Church, ed. pp. 125–144. Englewood Cliffs, NJ: Prentice Hall.

Beta-Adrenergic Agents

Dursch, I., H. H. D. Meyer, and H. Karg. 1995. Accumulation of the β-agonist clenbuterol by pigmented tissues in rat eye and hair of veal calves. *J. Anim. Sci.* 73:2050–2053.

Kim, Y. S., Y. B. Lee, W. N. Garrett, and R. H. Dalrymple. 1989. Effects of cimaterol on nitrogen retention and energy utilization in lambs. *J. Anim. Sci.* 67:674–681.

Mersmann, H. J. 1998. Overview of the effects of β-adrenergic receptor agonists on animal growth including mechanisms of action. *J. Anim. Sci.* 76:160–172.

Ricks, C. A., R. H. Dalrymple, P. K. Baker, and D. L. Ingle. 1984. Use of a β-agonist to alter fat and muscle deposition in steers. *J. Anim. Sci.* 59:1247–1255.

Smith, D. J., and G. D. Paulson. 1997. Distribution, elimination, and residues of (^{14}C) clenbuterol HCl in Holstein calves. *J. Anim. Sci.* 75:454–461.

Smith, D. J. 1998. The pharmacokinetics, metabolism, and tissue residues of β-adrenergic agonists in livestock. *J. Anim. Sci.* 76:173–194.

Antibiotics

American Society of Animal Science. 1986. Public health implications of the use of antibiotics in animal agriculture. *J. Anim. Sci.* 62(Suppl. 3).

CAST, 1989. *Antibiotics for Animals: The Antibiotic-Resistance Issue.* Ames, IA: Council for Agricultural Science and Technology.

Hays, V. W. and W. M. Muir. 1979. Efficacy and safety of feed additive use of antibacterial drugs in animal production. *Can. J. Anim. Sci.* 59:447–456.

Institute of Medicine. 1989. *Human Health Risks with the Subtherapeutic Use of Penicillin or Tetracyclines in Animal Feed.* Washington, DC: National Academy Press.

Kunin, C. M. 1993. Resistance to antimicrobial drugs: A worldwide calamity. *Ann. Intern. Med.* 118:557–561.

Mathew, A. G., W. G. Upchurch, and S. E. Chattin. 1998. Incidence of antibiotic resistance in fecal *Escherichia coli* isolated from commercial swine farms. *J. Anim. Sci.* 76:429–434.

Parker, D. S. and D. G. Armstrong. 1987. Antibiotic feed additives and livestock production. *Proc. Nutr. Soc.* 46:415–421.

Visek, W. J. 1978. The mode of growth promotion by antibiotics. *J. Anim. Sci.* 46:1447–1469.

Copper Sulfate and Other Chemotherapeutics

Banton, M. I., S. S. Nicholson, P. L. H. Jowett, M. B. Brantley, and C. L. Boudreaux. 1987. Copper toxicosis in cattle fed chicken litter. *J. Am. Vet. Med. Assoc.* 191:827–828.

Czarnecki, G. L., M. S. Edmonds, O. A. Izquierdo, and D. H. Baker. 1984. Effect of 3-nitro-4-hydroxyphenylarsonic acid on copper utilization by the pig, rat and chick. *J. Anim. Sci.* 59:997–1002.

Edmonds, M. S., O. A. Izquierdo, and D. H. Baker. 1985. Feed additive studies with newly weaned pigs: Efficacy of supplemental copper, antibiotics and organic acids. *J. Anim. Sci.* 60:462–469.

Hill, G. M., D. C. Mahan, S. D. Carter, G. L. Cromwell, R. C. Ewan, R. L. Harrold, A. J. Lewis, P. S. Miller, G. C. Shurson, and T. L. Veum. 2001. Effect of pharmacological concentrations of zinc oxide with or without the inclusion of an antibacterial agent on nursery pig performance. *J. Anim. Sci.* 79:934–941.

Roof, M. D. and D. C. Mahan. 1982. Effect of carbadox and various dietary copper levels for weanling swine. *J. Anim. Sci.* 55:1109–1117.

Stahly, T. S., G. L. Cromwell, and H. J. Monegue. 1980. Effects of the dietary inclusion of copper and (or) antibiotics on the performance of weanling pigs. *J. Anim. Sci.* 51:1347–1351.

Saponins and Environmentally Active Substances

Carlile, F. 1984. Ammonia in poultry houses: A literature review. *World Poult. Sci. Assoc. J.* 40:99–113.

Johnston, N. L., C. L. Quarles, D. J. Fagerberg, and D. D. Caveny. 1981. Evaluation of yucca saponin on broiler performance and ammonia suppression. *Poult. Sci.* 60:2289–2292.

Mader, T. L. and M. C. Brumm. 1987. Effect of feeding sarsaponin in cattle and swine diets. *J. Anim. Sci.* 65:9–15.

Nakaue, H. S., R. R. Lowry, P. R. Cheeke, and G. H. Arscott. 1980. The effect of dietary alfalfa of varying saponin content on egg cholesterol level and layer performance. *Poult. Sci.* 59:2744–2748.

Nakaue, H. S. and J. K. Koelliker. 1981. Studies with clinoptilolite in poultry. I. Effect of feeding varying levels of clinoptilolite (zeolite) to dwarf single comb White Leghorn pullets and ammonia production. *Poult. Sci.* 60:944–949.

Preston, R. L., S. J. Bartle, T. May, and S. R. Goodall. 1987. Influence of sarsaponin on growth, feed and nitrogen utilization in growing male rats fed diets with added urea or protein. *J. Anim. Sci.* 65:481–487.

Sim, J. S., W. D. Kills, and D. B. Bragg. 1984. Effect of dietary saponin on egg cholesterol level and laying hen performance. *Can. J. Anim. Sci.* 64:977–984.

Wallace, R. J., L. Arthaud, and C. J. Newbold. 1994. Influence of *Yucca shidigera* extract on ruminal ammonia concentrations and ruminal micro-organisms. *Appl. Environ. Micro.* 60:1762–1767.

Immunomodulators

Johnson, R. W. 1997. Inhibition of growth by pro-inflammatory cytokines: An integrated view. *J. Anim. Sci.* 75:1244–1255.

Klasing, K. C. 1988. Nutritional aspects of leukocytic cytokines. *J. Nutr.* 118:1436–1446.

Klasing, K. C. and D. M. Barnes. 1988. Decreased amino acid requirements of growing chickens due to immunologic stress. *J. Nutr.* 118:1158–1164.

Spurlock, M. E. 1997. Regulation of metabolism and growth during immune challenge: An overview of cytokine function. *J. Anim. Sci.* 75:1773–1783.

Suchner, U., K. S. Kuhn, and P. Furst. 2000. The scientific basis of immunonutrition. *Proc. Nutr. Soc.* 59:553–563.

Thaler, R. C., J. L. Nelssen, G. A. Anderson, F. Blecha, C. G. Chitko, S. K. Chapes, and E. R. Clough. 1989. Evaluation of a biological response modifier: Effects on starter pig performance. *J. Anim. Sci.* 67:2341–2346.

Xanthophylls

Bendich, A. and S. S. Shapiro. 1986. Effect of β-carotene and canthaxanthin on the immune responses of the rat. *J. Nutr.* 116:2254–2262.

Herber-McNeill, S. M. and M. E. Van Elswyk. 1998. Dietary marine algae maintains egg consumer acceptability while enhancing yolk color. *Poult. Sci.* 77:493–496.

PART II

APPLIED NUTRITION: FEEDING BEHAVIOR AND WATER REQUIREMENTS

Goats are intermediate feeders, having both grazing and browsing feeding behavior. This photograph illustrates several adaptations that expand the nutritional horizons of goats. They can maintain bipedal stance to increase their access to browse and nimbly climb low growing trees to browse upon them. The physical structure of their mouth allows them to consume vegetation that is heavily fortified with physical defenses, such as the spines and thorns visible here. (Courtesy of A. N. Bhattacharya.)

Two of the most important factors influencing feed utilization are the level of feed intake and satisfaction of water requirements. Factors regulating voluntary feed intake are described in some detail in Chapter 9. Maximizing feed intake is critical to maximizing animal performance. Ideally, animals are fed individually according to their metabolic individuality. This ideal is already a reality with dairy cattle and, with the use of electronically controlled feeding gates, may become feasible with other species.

Water is often an overlooked substance in animal nutrition, but it plays a critical role. Water utilization by livestock is becoming a major concern in many parts of the world, particularly in large confinement units, such as feedlots and dairies. Water concerns relate to competition between humans and livestock for scarce water resources, and groundwater contamination by effluent from confinement livestock units.

There are large differences among animal species in their ability to conserve and efficiently utilize water. Perhaps in arid areas suffering from desertification, the domestication and production of water-efficient game animals like the oryx might permit more efficient and ecologically sound resource use than would continued production of traditional domestic livestock. Cattle, for instance, are notoriously inefficient in water utilization. These ideas are discussed in Chapter 10.

CHAPTER 9

Feeding Behavior and Regulation of Feed Intake

Objectives

1. To discuss important factors regulating voluntary feed intake by animals:
   - palatability
   - diet caloric density
   - environmental factors
2. To relate gut anatomy of herbivores to feeding strategy:
   - generalist vs specialist feeders
   - concentrate selectors
   - intermediate feeders
   - bulk and roughage eaters
3. To relate plant factors, such as chemical defenses, to animal feeding behavior and strategy
4. To highlight areas of current interest:
   - aversive conditioning
   - modification of feed intake
   - plant-animal interactions on a coevolutionary basis

FEEDING BEHAVIOR

Feeding behavior and digestive tract physiology in animals are closely interrelated. The feeds normally consumed by animals are those they are capable of digesting. There are three major types of animal feeding behavior: carnivorous, omnivorous, and herbivorous. **Carnivores,** such as cats, are adapted to a meat-based diet and require a high-quality, highly digestible source of nutrients. Carnivores have simple digestive tracts with little microbial activity. They require certain nutrients, such as preformed vitamin A and taurine, which in the wild can be obtained only from consumption of meat. At the other end of the spectrum, **herbivores** are animals that normally consume only plant material and have a more complex digestive tract with symbiotic microbial activity that permits the digestion of plant fiber. Herbivores include the ruminants, such as cattle, and non-ruminant herbivores, such as horses and rabbits. **Omnivores** are animals that are not fastidious in their feeding behavior and consume a wide variety of animal and plant foods. Swine, poultry, and humans are examples.

Ruminant animals vary considerably in their feeding strategies. Hofmann (1989), in a classic study of the stomach anatomy and function of African wild ruminants, classified ruminants into three groups based on feeding strategy: concentrate selectors, bulk and roughage eaters, and intermediate feeders. **Concentrate selectors** select the more nutritious low-fiber parts of herbage

FIGURE 9.1 A small African ruminant, the red duiker, with a mature body weight of about 15 kg. There has been interest in domesticating small ruminants such as duikers and dik-diks for use as the "laboratory rat" of ruminant nutrition. However, digestive function in these concentrate-selector ruminants is substantially different than in the common domestic ruminants, which are grazers and intermediate feeders.

such as leaves, fruit, and other soft, succulent plant parts. This material can be rapidly fermented, and much of the plant cell contents fraction is absorbed directly without fermentation. Concentrate selectors (Fig. 9.1) have a stomach anatomy adapted to the use of low-fiber forage. They have a relatively small rumen and a small omasum that can become impacted readily with poor-quality fibrous herbage. The reticular groove is well-developed, allowing high-quality feed to bypass rumen fermentation. Examples of concentrate selectors are the deer, moose, and giraffe. These animals tend to nibble fastidiously on a wide variety of plants and plant parts and often occupy ecological niches unavailable to grazing ruminants.

In contrast, **bulk and roughage eaters,** such as cattle, are grazing animals, with large rumens adapted to the intake of large amounts of low-energy fibrous feeds. Rumination is pronounced. The omasum is highly developed, retaining fibrous feed in the rumen to maximize fiber digestion. **Intermediate feeders,** such as sheep and goats, share feeding and digestive strategies of both the concentrate selectors and grazers. They are very adaptable to varying environments and habitats. Sheep graze on grass, but they will also extensively utilize shrubs and forbs. Their feeding behavior is much more selective than that of cattle.

An appreciation of these differences is desirable when dealing with domestic ruminants and vital when dealing with wild animals. Differences in digestive physiology and feeding strategy explain why winter feeding of deer with hay may produce mortality, or why the proper selection of forage for feeding giraffes in a zoo will determine whether they will survive.

While the three anatomical categories of ruminants described by Hofmann (1989) correlate well with feeding behavior, it appears from studies to test Hofmann's hypotheses that the differences in nutritional ecology of ruminants relate more to feeding behavior and diet selection than to anatomical differences in stomach anatomy (Gordon and Illius, 1994, 1996; Robbins et al., 1995). However, testing of these hypotheses has led to much greater understanding of ruminant evolution, behavior, and ecology of herbivore communities (Robbins et al., 1995).

FACTORS AFFECTING FEED INTAKE

The regulation of feed intake and appetite is a complex subject. This section emphasizes the properties of feeds affecting feed intake. More detailed information can be obtained in the reviews of Forbes (1986a,b; 1996) and the National Research Council (1987). A symposium on controls of feeding in farm animals, with separate review articles on pigs, horses, sheep, and cattle, was published in the *Journal of Animal Science* (59:1345–1380, 1984). A series of review articles on the regulation of voluntary forage intake of ruminants was published in the *Journal of Animal Science* (74:3029–3081, 1996).

Maximum feed intake can be achieved only with free choice water available. When water is restricted, feed intake is reduced. If a group or pen of animals goes off feed, the first thing to check is the water supply. Large amounts of water are required to moisten food in the gut. Ingredients with high water-absorbing properties, such as bran, increase the water requirements.

Palatability and Feed Preference

Palatability is a determinant of feed intake. **Palatability** is the summation of the taste, olfactory, and textural characteristics of a feedstuff that determine its degree of acceptance. Taste is a major component of the palatability complex. The major taste responses are sweet, salty, bitter, and acid. Virtually all animals except strict carnivores show a preference for the sweet taste. Molasses and to a lesser extent sucrose are used in feeds as a source of sweet taste to increase palatability. Most animals also show a pronounced appetite for salt. This is particularly true of herbivores. Animals show quite divergent responses to bitterness. Generally the herbivorous species are quite tolerant of bitter substances because many of the secondary compounds in plants, such as alkaloids and glycosides, are bitter. Herbivores that are intolerant of bitterness are not likely to find anything acceptable to eat, which promotes either rapid evolution of tolerance to bitter substances or extinction!

Most domestic animals are **generalist feeders** and are quite tolerant of a wide array of divergent tastes and feed textures. Lack of palatability of feeds is not usually a major problem. Specialist feeders are animals that eat only a narrow range of feeds. Many insects are **specialist feeders** and feed only on a particular plant species. The koala is an example of a mammalian specialist feeder; it will eat only the leaves of a few species of eucalyptus.

Knowledge of species differences in feed preferences and taste responses can be useful in developing and employing feed flavors. For example, cats respond favorably to the herb catnip (*Nepeta cataria*), whereas rabbits are attracted to thyme (*Thymus vulgaris*). Anise seed or oil is often used as a feed flavor or attractant.

Feed preferences are often evaluated in a two-choice self-selection test. Results from such evaluations must be interpreted with caution. A particular ingredient may appear unpalatable to animals when it is presented in a choice situation, but, if they are not given a choice, they may consume it readily.

Secondary Compounds and Palatability

Secondary compounds in plants are those compounds that are not involved in the primary processes of cellular metabolism, but have secondary roles, such as in chemical defenses. Examples of secondary compounds include a wide variety of toxic substances such as alkaloids, cyanogenic compounds, phytoestrogens, tannins, toxic amino acids, and other plant toxins. These compounds help protect plants against being eaten and thus serve as a form of chemical defense against predators, which include bacteria, viruses, insects, and mammalian herbivores. Thus, in crop and forage plants, there must be a balance between having sufficient chemical defenses to survive attacks by insects and other pests but not be so strongly protected that they are highly unpalatable or toxic to livestock. Numerous examples of this type were encountered earlier, including saponins in alfalfa, tannins in grain sorghum and trefoil, alkaloids in reed canarygrass, and mimosine in leucaena. Differences in palatability of various forages can be explained largely by their contents of secondary compounds or, in some cases, by the presence of physical defenses, such as prickliness or coarseness. It is very difficult to maintain a mixture of highly palatable and less palatable forage species in a pasture; the palatable plants will be overgrazed and killed out while the underutilized plants will become mature and even more unpalatable.

As plants have evolved chemical and physical defenses, herbivores have developed means of surmounting them. This is most obvious in the insect world because of the short-generation interval. Many insects, having developed the enzymatic capacity to detoxify plant toxins, are specialist feeders and feed only on plants containing the particular toxin to which they are adapted. The toxins act as attractants and feeding stimulants. In some cases, the plant toxins actually become essential metabolites for the insect and may be used in the synthesis of pheromones. As herbivores surmount the plant chemical defenses, the plants often chemically modify the toxin to overcome the herbivores' metabolic adaptations. These interactions are known as **coevolution**. These continual changes in plant and herbivore enzymatic activity represent a coevolutionary arms race! Examples of mammalian adaptations to surmount plant defenses include the enzymatic resistance of sheep and goats to many poisonous plants (Cheeke, 1998), and the ability of the rhinoceros and the goat to munch contentedly on twigs that are armed with vicious thorns and spines (Cooper and Owen-Smith, 1986).

The effect of secondary compounds on diet selection and feed intake of wild animals is an active area of research (Cooper et al., 1988). Chemical factors influencing palatability are either nutrients, fiber, or secondary compounds. Soluble carbohydrates are probably the main nutrient category involved. Although there often is an association between protein content and palatability, Cooper et al. (1988) suggest that animals do not have taste receptors for protein, but because leaf protein concentration is largely chloroplastic, high concentrations of leaf protein are correlated with high photosynthetic rates and, thus, high soluble carbohydrates.

Many secondary compounds in plants are bitter or otherwise unpleasant tasting. Alkaloids and glycosides are often bitter, while phenolic compounds such as tannins are astringent. Sheep and goats and other concentrate-selector herbivores are more tolerant of bitter substances than are grazers such as cattle and horses. Many wild herbivores (e.g., deer) have **tannin-binding salivary proteins** rich in the amino acid proline that bind with tannins and reduce their toxic and astringent effects (Austin et al., 1989). This is a metabolic adaptation that permits many browsers to utilize chemically protected plants (McArthur et al., 1995).

Silica is deposited in the cell walls of many grass species. It may act as a "varnish" on cell walls to reduce accessibility of rumen microbes to the cell contents, and it contributes to the physical roughness or harshness of many grass species. This may have negative effects on palatability of grasses to grazing animals (Shewmaker et al., 1989). Silica in forages is also important as a cause of urinary calculi formation (see Chapter 15).

Preferences for certain forages may involve the absence of unpalatable secondary compounds, or the presence of palatable components such as sugars or amino acids. However, Mayland et al. (2000) found that grazing preferences of cattle for tall fescue cultivars was not related to malate, citrate or amino acid concentrations among cultivars.

Dietary Energy Level

If feeds are sufficiently palatable to be readily consumed, the main dietary factor that controls voluntary feed intake is dietary energy concentration. A longstanding maxim in animal nutrition is that animals eat to satisfy their energy requirements. The amount of feed consumed is regulated so as to provide a constant intake of digestible energy. If a low-energy diet is used, feed intake is high, whereas with a high-energy diet, feed intake is reduced. The energy intake is about the same in each case. This is illustrated in Fig. 9.2. The lower limit of this relationship is when **gut capacity** is reached and the animal cannot consume enough feed to meet its energy requirements; thus its performance is reduced.

Because of this relationship, there is no specific energy concentration per pound or kilogram of diet required in diets for livestock. The requirement is best expressed in kilocalories per animal per day. Above the energy concentration at which gut capacity becomes limiting, animals can adjust their voluntary intake to consume the required amount of kilocalories. This regulation is largely accomplished by concentrations of blood metabolites. In nonruminants, blood glucose is the principal regulator of feed intake. When glycogen reserves are depleted and blood glucose concentration begins to fall, the appetite center in the brain initiates neural activity to produce sensations of hunger, stimulating feed intake. In ruminants, blood concentrations of VFA fulfill this role (Illius and Jessop, 1996). Control of intake by effects of blood metabolites on the appetite center is known as **chemostatic regulation** of feed intake. At the molecular level, it appears that a number of neuropeptides are involved in the function of the appetite center (Baile and McLaughlin, 1987). The opioid peptides, such as

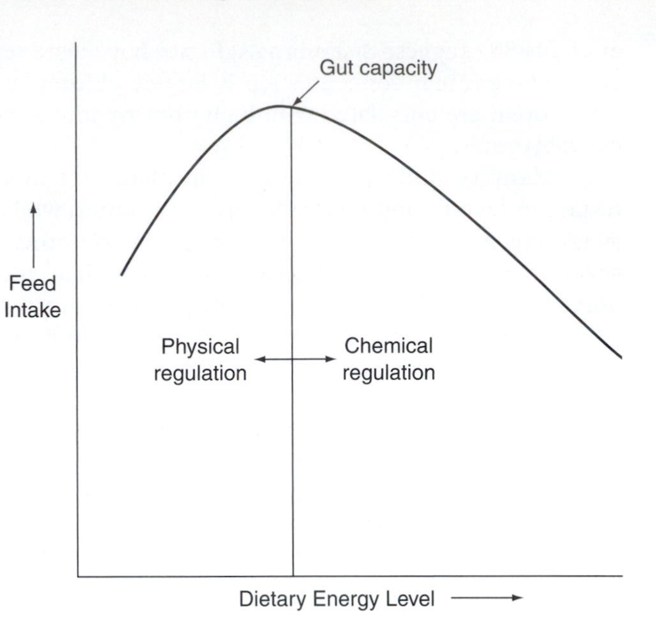

FIGURE 9.2 The relationship between feed intake and dietary energy level. Animals eat to meet their energy requirements. Why does feed intake decrease at very low diet energy levels? Would this part of the graph (physical regulation of feed intake portion) apply mainly to ruminants or nonruminants under practical conditions?

β-endorphin, stimulate feed intake, whereas cholecystokinin (CCK) inhibits feed intake. Because CCK is a gastrointestinal hormone as well as one that is released in the brain, there may be a connection between gastrointestinal tract release of CCK and development of satiety.

With most practical diets fed to swine and poultry in the industrialized countries, the energy concentration of the diet is above the level at which gut capacity limits feed intake. In developing countries, use of ingredients such as rice bran may result in diets that are too low in digestible energy to support maximum performance. With ruminant animals, forage-based diets are too low in digestible energy to meet the energy requirements for maximum growth and lactational performance. This is why the beef feedlot industry and high-concentrate feeding of dairy cattle have developed in North America and Europe. When feed resources are abundant, high-concentrate diets for ruminants are economically more attractive than forage-based systems because of the much higher productivity obtained.

Because of the relationship between dietary energy concentration and feed intake, the feed-conversion efficiency is largely dependent on the energy level of the feed. Feed efficiency is commonly expressed as the amount of feed required per unit of live weight gain. Also, because of the energy density-feed intake relationship, for maximum precision in diet formulation, it is desirable to express nutrient requirements on a per kilocalorie basis. Thus, regardless of the amount of feed required to meet the energy requirement, the requirements for other nutrients would automatically be met as well. On a percentage basis, the optimal percent of crude protein, for example, would be high in a high-energy diet and lower in a low-energy diet.

In ruminants, the break-off point—where gut capacity ceases to limit feed intake and intake is regulated by chemostatic means—is at an energy level approximately at the change from a roughage-based diet to a concentrate diet (Fig. 9.2). This corresponds to a dry-matter digestibility of approximately 66 to 69 percent. As forage quality declines, so does feed intake, largely as a result of a reduced rate of digestion leading to a longer retention time in the rumen for poor-quality roughages.

Ketelaars and Tolkamp (1996) have proposed that the unifying factor relating energy metabolism and feed intake is the efficiency of oxygen utilization. Voluntary energy intake corresponds to the feed consumption level at which oxygen efficiency (net energy yield per l of O_2 consumed) is maximum.

Protein and Amino Acid Concentrations

Dietary energy level has a much greater impact on voluntary feed intake than dietary protein status. Nevertheless, protein intake in growing animals is subject to some regulation (Henry, 1985). The growing pig, for example, has the ability to preferentially select diets adequate in essential amino acids over amino-acid-deficient diets (Henry, 1987). Plasma levels of certain amino acids (e.g., lysine and tryptophan) provide signals for release of neurotransmitters, such as serotonin in the brain, that play a role in feed-intake regulation.

The provision of bypass protein to ruminants fed low-quality roughages increases feed intake. Part of the response in growing animals may be that bypass protein provides amino acids in which microbial protein is deficient, thus increasing growth rate, which in turn will increase feed intake. However, there is a specific effect of bypass protein on feed intake in addition to effects it might have on improved absorbed nutrient balance. The effect of bypass protein on enhancing roughage intake seems to be facilitation of greater fill, perhaps affecting the set point at which rumen distention causes cessation of intake (Ndlovu and Buchanan-Smith, 1987). The effect of bypass protein on stimulating feed intake is greater with tropical than temperate roughages. Supplementation with bypass energy sources, such as rice polishings, improves growth performance of animals fed low-quality roughages without affecting roughage intake (Preston, 1982).

Ruminants may have the ability to select a diet of adequate nitrogen content when grazing. Sheep and goats graze more selectively than cattle and, under conditions of poor-quality pasture, may be better able to select a diet of adequate nitrogen content than cattle.

Forage Composition

In ruminants, forage quality and composition influence feed intake. A major factor regulating forage intake is the **neutral detergent fiber (NDF)** and its digestibility. The NDF seems to be the major component limiting rumen fill. Van Soest (1994) used the following analogy **(the skyscraper theory)** to illustrate this effect:

> The effect [of NDF breakdown] may be compared to the demolition of a tall city skyscraper; the contents of the rooms, non-bearing walls, plaster, doors and furniture can be expelled without altering the volume of the building. Only when the wrecking ball smashes the structure does its effective volume decrease.

In the rumen, only when the NDF is digested is its effect on rumen fill eliminated (Allen, 1996). Lippke (1986) confirmed this relationship with cattle; the intake of forage was regulated by the intake of indigestible NDF. When given a choice, cattle selected a diet that maximized digestible organic matter intake. The NDF content is highly correlated with rumination or chewing time. Both the initial volume of the feed and the amount of chewing required to reduce its volume have a role in the fill effect. Because both the chewing time and feed volume are correlated with NDF content, the NDF is probably the best estimator of forage quality. Some ingredients, such as finely ground cottonseed hulls and soybean hulls, have a high NDF content but stimulate little rumination because of their small particle size. Therefore, for these ingredients, NDF may not be a good estimator of potential feed intake.

In dairy cattle nutrition, there is an increasing tendency to use NDF requirement figures in ration formulation to optimize the roughage intake for proper rumen fermentation and milk fat production. Another approach is that of French researchers (Jarrige et al., 1986), who have developed a "fill unit" system to estimate the **ingestibility** of forages and forage-concentrate diets. Separate fill units were developed for sheep and cattle. The system proposed that the fill unit value of a forage is a unique characteristic, just like its protein or energy content, and that animals consume a fixed amount of fill units rather than a fixed amount of dry matter. Fill units can be entered into computer programs for diet formulation. Although undoubtedly numerous refinements will be needed, this or a similar system will be very useful in estimating feed intake and thus providing nutrients in the amount needed per animal per day, rather than simply as a percent of diet.

In the United States, the concept of **effective NDF (eNDF)** has been developed to measure NDF that contributes to rumen fill. The eNDF is determined by subtracting from total NDF the NDF fraction that passes through a 1 mm screen (Van Soest, 1994).

The water content of forages can sometimes influence intake. Gallavan et al. (1989) found that the low dry-matter content of lush wheat pasture limited the dry-matter intake of sheep to below maintenance. A minimum dry-matter content of 17 percent was required for sheep to consume their maintenance requirements, and a dry-matter content of 28 to 30 percent was required for optimal growth. Intracellular water of forages can limit intake through a bulk effect because of the time period required for its release in the rumen and subsequent absorption.

Environmental Temperature

The environmental temperature has an important influence on feed intake. The metabolic rate of an animal is at a minimum in the **comfort zone** (Fig. 9.3), also known as the *zone of thermoneutrality*. As environmental temperatures decrease below the comfort zone, an increase in metabolic rate is necessary to maintain body temperature. Thus feed intake increases as environmental temperature decreases. It is difficult to quantify these relationships (e.g., to indicate that with a 10°C decrease in environmental temperature the feed intake will increase by a certain percent) because of the involvement of other factors, such as humidity and wind velocity. Kennedy et al. (1986) discussed some of the complexities in ruminants. In cold environments, digestibility is reduced because passage rate through the rumen is increased. The **heat of rumen fermentation** is a negative factor for an animal in the comfort zone but is beneficial at low environmental temperatures. Roughages, which produce a high amount of heat of fermentation, have a higher net energy value at lower than at higher temperatures. This is due partly to the greater ruminal heat of fermentation and partly to the high heat increment of acetate, the VFA produced in greatest proportion with high-roughage diets.

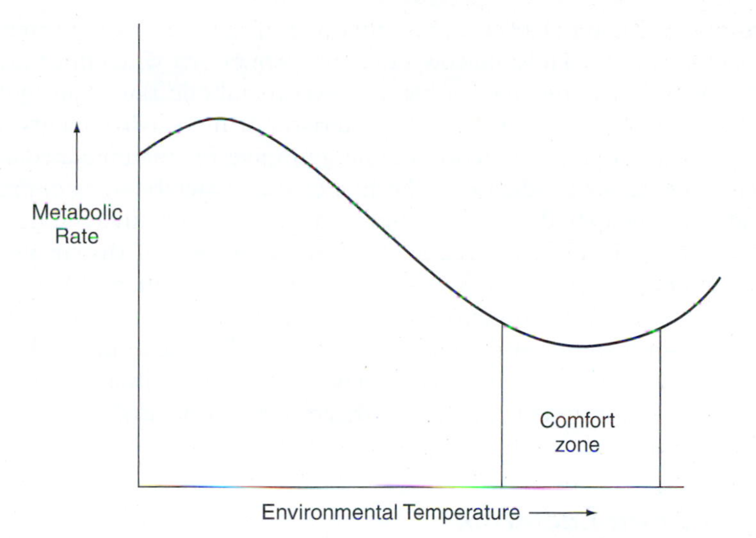

FIGURE 9.3 Metabolic rate (and consequently feed intake) is at a minimum in the environmental temperature range where the animal does not need to employ cooling or heating mechanisms (the comfort zone). At low temperatures, metabolic rate is increased to maintain body temperature, so feed intake is also increased.

Hormonal changes influence winter feed intake. During a period of exposure to cold, the comfort zone is lowered by alterations in the output of such hormones as thyroxin. Wild ungulates, such as deer, accumulate large amounts of body fat in the summer and have a very low feed intake in the winter, surviving on body-fat stores. Domestic ruminants appear to have vestiges of this photoperiod-controlled feed intake pattern (Kennedy et al., 1986). The winter hair coat also provides insulation to reduce energy requirements.

Such complexities make it difficult to quantify on a practical basis the relationships between environment, energy requirements, and feed intake. This problem is reflected in the NRC Nutrient Requirements publications, which provide no information on environmental temperature effects on requirements except to state that nutrient allowances are influenced by environmental temperature and animal housing conditions.

Under conditions of high environmental temperature, the **heat increment (HI)** of feeds becomes important. The HI is the extra heat production that arises from the metabolism of nutrients. It is highest for protein and corresponds to the extra metabolism (operation of the urea cycle in the liver) required to deaminate amino acids and convert the ammonia to urea when amino acids are used as an energy source. Thus, under conditions of high temperatures, the dietary protein level may need to be reduced. Under conditions of cold environmental temperatures, the percent of dietary protein can also be lowered because feed intake is increased. Because the HI is lowest for fat, it is desirable to increase the dietary fat content under conditions of high environmental temperatures.

The **heat increment** includes the heat of rumen fermentation. The heat increment is useful at temperatures below the comfort zone and detrimental at high environmental temperatures. The heat of fermentation is higher with high roughage diets than with concentrates.

Preston and Leng (1987) suggest that one of the reasons imported breeds of livestock, such as the Holstein cow, often perform poorly when introduced into the tropics is because they have a higher basal metabolic rate than indigenous breeds. With the lower feed intake characteristic of hot environments, the imported breeds use a greater fraction of energy intake for maintenance and thus have less available for production. The higher basal metabolic rate, producing more heat to be dissipated, also contributes to a greater sensitivity to heat stress.

Metabolic imbalance can reduce feed intake by way of **thermogenic effects.** In ruminants, an excess of C_2-energy sources (acetate and butyrate) relative to gluconeogenic precursor (propionate) leads to excessive heat production from disposal of the excess C_2 sources. The feed intake is reduced in response to this thermogenic effect, particularly in hot climates (Preston and Leng, 1987). As mentioned earlier, the thermogenic effect of acetate is useful under cold conditions.

Pregnancy and Lactation

Voluntary feed intake shows characteristic changes in pregnancy and lactation. In late gestation, feed intake decreases, perhaps an effect of the high levels of circulating estrogens at that time that may alter metabolism to reduce energy requirements (Forbes, 1986a). Also, the increasing size of the fetus or fetuses takes up an increasing volume of the abdominal cavity so that in ewes with multiple fetuses, for instance, the **rumen volume** is substantially reduced.

During **peak lactation,** feed intake is very high to support high levels of milk production. In dairy cattle, the peak in feed intake is reached after peak lactation occurs, so body reserves are mobilized during this period. In ruminants, rapid rates of utilization of VFA for synthesis of milk may reduce blood VFA levels, reducing the negative feedback on chemoreceptors in the appetite center, partially accounting for the increased feed intake during lactation.

Learning and Conditioning

Animals often exhibit **neophobia,** a reluctance to accept a new food. Such a reaction can have practical implications. In Australia, supplemental feeding of livestock is sometimes required under drought conditions; grain or pellets are often the least expensive source of emergency feed. However, sheep that have never eaten these feeds may not recognize them as feed and may starve. Chapple et al. (1987) found that it took sheep several weeks to overcome fear of the feed trough and supplementary wheat, and then a period was needed to learn to eat, chew, and swallow wheat. The learning process was accelerated if there were some experienced animals in the flock. Animals that have learned to eat a novel feed at a young age will readily accept it again even after a several-year lack of exposure to it. Livestock can also be trained to avoid poisonous plants, such as larkspur, by a process of **aversive conditioning** (Ralphs et al., 1988; Lane et al., 1990). If cattle are fed larkspur and then dosed with lithium chloride, they associate the temporary discomfort induced by the lithium chloride with the larkspur and avoid the plant in the future. Various feed additives such as monensin (Baile et al., 1979) and urea may cause a temporary feed aversion, which is overcome during an adaptation period.

Metabolic Body Size

Common experience indicates that the feed intake of animals does not vary directly with body size. The smallest mammal, the shrew, weighs approximately 1 gram and eats its body weight in food daily. It has an intense rate of metabolism and eats voraciously just to keep alive. It is evident that the larger an animal is, the less feed it consumes per unit of body weight. This relationship was one of great interest to animal scientists for much of the twentieth century. Two of the most active researchers in this area were Dr. Samuel Brody of the University of Missouri and Dr. Max Kleiber of the University of California. Their contributions were summarized in two classic books (Brody, 1964; Kleiber, 1961) dealing with bioenergetics and its relationship to body size.

Brody conducted calorimetric experiments with animals ranging in size from mice to elephants and found that the metabolic rate (hence energy requirements and feed intake) of animals per unit of body weight decreased as body size increased. When body weight doubled, metabolic rate increased by only 73 percent. Brody derived an equation relating **basal metabolic rate (BMR)** to body size:

$$\text{BMR} = 70.5\,\text{W}^{0.734}$$

The BMR is the lowest rate of metabolism of an animal under conditions of complete rest, in a thermoneutral environment, and in a fasting state.

Kleiber conducted similar research and arrived at the exponent of 0.75. Brody and Kleiber debated for years as to which exponent was correct! In reality, the relationship is influenced by other factors such as body surface area, species differences in endocrine regulation of metabolic rate, and age, so there is no "best" exponent. The exponent 0.75 has become accepted as the most suitable figure for general use; the body weight raised to the 3/4 power ($W^{0.75}$) is known as **metabolic body size.** Recent research (West et al., 2002) has shown that the 0.75 exponent applies to cold-blooded animals, plants, and even to intracellular organelles and molecules (mitochondria, respiratory chain complex). Kleiber, in his classic book *The Fire of Life,* illustrated the effect of body size on feed intake by comparing the utilization of 1 ton of hay by 1,300 lb of either cattle or rabbits (Fig. 9.4). The higher daily feed intake of smaller animals (e.g., sheep, rabbits) is compensated for by higher productivity (e.g., gain per day).

Metabolic body size is widely used in animal nutrition research, particularly for interspecies comparisons of feed intake (e.g., sheep vs cattle). A typical example is the data of Reid et al. (1988) shown in Table 9–1. On a metabolic size basis, the intake by cattle of temperate (C_3) grasses and legumes and tropical (C_4) grasses was greater than for sheep, indicating the greater capacity for forage intake of grazers (cattle) over intermediate feeders (sheep).

The relative efficiencies of ruminant and nonruminant digestion are influenced by body size and the abundance and quality of plant material. Among herbivores in general, the ruminants dominate in numbers in the intermediate body-size range. Very small and very large herbivores are mainly nonruminant herbivores. Demment and Van Soest (1985) considered this subject in detail. Gut volume among herbivores varies directly with body weight, whereas metabolic rate varies with a fractional power of body weight. Thus, small herbivores have a high metabolic requirement for energy and must consume high-quality diets to be

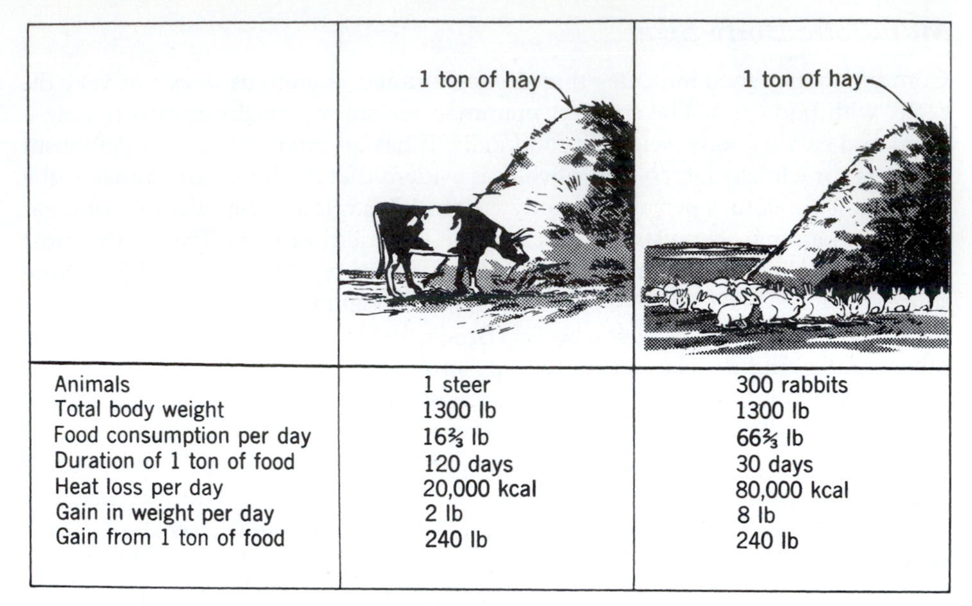

FIGURE 9.4 The effect of body size on feed utilization. Small animals have higher feed intake per unit of body weight than large animals. (From Kleiber, 1961; Courtesy of John Wiley and Sons, Inc.)

TABLE 9–1. Forage Intake of Sheep and Cattle Consuming Temperate and Tropical Forages, Expressed on a Metabolic Size Basis

Forage Fraction Intake ($g/kg\ W^{0.75}$)	*C_3 Grasses*	*C_3 Legumes*	*C_4 Grasses*
Dry matter			
Sheep	71.4	90.7	65.7
Cattle	89.0	94.8	90.0
Neutral detergent fiber (NDF)			
Sheep	40.8	43.7	48.6
Cattle	58.1	48.6	66.7

Adapted from Reid et al. (1988).

able to meet their energy requirements. For this reason, the smallest ruminants are concentrate selectors. Small nonruminant herbivores are better able to utilize high-quality forage than small ruminants because of direct digestion without fermentation, combined with coprophagy. As the body size of ruminants decreases, fermentation of forages can no longer meet the increasing (per unit of weight) energy requirements. With very large ruminants, however, insufficient fibrous feed can be ingested to meet the energy requirements of a large body mass through fermentation by-products. Large animals generally are not able to select sufficient amounts of the rapidly digested foods used by the smaller concentrate selectors because these foods are comparatively rare. Thus, large body size favors nonruminant herbivores, such as the elephant and rhinoceros, because intake of low-quality forage is not limited by retention time in the gut.

FIGURE 9.5 Uneaten grass growing on dung spots in a cattle pasture. Animals avoid grazing forage growing on dung spots from their own species but will consume it from dung spots from other species. Thus, mixed animal species grazing (e.g., sheep and cattle) can improve forage utilization.

Other Factors

Many other factors, often in combination with some of the preceding items, may influence feed intake. **Smell** is very important. Sometimes animals will reject a diet or a forage without even tasting it, suggesting that olfactory cues are involved. Grazing animals will not graze in the immediate area of manure droppings (Fig. 9.5), but if the grass around the dung is cut and taken to an animal, it will be readily consumed (Marten, 1978). This behavior is probably beneficial in reducing exposure to internal parasites. Specific volatile chemicals in cattle feces deter cattle from grazing forage near their feces for about 30 days (Dohi et al., 1991). Urine patches are avoided for only a few days (Sporndly, 1996). Some animals, such as horses, establish dunging areas in a pasture; such areas may occupy up to 25 percent of the total area (Archer, 1978). Harrowing pastures helps to spread out the dung piles and reduce the area of uneaten forage. Mixed grazing can also be used; animals avoid manure only from their own species.

Fatigue can limit feed intake. Ruminants consuming coarse forages may become fatigued in seeking, ingesting, chewing, and ruminating feed (Preston and Leng, 1987; McLeod and Smith, 1989). **Rumination time** is directly related to plant cell-wall content. Sheep and cattle have a maximum rumination time of approximately 10 hours per day and cannot be forced to ruminate beyond this limit (Van Soest, 1994). Small ruminants, such as sheep and goats, must ruminate fibrous feeds to a smaller particle size than cattle for the indigestible fiber to exit the rumen through the smaller omasal orifice. Therefore, rumination may be a more limiting factor for the smaller ruminants. Horses can, if necessary, graze almost continuously.

PREDICTING FEED INTAKE

Prediction of feed intake is very important in meeting nutrient requirements of animals. Ideally, diets are to be formulated and fed so as to precisely meet an individual animal's nutrient requirements. This is done with dairy cattle, where feed intake can be individually regulated and adjusted according to the lactational potential and performance of the cow. Computer modeling is used to account for factors affecting feed intake, and it provides equations for predicting intake and digestibility. Prediction of intake and digestibility for ruminants has been reviewed by Mertens (1987) and the NRC (1987). These sources should be consulted for more information.

Feed intake is readily measured in pen-fed animals. Feed intake measurements are a normal component of feeding trials. If the unconsumed feed (orts) is appreciably different from the offered feed, then in experimental work the orts should be analyzed and a correction of nutrient intake calculated. For example, if the animals are selectively sorting out an unpalatable component, then the measured nutrient intake will not be the same as the actual intake. Measurement of feed intake of grazing animals is more difficult. A common procedure is to surgically prepare selected animals with an esophageal fistula. When feed intake is to be measured, a collection bag is attached to the animal, with the consumed feed going into the bag rather than into the rumen. The bag contents (extrusa) can be analyzed for botanical composition and nutrient content.

QUESTIONS AND STUDY GUIDE

1. The dik-dik and duiker (Fig. 9.1) are small African antelopes, weighing approximately 3.7 to 4.5 kg and 12 to 16 kg, respectively. Since these animals are very much smaller than our domestic ruminants, what do you think of the idea that these animals could be domesticated and used as the "laboratory rat" for ruminant nutrition studies? Obviously there would be advantages of lower feed costs, the ability to use small amounts of a test forage, and so on. How well could results with these animals be extrapolated to sheep and cattle?
2. Wildlife specialists have noted that success with emergency feeding programs for starving deer and elk is much greater with elk than with deer. Elk fed alfalfa or alfalfagrass hay recover well from starvation, whereas deer may eat the hay and die. What explanations might there be to account for these differences?
3. In a feed preference study, two diets were offered free-choice to pigs, and the voluntary intake of each diet was measured. Of the total feed consumed, 97 percent consisted of Diet A and only 3 percent of Diet B. Based on these results, is Diet B of value as a pig feed? Explain.
4. Why do plants contain toxins?

5. According to the NRC (1987), daily requirements for maximum growth of a young (20–50 kg) pig are 6,460 kcal DE, 285 g crude protein, and 14.3 g lysine. Assume you have two diets: one with 2,700 kcal DE/kg and one with 3,800 kcal DE/kg. Calculate the following for each diet.
   a. Expected daily feed intake
   b. Optional percent of protein and lysine in each diet
   c. Optional protein/kcal and lysine/kcal ratios, expressed as mg/kcal

   Discuss the possible benefit of expressing these nutrients on a nutrient/kcal basis rather than on a percent basis.
6. What is the highest possible energy level you could achieve in a diet? For example, could you formulate a diet that contained 12,000 kcal DE/kg? What ingredients would you use to achieve this energy level?
7. When beef calves are taken off pasture and put in a feedlot and fed grain, their growth rate should be at least twice as high in the feedlot as on pasture. Why?
8. A sheepherder once said, "Silage is a hot feed. If you feed it to ewes, it'll burn 'em out." Explain what he probably meant. What is a hot feed, and how do you "burn out" an animal? Should these terms be used by animal nutritionists?
9. When the environmental temperature exceeds the comfort zone of an animal, why does its metabolic rate increase (Fig. 9.3)?
10. A pregnant ewe is "eating for two or more." Why, then, does her feed intake decline in late gestation?
11. What are some possible applications of aversive conditioning in animal production?
12. Look at the data in Table 9–1. Assume that the average body weights of the animals used were 50 kg for sheep and 500 kg for cattle. What is the average metabolic body size for each species (i.e., calculate $W^{0.75}$)? Calculate the average feed intake (kg) per animal for C_3 grass. Compare the feed intake of 10 sheep (500 kg total body weight) with one 500 kg bovine. Discuss.
13. In the data of Table 9–1, why was the NDF intake for sheep higher with C_4 grass than with the other forages, even though the feed intake was lowest with this forage?
14. Why are ruminants poorly represented in number among the smallest and largest herbivores? What are some examples of very small and very large nonruminant herbivores and very large and very small ruminants?
15. Suppose you are a range nutritionist and you wish to determine what plants, and how much of each, cows are consuming. You also wish to estimate their daily dry-matter and crude-protein intake. Describe how you would do this, with the animals being allowed to graze and behave in a normal manner.

REFERENCES

Allen, M. S. 1996. Physical constraints on voluntary intake of forages by ruminants. *J. Anim. Sci.* 74:3063–3075.

Archer, M. 1978. Further studies on palatability of grasses to horses. *J. Br. Grassland Soc.* 33:239–243.

Austin, P. J., L. A. Suchar, C. T. Robbins, and A. E. Hagerman. 1989. Tannin-binding proteins in saliva of deer and their absence in saliva of sheep and cattle. *J. Chem. Ecol.* 15:1335–1347.

Baile, C. A. and C. L. McLaughlin. 1987. Mechanisms controlling feed intake in ruminants: A review. *J. Anim. Sci.* 64:915–922.

Baile, C. A., C. L. McLaughlin, E. L. Potter, and W. Chalupa. 1979. Feeding behavior changes of cattle during introduction of monensin with roughage or concentrate diets. *J. Anim. Sci.* 48:1501–1508.

Beede, D. K. and R. J. Collier. 1989. Potential nutritional strategies for intensively managed cattle during thermal stress. *J. Anim. Sci.* 62:543–554.

Brody, S. 1964. *Bioenergetics and Growth.* New York: Hafner.

Chapple, R. S., M. Wodzicka-Tomaszewska, and J. J. Lynch. 1987. The learning behavior of sheep when introduced to wheat. II. Social transmission of wheat feeding and the role of the senses. *Appl. Anim. Behav. Sci.* 18:163–172.

Cheeke, P. R. 1998. *Natural Toxicants in Feeds, Forages, and Poisonous Plants.* Upper Saddle River, NJ: Prentice Hall, Inc.

Cooper, S. M. and N. Owen-Smith. 1986. Effects of plant spinescence on large mammalian herbivores. *Oecologia* 68:446–455.

Cooper, S. M., N. Owen-Smith, and J. P. Bryant. 1988. Foliage acceptability to browsing ruminants in relation to seasonal changes in the leaf chemistry of woody plants in a South African savanna. *Oecologia* 75:336–342.

Demment, M. W. and P. J. Van Soest. 1985. A nutritional explanation for body-size patterns of ruminant and non-ruminant herbivores. *Am. Nat.* 125:641–672.

Dohi, H., A. Yamada, and S. Entsu. 1991. Cattle feeding deterrents emitted from cattle feces. *J. Chem. Ecol.* 17:1197–1203.

Forbes, J. M. 1986a. The effects of sex hormones, pregnancy, and lactation on digestion, metabolism, and voluntary food intake. In *Control of Digestion and Metabolism in Ruminants*, L. P. Milligan, W. L. Grovum, and A. Dobson, eds. pp. 420–435. Englewood Cliffs, N. J. Reston, Prentice-Hall.

______ . 1986b. *The Voluntary Food Intake of Farm Animals.* London: Butterworths.

______ . 1996. Integration of regulatory signals controlling forage intake in ruminants. *J. Anim. Sci.* 74:3029–3035.

Gallavan, R. H., Jr., W. A. Phillips, and D. L. Von Tungeln. 1989. Forage intake and performance of yearling lambs fed harvested wheat forage. *Nutr. Rep. Int.* 39:643–648.

Gordon, I. J. and A. W. Illius. 1994. The functional significance of the browser-grazer dichotomy in African ruminants. *Oecologia* 98:167–175.

Gordon, I. J. and A. W. Illius. 1996. The nutritional ecology of African ruminants: A reinterpretation. *J. Anim. Ecol.* 65:18–28.

Henry, Y. 1985. Dietary factors involved in feed intake regulation in growing pigs: A review. *Livestock Prod. Sci.* 12:339–354.

______ . 1987. Self-selection by growing pigs of diets differing in lysine content. *J. Anim. Sci.* 65:1257–1265.

Hofmann, R. R. 1973. *The Ruminant Stomach.* Nairobi, Kenya: East African Literature Bureau.

______ . 1989. Evolutionary steps of ecophysiological adaptation and diversification of ruminants: A comparative view of their digestive system. *Oecologia* 78:443–457.

Illius, A. W. and N. S. Jessop. 1996. Metabolic constraints on voluntary intake in ruminants. *J. Anim. Sci.* 74:3052–3062.

Jarrige, R., C. Demarquilly, J. P. Dulphy, A. Hoden, J. Robelin, C. Beranger, Y. Geay, M. Journet, C. Malterre, D. Micol, and M. Petit. 1986. The INRA "fill unit" system for predicting the voluntary intake of forage-based diets in ruminants: A review. *J. Anim. Sci.* 63:1737–1758.

Kennedy, P. M., R. J. Christopherson, and L. P. Milligan. 1986. Digestive responses to cold. In *Control of Digestion and Metabolism in Ruminants*, L. P. Milligan, W. L. Grovum, and A. Dobson, eds. pp. 285–306. Englewood Cliffs, NJ: Prentice Hall.

Ketelaars, J. J. M. H. and B. J. Tolkamp. 1996. Oxygen efficiency and the control of energy flow in animals and humans. *J. Anim. Sci.* 74:3036–3051.

Kleiber, M. 1961. *The Fire of Life: An Introduction to Animal Energetics*. New York: Wiley.

Lane, M. A., M. H. Ralphs, J. D. Olsen, F. D. Provenza, and J. A. Pfister. 1990. Conditioned taste aversion: Potential for reducing cattle loss to larkspur. *J. Range Manage*. 43:127–131.

Lippke, H. 1986. Regulation of voluntary intake of ryegrass and sorghum forages in cattle by indigestible neutral detergent fiber. *J. Anim. Sci.* 63:1459–1468.

McArthur, C., G. D. Sanson, and A. M. Beal. 1995. Salivary proline-rich proteins in mammals: Roles in oral homeostasis and counteracting dietary tannin. *J. Chem. Ecol.* 21:663–691.

McLeod, M. N. and B. R. Smith. 1989. Eating and ruminating behaviour in cattle given forages differing in fibre content. *Anim. Prod.* 48:503–511.

Marten, G. C. 1978. The animal-plant complex in forage palatability phenomena. *J. Anim. Sci.* 46:1470–1477.

Mayland, H.F., S.A. Martin, J. Lee and G.E. Shewmaker. 2000. Malate, citrate, and amino acids in tall fescue cultivars: relationship to animal preference. Agron. J. 92: 206–210.

Mertens, D. R. 1987. Predicting intake and digestibility using mathematical models of ruminal function. *J. Anim. Sci.* 64:1548–1558.

Mills, C. F. and G. K. Davis. 1987. Molybdenum. In *Trace Elements in Human and Animal Nutrition*, W. Mertz, ed. pp. 429–463. Orlando, FL: Academic Press.

National Research Council. 1987. *Predicting Feed Intake of Food-Producing Animals*. Washington, DC: National Academy Press.

Ndlovu, L. R. and J. G. Buchanan-Smith. 1987. Alfalfa supplementation of corncob diets for sheep: Effect of ruminal or postruminal supply of protein on intake, digestibility, digesta passage and liveweight changes. *Can. J. Anim. Sci.* 67:1075–1082.

Preston, T. R. 1982. Nutritional limitations associated with the feeding of tropical forages. *J. Anim. Sci.* 54:877–884.

Preston, T. R. and R. A. Leng. 1987. *Matching Ruminant Production Systems with Available Resources in the Tropics and Sub-Tropics*. Armidale, Australia: Penambul Books.

Ralphs, M. H., J. D. Olsen, J. A. Pfister, and G. D. Manners. 1988. Plant-animal interactions in larkspur poisoning in cattle. *J. Anim. Sci.* 66:2334–2342.

Reid, R. L., G. A. Jung, and W. V. Thayne. 1988. Relationships between nutritive quality and fiber components of cool season and warm season forages: A retrospective study. *J. Anim. Sci.* 66:1275–1291.

Robbins, C. T., D. E. Spalinger, and W. van Hoven. 1995. Adaptation of ruminants to browse and grass diets: Are anatomical based browser-grazer interpretations valid? *Oecologia* 103:208–213.

Shewmaker, G. E., H. F. Mayland, R. C. Rosenau, and K. H. Asay. 1989. Silicon in C-3 grasses: Effects on forage quality and sheep preference. *J. Range Manage*. 42:122–127.

Sporndly, E. 1996. The effect of fouling on herbage intake of dairy cows on late season pasture. *Acta. Agric. Scand.* 46(A):144–153.

Van Soest, P. J. 1994. *Nutritional Ecology of the Ruminant*. Ithaca, NY: Cornell University Press.

West, G. B., W. H. Woodruff, and J. H. Brown. 2002. Allometric scaling of metabolic rate from molecules and mitochondria to cells and mammals. *Proc. Nat. Acad. Sci.* 99:2473–2478.

CHAPTER 10

Water

Objectives 1. To emphasize the unique role of water in metabolism of living organisms:
water intake of animal species
nutrients and toxic elements in water
adaptations to arid environments
ecological impacts of species differences in water requirements

> Water is the fluid matrix of the animal body in which are embedded the living cells with their contained protoplasm and the intercellular substance giving form and structure to the body and protection from environmental stress. The high solvent power of water permits the formation of a large variety of solutions, true and colloidal, within which the reactions of metabolism occur. Most, but not all, enzymes are soluble in water. The high dielectric constant of water promotes ionization of organic, and especially inorganic, solutes, thus facilitating the initiation and velocity of metabolic reactions. Water enters into many metabolic reactions (hydrolyses) and is produced in many others. . . . The low thermal conductivity of water and its high specific heat narrow the amplitude of variation in body temperature induced by warming or cooling of its environment. . . . The high heat of vaporization of water accounts in part for the high proportion of the body heat dissipated in the production of water vapor on the external surfaces of the body.
>
> H. H. Mitchell (1962)

Water is required in greater quantity than any other orally ingested substance. Sources of water include drinking water, water occurring in feedstuffs, and **metabolic water** arising from nutrient metabolism. Examples of metabolic water include that produced when fats, carbohydrates, and amino acids are metabolized to carbon dioxide and water. In some desert animals, metabolic water may make up the major or sole source of water. The kangaroo rat, for example, never drinks (Schmidt-Nielsen, 1964). It has a very low rate of water loss, achieved by having no sweat glands, excreting a highly concentrated urine, and having a low evaporation rate from the expired air because of a counter-current cooling mechanism in the nasal passages. Kangaroo rats eat a diet of dry seeds and plant material, with very little green or succulent matter. In contrast, another desert rodent, the pack rat, has a very high water requirement, which it meets by consuming succulent vegetation such as cactus. The behavioral and physiological adaptations of desert animals to a dry environment have been discussed in detail by Schmidt-Nielsen (1964) (Figure 10.1).

One of the factors influencing species differences in water requirements is the nature of the **nitrogenous end products** of protein metabolism excreted in the urine. Most mammals excrete **urea,** which is toxic to the tissues unless in

FIGURE 10.1
Livestock differ greatly in their efficiency of water utilization. Animals adapted to arid conditions have various mechanisms to conserve body water and survive dehydration. Camels have lower water requirements than other domestic animals. (Courtesy of A. N. Bhattacharya.)

TABLE 10–1. Expected Water Consumption of Various Classes of Adult Livestock in a Temperate Climate

Species	*Water (liters/day)*	*Species*	*Water (liters/day)*
Beef cattle	26–66	Sheep and goats	4–15
Dairy cattle	38–110	Chickens	0.2–0.4
Horses	30–45	Turkeys	0.4–0.6
Swine	11–19		

Adapted from National Research Council (1974).

dilute solution. Thus large amounts of water are required to dilute it. Birds have a lower water requirement than mammals because they excrete **uric acid** in a nearly solid form. Fish excrete **ammonia** directly from the gills, and many have such low water requirements that they never drink. With mammals, high-protein diets increase the amount of water required for dilution of urinary urea. The nature of the digestive tract and feeding strategy also influence water requirements. Ruminants require large quantities of water to form a suspension of ingesta in the rumen and, thus, have higher water requirements than nonruminant species. Feedstuffs with high water-absorbing characteristics, such as wheat bran and dried forage (e.g., alfalfa hay), increase the water requirements. Average water requirement figures for livestock are given in Table 10–1. Water requirements are increased in cold weather because feed intake is increased. Mature beef cattle and sheep can rely solely on snow as a water source, but more productive animals, such as feedlot cattle and dairy animals, must have free access to drinking water (Degen and Young, 1990).

NUTRIENTS AND TOXIC ELEMENTS IN WATER

The concentration of minerals in water is normally not of sufficient magnitude to be nutritionally significant. More commonly, water can be a source of mineral toxicity. Highly **saline water** may contain sufficient calcium, magnesium, sodium, bicarbonate, chloride, and sulfate ions to exert toxic effects. Sulfates are more injurious than chlorides; hydroxides are more harmful than carbonates, which are more harmful than bicarbonates. The NRC (1987) recommendations concerning saline water are given in Table 10–2. The upper safe limits for a number of toxic ions are given in Table 10–3.

One of the most common toxic substances in water is the **nitrate** ion. Contamination of water sources with nitrate is common in areas with a high concentration of animals (e.g., feedlots), with heavy fertilization of fields with manure or

TABLE 10–2. A Guide to the Use of Saline Waters for Livestock and Poultry

Total Soluble Salts Content of Waters (mg/liter)	*Comment*
Less than 1,000	These waters have a relatively low level of salinity and should present no serious burden to any class of livestock or poultry.
1,000–2,999	These waters should be satisfactory for all classes of livestock and poultry. They may cause temporary and mild diarrhea in livestock not accustomed to them or watery droppings in poultry (especially at the higher levels) but should not affect their health or performance.
3,000–4,999	These waters should be satisfactory for livestock, although they might very possibly cause temporary diarrhea or be refused at first by animals not accustomed to them. They are poor waters for poultry, often causing watery feces and (at the higher levels of salinity) increased mortality and decreased growth, especially in turkeys.
5,000–6,999	These waters can be used with reasonable safety for dairy and beef cattle, sheep, swine, and horses. It may be well to avoid the use of those approaching the higher levels for pregnant or lactating animals. They are not acceptable waters for poultry, almost always causing some type of problem, especially near the upper limit, where reduced growth and production, or increased mortality, will probably occur.
7,000–10,000	These waters are unfit for poultry and probably for swine. Considerable risk may exist in using them for pregnant or lactating cows, horses, sheep, the young of these species, or for any animals subjected to heavy heat stress or water loss. In general, their use should be avoided, although older ruminants, horses, and even poultry and swine may subsist on them for long periods of time under conditions of low stress.
More than 10,000	The risks with these highly saline waters are so great that they cannot be recommended for use under any conditions.

Adapted from National Research Council (1974).

TABLE 10–3. Recommended Limits of Concentration of Some Potentially Toxic Substances in Drinking Water for Livestock and Poultry

Item	*Safe Upper Limit of Concentration (mg/liter)*	*Item*	*Safe Upper Limit of Concentration (mg/liter)*
Arsenic	0.2	Lead	0.1
Barium	Not established	Manganese	Not established
Cadmium	0.05	Mercury	0.010
Chromium	1.0	Molybdenum	Not established
Cobalt	1.0	Nickel	1.0
Copper	0.5	Nitrate-N	100.0
Cyanide	Not established	Nitrite-N	10.0
Fluoride	2.0	Vanadium	0.1
Iron	Not established	Zinc	25.0

Adapted from National Research Council (1974).

nitrogenous fertilizers, or with contamination from septic tanks. Levels of 200 ppm nitrate in water are potentially hazardous; levels above 1,500 ppm may cause acute toxicity. Nitrate is converted to nitrite ion in the rumen; nitrite combines with the hemoglobin molecule to prevent it from transporting oxygen. Death is thus from anoxia. The blood of affected animals is chocolate in color due to the color of **methemoglobin,** formed when nitrite combines with hemoglobin.

Iron salts in groundwater sometimes cause problems with water quality. Iron in oxygen-free groundwater is normally in the soluble, reduced state (Fe^{2+}). When exposed to oxygen or chlorine, it is oxidized to the reddish, less soluble Fe^{3+} form. This leads to iron deposits in pipes, tanks, and fixtures. Wells may become contaminated with iron-utilizing bacteria; these cause foul odors and plug water lines. Treatment of the water system and well with bleach may control bacteria.

Pesticides and other **agricultural chemicals** may enter water supplies from agricultural runoff, accidental spills, or faulty waste-disposal systems. The organophosphates, such as malathion, are the most toxic. Insecticides of plant origin, such as pyrethrins and rotenones, are not considered a hazard.

Lakes and stock-watering ponds may have a heavy "bloom" of toxic **blue-green algae (cyanobacteria).** Many species of algae produce extremely potent toxins. Signs of blue-green algae toxicosis in pigs include vomiting, frothing at the mouth, coughing, muscle tremors, rapid breathing, and bloody diarrhea (Chengappa et al., 1989). Severe liver damage occurs. In dairy cattle, clinical signs include anorexia, reluctance to move, dementia, and ruminal atony. Serum enzymes characteristic of liver damage are elevated (Galey et al., 1987), and death is due to massive liver damage.

Blue-green algae in farm ponds can be controlled by dragging a porous bag containing copper sulfate through the pond. However, this sometimes leads to proliferation of copper-tolerant green algae. Though these algae are normally controlled by zooplankton, which feed on them, high levels of copper kill the zooplankton, allowing the green algae to proliferate. The green algae are not usually toxic, but they can cause severe water quality problems, leading to turbid, odiferous, bad-tasting water.

Water quality can influence development of **polioencephalomalacia (PEM)** in feedlot cattle. Water high in sulfates promotes PEM, apparently via a complex interaction involving copper, sulfur, and thiamin. Gooteratne et al. (1989) and Cummings et al. (1995) reviewed these interrelationships. When water is a major source of dietary sulfate, risk of PEM may increase in hot weather because of increased water consumption (McAllister et al., 1997).

ECOLOGICAL IMPLICATIONS OF WATER REQUIREMENTS OF ANIMALS

Humans and livestock are potentially competitive for several types of resources, including land, cereal grains, and water. In arid regions, including much of the western United States, rapidly expanding human populations are creating an ever-increasing demand on water resources. The use of large amounts of water for livestock, and especially cattle, is of concern (Tamminga, 1996; Van Horn et al., 1996). Besides the water actually used by animals, the amount of water excreted is also important as a potential source of ground water contamination. Cattle, in fact, are poor utilizers of water, and in semiarid regions, other kinds of animals might be better choices. Maloiy et al. (1979) classified large herbivores into three main physiological ecotypes:

1. Wet tropical and wet temperate. Herbivores with high rates of water and energy use and poor urine concentrating ability, for example, buffalo, cattle, pig, eland, water buck, elephant, horse, moose, and reindeer
2. Warm, dry savanna, semiarid. Herbivores with intermediate rates of water and energy use and good renal concentrating ability, for example, sheep, wildebeest, and donkey
3. Arid zone animals with low rates of energy and water turnover and medium to high urine-concentrating ability, for example, camel, goat, oryx, gazelle, and ostrich

Thus cattle are relatively unadapted to semiarid conditions. Zebu breeds are only slightly superior to European breeds in water utilization. Sheep are considerably more efficient in water utilization than cattle and, under temperate conditions, can often meet their water needs from forage alone. Cattle, however, virtually always require access to drinking water. Adaptations of sheep include a good urine-concentrating ability, the excretion of relatively dry fecal pellets, and the insulatory properties of the fleece, reducing the need for evaporative cooling (Silanikove, 1992). Of the common domestic ruminants, goats are clearly the best adapted to arid conditions, especially the Bedouin goat, raised by Bedouin nomads on the deserts of the Middle Eastern countries. These animals can lose as much as 30 percent of their body weight through dehydration and still continue to feed normally. When provided access to water, they can replenish this loss within 2 minutes (Figure 10.2). In these goats, the rumen acts as an osmotic barrier, preventing osmotic shock to the tissues after rapid rehydration (Louw, 1984; Shkolnik et al., 1980). Because the Bedouin goats need drink only once every 2

FIGURE 10.2 Desert goats need drink only once every 2 to 4 days. Thus they can use forage resources that are far removed from water supplies more effectively than most other domestic animals. (Courtesy of A. N. Bhattacharya.)

to 4 days, they can utilize forage resources that are far removed from water. In cattle rehydrated after water deprivation, about 20 percent of ingested water bypasses the rumen to the intestine (Cafe and Poppi, 1994).

Other herbivores well adapted to desert conditions include the oryx, gazelle (e.g., springbok), ostrich, and camel. In terms of conversion of the meager desert forage into meat or milk, the oryx and goat seem to have the most potential (Louw, 1984).

The **oryx** (Figure 10.3) is the only grazing ruminant native to East Africa that is adapted to extremely arid conditions without surface water. Efforts to domesticate the oryx are in progress (Stanley-Price, 1985a,b). Oryx can be quickly tamed and moved as a closely packed herd and are highly adapted to desert conditions. The oryx has an extremely low rate of evaporative water loss (Stanley-Price, 1985a) and a low metabolic rate. It ruminates at night when the heat of fermentation is least burdensome, and, in fact, this eliminates the need for increased thermogenesis at night, when temperatures are often quite cold in the desert.

Thus, the efficiency of conversion of forage and water to useful products in many arid and semiarid regions of the world might be enhanced by utilizing animals with superior biochemical and behavioral adaptations to an arid environment. These animals probably will not be the traditional domesticated ruminants, but perhaps newly domesticated species like the oryx.

FIGURE 10.3 The oryx has extremely low water requirements and is highly adapted to desert conditions. Efforts to domesticate the oryx are in progress.

QUESTIONS AND STUDY GUIDE

1. Can a case be made that water is not a nutrient? Discuss. Is oxygen a nutrient?
2. Would you expect the Dalmatian dog to have a different requirement for water than a similar-size dog of another breed? Why?
3. The water requirements of chickens fed barley or rye are different than the requirement for birds fed a corn-based diet. Discuss factors that might account for this. What management problems in a poultry house could develop if broilers on litter on the floor are fed barley instead of corn?
4. Why are cattle more sensitive to nitrate in water than are chickens? Provide two reasons.
5. Explain why camels and goats have lower water requirements than cattle.

REFERENCES

Beasley, V. R., W. O. Cook, A. M. Dahlem, S. B. Hoover, R. A. Lovell, and W. M. Valentine. 1989. Algae intoxication in livestock and waterfowl. *Vet. Clin. N. Am.: Food Anim. Pract.* 5:345–361.

Cafe, L. M. and D. P. Poppi. 1994. The fate and behavior of imbibed water in the rumen of cattle. *J. Agr. Sci.* 122:139–144.

Chengappa, M. M., L. W. Pace, and B. G. McLaughlin. 1989. Blue-green algae (*Anabaena spiroides*) toxicosis in pigs. *J. Am. Vet. Med. Assoc.* 194:1724–1725.

Clemens, E. T. and G. M. O. Maloiy. 1983. Digestive physiology of East African wild ruminants. *Comp. Biochem. Physiol.* 76:319–333.

Cummings, B. A., D. H. Gould, D. R. Caldwell, and D. W. Hamas. 1995. Ruminal microbial alterations associated with sulfide generation in steers with dietary sulfate-induced polioencephalomalacia. *Am. J. Vet. Res.* 56:1390–1395.

Degen, A. A. and B. A. Young. 1990. The performance of pregnant cows relying on snow as a water source. *Can. J. Anim. Sci.* 70:507–515.

Galey, F. D., V. R. Beasley, W. W. Carmichael, G. Kleppe, S. B. Hooser, and W. M. Haschek. 1987. Blue-green algae (*Microcystis aeruginosa*) hepatotoxicosis in dairy cows. *Am. J. Vet. Res.* 48:1415–1420.

Gooteratne, S. R., W. T. Buckley, and D. A. Christensen. 1989. Review of copper deficiency and metabolism in ruminants. *Can. J. Anim. Sci.* 69:819–845.

Louw, G. N. 1984. Water deprivation in herbivores under arid conditions. In *Herbivore Nutrition in the Subtropics and Tropics,* F. M. C. Gilchrist and R. I. Mackie, eds. pp. 106–126. Craighall, Republic of South Africa: The Science Press.

Maloiy, G. M. O., W. V. Macfarlane, and A. Shkolnik. 1979. Mammalian herbivores. In *Comparative Physiology of Osmoregulation in Animals,* Vol. 11, G. M. O. Maloiy, ed. pp. 185–209. London: Academic Press.

McAllister, M. M., D. H. Gould, M. F. Raisbeck, B. A. Cummings, and G. H. Loneragan. 1997. Evaluation of ruminal sulfide concentrations and seasonal outbreaks of polioencephalomalacia in beef cattle in a feedlot. *J. Am. Vet. Med. Assoc.* 211:1275–1279.

Mitchell, H. H. 1962. *Comparative Nutrition of Man and Domestic Animals.* New York: Academic Press.

National Research Council. 1974. *Nutrients and Toxic Substances in Water for Livestock and Poultry.* Washington, DC: National Academy Press.

Schmidt-Nielsen, K. 1964. *Desert Animals. Physiological Problems of Heat and Water.* London: Oxford University Press.

Shkolnik, A., E. Maltz, and I. Choshniak. 1980. The role of the ruminant's digestive tract as a water reservoir. In *Digestive Physiology and Metabolism in Ruminants,* Y. Ruckebusch and P. Thivend, eds. pp. 731–742. Westport, CT: AVI Publishing Co.

Silanikove, N. 1992. Effects of water scarcity and hot environment on appetite and digestion in ruminants: A review. *Livestock Prod. Sci.* 30:175–194.

Stanley-Price, M. R. 1985a. Game domestication for animal production in Kenya: Feeding trials with oryx, zebu cattle and sheep under controlled conditions. *J. Agr. Sci.* 104:367–374.

______ .1985b. Game domestication for animal production in Kenya: The nutritional ecology of oryx, zebu cattle and sheep under free-range conditions. *J. Agr. Sci.* 104:375–382.

Tamminga, S. 1996. A review on environmental impacts of nutritional strategies in ruminants. *J. Anim. Sci.* 74:3112–3124.

Van Horn, H. H., G. L. Newton, and W. E. Kunkle. 1996. Ruminant nutrition from an environmental perspective: Factors affecting whole-farm nutrient balance. *J. Anim. Sci.* 74:3082–3102.

PART III

FEED MANUFACTURING AND DIET FORMULATION

An example of a feed mill. The silos are used for storage of bulk ingredients. These are transferred to the mixer and pellet mill by the pipes shown in the superstructure. Weighing and transfer of ingredients is controlled by computer. (Courtesy of R.A. Swick.)

This part describes the fundamentals of feed processing, the preparation of manufactured feeds, and the techniques of diet formulation. Understanding the functional properties of feed ingredients in influencing pellet quality, for example, is important in understanding the nonnutritive properties of feedstuffs. Pelletability and other feed-processing characteristics of feeds are discussed in Chapter 12.

The basic principles of diet formulation are presented in Chapter 11. Because of the ever-increasing availability and sophistication of desktop computer technology, computer programs for least-cost diet formulation should be obtained from current sources.

CHAPTER 11

Diet Formulation

Objectives

1. To identify information needed for diet formulation:
   - nutritional requirements
   - composition of feedstuffs
   - bioavailability of nutrients
   - nonnutritive factors
   - costs of available feedstuffs
2. To describe the mathematical manipulations needed to formulate diets

The objectives in diet formulation are to design diets to meet the animal's nutrient requirements and provide the maximum or optimum economic return to the livestock producer. In the case of animals raised for exhibition purposes or as pets, the economic factor is not as critical. In general, though, people would like to feed their animals adequately at a reasonable cost. It is important to note the distinction between diets that provide the maximum economic return and diets that have the lowest cost. The lowest cost diet, even though it may be nutritionally adequate, may not be the most economical. It is the net return from the use of the diet that indicates the economic efficiency.

The almost universal use of computer technology has eliminated most of the need to formulate diets by hand calculation. Some of the most simple calculations will be briefly described, but it should be recognized that there is abundant computer software available for diet formulation.

INFORMATION NEEDED TO FORMULATE DIETS

The following information is required for the efficient formulation of diets.

1. Nutritional requirements of the animal
2. Nutrient composition of available feedstuffs
3. Nutrient availability (bioavailability)
4. Nonnutritive characteristics of the feedstuffs, such as palatability, pelleting properties, and associative effects
5. Costs of available feedstuffs
6. Expected daily feed intake

Nutritional Requirements

Standard sources in the United States for nutrient requirements of livestock are the NRC publications in the series *Nutrient Requirements of Domestic Animals.* Each of these publications contains tables of nutrient requirements of animals for various productive functions (growth, gestation, reproduction, etc.). The requirements are expressed as amount per unit of diet (%, mg/kg, etc.) or as amount per animal per day. These figures are the minimum requirements for maximum production and do not include a margin of safety. Many nutritionists use their judgment to provide a margin of safety by increasing the figures by 5 or 10 percent or more. Environmental conditions, stress, animal housing conditions, breed or strain of animal, disease incidence, and projected length and conditions of feed storage are factors that might influence selection of an appropriate margin of safety.

Traditionally, diets have been formulated on a percentage basis so that a diet for growing pigs might contain 15 percent protein, whereas one for finishing pigs might have 13 percent protein. In some cases, as with dairy cattle where feed intake can be very precisely controlled, formulation on the basis of nutrient intake per animal per day will allow greater efficiency in meeting nutrient needs. It has also been the traditional belief in the United States that a well-formulated diet is one that meets NRC requirements and will support maximum animal performance. Preston and Leng (1987) are proponents of the viewpoint that application of this philosophy in developing countries is inappropriate and that, instead, the objective should be to optimize the use of available resources and minimize the use of imported ingredients. Under these conditions, NRC requirements generally cannot be economically achieved and optimal production is less than maximal.

Nutrient requirement figures are continually "fine-tuned" as new research becomes available and as animal genetics and management systems change. The entire tables of NRC values can be entered in a computer program, and diets can be formulated so that for each nutrient the listed requirement is met or exceeded. The most recent NRC publications (since 1989) contain diskettes of the nutrient requirements.

Nutrient Composition of Feedstuffs

Extensive tables of composition of feedstuffs are available (NRC, 1982). As with requirement figures, the composition data can be placed in computer files. In the future, it is likely that standardized computer lists of ingredient composition will become available commercially. Most feed companies have proprietary files of ingredient composition for use in ration formulation.

It is important not to be dazzled or intimidated by lengthy computer printouts. Feedstuffs are inherently variable products; their composition is greatly influenced by harvesting conditions, environmental factors, fertilization and irrigation practices, and so on. Thus considerable judgment is needed in assessing whether an apparent deficiency of a particular nutrient is truly important.

Nutrient Availability

For most feedstuffs, the nutrients are not completely released during digestion. This is particularly true with the energy fraction, and, for this reason, either digestible energy, metabolizable energy, or net energy values should be used rather than gross energy. These values vary between animal species, especially for fibrous feeds. For instance, the DE value for alfalfa meal is higher in dairy cattle than in swine. Ideally, species-specific values for energy (DE, ME, or NE), protein, and amino acids should be used in ration formulation, along with the requirements expressed on the same basis. For poultry, ME values are used, whereas for swine, either DE or ME are satisfactory. For ruminants, ME or NE values should be used because of the variability of energy losses in rumen gases.

Nonnutritive Characteristics

One of the most obvious nonnutritive characteristics of importance is **palatability.** Certain ingredients, such as molasses and other sweet materials, are highly palatable. Others, such as feathermeal, are somewhat unpalatable. The experience and judgment of the nutritionist are important factors in the evaluation of palatability characteristics. Other nonnutritive qualities include pelletability, handling qualities (e.g., will the ingredient flow freely or bridge in bins), associative effects, and presence of toxins, such as gossypol in cottonseed meal. **Associative effects** refer to the interactions between feedstuffs that can influence their nutritive value. They may be positive or negative. The DE content of a forage, for instance, may be less if the forage is fed along with a concentrate than if fed alone. Thus there may be a negative associative effect between fibrous feeds and concentrates in ruminants. A positive associative effect is the improvement in utilization of low-quality roughage when it is supplemented with alfalfa. These types of associative effects are discussed in more detail in Chapter 6. Pelleting qualities of feed ingredients are very important and are discussed in more detail in Chapter 12.

Costs of Available Feedstuffs

Costs of ingredients vary according to a number of factors. The ingredient in largest supply will often dictate prices for other ingredients. For example, the plant protein market in North America is dominated by soybean meal. The cost of soybean meal is influenced by crop yields in the United States (which in turn are influenced by acreage planted, weather conditions, and so forth) and by supplies available on the world market. The cost of other protein sources, such as cottonseed and canola meal, will be pegged to the price of soybean meal. The canola marketing board, for instance, may price canola meal at a slightly cheaper rate per unit of protein than soybean meal to induce feed manufacturers and producers to use it.

In **least-cost diet formulation,** the prices of available ingredients are entered into the computer. These prices may change on a daily or weekly basis. The computer solves an array of simultaneous equations to provide the solution for the least expensive combination of ingredients to satisfy the requirement for each nutrient.

MATHEMATICS OF DIET FORMULATION

Diets may be formulated by hand or by computer. **Formulation by hand** refers to making simple algebraic calculations to determine how feedstuffs of known composition can be combined to meet specific nutrient requirements. It is difficult to include cost factors effectively in hand formulation, whereas a major advantage of **computer formulation** is that it calculates the least expensive way of meeting the nutrient requirements. Another major advantage of computer formulation is the speed and ease of checking different options. It is also easy to calculate **opportunity costs,** which refer to the price an ingredient would have to be for the computer to select it.

Diet Formulation by Hand

In hand calculation of diet formulae, the usual procedure is to select the main energy source and protein supplement to be used primarily on the basis of cost. In the case of grains, the cheapest source is that which has the lowest cost per kcal of digestible energy. Protein sources would be selected on the basis of cost per unit of protein. After the two major ingredients are selected, the quantity of each that is needed to make a mixture with the desired protein content is calculated. Then the content of other major nutrients in the diet mix is calculated, and any deficient nutrients are added in a supplement. The basic steps in hand formulation of diets are illustrated here for a swine grower diet.

1. Select the main sources of energy and protein. First select the grain and main protein supplement to be used. This selection is based on cost. The costs of available grains are compared, and the cheapest source of DE or ME selected. A similar process is used with protein supplements to find the cheapest source of good-quality protein. Suppose in this case that corn and soybean meal are the best sources of energy and protein, respectively.

2. Decide how much **slack space** is needed. The slack space will be used for salt, mineral, and vitamin supplements, and any other additives that will be used. Assume that all of these additives will fit in a slack space of 2.5 percent. Therefore, the corn and soybean meal will add up to 97.5 percent.

3. Look up the requirement for crude protein, which in the case of a swine grower diet is 15 percent. Then calculate the combination of corn and soybean meal that is necessary, in 97.5 percent of the diet, to provide 15 percent crude protein in 100 percent of the diet.

4. To perform this calculation, we need to know the percent of crude protein of the corn and soybean meal (SBM) we will use. Assume that these are 8.5 and 44 percent, respectively.

Algebraic solution:

For 100 lbs of mixed feed:

$$\begin{aligned}
\text{Let amount of corn} &= x \text{ lb} \\
\text{Then amount of SBM} &= 97.5 - x \\
\text{lb corn protein} &= 0.085x \\
\text{lb SBM protein} &= 0.44\,(97.5 - x) \\
\text{lb corn protein} + \text{lb SBM protein} &= 15 \text{ lb} \\
0.085\,x + 0.44(97.5 - x) &= 15 \\
0.085\,x + 42.9 - 0.44x &= 15
\end{aligned}$$

Multiply both sides of the equation by -1

$$
\begin{aligned}
(-1)(0.085x + 42.9 - 0.44x) &= (-1)(15) \\
-0.85x - 42.9 + 0.44x &= -15 \\
0.44x - 0.085x &= -15 + 42.9 \\
0.355x &= 27.9 \\
x &= 78.59 \\
97.5 - x &= 18.91
\end{aligned}
$$

Check the calculation

$$
\begin{aligned}
\text{78.59 lb corn @ 8.5\% CP} &= 6.68 \\
\text{18.91 lb SBM @ 44\% CP} &= \underline{8.32} \\
\text{Total percentage} &= 15.00
\end{aligned}
$$

5. The next step is to determine if this mixture meets the essential amino acid requirements. This can be done by using the percent essential amino acid values for corn and SBM. If any amino acids are deficient, then either crystalline amino acids or a protein supplement with a high content of the amino acid must be added. The value of a computer in making these calculations should be apparent. As an example, calculate the lysine content of our corn-SBM diet:

$$
\begin{aligned}
\text{\% lysine in corn} &= 0.3\% \\
\text{\% lysine in SBM} &= 2.8\%
\end{aligned}
$$

Pounds of lysine in our 100-lb diet are

$$
\begin{aligned}
\text{78.59 lb corn} \times \text{0.003 CP} &= 0.236 \\
\text{18.91 lb SBM} \times 0.028 &= \underline{0.530} \\
\text{Total lb lysine} &= 0.766
\end{aligned}
$$

The NRC requirement is 0.75 percent. Thus this mixture provides adequate lysine. All of the other amino acids could be checked in the same way. In practice, a nutritionist would know that lysine is the first limiting amino acid in a corn-SBM mixture for growing swine, so if it is adequate, then the others will be adequate also.

6. Similar calculations can be made for calcium and phosphorus. Any deficiency can be overcome with the appropriate amount of supplement, using the slack space. Further additions will be 0.25 percent salt, a vitamin premix, and a trace mineral premix. Any remaining slack space can be filled using corn.

In the NRC (1988) swine bulletin, a similar calculation is presented, except that instead of formulating the diet to provide 15 percent crude protein, it was formulated to provide 0.75 percent lysine. In that case, the proportions of corn and soybean meal are 78.4 and 19.1 percent, respectively, and the ration is slightly less than 15 percent crude protein. The NRC diet is shown in Table 11–1. The nutrient composition of a simple mixture of 79 percent corn and 19 percent soybean meal is shown in Table 11–2. The nutrient deficiencies, which must be made up by supplementation, include calcium, phosphorus, vitamin A, and some of the B-complex vitamins.

TABLE 11–1. Composition of NRC Swine Grower Diet

Nutrient	*Percent*
Corn	78.67
Soybean meal	19.10
Dicalcium phosphate	0.85
Ground limestone	0.83
Sodium chloride	0.25
Vitamin premix	0.10
Trace mineral premix	0.10
Antimicrobial premix	0.10
Total	100.00

Adapted from National Research Council (1988).

Rather than using the preceding algebraic equations, the **Pearson's square** can be used, which basically is a diagrammatic version of the equations. The use of the square is illustrated using the same swine diet as previously formulated

% CP of SBM = 44
% CP of corn = 8.5
% CP desired in ration = 15

where CP is crude protein.

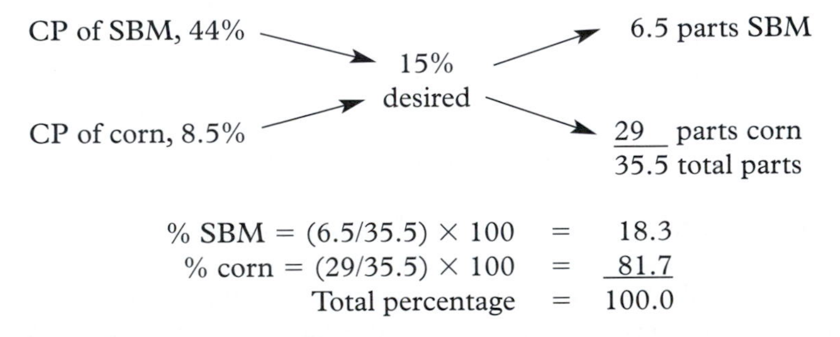

% SBM = (6.5/35.5) × 100 = 18.3
% corn = (29/35.5) × 100 = 81.7
Total percentage = 100.0

Note that this answer is slightly different because no slack space was used. To recalculate with a slack space, we would say that we want all of the 15 lbs of CP to be in the 97.5 lbs corn and SBM. Therefore, for a 100 percent mixture of corn and SBM, the percent of protein should be 15.39, so that when 97.5 percent of this mixture is used, it provides 15 percent CP.

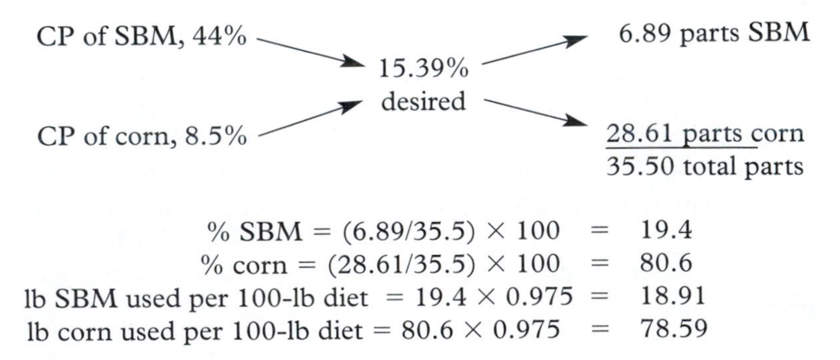

% SBM = (6.89/35.5) × 100 = 19.4
% corn = (28.61/35.5) × 100 = 80.6
lb SBM used per 100-lb diet = 19.4 × 0.975 = 18.91
lb corn used per 100-lb diet = 80.6 × 0.975 = 78.59

TABLE 11–2. Nutrients in Corn and Corn + Soybean Meal (44% protein) Compared with Nutrient Requirements of a Pig

Nutrient	*Corn*	*Corn + Soybean Meal (79%:19%)*	*Requirement (25-kg pig)*
Indispensable amino acids, %			
Arginine	0.43	0.95	0.25
Histidine	0.27	0.43	0.22
Isoleucine	0.35	0.66	0.46
Leucine	1.19	1.58	0.60
Lysine	0.25	0.75	0.75
Methionine + cystine	0.40	0.54	0.41
Phenylalanine + tyrosine	0.84	1.35	0.66
Threonine	0.36	0.61	0.48
Tryptophan	0.09	0.19	0.12
Valine	0.48	0.76	0.48
Linoleic acid, %	2.2	1.8	0.1
Mineral elements			
Calcium, %	0.03	0.08	0.60
Phosphorus, total, %	0.28	0.34	0.50
Phosphorus, available, %	0.04	0.08	0.23
Sodium, %	0.01	0.02	0.10
Chlorine, %	0.05	0.05	0.08
Magnesium, %	0.11	0.14	0.04
Potassium, %	0.33	0.66	0.23
Copper, %	3.5	7.2	4.0
Iodine, mg/kg	0.03	0.06	0.14
Iron, mg/kg	33	53	60
Manganese, mg/kg	5.7	10.3	2.0
Selenium, mg/kg	0.07	0.07	0.15
Sulfur, %	0.11	0.17	[a]
Zinc, mg/kg	19	25	60
Vitamins			
Vitamin A, IU/kg	200	158	1,300
Vitamin D, IU/kg			
Vitamin E, IU/kg	21	17	150
Vitamin K, mg/kg			11
Biotin, mg/kg	0.07	0.12	0.50[b]
Choline, g/kg	0.50	0.89	0.05
Folacin, mg/kg	0.30	0.35	0.30
Niacin, available, mg/kg	[c]	5.4	0.30
Pantothenic acid, mg/kg	5.1	7.1	10.0
Riboflavin, mg/kg	1.1	1.4	8.0
Thiamin, mg/kg	3.7	4.1	2.5
Vitamin B_6 mg/kg	6.2	6.0	1.0
Vitamin B_{12} μg/kg			1.0
Ascorbic acid[d]			10.0

Adapted from National Research Council (1988).

[a]The requirement is unknown but is met by the sulfur from methionine and cystine.

[b]The requirement is generally met by microbial synthesis.

[c]The niacin in cereal grain is unavailable.

[d]The requirement is met by metabolic synthesis.

Diet Formulation by Computer

Virtually all commercial diets are now formulated by computer using linear programming techniques. **Linear programming** involves the simultaneous solution of many linear equations to determine the optimum allocation of feed ingredients to meet an objective, which usually is to satisfy a set of nutrient requirements. Generally, the optimum allocation means the least expensive selection of ingredients. Thus most diets are formulated on a least-cost basis. In essence, the computer calculates the combination of available ingredients that will meet all the specified requirements at the lowest cost. Many computer programs for diet formulation are available from commercial sources. The selection of the best one to use is based on the computer available and the particular objectives desired.

ECOLOGICAL NUTRITION AND NUTRIENT MANAGEMENT

Traditionally, animal nutritionists have been concerned with meeting the nutrient requirements of livestock with optimal effects on economic returns. A new factor has come into play, however. With the increasing size of livestock enterprises and the concentration of large numbers of animals on small land areas, the environmental effects of nutrients excreted has become important, especially with regard to **nitrogen and phosphorus excretion.** Pollution of surface water and groundwater with these elements is of major concern. In many parts of the world, livestock production is being phased down by legislative edict to reduce pollution. An important aspect of diet formulation is the balancing of diets to minimize the excretion of nitrogen, phosphorus, potassium and copper. To minimize nitrogen excretion, diets should be balanced with respect to amino-acid requirements, and excess dietary protein should not be fed. Phosphorus excretion can be reduced by dietary addition of phytase (see Chapter 7) to increase bioavailability of organic phosphorus in feeds.

Aspects of "ecological nutrition" have been reviewed by Tamminga (1996) and Van Horn et al. (1996).

QUESTIONS AND STUDY GUIDE

1. The following grains are available for use at a feedmill. Their costs per ton are also shown. If you were formulating a ration for laying hens, which of these grains would be the best choice? Corn ($134), milo ($128), barley ($123), or wheat ($131). Calculate the cost per megacalorie of ME for each grain.
2. In the example of a swine grower diet that was formulated, it was specified that 15 percent crude protein was needed, but a requirement for energy was not specified. Why not?
3. The following ingredients, and their cost per ton, are available for use in formulating a grower diet for pigs: soybean meal ($158), cottonseed meal ($139), nondecorticated safflower meal ($65), and sesame meal ($116). Which protein supplement is the best buy? Defend your choice.

4. A rabbit raiser is attempting to formulate a 15 percent protein maintenance diet using alfalfa meal, wheat middlings, soybean meal, and cottonseed meal as the only sources of energy and protein. Each time he puts the information into his microcomputer, the computer displays "no answer possible." What is the problem?
5. Under conditions of very cold environmental temperatures, should the percent of protein in animal diets be increased, decreased, or unchanged for maximum efficiency of protein utilization? Why?
6. If a simple mixture of corn and soybean meal, as shown in Table 11–2, were fed to growing pigs, what deficiency signs would you expect to see? Why?
7. Gross energy values of feedstuffs are of virtually no use as indicators of the energy value of feeds. Explain why. Why are GE values of feedstuffs often measured, if they are not useful expressions of feed energy?

REFERENCES

National Research Council. 1982. *United States-Canadian Tables of Feed Composition.* Washington, DC: National Academy Press.

______ . 1988. *Nutrient Requirements of Swine.* Washington, DC: National Academy Press.

Preston, T.R. and R.A. Leng. 1987. *Matching Ruminant Production Systems with Available Resources in the Tropics and Sub-Tropics.* Armidale, Australia: Penambul Books.

Tamminga, S. 1996. A review on environmental impacts of nutritional strategies in ruminants. *J. Anim. Sci.* 74:3112–3124.

Van Horn, H.H., G.L. Newton, and W.E. Kunkle. 1996. Ruminant nutrition from an environmental perspective: Factors affecting whole-farm nutrient balance. *J. Anim. Sci.* 74:3082–3102.

CHAPTER 12

Feed Manufacturing and Processing

Objectives

1. To describe the general processes involved in feed manufacturing
2. To discuss feed quality and regulation of additives
3. To describe nonnutritive characteristics of ingredients that are important in feed milling

FEED MANUFACTURING

Feed manufacturing is the process of converting ingredient raw materials (i.e., feedstuffs) into balanced diets that are then sold to producers of livestock and other animals. In many countries, the terms "feed compounding" and "compound feeds" are used. Manufactured feeds are produced in feed mills that have equipment to process feedstuffs (e.g., grinding and extruding) for mixing in the desired proportions and for mixing the ingredients to produce the finished product. Often the mixed feed is pelleted, or it may be marketed as a meal-type (mash) feed. In modern animal agriculture, swine and poultry are usually fed manufactured feeds. With swine, it is a common practice for farmers to purchase a mixed supplement to be added to their own grain and mixed at the farm. Dairy cattle are fed manufactured concentrate feed along with hay and/or silage. Beef cows and sheep are primarily grazing animals and generally receive little if any manufactured feed. Calves and lambs may be fed supplementary commercial feed. Feedlot cattle are usually fed diets mixed at the feedlot and so are not major consumers of manufactured feed.

Specialty feeds are a significant item in the feed industry. These include horse, rabbit, llama, ratite, and pet foods, calf milk replacers, and feeds manufactured for sale in retail feedstores to people who raise a few animals as a hobby. For example, a pelleted calf grower ration can be purchased in bags at a local feedstore. A commercial beef producer would not generally find use of this feed economically feasible, whereas a part-time or hobby farmer may use it with less concern for or probably an unawareness of the fact that it may be uneconomical. Thus, this type of feed may be regarded as a specialty item also.

Modern feed mills are largely computer controlled (Fig. 12.1). The process begins with company nutritionists who computer-formulate diets, using NRC or other recognized nutrient requirement figures, tables of feed composition, and current prices of ingredients. Many diets are least-cost formulas, in which the ingredients are selected to meet the prescribed nutrient requirement figures at the lowest cost. A least-cost formula is not necessarily optimal. Even though it provides nutrients in adequate quantities, there may be other factors, such as palatability and physical texture, that reduce animal performance. The informed

FIGURE 12.1 The control panel of an automated computerized feed mill. (Courtesy of R.A. Swick.)

judgment of an experienced animal nutritionist is an essential component of diet formulation. A discussion of some methods of formulating optimal return and least-cost-gain diets is given by Hertzler et al. (1988). The computer-formulated diet formula is then put into the mill's computer system. The operator provides information as to the quantity of feed desired, mixing instructions, and so on. Ingredients are weighed out automatically, the feed mixed, pelleted (if necessary), and discharged into bulk trucks or bags. This automated process greatly reduces or eliminates human error, which can be a major problem in nonautomated mills.

Feed manufacturing is subject to numerous regulations for the protection of the customer and the consumer. **Feed labeling** requires various information, including a list of ingredients and a guaranteed analysis. The following information is required by law in the United States.

1. The net weight.
2. The product name and the brand name, if any, under which the commercial feed is distributed.
3. Guaranteed analysis as state regulations determine to advise the user of the composition of the feed or to support claims made in the labeling. In all cases, the substances or elements must be determinable by laboratory methods, such as the methods published by the Association of Official Analytical Chemists (AOAC).
4. The common or usual name of each ingredient used in the manufacture of the commercial feed. Some states may permit the use of a collective

term for a group of ingredients that perform similar function; or regulations may exempt commercial feeds, or any group thereof, from the requirement of an ingredient statement if the authorities find that such statement is not required in the interest of consumers.

5. The name and principal mailing address of the manufacturer or the person responsible for distributing the commercial feed.
6. Adequate directions for use for all commercial feeds containing drugs and for such other feeds as may be required by regulation as necessary for their safe and effective use.
7. Such precautionary statements as regulations require for the safe and effective use of the commercial feed.

The limitations of the information on a **feed tag** should be recognized. There is no information on protein quality (amino acid balance). The digestible or metabolizable energy value of the feed cannot be deduced from the tag, except that the crude-fiber level provides some indication if low-energy ingredients have been used.

The use of **drugs in feeds** is strictly regulated. In 1986, the Food and Drug Administration established regulations called the **Second Generation Medicated Feed Program.** Under Second Generation, the degree of regulation is based on the potential risk to humans from drug residues in animal products. Those drugs with the greatest potential for hazard are subjected to greater regulatory pressure than formerly, whereas those with little danger are less regulated. The Second Generation regulations provide feed companies with options as to the degree of regulation to which they will be subject. They can opt not to use additives that would require them to undergo more intense regulatory pressure.

Under Second Generation, all animal drugs have been placed in one of two categories. A **Category I drug** is one for which there is no withdrawal at the lowest approved use level for each species. Examples are zinc bacitracin, bambermycins, lasalocid, monensin, oxytetracycline, tylosin, and virginiamycin. A **Category II drug** is one in which a **withdrawal period** is required at the lowest use level for at least one species, or which is regulated on a no-residue or zero-tolerance basis because of carcinogenic concern. Examples include carbadox, furazolidone, lincomycin, nitrofurazone, sulfamethazine, and roxarsone. Current U.S. regulations concerning medicated feeds can be obtained from *The Feed Additive Compendium,* published annually by the Miller Publishing Company (see Chapter 1).

Increasing quantities of feed in the United States are being manufactured on farms rather than in commercial feed mills. A principal reason for this trend is the increasing size of livestock operations with large feedlots, dairies, confinement swine operations, and very large broiler and egg complexes. If grain is produced on the farm, on-farm mixing eliminates the cost of transporting it to and from a central mill location. Large livestock and poultry operations may realize cost savings through bulk purchase of commodities directly, rather than going through a commercial feed mill. On the other hand, feed quality is often more variable with farm-mixed feeds, and the economy of scale for efficient use of machinery may be lacking. However, as livestock enterprises continue to consolidate into fewer but larger operations, on-site feed preparation is likely to increase.

FEED PROCESSING

Feed Mixing

Various types of feed mixing equipment are used to mix the ingredients into a homogeneous batch of material. The raw materials should be ground to a similar particle size to avoid separation after mixing. The efficiency of mixing can be influenced by the order of adding ingredients. Ingredients that make up the major part of the feed should be added first, followed by liquid material. Small quantities, such as vitamin and trace mineral preparations, should be added as a **premix,** in which the material is premixed with some of the mixed feed before being added to the mixer. With vertical mixers, any ingredient making up less than 2.5 percent of the total batch should be premixed. A level of 1 percent is acceptable in a horizontal mixer.

Pelleting

Pelleting of feeds (Fig. 12.2) is accomplished by forcing the mixed feed ingredients through a chamber with holes (pellet die). As the extruded material leaves the die, it is cut off by knives to pellets of a predetermined length. There are a number of **advantages of pelleted feeds.** Pelleting increases the bulk density of feeds, which reduces the volume of storage and transportation space needed, and it may increase feed intake because more weight of feed occupies a given volume. Pelleting prevents animals from sorting ingredients; they must consume the entire mixed feed. Many animals prefer pelleted feeds over the same mixture fed in a mash form. Pelleting reduces dustiness, improving the feed's acceptability to animals. On the negative side, pelleting requires a large amount of electrical energy and so adds an appreciable cost (approximately 10 percent) to the feed.

One of the main concerns with pelleted feed is obtaining good **pellet quality.** If the pellets crumble readily or contain a lot of dust (fines), customers will complain. Pellet quality can be maintained by the use of **pellet binders** (see Chapter 8) and by the selection of ingredients that will form good pellets. The characteristics of an ingredient that control how it will react in pelleting are called its **functional properties.** The functional properties of starches and proteins are of major importance in determining the pelleting quality of feed ingredients (Wood, 1987), especially with regard to how they react under conditions of moist heat (steam).

Normally the feed is **preconditioned** with steam before entering the pellet mill. Steam releases natural adhesive properties in feeds to facilitate pelleting. It softens ingredient particles so that they can more readily bind with each other under pressure. Heat and moisture cause starch **gelatinization,** helping to bind the particles together, as well as improving starch digestion. In general, fine grinding of ingredients improves pellet quality by increasing the surface area to allow better moisture and heat penetration during conditioning.

Some ingredients such as wheat and wheat-processing by-products contain endosperm proteins with good functional properties for pelleting. **Wheat gluten,** when moist, has a gooey consistency, helping to bind the feed together. In turn, pelleting improves animal utilization of wheat by-products (Skoch et al., 1983a, b). The endosperm proteins of triticale, rye, and barley also react with

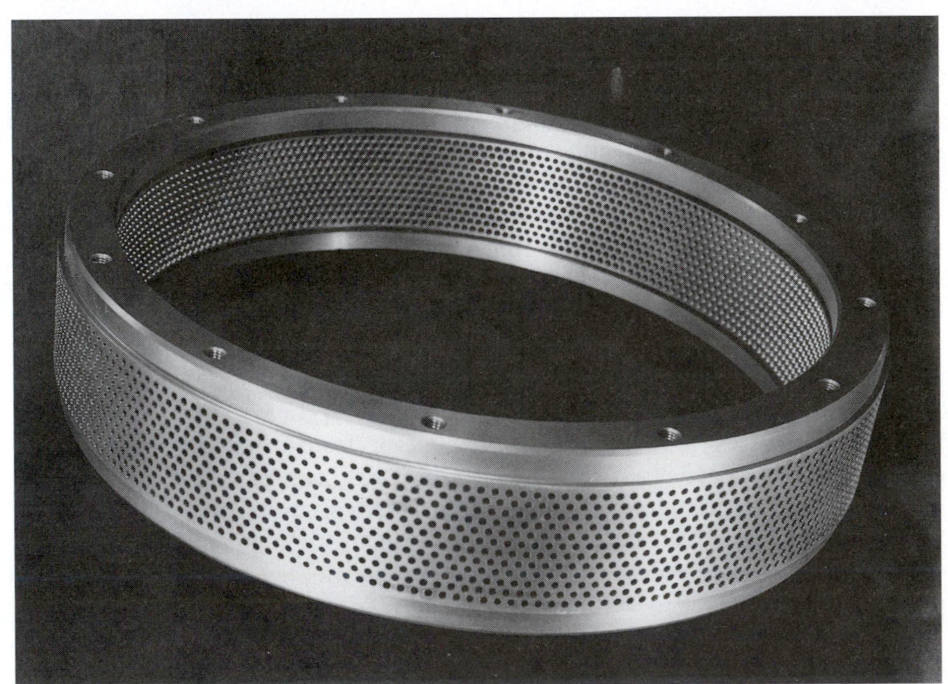

FIGURE 12.2 A commercial pellet mill (top). The cover of the die is swung open in this picture. The die (bottom) contains holes of a particular size through which the mixed feed is forced. Pellet diameter is determined by the size of the holes in the die, and pellet length can be adjusted according to the setting of the knives that slice the pellets off the die. During operation, the pellet die is rotating. (Courtesy of Sprout-Waldron Division, Koppers Company, Inc.)

water to increase viscosity, but those of corn, sorghum, millet, rice, and oats do not. The glucans and pentosans in barley, rye, and oats have viscous properties when wet, improving pellet quality. Thus endosperm proteins, starch, and soluble fiber have independent and additive properties in improving pellet stability. The pelleting qualities of various feedstuffs have been reviewed in depth by Thomas et al. (1998).

High levels of added **fat** coat the feed particles and interfere with the formation of hydrophilic attachments between ingredients, thus reducing pellet quality. Added fats at levels of more than 5 percent tend to cause pellet crumbling. **Molasses** increases the pelletability of feeds. It is often injected directly into the conditioning chamber; molasses is less likely to cause a buildup on the equipment or to cause plugging when it is warm. The best way to get an even distribution of molasses is to inject it with steam in the form of very fine droplets.

Some ingredients are heat sensitive and may caramelize and harden in the mill. Sucrose, milk sugar, whey, and dried milk powder will begin to caramelize at approximately 140°F. They become hard and glassy and plug the equipment.

As pellets leave the pellet mill (the die) they are moved over a shaker to remove fines and through a dryer to remove excess moisture. They may be sprayed with molasses or fat at this point also.

Pelleting may improve the utilization of diets containing bulky, low-density ingredients, such as wheat milling by-products. Patience et al. (1977) noted that pelleting improved feed conversion and energy digestibility of swine diets containing wheat millfeed (shorts). The increased bulk density of the pelleted diets was advantageous in feed handling and storage. Skoch et al. (1983b) observed similar results with swine diets containing wheat middlings, whereas only slight benefits occurred with pelleting of a corn-soy diet (Skoch et al., 1983a).

Bioavailability of nutrients can be influenced by pelleting. For example, the availability of phytate phosphorus in grains is increased by steam pelleting (Bayley et al., 1975). On the other hand, there may be destruction of heat-labile nutrients, such as vitamin A.

With animals that readily consume a diet in either the mash or pelleted form, pelleting is generally not economically advantageous but does offer some other benefits, such as less dust and easier handling. For animals that do not readily consume a mash diet, such as rabbits, pelleting the feed is a necessity.

Other Processing Methods

Grains and Plant Protein Sources Most feedstuffs are subject to processing methods of some sort. Rarely are whole grains or seeds fed, except in a few instances, such as with seed for caged birds (canaries, parrots, finches). Common processes include grinding in a hammermill, dry or steam rolling, and flaking. Breakdown of the grain structure aids in digestion, particularly in species that do not masticate their feed finely. If horses or cattle are fed whole grains, for example, a significant portion will pass through the gut undigested, whereas sheep masticate their feed more finely so most grains and seeds are subject to digestion. As a general rule, coarse grinding of plant energy and protein sources is preferable to fine grinding. Although finely ground particles have a high surface area exposed to digestive enzymes, there may be digestive disturbances with finely ground feed particles, as well as respiratory problems from dust. In swine, finely ground grains promote **stomach ulcers.** Large particles in the digesta are useful in promoting normal gut motility.

A **particle size** of 600 μm or slightly less is optimal for corn in diets for swine (Wondra et al., 1995). Owens et al. (1997) have reviewed grain processing for feeding feedlot cattle. **Steam rolling** or flaking of corn, milo, and wheat improves feed efficiency compared to the dry, rolled grains. **Steam flaking** of grains tends to give higher daily gains. Coarse flaking is preferable to fine flaking.

Popped, Micronized, and Extruded Grains Popped corn for feedlot use is produced by use of dry heat, causing moisture in the grain to expand, rupturing the endosperm. **Popping** increases starch digestibility in the rumen. **Micronized grains** are produced in a similar manner as popping, with the heat provided as infrared energy in what are essentially huge microwave ovens.

Grain **tempering** (conditioning) involves pretreatment with added moisture prior to further processing such as rolling or flaking. Surfactants such as yucca extract are often added to improve the dispersion of moisture throughout the grain (Wang et al., 2003). Tempering increases the efficiency (increased output with lower energy requirements) of the subsequent processing procedure.

Most plant protein sources contain heat-labile toxic factors that must be destroyed by heat treatment, as discussed in Chapter 4. Usually this is done at the site of production of the commercial product. Soybean meal and cottonseed meal, for example, are heated at the oil extraction plant. In feed mills and on the farm, oilseeds can be processed by extrusion. **Extruders** are machines in which soybeans or other oilseeds are forced through a tapered die (Fig. 12.3). The frictional pressure causes sufficient heating to inactivate many toxins. Extruded (full-fat) soybeans are a common high-energy ingredient in the feed trade.

When proteins are heat-treated, protein and amino acid bioavailability may be reduced. Heating may cause a substantial loss of available lysine as a result of the **browning** or **Maillard reaction.** The browning reaction involves a reaction between the free amino group on the side chain of lysine with a reducing sugar to form brown, indigestible polymers. Under controlled conditions, browning can be used to treat proteins advantageously to reduce their rumen degradability (Cleale, 1987).

Expander conditioning, or **high-shear conditioning,** is a relatively new concept in feed manufacturing. Expanders are similar to extruders and provide high-temperature, short-time cooking of the feed before it is pelleted. Expanding increases digestibility of carbohydrates, hydrolyzes starch, and inactivates heat-labile toxins and lipases that cause fat oxidation and rancidity. The process kills most pathogenic and spoilage microbes. Expander-pelleting increases pellet durability and pellet hardness (van der Poel et al., 1997, 1998).

Roughages Roughages are processed in various ways. Most conserved forage is in the form of hay. **Chopping** or **grinding hay** may improve its utilization by reducing feed wastage. Chopped hay is often more palatable than long hay. With ground hay, the loss of undigested small particles from the rumen may cancel out any benefits from the processing.

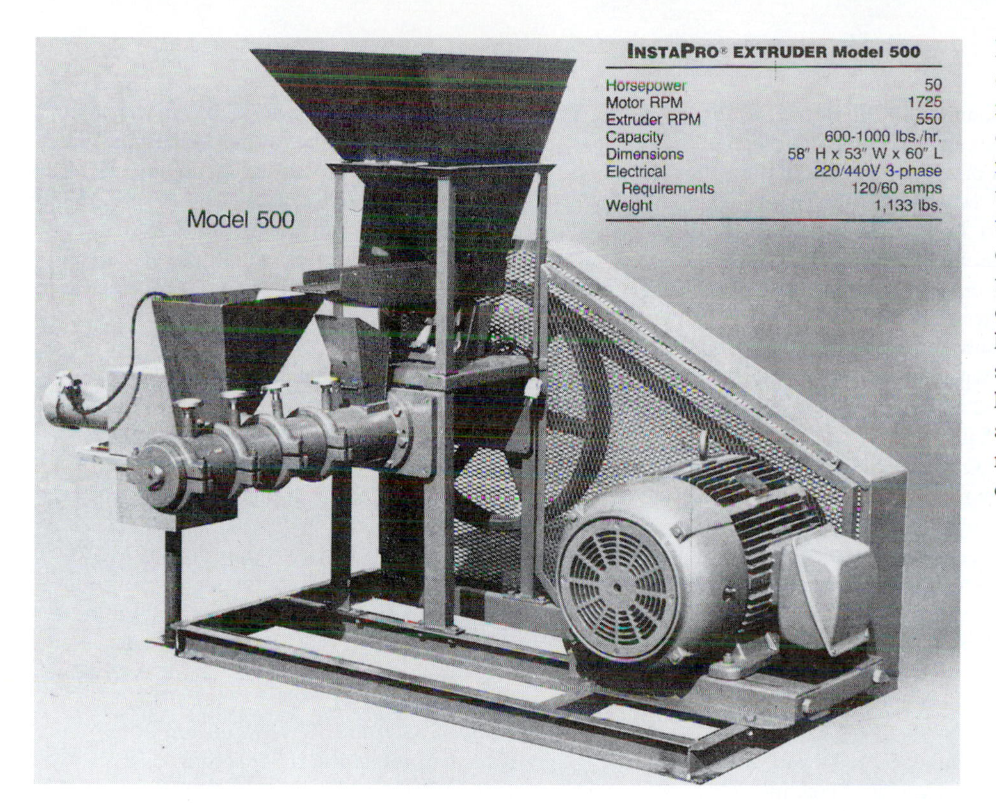

FIGURE 12.3 An extruder used to process raw soybeans and other oilseeds. The seeds are fed into the machine through the hopper at the top and are extruded through a tapered die at the center. The frictional heat generated is sufficient to destroy heat-labile toxins such as trypsin inhibitors in raw soybeans. (Courtesy of Triple "F" Feeds.)

Other processes used to some extent with roughages include the preparation of **cubes** and **wafers.** Hay is compressed through a machine to compact it into large cubes or wafers. The process is expensive and appears to offer little advantage over traditional means of forage handling.

QUESTIONS AND STUDY GUIDE

1. Look up the paper by Hertzler et al. (1988). Compare the concepts of linear versus nonlinear programming for diet formulation, as discussed in this paper.
2. What does the term "withdrawal period" mean with respect to feed additives?
3. What are functional properties of feeds? Which do you think would make a better-quality pellet: a wheat-based or a corn-based diet? Why?
4. Why is steam used in feed manufacturing?
5. What effects would you expect low (5 percent) and high (30 percent) levels of dried whey (70 percent lactose) in a feed to have on the pelletability and pellet quality of a feed?
6. How might the Maillard reaction be usefully employed in ruminant nutrition?

REFERENCES

Bayley, H.S., J. Pos, and R.G. Thomson. 1975. Influence of steam pelleting and dietary calcium level on the utilization of phosphorus by the pig. *J. Anim. Sci.* 46:857–863.

Cleale, R.M., IV, T.J. Klopfenstein, R.A. Britton, L.D. Satterlee, and S.R. Lowry. 1987. Induced non-enzymatic browning of soybean meal. III. Digestibility and efficiency of protein utilization by ruminants of soybean meal treated with xylose or glucose. *J. Anim. Sci.* 65:1327–1335.

Hertzler, G., D.E. Wilson, D.D. Loy, and G.H. Rouse. 1988. Optimal beef cattle diets formulated by nonlinear programming. *J. Anim. Sci.* 66:1115–1123.

Owens, F.N., D.S. Secrist, W.J. Hill, and D.R. Gill. 1997. The effect of grain source and grain processing on performance of feedlot cattle: A review. *J. Anim. Sci.* 75:868–879.

Patience, J.F., L.G. Young, and I. McMillan. 1977. Utilization of wheat shorts in swine diets. *J. Anim. Sci.* 45:1294–1301.

Skoch, E.R., K.C. Behnke, C.W. Deyoe, and S.F. Binder. 1981. The effect of steam-conditioning rate on the pelleting process. *Anim. Feed Sci. Tech.* 6:83–90.

Skoch, E.R., S.F. Binder, C.W. Deyoe, G.L. Allee, and K.C. Behnke. 1983a. Effects of pelleting conditions on performance of pigs fed a corn-soybean meal diet. *J. Anim. Sci.* 57:922–928.

Skoch, E.R., S.F. Binder, C.W. Deyoe, G.L. Allee, and K.C. Behnke. 1983b. Effects of steam pelleting conditions and extrusion cooking on a swine diet containing wheat middlings. *J. Anim. Sci.* 57:929–935.

Thomas, M. and A.F.B. van der Poel. 1996. Physical quality of pelleted animal feed. 1. Criteria for pellet quality. *Anim. Feed Sci. Tech.* 61:89–112.

Thomas, M., T. van Vliet, and A.F.B. van der Poel. 1998. Physical quality of pelleted animal feed. 3. Contribution of feedstuff components. *Anim. Feed Sci. Tech.* 70:59–78.

Thomas, M., D.J. van Zuilichem, and A.F.B. van der Poel. 1997. Physical quality of pelleted animal feed. 2. Contribution of processes and its conditions. *Anim. Feed. Sci. Tech.* 64:173–192.

van der Poel, A.F.B., A. Schoterman, and M.W. Bosch. 1998. Effect of expander conditioning and/or pelleting of a diet on the ileal digestibility of nutrients and on feed intake after choice feeding of pigs. *J. Sci. Food Agric.* 76:87–90.

van der Poel, A.F.B., H.M.P. Fransen, and M.W. Bosch. 1997. Effect of expander conditioning and/or pelleting of a diet containing tapioca, pea and soybean meal on the total tract digestibility in growing pigs. *Anim. Feed Sci. Tech.* 66:289–295.

Wang, Y., D. Greer, and T.A. McAllister. 2003. Effects of moisture, roller setting, and saponin-based surfactant on barley processing, ruminal degradation of barley, and growth performance by feedlot steers. *J. Anim. Sci.* 81:2145–2154.

Wondra, K.J., J.D. Hancock, K.C. Behnke, R.H. Hines, and C.R. Stark. 1995. Effects of particle size and pelleting on growth performance, nutrient digestibility, and stomach morphology in finishing pigs. *J. Anim. Sci.* 73:757–763.

Wood, J.F. 1987. The functional properties of feed raw materials and their effect on the production and quality of feed pellets. *Anim. Feed Sci. Tech.* 18:1–17.

PART IV

FEEDING AND NUTRITION OF LIVESTOCK AND SPECIALTY ANIMALS

Production of champion animals and profitable livestock requires a sound nutritional program. (Courtesy of L. A. Lawrence.)

For most of the species discussed in this section, entire books are available on feeding and nutrition. The intent of these chapters is to distill the basics of unique production needs, nutritional requirements, and nutritional problems into a discussion that will provide a good general background on the basic principles of adequate feeding of various animal species. Thus, it is not intended that the specialist should seek no further, but it is hoped that these chapters will point the way to more comprehensive information.

This section deals with virtually all of the major and minor domestic livestock and poultry species. Newer areas have received more emphasis than might be deemed necessary by their present economic importance. It is relatively easy to find information on nutritional needs of dairy cattle, turkeys, beef cattle, and pigs, but it is much more difficult to find readily accessible information on the feeding and nutrition of fish, exotic and zoo animals, ostriches, llamas, and rabbits, for example.

The National Research Council series on the *Nutrient Requirements of Domestic Animals* covers all of the major species. The most recent NRC publications (e.g., dairy cattle, horse) have computer diskettes containing software programs that allow the calculation of nutrient requirements and intakes for animals of different size, production levels, or activity function. As revised editions of the NRC publications are released, they have become larger and more complex. The approach taken here is to present representative tables that illustrate the type of information available and to provide requirement data for use in student formulation of diets. Therefore, rather than attempting the impossible task of keeping the book current with the diet formulation and feed composition software continually becoming available, background information on each species is provided to establish a foundation upon which to build as dictated by personal interest or class orientation.

Part Objectives

1. To understand the nutrition and feeding of each species as follows: nutrient requirements for specific productive stages, feeding programs, and special or unique nutritional problems

CHAPTER 13

Feeding and Nutrition of Swine

T. L. Veum and P. R. Cheeke

Objectives

1. To describe nutritional needs of swine
2. To describe common types of swine diets
3. To describe special nutritional problems of swine:
   - gastric ulcers
   - leg disorders
   - mycotoxicoses

In the first half of the twentieth century, most swine in the United States were raised in outdoor systems on pasture and supplemented with concentrate feeds. Pasture and insects made a significant contribution to their nutritional needs, and sunshine provided adequate vitamin D activity. During the 1960s and 1970s there was a rapid shift to "complete or **total confinement" production** of swine, mainly for size economics and management considerations. Most swine are now raised in complete confinement systems, except for a recent interest by a number of producers to produce "organically grown pigs" outdoors for a niche market. Confinement reared pigs are totally dependent on the rations provided without any opportunity to obtain nutrients from pasture or other sources for their total life-cycle nutrition. In swine diet formulation, the nutrients or nutrient categories of most concern are energy (DE or ME), the limiting essential amino acids, calcium, phosphorus, trace minerals, and vitamins. High-energy diets based on grains (usually corn) are used. Because nutrient requirements vary with age, gender, and productive function, it is desirable to alter diet composition for different stages of the life-cycle from weaning to slaughter weight. The diet changes made about every 1 or 2 weeks for increasing body weight are called "**phase feeding.**" At about 50 kg body weight, barrows are usually fed separately from gilts (split-sex feeding) because gilts are leaner and have higher amino acid requirements than those of barrows. Baby pigs have the highest nutritional requirements, which decrease as growth proceeds. Lactation is also a nutritionally demanding process for a good milking sow nursing a large litter. Adult animals under maintenance conditions have the lowest nutrient requirements.

NUTRIENT REQUIREMENTS

Energy Requirements

Energy requirements for swine are expressed as digestible energy (DE) or metabolizable energy (ME), and the requirements for pigs from weaning to slaughter weight are shown in Table 13–1 (NRC, 1998), and the requirements for barrows and gilts with split-sex feeding are shown in Table 13–2. Values for ME

TABLE 13–1. Dietary Energy and Amino Acid Requirements of Growing Pigs Fed Ad Libitum (90% dry matter)[a]

	Body Weight (kg)					
	3–5	*5–10*	*10–20*	*20–50*	*50–80*	*80–120*
Average weight in range (kg)	4	7.5	15	35	65	100
DE content of diet (kcal/kg)	3,400	3,400	3,400	3,400	3,400	3,400
ME content of diet (kcal/kg)[b]	3,265	3,265	3,265	3,265	3,265	3,265
Estimated DE intake (kcal/day)	855	1,690	3,400	6,305	8,760	10,450
Estimated ME intake (kcal/day)[b]	820	1,620	3,265	6,050	8,410	10,030
Estimated feed intake (g/day)	250	500	1,000	1,855	2,575	3,075
Crude protein (%)[c]	26.0	23.7	20.9	18.0	15.5	13.2
	Amino acid requirements[d]: Apparent ileal digestible basis (%)					
Arginine	0.51	0.46	0.39	0.31	0.22	0.14
Histidine	0.40	0.36	0.31	0.25	0.20	0.16
Isoleucine	0.69	0.61	0.52	0.42	0.34	0.26
Leucine	1.29	1.15	0.98	0.80	0.64	0.50
Lysine	1.26	1.11	0.94	0.77	0.61	0.47
Methionine	0.34	0.30	0.26	0.21	0.17	0.13
Methionine + cystine	0.71	0.63	0.53	0.44	0.36	0.29
Phenylalanine	0.75	0.66	0.56	0.46	0.37	0.28
Phenylalanine + tyrosine	1.18	1.05	0.89	0.72	0.58	0.45
Threonine	0.75	0.66	0.56	0.46	0.37	0.30
Tryptophan	0.22	0.19	0.16	0.13	0.10	0.08
Valine	0.84	0.74	0.63	0.51	0.41	0.32
	Amino acid requirements[d]: Total basis (%)					
Arginine	0.59	0.54	0.46	0.37	0.27	0.19
Histidine	0.48	0.43	0.36	0.30	0.24	0.19
Isoleucine	0.83	0.73	0.63	0.51	0.42	0.33
Leucine	1.50	1.32	1.12	0.90	0.71	0.54
Lysine	1.50	1.35	1.15	0.95	0.75	0.60
Methionine	0.40	0.35	0.30	0.25	0.20	0.16
Methionine + cystine	0.86	0.76	0.65	0.54	0.44	0.35
Phenylalanine	0.90	0.80	0.68	0.55	0.44	0.34
Phenylalanine + tyrosine	1.41	1.25	1.06	0.87	0.70	0.55
Threonine	0.98	0.86	0.74	0.61	0.51	0.41
Tryptophan	0.27	0.24	0.21	0.17	0.14	0.11
Valine	1.04	0.92	0.79	0.64	0.52	0.40

Adapted from NRC, *Nutrient Requirement of Swine* (1998).

[a]Mixed gender (1:1 ratio of barrows to gilts) of pigs with high-medium lean growth rate (325 g/day of carcass fat-free lean) from 20 to 120 kg body weight.

[b]Assumes that ME is 96% of DE. In corn-soybean meal diets of these crude-protein levels, ME is 94–96% of DE.

[c]Crude protein levels apply to corn-soybean meal diets. In 3–10 kg pigs fed diets with dried plasma and/or dried milk products, protein levels will be 2–3% less than shown.

[d]Total amino acid requirements are based on the following types of diets: 3–5 kg pigs, corn-soybean meal diet that includes 5% dried plasma and 25–50% dried milk products; 5–10 kg pigs, corn-soybean meal diet that includes 5 to 25% dried milk products; 10–120 kg pigs, corn-soybean meal diet.

TABLE 13–2. Dietary Energy and Amino Acid Requirements of Barrows and Gilts of Different Lean Growth Rates Fed Ad Libitum (90% dry matter)[a]

Body weight range	*50–80 kg Body weight*						*80–120 kg Body Weight*					
Lean gain (g/day) ***Gender***	***300 Barrow***	***300 Gilt***	***325 Barrow***	***325 Gilt***	***350 Barrow***	***350 Gilt***	***300 Barrow***	***300 Gilt***	***325 Barrow***	***325 Gilt***	***350 Barrow***	***350 Gilt***
Average weight in range (kg)	65	65	65	65	65	65	100	100	100	100	100	100
DE content of diet (kcal/kg)	3,400	3,400	3,400	3,400	3,400	3,400	3,400	3,400	3,400	3,400	3,400	3,400
ME content of diet (kcal/kg)[b]	3,265	3,265	3,265	3,265	3,265	3,265	3,265	3,265	3,265	3,265	3,265	3,265
Estimated DE intake (kcal/day)	9,360	8,165	9,360	8,165	9,360	8,165	11,150	9,750	11,150	9,750	11,150	9,750
Estimated ME intake (kcal/day)[b]	8,985	7,840	8,985	7,840	8,985	7,840	10,705	9,360	10,705	9,360	10,705	9,360
Estimated feed intake (g/day)	2,750	2,400	2,755	2,400	2,755	2,400	3,280	2,865	3,280	2,865	3,280	2,865
Crude protein (%)[c]	14.2	15.5	14.9	16.3	15.6	17.1	12.2	13.2	12.7	13.8	13.2	14.4
				Amino acid requirements[d]: Apparent ileal digestible basis (%)								
Arginine	0.19	0.21	0.21	0.24	0.23	0.26	0.12	0.13	0.13	0.15	0.15	0.17
Histidine	0.17	0.20	0.19	0.21	0.20	0.23	0.14	0.15	0.15	0.17	0.16	0.18
Isoleucine	0.29	0.34	0.31	0.36	0.34	0.39	0.23	0.26	0.24	0.28	0.26	0.30
Leucine	0.56	0.64	0.60	0.69	0.65	0.74	0.43	0.50	0.47	0.53	0.50	0.57
Lysine	0.53	0.61	0.57	0.66	0.61	0.71	0.41	0.47	0.44	0.51	0.47	0.54
Methionine	0.15	0.17	0.16	0.18	0.17	0.20	0.12	0.13	0.13	0.14	0.13	0.15
Methionine + cystine	0.31	0.36	0.34	0.39	0.36	0.41	0.25	0.29	0.27	0.31	0.29	0.33
Phenylalanine	0.32	0.36	0.34	0.39	0.37	0.42	0.24	0.28	0.26	0.30	0.28	0.32
Phenylalanine + tyrosine	0.50	0.58	0.54	0.62	0.58	0.67	0.39	0.45	0.42	0.49	0.45	0.52
Threonine	0.32	0.37	0.35	0.40	0.37	0.43	0.26	0.30	0.28	0.32	0.30	0.34
Tryptophan	0.09	0.10	0.10	0.11	0.10	0.12	0.07	0.08	0.07	0.09	0.08	0.09
Valine	0.36	0.41	0.38	0.44	0.41	0.47	0.28	0.32	0.30	0.34	0.32	0.37
				Amino acid requirements[d]: Total basis (%)								
Arginine	0.24	0.27	0.26	0.29	0.28	0.32	0.16	0.18	0.18	0.20	0.19	0.22
Histidine	0.21	0.24	0.23	0.26	0.24	0.28	0.17	0.19	0.18	0.20	0.19	0.22
Isoleucine	0.36	0.41	0.39	0.45	0.42	0.48	0.29	0.33	0.31	0.35	0.33	0.37
Leucine	0.61	0.71	0.67	0.77	0.72	0.83	0.46	0.54	0.50	0.58	0.54	0.63
Lysine	0.67	0.76	0.72	0.82	0.77	0.88	0.53	0.60	0.57	0.64	0.60	0.69
Methionine	0.17	0.20	0.19	0.21	0.20	0.23	0.14	0.15	0.15	0.17	0.16	0.18
Methionine + cystine	0.38	0.44	0.41	0.47	0.44	0.50	0.31	0.35	0.33	0.38	0.35	0.40
Phenylalanine	0.38	0.44	0.41	0.47	0.44	0.51	0.29	0.34	0.32	0.36	0.34	0.39
Phenylalanine + tyrosine	0.61	0.70	0.65	0.75	0.70	0.80	0.48	0.54	0.51	0.59	0.55	0.63
Threonine	0.44	0.50	0.47	0.54	0.51	0.58	0.36	0.41	0.38	0.44	0.41	0.46
Tryptophan	0.12	0.14	0.13	0.15	0.14	0.16	0.10	0.11	0.10	0.12	0.11	0.13
Valine	0.45	0.51	0.48	0.55	0.52	0.59	0.35	0.40	0.38	0.43	0.40	0.46

Adapted from NRC, *Nutrient Requirements of Swine* (1998).

[a]Average lean growth rates of 300, 325, and 350 g/day of carcass fat-free lean represent pigs with medium, high-medium, and high lean growth rates from 20 to 120 kg body weight.

[b]Assumes that ME is 96% of DE.

[c]Crude-protein and total amino acid requirements are based on a corn-soybean meal diet.

[d]Estimated from the growth model.

are usually 94 to 97 percent of the DE figures. The main energy source in growing-finishing swine and sow diets is usually starch from cereal grains. Except for the neonatal pig, starch is completely digested in the small intestine. Intestinal lactase production is high at birth, but pigs younger than 3 weeks have insufficient intestinal sucrase and pancreatic amylase secretion to digest sucrose or starch at birth. Swine produced in industrialized countries are usually fed diets high in energy and low in fiber. Diet formulations are based on cereal grains (see Chapter 2), primarily corn in the United States, as the main dietary energy source. In countries where the growing season is too short for corn (e.g., Scandinavia), other grains such as sorghum, barley, wheat, and rye may be used. The commercial availability of carbohydrate enzymes (e.g., β-glucanase) that increase the dry matter and energy digestibilities of barley and oats has made those grains more competitive for swine feeding. In tropical countries, cassava meal and sweet potato meal are high-energy feedstuffs available for swine feeding (see Chapter 3)

In many developing countries, swine are fed grass and aquatic weeds (e.g., in China) and fibrous agricultural by-products. The hindgut of the pig is relatively large, and microbial digestion of **fibrous feedstuffs** can be of significance. When the crude-fiber content of the diet exceeds 10 to 15 percent, caloric intake may be reduced because of excessive bulk or reduced palatability (NRC, 1998). In general, pigs and other nonruminants respond to diet dilution with fiber by increasing feed intake sufficient to meet their caloric requirements until gut capacity becomes limiting. With some feedstuffs such as alfalfa, palatability is reduced and pigs do not consume sufficient feed to compensate for the caloric dilution by fiber. Alfalfa contains constituents (e.g., saponins) that are bitter and unpalatable to the pig (LeaMaster and Cheeke, 1979). Alfalfa meal fiber and other sources of fiber are fermented in the hindgut, producing VFA (Kass et al., 1980), which can provide up to 30 percent of the maintenance energy requirement.

Swine make efficient use of dietary lipids. **Linoleic acid** is a dietary essential (Table 13–3) from which other unsaturated fatty acids (e.g., arachidonic acid) can be synthesized. Theoretically, if pigs eat to meet their energy requirements, substitution of fat for carbohydrate should not increase performance or energy intake. In general, so long as the diet is adequate in protein content, addition of fat tends to improve gains and reduce the ME required per unit of gain (NRC, 1998). Fat metabolism in pigs has been reviewed by Farnworth and Kramer (1987). As discussed in Chapter 3, addition of fat to sow diets prior to farrowing may increase baby pig survival. High-quality animal and vegetable fat by-products are excellent high-energy supplements, especially for growing-finishing pigs and lactating sows. Fat increases diet palability and energy intake, reduces dust in the swine buildings, and reduces wear on automated feeding conveyors. The amount of fat added is limited, usually 1 to 4 percent, to prevent bridging in the storage tanks and feeders. Animal fats, generally the most economical source of fat, are a solid at room temperature and must be heated before mixing in a feed. However, vegetable oils such as soy or corn oils are a liquid at room temperature because of their high content of unsaturated fatty acids. Fat may also be sprayed on pelleted feeds after pelleting, increasing the total amount of fat supplementation that can be used.

Protein and Amino Acid Requirements

Protein and essential amino acid requirements for growing pigs are listed in Table 13–1 as a percentage of the diet, and for split-sex feeding from 50 to 20 kg in Table 13–2. Under practical conditions, lysine, methionine, tryptophan, and threonine are the main amino acids of concern. However, valine may also become limiting in very low protein diets. Metabolically, the L-isomers are required, although in some cases the D-isomer is used by being converted in the liver to the L form. Thus, the relative potencies are: D-methionine, 100 percent of the L form; D-tryptophan, 60 to 70 percent; D-lysine, 0 percent; and D-threonine, 0 percent.

Lysine is usually the first limiting amino acid in diets for swine of all ages. Methionine and threonine may also become limiting in Phase 1 and 2 nursery diets that contain supplemental blood byproducts as protein sources. When crude protein levels are reduced for the purpose of reducing nitrogen excretion in swine urine and feces, and crystalline amino acid supplementation is used to balance the diet for limiting amino acids, tryptophan may also become a limiting amino acid. Lysine, methionine, threonine, and tryptophan are all available as feed supplements, with their use determined by cost relative to protein and (or) environmental (pollution) requirements. Practical lactation diets, however, differ from other swine diets in that valine is the second limiting amino acid after lysine. (Valine currently is not cost effective as a feed supplement, so plant or animal proteins are used to meet the valine requirement. This limits the reduction in crude protein that can be achieved in lactating sow diets for the purpose of reducing nitrogen excretion in sow urine and feces.) See the footnotes in example diet Table 13–4 and 13–8 concerning the limitations in percentage protein reduction in relation to the amino acid supplementation required.

Soybean meal is the plant protein supplement of choice in the United States. The simple corn-soybean meal diet supplemented with minerals and vitamins has become the standard against which all other more complex diet formulations for growing-finishing pigs and sows are evaluated by economics and performance. However, other plant oilseed meal supplements (e.g., canola meal, cottonseed meal, peanut meal) can be used when the economics for their use is favorable and their undesirable characteristics are considered in diet formulation. **Animal protein sources,** such as meat and bone meal and blood meal, are excellent protein (amino acid) sources for swine in the United States where their use has not been banned because of Bovine Spongiform Encephalopathy (BSE, also called *mad cow disease*).

For pigs at weaning, **spray-dried animal plasma** is an expensive supplement that is generally added at 4 to 6 percent of the first (Phase 1) postweaning diet because it stimulates feed intake, enhances immunity, and improves growth performance of pigs weaned at 14 to 17 days of age (Jiang et al., 2000; Jensen et al., 2001: Owusu-Asiedu et al., 2002), the common weaning age on confinement swine farms. Small additions of **blood meal** or blood cells (blood meal minus the plasma) ranging from 2 to 3 percent of the diet, or fish meal at 3 to 5 percent, usually replace the animal plasma in the second postweaning (Phase 2) diet (Kim and Easter, 2001; DeRouchey et al., 2002). In Europe the use of animal by-products in animal feeds has been banned because of several outbreaks of BSE in the late 1990s. However, pigs generally are older when weaned in Europe, reducing the need for high-quality animal proteins at weaning. It has also been demonstrated in the United States that both higher birth weights and higher weaning weights improve postweaning growth performance of pigs (Mahan et al., 1998; Wolter et al., 2002) Therefore, the ingredients selected and the nutrient composition of the protein supplement needed are determined both by age (nutrient requirements) and ingredient cost (least-cost linear programming), resulting in a "best-cost" diet formulation (see Chapter 11).

Improvements in grain genetics, such as high lysine corn and barley and high-oil corn (see Chapter 2), have made these grains attractive for use in swine feeding. There is a significant environmental trend to increase the use of **synthetic amino acids** and lower the level of supplemental protein, especially for growing-finishing pigs. Use of synthetic amino acids will require greater consideration of electrolyte balance in diet formulation (see Chapter 7).

Mineral Requirements

The mineral requirements for growing swine are listed in Table 13–3. For macrominerals, practical grain-oilseed meal diets are deficient in calcium (Ca), phosphorus (P), sodium (Na), and chloride (Cl). **Calcium and P** are essential for normal skeletal growth, bone mineralization, and bone strength. All plant feedstuffs used in swine diets are low in Ca and P. The bioavailability of P in plant sources is very low because of the presence of phytate. Common dietary sources of P, which also provide Ca, include dicalcium phosphate, bone meal, and defluorinated rock phosphate. Once the P requirement is met, then the remaining Ca requirement needs to be met. Ground limestone is an economical source of Ca, and it does not contain P. The requirements for **salt** (NaCl) range from 0.40 percent of the diet for sows and weaning pigs to 0.20 percent for finishing pigs. However, NaCl supplementation usually ranges from 0.25 to 0.50 percent in practice. Even though animal plasma and dried whey contribute Na and Cl to the weaning diet, pig performance is improved when salt is added to increase dietary Cl to 0.38 percent (Mahan et al., 1999).

Micro- or trace minerals are provided using a trace mineral premix that is added to the diet at a small, specified percentage ranging from 0.10 to about 0.30 percent. This premix should contain high-quality salts (sulfates, carbonates, chlorides, etc.) of zinc, iron, copper, and manganese, plus appropriate forms of iodine and selenium. Deficiencies of these nutrients lead to smaller litter size and greater baby pig mortality. The **phytates** in plant ingredients greatly reduce the bioavailability of the natural trace minerals, especially zinc in the presence of soy protein. The **iron requirement** of the baby pig is also increased because of its rapid growth rate, high rate of hemoglobin synthesis, low placental transfer or iron, and low transfer of iron into sow milk. Attempts to improve the iron status of baby pigs by supplementation of gestation and lactation diets with iron have been unsuccessful. Therefore, swine producers routinely administer iron by injection (iron dextran) into the neck muscle of newborn pigs. High dietary copper as **copper sulfate** (100 to 250 ppm copper) is widely used to reduce enteritis and increase daily gain in weanling and growing swine. This use is described in Chapter 8. Zinc oxide is also used as a growth promotant (see Chapter 8). Mineral requirements are reviewed by Mahan (1990) and the NRC (1998).

TABLE 13–3. Dietary Mineral, Vitamin, and Fatty Acid Requirements of Growing Pigs Fed Ad Libitum (90% dry matter)[a]

	Body Weight (kg)					
	3–5	*5–10*	*10–20*	*20–50*	*50–80*	*80–120*
Average weight in range (kg)	4	7.5	15	35	65	100
DE content of diet (kcal/kg)	3,400	3,400	3,400	3,400	3,400	3,400
ME content of diet (kcal/kg)[b]	3,265	3,265	3,265	3,265	3,265	3,265
Estimated DE intake (kcal/day)	855	1,690	3,400	6,305	8,760	10,450
Estimated ME intake (kcal/day)[b]	820	1,620	3,265	6,050	8,410	10,030
Estimated feed intake (g/day)	250	500	1,000	1,855	2,575	3,075
	Requirements (% or amount/kg of diet)					
Mineral elements						
Calcium (%)[c]	0.90	0.80	0.70	0.60	0.50	0.45
Phosphorus, total (%)[c]	0.70	0.65	0.60	0.50	0.45	0.40
Phosphorus, available (%)[c]	0.55	0.40	0.32	0.23	0.19	0.15
Sodium (%)	0.25	0.20	0.15	0.10	0.10	0.10
Chlorine (%)	0.25	0.20	0.15	0.08	0.08	0.08
Magnesium (%)	0.04	0.04	0.04	0.04	0.04	0.04
Potassium (%)	0.30	0.28	0.26	0.23	0.19	0.17
Copper (mg)	6.00	6.00	5.00	4.00	3.50	3.00
Iodine (mg)	0.14	0.14	0.14	0.14	0.14	0.14
Iron (mg)	100	100	80	60	30	40
Manganese (mg)	4.00	4.00	3.00	2.00	2.00	2.00
Selenium (mg)	0.30	0.30	0.25	0.15	0.15	0.15
Zinc (mg)	100	100	80	60	50	50
Vitamins						
Vitamin A (IU)[d]	2,200	2,200	1,750	1,300	1,300	1,300
Vitamin D_3 (IU)[d]	220	220	200	150	150	150
Vitamin E (IU)[d]	16	16	11	11	11	11
Vitamin K (menadione) (mg)	0.50	0.50	0.50	0.50	0.50	0.50
Biotin (mg)	0.08	0.05	0.05	0.05	0.05	0.05
Choline (g)	0.60	0.50	0.40	0.30	0.30	0.30
Folacin (mg)	0.30	0.30	0.30	0.30	0.30	0.30
Niacin, available (mg)[e]	20.00	15.00	12.50	10.00	7.00	7.00
Pantothenic acid (mg)	12.00	10.00	9.00	8.00	7.00	7.00
Riboflavin (mg)	4.00	3.50	3.00	2.50	2.00	2.00
Thiamin (mg)	1.50	1.00	1.00	1.00	1.00	1.00
Vitamin B_6 (mg)	2.00	1.50	1.50	1.00	1.00	1.00
Vitamin B_{12} (μg)	20.00	17.50	15.00	10.00	5.00	5.00
Linoleic acid (%)	0.10	0.10	0.10	0.10	0.10	0.10

Adapted from NRC, *Nutrient Requirements of Swine* (1998).

[a] Pigs of mixed gender (1:1 ratio of barrows to gilts). The requirements of certain minerals and vitamins may be slightly higher for pigs having high lean growth rates (>325 g/day of carcass fat-free lean), but no distinction is made.

[b] Assumes that ME is 96% of DE. In corn-soybean meal diets, ME is 94–96% of DE, depending on crude protein level of the diet.

[c] The percentages of calcium, phosphorus, and available phosphorus should be increased by 0.05 to 0.1 percentage points for developing boars and replacement gilts from 50 to 120 kg body weight.

[d] Conversions: 1 IU vitamin A = 0.344 μg (g retinyl acetate; 1 IU vitamin D_3 = 0.025 μg cholecalciferol; 1 IU vitamin E = 0.67 mg of D-α-tocopherol or 1 mg of DL-α-tocopheryl acetate.

[e] The niacin in corn, grain sorghum, wheat, and barley is unavailable. Similarly, the niacin in by-products made from these cereal grains is poorly available unless the by-products have undergone a fermentation or wet-milling process.

Mineral Utilization Because the minerals in cereal grain oilseed meal-based diets are poorly digested by the pig, the swine industry is using available nutrition technology to increase mineral digestion (absorption) and consequently, to reduce the excretion in manure. **Phytase** enzymes have been developed commercially that will hydrolyze some of the phytate P in low-P grain-soybean meal diets for growing-finishing swine. The addition of 500 units of phytase per kg of diet releases, or makes available, about 0.1 percent P (% on a diet basis) that was bound as phytate in the feed ingredients (Harper et al., 1997; Liu et al., 1997, 1998). Phytase effectiveness is also enhanced by lowering the Ca to total P ratio to 1:1 (Liu et al., 2000). The availability of trace minerals is also increased by phytase supplementation (Adeola et al., 1995; Stahl et al., 1999).

The development of nontransgenic (non-GMO) low-phytic acid grains by USDA-ARS plant geneticists will significantly reduce P excretion by nonruminant animals when this technology reaches the commercial market (Raboy et al., 2000, 2001). Phosphorus excretion by growing swine was reduced 16 to 19 percent by feeding low-phytic acid grains (corn or barley) in diets containing soybean meal (Veum et al., 2001, 2002). The development of a low-phytic acid soybean seed (Wilcox et al., 2000) will greatly reduce the excretion of P and other minerals when fed in combination with a low-phytic acid grain, because regular soybean meal contains a high level of phytic acid. Of course, phytase may be used in combination with one or more low-phytic acid ingredients to increase the available P in the diet until the requirement is met.

Vitamin Requirements

Vitamin requirements expressed as a percentage or amount per kilogram of diet for growing swine are listed in Table 13–3. Commercial vitamin premixes are formulated with crystalline or synthetic vitamins to provide specific amounts of the vitamins known to be limiting or deficient in the feed ingredients. Premixes for growing-finishing swine fed grain-oilseed meal diets in confinement usually contain the fat-soluble vitamins A, D, E, and possibly K. Vitamin K may be added because of the frequent occurrence of mold toxins in cereal grains that interfere with the blood clotting mechanism in swine. The water-soluble vitamins required by growing-finishing pigs include riboflavin, niacin, pantothenic acid, and vitamin B_{12}. Vitamin B_{12} is provided because animal protein ingredients, a natural source of B_{12}, may not be used in growing-finishing diets.

A common belief is that pigs exposed to their feces meet part of the B-complex vitamin requirements by **coprophagy** and have lower vitamin requirements than those raised on slatted floors. However, De Passille et al. (1989) and Bilodeau et al. (1989), in extensive studies of coprophagy, found that floor type and coprophagy had very little influence on the B-complex vitamin requirements of swine.

In addition to the vitamins required for growing-finishing pigs, the premix for gestating and lactating sows should contain choline and folic acid because those vitamins improve reproductive performance. Choline is usually adequate in practical diets fed to growing-finishing pigs, whereas weanling pigs usually require supplemental choline because of their higher requirements.

Folic acid is particularly important in sow nutrition. Folates function as coenzymes in nucleic acid synthesis, so the requirement is highest during periods of rapid cell division, such as during rapid growth of the fetuses. Supplementation of corn-soy diets for sows increases the conception rate and number of pigs born (Lindemann and Kornegay, 1989; Thaler et al., 1989) and reduces embryonic mortality in the first month of gestation, particularly when ovulation rate is high (Tremblay et al., 1989).

Water

Water is required in a larger quantity by swine than any other nutrient. Water functions as a structural element by giving form to the body through cell turgidity, in temperature regulation, transport of nutrients from the site of absorption to the cells of the body tissues, and removal of waste products from cells. Water has a role in facilitating virtually every chemical reaction that takes place in the body, plus lubrication of the joints as a component of synovial fluid, and cushioning for the nerves as cerebrospinal fluid (Thacker, 2001). The water content of the body tissues is highest at birth and declines with age as the percentage of body fat increases, with a significant inverse relationship between body water and fat content (Schmidt et al., 1973). The body water content of the pig remains remarkably constant at a given body weight and fat content. Swine lose body water mainly by urination, followed by defecation, respiration and by evaporation from the skin even though the sweat glands of the pig remain dormant. Therefore, it is essential that swine are provided *ad libitum* access to fresh, high quality drinking water to maximize feed consumption and growth rate (Thacker, 2001). The water requirements for pigs of all ages, and the factors affecting water consumption by pigs have been reviewed by Thacker (2001).

Suckling pigs may consume water even though milk contains 80% water. Water consumption by nursing pigs varies considerably and increases as the temperature in the farrowing rooms increases (Fraser et al., 1988). Pigs weaned at 3 weeks of age consume about 0.5, 1.0 and 1.5 liters daily during weeks 1, 2 and 3 postweaning, respectively (Gill et al., 1986). Feed consumption by weanling and growing-finishing pigs has a positive relationship with water consumption (Barber et al., 1989; Thacker, 2001). The voluntary water intake of growing pigs consuming feed *ad libitum* is about 2.5 units (weight) of water per 1.0 unit of feed (Braude et al., 1957). However, when feed intake was restricted the water intake increased to 3.7 units of water per unit of feed (Cumby, 1986), most likely for "gut fill" to compensate for the low feed allowance. Water consumption increases considerably at high temperatures of 30°C or higher, and pigs may attempt to wet the pen floor with spilled water, urine or feces; and then lay on the wet surface to cool their skin (Close et al., 1971).

Most growing-finishing swine in North America are fed dry feed in self feeders, either as a meal or in pelleted form. However, there is some interest in various low-cost wet feeding systems that usually consist of a feeder with one or more watering devices attached that allows the pigs to determine the amount of water to either consume directly or to run on the feed. These 'wet/dry feeders' are usually designed to limit feed intake to assure that wet feed it not left in the feeder, so adequate feeder space is required to accommodate all the pigs in the pen at one time when the feed is dispensed at regular, controlled intervals. The feeder design must also prevent wicking of water back into the feed storage area of the feeder (Brumm and Gonyou, 2001). The type of drinking device used and the adjustment will also affect water wastage for wet or dry feeding systems (Gill and Barber, 1990). The effect of wet versus dry feeding on feed efficiency has been variable (Brumm and Gonyou, 2001). However, Liu et al. (1997) found that mixing 1 part of a low-phosphorus diet containing the enzyme phytase with 2 parts water, and 'soaking' the slurry for 2 hours prior to feeding doubled the efficacy of the phytase compared with dry feeding for growing pigs.

The water intake of nonpregnant gilts is about 11.5 kg daily. During pregnancy water intake will increase, and may reach 20 kg daily. Lactating sows require a considerable amount of water to produce 8 to 12 kg of milk per day, and to excrete the metabolic acid products in urine. Daily consumption will vary from 12 to 40 liters per day, with a mean of 18 liters daily (Lightfoot, 1978).

SWINE DIETS

Many different diet formulations are required in a complete farrow-to-finish swine production system that includes nursery, grower, finisher, gestation, and lactation diets. Intensive confinement systems generally wean pigs at 14 to 17 days of age, and generally do not provide creep-feed for nursing pigs. Thus, the first exposure to feed, except for sow feed while nursing, is the first weaning diet. The advantages of early weaning are an increase in the number of litters that can be farrowed in the facility annually and an increase in sow productivity (pigs produced per sow annually). Nutrient efficiency is increased by feeding the baby pig directly compared with feeding the sow to produce milk for the pigs. Successful methods of weaning pigs at 3 to 5 days of age and feeding a synthetic milk diet have been developed, but are not widely applied on a practical scale because of the expensive facilities and high degree of management skill required. Adequate sanitation is critical. Early weaning of pigs allows for greater reproductive efficiency in a swine herd because the sows can be rebred sooner, and total production per sow per year is increased. Refer to the text *Swine Nutrition* by Lewis and Southern (2001) for a comprehensive review of all aspects of swine nutrition.

Preweaning Diets

A **prestarter diet** is needed when baby pigs need to be removed from the sow and reared artificially because of sow lactation failure. Prestarter diets have a high protein content (22 to 24 percent) and must contain high-quality readily digestible feedstuffs such as spray-dried animal plasma, spray-dried skim milk, spray-dried whey, lactose as the carbohydrate source, and 1 or 2 percent fat as corn or soybean oil. These ingredients are very palatable and digestible by the baby pig. Because the baby pig lacks intestinal sucrase and pancreatic amylase activities at birth, sucrose (table sugar) and starch should not be added to a prestartar or the first postweaning diet to prevent digestive disturbances and diarrhea (Veum and Odle, 2001). A **starter diet** (20 to 22 percent crude protein) may be used as a creep-feed before weaning when pigs are not weaned until 3 or more weeks of age. Starter diets contain grain (usually corn) and soybean protein (usually soybean meal), a large amount of lactose, and spray-dried animal plasma or spray-dried skim milk. **Creep-feeding** has generally been eliminated in highly intensive production systems that wean pigs at 14–17 days of age because with high milking sows very little creep-feed is consumed before weaning.

Weaning Diets

Weaning is a critical period for the young pig. The stress of weaning accompanied by dietary changes makes the animal very susceptible to **postweaning diarrhea** (enteritis). Starter diets usually are medicated with antibiotics to help control enteritis. As discussed in Chapter 8, other feed additives such as probiotics, organic acids, copper sulfate, and zinc oxide may also be efficacious in controlling postweaning enteric problems. Pigs normally eat very little feed the first 1 to 2 days after weaning. Then when the pigs start eating they may overeat and overload the gut with readily fermented substrate, causing the proliferation of undesirable organisms such as pathogenic strains of *E. coli.* Simple diets composed of only a few ingredients, as opposed to complex diets, reduce enteric problems by reducing feed intake (Ball and Aherne, 1982).

Segregated Early Weaning (SEW) The SEW system is based on weaning the pigs while maternal antibody protection is at its peak, and then moving the pigs to an isolated housing building to prevent disease exposure and activation of the immune system. For the first 2 to 3 weeks of life, the baby pig has passive immunity obtained via maternal antibodies in the mother's colostrum. However, this immunity is limited only to the antigens to which the mother was exposed prior to farrowing. In the SEW system, pigs are weaned at about 14 to 17 days of age and moved to isolated, clean facilities without exposure to any other swine. This weaning method has been adopted as a standard practice by most of the swine industry, combined with the **all-in and all-out facility management concept.** This means the buildings are completely emptied to allow cleanup, sanitizing, and (preferably) 1 day of vacancy prior to refilling the building with same-source and same-age pigs. This SEW system minimizes the young pigs' exposure to diseases (antigens), which also minimizes the activation of their immune system.

Activation of the immune system by exposure to antigens reduces growth performance of the pig, because nutrients are diverted from growth to the synthesis of **cytokines** (see Chapter 8) and other components of the immune system (Dritz et al., 1996; Williams et al., 1997a, b, c). Thus, preventing the activation of a pig's immune system by minimizing its exposure to antigens results in more efficient nutrient utilization for growth. Pigs raised in the SEW system have higher amino acid requirements than conventionally reared pigs (Williams et al., 1997a, b), and the diets should contain protein sources with highly digestible amino acids (Bergstrom et al., 1997). The advantages of the SEW system are maintained throughout the entire growth period (Williams et al., 1997c) with increased growth rate, feed conversion efficiency, and carcass leanness. Therefore, the pigs' level of chronic immune system activation will affect nutrient requirements at all stages of growth.

Nursery Diets

Pigs in the nursery (postweaning) are fed a series of diets called **phase feeding.** This consists of reducing or replacing very expensive, high-quality ingredients with less expensive ingredients by reformulating the diets every 1 or 2 weeks until the final result is a simple and practical corn-soybean meal (grain-oilseed meal) diet. However, as mentioned previously, best-cost formulation may bring in other ingredients to either complement or replace corn or soybean meal on a cost basis.

The first, or Phase 1, diet requires at least one or more very high quality protein sources. Spray-dried animal plasma is usually the protein of choice because of its increase in feed intake and pig performance the first week postweaning compared with that of other proteins. However, because animal plasma is more expensive than other protein sources, its use is generally limited to about 4 to 6 percent of the diet. A whey protein product that is high in protein can either be fed in combination with or completely replace animal plasma in weanling pig diets (Grinstead et al., 2000). Soybean meal added at 20 percent of the Phase 1 diet does not reduce pig performance (Nessmith et al., 1997b). Other good protein sources include spray-dried skim milk, spray-dried whey protein concentrate, and fish meal. Further-processed soy products (soy protein concentrate, isolated soy protein) are generally not better than soybean meal in weanling pig diets. Spray-dried whole whey contains about 70 percent lactose, an essential carbohydrate in the weaning diet, as well as a high-quality whey protein. The first postweaning diet should contain a high amount of lactose (Mahan, 1992). Lactose is reduced in the second diet and eliminated in the third postweaning diet. Crystalline lactose is also available for use as a feed ingredient and can be used instead of dried whey (Mahan, 1993: Nussmith et al., 1997a). More ingredients are used initially at weaning to provide the quality of nutrients (amino acids and carbohydrate) that encourage the consumption of dry feed. The complexity of the diet is reduced in subsequent dietary phases until the pigs are consuming a typical corn-soy grower diet. Example weaning diets are shown in Table 13–4.

During the weaning period, various stresses induce morphological changes in the mucosa of the small intestine. The villi may be blunted with areas of disruption and erosion of microvilli (Fig. 13.1). These changes impair nutrient absorption and increase substrate available for microbial growth. British workers have suggested that the morphological changes in the intestinal mucosa after weaning are a result of transient **hypersensitivity to food antigens,** caused by exposure to the antigens by consumption of small amounts of creep feed before weaning (Miller et al., 1984a, b). They suggested that abrupt weaning without prior consumption of creep feed would aid in preventing postweaning diarrhea. However, subsequent work (Hampson, 1986; Hampson and Smith, 1986; Hampson et al., 1988) has failed to confirm this interesting hypothesis. Cera et al. (1989) suggested that a highly digestible and absorbable diet should be used for approximately 2 weeks postweaning.

The cells lining the villi, the **enterocytes,** are formed in the crypts of Lieberkuhn at the base of the villi (Fig. 13.1; also see Fig. 1.3). The formation and growth of enterocytes consumes about 20 percent of an animal's energy intake (Cant et al., 1996), and probably more in the very young animal. Milk contains **bioactive peptides** such as epidermal growth factor (EGF), and insulin-like growth factors (IGF-I and IGF-II) that regulate growth and development of enterocytes (Odle et al., 1996; Pluske et al., 1997). Cytokines produced by the enterocytes also impact the intestinal immune system and play a role in maintaining mucosal integrity. Future possibilities include the enhancement of bioactive peptides in sow's milk using transgenic methodology (Odle et al., 1996). As the understanding of the factors regulating intestinal cell growth and development increases, prevention of postweaning diarrhea by dietary manipulation should become more successful.

TABLE 13–4. Example Weaning, Grower, and Finisher Diets on a Total Amino Acid Basis[a]

Gender:	*Mixed (barrows and gilts)*				*Barrows*		*Gilts[b]*	
Body wt. range, kg:	*4-8*	*8–14*	*14–20*	*20–50*	*50–80*	*80–120*	*50–80*	*80–120*
Diet description:	*Phase 1*	*Phase 2*	*Phase 3*	*Grower*	*Early finisher*	*Late finisher*	*Early finisher*	*Late finisher*
Ingredients, % of diet								
Ground corn	40.35	49.70	65.03	70.67	80.81	85.42	76.46	81.85
Soybean meal, 49%	20.00	24.01	30.24	24.79	14.97	10.58	19.37	14.20
Spray-dried (whole) whey	20.00	20.00	—	—	—	—	—	—
Lactose	10.00	—	—	—	—	—	—	—
Spray-dried animal plasma	6.00	—	—	—	—	—	—	—
Spray-dried blood cells	—	2.00	—	—	—	—	—	—
Dicalcium phosphate (22% Ca, 18.5%P)	0.49	0.51	1.01	0.84	0.67	0.49	0.68	0.46
Ground limestone (39% Ca)	1.04	0.75	0.69	0.70	0.65	0.66	0.59	0.65
Fat source[c]	1.00	2.00	2.00	2.00	2.00	2.00	2.00	2.00
White salt	0.35	0.40	0.50	0.50	0.50	0.50	0.50	0.50
Trace mineral premix[d]	0.15	0.15	0.15	0.15	0.10	0.10	0.10	0.10
Vitamin premix[d]	0.25	0.25	0.25	0.20	0.15	0.15	0.15	0.15
L-lysine — HCl	0.20	0.12	0.15	0.15	0.15	0.10	0.15	0.10
DL methionine	0.16	0.10	0.04	—	—	—	—	—
Chemical composition, % (based on NRC, 1998)								
Crude protein	20.29	19.97	19.92	17.78	13.96	12.21	15.68	13.63
Lysine	1.46	1.30	1.20	1.05	0.78	0.62	0.90	0.72
Methionine	0.44	0.39	0.35	0.29	0.24	0.22	0.26	0.23
Threonine	0.91	0.80	0.75	0.66	0.51	0.44	0.58	0.50
Tryptophan	0.27	0.25	0.24	0.20	0.15	0.12	0.17	0.14
Calcium	0.80	0.70	0.65	0.60	0.50	0.45	0.50	0.45
Total P	0.61	0.57	0.60	0.54	0.46	0.41	0.48	0.42
Available P	0.40	0.32	0.28	0.24	0.19	0.15	0.20	0.15

Diets formulated using Brill Software and NRC (1998) feedstuff nutrient values.

[a]In practice, more grower and finisher phases are used than are shown here, mainly to more precisely meet the requirements and to reduce production cost. Crude protein may be reduced 3.0% in this grower diet, and 2.5 to 3.0% in these finisher diets when L-lysine, DL methionine, L-threonine, and L-tryptophan are used in the formulation. Greater reductions in crude protein will result in deficiencies of isoleucine and (or) valine.

[b]Finishing diets for barrows and gilts are based on the requirements for 350 g of lean gain/day.

[c]For Phase 1 and 2 diets, a high-quality fresh oil (e.g., corn or soybean oil) or animal fat may be used. For grower and finisher diets, high-quality animal or animal-vegetable blends are adequate. These grower and finisher diets average 3,560 kcal of DE and 3,420 kcal of ME per kg with the 2% added fat. Therefore, fat may be deleted and the DE and ME per kg will still be adequate.

[d]The amounts of trace mineral or vitamin premixes used will vary, depending on the concentration of the premix.

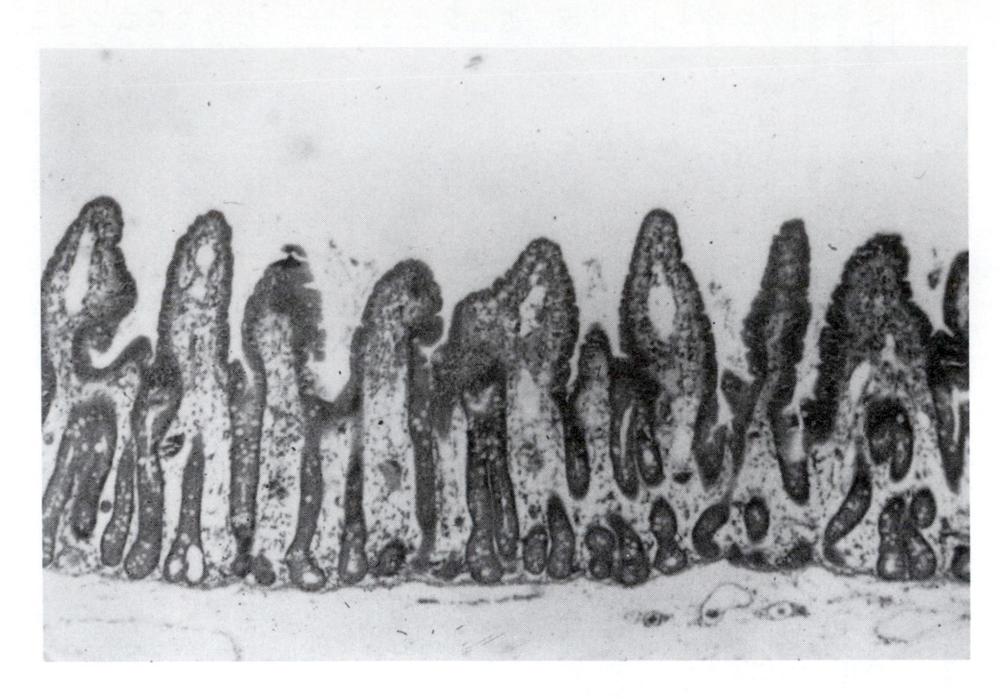

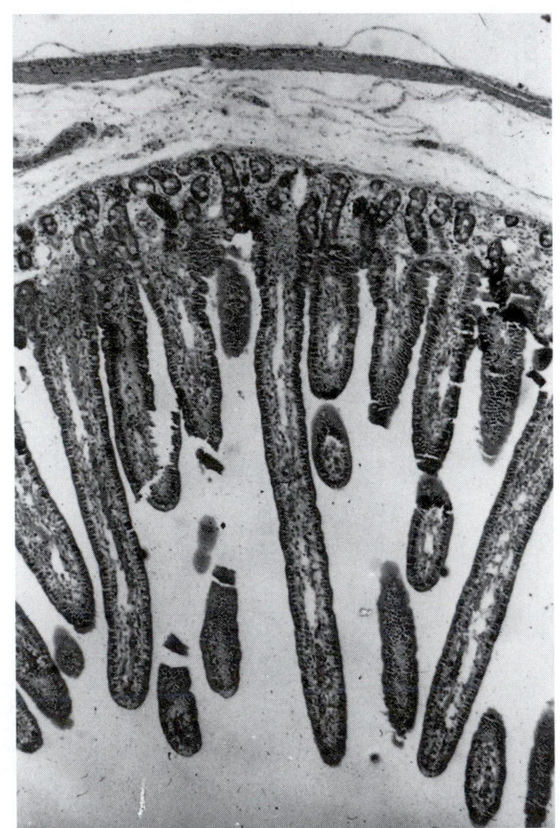

FIGURE 13.1 An electron microscopic view of the normal intestinal mucosa of a preweaning pig (top). Note the elongated, narrow villi. (The midsections of some villi are missing because of the plane of the tissue slicing.) The individual cells lining the villi, the enterocytes, are visible. These are formed in the **crypts of Lieberkuhn** at the base of the villi, and move up the villi as they elongate. In contrast, the mucosa of a newly weaned pig (bottom) reveals shortened, flattened villi. Nutrient absorption is impaired, and the pig is susceptible to postweaning diarrhea or enteritis. (Courtesy of D. J. Hampson.)

The incidence of postweaning diarrhea is influenced by the feed intake pattern. Initial postweaning feed intake is erratic; pigs tend not to eat for a period following weaning and then may consume a large meal, overloading the gut and causing diarrhea (Ball and Aherne, 1987a). Large single meals tend to cause temporary gut stasis followed by fluid accumulation and rapid peristaltic activity. Gut stasis may allow rapid proliferation of hemolytic *E. coli* (Ball and Aherne, 1987a). For pigs weaned at 4 weeks of age, a moderate degree of feed restriction (85 percent of unrestricted DE intake) may be helpful in reducing diarrhea. The energy intake is more important than is the total volume of feed consumed in predisposing pigs to diarrhea (Ball and Aherne, 1987b).

Large amounts of water are recycled in the intestinal tract. Miller et al. (1984b) describe how this could influence diarrhea in baby pigs. Water and electrolytes are continuously secreted into the small intestine, mainly from crypt enterocytes of the duodenum and jejunum, and are reabsorbed by villus enterocytes of the ileum. If this fluid recycling is not carefully balanced, the hindgut will be overloaded with fluid. This balance can be disturbed by **enterotoxins,** which are hypersecretory agents. At weaning, with villus atrophy, there may be a dominance of secretory cells, thus leading to fluid overload of the hindgut and diarrhea.

Growing-Finishing Diets

Pigs usually leave the nursery at a body weight of approximately 50 to 60 lb (23 to 27 kg), and are moved to a grower or grower-finisher building. A grower diet (16 to 18 percent crude protein) is fed until a weight of approximately 110 lb (50 kg) is reached. From that weight until market, a finisher diet (13 to 15 percent crude protein) is used. The grower-finisher diets are based on corn or another cereal grain and supplemented with a protein source such as soybean meal. There is generally no advantage to using a mixture of grains; the grain that is the least expensive source of DE is usually the best choice. Examples of typical diets for growing pigs are shown in Table 13–4. Commercial feed manufacturers and most large integrators usually pellet these diets because pelleted feeds are more palatable and feed waste is reduced. Feed prepared on the farm is fed in the meal (mash) form.

Breeding Herd Diets

Replacement boars and gilts should be fed in the same manner as the market pigs until a weight of approximately 200 lb (90 kg) is reached. Then they should be fed at approximately 75 percent of *ad libitum* intake, to allow adequate growth but to prevent excessive fat deposition. With most diets, 5 to 6 lb of feed per day will be adequate. “Flushing” gilts for 1 to 2 weeks before breeding by putting them on full feed may increase litter size. They should be returned to a restricted feeding regime immediately after mating, or conception rate may be impaired.

The NRC (1998) energy and amino acid requirements expressed as a percentage of the diet for gestating and lactating sows are shown in Tables 13–5 and 13–6, respectively. During gestation, feed intake should be restricted to prevent **obesity** because excessively fat sows may have smaller, less viable litters. Underfeeding will have the same effects. Underfeeding will reduce pig birth weight and survivability. Feed intake should be regulated to maintain the animals in

TABLE 13–5. Dietary Energy and Amino Acid Requirements of Gestating Sows (90% dry matter)[a]

	Body Weight at Breeding (kg)					
	125	*150*	*175*	*200*	*200*	*200*
	Gestation Weight Gain (kg)[b]					
	55	*45*	*40*	*35*	*30*	*35*
	Anticipated Pigs in Litter					
	11	*12*	*12*	*12*	*12*	*14*
DE content of diet (kcal/kg)	3,400	3,400	3,400	3,400	3,400	3,400
ME content of diet (kcal/kg)[c]	3,265	3,265	3,265	3,265	3,265	3,265
Estimated DE intake (kcal/day)	6,660	6,265	6,405	6,535	6,115	6,275
Estimated ME intake (kcal/day)[c]	6,395	6,015	6,150	6,275	5,870	6,025
Estimated feed intake (kg/day)	1.96	1.84	1.88	1.92	1.80	1.85
Crude protein (%)[d]	12.9	12.8	12.4	12.0	12.1	12.4
	Amino acid requirements: Apparent ileal digestible basis (%)					
Arginine	0.03	0.00	0.00	0.00	0.00	0.00
Histidine	0.15	0.15	0.14	0.13	0.13	0.14
Isoleucine	0.26	0.26	0.25	0.24	0.24	0.25
Leucine	0.47	0.46	0.43	0.40	0.40	0.43
Lysine	0.45	0.45	0.42	0.40	0.40	0.42
Methionine	0.13	0.13	0.12	0.11	0.12	0.12
Methionine + cystine	0.30	0.31	0.30	0.29	0.30	0.31
Phenylalanine	0.27	0.26	0.24	0.23	0.23	0.24
Phenylalanine + tyrosine	0.45	0.44	0.42	0.40	0.41	0.43
Threonine	0.32	0.33	0.32	0.31	0.32	0.33
Tryptophan	0.08	0.08	0.08	0.07	0.07	0.08
Valine	0.31	0.30	0.28	0.27	0.27	0.28
	Amino acid requirements: Total basis (%)[d]					
Arginine	0.06	0.03	0.00	0.00	0.00	0.00
Histidine	0.19	0.18	0.17	0.16	0.17	0.17
Isoleucine	0.33	0.32	0.31	0.30	0.30	0.31
Leucine	0.50	0.49	0.46	0.42	0.43	0.45
Lysine	0.58	0.57	0.54	0.52	0.52	0.54
Methionine	0.15	0.15	0.14	0.13	0.13	0.14
Methionine + cystine	0.37	0.38	0.37	0.36	0.36	0.37
Phenylalanine	0.32	0.32	0.30	0.28	0.28	0.30
Phenylalanine + tyrosine	0.54	0.54	0.51	0.49	0.49	0.51
Threonine	0.44	0.45	0.44	0.43	0.44	0.45
Tryptophan	0.11	0.11	0.11	0.10	0.10	0.11
Valine	0.39	0.38	0.36	0.34	0.34	0.36

Adapted from NRC, *Nutrient Requirements of Swine* (1998).

[a]Daily intakes of DE and feed and the amino acid requirements are estimated by the gestation model.

[b]Weight gain includes maternal tissue and products of conception.

[c]Assumes that ME is 96% of DE.

[d]Crude protein and total amino acid requirements are based on a corn-soybean meal diet.

TABLE 13–6. Dietary Amino Acid Requirements of Lactating Sows (90% dry matter)[a]

	Sow Postfarrowing Weight (kg)					
	175	*175*	*175*	*175*	*175*	*175*
	Anticipated Lactational Weight Change (kg)[b]					
	0	*0*	*0*	*−10*	*−10*	*−10*
	Daily Weight Gain of Pigs (g)[b]					
	150	**200**	**250**	**150**	**200**	**250**
DE content of diet (kcal/kg)	3,400	3,400	3,400	3,400	3,400	3,400
ME content of diet (kcal/kg)[c]	3,265	3,265	3,265	3,265	3,265	3,265
Estimated DE intake (kcal/day)	14,645	18,205	21,765	12,120	15,680	19,240
Estimated ME intake (kcal/day)[c]	14,060	17,475	20,895	11,635	15,055	18,470
Estimated feed intake (kg/day)	4.31	5.35	6.40	3.56	4.61	5.66
Crude protein (%)[d]	16.3	17.5	18.4	17.2	18.5	19.2
	Amino acid requirements: Apparent ileal digestible basis (%)					
Arginine	0.34	0.41	0.46	0.33	0.41	0.47
Histidine	0.27	0.30	0.32	0.29	0.32	0.34
Isoleucine	0.37	0.41	0.44	0.41	0.44	0.47
Leucine	0.77	0.86	0.92	0.83	0.92	0.98
Lysine	0.66	0.73	0.79	0.72	0.79	0.84
Methionine	0.18	0.20	0.21	0.19	0.21	0.22
Methionine + cystine	0.33	0.36	0.38	0.36	0.39	0.40
Phenylalanine	0.36	0.40	0.43	0.39	0.43	0.46
Phenylalanine + tyrosine	0.75	0.83	0.89	0.82	0.90	0.96
Threonine	0.40	0.43	0.46	0.44	0.47	0.49
Tryptophan	0.11	0.12	0.13	0.13	0.14	0.14
Valine	0.55	0.61	0.66	0.61	0.67	0.71
	Amino acid requirements: Total basis (%)[d]					
Arginine	0.40	0.48	0.54	0.39	0.49	0.55
Histidine	0.32	0.36	0.38	0.34	0.38	0.40
Isoleucine	0.45	0.50	0.53	0.50	0.54	0.57
Leucine	0.86	0.97	1.05	0.95	1.05	1.12
Lysine	0.82	0.91	0.97	0.89	0.97	1.03
Methionine	0.21	0.23	0.24	0.22	0.24	0.26
Methionine + cystine	0.40	0.44	0.46	0.44	0.47	0.49
Phenylalanine	0.43	0.48	0.52	0.47	0.52	0.55
Phenylalanine + tyrosine	0.90	1.00	1.07	0.98	1.08	1.14
Threonine	0.54	0.58	0.61	0.58	0.63	0.65
Tryptophan	0.15	0.16	0.17	0.17	0.18	0.19
Valine	0.68	0.76	0.82	0.76	0.83	0.88

Adapted from NRC, *Nutrient Requirements of Swine* (1998).
[a]Daily intakes of DE and feed and the amino acid requirements are estimated by the lactation model.
[b]Assumes 10 pigs per litter and a 21-day lactation period.
[c]Assumes that ME is 96% of DE. In corn-soybean meal diets of these crude protein levels, ME is 95–96% of DE.
[d]Crude protein and total amino acid requirements are based on a corn-soybean meal diet.

good body condition without excessive fat accumulation. A high plane of nutrition reduces early embryonic survival, so feed intake should be restricted during the first month of gestation. During midgestation, intake should be regulated on an individual sow basis to maintain good body condition. During late gestation, when most of the piglet growth occurs, feed intake should be increased. Supplementation with fat in late gestation may improve baby pig survival and performance (see Chapter 3). The **gestation diet** may contain 10 to 30 percent alfalfa meal or other low-energy ingredients to reduce the energy level and maintain body condition. Intensive swine farms, however, generally use an automated feeding system with individual feeding adjustments for each sow. This allows a simple, fortified corn-soybean meal type diet to be fed according to each individual's needs. Sows and gilts should gain a minimum of about 35 and 46 kg, respectively, during gestation to maintain body condition and allow for piglet growth.

Sows are susceptible to **constipation** when put in the farrowing crates a few days before parturition. In some management situations, it may be advisable to feed a laxative diet at that time. Wheat bran or beet pulp (10 to 15 percent of diet) or 0.75 to 1.0 percent potassium chloride are effective laxatives. The inclusion of added fat (see Chapter 3) in the diet prior to farrowing may increase the energy stores of the baby pigs at birth and improve neonatal survival. After parturition, sows should be fed a nutrient-dense **lactation diet** *ad libitum* to minimize weight loss during lactation.

The nutritional requirements of sows have been reviewed for protein and amino acids (Speer, 1990), energy (Noblet et al., 1990), and minerals (Mahan, 1990). The NRC vitamin and mineral requirements for breeding swine are shown in Table 13–7.

Mature boars should be fed a restricted quantity of the gestation diet. They should be fed to maintain adequate condition without becoming obese. Examples of conventional diet formulas for gestation and lactation diets are given in Table 13–8.

SPECIAL NUTRITIONAL CONCERNS WITH SWINE

Gastric Ulcers **Gastric ulcers** (esophagogastric parakeratosis) can be a significant problem in growing-finishing pigs. Damage to the stomach lining may impair digestive function or, in severe cases, lead to perforation of the stomach, peritonitis, and death. Contributing factors seem to be stress (e.g., overcrowding) and the use of finely ground high-energy diets (Potkins et al., 1989). Such diets promote bacterial fermentation in the esophageal region of the stomach, and the organic acids produced cause irritation to the mucosa. Oats and oat hulls tend to have protective effects by inhibiting gastric acid secretion. Fibrous feedstuffs in general seem to have favorable effects in reducing gastric ulcers in swine (Lee and Close, 1987).

Leg Disorders Leg disorders and lameness are common problems in pigs raised in confinement, particularly on concrete or slotted floors. The balance of calcium, phosphorus, and vitamin D may be involved. The calcium and phosphorus levels adequate for maximum growth are not necessarily adequate for

TABLE 13–7. Dietary Mineral, Vitamin, and Fatty Acid Requirements of Gestating and Lactating Sows (90% dry matter)[a]

	Gestation	*Lactation*
DE content of diet (kcal/kg)	3,400	3,400
ME content of diet (kcal/kg)[b]	3,265	3,265
DE intake (kcal/day)	6,290	17,850
ME intake (kcal/day)[b]	6,040	17,135
Feed intake (kg/day)	1.85	5.25
	Requirements (% or amount/kg of diet)	
Mineral elements		
Calcium (%)	0.75	0.75
Phosphorus, total (%)	0.60	0.60
Phosphorus, available (%)	0.35	0.35
Sodium (%)	0.15	0.20
Chlorine (%)	0.12	0.16
Magnesium (%)	0.04	0.04
Potassium (%)	0.20	0.20
Copper (mg)	5.00	5.00
Iodine (mg)	0.14	0.14
Iron (mg)	80	80
Manganese (mg)	20	20
Selenium (mg)	0.15	0.15
Zinc (mg)	50	50
Vitamins		
Vitamin A (IU)[c]	4,000	2,000
Vitamin D_3 (IU)[c]	200	200
Vitamin E (IU)[c]	44	44
Vitamin K (menadione) (mg)	0.50	0.50
Biotin (mg)	0.20	0.20
Choline (g)	1.25	1.00
Folacin (mg)	1.30	1.30
Niacin, available (mg)[d]	10	10
Pantothenic acid (mg)	12	12
Riboflavin (mg)	3.75	3.75
Thiamin (mg)	1.00	1.00
Vitamin B_6 (mg)	1.00	1.00
Vitamin B_{12} (μg)	15	15
Linoleic acid (%)	0.10	0.10

Adapted from NRC, *Nutrient Requirements of Swine* (1998).

[a]The requirements are based on the daily consumption of 1.85 and 5.25 kg of feed, respectively. If lower amounts of feed are consumed, the dietary percentage may need to be increased.

[b]Assumes that ME is 96% of DE.

[c]Conversions: 1 IU vitamin A = 0.344 μg retinyl acetate; 1 IU vitamin D_3 = 0.025 μg cholecalciferol; 1 IU vitamin E = 0.67 mg of D-α-tocopherol or 1 mg of DL-α-tocopheryl acetate.

[d]The niacin in corn, grain sorghum, wheat, and barley is unavailable. Similarly, the niacin in by-products made from these cereal grains is poorly available unless the by-products have undergone a fermentation or wet-milling process.

TABLE 13–8. Example Gestation and Lactation Diets on a Total Amino Acid Basis

	Gestation[a]	*Lactation*[b]
Ingredients, % of diet		
Ground corn	83.38	67.69
Soybean meal, 49%	11.25	27.14
Fat source	2.00	2.00
Dicalcium phosphate (22% Ca, 18.5% P)	1.62	1.47
Ground limestone	0.73	0.69
White salt	0.51	0.51
Trace mineral premix[c]	0.15	0.15
Vitamin premix[c]	0.25	0.25
L-lysine - HCL	0.10	0.11
Chemical composition, % (based on NRC, 1998)		
Crude protein	12.36	18.61
Lysine	0.64	1.08
Methionine	0.22	0.30
Threonine	0.45	0.70
Tryptophan	0.12	0.22
Valine	0.58	0.88
Calcium	0.75	0.75
Total P	0.62	0.66
Available P	0.36	0.36
DE, kcal/kg	3,509	3,540
ME, kcal/kg	3,382	3,382

Diets formulated using Brill Software and NRC (1998) feedstuff nutrient values.

[a]Gestation requirements were based on a 200 kg body weight at breeding, a 35 kg gestation weight gain, and 14 piglets expected in the litter. Crude protein may be reduced 3.27% by using L-lysine, L-threonine, L-tryptophan, and DL-methionine in this formulation. Further reduction in crude protein will result in an isoleucine deficiency. A feed ingredient higher in fiber (e.g., alfalfa meal) may be included in the formulation when needed to reduce constipation.

[b]Lactation requirements are based on a 175 kg postfarrowing sow weight, an expected total lactation body weight loss of 10 kg, and an expected daily weight gain of 250 g for each nursing pig. After lysine, valine is the second limiting amino acid in this formulation. Crystalline L-valine addition is not economically practical at present. However, if valine, lysine, threonine, tryptophan (L-forms), and DL methionine were used in this formulation, crude protein could be reduced 3.76%. Further reduction in crude protein will result in an isoleucine deficiency.

[c]The amounts of trace mineral and vitamin premixes used will vary, depending on the concentration of the premix. In addition to the vitamins required in a growing-finishing swine premix, the sow premix should also contain choline and folic acid.

maximum bone mineralization (NRC), although it has not been proven that maximum bone mineralization improves structural soundness of bone. Brennan and Aherne (1986) found that elevated calcium-phosphorus in excess of NRC recommendations did not reduce the incidence and severity of joint lesions or affect leg soundness scores.

Stress-Susceptible Syndrome Some breeds of swine have an inherited tendency to develop a rapid, fatal increase in body temperature (malignant hyperthermia) when subjected to normal management procedures such as transportation, exercise, mating, and parturition (Hoppe et al., 1989). The meat from these pigs is pale, soft, and exudative **(PSE syndrome).** The stress syndrome can be triggered by halothane anaesthesia, which can be used to detect stress-sensitivity in breeding animals. Higher than NRC-recommended levels of vitamin E appear to have protective effects against the PSE-stress-sensitivity syndrome (Hoppe et al., 1989).

Mycotoxins The grains and protein supplements used in swine feeding are often contaminated with mycotoxins such as **aflatoxin.** Swine are highly susceptible to acute aflatoxin poisoning, with severe liver damage a result. Chronic toxicity with feeds containing 0.4 ppm or more aflatoxin is characterized by reduced gain and feed efficiency in mild cases, and severe liver and kidney damage with higher aflatoxin levels. Systemic hemorrhaging is a common sign. Increased dietary levels of methionine and protein help to protect against aflatoxin toxicity (Coffey et al., 1989).

Swine are highly sensitive to **zearalenone,** a *Fusarium* mycotoxin with estrogen activity. Corn often contains zearalenone. Hyperestrogenic effects are seen, including swelling of the vulva and mammary glands, infertility, and vaginal prolapse. Other common mycotoxins to which swine are often exposed are the **trichothecenes,** including vomitoxin, T-2 toxin, and diacetoxyscirpenol (DAS). These toxins cause feed refusal and vomiting, and a number of other signs including diarrhea, hemorrhage, abortion, and neurological effects. The biological effects of these and other mycotoxins have been reviewed by Cheeke (1998).

Odor Control and Ecological Nutrition A major concern with intensive swine production is **air and water pollution.** The problems involve excretion of nitrogen, phosphorus, and other minerals that may pollute surface and ground waters, and pollution of the air with ammonia and swine odors. Aspects of ecological nutrition to reduce excretion of nutrients have been dealt with in Chapters 8 and 11. For example, use of phytases as feed additives can increase phosphorus bioavailability and reduce phosphorus excretion. Nitrogen excretion in the urine can be minimized by avoiding the feeding of excess protein and by using well-balanced diets that precisely meet amino acid requirements.

The odor associated with intensive swine production causes tremendous environmental problems. **Swine odors** are offensive to people and can cause health problems. More than 200 odorous and offensive compounds have been identified in swine manure and lagoons. Many of the offensive compounds in swine excreta are derived from protein. Degradation of amino acids in the hindgut by microbial action produces a number of foul-smelling compounds. The indole ring of tryptophan gives rise to skatole (3-methyl indole) and various other indolic compounds. Sulfur amino acids can be converted to odorants such as dimethyl sulfide. Other compounds produced from amino acids have the descriptive names of *putrescine* and *cadaverine*. The production of undesirable compounds such as these can be minimized by fine-tuning diets to provide an optimal balance of amino acids (Hobbs et al., 1996).

QUESTIONS AND STUDY GUIDE

1. Why are baby pigs susceptible to developing diarrhea? Find a recent paper (within the last 2 years) dealing with postweaning diarrhea or enteritis in pigs, and discuss its major findings.
2. By what mechanisms does wheat bran act as a laxative in sows?
3. Why do diets containing oat hulls and alfalfa meal tend to prevent gastric ulcers in pigs?
4. Is it nutritionally more efficient to feed a baby pig directly (i.e., by early weaning) than to feed the sow to produce milk? Explain.
5. Discuss the SEW system and its implications in swine nutrition and animal welfare.
6. When formulating Phase 1 weaning diets, which supplemental carbohydrate sources should be used, and which sources should not be used? Explain.
7. Why are barrows and gilts usually fed separately (split-sex feeding) during the finishing phases?
8. Why has the swine industry gradually moved to fewer swine farms that are much larger, with pigs produced in multiple confinement units?
9. When grain-plant protein based diets are fed to growing-finishing swine, which macrominerals must be provided as supplements?
10. Which trace minerals are normally included in a swine "trace mineral" premix?
11. How will low phytic acid grains benefit agriculture and the environment when they become available economically?
12. Which "fat soluble" vitamins should be included in a swine vitamin premix for growing pigs?
13. Which "water soluble" vitamins should be included in a swine vitamin premix for growing pigs?
14. Which additional vitamins are beneficial for gestating and lactating sows?
15. **a.** Which amino acids become the "limiting amino acids" in practical grain-plant protein diets?
    b. Are crystalline forms of these amino acids available as feed supplements? Do the D and L forms have equivalent feeding values?

REFERENCES

Adeola, O., B. V. Lawrence, A. L. Sutton, and T. R. Cline. 1995. Phytase-induced changes in mineral utilization in zinc-supplemented diets for pigs. *J. Anim. Sci.* 73:3384–3391.

Ball, R. O. and F. X. Aherne. 1982. Effect of diet complexity and feed restriction on the incidence and severity of diarrhea in early-weaned pigs. *Can. J. Anim. Sci.* 62:907–913.

______ . 1987a. Influence of dietary nutrient density, level of feed intake and weaning age on young pigs. I. Performance and body composition. *Can. J. Anim. Sci.* 67:1093–1103.

______ . 1987b. Influence of dietary nutrient density, level of feed intake and weaning age on young pigs. II. Apparent nutrient digestibility and incidence and severity of diarrhea. *Can. J. Anim. Sci.* 67:1105–1115.

Barber, J., P. H. Brooks, and J. L. Carpenter. 1989. The effects of water delivery rate on the voluntary feed intake, water use and performance of early weaned pigs from 3–6 weeks of age. In *The Voluntary Feed Intake of Pigs,* Forbes, J. M., M. A. Varley, and T. L. J. Lawrence, Eds., British Society of Animal Production, Edinburgh, 103–104.

Bergstrom, J. R., J. L. Nelssen, M. D. Tokach, R. D. Goodband, S. S. Dritz, K. Q. Owen, and W. B. Nessmith, Jr. 1997. Evaluation of spray-dried animal plasma and select menhaden fish meal in transition diets of pigs weaned at 12 to 14 days of age and reared in different production systems. *J. Anim. Sci.* 75:3004–3009.

Bilodeau, R., J. J. Matte, A. M. B. De Passille, C. L. Girard, and G. J. Brisson. 1989. Effects of floor type on serum folates, serum vitamin B_{12}, plasma biotin and on growth performances of pigs. *Can. J. Anim. Sci.* 69:779–788.

Braude, R., P. M. Clarke, K. G. Mitchell, A. S. Cray, A. Franke, and P. H. Sedgwick. 1957. Unrestricted whey for fattening pigs. J. Agric. Sci. (Cambridge), 49:347–356.

Brennan, J. J. and F. X. Aherne. 1986. Effect of dietary calcium and phosphorus levels on performance, bone bending moment and the severity of osteochondrosis and lameness in boars and gilts slaughtered at 100 or 130 kg body weight. *Can. J. Anim. Sci.* 66:777–790.

Brumm, M. C. and J. M. Dahlquist, 1997. Impact on feeder and drinker designs on pig performance, water use, and manure production, Nebraska Swine Report EC97-219, University of Nebraska Coop. Ext., Lincoln, 34–35.

Brumm, M. C. and H. W. Gonyou. 2001. Effects of facility design on behavior and feed and water intake. Pages 499–518 in *Swine Nutrition.* 2nd ed. A. J. Lewis and L. L. Southern, Ed., CRC Press, New York, NY.

Cant, J. P., B. W. McBride, and W. G. Croom, Jr. 1996. The regulation of intestinal metabolism and its impact on whole animal energetics. *J. Anim. Sci.* 74:2541–2553.

Cera, K. R., D. C. Mahan, R. F. Cross, G. A. Reinhart, and R. E. Whitmoyer. 1989. Effect of age, weaning and postweaning diet on small intestinal growth and jejunal morphology in young swine. *J. Anim. Sci.* 66:574–584.

Cheeke, P. R. 1998. *Natural Toxicants in Feeds, Forages, and Poisonous Plants.* Upper Saddle River, NJ: Prentice Hall, Inc.

Close, W. H., L. E. Mount, and I. B. Start. 1971. The influence of environmental temperature and plane of nutrition on heat losses from groups of growing pigs. Anim. Prod. 13:285–294.

Coffey, M. T., W. M. Hagler, Jr., and J. M. Cullen. 1989. Influence of dietary protein, fat or amino acids on the response of weanling swine to aflatoxin B_1. *J. Anim. Sci.* 7:465–472.

Cumby, T. R. 1986. Design requirements of liquid feed systems for pigs: a review. J. Agric. Eng. Res. 34: 153–172.

De Passille, A. M. B., R. R. Bilodeau, C. L. Girard, and J. J. Matte. 1989. A study on the occurrence of coprophagy behavior and its relationship to B-vitamin status in growing-finishing pigs. *Can. J. Anim. Sci.* 69:299–306.

DeRouchey, J. M., M. D. Tokach, J. L. Nelssen, R. D. Goodband, S. S. Dritz, J. C. Woodworth, and B. W. James. 2002. Comparison of spray-dried blood meal and blood cells in diets for nursery pigs. *J. Anim. Sci.* 80:2879–2886.

Dritz, S. S., K. Q. Owen, R. D. Goodband, J. L. Nelssen, M. D. Tokach, M. M. Chengappa, and F. Blecha. 1996. Influence of lipopolysaccharide-induced immune challenge and diet complexity on growth performance and acute-phase protein production in segregated early-weaned pigs. *J. Anim. Sci.* 74:1620–1628.

Farnworth, E. R. and J. K. G. Kramer. 1987. Fat metabolism in growing swine: A review. *Can. J. Anim. Sci.* 67:301–318.

Fraser, D., P. A. Phillips, B. K. Thompson, and W. B. Peeters Weem. 1988. Use of water by piglets in the first days after birth. *Can. J. Anim. Sci.* 68:603–610.

Gill, B. P., and J. Barber. 1990. Water delivery systems for growing pigs. Farm Building Progr., 102:19–22.

Gill, B. P., P. H. Brooks, and J. L. Carpenter. 1986. The water intake of weaned pigs from 3 to 6 weeks of age. Anim. Prod. 42:270 (Abstr.).

Gonyou, H. W. and Z. Lou. 2000. Effects of eating space and availability of water in feeders on productivity and eating behavior of grower-finisher pigs. J. Anim. Sci. 78:865–870.

Grinstead, G. S., R. D. Goodband, S. S. Dritz, M. D. Tokach, J. L. Nelssen, J. C. Woodworth, and M. Molitor. 2000. Effects of a whey protein product and spray-dried animal plasma on growth performance of weanling pigs. *J. Anim. Sci.* 78:647–657.

Hampson, D. J. 1986. Alterations in piglet small intestinal structure at weaning. *Res. Vet. Sci.* 40:32–40.

Hampson, D. J. and W. C. Smith. 1986. Influence of creep feeding and dietary intake after weaning on malabsorption and occurrence of diarrhoea in the newly weaned pig. *Res. Vet. Sci.* 41:63–69.

Hampson, D. J., Z. F. Fu, and W. C. Smith. 1988. Pre-weaning supplementary feed and porcine post-weaning diarrhoea. *Res. Vet. Sci.* 44:309–314.

Harper, A. F., E. T. Kornegay, and T. C. Schell. 1997. Phytase supplementation of low-phosphorus growing-finishing pig diets improves performance, phosphorus digestibility, and bone mineralization and reduces phosphorus excretion. *J. Anim. Sci.* 3174–3186.

Hobbs, P. J., B. F. Pain, R. M. Kay, and P. A. Lee. 1996. Reduction of odorous compounds in fresh pig slurry by dietary control of crude protein. *J. Sci. Food Agric.* 71:508–514.

Hoppe, P. P., G. G. Duthie, J. R. Arthur, F. J. Schoner, and H. Wiesche. 1989. Vitamin E and vitamin C supplementation and stress-susceptible pigs; effects of halothane and drug-induced muscle contraction. *Livestock Prod. Sci.* 22:341–350.

Jensen, A. R., J. Elnif, D. G. Burrin, and P. T. Sangild. 2001. Development of intestinal immunoglobin absorption and enzyme activities in neonatal pigs is diet dependent. *J. Nutr.* 131:3259–3265.

Jiang, R., X. Chang, B. Stoll, K. J. Ellis, R. J. Shypailo, E. Weaver, J. Campbell, and D. G. Burrin. 2000. Dietary plasma protein is used more efficiently than extruded soy protein for lean tissue growth in early-weaned pigs. *J. Nutr.* 130:2016–2019.

Kass, M. L., P. J. Van Soest, and W. G. Pond. 1980. Utilization of dietary fiber from alfalfa by growing swine. II. Volatile fatty acid concentrations in and disappearance from the gastrointestinal tract. *J. Anim. Sci.* 50:192–197.

Kim, S. W. and R. A. Easter. 2001. Nutritional value of fish meals in the diet for young pigs. 2001. *J. Anim. Sci.* 79:1829–1839.

LeaMaster, B. R. and P. R. Cheeke. 1979. Feed preferences of swine: Alfalfa meal, high and low saponin alfalfa, and quinine sulfate. *Can. J. Anim. Sci.* 59:467–469.

Lee, P. A. and W. H. Close. 1987. Bulky feeds for pigs: A consideration of some non-nutritional aspects. *Livestock Prod. Sci.* 16:395–405.

Lewis, A. J. and L. L. Southern, eds. 2001. *Swine Nutrition,* 2d ed. Boca Raton, FL: CRC Press.

Lightfoot, A. L. 1978. Water consumption of lactating sows. Anim. Prod. 26:386 (Abstr.).

Lindemann, M. D. and E. T. Kornegay. 1989. Folic acid supplementation to diets of gestating-lactating swine over multiple parities. *J. Anim. Sci.* 67:459–464.

Liu, J., D. Bollinger, D. R. Ledoux, M. R. Ellersieck, and T. L. Veum. 1997. Soaking increases the efficacy of supplemental microbial phytase in a low-phosphorus corn-soybean meal diet for growing pigs. *J. Anim. Sci.* 75:1292–1298.

Liu, J., D. Bollinger, D. R. Ledoux, and T. L. Veum. 1998. Lowering the dietary calcium to total phosphorus ratio increases phosphorus utilization in low-phosphorus corn-soybean meal diets supplemented with microbial phytase for growing-finishing pigs. *J. Anim. Sci.* 76:808–813.

______ . 2000. Effects of dietary calcium:phosphorus ratios on apparent absorption of calcium and phosphorus in the small intestine, cecum, and colon of pigs. *J. Anim. Sci.* 78:106–109.

Mahan, D. C. 1990. Mineral nutrition of the sow: A review. *J. Anim. Sci.* 68:573–582.

______ . 1992. Efficacy of dried whey and its lactalbumin and lactose components at two dietary lysine levels on postweaning pig performance and nitrogen balance. *J. Anim. Sci.* 70:2182–2187.

______ . 1993. Evaluating two sources of dried whey and the effects of replacing the corn and dried whey component with corn gluten meal and lactose in the diets of weanling swine. *J. Anim. Sci.* 71:2860–2866.

Mahan, D. C., G. L. Cromwell, R. C. Ewan, C. R. Hamilton, and J. T. Yen. 1998. Evaluation of the feeding duration of a phase 1 nursery diet to three-week-old pigs of two weaning weights. *J. Anim. Sci.* 76:578–583.

Mahan, D. C., T. D. Wiseman, E. Weaver, and L. Russell. 1999. Effect of supplemental sodium chloride and hydrochloride acid added to initial starter diets containing spray-dried blood plasma and lactose on resulting performance and nitrogen digestibility of 3-week-old weaned pigs. *J. Anim. Sci.* 77:3016–3021.

Miller, B. G., T. J. Newby, C. R. Stokes, and F. J. Bourne. 1984a. Creep feeding and post weaning diarrhoea in piglets. *Vet. Rec.* 114:296–297.

______ . 1984b. Influence of diet on postweaning malabsorption and diarrhoea in the pig. *Res. Vet. Sci.* 36:187–193.

National Research Council. 1988, 1998. *Nutrient Requirements of Swine*. Washington, DC: National Academy Press.

Nessmith, W. B. Jr., J. L. Nelssen, M. D. Tokach, R. D. Goodband, and J. R. Bergstrom. 1997a. Effects of substituting deproteinized whey and(or) crystalline lactose for dried whey on weaning pig performance. *J. Anim. Sci.* 75:3222–3228.

Nessmith, W. B. Jr., J. L. Nelssen, M. D. Tokach, R. D. Goodband, J. R. Bergstrom, Dritz, S. S., and B. T. Richert. 1997b. Evaluation of the interrelationships among lactose and protein sources in diets for segregated early-weaned pigs. *J. Anim. Sci.* 75:3214–3221.

Noblet, J., J. Y. Dourmad, and M. Etienne. 1990. Energy utilization in pregnant and lactating sows: Modeling of energy requirements. *J. Anim. Sci.* 68:562–572.

Odle, J., R. T. Zijlstra, and S. M. Donovan. 1996. Intestinal effects of milkborne growth factors in neonates of agricultural importance. *J. Anim. Sci.* 74:2509–2522.

Owusu-Asiedu, A., S. K. Baidoo, C. M. Nyachoti, and R. R. Marquardt. 2002. Response of early-weaned pigs to spray-dried porcine or animal plasma-based diets supplemented with egg-yolk antibodies against enterotoxigenic *Escherichia coli*. *J. Anim. Sci.* 80:2895–2903.

Pluske, J. R., D. J. Hampson, and I. H. Williams. 1997. Factors influencing the structure and function of the small intestine in the weaned pig: A review. *Livestock Prod. Sci.* 51:215–236.

Pond, W. G. and J. H. Maner. 1984. *Swine Production and Nutrition*. Westport, CT: AVI Publishing Co.

Potkins, Z. V., T. L. J. Lawrence, and J. R. Thomlinson. 1989. Oesophagogastric parakeratosis in the growing pig: Effects of the physical form of barley-based diets and added fibre. *Res. Vet. Sci.* 47:60–67.

Powley, J. S., P. R. Cheeke, D. C. England, T. P. Davidson, and W. H. Kennick. 1981. Performance of growing-finishing swine fed high levels of alfalfa meal: Effects of alfalfa level, dietary additives and antibiotics. *J. Anim. Sci.* 53:308–316.

Raboy, V., P. E. Gerbasi, K. A. Young, S. D. Stoneberg, S. G. Pickett, A. T. Bauman, P. P. N. Murthy, W. F. Sheridan, and D. S. Ertl. 2000. Origin and seed phenotype of maize low phytic acid 1–1 and low phytic acid 2–1. *Plant Physiol.* 124:355–368.

Raboy, V., K. A. Young, J. A. Dorsch, and A. Cook. 2001. Genetics and breeding of seed phosphorus and phytic acid. *J. Plant Physiol.* 158:489–497.

Schmidt, M. K., T. L. Veum, J. L. Clark, and G. F. Krause. 1973. Chemical composition of crossbred swine from birth to 136 kg with two planes of nutrition from 53 to 136 kilograms. J. Anim. Sci. 37:683–687.

Speer, V. C. 1990. Partitioning nitrogen and amino acids for pregnancy and lactation in swine: A review. *J. Anim. Sci.* 68:553–561.

Stahl, C. H., Y. M. Han, K. R. Roneker, N. A. House, and X. G. Lei. 1999. Phytase improves iron bioavailability for hemoglobin synthesis in young pigs. *J. Anim. Sci.* 77:2135–2142.

Thacker, P. A. 2001. Water in swine nutrition. Pages 381–398 in *Swine Nutrition*. 2nd ed. A. J. Lewis and L. L. Southern, Ed. CRC Press, New York, NY.

Thacker, P. A. and R. N. Kirkwood, eds. 1990. *Nontraditional Feed Sources for Use in Swine Production*. Stoneham, MA: Butterworths.

Thaler, R. C., J. L. Nelssen, R. D. Goodband, and G. L. Allee. 1989. Effect of dietary folic acid supplementation on sow performance through two parities. *J. Anim. Sci.* 67:3360–3369.

Tremblay, G. F., J. J. Matte, J. J. Dufour, and G. J. Brisson. 1989. Survival rate and development of fetuses during the first 30 days of gestation after folic acid addition to a swine diet. *J. Anim. Sci.* 67:724–732.

Veum, T. L., D. R. Ledoux, D. W. Bollinger, V. Raboy, and A. Cook. 2002. Low-phytic acid barley improves calcium and phosphorus utilization and growth performance in growing pigs. *J. Anim. Sci.* 80:2663–2670.

Veum, T. L., D. R. Ledoux, V. Raboy, and D. S. Ertl. 2001. Low-phytic acid corn improves nutrient utilization for growing pigs. *J. Anim. Sci.* 79:2873–2880.

Veum, T. L. and J. Odle. 2001. Feeding neonatal pigs. In *Swine Nutrition,* 2d ed., A. J. Lewis and L. L. Southern, eds. pp. 671–690. Boca Raton, FL: CRC Press.

Wilcox, J. R., G. S. Premachandra, K. A. Young, and V. Raboy. 2000. Isolation of high-seed inorganic P, low-phytate soybean mutants. *Crop Sci.* 40:1601–1605.

Williams, N. H., T. S. Stahly, and D. R. Zimmerman. 1997a. Effect of chronic immune system activation on the rate, efficiency, and composition of growth and lysine needs of pigs fed from 6 to 27 kg. *J. Anim. Sci.* 75:2463–2471.

______ . 1997b. Effect of chronic immune system activation on body nitrogen retention, partial efficiency of lysine utilization, and lysine needs of pigs. *J. Anim. Sci.* 75:2472–2480.

______ . 1997c. Effect of level of chronic immune system activation on the growth and dietary lysine needs of pigs fed from 6 to 112 kg. *J. Anim. Sci.* 75:2481–2496.

Wolter, B. F., M. Ellis, B. P. Corrigan, and J. M. DeDecker. 2002. The effect of birth weight and feeding of supplemental milk replacer to piglets during lactation on preweaning and postweaning growth performance and carcass characteristics. *J. Anim. Sci.* 80:301–308.

CHAPTER 14

Feeding and Nutrition of Avian Species

D. R. Ledoux and P. R. Cheeke

Objectives

1. To describe the major nutritional requirements of poultry
2. To briefly discuss specific aspects of nutrition of waterfowl, game birds, and exotic birds
3. To describe some of the major nutritional problems encountered in the poultry industry

Diets for commercial production of poultry in the industrialized countries have become very sophisticated, with precise formulation to meet a lengthy list of required nutrients. High-energy diets are used to maximize feed-conversion efficiency. On large enterprises, use of low-energy diets could result in the need to handle much larger quantities of feed and increased labor, transportation, handling, storage, and feed hopper requirements. Thus, unless substitution of lower-energy ingredients will have a major favorable impact on net income from the enterprise, it is generally most economically efficient to strive for the best feed conversion possible and to minimize the feed input per unit of product (kilogram of gain or per dozen eggs).

The major types of commercial poultry raised in North America are chickens (layers and broilers) and turkeys. Other poultry, of much less economic importance, include waterfowl (ducks and geese), guinea fowl, ostrich, and game birds, such as pheasants and quail. These minor species in North America are more significant in other parts of the world. Ducks are particularly important in Southeast Asia (e.g., Indonesia) and China, and geese are raised in large numbers in France and Eastern European countries (e.g., Poland, Hungary). Guinea fowl (keets) are raised as meat birds in France, as are pigeons (squab). Ostrich are raised commercially in South Africa for their feathers; they are becoming more common in the United States as meat producers, pets, and novelties. Caged birds (finches, canaries, parrots, and so on) are raised as pets; some feed companies produce special diets for these species.

NUTRIENT REQUIREMENTS OF POULTRY

Energy Requirements

Energy requirements of poultry are usually expressed in terms of ME. There has been considerable interest in **true metabolizable energy (TME).** Sibbald (1982) has developed a technique in which TME is determined in roosters by force-feeding a given quantity of the test feed, collecting the excreta, and determining GE of feed and excreta. A correction is made for endogenous losses of energy of a fasting bird. This endogenous correction yields similar ME results to other more difficult assays. Although the intricacies of the technique are controversial, TME values are useful in diet formulation.

High-energy cereal grains (corn or sorghum) are the principal energy sources used in modern poultry diets. Fat is often added at levels of 3 to 8 percent to increase the dietary energy concentration.

Protein and Amino Acid Requirements

Amino acid requirements for poultry are somewhat different than for swine in that glycine or serine, which can be used interchangeably, are dietary requirements, and the arginine requirement is higher than for swine. Both of these differences are related to mechanisms of nitrogen metabolism. The bird lacks the full complement of urea cycle enzymes involved in urea synthesis and excretes **uric acid** as an end product of protein metabolism. In mammals, urea for excretion is formed in the liver; in the final step of the urea cycle, a molecule of urea and a molecule of arginine are formed. Conversely, in the formation of uric acid in the bird, glycine is required as it contributes part of the uric acid molecule, and the bird is unable to synthesize arginine.

Soybean meal is the major plant protein supplement used in poultry diets. Cottonseed meal should be avoided in poultry diets, especially for layers, because of gossypol toxicity (see Chapter 4). Because the protein and amino acid requirements as a percent of diet are higher for poultry than for swine, high-protein supplements such as fish meal are more commonly used in poultry feeding. Methionine is the first-limiting amino acid in corn-soy diets for poultry. Synthetic methionine is widely used as a supplement, either as DL-methionine or methionine hydroxy analog.

Whereas in the past most diets were formulated on the actual content of the amino acids in the feedstuff, there has been a shift toward formulation of diets on a digestible amino acid basis. Digestible amino acids refer to that portion of the protein that is actually absorbed by the birds rather than excreted. This is a more accurate form of expressing requirements and should be used when sufficient data are available.

Mineral Requirements

Poultry have high requirements for mineral elements. In the case of broilers, the rapid growth rate increases requirements. Also, modern strains of broilers have been selected for a very meaty conformation, putting a strain on the legs and other support systems. Proper balance of various minerals, including calcium, phosphorus, zinc, copper, and manganese, and the electrolytes sodium, potassium, and chlorine, is necessary to prevent leg disorders such as perosis and tibial dyschrondroplasia.

The **calcium** requirement of laying hens is very high (3.4 percent of the diet) for eggshell formation (Table 14–1). Good eggshell quality is very important to egg producers; their profit margin can be lost with a high rate of broken or cracked eggs. Thus for layers, close monitoring of eggshell quality is necessary, with changes in formulation needed if shell quality starts to fall. Excess levels of dietary calcium may induce deficiencies of other minerals, particularly manganese and zinc.

The organic or phytin **phosphorus** in grains and plant protein supplements has a very low bioavailability in poultry, especially in chicks. Proper supplementation with calcium and phosphorus is critical. Phosphorus sources are expensive and sometimes difficult to obtain because phosphate deposits are mined in relatively few parts of the world. Raw rock phosphate may contain toxic concentrations of other elements such as fluorine and vanadium; it usually must be heat-processed to produce a safe product (e.g., defluorinated rock phosphate). The use of enzyme or microbial feed supplements to provide phytase activity in the gut has been shown to be an effective approach for increasing the utilization of plant phosphorus and reducing the amount of inorganic phosphate supplementation necessary. The use of genetically modified grains containing lower levels of phytic acid (but same amount of total P) also shows promise for increasing the utilization of plant phosphorus and reducing the amount of inorganic phosphate supplementation necessary (Li et al. 2000; Jang et al. 2003).

Vitamin Requirements

Because of the simple digestive tract and rapid rate of passage of digesta in poultry, there is little intestinal synthesis of B-complex vitamins. Birds do not utilize vitamin D_2 efficiently, so poultry diets should be supplemented with vitamin D_3. Requirements of poultry for **vitamin D** are expressed in International Chick Units (ICU), based on the activity in chick bioassays. One ICU is defined as 0.025 mg of vitamin D_3. Vitamin D is also available commercially in the hydroxylated form, 25 OH D_3. It is absorbed more efficiently in this form than as vitamin D_3. Use of the 25 OH D_3 may help prevent skeletal deformities in young chicks induced by the fat malabsorption syndrome (see Fat Disorders, this chapter).

Vitamin E deficiency signs in poultry include nutritional muscular dystrophy, exudative diathesis, and encephalomalacia (crazy chick disease). As in other species, the requirement for vitamin E depends upon the dietary selenium level and the presence of synthetic antioxidants, which have a sparing effect on vitamin E. Diets high in unsaturated fatty acids increase vitamin E requirements and enhance its destruction in the diet.

Vitamin K is important in poultry nutrition because of the lack of microbial vitamin synthesis in the gut. Signs of vitamin K deficiency are related to its role in blood clotting. Mild deficiency may result in small internal hemorrhages that cause blemishes on the carcass. Numerous outbreaks of **hemorrhagic syndrome** have occurred in commercial flocks. Therapeutic doses of menadione (water-soluble vitamin K) in the water are effective therapy. Some cases of field hemorrhagic syndrome are caused by mycotoxins and do not respond to vitamin K.

Vitamin requirements are higher for hatchability than for egg production. Therefore, breeder birds producing eggs for hatching usually are fed a diet with a higher level of vitamin supplementation than are birds used for commercial egg production.

Xanthophyll pigments, which are chemically similar to β-carotene, are important in poultry nutrition. Individual **xanthophylls** vary in color; lutein in alfalfa is yellow, whereas zeaxanthin in corn is orange-red. Marigold meal is the

TABLE 14–1. Nutrient Requirements of Leghorn-Type Laying Hens as Percentages or Units per Kilogram of Diet (90 percent dry matter)

		Dietary Concentrations Required by White-Egg Layers at Different Feed Intakes			*Amounts Required per Hen Daily (mg or IU)*		
Nutrient	***Unit***	***80[a,b]***	***100[a,b]***	***120[a,b]***	***White-Egg Breeders at 100 g of Feed per Hen Daily[b]***	***White-Egg Layers at 100 g of Feed per Hen Daily***	***Brown-Egg Layers at 110 g of Feed per Hen Daily[c]***
Protein and amino acids							
Crude protein[d]	%	18.8	15.0	12.5	15,000	15,000	16,500
Arginine[e]	%	***0.88***	***0.70***	***0.58***	***700***	***700***	*770*
Histidine	%	***0.21***	***0.17***	***0.14***	***170***	***170***	***190***
Isoleucine	%	0.81	0.65	0.54	650	650	***715***
Leucine	%	***1.03***	***0.82***	***0.68***	***820***	***820***	***900***
Lysine	%	0.86	0.69	0.58	690	690	***760***
Methionine	%	0.38	0.30	0.25	300	300	***330***
Methionine + cystine	%	0.73	0.58	0.48	580	580	***645***
Phenylalanine	%	***0.59***	***0.47***	***0.39***	***470***	***470***	***520***
Phenylalanine + tyrosine	%	***1.04***	***0.83***	***0.69***	***830***	***830***	***910***
Threonine	%	***0.59***	***0.47***	***0.39***	***470***	***470***	***520***
Tryptophan	%	0.20	0.16	0.13	160	160	175
Valine	%	***0.88***	***0.70***	***0.58***	***700***	***700***	*770*
Fat							
Linoleic acid	%	1.25	1.0	0.83	1,000	1,000	***1,100***
Macrominerals							
Calcium[f]	%	**4.06**	3.25	2.71	3,250	3,250	3,600
Chloride	%	0.16	0.13	0.11	130	130	***145***
Magnesium	mg	625	**500**	420	50	50	***55***
Nonphytate phosphorus[g]	%	0.31	0.25	0.21	250	250	***275***
Potassium	%	0.19	0.15	0.13	150	150	***165***
Sodium	%	0.19	0.15	0.13	150	150	***165***

Trace minerals							
Copper	mg	?	?	?	?	?	?
Iodine	mg	0.044	0.035	0.029	0.010	0.004	***0.004***
Iron	mg	56	45	38	6.0	4.5	***5.0***
Manganese	mg	25	20	17	2.0	2.0	***2.2***
Selenium	mg	0.08	0.06	0.05	0.006	0.006	***0.006***
Zinc	mg	44	35	29	4.5	3.5	***3.9***
Fat soluble vitamins							
A	IU	3,750	3,000	2,500	300	300	***330***
D_3	IU	375	300	250	30	30	***33***
E	IU	***6***	***5***	***4***	***1.0***	0.5	***0.55***
K	mg	0.6	0.5	0.4	0.1	***0.05***	***0.055***
Water soluble vitamins							
B_{12}	mg	0.004	***0.004***	***0.004***	0.008	***0.0004***	***0.0004***
Biotin	mg	***0.13***	***0.10***	***0.08***	***0.01***	***0.01***	***0.011***
Choline	mg	1,310	1,050	875	105	105	***115***
Folacin	mg	0.31	0.25	0.21	0.035	0.025	***0.028***
Niacin	mg	12.5	10.0	8.3	1.0	1.0	***1.1***
Pantothenic acid	mg	2.5	2.0	1.7	0.7	0.20	***0.22***
Pyridoxine	mg	3.1	2.5	2.1	0.45	0.25	***0.28***
Riboflavin	mg	3.1	2.5	2.1	0.36	0.25	***0.28***
Thiamin	mg	0.88	0.70	0.60	0.07	***0.07***	***0.08***

Adapted from National Research Council (1994).

NOTE: Where experimental data are lacking, values typeset in bold italics represent an estimate based on values obtained for other ages or related species.

[a]Grams feed intake per 100 hen daily.

[b]Based on dietary ME_n concentrations of approximately 2,900 kcal/kg and an assumed rate of egg production of 90 percent (90 eggs per 100 hens daily).

[c]Italicized values are based on those from white-egg layers, but were increased 10 percent because of larger body weight and possibly more egg mass per day.

[d]Laying hens do not have a requirement for crude protein per se. However, there should be sufficient crude protein to ensure an adequate supply of nonessential amino acids. Suggested requirements for crude protein are typical of those derived with corn-soybean meal diets, and levels can be reduced somewhat when synthetic amino acids are used.

[e]Italicized amino acid values for white-egg-laying chickens were estimated by using Model B (Hurwitz and Bornstein, 1973), assuming a body weight of 1,800 g and 47 g of egg mass per day.

[f]The requirement may be higher for maximum eggshell thickness.

[g]The requirement may be higher in very hot temperatures.

richest available source of xanthophylls. Synthetic xanthophylls such as **canthaxanthin** are also used as pigmenting agents. Aflatoxin contamination of feeds may impair xanthophyll absorption, resulting in pale bird syndrome (Schaeffer and Hamilton, 1990).

NUTRIENT REQUIREMENTS OF CHICKENS

Broiler Nutrient Requirements

Broilers are young, rapidly growing birds raised for meat. The broiler industry has become very sophisticated in all aspects, with the result that poultry meat is the least expensive meat available in the United States. This is one of the reasons that per capita consumption of chicken has been steadily increasing. Genetic programs have produced broilers with a very rapid growth rate. These birds require nutrient-rich diets to meet genetic potential for rapid growth. Broilers now reach market weight at less than 7 weeks of age. Typically, broilers are raised by contract growers, who produce birds for vertically integrated poultry companies (for discussion, see Cheeke, 2004). Broilers are raised at high stocking density (Fig. 14.1) on the floor, as contrasted with layers, which are kept in cages (Fig. 14.2).

Broilers are fed **multiple-stage diets**, meaning that the composition is changed as age increases. The starter diet is fed for the first 3 weeks and contains 23 to 24 percent crude protein. A grower diet is fed from 3 to 6 weeks, and a finisher diet past 6 weeks. Alternatively, a two-stage system with a starter and finisher diet may be used. Feed is provided *ad libitum* to ensure maximum growth rate. Broiler diets generally contain a coccidiostat to control coccidiosis and a growth-promoting antibiotic or copper sulfate (see Chapter 8). Diets are usually fed as mash, although **crumbles** (rolled pellets) are sometimes used.

Broiler breeders become excessively fat and heavy, causing leg problems and low egg production, if full-fed on high energy feed. Broiler breeder diets are formulated to provide about 120 percent of the requirements for all nutrients (other than energy), and their consumption is limited to 80 percent of what *ad libitum* intake would be.

Nutrient requirements of broilers are given in Table 14–2 and examples of typical broiler diets in Table 14–3.

FIGURE 14.1 Intensive production of broiler chickens. (Courtesy of R. A. Swick.)

FIGURE 14.2 A modern egg farm, with layers kept in small cages. The eggs roll to a conveyor belt in front to be moved automatically to the egg processing facility. Feed and water are provided automatically. There is little human labor involved.

TABLE 14–2. Nutrient Requirements of Broilers as Percentages or Units per Kilogram of Diet (90 percent dry matter)

Nutrient	*Unit*	*0 to 3 Weeks*[a] *3, 200*[b]	*3 to 6 Weeks*[a] *3, 200*[b]	*6 to 8 Weeks*[a] *3, 200*[b]
Protein and amino acids				
Crude protein[c]	%	23.00	20.00	18.00
Arginine	%	1.25	1.10	1.00
Glycine + serine	%	1.25	***1.14***	***0.97***
Histidine	%	0.35	***0.32***	***0.27***
Isoleucine	%	0.80	***0.73***	***0.62***
Leucine	%	1.20	***1.09***	***0.93***
Lysine	%	1.10	1.00	0.85
Methionine	%	0.50	0.38	0.32
Methionine + cystine	%	0.90	0.72	0.60
Phenylalanine	%	0.72	***0.65***	***0.56***
Phenylalanine + tyrosine	%	1.34	***1.22***	***1.04***
Proline	%	0.60	***0.55***	***0.46***
Threonine	%	0.80	0.74	0.68
Trytophan	%	0.20	0.18	0.16
Valine	%	0.90	***0.82***	0.70

(continues)

TABLE 14–2. Nutrient Requirements of Broilers as Percentages or Units per Kilogram of Diet (90 percent dry matter) (*Continued*)

Nutrient	*Unit*	*0 to 3 Weeks*[a] *3, 200*[b]	*3 to 6 Weeks*[a] *3, 200*[b]	*6 to 8 Weeks*[a] *3, 200*[b]
Fat				
Linoleic acid	%	1.00	1.00	1.00
Macrominerals				
Calcium[d]	%	1.00	0.90	0.80
Chlorine	%	0.20	0.15	0.12
Magnesium	mg	600	***600***	***600***
Nonphytate phosphorus	%	0.45	0.35	0.30
Potassium	%	0.30	0.30	0.30
Sodium	%	0.20	0.15	0.12
Trace minerals				
Copper	mg	8	***8***	***8***
Iodine	mg	***0.35***	0.35	0.35
Iron	mg	80	***80***	***80***
Manganese	mg	***60***	***60***	***60***
Selenium	mg	0.15	0.15	0.15
Zinc	mg	40	***40***	***40***
Fat soluble vitamins				
A	IU	***1,500***	***1,500***	***1,500***
D_3	ICU	200	200	200
E	IU	***10***	***10***	***10***
K	mg	0.50	0.50	0.50
Water soluble vitamins				
B_{12}	mg	0.01	***0.01***	***0.007***
Biotin	mg	0.15	0.15	***0.12***
Choline	mg	1,300	***1,000***	***750***
Folacin	mg	0.55	0.55	***0.50***
Niacin	mg	35	30	25
Pantothenic acid	mg	***10***	***10***	***10***
Pyridoxine	mg	3.5	3.5	***3.0***
Riboflavin	mg	3.6	3.6	3
Thiamin	mg	***1.80***	***1.80***	***1.80***

Adapted from National Research Council (1994).

NOTE: Where experimental data are lacking, values typeset in bold italics represent an estimate based on values obtained for other ages or related species.

[a] The 0- to 3-, 3- to 6-, and 6- to 8- week intervals for nutrient requirements are based on chronology for which research data were available; however, these nutrient requirements are often implemented at younger age intervals or on a weight-of-feed consumed basis.

[b] These are typical dietary energy concentrations, expressed in kcal ME_m/kg diet. Different energy values may be appropriate depending on local ingredient prices and availability.

[c] Broiler chickens do not have a requirement for crude protein per se. There, however, should be sufficient crude protein to ensure an adequate nitrogen supply for synthesis of nonessential amino acids. Suggested requirements for crude protein are typical of those derived with corn-soybean meal diets, and levels can be reduced when synthetic amino acids are used.

[d] The calcium requirement may be increased when diets contain high levels of phytate phosphorus (Nelson, 1984).

TABLE 14–3. Percentage Composition of Typical Broiler Diets

Ingredient	*Starter, 0–3 wks*	*Grower, 3–6 wks*	*Finisher, 6–9 wks*
Ground yellow corn	55.10	62.75	67.50
Soybean meal	29.00	25.00	20.25
Meat and bone meal	5.00	4.00	4.00
Fish meal (menhaden)	2.00	—	—
Dehydrated alfalfa meal	2.00	2.00	2.00
Limestone	0.25	0.70	0.60
Dicalcium phosphate	0.30	0.70	0.80
Stabilized fat	5.00	3.50	3.50
Trace mineralized salt	0.25	0.25	0.25
Vitamin premix	1.00	1.00	1.00
Methionine	0.10	0.10	0.10

Adapted from Jurgens (1993).

Layer Nutrient Requirements

Layers are chickens raised for egg production; most are of the Leghorn breed. They are smaller bodied than the meat breeds and have a slower growth rate and, therefore, lower nutritional requirements than broilers during the early growth stages. Because of the high nutrient content of eggs and the high level of production of modern layers, the nutritional requirements of birds in full production are very high, particularly for protein, energy, and calcium.

Flock replacement chicks are fed a 20 percent crude protein all-mash starter diet for the first 6 weeks, followed by a 15 to 16 percent protein grower feed to 12 weeks, and a 12 to 13 percent protein developer feed until sexual maturity (age at first egg laying, approximately 18 to 20 weeks). Examples of typical diets are given in Table 14–4. Feed may be restricted during the developer period to delay onset of sexual maturity. This is desirable to reduce the number of small eggs produced when the birds begin laying. Layers have a high calcium requirement for eggshell formation. This is met by providing adequate calcium in the diet by the addition of crushed limestone (calcium carbonate) or crushed oyster shell. When adequate calcium (3.75 g or more per day) is provided, the hen should be able to absorb the 2 grams of calcium needed daily for consistent egg production. The large particle size of oyster shell may be more beneficial in metering calcium into the blood from the gut over a more prolonged time period than when calcium is provided in the diet in a finely ground form such as limestone flour. However, when adequate calcium (3.75 g or more per day) is provided, there is little difference in response to oyster shell and limestone flour.

Breeder hens should receive a diet containing adequate levels of minerals and vitamins because borderline deficiencies of some of these nutrients impair fertility and hatchability and may cause developmental defects. Examples of typical layer and breeder diets are given in Table 14–5. Layers are often "phase-fed," with an increase in the calcium level at approximately 40 weeks of age to maintain eggshell quality.

TABLE 14–4. Percentage Composition of Typical Diets for Replacement Pullets

Ingredient	*Starter, 0–6 wks*	*Grower, 6–14 wks*	*Developer, 14–20 wks*
Ground yellow corn	61.25	59.00	61.60
Wheat middlings	4.00	3.00	6.80
Oats	8.00	18.00	20.00
Dehydrated alfalfa meal	2.50	2.50	2.50
Fish meal, menhaden	2.50	—	—
Meat and bone meal	2.50	—	—
Soybean meal	16.30	14.40	6.00
Dicalcium phosphate	1.00	1.00	1.00
Limestone	0.50	0.75	0.7
Salt	0.25	0.25	0.25
Trace mineral premix	0.10	0.10	0.10
Vitamin premix	1.00	1.00	1.00
Methionine	0.10	—	—

Adapted from Jurgens (1988).

TABLE 14–5. Percentage Composition of Complete Layer and Breeder Diets

Ingredient	*Phase 1 Layer Diet, 22–40 wks*		*Phase 1 Breeder Diet*	*Phase 2 Layer Diet (after 40 wks)*		*Phase 2 Breeder Diet*
Ground yellow corn	60.2	66.75	55.3	58.7	26.7	66.3
Wheat shorts	—	—	—	5.0	—	—
Ground sorghum	—	—	—	—	38.0	—
Hominy feed	2.5	—	10.0	—	—	—
Stabilized grease	2.0	—	1.5	2.5	2.5	—
Soybean meal, 49% CP	21.0	12.5	15.0	20.5	14.5	14.5
Fish meal, 60% CP	—	1.5	3.5	—	—	2.5
Dried fish solubles	0.5	—	1.0	—	—	—
Meat and bone meal, 50% CP	—	5.0	—	—	5.0	2.5
Corn distillers dried grains	2.5	5.0	5.0	—	—	2.5
Alfalfa meal, 17% CP	1.5	1.5	2.0	2.5	4.0	2.5
Dicalcium phosphate	1.5	—	1.0	1.5	—	0.5
Limestone	5.0	5.0	4.5	4.5	4.5	4.0
Oyster shell	2.0	2.0	2.0	4.0	4.0	4.0
Trace mineralized salt	0.25	0.25	0.25	0.25	0.25	0.25
DL-methionine	0.05	0.05	0.025	0.05	50.08	0.025
Vitamin premix	0.5	0.5	0.5	0.5	0.5	0.5

Adapted from Scott et al. (1982).

NUTRIENT REQUIREMENTS OF TURKEYS

Turkeys are raised for meat. Strains have been selected for rapid growth and efficient meat production. Essentially all of the commercially produced birds are white, whereas in the past the bronze-colored breeds predominated. There is a niche market for so-called **heritage breeds** of turkeys, which are traditional breeds that have fallen by the wayside with the development of "industrial" white birds (Christman and Haives, 1999). Turkey meat has changed from being a seasonal dish (Thanksgiving and Christmas) to having year-round consumption, with new products such as turkey ham contributing to this trend. The consumer demand for white meat has also increased turkey consumption.

Young turkey poults have a higher **protein** requirement than chickens. Starter diets (zero to 4 weeks) should contain 28 percent crude protein. The protein content can be reduced by approximately 2 percent during each 4-week period, up to a market age of approximately 20 weeks. Turkeys are sometimes raised outside on ranges, but the nutritional contribution of grass, insects, and so on that they obtain is usually of little significance. Nutrient requirements of turkeys are given in Table 14–6 and examples of typical diets are shown in Table 14–7 and 14–8. The nutrition and feeding of turkeys has been reviewed in depth by Scott (1987).

Turkeys are affected by a number of disorders that have a nutritional component. These include leg weakness, hock joint disorder, footpad dermatitis, pendulous crop, ascites, and aortic rupture. **Leg weakness disorders** are associated with deficiencies of nutrients involved in bone formation, including calcium, phosphorus, vitamin D_3, choline, biotin, folic acid, manganese, and zinc. Enlargement of the hock joint may occur with deficiencies of niacin, biotin, vitamin E, or zinc. **Footpad dermatitis** is typical of biotin deficiency. Diets causing sticky droppings, such as with barley or poorly processed soybean meal, may cause droppings to adhere to the feet and cause dermatitis. **Pendulous crop,** caused by yeast proliferation in the crop, may occur with diets high in fermentable carbohydrate. The distention of the crop associated with gas production from the fermentation may interfere with passage of ingesta from the crop to the proventriculus, causing the pendulous crop condition. The disorder can be treated with a fungal inhibiting antibiotic. **Ascites,** or fluid accumulation in body cavities, has been observed with high salt intakes. Selenium deficiency also causes an edematous condition, **exudative diathesis.** A selenium deficiency seen in turkeys is **gizzard myopathy,** or degeneration of the gizzard. However, it should be noted that with the advent of computer diet formulation, these problems are rare in commercially raised birds.

Aortic rupture occasionally occurs in market turkeys. Although aortic aneurysm can be induced experimentally with copper deficiency and by feeding lathyrogenic substances (lathyrogens are toxic amino acids in seeds of *Lathyrus* species), the causes of the problem under field conditions have not been identified. Scott (1987) has suggested that gizzerosine, the gizzard erosion factor, may play a role in aortic aneurysm and rupture.

TABLE 14–6. Nutrient Requirements of Turkeys as Percentages or Units per Kilogram of Diet (90 percent dry matter)

Nutrient	Unit	Growing Turkeys, Males and Females						Breeders	
		0 to 4 Weeks[a], 0 to 4 Weeks[b], 2,800[c]	4 to 8 Weeks[a], 0 to 8 Weeks[b], 2,900[c]	8 to 12 Weeks[a], 8 to 11 Weeks[b], 3,000[c]	12 to 16 Weeks[a], 11 to 14 Weeks[b], 3,100[c]	16 to 20 Weeks[a], 14 to 17 Weeks[b], 3,200[c]	20 to 24 Weeks[a], 17 to 20 Weeks[b], 3,300[c]	Holding; 2,900[c]	Laying Hens; 2,900[c]
Protein and amino acids									
Protein[d]	%	28.0	26	22	19	16.5	14	12	14
Arginine	%	1.6	*1.4*	*1.1*	*0.9*	*0.75*	*0.6*	*0.5*	*0.6*
Glycine + serine	%	1.0	*0.9*	*0.8*	*0.7*	*0.6*	*0.5*	*0.4*	*0.5*
Histidine	%	0.58	*0.5*	*0.4*	*0.3*	*0.25*	*0.2*	*0.2*	*0.3*
Isoleucine	%	1.1	*1.0*	*0.8*	*0.6*	*0.5*	*0.45*	*0.4*	*0.5*
Leucine	%	1.9	*1.75*	*1.5*	*1.25*	*1.0*	*0.8*	*0.5*	*0.5*
Lysine	%	1.6	1.5	1.3	1.0	0.8	0.65	*0.5*	*0.6*
Methionine	%	0.55	0.45	0.4	0.35	0.25	*0.25*	*0.2*	*0.2*
Methionine + cystine	%	1.05	0.95	0.8	0.65	0.55	*0.45*	*0.4*	*0.4*
Phenylalanine	%	1.0	*0.9*	*0.8*	*0.7*	*0.6*	*0.5*	*0.4*	*0.55*
Phenylalanine + tyrosine	%	1.8	*1.6*	*1.2*	*1.0*	*0.9*	*0.9*	*0.8*	*1.0*
Threonine	%	1.0	*0.95*	*0.8*	*0.75*	*0.6*	*0.5*	*0.4*	*0.45*
Tryptophan	%	0.26	*0.24*	*0.2*	*0.18*	*0.15*	*0.13*	*0.1*	*0.13*
Valine	%	1.2	*1.2*	*0.9*	*0.8*	*0.7*	*0.6*	*0.5*	*0.58*
Fat									
Linoleic acid	%	1.0	*1.0*	*0.8*	*0.8*	*0.8*	*0.8*	*0.8*	1.1
Macrominerals									
Calcium[e]	%	1.2	1.0	0.85	0.75	0.65	0.55	0.5	2.25
Nonphytate Phosphorus[f]	%	0.6	0.5	0.42	0.38	0.32	0.28	0.25	0.35
Potassium	%	0.7	*0.6*	*0.5*	*0.5*	*0.4*	*0.4*	*0.4*	*0.6*
Sodium	%	0.17	*0.15*	*0.12*	*0.12*	*0.12*	*0.12*	*0.12*	*0.12*
Chlorine	%	0.15	*0.14*	*0.14*	*0.12*	*0.12*	*0.12*	*0.12*	*0.12*
Magnesium	mg	500	*500*	*500*	*500*	*500*	*500*	*500*	*500*

Trace minerals									
Manganese	mg	60	***60***	***60***	***60***	***60***	***60***	***60***	***60***
Zinc	mg	70	***65***	***50***	***40***	***40***	***40***	***40***	***65***
Iron	mg	***80***	***60***	***60***	***60***	***50***	***50***	***50***	***60***
Copper	mg	***8***	***8***	***6***	***6***	***6***	***6***	***6***	***8***
Iodine	mg	***0.4***	***0.4***	***0.4***	***0.4***	***0.4***	***0.4***	***0.4***	***0.4***
Selenium	mg	0.2	***0.2***	***0.2***	***0.2***	***0.2***	***0.2***	***0.2***	***0.2***
Fat soluble vitamins									
A	IU	5,000	5,000	***5,000***	***5,000***	***5,000***	***5,000***	***5,000***	***5,000***
D[g] $_3$	ICU	1,100	***1,100***	***1,100***	***1,100***	***1,100***	***1,100***	***1,100***	1,100
E	IU	12	***12***	***10***	***10***	***10***	***10***	***10***	25
K	mg	1.75	***1.5***	***1.0***	***0.75***	***0.75***	***0.50***	***0.5***	***1.0***
Water soluble vitamins									
B_{12}	mg	0.003	0.003	***0.003***	***0.003***	***0.003***	***0.003***	***0.003***	***0.003***
Biotin[h]	mg	0.25	0.2	***0.125***	***0.125***	***0.100***	***0.100***	***0.100***	***0.20***
Choline	mg	1,600	1,400	***1,100***	***1,100***	***950***	***800***	***800***	***1,000***
Folacin	mg	1.0	1.0	***0.8***	***0.8***	***0.7***	***0.7***	***0.7***	***1.0***
Niacin	mg	60.0	60.0	***50.0***	***50.0***	***40.0***	***40.0***	***40.0***	***40.0***
Pantothenic acid	mg	10.0	9.0	***9.0***	***9.0***	***9.0***	***9.0***	***9.0***	***16.0***
Pyridoxine	mg	***4.5***	***4.5***	***3.5***	***3.5***	***3.0***	***3.0***	***3.0***	***4.0***
Riboflavin	mg	4.0	3.6	***3.0***	***3.0***	***2.5***	***2.5***	***2.5***	***4.0***
Thiamin	mg	2.0	2.0	***2.0***	***2.0***	***2.0***	***2.0***	***2.0***	***2.0***

Adapted from National Research Council (1994).

Note: Where experimental data are lacking, values typset in bold italics represent estimates based on values obtained from other ages or relate species or from modeling experiments.

[a]The age intervals for nutrient requirements of males are based on actual chronology from previous research. Genetic improvements in body weight gain have led to an earlier implementation of these levels, at 0 to 3, 3 to 6, 6 to 9, 9 to 12, 12 to 15, and 15 to 18 weeks, respectively, by the industry at large.

[b]The age intervals for nutrient requirements of males are based on actual chronology from previous research. Genetic improvements in body weight gain have led to an earlier implement of these levels, at 0 to 3, 3 to 6, 6 to 9, 9 to 12, 12 to 14, and 14 to 16 weeks, respectively, by the industry at large.

[c]These are approximate metabolizable energy (*ME*) values provided with typical corn-soybean-meal-based feeds, expressed in kcal ME_n/kg diet. Such energy, when accompanied by the nutrient levels suggested, is expected to proide near-maximum growth, particularly with pelleted feed.

[d]Turkeys do not have a requirement for crude protein per se. However, there should be sufficient crude protein to ensure an adequate nitrogen supply for synthesis of nonessential amino acids. Suggested requirements for crude protein are typical of those derived with corn-soybean meal diets, and levels can be reduced when synthetic amino acids are used.

[e]The calcium requirement may be increased when diets contain high levels of phytate phosphorus (Nelson, 1984).

[f]Organic phosphorus is generally considered to be associated with phytin and is of limited availability.

[g]These concentrations of vitamin D are considered satisfactory when the associated calcium and phosphorus levels are used.

[h]Requirement may increase with wheat-based diets.

TABLE 14–7. Percentage Composition of Turkey Grower Diets

	Age of Birds					
Ingredient	***0–4 wks***	***5–8 wks***	***9–12 wks***	***13–16 wks***	***17–20 wks***	***20+ wks***
Ground yellow corn	48.15	51.00	62.30	70.50	75.95	80.65
Soybean meal	35.65	37.00	26.60	18.90	13.75	10.00
Meat and bone meal	7.50	7.50	7.50	7.50	6.45	4.00
Fish meal, menhaden	4.00	—	—	—	—	—
Dehydrated alfalfa meal	2.00	—	—	—	—	—
Stabilized fat	1.00	2.00	1.50	1.50	2.00	2.50
Dicalcium phosphate	—	0.35	0.50	0.20	0.50	1.20
Limestone	0.25	0.70	0.15	—	—	0.30
Salt	0.25	0.25	0.25	0.25	0.25	0.25
Trace mineral mix	0.10	0.10	0.10	0.10	0.10	0.10
DL methionine	0.10	0.10	0.10	0.05	—	—
Vitamin premix	1.00	1.00	1.00	1.00	1.00	1.00

Adapted from Jurgens (1993).

TABLE 14–8. Percentage Composition of Representative Turkey Breeder Diets

Ingredient	***Holding Diet***	***Breeder Diet***
Ground yellow corn	58.20	71.00
Ground oats	25.00	—
Soybean meal	6.00	11.10
Fish meal, menhaden	2.00	4.00
Dehydrated alfalfa meal	5.00	4.00
Dicalcium phosphate	2.00	2.00
Limestone	0.40	4.50
Salt	0.30	0.30
Trace mineral mix	0.10	0.10
Vitamin premix	1.00	1.00

Adapted from Jurgens (1993).

NUTRIENT REQUIREMENTS OF WATERFOWL, RATITES, AND GAME BIRDS

Ducks are raised commercially in the United States to produce ducklings for the restaurant trade. The White Pekin breed is the principal meat producer. In China and Southeast Asian countries, such as Indonesia and Thailand, ducks are raised extensively for meat and eggs. Egg-producing breeds, such as the Indian Runner, can achieve egg-production rates equal to or greater than those of chickens. In Asia, duck raising is often integrated into crop production by herding the ducks through rice fields to control insects and to harvest waste grain. **Geese** are the other major waterfowl raised commercially. Goose production is a minor activity in North America, but large numbers are raised in European countries such as France, Hungary, Poland, and Russia. They are raised to produce meat, feathers, down, and in France, liver pate (*pâté de foie gras*). Fatty livers are produced by force-feeding the birds with a high-energy diet.

Ducks and geese are fed diets similar to those for chickens. **Game birds,** such as pheasants and quail, have protein requirements similar to those of turkeys and can be fed turkey diets. The vitamin requirements, especially for niacin, tend to be higher for waterfowl and game birds than for chickens. Leg weakness is a major problem in waterfowl because of their rapid growth rate; adequate choline and niacin are necessary to prevent these disorders. Geese are grazing birds and can be fed on range. The amount of concentrate feed for geese can be limited to 1 to 2 pounds of feed per bird per week. If full fed on grain, they will reduce their forage intake. For several weeks prior to marketing, they should have free access to concentrate feed to develop adequate finish. Although goslings have a very rapid growth rate, a large amount of the gain is skin, fat, and feathers. They are less efficient than chickens in the accretion of lean tissue (Summers et al., 1987). Their ability to utilize forages may be of value in some countries where feed grains are expensive.

Although **geese** are herbivorous and accept high-fiber diets, the digestibility of fiber is low in this species. The ability of geese to utilize fibrous feeds results from a fast passage rate that allows a high feed intake, and the efficient manner in which the gizzard breaks down plant cell walls, which allows digestion of the cell contents (Stevenson, 1989). Hollister et al. (1984) found that the growth rate of goslings fed diets containing either 20 or 40 percent dehydrated alfalfa meal or Kentucky bluegrass was only slightly less than with a high-energy corn-soy diet. Growth on the fibrous diets was improved when the diets were pelleted. Geese are sensitive to bitter substances in the diet (Cheeke et al., 1983) and when grazing prefer grasses and weeds to alfalfa. One reason may be the presence of bitter substances, such as saponins, in alfalfa.

Nutrient requirements are listed for ducks (Table 14–9), geese (Table 14–10), pheasants (Table 14–11), and quail (Table 14–12).

Other birds raised commercially for meat production, especially in Europe, are **pigeons** and **guinea fowl.** The young birds, known as squabs (pigeons) and keets (guinea fowl) are considered delicacies. Diets adequate for chickens can be used for these birds. The starter diet for keets should contain at least 21 percent protein and the finisher diet at least 17 percent (Hughes and Jones, 1980). Pelleting the feed improves growth performance (Oguntona and Hughes, 1988).

Ostriches and other **ratite birds** (emu, kiwi, cassowary, rhea) are large, flightless birds that have been separated from the main lines of avian evolution for 80 to 90 million years, since the Middle Cretaceous period. Their occurrence in Africa, South America, and Australia stems from having common ancestors that lived when these continents were joined. These birds have enlarged ceca with microbial fermentation and utilize fibrous vegetation to a considerable extent. There is evidence of nitrogen recycling from the cloaca to the cecum, increasing the efficiency of protein utilization. The metabolic rate is lower than for other avian species, thus contributing to their ability to survive on low-quality diets (Dawson and Herd, 1983). To a limited extent, ratites such as ostriches and emus are produced commercially in North America (Fig. 14.3).

The ME values of alfalfa and high-fiber grains, such as oats and barley, are higher for ostriches than for chickens because of fiber fermentation in the ceca (Cilliers et al., 1994, 1997a). There is a long retention time of fiber in the hindgut,

TABLE 14–9. Nutrient Requirements of White Pekin Ducks as Percentages or Units per Kilogram of Diet (90 percent dry matter)

Nutrient	*Unit*	*0 to 2 Weeks; 2,900*[a]	*2 to 7 Weeks; 3,000*[a]	*Breeding; 2,900*[a]
Protein and amino acids				
Protein	%	***22***	16	***15***
Arginine	%	***1.1***	***1.0***	
Isoleucine	%	***0.63***	***0.46***	***0.38***
Leucine	%	***1.26***	***0.91***	***0.76***
Lysine	%	***0.90***	***0.65***	***0.60***
Methionine	%	0.40	***0.30***	***0.27***
Methionine + cystine	%	0.70	0.55	***0.50***
Tryptophan	%	***0.23***	***0.17***	***0.14***
Valine	%	***0.78***	***0.56***	***0.47***
Macrominerals				
Calcium	%	***0.65***	***0.60***	***2.75***
Chloride	%	0.12	***0.12***	***0.12***
Magnesium	mg	***500***	***500***	***500***
Nonphytate phosphorus	%	***0.40***	***0.30***	
Sodium	%	0.15	***0.15***	***0.15***
Trace minerals				
Manganese	mg	***50***	?[b]	?
Selenium	mg	0.20	?	?
Zinc	mg	***60***	?	?
Fat soluble vitamins				
A	IU	***2,500***	***2,500***	***4,000***
D_3	IU	***400***	***400***	***900***
E	IU	***10***	***10***	***10***
K	mg	***0.5***	***0.5***	***0.5***
Water soluble vitamins				
Niacin	mg	***55***	***55***	***55***
Pantothenic acid	mg	***11.0***	***11.0***	***11.0***
Pyridoxine	mg	***2.5***	***2.5***	***3.0***
Riboflavin	mg	***4.0***	***4.0***	***4.0***

Adapted from National Research Council (1994).
NOTE: For nutrients not listed or those for which no values are given, see requirements of broiler chickens (Table 14–2) as a guide. Where experimental data are lacking, values typeset in bold italics represent an estimate based on values obtained for other ages or species.
[a]These are typical dietary energy concentrations as expressed in kcal ME_n/kg diet.
[b]Question marks indicate that no estimates are available.

with production of VFA that provides a significant portion of absorbed ME (Swart et al., 1993a, b). In contrast, the **emu** has a poorly developed hindgut and is not well adapted to high fiber diets.

The digestibility of crude protein and amino acids is higher in ostriches than in chickens (Cilliers et al., 1997b). Therefore, for both protein and energy, specific bioavailable nutrient values for ostriches are needed for optimal diet formulation. Extrapolation from chicken values is not appropriate.

TABLE 14–10. Nutrient Requirements of Geese as Percentages or Units per Kilogram of Diet (90 percent dry matter)

Nutrient	*Unit*	*0 to 4 Weeks; 2,900*[a]	*After 4 Weeks; 3,000*[a]	*Breeding; 2,900*[a]
Protein and amino acids				
Protein	%	***20***	***15***	***15***
Lysine	%	1.0	0.85	***0.6***
Methionine+ cystine	%	0.60	0.50	***0.50***
Macrominerals				
Calcium	%	***0.65***	***0.60***	***2.25***
Nonphytate phosphorus	%	***0.30***	***0.3***	***0.3***
Fat soluble vitamins				
A	IU	***1,500***	***1,500***	***4,000***
D_3	IU	***200***	***200***	***200***
Water soluble vitamins				
Choline	mg	***1,500***	***1,000***	?
Niacin	mg	***65.0***	***35.0***	***20.0***
Pantothenic acid	mg	***15.0***	***10.0***	***10.0***
Riboflavin	mg	3.8	***2.5***	***4.0***

Adapted from National Research Council (1994).
NOTE: For nutrients not listed or those for which no values are given, see requirements of chickens (Table 14–2) as a guide. Where experimental data are lacking, values typeset in bold italics represent an estimate based on values obtained for other ages or species.
[a]These are typical dietary energy concentrations, expressed in kcal ME_n/kg diet.

The following recommendations for feeding **ostriches** are provided by Thornberry (1988). Young chicks can be started on turkey or game bird starter ration containing at least 26 percent protein. They should have continuous access to the starter diet for the first 3 weeks, after which they can be fed all the starter diet they will consume in two 30-minute feeding periods per day. At this time, they should have free access to alfalfa pellets. This feeding program will help to prevent excessive weight gain that causes **leg weakness** problems. At 8 weeks of age, the ostrich chicks can be fed a game bird or turkey grower ration for two feeding periods daily, with continuous access to alfalfa pellets. If good-quality pasture is available, they can be ranged on pasture with the supplemental ration provided twice daily. Forages should be young and succulent. Adult birds can be kept on pasture, with daily supplements of a turkey breeder diet during the breeding season. Birds in dry-lots can be fed alfalfa pellets or good-quality hay.

Ratites have a tendency to consume foreign objects, including sand and pebbles. **Impaction** of the gut with sand can occur. Morrow et al. (1997) reported a case of hypophosphatemic **rickets** and nephrocalcinosis in ostrich chicks reared on limestone sand. The birds ingested large quantities of the sand and the absorbed calcium caused a calcium: phosphorus imbalance. Urinary excretion of excess calcium resulted in excessive phosphorus loss (see action of PTH, Chapter 2), calcification of the kidneys, and permanent stunting of the birds. Calcium: phosphorus ratios in excess of 2:1 may lead to hypophosphatemic rickets in rheas (Grone et al., 1995).

TABLE 14–11. Nutrient Requirements of Ring-Necked Pheasants as Percentages or Units per Kilogram of Diet (90 percent dry matter)

Nutrient	*Unit*	*0 to 4 Weeks; 2,800*[a]	*4 to 8 Weeks; 2,800*[a]	*9 to 17 Weeks; 2,700*[a]	*Breeding; 2,800*[a]
Protein and amino acids					
Protein	%	28	24	18	15
Glycine + serine	%	***1.8***	***1.55***	***1.0***	***0.50***
Linoleic Acid	%	***1.0***	***1.0***	***1.0***	***1.0***
Lysine	%	***1.5***	***1.40***	***0.8***	***0.68***
Methionine	%	***0.50***	***0.47***	***0.30***	***0.30***
Methionine + cystine	%	***1.0***	***0.93***	***0.6***	***0.60***
Protein	%	28	***24***	***18***	15
Macrominerals					
Calcium	%	1.0	***0.85***	***0.53***	***2.5***
Chlorine	%	***0.11***	***0.11***	***0.11***	***0.11***
Nonphytate phosphorus	%	0.55	***0.50***	***0.45***	***0.40***
Sodium	%	***0.15***	***0.15***	***0.15***	***0.15***
Trace minerals					
Manganese	mg	***70***	***70***	***60***	***60***
Zinc	mg	60	***60***	***60***	***60***
Water soluble vitamins					
Choline	mg	***1,430***	***1,300***	***1,000***	***1,000***
Niacin	mg	70.0	***70***	***40.0***	***30.0***
Pantothenic acid mg	mg	10.0	***10.0***	***10.0***	***16.0***
Riboflavin	mg	3.4	***3.4***	***3.0***	***4.0***

Adapted from National Research Council (1994).

NOTE: Where experimental data are lacking, values typeset in bold italics represent an estimate based on values obtained for other ages or species. For nutrients not listed or those for which no values are given, see requirements of turkeys (Table 14–6) as a guide.

[a]These are typical dietary energy concentrations, expressed in kcal ME_n/kg diet.

FIGURE 14.3 Ostriches have been domesticated in South Africa, where the birds are raised as sources of plumes (feathers), meat, and leather. There is a small commercial ostrich industry in North America.

TABLE 14–12. Nutrient Requirements of Bobwhite Quail as Percentages or Units per Kilogram of Diet (90 percent dry matter)

Nutrient	*Unit*	*0 to 6 Weeks; 2,800*[a]	*After 6 Weeks; 2,800*[a]	*Breeding; 2,800*[a]
Protein and amino acids				
Protein	%	***26***	***20.0***	24.0
Methionine + cystine	%	***1.0***	***0.75***	***0.90***
Fat				
Linoleic acid	%	***1.0***	***1.0***	***1.0***
Macrominerals				
Calcium	%	0.65	***0.65***	***2.4***
Nonphytate phosphorus	%	***0.45***	***0.30***	***0.70***
Sodium	%	***0.15***	***0.15***	***0.15***
Trace minerals				
Chlorine	%	***0.11***	***0.11***	***0.11***
Iodine	mg	0.30	***0.30***	***0.30***
Water soluble vitamins				
Choline	mg	1,500.0	***1,500.0***	***1,000.0***
Niacin	mg	30.0	***30.0***	***20.0***
Pantothenic acid	mg	12.0	***9.0***	***15.0***
Riboflavin	mg	3.8	***3.0***	***4.0***

Source: Adapted from National Research Council (1994).

NOTE: Where experimental data are lacking, values typeset in bold italics represent an estimate based on values obtained for other ages or species. For values not listed for the starting-growing periods, see requirements for turkeys as a guide.

[a]These are typical dietary energy concentrations, expressed in kcal ME_n/kg diet.

NUTRIENT REQUIREMENTS OF EXOTIC BIRDS

A variety of exotic birds such as canaries, finches, parrots, and cockatiels are raised by bird fanciers. With the widespread destruction of the native tropical rain forest habitat of many of these birds, an interest in raising and propagating them in aviaries is developing. As a result, there is an increasing demand for commercial feed for cage birds, and some feed companies produce special feed formulae for this purpose. As with the production of all other types of feeds, knowledge of the nutritional requirements and digestive abilities of the birds is necessary for optimal formulation of cage-bird diets. This is especially true when reproduction is desired, as nutrient requirements for reproduction are much higher than for an adult kept under maintenance conditions, as many pet birds are. Relatively little research has been conducted in this area. The cage-bird program of the Department of Avian Sciences, University of California, Davis, has been one of the major sources of information.

The majority of cage birds are psittacines and passerines. **Psittacines** are members of the parrot family (family Psittacidae) and include parrots, macaws, parakeets (budgerigars), lories, cockatiels, and cockatoos. **Passerines** are members of the order Passeriformes or perching birds, and include finches and canaries. Many of the psittacines are **altricial birds,** meaning that they are physiologically immature at hatching and require a long period of intensive care by the parent birds or aviculturist. The newly hatched young of altricial birds "look like embryos prematurely escaped from their shells" (Roudybush and Grau, 1986) and are completely dependent on the parent birds for food, water, warmth, and care. In many cases, partially digested crop contents (e.g., pigeon "milk") are regurgitated by the parent and fed to the offspring. Artificial rearing of these birds involves hand-feeding a liquid concoction resembling the crop contents of the adult bird in composition.

Adult psittacines and passerines are seed-eating birds. They are commonly fed mixtures of "bird seed" such as sunflower, millet, sorghum, oat, corn, canary grass (*Phalaris canariensis*), pumpkin, rape, buckwheat, flax, and safflower seeds. Fresh vegetable supplements and mineral-vitamin mixtures are often given. According to Harrison and Harrison (1986), the typical caged bird fed on a diet consisting primarily of seeds has a characteristic appearance. The feathering is lackluster, and the feathers lack a smooth, interlocking, tight look. The epithelial tissues have a dry, flaky texture; the beak and nails may be long and rough. Varying the diet by including supplemental protein, fresh green or yellow vegetables, and fruits to an all-seed diet is beneficial in overcoming these signs of mild, generalized malnutrition (Harrison, 1998).

Obesity from overnutrition is common in budgerigars and cockatoos (Harrison and Harrison, 1986). Often cockatoos will resist eating anything but sunflower seeds and develop lipomas (fat deposits) as a result. These may require surgical removal.

In developing **pelleted feeds** to take the place of seed mixtures, factors that should be considered include palatability, moisture content, odor of the excreta, and bird performance. Feeding excess protein may cause moist, odoriferous droppings because of the extra uric acid being excreted. Roudybush and Grau (1986) found that 20 percent dietary crude protein was optimal for growing cockatiels. An example of a satisfactory diet developed by these workers for cockatiels is presented in Table 14–13.

The feeding and nutrition of psittacine and passerine birds has been reviewed by Harrison (1986) and Ullrey et al. (1991). Digestibility studies have been conducted with canaries (Taylor et al., 1994) and budgerigars (Earle and Clarke, 1991). Information on diseases, nutrition, surgical techniques, and management of exotic birds is provided by Harrison and Harrison (1986).

TABLE 14–13. Example of a Diet Satisfactory for Cockatiels

Ingredient	%	*Ingredient*	%
Ground yellow corn	81.77	DL-methionine	0.26
Isolated soy protein	14.46	Mineral premix	0.12
$CaCO_3$	0.64	Vitamin premix	0.18
Dicalcium phosphate	2.57		

From Roudybush et al. (1984).

SPECIAL NUTRITIONAL PROBLEMS OF POULTRY

Leg Disorders

Leg abnormalities are a problem in poultry, especially in birds kept on wire. **Perosis** or slipped tendon (Fig. 14.4) is caused by deficiencies of choline and manganese. Perosis is due to an abnormality of the joint in the long bones of the leg, causing the tendon to slip and pull the leg sideways. **Cage layer fatigue** is a type of osteoporosis that involves excessive mobilization of calcium from the leg bones, which causes the birds to have difficulty standing and broken bones. It is primarily a problem with layers kept in wire cages at high stocking density. Lack of exercise is probably a contributing factor. At the end of the laying cycle, many layers have broken bones. Nutritional factors are no doubt involved, relating to calcium, phosphorus, and vitamin D. However, supplementation with these nutrients does not prevent osteoporosis (Rennie et al., 1997). Genetic factors may be involved as well.

Another leg disorder seen in broilers is **tibial dyschondroplasia** (TD) in which there is abnormal formation of cartilage in the long bones. The cartilage forms a thickened layer below the epiphyseal plate. The lesion arises from the failure of growth plate chondrocytes to differentiate (Rennie and Whitehead, 1996). Dietary electrolyte balance and dietary alterations that affect acid-base or cation-anion balance have a role in TD (Hulan et al., 1987). High dietary chloride levels provoke an increased incidence (Edwards and Veltmann, 1983). Dietary zeolite has been shown to reduce the incidence and severity of TD (Ballard and Edwards, 1988), apparently by facilitating calcium utilization. High levels of

FIGURE 14.4 Perosis, or slipped tendon, can be caused by deficiencies of choline and/or manganese.

vitamin A were reported by Jensen et al. (1983) to increase leg disorders, but this result was not confirmed by Ballard and Edwards (1988). Edwards (1984) found that a diet low in calcium, and high in phosphorus and chloride, produced a high incidence of TD in broilers. Supplementation of high phosphorus diets with calcium reduces the incidence of TD (Edwards and Veltmann, 1983).

Edwards (1990) suggested that TD may be a manifestation of the inability of the rapidly growing broiler to convert vitamin D_3 to 1,25-$(OH)_2D_3$ adequately for maximum calcium absorption and bone formation. Dietary supplementation with active **vitamin D metabolites** such as 1-$(OH)D_3$, 25-(OH) D_3, and 1,25-$(OH)_2$ D_3 is effective in preventing TD (Elliot et al., 1995; Rennie and Whitehead, 1996). The 25(OH) D_3 metabolite is about ten times as toxic as vitamin D_3 to broilers (Yarger et al., 1995), and should, therefore, be used prudently in diet formulation. The 25 (OH) D_3 is especially effective in preventing TD when the dietary calcium level is below 0.85% (Ledwaba and Roberson, 2003).

Fat Disorders

Excessively fat broilers are a problem in the United States poultry industry. Selection of birds for very rapid growth rate and the use of high-energy diets are implicated. Maintaining a proper balance of protein to ME helps prevent excessive fat deposition.

Fatty liver syndrome (also known as *fatty liver and kidney syndrome* and *fatty liver hemorrhagic syndrome*) affects layers, with the deposition of large amounts of fat in the liver. The cause is not completely understood, but **lipotropic agents** (substances that reduce fat deposition in the liver or hasten its removal) such as choline, vitamin B_{12}, and methionine show evidence of protective activity (Whitehead and Randall, 1982).

Antioxidants such as vitamin E and ascorbic acid do not affect fatty liver syndrome, but dietary cysteine supplementation has a protective effect (Diaz et al., 1994). Cysteine promotes lipoprotein synthesis, which facilitates transport of triglycerides from the liver in association with serum lipoproteins.

Oily bird syndrome is sometimes a problem with broilers, observable only after the birds have been processed, and involves a greasy appearance of the fat and skin. Nutritional factors such as the protein-to-calorie ratio may have a role in the syndrome (Jensen et al., 1980).

Another nutrition-related problem with broilers is **sudden death syndrome (SDS)** or flip-over disease. This appears to involve a heart failure (heart attack). Birds that develop SDS have altered fatty acid content of tissues (Rotter et al., 1988), with decreased arachidonic acid levels. This could result in reduced prostaglandin synthesis, leading to impaired heart function. Biotin deficiency has also been implicated in SDS. Biotin is a cofactor for enzymes involved in fatty acid metabolism. Feeding high-energy diets increases incidence of SDS (Proudfoot et al., 1984). Impaired absorption of fat or **fat malabsorption syndrome (FMS)** can be caused by a number of factors. In newly hatched chicks and poults, adequate lipase secretion for fat digestion may not occur until 10–20 days post-hatching (Jin, et al 1998; Sklan and Noy, 2003). Because the absorption of fat-soluble vitamins is linked to fat absorption, vitamin D deficiency-induced rickets and skeletal deformities may occur. Provision of vitamin D in the form of

25-OH D_3 (see Chapter 7) in the diet may prevent these problems because the vitamin is more effectively absorbed in the hydroxylated form.

Impaired bile secretion can also be a cause of FMS because bile is essential for fat absorption. For example, liver damage caused by aflatoxin in the diet may result in vitamin D deficiency.

As discussed in Chapter 8, dietary non-starch polysaccharides can cause FMS in poultry.

Gizzard Erosion

Gizzard erosion is a disorder that has been observed in poultry for many years. It is characterized by erosion or necrosis of the gizzard lining, often with ulceration into the muscular gizzard wall. The entire lining of the gizzard may have a dry, scaly, brownish color rather than the normal greenish-yellow color. The upper part of the small intestine may be ulcerated as well. The condition is related to dietary ingredients, particularly fish meal or grains contaminated with mycotoxins.

The gizzard erosion factor in fish meal has been identified by Japanese scientists in studies of black vomit disease as a dipeptidelike compound composed of histamine and lysine (Masumura et al., 1985). This compound, called **gizzerosine,** is formed by a Maillard (browning) reaction when fish meal is heated, especially when high temperatures are used (Sugahara, 1995). Gizzerosine induces an abnormally high secretion of hydrochloric acid in the proventriculus, apparently acting as an analog of histamine, which is the physiological regulator of gastric acid secretion. It is approximately ten times as potent as histamine in stimulating gastric acid secretion. Gizzerosine is metabolized much more slowly than histamine and has a higher affinity than histamine for gastric histamine receptors (Ito et al., 1988). Sugahara et al. (1988) suggest a maximum concentration of 0.4 ppm gizzerosine in practical poultry diets. Higher levels cause retarded growth and mortality.

Mycotoxins also cause gizzard erosion. For example, aflatoxin has a synergistic effect in increasing the toxicity of gizzerosine (Diaz and Sugahara, 1995). Some mycotoxins cause damage to the liver and bile duct, decreasing the normal secretion of bile. Bile is periodically regurgitated into the gizzard, neutralizing gastric acid. Thus, absence of normal bile levels results in hyperacidity of the gizzard and mucosal damage. In addition, fish oils not adequately protected with antioxidants may cause gizzard erosion, probably due to peroxidation of membrane tissue. Vitamin E protects against this damage. The synthetic antioxidant ethyoxyquin is also very effective in protecting against peroxide damage (Cabel et al., 1988).

Ascites Disorders

Ascites is fluid accumulation in body cavities. Significant occurrence of abdominal ascites is noted in broilers in many countries. In most cases, ascites is associated with liver, heart, or lung damage. The condition is also known as *pulmonary hypertension syndrome*. Fast-growing broilers are most susceptible; the mortality rate in affected flocks can be reduced by restricting feed to reduce growth rate. The problem may be that the extremely rapid rate of muscle growth in modern broilers outgrows the cardiovascular support systems. Another cause of ascites is high altitude, which intensifies the condition because of the lower oxygen content of the air. Mycotoxins and other hepatotoxic agents increase susceptibility. Excess sodium in the feed or water also may cause ascites, as do poor-quality dietary fats containing oxidized fatty acids.

Dietary antioxidants such as vitamin E are effective in reducing the incidence of ascites in broilers (Bottje et al., 1995).

Heat Stress

Heat stress is a major problem in poultry production in the southern United States and in tropical countries. Dietary management may aid in minimizing the detrimental effects. Heat stress has various physiological effects; for example, increased respiration rate decreases the blood CO_2 level, which increases blood pH (Teeter et al., 1985). The consequences of decreased blood CO_2 and elevated pH are not fully known. Increased weight gain in heat-stressed broilers given carbonated water (Bottje and Harrison, 1985) or water supplemented with NH_4Cl and HCl (Teeter et al., 1985) has been noted. On the other hand, Teeter and Smith (1986) observed that addition of 0.48 percent KCl to drinking water increased the growth rate of heat-stressed broilers by 20 percent even though the blood had an alkaline pH, and blood CO_2 was depressed, indicating that factors other than blood CO_2 and pH are involved.

Egg Production Disorders

With layers, maintenance of good **egg shell quality** can be a problem. The major concerns are with shell thickness and shell structure. It is desirable that egg shells not break during handling and that the shell resist penetration by microorganisms. Shells should be attractive in appearance, that is, not misshapen, bumpy, or blotchy. The shell is virtually 100 percent calcium carbonate. Thus adequate dietary levels of calcium are essential for good shell quality; adequate vitamin D is necessary for calcium absorption and shell formation. Dietary zeolites improve eggshell quality (Roland et al., 1985), presumably by facilitating calcium utilization, whereas excesses of phosphorus and deficiencies of manganese can cause thin eggshells.

Besides calcium, a source of carbonate ion is needed to form the calcium carbonate in the shell. The blood bicarbonate is the source of carbonate ion used in shell formation. Blood acid-base and electrolyte balance influence availability of carbonate ion. In hot weather, the respiration rate of birds increases markedly. This can so increase carbon dioxide loss that blood bicarbonate levels are too low to support proper eggshell formation. Providing a dietary source, such as sodium bicarbonate, may be useful under these conditions. There is some evidence that feeding ascorbic acid (vitamin C) may also improve eggshell quality under stress conditions.

Egg size is another important economic consideration for the poultry producer. Energy intake is a major nutritional determinant of egg size. Some feedstuffs, such as canola meal, may cause a reduction in egg size, probably due to a reduction in feed intake (Summers et al., 1988).

Egg size is affected by the lipid content and composition of the diet. Corn and corn oil in the diet stimulate production of maximum egg size due to their content of linoleic acid. March and MacMillan (1990) reviewed the effects of linoleic acid on egg size and concluded that **linoleic acid** is necessary for the synthesis of lipoproteins in the liver that can be transported to the ovary for uptake by the developing ova. Egg size cannot be maximized if the amount of linoleic acid present is not adequate to support maximal lipoprotein synthesis.

Other Nutritional Problems

Urolithiasis in poultry is a syndrome characterized by mineralization of the kidney (kidney stones), resulting in acute uricemia (elevated blood uric acid), visceral gout, and mortality (Wideman et al., 1989). The condition may be induced by excessive dietary calcium. Urinary acidification by feeding ammonium chloride may aid in reducing renal urolith formation. Dietary methionine hydroxy analog has also been shown to reduce calcium-induced kidney damage (Wideman et al., 1989). These authors provide a good review of dietary factors influencing urolithiasis.

Mycotoxins can cause important losses in poultry production. Some of the important mycotoxins such as aflatoxin, zearalenone, trichothecenes, citrinin, and ochratoxin have been mentioned earlier. Moldy feed ingredients may impair performance even when mycotoxins are not detectable. Aflatoxins affect the immune system, increasing susceptibility to disease. Mycotoxins have a role in field hemorrhagic syndromes in which small hemorrhages may blemish the carcass. Aflatoxins and ochratoxin can cause pale bird syndrome by impairing pigment absorption (see Chapter 8).

QUESTIONS AND STUDY GUIDE

1. Why are energy requirements for poultry expressed as ME rather than DE, whereas for swine both measures are used?
2. What types of diets for poultry would generally be supplemented with a source of xanthophyll?
3. How are the nutritional requirements of turkeys different from those of broiler chickens?
4. Suppose you are working for a "wildlife safari" park, and you are responsible for the management of avian species. You have incubated emu eggs, which have hatched. You are feeding a commercial turkey feed to the young emus. After a few weeks, they have pronounced leg problems, with leg weakness and deformities. What is the problem, and how would you prevent it in the future?
5. Imagine that you are working as a summer intern student in an extension office. A person calls in and wants to know why her ducklings are not doing well. You ask about what feed she is using, and she says she is feeding only cracked wheat. What nutrient deficiencies and deficiency signs would you expect the ducklings to have? What would be your advice?
6. What causes perosis in avian species, and how can it be prevented?

REFERENCES

Ballard, R. and H.M. Edwards, Jr. 1988. Effects of dietary zeolite and vitamin A on tibial dyschondroplasia in chickens. *Poult. Sci.* 67:113–119.

Bottje, W., B. Enkvetchakul, R. Moore, and R. McNew. 1995. Effect of a-tocopherol on antioxidants, lipid peroxidation, and the incidence of pulmonary hypertension syndrome (ascites) in broilers. *Poult. Sci.* 74:1356–1369.

Bottje, W.G. and P.C. Harrison. 1985. The effect of tap water, carbonated water, sodium bicarbonate and calcium chloride on blood acid-base balance in cockerels subjected to heat stress. *Poult. Sci.* 64:107–113.

Cabel, M.C., T.L. Goodwin, and P.W. Waldroup. 1988. Feather meal as a nonspecific nitrogen source for abdominal fat reduction in broilers during the finishing period. *Poult. Sci.* 67:300–306.

Cabel, M.C., P.W. Waldroup, W.D. Shermer, and D.F. Calabotta. 1988. Effects of ethoxyquin feed preservative and peroxide level on broiler performance. *Poult. Sci.* 67:1725–1730.

Cheeke, P.R. 2004. *Contemporary Issues in Animal Agriculture,* Upper Saddle River, NJ: Prentice Hall.

Cheeke, P.R., J.S. Powley, H.S. Nakaue, and G.H. Arscott. 1983. Feed preference responses of several avian species fed alfalfa meal, high-and low-saponin alfalfa, and quinine sulfate. *Can. J. Anim. Sci.* 63:707–710.

Christman, C.J. and R.O. Hawes 1999. *Birds of a Feather, Saving Rare Turkeys from Extinction.* Pittsboro, NC: The American Livestock Breeds Conservancy.

Cilliers, S.C., J.P. Hayes, A. Chwalibog, J. J. du Preez, and J. Sales. 1997a. A comparative study between mature ostriches (*Struthio camelus*) and adult cockerels with respect to true and apparent metabolisable energy values for maize, barley, oats and triticale. *Brit. Poult. Sci.* 38:96–100.

______ . 1997b. A comparative study between mature ostriches (*Struthio camelus*) and adult cockerels with regard to the true and apparent digestibilities of amino acids. *Brit. Poult. Sci.* 38:311–313.

Cilliers, S.C., J.P. Hayes, J.S. Maritz, A. Chwalibog, and J.J. du Preez. 1994. True and apparent metabolizable energy values of lucerne and yellow maize in adult roosters and mature ostriches (*Struthio camelus*), *Anim. Prod.* 59:309–313.

Dawson, T.J. and R.M. Herd. 1983. Digestion in the emu: Low energy and nitrogen requirements of this large ratite bird. *Comp. Biochem. Physiol.* 75:41–45.

Diaz, G.J., E.J. Squires, and R.J. Julian. 1994. Effect of selected dietary antioxidants on fatty liver-haemorrhagic syndrome in laying hens. *Brit. Poult. Sci.* 35:621–629.

Diaz, G.J. and M. Sugahara. 1995. Individual and combined effects of aflatoxin and gizzerosine in broiler chickens. *Brit. Poult. Sci.* 36:729–736.

Earle, K.E. and N.R. Clarke. 1991. The nutrition of the budgerigar (*Melopsittacus undulatus*). *J. Nutr.* 121:S186–S192.

Edwards, H.M., Jr. 1984. Studies on the etiology of tibial dyschondroplasia in chickens. *J. Nutr.* 114:1001–1013.

______ . 1990. Efficacy of several vitamin D compounds in the prevention of tibial dyschondroplasia in broiler chickens. *J. Nutr.* 120:1054–1061.

Edwards, H.M., Jr., and J.R. Veltmann, Jr. 1983. The role of calcium and phosphorus in the etiology of tibial dyschondroplasia in young chicks. *J. Nutr.* 113:1568–1575.

Elliot, M.A., K.D. Roberson, G.N. Rowland III, and H.M. Edwards, Jr. 1995. Effect of dietary calcium and 1,25- dihydroxycholecalciferol on the development of tibial dyschondroplasia in broilers during the starter and grower periods. *Poult. Sci.* 74:1495–1505.

Grone, A., D.E. Swayne, and L.A. Nagode. 1995. Hypophosphatemic rickets in rheas (*Rhea americana*). *Vet. Pathol.* 32:324–327.

Harrison, G.J. 1986. Feeding psittacine and passerine birds. In *Zoo and Wild Animal Medicine,* M.E. Fowler, ed. pp. 479–488. Philadelphia: W.B. Saunders.

Harrison, G.J. 1998. Twenty years of progress in pet bird nutrition. *J. Amer. Vet. Med. Assn.* 212: 1226–1230.

Harrison, G.J. and L.A. Harrison. 1986. *Clinical Avian Medicine and Surgery Including Aviculture.* Philadelphia: W.B. Saunders.

Hollister, A.G., H.S. Nakaue, and G.H. Arscott. 1984. Studies with confinement-reared goslings. 1. Effects of feeding high levels of dehydrated alfalfa and Kentucky bluegrass to growing goslings. *Poult. Sci.* 63:532–537.

Hughes, B.L. and J.E. Jones. 1980. Diet regimes for growing guineas as meat birds. *Poult. Sci.* 59:582–584.

Hulan, H.W., R.G. Ackman, W.M.N. Ratnayake, and F.G. Proudfoot. 1988. Omega–3 fatty acid levels and performance of broiler chickens fed redfish meal or redfish oil. *Can. J. Anim. Sci.* 68:533–547.

______. 1989. Omega–3 fatty acid levels and general performance of commercial broilers fed practical levels of redfish meal. *Poult. Sci.* 68:153–162.

Hulan, H.W., P.C.M. Simons, P.J.W. Van Schagen, K.B. McRae, and F.G. Proudfoot. 1987. Effect of dietary cation-anion balance and calcium content on general performance and incidence of leg abnormalities of broiler chickens. *Can. J. Anim. Sci.* 67:165–177.

Ito, Y., H. Terao, T. Noguchi, and H. Naito. 1988. Gizzerosine raises the intracellular cyclic adenosine–3′, 5′-monophosphate level in isolated chicken proventriculus. *Poult. Sci.* 67:1290–1294.

Jang, D. A., J. G. Fadel, K. C. Klasing, A. J. Mireles, Jr., R. A. Ernst, K. A. Young, A. Cook, and V. Raboy, 2003. Evaluation of low-phytate corn and barley on broiler chick performance. *Poultry Sci.* 82:1914–1924.

Jensen, L.S., I. Bartov, M.J. Beirne, J.R. Veltman, Jr., and D.L. Fletcher. 1980. Reproduction of the oily bird syndrome in broilers. *Poult. Sci.* 59:2256–2266.

Jensen, L.S., D.L. Fletcher, M.S. Lilburn, and Y. Akiba. 1983. Growth depression in broiler chicks fed high vitamin A levels. *Nutr. Rep. Int.* 28:171–179.

Jin, S-H. Corless and J. L. Sell. 1998. Digestive system development in post-hatch poultry. *World's Poult Sci.J.* 54:335–345.

Jurgens, M.H. 1993. *Animal Feeding and Nutrition.* Dubuque, IA: Kendall/Hunt Publishing Company.

Ledwaba, M.F. and K.D. Roberson. 2003. Effectiveness of twenty-five-hydroxycholecalciferol in the prevention of fibial dyschondroplasia in Ross cockerals depends on dietary calcium level. *Poult. Sci.* 82:1769–1777.

Leeson, S. 1986. Nutritional considerations of poultry during heat stress. *World Poult. Sci. J.* 42:69–81.

Leeson, S. and J.D. Summers. 1988. Some nutritional implications of leg problems with poultry. *Brit. Vet J.* 144:81–92.

Li, Y. C., D. R. Ledoux, T. L. Veum, V. Raboy, and D. S. Ertl, 2000. Effects of low phytic acid corn on phosphorus utilization, performance, and bone mineralization in broiler chicks. *Poultry Sci.* 79:1444–1450.

March, B.E. and C. MacMillan. 1990. Linoleic acid as a mediator of egg size. *Poult. Sci.* 69:634–639.

Martosiswoyo, A.W. and L.S. Jensen. 1988. Effect of formulating diets using different meat and bone meal energy data on broiler performance and abdominal fat content. *Poult. Sci.* 67:294–299.

Masumura, T., M. Sugahara, T. Noguchi, K. Mori, and H. Naito. 1985. The effect of gizzerosine, a recently discovered compound in overheated fish meal, on the gastric acid secretion in the chicken. *Poult. Sci.* 64:356–361.

Morrow, C.J., A.P. Browne, C.J. O'Donnell, and B.H. Thorp. 1997. Hypophosphataemic rickets and nephrocalcinosis in ostrich chicks brooded and reared on limestone sand. *Vet. Rec.* 140:531–532.

National Research Council. 1984 and 1994. *Nutrient Requirements of Poultry.* Washington, DC: National Academy Press.

Oguntona, T. and B.L. Hughes. 1988. Effect of energy levels and feed pelleting on growth and feed conversion of guinea fowl keets. *Nutr. Rep. Int.* 38:1283–1288.

Proudfoot, F.G., H.W. Hulan, and K.B. McRae. 1984. The effects of dietary micronutrient, fat and protein components in pelleted feeds on the incidence of sudden death syndrome and other traits among male broiler chickens. *Can. J. Anim. Sci.* 64:159–164.

Rennie, J.S., R.H. Fleming, H.A. McCormack, C.C. McCorquodale, and C.C. Whitehead. 1997. Studies on effects of nutritional factors on bone structure and osteoporosis in laying hens. *Brit. Poult. Sci.* 38:417–424.

Rennie, J.S. and C.C. Whitehead. 1996. Effectiveness of dietary 25- and 1- hydroxycholecalciferol in combatting tibial dyschondroplasia in broiler chickens. *Brit. Poult. Sci.* 37:413–421.

Roland, D.A., Sr., S.M. Laurent, and H.D. Orloff. 1985. Shell quality as influenced by zeolite with high ion-exchange capability. *Poult. Sci.* 64:1177–1187.

Rotter, B., W. Guenter, and B.R. Boycott. 1988. Sudden death syndrome in broilers: Influence of dietary fats on the incidence and tissue composition. *Nutr. Rep. Int.* 38:369–380.

Roudybush, T.E., C.R. Grau, T. Jermin, and D. Nearenberg. 1984. Pelleted and crumbled diets for cockatiels. *Feedstuffs* 56(43):18–20.

Roudybush, T.E. and C.R. Grau. 1986. Food and water interrelations and the protein requirement for growth of an altricial bird, the cockatiel (*Nymphicus hollandicus*). *J. Nutr.* 116:552–559.

Schaeffer, J.L. and P.B. Hamilton. 1990. Effect of dietary lipid on lutein metabolism during aflatoxicosis in young broiler chickens. *Poult. Sci.* 69:53–59.

Scott, M.L., M.C. Nesheim, and R.J. Young. 1982. *Nutrition of the Chicken.* Ithaca, NY: M.L. Scott of Ithaca.

Scott, M.L. 1987. *Nutrition of the Turkey.* Ithaca, NY: M.L. Scott of Ithaca.

Sibbald, I.R. 1982. Measurement of bioavailable energy in poultry feedingstuffs: A review. *Can. J. Anim. Sci.* 62:983–1048.

Sibbald, I.R. 1987. Estimation of bioavailable amino acids in feedingstuffs for poultry and pigs: A review with emphasis on balance experiments. *Can J. Anim. Sci.* 67:221–300.

Sklan, D. and Y. Noy. 2003. Functional development and intestinal absorption in the young poult. *Brit Poult. Sci.* 44:651–658.

Stevenson, M.H. 1989. Nutrition of domestic geese. *Proc. Nutr. Soc.* 48:103–111.

Sugahara, M. 1995. Black vomit, gizzard erosion and gizzerosine. *World's Poult. Sci. J.* 51:293–306.

Sugahara, M., T. Hattori, and T. Nakajima. 1988. Effect of synthetic gizzerosine on growth, mortality and gizzard erosion in broiler chicks. *Poult. Sci.* 67:1580–1584.

Summers, J.D., G. Hurnik, and S. Leeson. 1987. Carcass composition and protein utilization of Embden geese fed varying levels of dietary protein supplemented with lysine and methionine. *Can. J. Anim. Sci.* 67:159–164.

Summers, J.D., S. Leeson, and D. Spratt. 1988. Canola meal and egg size. *Can. J. Anim. Sci.* 68:907–913.

Swart, D., R.I. Mackie, and J.P. Hayes. 1993a. Influence of live mass, rate of passage and site of digestion on energy metabolism and fibre digestion in the ostrich (*Struthio camelus* var. *domesticus*). *S. Afr. J. Anim. Sci.* 23:119–126.

______ . 1993b. Fermentative digestion in the ostrich (*Struthio camelus* var. *domesticus*), a large avian species that utilizes cellulose. *S. Afr. J. Anim. Sci.* 23:127–135.

Taylor, E.J., H.M.R. Nott, and K.E. Earle. 1994. The nutrition of the canary (*Serinus canarius*). *J. Nutr.* 124:2636S–2637S.

Teeter, R.G. and M.O. Smith. 1986. High chronic ambient temperature stress effects on broiler acid-base balance and their response to supplemental ammonium chloride, potassium chloride and potassium carbonate. *Poult. Sci.* 65:1777–1781.

Teeter, R.G., M.O. Smith, F.N. Owens, S.C. Arp, S.S. Sangiah, and J.E. Breazile. 1985. Chronic heat stress and respiratory alkalosis: Occurrence and treatment in broiler chicks. *Poult. Sci.* 64:1060–1064.

Thornberry, F.D. 1988. *Ostrich Production,* Special report. College Station, TX: Texas A & M University.

Ullrey, D.E., M.E. Allen, and D.J. Baer. 1991. Formulated diets versus seed mixtures for psittacines. *J. Nutr.* 121:S193–S205.

Whitehead, C.C. and C.J. Randall. 1982. Interrelationships between biotin, choline and other B-vitamins and the occurrence of fatty liver and kidney syndrome and sudden death syndrome in broiler chickens. *Brit. J. Nutr.* 48:177–184.

Wideman, R.F., W.B. Roush, J.L. Satnick, R.P. Glahn, and N.O. Oldroyd. 1989. Methionine hydroxy analog (free acid) reduces avian kidney damage and urolithiasis induced by excess dietary calcium. *J. Nutr.* 119:818–828.

Yarger, J.G., C.L. Quarles, B.W. Hollis, and R.W. Gray. 1995. Safety of 25–hydroxycholecalciferol as a source of cholecalciferol in poultry rations. *Poult. Sci.* 74:1437–1446.

CHAPTER 15

Feeding and Nutrition of Beef Cattle

T. DelCurto and P.R. Cheeke

Objectives

1. To describe the major nutrient requirements of beef cattle and factors affecting requirements
2. To describe feeding systems for beef cattle production and supplementation programs for forage-based systems
3. To discuss the feeding of stocker and feedlot cattle
4. To describe the use of crop residues in beef cattle feeding systems

Beef cattle are well adapted to the utilization of forages. In the developed countries, there has been a trend in recent years to feed diets high in grain to ruminant animals. Most beef cattle in the United States are finished in feedlots where they are fed diets high in grain and low in roughage. Grains are usually the least expensive sources of DE for livestock in the developed countries, most of which have large grain surpluses. In addition, because high-concentrate feeds allow full expression of genetic potential for growth, the time taken to reach market weight is much less in grain-fed animals than in those fed forage. Grain-fed cattle can reach slaughter weight at approximately 1 year of age or slightly more, whereas cattle raised on forage may take 1.5 to 3 years (or more on tropical pastures). Therefore, the nutrient expenditure for maintenance needs is much less over the lifetime of the concentrate-fed animal, and the total number of animals that can be produced is greater. Grain feeding is also used because of the widespread belief that the eating quality of the beef is superior with grain-fed animals. Concerns about the effects of animal fat on human health may result in a reduction in the amount of grain fed to beef cattle. In the United States, numerous cattle producers are marketing “light” or “natural” beef from animals fed little or no grain prior to slaughter.

Although raising ruminants on high-concentrate diets makes economic sense at the present time, it may not make sense in terms of optimal biological efficiency. Nonruminants, such as swine and poultry, can produce meat more efficiently from grain than ruminants can because they do not have the energy losses associated with rumen fermentation. Therefore, to maximize meat production from a given amount of grain, it should be fed to broilers or swine rather than to beef cattle. Second, the ruminant digestive tract has evolved to utilize fibrous feeds. It seems intuitively logical to feed ruminants the types of feeds to which they are best adapted, rather than the concentrate feeds that may induce a number of metabolic problems (e.g., displaced abomasum and lactic acidosis).

NUTRIENT REQUIREMENTS

Energy Requirements

Expressing energy requirements and energy values of feedstuffs for ruminants is somewhat more complex than for swine and poultry because of rumen fermentation and the complexity of interactions between diet and fermentation end products. One example of this complexity is the effect of balance of absorbed VFA on metabolic efficiency. If there is a surplus of acetate (C_2) or a deficiency of propionate (C_3), the C_2 energy cannot be utilized in the citric acid cycle reactions of metabolism. An extra loss of heat energy associated with conversion of amino acids to citric acid cycle intermediates will result (see **Heat Increment,** Chapter 9). This extra loss of heat energy reduces the net energy value of the diet.

For ruminants, DE is less suitable as a measure of feed energy than for swine because of the variable losses of rumen gas energy (mainly methane). The major weakness of DE as a basis for feeding systems for ruminants is that it overestimates the available energy of high-fiber feedstuffs relative to concentrates (NRC, 1984). The efficient use of roughage ME for both maintenance and gain is lower for diets with a high roughage-to-concentrate ratio. The ME category, which corrects for rumen gas and urinary losses, shares the same fault. For these reasons, net energy has become widely used with beef cattle, particularly for feedlot animals.

The use of NE is complicated by the fact that energy is used with different efficiency for maintenance and productive functions. To account for these differences, in the NE system two NE values are assigned to each feedstuff, one giving its caloric value for maintenance (NE_m) and one for gain (NE_g). Advantages of this system are that first, these NE values are independent of diet and do not have to be adjusted for varying roughage-to-concentrate ratios and second, that feed requirements for maintenance are estimated separately from those for production.

Net energy values for most feedstuffs commonly used in feedlot diets have been determined using the **comparative slaughter technique.** This is an alternative to determination of NE by measuring all energy losses and correcting the gross energy (GE) for these losses. In ruminants, these energy losses include the contribution of feces, urine, rumen gases, and heat production. With the comparative slaughter technique, the utilizable portion of the GE is determined by measuring energy gain of the tissues.

Briefly, this method involves establishing a statistical relationship between the energy content of a carcass and its **specific gravity.** The higher the fat content, and, therefore, energy content, the lower the density of the carcass, and the lower its specific gravity. The specific gravity is the weight of the carcass divided by the weight of an equal volume of water, determined by measuring the amount of water displaced when the carcass is submerged in a tank of water. The energy content of the carcass is determined by grinding and thoroughly mixing it and burning a sample in a bomb calorimeter. After the relationship between carcass energy content and specific gravity has been experimentally determined, the NE value of a feedstuff is measured by feeding it to a group of feedlot cattle, slaughtering some at the start and the rest at the end of the feeding period, measuring specific gravity of the carcasses, and calculating total energy gain during the period of consumption of the feedstuff. From the total GE intake and carcass energy gain, the NE_g value is calculated. The development of this technique and its use to determine NE values for feedstuffs used in cattle feeding in the United States was pioneered by University of California researchers Lofgreen and Garrett (1968).

The NE_m value is useful for estimating energy and feed requirements for mature beef animals, such as bulls and dry cows. For growing animals, the **NE_m** value is used to estimate that part of the total feed intake needed for satisfying maintenance requirements that must be met before productive functions take place. The **NE_g** requirements vary according to the type of tissue being formed by the animal. Tissue growth is mostly protein (muscle) or fat. The caloric value of fat is 9.4 kcal/g and for the fat-free tissue (mostly protein) approximately 5.6 kcal/g. The proportions of protein and fat that are synthesized are related to the level of energy intake above maintenance and to the stage of growth. Fat is deposited in growing animals mainly when the energy intake above maintenance is high. The stage of growth will be different at the same body weights for different breeds of cattle, depending on mature body size, body type (large vs small frame), sex, and so on. For this reason, the NRC (1984) published separate requirements for medium- and large-frame steers, and for heifers and bulls of various body weights, and discussed in detail the background data in support of these adjustments.

Energy requirements are modified by environmental conditions. Beef cattle are particularly likely to be exposed to extremes in environment as opposed to confinement-housed animals, such as swine and poultry. These conditions include very hot or cold weather, wind velocity (wind chill factor), precipitation, and animal factors such as age, breed, hair coat, body condition, period of adaptation, and diet. **Environmental temperature** affects net energy values. In the thermal neutral or comfort zone, or above the comfort zone, the heat of rumen fermentation is detrimental, representing excess heat that must be eliminated. Under cold conditions, the **heat of rumen fermentation** and the heat of nutrient metabolism (the heat increment) are beneficial in helping to maintain body temperature and, therefore, contribute to the NE of a feed. Low-quality roughages have a high heat of rumen fermentation, so straw and other coarse feeds are of greater value in cold environments than under moderate conditions. Animals adapted to cold make the adaptation through changes in hormone secretion; in cold-adapted animals the comfort zone can be shifted significantly to lower temperatures.

Energy intake of beef cows is a major determinant of **reproductive efficiency.** It is of particular concern in relation to estrous cycles and initiation of pregnancy. After calving, beef cows should be rebred within 60 to 90 days to maintain a once-per-year calving interval. This postpartum period is one of high energy requirements to support lactation. If a cow is not able to consume sufficient feed to maintain a positive energy balance, initiation of estrous cycling may not occur, resulting in a **prolonged calving interval.** Grazing cattle may not be able to consume a sufficient amount of forage to maintain a positive energy balance and may have a prolonged **postpartum anestrus** period. Cows with a high level of milk production are particularly susceptible. Short and Adams (1988) reviewed the effects of nutrition on beef-cow reproduction. The effects of energy intake appear to be mediated by blood glucose level, which in turn influences the release of gonadotropins that control the estrous cycle (Richards et al., 1989).

Nutrition has an effect on **dystocia** (difficult calving), although the effects are not always consistent and predictable. Very high and very low planes of nutrition for cows in late gestation are undesirable. High levels of feeding cause obesity and deposition of fat in the birth canal, reducing the pelvic opening and causing dystocia. Low levels of nutrition do not necessarily reduce dystocia incidence, and result in weak, less active calves and in impairment of subsequent reproduction potential of the cow.

With young breeding stock, particularly animals used in exhibition and showing, grain should be fed cautiously to avoid disorders of **grain overload** (see Chapter 2). Soundness of feet and legs is very important in cattle exhibited in shows and in cattle expected to perform well under rugged conditions. **Laminitis** is a common problem in intensively fed cattle, particularly calves (Greenough et al., 1990). Animals that have experienced laminitis will often have permanently distorted feet and legs.

Protein Requirements

Ruminants require dietary nitrogen to support rumen microbial growth, and absorbed amino acids to satisfy metabolic requirements for protein synthesis. The minimum concentration of **rumen ammonia** necessary for optimal microbial growth and rumen fermentation in beef cattle has been estimated at 20 mg/100 ml of rumen fluid (NRC, 1984). The addition of a source of NPN to the diet is beneficial only when rumen ammonia concentrations are below this level. Various techniques have been developed to estimate the degree to which dietary nitrogen sources are partitioned between fermentable nitrogen and bypass protein so that the amount of NPN, such as urea, to be added to the diet to provide useful fermentable nitrogen can be calculated. The main problem with such calculations is that the degree of ruminal degradation of a specific protein is difficult to predict, as it is influenced by numerous factors such as roughage level, feed intake level, particle size, and feed processing. Therefore, systems that attempt to partition dietary nitrogen according to its fate in the rumen are as yet of limited practical value (NRC, 1984).

The 1996 NRC publication on *Nutrient Requirements of Beef Cattle* has listed protein requirements in terms of rumen versus intestinal digestion of dietary protein, using the following categories.

Metabolizable protein (MP): True protein absorbed in the intestine, supplied by microbial protein and undegraded intake protein.

Undegraded intake protein (UIP): Dietary protein that is not fermented in the rumen, also known as *bypass* or *escape protein.*

Degraded intake protein (DIP): Dietary protein that is fermented in the rumen and converted to microbial protein.

Bacterial crude protein (BCP): Bacterial protein synthesized in the rumen. Because protozoa are retained in the rumen, BCP is more or less the same as microbial protein.

Unfortunately, these changes have introduced a degree of complexity that makes the nutrient requirement tables incomprehensible to many potential users. For many (or most) beef cattle producers, application of fermentable versus nonfermentable protein values is not practical. In addition, on low-quality, high-fiber diets, the accuracy of the 1996 beef NRC models is highly variable and strongly influenced by the value used to estimate microbial efficiency.

Mineral Requirements

Calcium and **phosphorus** are the minerals required in greatest quantity by beef cattle. Forage legumes are rich sources of calcium and moderately good sources of phosphorus, whereas grains and grain by-products are good sources of phosphorus but very poor sources of calcium. Grasses are moderate to low in both calcium and phosphorus content. With pasture- or hay-fed cattle, calcium supplementation is usually necessary only for lactating cows. Phosphorus is particularly important in the postpartum period; cows with borderline phosphorus deficiency may not have normal estrous cycles, and calving interval may be lengthened. Magnesium is of major concern in areas where grass tetany is a problem (see Chapter 5).

Specific **trace mineral deficiencies** occur in many areas, depending upon local soil conditions (see Chapter 7). Selenium, copper, and cobalt deficiencies are practical problems in many areas. Information on the trace mineral status of a particular geographical area must be obtained from local extension agents or other sources.

Grazing cattle should have access to a mineral mix containing calcium, phosphorus, salt, and trace elements. In areas with widely recognized trace-element deficiencies, a **custom mineral mix** with appropriate adjustments may be needed. In some cases, mineral toxicities (e.g., molybdenum) may be of concern. Because grains are much lower than forages in potassium, deficiency of this element is sometimes encountered with high-concentrate diets. Feedlot cattle sometimes respond to potassium supplementation (see Chapter 7).

Cattle and sheep grazing semiarid rangelands in North America, Australia, and Russia frequently develop **urinary calculi,** composed mainly of silica. This condition is often referred to as **silica urolithiasis.** The calculi are deposits of insoluble silicates in the urinary tract. Displacement of large calculi from the bladder to the urethra causes an obstruction to the normal flow of urine. The blockage may lead to rupture of the bladder or urethra, a fatal condition unless relieved by surgical intervention. Calculi are generally only a problem in males because of their smaller urethra diameter.

The **silica content of range grasses** becomes progressively greater with plant maturity and is greatest in weathered native plants in which the soluble components have been washed out. Weathered range grasses may contain as much as 6.5 percent silica, occurring in the cell wall material. Silica is solubilized in the rumen, keeping the rumen saturated with silicic acid. Silicic acid is absorbed. Because of the low intakes of protein and minerals in ruminants grazing mature weathered grasses, water resorption in the kidney is high and urine output low. This, along with the low water intake common under range conditions, concentrates silicic acid in the urine, where it may precipitate out as insoluble silicates, forming the calculi. Dietary or management manipulations to increase water consumption, such as feeding supplements high in salt, are beneficial. Stewart et al. (1990) found that diets with high calcium-to-phosphorus ratios and alkali-forming potential (see Chapter 7) contribute to silica urolithiasis. They suggested that supplementation with a phosphorus source having an acid-forming potential, such as phosphoric acid in a liquid supplement, would be beneficial.

Urinary calculi in feedlot cattle and sheep generally involve deposition of calcium and magnesium phosphates in association with proteinaceous matrix material. These phosphatic calculi are promoted by diets that promote an alkaline urine high in phosphates. Concentrates such as grains and plant protein supplements have excess phosphorus for feedlot animals and inadequate calcium. Careful diet formulation is necessary to obtain optimal calcium:phosphorus ratios of 1.5–2.0:1. If problems with calculi persist, a urine acidifying agent such as 0.5 percent ammonium chloride can be added to the diet.

Vitamin Requirements

The B-complex vitamins and vitamin K are usually synthesized in adequate amounts in the rumen, and vitamin D is obtained with exposure to sunlight. Therefore, vitamins A and E are the major vitamins of concern. Green forage is a good source of β-carotene. Dietary **vitamin A deficiency** is most likely when cattle are fed high-concentrate diets, pasture or hay that is dry and sun bleached, or feeds that have been stored for extensive periods. Vitamin A is often included in trace mineral mixtures, although some elements such as copper promote destruction (oxidation) of the vitamin. The use of stable forms (vitamin A esters, e.g., retinyl acetate or palmitate) will help to reduce these losses.

Vitamin E has been shown to be valuable in regions where selenium may be deficient. Supplementation with vitamin E reduces selenium needs and shows additive responses when supplemented with selenium in selenium-deficient regions such as the Pacific Northwest. Vitamin E can be provided to cattle by intermuscular injection (i.e., 3,000 IU) or in a trace mineral mixture (i.e., 1,000 IU per day). Trace mineral mixture supplementation is more practical, however, because the serum benefits from an injection usually lasts fewer than 14 days (Hidiroglou et al., 1994). A recent innovation is the use of high doses of **vitamin E** in beef cattle to improve the **shelf life** of the meat (Liu et al., 1995). Vitamin E is an antioxidant that delays the oxidation of lipids and hemoglobin in meat, preserving its red color for a longer period. Supplementation of beef animals with 500 IU of vitamin E per day for about 120 days is effective in increasing the concentration of vitamin E in muscle tissue and improving the appearance and shelf life of beef (Liu et al., 1995; Faustman et al., 1998).

FEEDING BEEF CATTLE

Grazing Management

The general nature of the beef cattle industry in North America is that cows are raised on farms and ranches, primarily consume pasture and crop residues (stubble, corn stover), and are fed hay or silage during the winter feeding season. The calving season is generally in the spring or less commonly in the autumn. Calves suckle the cows until weaning (3 to 6 months of age) at which time they continue to be raised on forage or are sold to a feedlot for concentrate feeding and fattening (finishing).

For the cow herd, the objective should generally be to maximize the **utilization of roughages.** It is not economical to raise beef cows on valuable high-fertility cropland that can be better used to produce crops. Beef cattle production usually takes place in areas with soil, topography, or climatic characteristics that result in land that is reasonably inexpensive and unsuitable for intensive crop production, such as hilly or swampy areas, and arid and semiarid rangelands. The most feasible way to produce human food from these areas is to harvest forage with the ruminant animal. During much of the grazing season, beef cattle require little in the way of supplements. Salt and a mineral mixture should be provided free choice. Rotational grazing (see Chapter 6) using high-voltage electric fences increases stocking capacity and efficiency of forage utilization. If an extensive period of grazing on poor-quality, mature, dry forage, or crop residue will occur, a vitamin supplement with vitamins A and E should be provided. The ability of the forage resources to meet the nutritional demands of beef cattle is influenced by a number of factors (Fig. 15.1). In general, the role of the beef manager is to select cattle (breed and type) that can use forage resources with a minimal amount of supplemental input. Forage quality and production are strongly influenced by growing season, amount and distribution of precipitation, elevation, and composition of vegetation.

In many areas, poisonous plants are of concern. In the United States, **tall fescue toxicosis** (see Chapter 5) is the most important natural toxicity problem in the eastern part of the country, while in the west, larkspur poisoning causes more losses than any other poisonous plant. Other significant problems include birth defects in calves caused by plant toxins, such as crooked calf disease caused by ingestion of lupins, or poison hemlock ingested by gestating cows. Abortion caused by consumption of Ponderosa pine needles is a problem in many western states. More information on **poisonous plants** is provided by Cheeke (1998).

FIGURE 15.1 Beef production systems have to be adapted to available forage resources. The above cows are grazing dormant or stockpiled forage in the Great Basin region. This region spans major portions of Oregon, Idaho, Nevada, and Utah. This arid, high elevation environment typically has growing seasons (periods of green vegetative growth) of fewer than 100 days. As a result, the beef cattle producer has to base his or her production decisions on limited forage resources in terms of nutritional potential.

Winter Feeding

For the nongrazing feeding periods (winter feeding in most areas), cows can be fed conserved forage, such as standing forage, which is "stockpiled" for later grazing (Hitz and Russell, 1998). Hay, crop residues, or silage are other possibilities. If these forages are of reasonable quality (green in color, crude protein of 7 to 8 percent or more), they are adequate as the main or sole feed. Legume hays such as alfalfa contain higher energy and protein levels than needed by wintering cows and should not be fed as the main feed if maximum economic efficiency is desired. Good-quality legume hay can be fed several times weekly as a supplement to provide fat-soluble vitamins, minerals, bypass protein, rumen fermentable nitrogen, and readily fermentable carbohydrate. Supplementation with energy or protein sources may be needed with poor-quality roughages, with protein usually being the supplement of choice if the goal is to optimize the use of the low-quality forage (DelCurto et al., 2000b). Quality must be ascertained by obtaining a forage analysis, estimating forage intake, and comparing nutrient intake with NRC requirements (Table 15–1 on page 382).

Because of the myriad activities of the rumen, the mature beef cow is a nutritionally resilient animal and precise attention to ration formulation is not usually necessary. The old adage "The eye of the master fattens the cattle" applies. The livestock producer should be pragmatic and observant. Cows should be managed to keep them in good condition without being overly fat or thin, readily accomplished over a wide range of available feedstuffs.

Winter feeding of beef cows on corn stover and cereal crop residues is discussed elsewhere in this chapter (Alternative Nutritional Management Strategies).

Feeding During Lactation

The period of greatest nutritional stress for the beef cow is during early lactation, when milk production is highest and preparation for the next breeding season begins (Fig. 15.2). Cows will not rebreed unless energy intake is adequate to maintain good body condition. Good-quality pasture is usually adequate for this purpose. If cows are not able to maintain condition, an **energy supplement** of grain or molasses (as a liquid supplement or as molasses blocks) may be provided. The objective normally should be to maximize the contribution of pasture or other low-cost roughage and minimize the use of supplements.

Weaned Calves and Feedlot Feeding

After weaning, calves may be continued on good-quality pasture (stocker cattle) or moved to a feedlot (Fig. 15.3). In the United States, a very high percentage of cattle are finished on concentrate rather than on forage. The concentrate feeding period varies greatly in length depending on many circumstances, such as the age at which a particular cow-calf operator sells calves and the availability and price of feeder cattle of various ages. As discussed in Chapter 9, full feeding on a forage diet cannot provide sufficient DE intake for maximum growth rate. Feeding a concentrate diet, on the other hand, does allow the animal to grow at its genetic potential. Therefore, when rapid growth is desired, growing-finishing cattle must be fed high-concentrate diets.

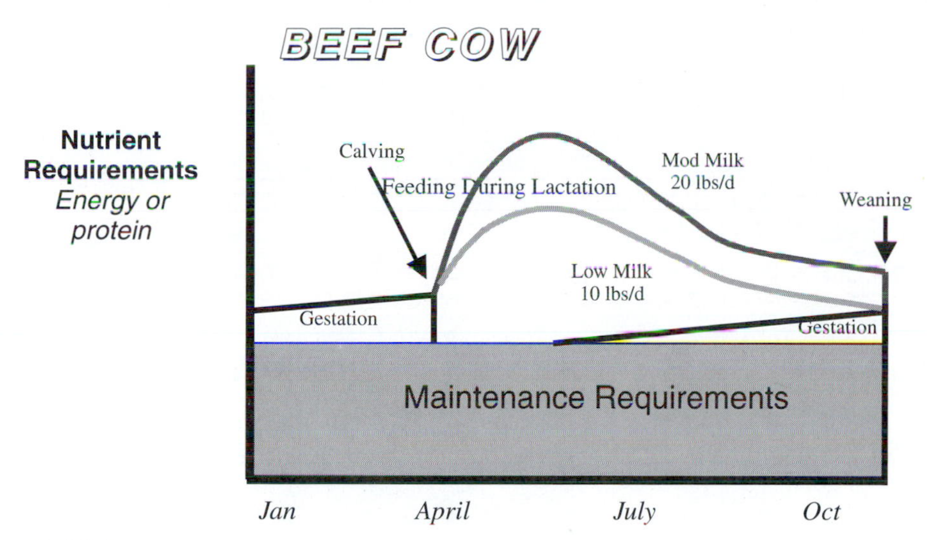

FIGURE 15.2 Stage of beef cow production determines the nutrient requirements of the animal. This figure partitions beef cattle requirements by maintenance, gestation, and milk production. Timing of calving, timing of weaning, and milk production have dramatic influences on nutrient requirements.

FIGURE 15.3 Cattle are fed concentrate diets in feedlots. Feedlot feeding produces rapid and efficient weight gains. The feed ingredients, which often consist of by-products such as potato processing wastes, are mixed in a feed truck and dispensed in the feeders. (Courtesy of Wade Nichols.)

TABLE 15–1. Nutrient Requirements of Breeding Cattle

			Energy								Total Protein		Calcium		Phosphorus		Vitamin A[d]
			Daily				In Diet DM										
Weight[a] kg	Daily Gain[b] kg	Daily DM[c] kg	ME, Mcal	TDN, kg	NE_m, Mcal	NE_g, Mcal	ME, Mcal/kg	TDN, %	NE_m, Mcal/kg	NE_g, Mcal/kg	Daily, g	In Diet DM, %	Daily, g	In Diet DM, %	Daily, g	In Diet DM, %	Daily, 1000's IU
Pregnant Yearling Heifers—Last Third of Pregnancy																	
325	0.4	7.1	14.2	3.9	8.04	NA[e]	2.00	55.2	1.15	NA	591	8.4	19	0.27	14	0.20	20
325	0.6	7.3	15.7	4.3	8.04	0.77	2.15	59.3	1.29	0.72	649	8.9	23	0.32	15	0.12	20
325	0.8	7.3	17.2	4.8	8.04	1.67	2.35	64.9	1.47	0.88	697	9.5	27	0.37	16	0.22	20
350	0.4	7.5	14.8	4.1	8.38	NA	1.99	55.0	1.14	NA	616	8.3	20	0.27	15	0.21	21
350	0.6	7.7	16.5	4.6	8.38	0.81	2.14	59.1	1.28	0.71	674	8.8	24	0.32	16	0.21	22
350	0.8	7.8	18.1	5.0	8.38	1.76	2.34	64.6	1.46	0.88	720	9.3	27	0.35	17	0.22	22
375	0.4	7.8	15.5	4.3	8.71	NA	1.98	54.7	1.13	NA	641	8.2	21	0.27	15	0.19	22
375	0.6	8.1	17.2	4.8	8.71	0.86	2.13	58.8	1.27	0.70	697	8.6	25	0.31	17	0.21	23
375	0.8	8.2	19.0	5.2	8.71	1.86	2.32	64.1	1.45	0.86	743	9.1	27	0.33	18	0.22	23
400	0.4	8.2	16.1	4.5	9.04	NA	1.97	54.4	1.12	NA	664	8.1	22	0.27	16	0.20	23
400	0.6	8.5	18.0	5.0	9.04	0.90	2.12	58.6	1.26	0.69	721	8.5	25	0.30	18	0.21	24
400	0.8	8.6	19.8	5.5	9.04	1.95	2.31	63.8	1.44	0.85	764	8.9	28	0.33	18	0.20	24
425	0.4	8.6	16.8	4.6	9.36	NA	1.96	54.1	1.11	NA	687	8.0	23	0.27	17	0.20	24
425	0.6	8.9	18.7	5.2	9.36	0.94	2.11	58.3	1.25	0.69	743	8.4	26	0.30	18	0.20	25
425	0.8	9.0	20.7	5.7	9.36	2.04	2.30	63.5	1.43	0.84	786	8.8	28	0.31	19	0.21	25
450	0.4	8.9	17.3	4.8	9.67	NA	1.95	53.9	1.10	NA	710	8.0	23	0.26	18	0.20	25
450	0.6	9.2	19.4	5.4	9.67	0.98	2.10	58.0	1.25	0.68	765	8.3	26	0.29	19	0.21	26
450	0.8	9.4	21.5	5.9	9.67	2.13	2.29	63.3	1.42	0.84	807	8.6	28	0.30	20	0.21	26
Dry Pregnant Mature Cows—Middle Third of Pregnancy																	
350	0.0	6.8	11.9	3.3	6.23	NA	1.76	48.6	0.92	NA	478	7.1	12	0.16	12	0.18	19
400	0.0	7.5	13.1	3.6	6.89	NA	1.76	48.6	0.92	NA	525	7.0	13	0.17	13	0.17	21
450	0.0	8.2	14.3	4.0	7.52	NA	1.76	48.6	0.92	NA	570	7.0	15	0.17	15	0.18	23
500	0.0	8.8	15.5	4.3	8.14	NA	1.76	48.6	0.92	NA	614	7.0	17	0.19	17	0.19	25
550	0.0	9.5	16.7	4.6	8.75	NA	1.76	48.6	0.92	NA	657	6.9	18	0.19	18	0.19	27
600	0.0	10.1	17.8	4.9	9.33	NA	1.76	48.6	0.92	NA	698	6.9	20	0.20	20	0.20	28
650	0.0	10.7	18.9	5.2	9.91	NA	1.76	48.6	0.92	NA	739	6.9	22	0.21	22	0.21	30
Dry Pregnant Mature Cows—Last Third of Pregnancy																	
350	0.4	7.4	14.7	4.1	8.38	NA	1.98	54.7	1.13	NA	609	8.2	20	0.27	15	0.20	21
400	0.4	8.2	16.0	4.4	9.04	NA	1.96	54.1	1.11	NA	657	8.0	22	0.27	16	0.20	23
450	0.4	8.9	17.2	4.8	9.67	NA	1.94	53.6	1.10	NA	703	7.9	23	0.26	18	0.21	24
500	0.4	9.5	18.3	5.1	10.29	NA	1.92	53.1	1.08	NA	746	7.8	25	0.26	20	0.21	27
550	0.4	10.2	19.5	5.4	10.90	NA	1.91	52.8	1.07	NA	790	7.8	26	0.25	21	0.21	29
600	0.4	10.8	20.6	5.7	11.48	NA	1.90	52.5	1.06	NA	832	7.7	28	0.26	23	0.21	30
650	0.4	11.5	21.7	6.0	12.06	NA	1.89	52.2	1.05	NA	872	7.6	30	0.26	25	0.22	32

Two-Year-Old Heifers Nursing Calves—First 3–4 Months Postpartum–5.0 kg Milk/Day																	
300	0.2	6.9	16.6	4.6	9.30[f]	0.72	2.41	66.6	1.53	0.93	814[g]	11.8	26	0.38	17	0.25	27
325	0.2	7.3	17.4	4.8	9.64[f]	0.77	2.37	65.5	1.49	0.90	841[g]	11.5	27	0.37	18	0.25	28
350	0.2	7.8	18.1	5.0	9.98[f]	0.81	2.34	64.6	1.46	0.88	866[g]	11.2	27	0.35	19	0.24	30
375	0.2	8.2	18.9	5.2	10.31[f]	0.86	2.31	63.8	1.44	0.85	892[g]	10.9	28	0.34	19	0.23	32
400	0.2	8.6	19.7	5.4	10.64[f]	0.90	2.29	63.3	1.42	0.84	916[g]	10.7	28	0.33	20	0.23	34
425	0.2	9.0	20.4	5.6	10.96[f]	0.94	2.27	62.7	1.40	0.82	939[g]	10.5	29	0.32	21	0.23	35
450	0.2	9.4	21.1	5.8	11.27[f]	0.98	2.25	62.2	1.38	0.80	963[g]	10.3	29	0.31	22	0.23	37
Cows Nursing Calves—Average Milking Ability—First 3–4 Months Postpartum—5.0 kg Milk/Day																	
350	0.0	7.7	16.6	4.6	9.98[f]	NA	2.15	59.4	1.29	NA	814[g]	10.6	23	0.30	18	0.23	30
400	0.0	8.5	17.9	4.9	10.64[f]	NA	2.11	58.3	1.25	NA	864[g]	10.2	25	0.29	19	0.22	33
450	0.0	9.2	19.1	5.3	11.27[f]	NA	2.08	57.5	1.23	NA	911[g]	9.9	26	0.28	21	0.23	36
500	0.0	9.9	20.3	5.6	11.89[f]	NA	2.05	56.6	1.20	NA	957[g]	9.7	28	0.28	22	0.22	38
550	0.0	10.6	21.5	5.9	12.50[f]	NA	2.03	56.1	1.18	NA	1001[g]	9.5	29	0.27	24	0.23	42
600	0.0	11.2	22.6	6.2	13.08[f]	NA	2.01	55.5	1.16	NA	1044[g]	9.3	31	0.28	26	0.23	44
650	0.0	11.9	23.9	6.6	13.66[f]	NA	2.00	55.3	1.15	NA	1086[g]	9.1	33	0.28	27	0.23	46
Cows Nursing Calves—Superior Milking Ability—First 3–4 Months Postpartum—10.0 kg Milk/Day																	
350	0.0	6.2	18.5	5.1	13.73[f]	NA	3.00	82.9	2.03	NA	1009[g]	16.4	36	0.58	24	0.39	24
400	0.0	7.6	21.4	5.9	14.39[f]	NA	2.80	77.4	1.86	NA	1099[g]	14.4	37	0.49	25	0.33	30
450	0.0	9.1	23.2	6.4	15.02[f]	NA	2.56	70.7	1.66	NA	1186[g]	13.1	39	0.43	26	0.29	35
500	0.0	10.0	24.6	6.8	15.64[f]	NA	2.45	67.7	1.56	NA	1246[g]	12.4	40	0.40	28	0.28	39
550	0.0	10.9	25.8	7.1	16.25[f]	NA	2.38	65.8	1.50	NA	1299[g]	12.0	42	0.39	30	0.27	45
600	0.0	11.6	27.0	7.5	16.83[f]	NA	2.32	64.1	1.45	NA	1348[g]	11.6	43	0.37	31	0.27	45
650	0.0	12.4	28.2	7.8	17.41[f]	NA	2.28	63.0	1.41	NA	1394[g]	11.3	45	0.36	33	0.26	48
Bulls, Maintenance, and Regaining Body Condition																	
650	0.4	12.3	24.3	6.7	9.91	2.06	1.98	54.8	1.13	0.57	904	7.4	25	0.20	23	0.19	48
650	0.6	12.6	26.7	7.4	9.91	3.21	2.11	58.4	1.25	0.69	957	7.6	27	0.21	24	0.19	49
650	0.8	12.8	28.7	7.9	9.91	4.40	2.24	62.0	1.37	0.79	998	7.8	29	0.23	25	0.20	50
700	0.4	13.0	25.7	7.1	10.48	2.18	1.98	54.8	1.13	0.57	942	7.3	26	0.20	25	0.20	51
700	0.6	13.4	28.2	7.8	10.48	3.40	2.11	58.4	1.25	0.69	994	7.4	29	0.22	26	0.20	52
700	0.8	13.5	30.3	8.4	10.48	4.66	2.24	62.0	1.37	0.79	1032	7.6	30	0.22	26	0.19	53
800	0.0	12.9	22.6	6.3	11.58	NA	1.75	48.4	0.91	NA	882	6.8	27	0.21	27	0.21	50
800	0.2	13.7	25.5	7.1	11.58	1.12	1.86	51.5	1.02	0.47	956	7.0	27	0.20	27	0.20	53
900	0.0	14.1	24.7	6.8	12.65	NA	1.75	48.4	0.91	NA	958	6.8	30	0.21	30	0.21	55
900	0.2	15.0	27.9	7.7	12.65	1.23	1.86	51.5	1.02	0.47	1031	6.9	31	0.21	31	0.21	58
1000	0.0	15.3	26.8	7.4	13.69	NA	1.75	48.4	0.91	NA	1032	6.8	33	0.22	33	0.22	60

Adapted from National Research Council (1984).

[a] Average weight for a feeding period.

[b] Approximately 0.4 ± 0.1 kg of weight gain/day over the last third of pregnancy is accounted for by the products of conception. Daily 2.15 Mcal of NE_m and 55 g of protein are provided for this requirement for a calf with a birth weight of 36 kg.

[c] Dry matter consumption should vary depending on the energy concentration of the diet and environmental conditions. These intakes are based on the energy concentration shown in the table and assuming a thermoneutral environment without snow or mud conditions. If the energy concentrations of the diet to be fed exceed the tabular value, limit feeding may be required.

[d] Vitamin A requirements per kilogram of diet are 2800 IU for pregnant heifers and cows and 3900 IU for lactating cows and breeding bulls.

[e] Not applicable.

[f] Includes 75 Mcal NE_m/kg of milk produced.

[g] Includes 33.5 g protein/kg of milk produced.

Feedlot rations are usually based on corn, milo, wheat, or barley, with supplementary protein and a minimum quantity of roughage. In many areas, by-products such as potato waste may constitute the base of the ration. The roughage is provided to maintain the health of the rumen and to promote rumination. Because of the negative effect of high-concentrate diets on fiber digestibility, the forage component of high-concentrate diets makes little if any nutritional contribution. With no roughage in the diet, **parakeratosis** or erosion of the rumen wall can occur. The ratio of roughage to concentrate can be important. If the roughage component constitutes more than approximately 20 percent of the diet, the negative **associative effects** of concentrate on fiber digestion (see Chapter 5) may reduce dietary DE.

When cattle are brought into a feedlot, they should be introduced gradually to a high-concentrate diet to avoid **lactic acidosis.** They should begin on a diet of hay, with concentrate introduced in increments of 10 percent of the diet every 4 to 5 days, beginning with a 50:50 roughage:concentrate proportion. The feeding of **ionophores** (e.g., monensin) during the adaptation period to high-concentrate diets can aid in preventing subclinical acidosis, which is characterized by reductions in rumen pH, feed intake, and animal performance (Burrin et al., 1988). The beneficial effect may be due in part to a suppression of feed intake when animals are first exposed to ionophores.

Adaptation to high-concentrate diets is also influenced by the grain used. The more rapid the rate of ruminal fermentation of a grain, the more likely it is to induce acidosis. Wheat is much more likely to induce acidosis than corn (Fulton et al., 1979); wheat-based diets cause lower rumen pH and reduced feed intake. Sorghum grain is less rapidly fermentable than corn, wheat, or barley (Spicer et al., 1986). High-moisture corn grain has a rapid rate of ruminal fermentation and is more likely to induce acidosis problems than dry corn (Stock et al., 1987).

Carbohydrate overload disorders such as lactic acidosis, laminitis, feedlot bloat, liver abscesses and polioencephalomalacia are discussed in Chapter 2.

The NRC requirements for energy and protein for growing-finishing beef cattle are listed in Tables 15–2 and 15–3. These tables can be used to formulate diets for a given class and weight of cattle to achieve various rates of gain. Computer programs can be helpful in establishing the optimal rate of gain under existing price conditions of cattle and feedstuffs.

The following brief outline summarizes the major facets of a beef cattle feeding program.

BASIC FEEDING PROGRAM FOR BEEF CATTLE

Cow Herd Feed good-quality forage (pasture in summer and grass-clover hay in winter). Based on forage analysis, supplement with sources of protein (e.g., soybean meal or cottonseed meal) or energy (grain, molasses, or high-quality hay) as needed, particularly for cows in early lactation. Trace-mineralized salt should be available free choice, and a calcium-phosphorus supplement offered, depending upon forage quality. Cattle that are grazing productive, well-fertilized grass-legume pastures may not require supplementation with anything but salt. With low-quality forage, **phosphorus supplementation** is particularly important because of its role in reducing calving interval (see Chapter 7). Range cows should

TABLE 15–2. Net Energy (NEg) Requirements of Growing and Finishing Beef Cattle (Mcal/day)[a]

	Body Weight, kg and NE_m Required									
Daily Gain, kg	*150* *3.30*	*200* *4.10*	*250* *4.84*	*300* *5.55*	*350* *6.24*	*400* *6.89*	*450* *7.52*	*500* *8.14*	*550* *8.75*	*600* *9.33*
Medium-Frame Steer Calves										
0.2	0.41	0.50	0.60	0.69	0.77	0.85	0.93	1.01	1.08	
0.4	0.87	1.08	1.28	1.47	1.65	1.82	1.99	2.16	2.32	
0.6	1.36	1.69	2.00	2.29	2.57	2.84	3.11	3.36	3.61	
0.8	1.87	2.32	2.74	3.14	3.53	3.90	4.26	4.61	4.95	
1.0	2.39	2.96	3.50	4.02	4.51	4.98	5.44	5.89	6.23	
1.2	2.91	3.62	4.28	4.90	5.50	6.69	6.65	7.19	7.73	
Large-Frame Steers, Compensating Medium-Frame Yearling Steers, and Medium-Frame Bulls										
0.2	0.36	0.45	0.53	0.61	0.68	0.75	0.82	0.89	0.96	1.02
0.4	0.77	0.96	1.13	1.30	1.46	1.61	1.76	1.91	2.05	2.19
0.6	1.21	1.50	1.77	2.03	2.28	2.52	2.75	2.98	3.20	3.41
0.8	1.65	2.06	2.43	2.78	3.12	3.45	3.77	4.08	4.38	4.68
1.0	2.11	2.62	3.10	3.55	3.99	4.41	4.81	5.21	5.60	5.98
1.2	2.58	3.20	3.78	4.34	4.87	5.38	5.88	6.37	6.84	7.30
1.4	3.06	3.79	4.48	5.14	5.77	6.38	6.97	7.54	8.10	8.64
1.6	3.53	4.39	5.19	5.95	6.68	7.38	8.07	8.73	9.38	10.01
Large-Frame Bull Calves and Compensating Large-Frame Yearling Steers										
0.2	0.32	0.40	0.47	0.54	0.60	0.67	0.73	0.79	0.85	0.91
0.4	0.69	0.85	1.01	1.15	1.29	1.43	1.56	1.69	1.82	1.94
0.6	1.07	1.33	1.57	1.80	2.02	2.23	2.44	2.64	2.83	3.02
0.8	1.47	1.82	2.15	2.47	2.77	3.06	3.34	3.62	3.88	4.15
1.0	1.87	2.32	2.75	3.15	3.54	3.91	4.27	4.62	4.96	5.30
1.2	2.29	2.84	3.36	3.85	4.32	4.77	5.21	5.64	6.06	6.47
1.4	2.71	3.36	3.97	4.56	5.11	5.65	6.18	6.68	7.18	7.66
1.6	3.14	3.89	4.60	5.28	5.92	6.55	7.15	7.74	8.31	8.87
1.8	3.56	4.43	5.23	6.00	6.74	7.45	8.13	8.80	9.46	10.10
Medium-Frame Heifer Calves										
0.2	0.49	0.60	0.71	0.82	0.92	1.01	1.11	1.20	1.29	
0.4	1.05	1.31	1.55	1.77	1.99	2.20	2.40	2.60	2.79	
0.6	1.66	2.06	2.44	2.79	3.13	3.46	3.78	4.10	4.40	
0.8	2.29	2.84	3.36	3.85	4.32	4.78	5.22	5.65	6.07	
1.0	2.94	3.65	4.31	4.94	5.55	6.14	6.70	7.25	7.79	
Large-Frame Heifer Calves and Compensating Medium-Frame Yearling Heifers										
0.2	0.43	0.53	0.63	0.72	0.81	0.90	0.98	1.06	1.14	1.21
0.4	0.93	1.16	1.37	1.57	1.76	1.95	2.13	2.31	2.47	2.64
0.6	1.47	1.83	2.16	2.47	2.78	3.07	3.35	3.63	3.90	4.16
0.8	2.03	2.62	2.98	3.41	3.83	4.24	4.63	5.01	5.38	5.74
1.0	2.61	3.23	3.82	4.38	4.92	5.44	5.94	6.43	6.91	7.37
1.2	3.19	3.97	4.69	5.37	5.03	6.67	7.28	7.88	8.47	9.03

Adapted from National Research Council (1984).
[a]Shrunk liveweight basis.

TABLE 15–3. Protein Requirements of Growing and Finishing Cattle (g/day)[a]

Daily Gain, kg	Body Weight Gain, kg									
	150	200	250	300	350	400	450	500	550	600
Medium-Frame Steer Calves										
0.2	343	399	450	499	545	590	633	675	715	
0.4	428	482	532	580	625	668	710	751	790	
0.6	503	554	601	646	688	728	767	805	842	
0.8	575	621	664	704	743	780	815	849	883	
1.0	642	682	720	755	789	821	852	882	911	
1.2	702	735	766	794	822	848	873	897	921	
Large-Frame Steer Calves and Compensating Medium-Frame Yearling Steers										
0.2	361	421	476	529	579	627	673	719	762	805
0.4	441	499	552	603	651	697	742	785	827	867
0.6	522	576	628	676	722	766	809	850	890	930
0.8	598	650	698	743	786	828	867	906	944	980
1.0	671	718	762	804	843	881	918	953	988	1021
1.2	740	782	822	859	895	929	961	993	1023	1053
1.4	806	842	877	908	938	967	995	1022	1048	1073
1.6	863	892	919	943	967	989	1011	1031	1052	1071
Medium-Frame Bulls										
0.2	345	401	454	503	550	595	638	680	721	761
0.4	430	485	536	584	629	673	716	757	797	835
0.6	509	561	609	655	698	740	780	819	856	893
0.8	583	632	677	719	759	798	835	871	906	940
1.0	655	698	739	777	813	849	881	914	945	976
1.2	722	760	795	828	860	890	919	947	974	1001
1.4	782	813	841	868	893	917	941	963	985	1006
Large-Frame Bull Calves and Compensating Large-Frame Yearling Steers										
0.2	355	414	468	519	568	615	661	705	747	789
0.4	438	494	547	597	644	689	733	776	817	857
0.6	519	574	624	672	718	761	803	844	884	923
0.8	597	649	697	741	795	826	866	905	942	979
1.0	673	721	765	807	847	885	922	958	994	1027
1.2	745	789	830	868	904	939	973	1005	1037	1067
1.4	815	854	890	924	956	986	1016	1045	1072	1099
1.6	880	912	943	971	998	1024	1048	1072	1095	1117
1.8	922	942	962	980	997	1013	1028	1043	1057	1071
Medium-Frame Heifer Calves										
0.2	323	374	421	465	508	549	588	626	662	
0.4	409	459	505	549	591	630	669	706	742	
0.6	477	522	563	602	638	674	708	741	773	
0.8	537	574	608	640	670	700	728	755	781	
1.0	562	583	603	621	638	654	670	685	700	
Large-Frame Heifer Calves and Compensating Medium-Frame Yearling Heifers										
0.2	342	397	449	497	543	588	631	672	712	751
0.4	426	480	530	577	622	665	707	747	787	825
0.6	500	549	596	639	681	721	759	796	832	867
0.8	568	613	654	693	730	765	799	833	865	896
1.0	630	668	703	735	767	797	826	854	881	907
1.2	680	708	734	758	781	803	824	844	864	883

Adapted from National Research Council (1984).
[a]Shrunk liveweight basis.

be managed to avoid pulmonary emphysema (see Chapter 5) associated with a rapid change from a sparse pasture to lush, succulent feed.

The cow herd is often grouped according to age, with first-calf heifers grouped separately from older cows. Since they are still growing, as well as producing a calf, first-calf heifers should receive high-quality forage and a protein-energy supplement.

Calves Calves graze forage and suckle cows for several months. At 3 to 4 weeks of age, they begin to graze forage, which during the next few months becomes their major nutrient source. **Creep feed** may be used to increase growth rate but may not be economical. The creep feed should be based on a grain (e.g., rolled corn) and a protein supplement (e.g., cottonseed meal). Wean calves at 7 to 8 months of age and sell (cow-calf operation) or overwinter the calves.

Postweaning Calves and Replacement Heifers Feed good-quality forage free choice. Supplement with grain and protein supplement as necessary to produce desired level of body weight gain (see nutrient requirement tables).

Market Cattle Weaned calves may be raised on roughage for a year or more before entering the feedlot, or they may enter the feedlot directly after weaning. Calves may be **preconditioned** by starting on grain for at least 30 days prior to sale as feeder calves as well as by receiving vaccinations. **Stocker cattle** are weaned calves that are forage-fed for a period of time before being sold as feeder calves to enter a feedlot. Stocker calves are usually purchased when grazing conditions are good, such as over the winter on wheat pasture in the United States and during the spring flush of grass growth in most other areas.

Feedlot Management Because of their high-energy diet, feedlot cattle have several nutritional problems not usually encountered with beef cows. Feedlot cattle are susceptible to a number of **grain overload disorders,** discussed in Chapter 2, that include acidosis, rumenitis, feedlot bloat, laminitis, liver abscess, polioencephalomalacia, and enterotoxemia. These disorders are all related to the rapid fermentation of soluble carbohydrate.

Start cattle on a complete, self-fed diet formulated to meet NRC requirements in a "warm-up" or "step-up" program. Begin with a 50:50 concentrate: roughage ratio and work up to a 90:10 ratio. The cattle can continue on the 90:10 concentrate:roughage diet for the remainder of the feeding period. Diets with no supplementary roughage have been successfully used in some instances; an ionophore (to reduce acidosis) should be fed in this case. This program is more successful with grains with a slower rate of rumen fermentation (corn, sorghum) than with a high rate of rumen fermentation (wheat).

MANAGEMENT TO REDUCE NUTRITIONAL INPUTS

One of the most fundamental objectives of economically sustainable beef cattle production is to not provide nutritional inputs such as harvested feeds and(or) supplements unless it is necessary. Therefore, the first goal of manager should be to match the biological cycle and nutritional demands of the cow herd with the forage resources available.

When is the best time to calve? One of the most fundamental management decisions that has profound effects on beef cattle nutritional requirements is calving date. Calving date sets the biological cycle that, in turn, determines the nutritional cycle of the cow herd and associated relationship to ranch resources. The North American beef cattle industry is dominated by spring-calving cattle. In addition, time of calving has generally been related to the "55 days to grass" philosophy. This traditional management strategy has gained popularity for a variety of reasons. First, gestation length in beef cattle is approximately 284 days. Therefore, if the cow herd calves approximately 55 days before the onset of green forage, the cows will be exposed to green, highly nutritious forage for approximately 25 days before they must conceive to stay on a 365-day calving interval. In a sense, the 25 days of high-quality forage is a natural "flushing" mechanism that normally prompts a cow to begin cycling, provided she was in adequate body condition initially. Obviously, if the goal is to match the cow nutritional requirements with available forage quality, a producer might coincide calving with the onset of green forage (DelCurto et al., 2000a, b). However, the "55 days to grass" philosophy has another advantage: the calf. A typical beef calf does not develop a fully functioning rumen until approximately 90 to 120 days of age. This normally coincides with the time a cow has passed its peak lactation period (day 70 to 90), and as a result, calf performance will depend largely on the quality of available forage. Thus, a calf born March 1 will be effectively utilizing the high-quality forage available in June. In contrast, a calf born May 1 will not be effectively utilizing forage resources until August, when forage quality is normally low in temperate regions of North America. Because of the vast difference in calf nutrition from day 90 to weaning, the earlier-born calf will have weaning weight advantages that greatly outweigh the 60-day difference in age. If higher weaning weights are a measure of economic importance (calves are marketed in the fall), then the "55 days to grass" philosophy may be the best approach.

Are weaning weights really important? The beef cattle industry in the United States has seen dramatic changes in production efficiencies over the last 30 years. In particular, weaning weights have increased from approximately 400 pounds in 1967 to greater than 600 pounds in 2002. The increase in weaning weights are related to increased use of continental breeds, greater selection for growth traits, and general improvements in management efficiency. Heavier weaning weights could increase the potential for profitability if the producer's goal is to market calves in the spring. This scenario would only be true if the increased income of heavier weaning weights outweighs the added cost of attaining that increase in weight. However, the increase in weaning weights is an improvement in production efficiency that has some indirect problems. First, the target slaughter weight of market cattle has not changed dramatically during this time period. As a result, the opportunities to put on postweaning weight have become limited with the heavier weaning weight cattle. For example, if a spring-calving cow/calf producer weans his or her cattle in late October at 600 pounds, he/she may choose to sell in the fall market or retain calves over the winter feeding period. However, because of the bigger calves, options are reduced. With only marginal gains of 1 to 1.5 pounds per head per day, this producer will come out of the winter feeding period (120 to 150 days) with 700- to 800-pound yearlings. The opportunities to place these animals on spring grass have now become very restricted. To fit market standards, yearlings need to be placed in the feedlot (avg.

90 days) with an expected gain of 300 to 350 pounds in order to meet a target end weight of 1,200 to 1,300 pounds. Therefore, heavy-weaning, spring-born calves have limited opportunities as stocker cattle on grass markets.

Another change in the beef cattle industry in recent years is the trend toward **retained ownership** and/or branded markets. These changes have indirectly led producers to reevaluate weaning weight goals because of opportunites to capture weight gains on yearlings and the need to provide cattle at finished weights on a yearly time frame. For producers who wish to retain ownership of cattle after weaning, weaning weight takes on less significance. In fact, these producers are the ones who should consider calving dates strongly. If a producer wishes to decrease costs per cow, moving the calving date to coincide cow nutrient demands with range/pasture forage quality may effectively reduce costs associated with supplementing cows during nutrient deficiencies. Weaning weight advantages are reduced, but the producer has more opportunites to capture gains in the stocker, backgrounding, and finishing phases.

PREPARING THE COW HERD FOR THE WINTER PERIOD

Because the winter period represents a time of high feed costs for beef cow-calf production, management strategies should emphasize decreasing the inputs. Getting the cow herd in good, fleshy condition going into the winter period should be a year-round management goal. Obviously, this involves monitoring range and/or pasture forage conditions, with particular attention to the quantity and quality of late summer and early fall forage. For most regions of North America, some period of forage dormancy is common, and for many areas this period is usually during the winter months and often requires feeding harvested and/or conserved forages.

A manager should monitor cow body condition and calf performance in late summer. Body condition refers to a visual appraisal and estimate of body fat or reserves. The most common approach relates cow condition to a 9-point scale, with 1 being extremely emaciated to 9 being extremely fat. For most production conditions, optimal cow condition should be maintained between body condition scores (BCS) of 3 and 6 (Figure 15.4). This approach insures that the cattle are neither too thin nor too fat to achieve optimal reproductive success.

When cows start losing body condition and/or calf performance begins to decline due to poor forage conditions, the producer should consider nutritional management strategies to optimize cow condition going into the winter period. A cow in good condition (5 or better) going into the winter period will be easier to feed and can lose some body condition without adversely affecting subsequent calving and rebreeding potential (Houghton et al., 1990). In contrast, cows that are thin going into the winter period are often thin at calving and require increased postpartum energy level to rebreed in a timely manner. Cows in thin condition at breeding (less the 4 BCS) usually have reduced reproductive efficiency. In addition, cattle on an increasing plane of nutrition at a BCS of 4 have higher conception rates than cattle on a level plane of nutrition at the same body condition (Wiltbank, 1982).

A cow going into the winter period with a body condition score of 5 will have approximately 140 lb of additional body weight (reserves) as compared to the

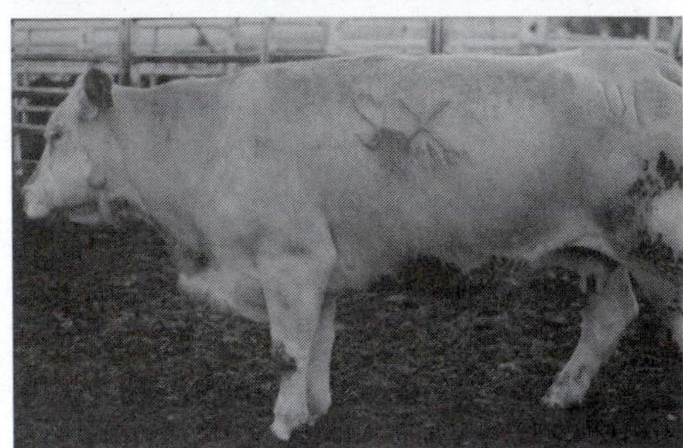

FIGURE 15.4 Managing the body condition of beef cattle is critical to reducing nutritional inputs and optimizing reproductive efficiency. The upper photos are of a cow with a BCS of 3 (body weight = 860 lb). The same cow in the lower photos is in a BCS of 6 (1,086 lb). The additional body reserves are important for cattle in higher latitudes and elevations. (Courtesy of Patrick Momont.)

same cow with a BCS of 3 (Fig. 15.4). Managing the condition of the cow to optimize condition going into the period when forage quality is limited helps the producer reduce the needs for supplemental inputs. The relative success of producers in managing cow body condition relates to a number factors but is largely influenced by the nutrient requirements of the cow. Cows with low to moderate milk production are easier to manage in terms of body condition than cows with higher milk producing capabilities. In environments where forage quality is limited for extended periods of time, the selection of cattle with lower requirements is essential if the producer wants to minimize nutritional inputs.

Early Weaning as a Management Tool

Tradionally, beef producers have weaned calves at approximately 7 months of age, which is usually late October or November for spring-calving herds. For many regions of North America, gains by both calves and cows are often poor by late August, particularly during years of poor forage quality/quantity. Early weaning of calves will allow the producer to provide them higher-quality sources of feed and potentially have greater gains at a similar cost to later weaning (Fig. 15.5). The cows, in turn, can be left on range, and, due to the removal of the nutrient demands associated with lactation, these dry cows will do well on range forage during the fall and, without a suckling calf, will come into winter in better condition. Improved body condition, in turn, translates into a cow that will be easier to maintain during the winter period and have a higher chance of breeding back and maintaining a 365-day calving interval.

A number of factors need to be considered when deciding if early weaning is appropriate. First, forage quality must be limiting to the point that calf gain will be reduced and cows will likely lose body condition from late August to the October or November weaning date. If forage quality and(or) quantity are not lim-

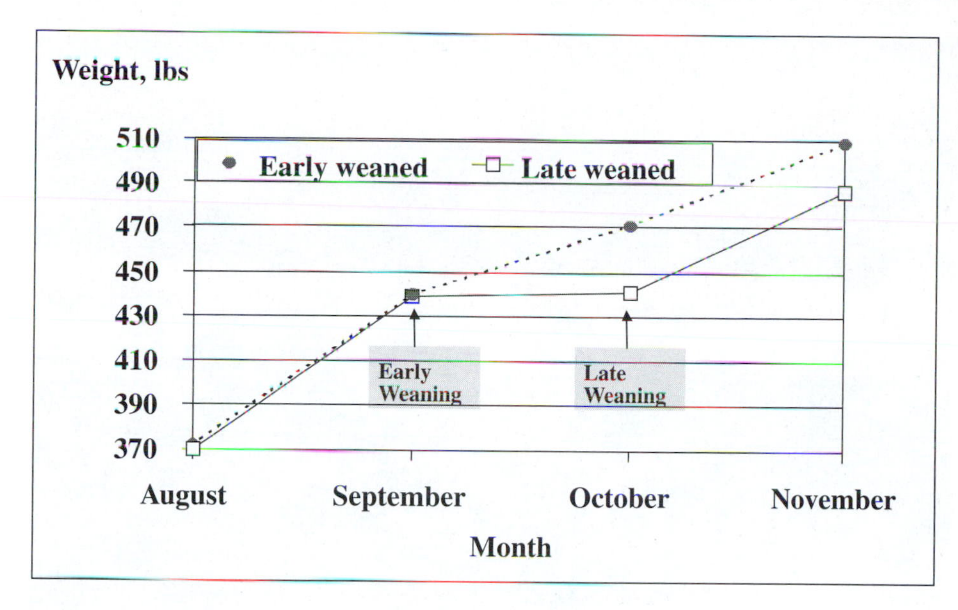

FIGURE 15.5 Influence of time of weaning on calf weights. (Courtesy of Turner and DelCurto, 1991.)

iting, there is really no advantage to early weaning. The real advantage of early weaning is to improve the weight and body condition of the cows from late summer to the beginning of the winter feeding period. In addition, the producer must provide adequate forage/nutrition to the early-weaned calf. For producers with limited nutritional options during the late summer and fall period, however, early weaning may provide an alternative that allows for more efficient management of mature cows' body condition. In turn, improved body condition of cattle going into the winter period will facilitate reduced supplemental feed requirements during the winter period.

ALTERNATIVE NUTRITIONAL MANAGEMENT STRATEGIES

Several **crop residues** such as wheat straw and corn stover are available in tremendous quantities and could support a large population of ruminant animals. For example, Van Soest (1994) states that if all of the arable land in the United States were used for cereal production, sufficient straw and stover would be produced to support approximately threefold the present ruminant population. Males (1987) calculated that if one-half of the available crop aftermath were left on the field for erosion control and the remainder were harvested and fed to beef cows as a wintering feed, 17.5 million brood cows (about 5 percent of the U.S. cow herd) could be wintered on cereal straw. However, the challenge is to develop economically viable beef production systems based on crop residues. The low content and bioavailability of nutrients in crop residues limit their use in diets for high-producing animals. Their low nutritional value generally precludes transporting them any distance.

FIGURE 15.6 Large numbers of beef cows in the United States are wintered on corn stover. (Courtesy of J. L. Johnson.)

Feeding Systems Based on Corn Stover

The residues from corn grain production are equal to or greater in quantity than the grain itself. Nebraska researchers (Klopfenstein et al., 1987) have pioneered development of beef-production systems based on **corn residues.** Grazing is the most economical means of utilizing corn residues and allows the most nutrients to remain in the soil. In many parts of the U.S. Corn Belt and midplains, cows are turned out into cornfields after the harvest and wintered on corn stalks with only modest inputs of supplementary feed (Fig. 15.6). Klopfenstein et al. (1987) suggest that with proper supplementation, stalk grazing could be an effective means of wintering weaned calves. They would then be suitable for turning out on grass in the spring and would be capable of making rapid gains. These authors present a cost-and-performance analysis for a corn residue-based system for wintering calves to produce yearlings ready for grass. Supplementation with a source of bypass protein greatly improves animal performance (Fernandez-Rivera and Klopfenstein, 1989a). A strong response to supplementation with alfalfa hay is also observed, probably because of its favorable effect on the rumen environment (see Chapter 6).

Fernandez-Rivera and Klopfenstein (1989b) developed a simulation model to predict performance of cattle grazing **cornstalks,** which should be useful in planning management systems to optimize utilization of corn crop residues. The first limiting nutrient in cornstalks for growing cattle is protein (Fernandez-Rivera and Klopfenstein, 1989a). Cornstalks from dryland fields support a higher growth rate of cattle than do stalks from irrigated fields (Fernandez-Rivera and Klopfenstein, 1989b), possibly because of greater digestibility and higher intake. Those factors that decrease grain yield tend to increase residue quality (Klopfenstein et al., 1987). When cattle are first turned into a cornfield, the availability of nutrients is initially high in the form of residual grain and leaves. As the higher-quality material is eaten, the residual stalks provide a lower-quality diet, and more supplementation may be necessary. By mid to late winter, weathering may result in substantial nutrient losses from corn residues (Russell et al., 1993). When cows are first introduced to a field, there is a danger of grain overload and acidosis. The field should be carefully checked for piles of spilled corn, and the cows given a full feeding of hay or other dry roughage before being turned out. **Nitrate poisoning** is a possibility if the corn has been heavily fertilized with nitrogen. There should be ample access to water; runoff water accumulating in depressions in the field may contain toxic levels of nitrate.

Wahlberg et al. (1988) described another system using ensiled whole corn plant for utilizing corn in beef cattle production. Harvesting corn as whole plant silage can result in the highest energy yield per unit land because all plant tissues are recovered. Carcass beef production per hectare was 6 to 12 percent greater with whole plant silage than with high-moisture shelled corn.

Feeding Systems Based on Cereal Straw

Residues from cereal-grain production represent a tremendous potential feedstuff resource, at least in terms of biomass available. Males (1987) and Anderson (1978) reviewed methods of optimizing the use of **cereal straw** in beef-production systems. Cereal-grain residues are low in protein (less than 6 percent), and high in NDF (greater than 60 percent) and ADF (greater than 40 percent). Cereal straw dry-matter digestibility and voluntary intake are low. Various treatments, discussed in Chapter 6, can be used to increase the nutritive value of straw. These include alkali treatment, ammoniation, supplementation with bypass protein, and nutrient supplementation to optimize rumen fermentation.

Representative trials reporting favorable effects of **ammoniation** include those of Herrera-Saldana et al. (1982, 1983) and Zorrilla-Rios et al. (1985). Ammoniation alters the physical structure of wheat straw by increasing its fragility (Zorrilla-Rios et al., 1985). This increases its physical breakdown during mastication and rumination and accelerates ruminal degradation. These factors contribute to the increased intake of ammoniated straw relative to untreated material. The beards of barley and common varieties of soft wheat can cause eye irritations and predispose cows to pink eye, other eye disorders, and sore mouths (Anderson, 1978). Presumably the favorable effects of ammoniation on physical structure would help to prevent these problems.

The physical form of straw also influences intake and utilization. Weisenburger and Mathison (1977) compared pelleted, ground, and chopped barley straw in diets for wintering cows. Chopped straw had the highest digestibility, whereas daily intake was highest for pelleted straw. Neither pelleting nor grinding was considered cost effective.

Optimal protein supplementation of **straw-based diets** is important. Males et al. (1982) and Wiedmeier et al. (1983) reported that when straw-based diets are fed, the protein level should be increased to exceed NRC requirement values by 30 percent. With diets based on ammoniated wheat straw, Males (1987) suggested that only half of the added nitrogen in ammonia-treated straw was effectively utilized in the rumen. Therefore, he suggests using 50 percent of the analyzed nitrogen (N) value as the N content of treated straw for ration formulation purposes. Several studies indicate a positive response to supplementation with rumen nondegradable protein, including the work of Church and Santos (1981) and Nelson et al. (1985). When **ammoniated straw** is used, the residual ammonia is sufficient to maintain optimal rumen fermentable N (Nelson et al., 1985), so a protein supplement with high bypass potential should be used. With untreated straw, the first-limiting factor may be rumen fermentable N (Dixon et al., 1981; Nelson et al., 1985). When this requirement has been met, a further response to a supplement of bypass protein will be noted. Fish meal is a very effective source of bypass protein in roughage-based diets (Laflamme, 1988).

Liquid supplements can be effectively used for cattle that are on pasture or fed low-quality roughages. They are usually based on molasses, with added sources of NPN, protein, phosphorus, salt, trace elements, and vitamin A. Urea can be used as a source of NPN; ammoniated molasses should not be used because of its imidazole toxicity (see Chapter 6), which causes "bovine bonkers" or "crazy cow disease." Natural proteins such as cottonseed meal give better results than urea (Pate et al., 1990) because of their contribution of bypass protein. Phosphoric acid is often used as the phosphorus source in liquid supplements. Liquid supplements may be fed in open troughs or in tanks with a lick wheel. Cattle lick the wheel, which rotates in the supplement and is covered continuously with the material.

Feeding systems using straw with appropriate treatment and supplementation can be developed for wintering beef cows and yearling beef animals. Males (1987) calculated that ammoniated straw was economically competitive with 10 percent protein grass hay valued at $60 U.S. per ton. Hay prices at this level or higher for hay of this quality were typical in the period from 1978 to 1987, suggesting that straw-based diets can be economically viable in the United States.

In developing countries, **rice straw** is widely used as a feedstuff for ruminants. Rice straw contains high levels of silica and oxalate in the leaves but is lower in lignin than other cereal straws. **Ammoniation** is effective in improving the digestibility and intake of rice straw. Numerous other fibrous crop residues, such as cotton gin trash, can be similarly used (Conner and Richardson, 1987).

Straw can also be harvested by grazing. In many areas, it is a common practice to graze beef cows on wheat stubble after the harvest. Older cows become quite skillful in picking up wheat kernels on the ground, so in the first week or two on stubble, they may obtain quite a high-energy diet. Anderson (1978) has provided a good review of straw feeding systems including grazing. To facilitate grazing, some farmers may use machinery to bunch the straw in piles. This bunched material is more easily found by cattle when the field is snow-covered; however, it may mold and become toxic (Bagley et al., 1983).

In the Pacific Northwest parts of the United States and parts of Canada, grass-seed production is an important industry. Large amounts of grass-seed straw are a by-product of seed production; traditionally, these residues have been burned to rid the fields of insect pests and pathogens and to dispose of the straw. Open-field burning causes extensive air pollution and is under increasing regulatory pressures. Development of alternate uses for the straw, such as livestock feeding, is urgently needed. As with cereal straws, **grass-seed straw** is low in protein, high in fiber, and of low digestibility, and the nutritive value is increased by chemical treatment (Anderson and Ralston, 1973; Schultz et al., 1974; Church and Champe, 1980). Use of equipment to spray chemical treatments (e.g., urea) on straw as it is being baled, particularly in large bales, is a practical method of enhancing grass-straw utilization. A potential hazard with grass-seed straws, particularly if open-field burning is banned, is the presence of mycotoxins, which cause conditions such as annual ryegrass toxicity and perennial ryegrass staggers. These and other mycotoxicoses associated with roughage feeding are reviewed by Cheeke (1998).

BISON FEEDING AND NUTRITION

There is a small niche market for bison meat in North America. Compared to beef, bison meat tends to be lower in fat, similar in cholesterol content, more tender, and has a more intense "gamey" taste (Koch et al., 1995). Bison are closely related to cattle, and can be crossed with cattle to produce fertile hybrids (cattalo, beefalo). Bison differ from cattle in their grazing behavior. They preferentially graze on grasses rather than broad-leaved plants (forbs), and their grazing pattern leaves patches of heavily and lightly grazed areas (Knapp et al., 1999). It is a widely held view that bison utilize low-quality forages better than cattle do. Koch et al. (1995) observed higher digestibilty coefficients with bison than with cattle, regardless of fiber or roughage content of the diet.

In general, bison can be fed with diets that are suitable for beef cattle. The same applies to other exotic bovids such as the yak.

QUESTIONS AND STUDY GUIDE

1. Diets for feedlot steers are being formulated, using protein supplements of either soybean meal or feathermeal. Which diet could use a larger percentage of the total nitrogen as supplementary urea: one with soybean meal or one with feathermeal? Why?
2. Wheat straw will have a higher net energy value for beef cows when fed as a major portion of the diet in a very cold environment than when fed in a warm environment. What are two reasons, both related to heat increment, that explain this?
3. Cows that are losing body condition after calving may not start cycling (estrous cycle) as soon as cows on a high plane of nutrition. Explain the effects of nutrition on calving interval.
4. Grains may have a negative associative effect on the nutritive value of grass hay for beef cattle. Explain.
5. Under what conditions might vitamin A deficiency in beef cattle be expected to occur? What symptoms would you expect to observe with vitamin A deficiency?

REFERENCES

Anderson, D.C. 1978. Use of cereal residues in beef cattle production systems. *J. Anim. Sci.* 46:849–861.

Anderson, D.C. and A.T. Ralston. 1973. Chemical treatment of ryegrass straw: *In vitro* dry matter digestibility and compositional changes. *J. Anim. Sci.* 37:148–152.

Bagley, C.V., J.B. McKinnon, and C.S. Asay. 1983. Photosensitization associated with exposure of cattle to moldy straw. *J. Am. Vet. Med. Assoc.* 183:802.

Burrin, D.G., R.A. Stock, and R.A. Britton. 1988. Monensin level during grain adaptation and finishing performance in cattle. *J. Anim. Sci.* 66:513–521.

Cheeke, P.R. 1998. *Natural Toxicants in Feeds, Forages, and Poisonous Plants.* Upper Saddle River, NJ: Prentice Hall, Inc.

Church, D.C. and K.A. Champe. 1980. Digestibility of untreated and hydroxide-treated annual ryegrass straw. *J. Anim. Sci.* 51:20–24.

Church, D.C. and A. Santos. 1981. Effect of graded levels of soybean meal and of a nonprotein nitrogen-molasses supplement on consumption and digestibility of wheat straw. *J. Anim. Sci.* 53:1609–1615.

Conner, M.C. and C.R. Richardson. 1987. Utilization of cotton plant residues by ruminants. *J. Anim. Sci.* 65:1131–1138.

DelCurto, T., D. Bohnert, and C. Ackerman. 2002a. Characteristics and challenges of sustainable beef production in western US. In *Strategic Supplementation of Beef Cattle Consuming Low-Quality Roughages in the Western United States,* pp. 6–18. Oregon State Univ. Agri. Exp. Sta. SB 683.

DelCurto, T., K. C. Olson, B. Hess, and E. Huston. 2002b. Optimal supplementation strategies for beef cattle consuming low-quality forages in the western United States. *J. Anim. Sci.* Symposium Paper Web Site.

Dixon, R.M., G.W. Mathison, and L.P. Milligan. 1981. Effect of rumen degradability of supplements on intake of barley straw by steers. *Can. J. Anim. Sci.* 61:1055–1058.

Faustman, C., W. K. M. Chan, D. M. Schaefer, and A. Havens. 1998. Beef color update: The role for vitamin E. *J. Anim. Sci.* 76:1019–1026.

Fernandez-Rivera, S. and T.J. Klopfenstein. 1989a. Diet composition and daily gain of growing cattle grazing dryland and irrigated cornstalks at several stocking rates. *J. Anim. Sci.* 67:590–596.

_______ . 1989b. Yield and quality components of corn crop residues and utilization of these residues by grazing cattle. *J. Anim. Sci.* 67:597–605.

Fernandez-Rivera, S., T.J. Klopfenstein, and R.A. Britton. 1989a. Growth response to escape protein and forage intake by growing cattle grazing cornstalks. *J. Anim. Sci.* 67:574–580.

Fernandez-Rivera, S., M. Lewis, T.J. Klopfenstein, and T.L. Thompson. 1989b. A simulation model of forage yield, quality and intake and growth of growing cattle grazing cornstalks. *J. Anim. Sci.* 67:581–589.

Fox, D.G., C.J. Sniffen, and J.D. O'Connor. 1988. Adjusting nutrient requirements of beef cattle for animal and environmental variations. *J. Anim. Sci.* 66:1475–1495.

Fulton, W.R., T.J. Klopfenstein, and R.A. Britton. 1979. Adaptation to high concentrate diets by beef cattle. I. Adaptation to corn and wheat diets. *J. Anim. Sci.* 49:775–784.

Greenough, P.R., J.J. Vermunt, J.J. McKinnon, F.A. Fathy, P.A. Berg, and R.D.H. Cohen. 1990. Laminitis-like changes in the claws of feedlot cattle. *Can. Vet. J.* 31:202–208.

Herrera-Saldana, R., D.C. Church, and R.O. Kellems. 1982. The effect of ammoniation treatment on intake and nutritive value of wheat straw. *J. Anim. Sci.* 54:603–608.

_______ . 1983. Effect of ammoniation treatment of wheat straw on *in vitro* and *in vivo* digestibility. *J. Anim. Sci.* 56:938–942.

Hidiroglou, M., R.R. Batra, and G. L. Roy. 1994. Changes in plasma α-tocopherol and selenium of gestation cows fed hay or silage. *J. Dairy Sci.* 77:190.

Hitz, A.C. and J.R. Russell. 1998. Potential of stockpiled perennial forages in winter grazing systems for pregnant beef cows. *J. Anim. Sci.* 76:404–415.

Houghton, P. L., R. P. Lemenager, L. A. Horstman, K. S. Hendrix, and G. E. Moss. 1990. Effects of body composition, pre- and postpartum energy level and early weaning on reproductive performance of beef cows and preweaning calf gain. *J. Anim. Sci.* 68:1438–1446.

Jurgens, M.H. 1993. *Animal Feeding and Nutrition.* Dubuque, IA: Kendall/Hunt.

Klopfenstein, T., L. Roth, S. Fernandez-Rivera, and M. Lewis. 1987. Corn residues in beef production systems. *J. Anim. Sci.* 65:1139–1148.

Knapp, A. K., J. M. Blair, J. M. Briggs, S. L. Collins, D. C. Hartnett, L. C. Johnson, and E. G. Towne. 1999. The keystone role of bison in North America tallgrass prairie. *BioScience* 49:39–50.

Koch, R. M, H. G. Jung, J. D. Crouse, V. H. Varel, and L. V. Cundiff. 1995. Growth, digestive capability, carcass, and meat characteristics of *Bison bison, Bos taurus,* and *Bos* × *Bison. J. Anim. Sci.* 73:1271–1281.

Laflamme, L.F. 1988. Fish meal as a supplement to hay-based diets. *Can. J. Anim. Sci.* 68:1323 (Abst.).

Liu, Q., M.C. Lanari, and D.M. Schaefer. 1995. A review of dietary vitamin E supplementation for improvement of beef quality. *J. Anim. Sci.* 73:3131–3140.

Lofgreen, G.P. and W.N. Garrett. 1968. A system for expressing the net energy requirements and feed values for growing and finishing cattle. *J. Anim. Sci.* 27:793–806.

Males, J.R. 1987. Optimizing the utilization of cereal crop residues for beef cattle. *J. Anim. Sci.* 65:1124–1130.

Males, J.R., W.E. McReynolds, C.T. Gaskins, and R.L. Preston. 1982. Supplementation of wheat straw diets to optimize performance of wintering beef cows. *J. Anim. Sci.* 54:384–390.

National Research Council. 1984. *Nutrient Requirements of Beef Cattle*. Washington, DC: National Academy Press.

_______ . 1996. *Nutrient Requirements of Beef Cattle*. Washington, DC: National Academy Press.

Nelson, M.L., I.G. Rush, and T.J. Klopfenstein. 1985. Protein supplementation of ammoniated roughages. II. Wheat straw supplemented with alfalfa, blood meal or soybean meal fed to wintering steers. *J. Anim. Sci.* 61:245–251.

Pate, F.M., D.W. Sanson, and R.V. Machen. 1990. Value of a molasses mixture containing natural protein as a supplement to brood cows offered low quality forages. *J. Anim. Sci.* 68:618–623.

Randel, R.D. 1990. Nutrition and postpartum rebreeding in cattle. *J. Anim. Sci.* 68:853–862.

Richards, M.W., R.P. Wettemann, and H.M. Schoenemann. 1989. Nutritional anestrus in beef cows: Concentrations of glucose and nonesterified fatty acids in plasma and insulin in serum. *J. Anim. Sci.* 67:2354–2362.

Russell, J.R., M.R. Brasche, and A.M. Cowen. 1993. Effects of grazing allowance and system on the use of corn-crop residues by gestating beef cows. *J. Anim. Sci.* 71:1256–1265.

Schultz, T.A., A.T. Raston, and E. Schultz. 1974. Effect of various additives on nutritive value of ryegrass straw silage. I. Laboratory silo and *in vitro* dry matter digestion observations. *J. Anim. Sci.* 39:920–930.

Short, R.E. and D.C. Adams. 1988. Nutritional and hormonal interrelationships in beef cattle reproduction. *Can. J. Anim. Sci.* 68:29–39.

Spicer, L.A., C.B. Theurer, J. Sowe, and T.H. Noon. 1986. Ruminal and post-ruminal utilization of nitrogen and starch from sorghum grain-, corn- and barley-based diets by beef steers. *J. Anim. Sci.* 62:521–530.

Stewart, S.R., R.J. Emerick, and R.H. Pritchard. 1990. High dietary calcium to phosphorus ratio and alkali-forming potential as factors promoting silica urolithiasis in sheep. *J. Anim. Sci.* 68:498–503.

Stock, R.A., D.R. Brink, R.T. Brandt, J.K. Merrill, and K.K. Smith. 1987. Feeding combinations of high moisture corn and dry corn to finishing cattle. *J. Anim. Sci.* 65:282–289.

Turner, H. A. and T. DelCurto. 1991. Nutritional and managerial considerations for range beef cattle production. In: *Veterinary Clinics of North America.* John Maas (ed.). W. B. Saunders, Co. pp. 95–126.

Van Soest, P.J. 1994. *Nutritional Ecology of the Ruminant.* Ithaca, NY: Cornell University Press.

Wahlberg, M.L., H.W. Harpster, and E.H. Cash. 1988. Nutrient utilization, efficiency and tissue gain in steers fed ensiled feedstuffs from the corn plant. *J. Anim. Sci.* 66:3021–3032.

Wiedmeier, R.D., J.R. Males, and C.T. Gaskins. 1983. Effect of dietary crude protein on the dry matter digestibility of wheat straw diets in cattle. *J. Anim. Sci.* 57:1568–1575.

Weisenburger, R.D. and G.W. Mathison. 1977. Protein requirements of beef cows fed pelleted, ground or chopped barley straw in the winter. *Can. J. Anim. Sci.* 57:719–725.

Wiltbank, J. N. 1982. Nutrition and reproduction in the beef females. In: W. R. Woods, (ed.). Proc. Symposium on Management of Food Producing Animals, Vol. II:770.

Zorrilla-Rios, J., F.N. Owens, G.W. Horn, and R.W. McNew. 1985. Effect of ammoniation of wheat straw on performance and digestion kinetics of cattle. *J. Anim. Sci.* 60:814–821.

CHAPTER 16

Feeding and Nutrition of Small Ruminants: Sheep, Goats, and Llamas

J.M. Thompson and P.R. Cheeke

Objectives

1. To describe feeding systems used in sheep and goat production
2. To discuss nutritional requirements for wool production
3. To discuss forage utilization and browsing behavior by goats, with particular reference to goat production in developing countries
4. To describe the nutritional peculiarities of camelids (camels and llamas)

Sheep and goats share a number of nutritional similarities and are often discussed together under the term **"small ruminants."** These animals are very important in developing countries because of their small size, feeding behavior characteristics, and low water requirements (Fig. 10.2). They are intermediate feeders (see Chapter 9), are more selective in their feeding habits than cattle, and more likely to consume browse. Most of the information on nutrient requirements for beef cattle (Chapter 15) applies as well to the small ruminants. Only those aspects that are unique to sheep, goats, and llamas will be discussed in this chapter.

FEEDING AND NUTRITION OF SHEEP

A high proportion of the diet of sheep is normally made up of forage. One viewpoint of some sheep farmers is that they are actually forage farmers, and the sheep are merely a tool for harvesting and marketing the forage. Lambs can be finished to a consumer acceptable market grade or end point without the need for a grain-feeding period. Therefore, more than with other species, there is often an effort to maximize **forage utilization** by sheep and minimize the use of concentrate feed. In many areas, the grazing season can be extended by using temporary pastures, such as brassica species (see Chapter 5), that can be seeded into dormant grass pasture or grain stubble (Koch et al., 1987). In North America, sheep production is a fairly minor economic activity. Sheep numbers have declined markedly since the 1940s for a variety of reasons, including predator problems, lack of profitability compared to beef cattle production, decline in the value of wool, and decreased market demand for lamb. New production techniques could reverse this pattern. Most North American sheep production involves either farm flocks or range sheep operations. Range sheep production is located in the western regions. **Range sheep** are generally under the control of a herder and are grazed on semiarid desert lands and on high mountain ranges.

Typically, a band of sheep will consist of 1,000 to 2,000 ewes and their lambs. Large sheep ranches may have numerous bands. **Farm flocks** are kept in fenced pastures and generally involve no more than a few hundred ewes. The ewes are kept on pasture or conserved forage (hay, silage) for most of the year. Supplemental feeds are provided only in periods of nutritional stress, primarily in the period immediately before and after lambing. With farm flocks, the lambs are usually fed on the farm until market weight is reached. Range lambs are often marketed as feeders (lightweight lambs), which are fattened on concentrate feeds in feedlots or on irrigated pastures and crop residues.

NUTRIENT REQUIREMENTS

Energy and Protein

The requirements for energy and protein are discussed by NRC (1985); examples of NRC requirements are given in Tables 16–1 and 16–2. Utilization of protein and energy is similar to that described for beef cattle (Chapter 15). Because their browsing ability allows them to select a more nutritious diet when availability of feed is not limiting, sheep and goats may perform better on a "rough" pasture than cattle. However, this should not be construed to mean that they have lower nutrient requirements and can survive on poorer quality feed. The opposite is true: these smaller ruminants have higher nutrient requirements and require higher-quality feed than cattle. Therefore, when they are fed mixed or pelleted feeds or conserved forage (hay or silage), which reduces their ability to select the more nutritious components, they require higher quality feed than cattle. Mature beef cows can be maintained on poorer quality hay than can ewes or goats.

Ewes respond well to **flushing** or an improvement in nutritional status just prior to breeding. Although ewe response to flushing is variable, producers can expect to improve lambing rates by 10 to 20 percent from flushing. About 2 weeks before the rams are turned in, the ewes should be put on lush grass pasture or other good-quality feed, or fed about 0.5 to 1.0 pounds of grain per ewe for 2 weeks before and 2 to 3 weeks after breeding begins. Legume pastures, such as red clover, should not be used for flushing ewes because of the potential for **phytoestrogen-induced infertility** (see Chapter 5).

After breeding, ewes in the early stages of pregnancy are fed just above maintenance conditions for approximately 3½ months on good pasture or hay. Ewe lambs bred to lamb at 12 to 13 months in age need to be fed at higher levels to allow for continued growth as well as fetal development. For the last third of pregnancy, the rapidly increasing size of the fetus(es) and the decreased rumen volume (see Chapter 9) necessitate that the nutrient density of the diet be increased. During the last trimester of pregnancy, ewes are quite susceptible to ketosis, which may result in pregnancy disease, pregnancy paralysis, or twin-lamb disease. **Ketosis** is caused by a metabolic shortage of glucose and excessive reliance on mobilized body fat to meet energy requirements. These events reflect the increased metabolic requirements of the fetuses for glucose and the inability of a roughage diet to supply sufficient gluconeogenic precursor

TABLE 16–1. Daily Nutrient Requirements of Sheep

Body Weight		Weight Change/Day		Dry Matter per Animal[a]			Nutrients per Animal									
							Energy[b]				Crude Protein		Ca,	P,	Vitamin A Activity	Vitamin E Activity
						% body	TDN		DE,	ME,						
kg	lb	g	lb	kg	lb	weight	kg	lb	Mcal	Mcal	g	lb	g	g	IU	IU
Ewes[c]																
Maintenance																
50	110	10	0.02	1.0	2.2	2.0	0.55	1.2	2.4	2.0	95	0.21	2.0	1.8	2,350	15
60	132	10	0.02	1.1	2.4	1.8	0.61	1.3	2.7	2.2	104	0.23	2.3	2.1	2,820	16
70	154	10	0.02	1.2	2.6	1.7	0.66	1.5	2.9	2.4	113	0.25	2.5	2.4	3,290	18
80	176	10	0.02	1.3	2.9	1.6	0.72	1.6	3.2	2.6	122	0.27	2.7	2.8	3,760	20
90	198	10	0.02	1.4	3.1	1.5	0.78	1.7	3.4	2.8	131	0.29	2.9	3.1	4,230	21
Flushing—2 weeks prebreeding and first 3 weeks of breeding																
50	110	100	0.22	1.6	3.5	3.2	0.94	2.1	4.1	3.4	150	0.33	5.3	2.6	2,350	24
60	132	100	0.22	1.7	3.7	2.8	1.00	2.2	4.4	3.6	157	0.34	5.5	2.9	2,820	26
70	154	100	0.22	1.8	4.0	2.6	1.06	2.3	4.7	3.8	164	0.36	5.7	3.2	3,290	27
80	176	100	0.22	1.9	4.2	2.4	1.12	2.5	4.9	4.0	171	0.38	5.9	3.6	3,760	28
90	198	100	0.22	2.0	4.4	2.2	1.18	2.6	5.1	4.2	177	0.39	6.1	3.9	4,230	30
Nonlactating—First 15 weeks gestation																
50	110	30	0.07	1.2	2.6	2.4	0.67	1.5	3.0	2.4	112	0.25	2.9	2.1	2,350	18
60	132	30	0.07	1.3	2.9	2.2	0.72	1.6	3.2	2.6	121	0.27	3.2	2.5	2,820	20
70	154	30	0.07	1.4	3.1	2.0	0.77	1.7	3.4	2.8	130	0.29	3.5	2.9	3,290	21
80	176	30	0.07	1.5	3.3	1.9	0.82	1.8	3.6	3.0	139	0.31	3.8	3.3	3,760	22
90	198	30	0.07	1.6	3.5	1.8	0.87	1.9	3.8	3.2	148	0.33	4.1	3.6	4,230	24
Last 4 weeks gestation—Lambing rate expected: 130–150% or last 4–6 weeks lactation suckling single[d]																
50	110	180 (45)	0.40 (0.10)	1.6	3.5	3.2	0.94	2.1	4.1	3.4	175	0.38	5.9	4.8	4,250	24
60	132	180 (45)	0.40 (0.10)	1.7	3.7	2.8	1.00	2.2	4.4	3.6	184	0.40	6.0	5.2	5,100	26
70	154	180 (45)	0.40 (0.10)	1.8	4.0	2.6	1.06	2.3	4.7	3.8	193	0.42	6.2	5.6	5,950	27
80	176	180 (45)	0.40 (0.10)	1.9	4.2	2.4	1.12	2.4	4.9	4.0	202	0.44	6.3	6.1	6,800	28
90	198	180 (45)	0.40 (0.10)	2.0	4.4	2.2	1.18	2.5	5.1	4.2	212	0.47	6.4	6.5	7,650	30
Last 4 weeks gestation—Lambing rate expected: 180–225%																
50	110	225	0.50	1.7	3.7	3.4	1.10	2.4	4.8	4.0	196	0.43	6.2	3.4	4,250	26
60	132	225	0.50	1.8	4.0	3.0	1.17	2.6	5.1	4.2	205	0.45	6.9	4.0	5,100	27
70	154	225	0.50	1.9	4.2	2.7	1.24	2.8	5.4	4.4	214	0.47	7.6	4.5	5,950	28
80	176	225	0.50	2.0	4.4	2.5	1.30	2.9	5.7	4.7	223	0.49	8.3	5.1	6,800	30
90	198	225	0.50	2.1	46	2.3	1.37	3.0	6.0	5.0	232	0.51	8.9	5.7	7,650	32
First 6–8 weeks lactation suckling singles or last 4–6 weeks lactation suckling twins[d]																
50	110	-25 (90)	-0.06(0.20)	2.1	4.6	4.2	1.36	3.0	6.0	4.9	304	0.67	8.9	6.1	4,250	32
60	132	-25 (90)	-0.06 (0.20)	2.3	5.1	3.8	1.50	3.3	6.6	5.4	319	0.70	9.1	6.6	5,100	34

70	154	-25 (90)	-0.06 (0.20)	2.5	5.5	3.6	1.63	3.6	7.2	5.9	334	0.73	9.3	7.0	5,950	38
80	176	-25 (90)	-0.06 (0.20)	2.6	5.7	3.2	1.69	3.7	7.4	6.1	344	0.76	9.5	7.4	6,800	39
90	198	-25 (90)	-0.06 (0.20)	2.7	5.9	3.0	1.75	3.8	7.6	6.3	33	0.78	9.6	7.8	7,650	40
First 6–8 weeks lactation suckling twins																
50	110	-60	-0.13	2.4	5.3	4.8	1.56	3.4	6.9	5.6	389	0.86	10.5	7.3	5,000	36
60	132	-60	-0.13	2.6	5.7	4.3	1.69	3.7	7.4	6.1	405	0.89	10.7	7.7	6,000	39
70	154	-60	-0.13	2.8	6.2	4.0	1.82	4.0	8.0	6.6	420	0.92	11.0	8.1	7,000	42
80	176	-60	-0.13	3.0	6.6	3.8	1.95	4.3	8.6	7.0	435	0.96	11.2	8.6	8,000	45
90	198	-60	-0.13	3.2	7.0	3.6	2.08	4.6	9.2	7.5	450	0.99	11.4	9.0	9,000	48
Ewe Lambs																
Nonlactating—First 15 weeks gestation																
40	88	160	0.35	1.4	3.1	3.5	0.83	1.8	3.6	3.0	156	0.34	5.5	3.0	1,880	21
50	110	135	0.30	1.5	3.3	3.0	0.88	1.9	3.9	3.2	159	0.35	5.2	3.1	2,350	22
60	132	135	0.30	1.6	3.5	2.7	0.94	2.0	4.1	3.4	161	0.35	5.5	3.4	2,80	24
70	154	125	0.28	1.7	3.7	2.4	1.00	2.2	4.4	3.6	164	0.36	5.5	3.7	3,290	26
Last 4 weeks gestation—Lambing rate expected: 100–120%																
40	88	180	0.40	1.5	3.3	3.8	0.94	2.1	4.1	3.4	187	0.41	6.4	3.1	3,400	22
50	110	160	0.35	1.6	3.5	3.2	1.00	2.2	4.4	3.6	189	0.42	6.3	3.4	4,250	24
60	132	160	0.35	1.7	3.7	2.8	1.07	2.4	4.7	3.9	192	0.42	6.6	3.8	5,100	26
70	154	150	0.33	1.8	4.0	2.6	1.14	2.5	5.0	4.1	194	0.43	6.8	4.2	5,950	27
Last 4 weeks gestation—Lambing rate expected: 130–175%																
40	88	225	0.50	1.5	3.3	3.8	0.99	2.2	4.4	3.6	202	0.44	7.4	3.5	3,400	22
50	110	225	0.50	1.6	3.5	3.2	1.06	2.3	4.7	3.8	204	0.45	7.8	3.9	4,250	24
60	132	225	0.50	1.7	3.7	2.8	1.12	2.5	4.9	4.0	207	0.46	8.1	4.3	5,100	26
70	154	215	0.47	1.8	4.0	2.6	1.14	2.5	5.0	4.1	210	0.46	8.2	4.7	5,950	27
First 6–8 weeks lactation suckling singles (wean by 8 weeks)																
40	88	-50	-0.11	1.7	3.7	4.2	1.12	2.5	4.9	4.0	257	0.56	6.0	4.3	3,400	26
50	110	-50	-0.11	2.1	4.6	4.2	1.39	3.1	6.1	5.0	282	0.62	6.5	4.7	4,250	32
60	132	-50	-0.11	2.3	5.1	3.8	1.52	3.4	6.7	5.5	295	0.65	6.8	5.1	5,100	34
70	154	-50	-0.11	2.5	5.5	3.6	1.65	3.6	7.3	6.0	301	0.68	7.1	5.6	5,450	38
First 6–8 weeks lactation suckling twins (wean by 8 weeks)																
40	88	-100	-0.22	2.1	4.6	5.2	1.45	3.2	6.4	5.2	306	0.67	8.4	5.6	4,000	32
50	110	-100	-0.22	2.3	5.1	4.6	1.59	3.5	7.0	5.7	321	0.71	8.7	6.0	5,000	34
60	132	-100	-0.22	2.5	5.5	4.2	1.72	3.8	7.6	6.2	336	0.74	9.0	6.4	6,000	38
70	154	-100	-0.22	2.7	6.0	3.9	1.85	4.1	8.1	6.6	351	0.77	9.3	6.9	7,000	40
Replacement Ewe Lambs[e]																
30	66	227	0.50	1.2	2.6	4.0	0.78	1.7	3.4	2.8	185	0.41	6.4	2.6	1,410	18
40	88	182	0.40	1.4	3.1	3.5	0.91	2.0	4.0	3.3	176	0.39	5.9	2.6	1,880	21
50	110	120	0.26	1.5	3.3	3.0	0.88	1.9	3.9	3.2	136	0.30	4.8	2.4	2,350	22

(continued on next page)

TABLE 16–1. Daily Nutrient Requirements of Sheep (*continued*)

Body Weight		*Weight Change/Day*		*Dry Matter per Animal*[a]			*Nutrients per Animal*									
							Energy[b]				*Crude Protein*					
						% body	*TDN*		*DE,*	*ME,*			*Ca,*	*P,*	*Vitamin A Activity*	*Vitamin E Activity*
kg	*lb*	*g*	*lb*	*kg*	*lb*	*weight*	*kg*	*lb*	*Mcal*	*Mcal*	*g*	*lb*	*g*	*g*	*IU*	*IU*
Replacement Ewe Lamb[e] *(continued)*																
60	132	100	0.22	1.5	3.3	2.5	0.88	1.9	3.9	3.2	134	0.30	4.5	2.5	2,820	22
70	154	100	0.22	1.5	3.3	2.1	0.88	1.9	3.9	3.2	132	0.29	4.6	2.8	3,290	22
Replacement Ram Lambs[e]																
40	88	330	0.73	1.8	4.0	4.5	1.1	2.5	5.0	4.1	243	0.54	7.8	3.7	1,880	24
60	132	320	0.70	2.4	5.3	4.0	1.5	3.4	6.7	5.5	263	0.58	8.4	4.2	2,820	26
80	176	290	0.64	2.8	6.2	3.5	1.8	3.9	7.8	6.4	268	0.59	8.5	4.6	3,760	28
100	220	250	0.55	3.0	6.6	3.0	1.9	4.2	8.4	6.9	264	0.58	8.2	4.8	4,700	30
Lambs Finishing—4 to 7 Months Old[f]																
30	66	295	0.65	1.3	2.9	4.3	0.94	2.1	4.1	3.4	191	0.42	6.6	3.2	1,410	20
40	88	275	0.60	1.6	3.5	4.0	1.22	2.7	5.4	4.4	185	0.41	6.6	3.3	1,880	24
50	110	205	0.45	1.6	3.5	3.2	1.23	2.7	5.4	4.4	160	0.35	5.6	3.0	2,350	24
Early Weaned Lambs—Moderate Growth Potential[f]																
10	22	200	0.44	0.5	1.1	5.0	0.40	0.9	1.8	1.4	127	0.38	4.0	1.9	470	10
20	44	250	0.55	1.0	2.2	5.0	0.80	1.8	3.5	2.9	167	0.37	5.4	2.5	940	20
30	66	300	0.66	1.3	2.9	4.3	1.00	2.2	4.4	3.6	191	0.42	6.7	3.2	1,410	20
40	88	345	0.76	1.5	3.3	3.8	1.16	2.6	5.1	4.2	202	0.44	7.7	3.9	1,880	22
50	110	300	0.66	1.5	3.3	3.0	1.16	2.6	5.1	4.2	181	0.40	7.0	3.8	2,350	22
Early Weaned Lambs—Rapid Growth Potential[f]																
10	22	250	0.55	0.6	1.3	6.0	0.48	1.1	2.1	1.7	157	0.35	4.9	2.2	470	12
20	44	300	0.66	1.2	2.6	6.0	0.92	2.0	4.0	3.3	205	0.45	6.5	2.9	940	24
30	66	325	0.72	1.4	3.1	4.7	1.10	2.4	4.8	4.0	216	0.48	7.2	3.4	1,410	21
40	88	400	0.88	1.5	3.3	3.8	1.14	2.5	5.0	4.1	234	0.51	8.6	4.3	1,880	22
50	110	425	0.94	1.7	3.7	3.4	1.29	2.8	5.7	4.7	240	0.53	9.4	4.8	2,350	25
60	132	350	0.77	1.7	3.7	2.8	1.29	2.8	5.7	4.7	240	0.53	8.2	4.5	2,820	25

Adapted from National Research Council (1985).

[a]To convert dry matter to an as-fed basis, divide dry matter values by the percentage of dry matter in the particular feed.

[b]One kilogram TDN (total digestible nutrients) = 4.4 Mcal DE (digestible energy), ME (metabolizable energy) = 82% of DE. Because of rounding errors, values in Tables 16–1 and 16–2 may differ.

[c]Values are applicable for ewes in moderate condition. Fat ewes should be fed according to the next lower weight category and thin ewes at the next higher weight category. Once desired or moderate weight condition is attained, use that weight category through all production stages.

[d]Values in parentheses are for ewes suckling lambs the last 4–6 weeks of lactation.

[e]Lambs intended for breeding, thus, maximum weight gains and finish are of secondary importance.

[f]Maximum weight gains expected.

(propionate). Mobilization of body fat produces ketogenic acetate, rather than allowing for glucose biosynthesis. Symptoms of ketosis, such as coma, are due to inadequate blood glucose to support brain metabolism, while excess acetate from fat metabolism is converted to ketones (e.g., acetone).

There is some indication that mild feed restriction in midpregnancy might be desirable because overfat ewes are susceptible to feed intake reduction in late pregnancy, making them more likely to experience ketosis (Wilkinson and Chestnutt, 1988). For the last trimester of pregnancy, nutrient restrictions should be avoided since inadequate levels at this time will result in lighter lambs at birth, unequal birth weights in twins, increased perinatal lamb losses, reduced mothering instincts, and lowered milk production (SID, 2002).

A supplement of 0.5 to 2 lb of grain or concentrate mix per ewe per day should be provided in late gestation to farm flock ewes. Range ewes are often on dry, mature range forage prior to lambing and should receive a protein supplement, such as cottonseed meal. This may also improve energy status by stimulating feed intake via the bypass protein effect (see Chapter 9). Protein-molasses blocks may also be used. Vitamin A supplementation should be given if green forage is not available.

Lambing is a critical period for the sheep producer. The entire annual economic return from a ewe (except for wool) depends on whether her lamb(s) survives. It is critical that the newborn lamb receives colostrum in the first hour of life. If it does not, it becomes progressively weaker, is unable to suckle, and dies of starvation. If a lamb is too weak to nurse, it should be given colostrum by stomach-tube.

In order to provide for the needs of the suckling lambs, proper feeding during lactation is important. Milk production in the ewe peaks at 3 to 4 weeks following lambing. Ewes nursing singles produce more milk than the lamb will consume so they adjust production downward (SID, 2002). Ewes suckling twins produce 30 to 50 percent more than ewes with singles (Treacher and Caja, 2002). Ewes with triplets produce 30 percent more than those raising twins (Loerch et al., 1985). Ewes suckling more than one lamb during early lactation have the greatest nutritional needs of any time during the production cycle. During this time, ewes need high-quality forages along with energy and protein supplementation. Even when productive ewes are fed at the recommended levels, they are expected to utilize body fat reserves and lose body weight and condition especially during the early lactation period.

A subjective body condition scoring system can be useful in assessing the nutritional status of ewes. A body condition score is based on estimated external fat, and recommended scores have been established for the various stages of production. The most common scoring system uses body condition scores ranging from 1 to 5 (Russell et al., 1969). The lowest score indicates an extremely emaciated sheep, and 5 represents excessive obesity. A score of 3 indicates average body condition. This would be the recommended target score for a ewe at breeding, late gestation, and early lactation. Ewes that are expected to give birth to or that are nursing twins should be one-half a condition score higher (3.5) for the respective stages of production (SID, 2002).

TABLE 16–2. Nutrient Concentration in Diets for Sheep (expressed on 100 Percent Dry-Matter Basis[a])

Body Weight		*Weight Change/Day*		*Energy[b]*			*Example Diet Proportions*						
kg	*lb*	*g*	*lb*	*TDN,[c] %*	*DE, Mcal/kg*	*ME, Mcal/kg*	*Concentrate, %*	*Forage, %*	*Crude Protein, %*	*Ca, %*	*Ph, %*	*Vitamin A Activity, IU/kg*	*Vitamin E Activity, IU/kg*
Ewes[d]													
Maintenance													
70	154	10	0.02	55	2.4	2.0	0	100	9.4	0.20	0.20	2,742	15
Flushing—2 weeks prebreeding and first 3 weeks of breeding													
70	154	100	0.22	59	2.6	2.1	15	85	9.1	0.32	0.18	1,828	15
Nonlactating—First 15 weeks gestation													
70	154	30	0.07	55	2.4	2.0	0	100	9.3	0.25	0.20	2,350	15
Last 4 weeks gestation—Lambing rate expected (130–150% or Last 4–6 weeks lactation suckling singles[e])													
70	154	180 (0.45)	0.40 (0.10)	59	2.6	2.1	15	85	10.7	0.35	0.23	3,306	15
Last 4 weeks gestation—Lambing rate expected (180–225%)													
70	154	225	0.50	65	2.9	2.3	35	65	11.3	0.40	0.24	3,132	15
First 6–8 weeks lactation suckling singles or last 4–6 weeks, lactation suckling twins[e]													
70	154	–25 (90)	–0.06 (0.20)		2.9	2.4	35	65	13.4	0.32	0.26	2.380	15
First 6–8 weeks lactation suckling twins													
70	154	–60	–0.13	65	2.9	2.4	35	65	15.0	0.39	0.29	2,500	15
Ewe Lambs													
Nonlactating—First 15 weeks gestation													
55	121	135	0.30	59	2.6	2.1	15	85	10.6	0.35	0.22	1,668	15
Last 4 weeks gestation—Lambing rate expected (100–120%)													
55	121	160	0.35	63	2.8	2.3	30	70	11.8	0.39	0.22	2,833	15
Last 4 weeks gestation—Lambing rate expected (130–175%)													
55	121	225	0.50	66	2.9	2.4	40	60	12.8	0.48	0.25	2,833	15
First 6–8 weeks lactation suckling singles (wean by 8 weeks)													
55	121	–50	–0.22	66	2.9	2.4	40	60	13.1	0.30	0.22	2,125	15
First 6–8 weeks lactation twins (wean by 8 weeks)													
55	121	–100	–0.22	69	3.0	2.5	50	50	13.7	0.37	0.26	2,292	15

Replacement Ewe Lambs[f]													
30	66	227	0.50	65	2.9	2.4	35	65	12.8	0.53	0.22	1,175	15
40	88	182	0.40	65	2.9	2.4	35	65	10.2	0.42	0.18	1,343	15
50–70	110–154	115	0.25	59	2.6	2.1	15	85	9.1	0.31	0.17	1,567	15
Replacement Ram Lambs[f]													
40	88	330	0.73	63	2.8	2.3	30	70	13.5	0.43	0.21	1,175	15
60	132	320	0.70	63	2.8	2.3	30	70	11.0	0.35	0.18	1,659	15
80–100	176–220	270	0.60	63	2.8	2.3	30	70	9.6	0.30	0.16	1,979	15
Lambs Finishing–4 to 7 Months Old[g]													
30	66	295	0.65	72	3.2	2.5	60	40	14.7	0.51	0.24	1,085	15
40	88	275	0.60	76	3.3	2.7	75	25	11.6	0.42	0.21	1,175	15
50	110	205	0.45	77	3.4	2.8	80	20	10.0	0.35	0.19	1,469	15
Early Weaned Lambs—Moderate and Rapid Growth Potential[g]													
10	22	250	0.55	80	3.5	2.9	90	10	26.2	0.82	0.38	940	20
20	44	300	0.66	78	3.4	2.8	85	15	16.9	0.54	0.24	940	20
30	66	325	0.72	78	3.3	2.7	85	15	15.1	0.51	0.24	1,085	15
40–60	88–32	400	0.88	78	3.3	2.7	85	15	14.5	0.55	0.28	1,253	15

Adapted from National Research Council (1985).

[a]Values are calculated from daily requirements listed in Table 16–2 divided by DM intake. The exception, vitamin E daily requirements/head, are calculated from vitamin E/kg diet × DM intake.

[b]One kilogram TDN = 4.4 Mcal DE (digestible energy); ME (metabolizable energy) = 82% of DE. Because of rounding errors, values in Tables 16–1 and 16–2 may differ.

[c]TDN calculated on following basis: hay DM, 55% TDN and on as-fed basis 50% TDN; grain DM, 83% TDN and on as-fed basis 75% TDN.

[d]Values are for ewes in moderate condition. Fat ewes should be fed according to the next lower weight category and thin ewes at the next higher weight category. Once desired or moderate weight condition is attained, use that weight category through all production stages.

[e]Values in parentheses are for ewes suckling lambs the last 4–6 weeks of lactation.

[f]Lambs intended for breeding; thus, maximum weight gains and finish are of secondary importance.

[g]Maximum weight gains expected.

Artificial Rearing of Lambs For a variety of reasons, lambs may be raised on **milk replacer.** Orphan lambs (bummers) or excess lambs in multiple births may have to be raised artificially. **Milk replacers** for lambs should contain 24 to 30 percent fat, 20 to 25 percent milk protein, and 30 to 35 percent lactose. Calf milk replacers do not contain sufficient fat for lambs and should not be used. Heaney et al. (1982) found that a fat level of 24 percent was adequate in lamb milk replacer. Up to one-third of the milk protein can be replaced with soybean protein (soyflour) without reducing performance (Heaney and Shrestha, 1987). Nonmilk proteins, such as soy products, fish meal, and whey, do not form a curd in the abomasum; curd formation is not necessary for efficient utilization of milk replacers by lambs (Delisle et al., 1988). To minimize labor, milk replacer should be fed free choice. One problem with free-choice availability of milk is **abomasal bloat** of young lambs, resulting in death. This is caused largely by overdrinking, followed by fermentation of milk in the abomasum, and gas production. Keeping the milk replacer refrigerated will reduce voluntary consumption and help to prevent bloat (Heaney and Shrestha, 1985).

Because of the expense of milk replacers, artificially reared lambs should be weaned as soon as possible. They should be eating concentrate feed adequately by 3 weeks of age and can be weaned at that time. An example of a suitable postweaning concentrate diet is given in Table 16–3. Good-quality hay should be provided free choice to reduce postweaning suckling of other lambs, which may cause mortality of male lambs due to urinary tract damage (Heaney and Shrestha, 1985). Lambs should be vaccinated for prevention of **enterotoxemia** caused by *Clostridium perfringens* (Type C and D). Enterotoxemia (overeating disease, pulpy kidney disease) is caused by the clostridial toxins produced when organisms proliferate in the intestine after a rich diet is consumed.

Wool Production Nutritional status affects the wool growth of sheep. Although wool is essentially pure protein, the daily protein requirement for wool growth is only a small component of the total protein requirement. However, microbial protein does not contain an adequate amino acid balance to support maximal wool growth. Australian researchers demonstrated that postruminal administration of sources of methionine or high-quality proteins to sheep markedly stimulated wool production (Colebrook and Reis, 1969). Utilization of sulfur amino acid-rich sources of nondegradable protein in sheep feeding might be a feasible way of increasing wool production. The main nutritional factor limiting wool growth is energy intake (Ryder and Stephenson, 1968). Limitations to

TABLE 16–3. Example of a Postweaning Ration for Early Weaned Lambs

Ingredient	*%*	*Ingredient*	*%*
Coarse-ground corn	59.15	Limestone	1.45
Soybean meal	25	Salt	0.8
Chopped, ground hay	8	Trace minerals	0.2
Molasses	5	Vitamins	*
Dicalcium phosphate	0.4		

Adapted from Heaney and Shrestha (1985).
*4 million IU vitamin A, 340,000 IU vitamin D_3, 4000 IU vitamin E per ton.

feed intake or increased energy requirements for productive functions (e.g., lactation) reduce the rate of wool growth. Severe stress, such as a short period of starvation, may cause cessation of wool growth and a break in the fiber after wool growth resumes. This results in fleece shedding.

The physiology of the growth of hair fibers is described in Chapter 21. These comments also apply to the growth of hair and wool fibers in sheep and goats.

Minerals and Vitamins

Mineral and vitamin requirements and metabolism of sheep are, in general, similar to those of beef cattle. Braithwaite (1983) found that ewes are unable to absorb sufficient calcium in late pregnancy and early lactation to meet the high demands for milk secretion and are in negative calcium balance. Ewes fed adequate calcium in mid and late lactation are able to replenish the depleted calcium stores. Therefore, a mineral mixture, balanced with calcium and phosphorus, should be available to ewes throughout gestation and lactation. **Hypocalcemia** (milk fever) is a significant and sometimes fatal disorder of ewes during late pregnancy and early lactation. Signs include ataxia (incoordination) and muscular tremors, depression, and rumen stasis. In areas of selenium deficiency, adequate supplementation with selenium is very important. Sheep are more susceptible than beef cattle to **selenium deficiency,** commonly called **white muscle disease.** Symptoms include weakness and lameness of young lambs, and sudden death. These signs are due to degeneration of skeletal and cardiac muscle, with heart failure often the cause of death. Areas of muscle damage are calcified (Fig. 16.1), accounting for the common name of "white muscle disease." Affected animals respond rapidly to injected selenium. Prevention is achieved by injecting ewes with selenium before lambing or by providing free-choice selenized salt.

Sheep are the most susceptible livestock to **copper toxicity.** Contamination of sheep feed with copper (from poultry or swine feed with high copper added as a growth promotant—see Chapter 8), or grazing sheep on pastures fertilized with manure from swine or poultry fed high copper diets, may lead to outbreaks of copper toxicity. Copper accumulates in the liver until the liver cells are saturated with the element. This is followed by oxidative damage to the liver, breakdown of liver cells, and release of large amounts of copper into the blood. Hemolysis of red blood cells occurs, causing anemia, jaundice, and red urine due to excretion of hemoglobin pigments. These and other metabolic changes, and kidney and liver damage, generally result in death within two to four days. Copper toxicity can occur whenever the dietary copper level is higher than 10 ppm for an extended period.

General Nutritional Factors in Sheep Production

In other sections of the text, reference has been made to numerous nutritional problems affecting sheep not discussed here. A brief listing of some of these, and the chapters where more information can be obtained, are given below.

Pasture Bloat Sheep are less susceptible than cattle to pasture bloat. The causative factors and preventative measures for bloat are discussed in Chapter 5.

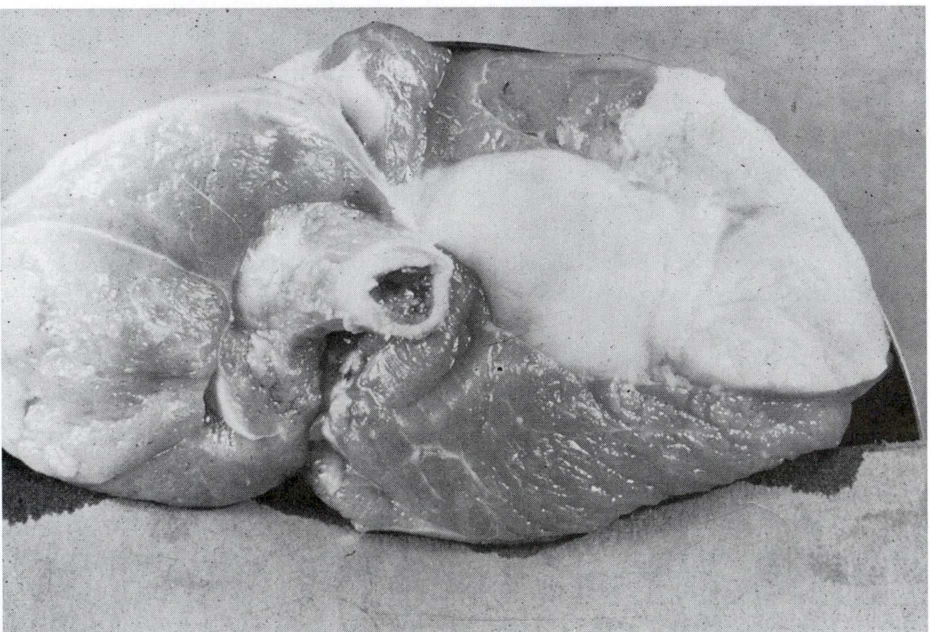

FIGURE 16.1 White muscle disease in a lamb caused by selenium deficiency (top). Degeneration of skeletal muscles occurs, causing lameness. Degeneration of the heart muscle also occurs, causing heart attack and death. The muscle tissue (bottom) from a selenium-deficient lamb shows extensive areas of calcification, accounting for the name *white muscle disease*. (Courtesy of J. E. Oldfield.)

Urinary Calculi Urinary calculi under range and feedlot conditions are important in sheep. Causes and prevention in sheep are the same as for beef cattle, as discussed in Chapter 15.

Polioencephalomalacia This disorder, occurring mainly in feedlot lambs fed high-concentrate diets, is discussed in Chapter 2.

Phytoestrogens Sheep are more affected than other livestock by reproductive problems caused by ingestion of plant estrogens, which are discussed in Chapter 5.

Poisonous Plants and Natural Toxins The interaction between the chemical defenses of plants (toxins) and herbivores is discussed in Chapter 9. As discussed there, sheep and goats and other small herbivores tend to be more resistant to poisonous plants than are cattle and other grazing animals. Therefore, both sheep and goats are used in the biological control of numerous poisonous plants, including leafy spurge, *Senecio* species, and larkspur. Sheep and goats are also more tolerant of toxins that affect palatability, so sheep and goats are often used for control of brush and chaparral not consumed by cattle. Sheep are less affected by toxins in *Brassica* species than cattle. Kale, rape, and related species are well utilized by sheep, generally without toxicity problems. Cull onions, which contain hemolytic factors similar to the brassica anemia factor (Chapter 5), are much less toxic to sheep than to cattle.

FEEDING AND NUTRITION OF GOATS

Goats are raised to only a limited extent in North America. They have a low status, often being considered a "poor man's cow." Modest numbers of dairy goats are raised in the United States (Fig. 16.2), and a considerable number of Angora goats are produced in Texas for **mohair.** Meat-producing goats are raised in the southwest United States; the meat is consumed primarily by Spanish-speaking Americans. Since this population is increasing, the demand for goat meat has also increased in the past decade. Some meat-producing (Spanish) and Angora goats are used in brush control programs in California to control chaparral, a fire-susceptible shrub ecotype. With increasing restrictions on the use of herbicides, employment of goats for brush control may increase.

In developing countries, goats are important meat and dairy animals and, along with sheep, constitute the major "small ruminant" species currently enjoying the attention of aid and development programs (Johnson et al., 1986; Devendra, 1981). Goats are often blamed for contributing to desertification in many parts of the world. Certainly, if animal management is not adequate, goats can cause immense ecological damage. On the other hand, if properly managed, they can make a very important contribution to human welfare in developing countries.

Feeding and Digestive Strategies

Goats are intermediate feeders with a propensity to browsing. They are quite fastidious in their eating behavior and select succulent and nutritious vegetation. In comparison to cattle and sheep, goats derive a much larger portion of their feed

FIGURE 16.2 Modern techniques of dairy-goat milk production are illustrated here. In localized areas of the United States, dairy goats are of economic importance. (Courtesy of G.F.W. Haenlein, University of Delaware.)

from browse if given the opportunity, although they will graze grass as well. Goats prefer grasses over legume forages such as clover and alfalfa (De Rosa et al., 1997). They have the ability to stand on their hind legs to browse (bipedal stance, see Part II opener figure) and are very nimble and can even climb some trees to reach the leafy vegetation. Goats are well adapted to arid climates, being heat tolerant and having low water requirements. They can go several days without drinking and are capable of traveling large distances for feed.

The **digestive efficiency** of goats is similar to that of sheep and cattle. Some reports suggest that fiber digestibility may be higher in goats, but most controlled studies have not shown any significant differences among the domestic ruminants (NRC, 1981; Quick and Dehority, 1986; Pfister and Malechek, 1986), although Brown and Johnson (1985) found a higher digestibility of high-fiber diets in sheep than in goats kept under identical conditions. There is substantial evidence that goats can digest **browse** more efficiently than can cattle and sheep. In New Zealand, Howe et al. (1988) found that goats digested gorse (*Ulex europaeus*) more efficiently than sheep. Gorse is a leguminous shrub widely grown in England as a hedge because of its vicious thorns. It is an introduced pest in New Zealand and the coastal areas of northwestern North America (e.g., Oregon, Washington, and southern British Columbia). The ability of goats to surmount physical defenses of plants is exemplified by their willingness to eat plants such as gorse and black locust. As Kingsbury (1983) aptly states, "Thorns are not always an effective defense. Anyone who has watched a goat avidly consume black locust (*Robinia pseudoacacia*), branches, thorns, and all, will be forced to that conclusion."

Howe et al. (1988) reported that as the lignin content of browse increased above 12 percent, goats were superior to sheep in digestion and intake of highly lignified material. They suggested three possible mechanisms: differences in rumen environment, in mean rumen retention time, and in chewing and rumination behavior. On low N diets, goats maintain higher rumen ammonia concentrations, perhaps by greater nitrogen recycling. Higher rumen ammonia would presumably result in greater microbial activity on low N diets. Goats have a longer mean rumen retention time than sheep, possibly increasing digestibility due to a longer exposure of lignified material to microbial action. Goats also have a longer rumination cycle than sheep, which could result in greater physical degradation of lignified browse and increased digestibility. McCabe and Barry (1988) found that goats were superior to sheep in utilizing willow browse, which is highly lignified and also high in tannins (3 to 6 percent). Voluntary ME intake was about twice as high for goats as for sheep. Wilson (1977) also reported that goats are superior to sheep in utilization of browse.

The preference of goats for browse and shrubby species, and their superior ability to consume and digest highly lignified browse, suggest useful applications in **mixed grazing.** In parts of the United States, such as the Edwards Plateau of Texas, it is common to use mixed grazing systems involving sheep, goats, and cattle. Goats effectively utilize various oak (*Quercus*) species such as Gambel oak (*Q. gambelii*), which is very common on rangelands of the Southwest. Oak foliage is high in tannin content, to which goats seem to be tolerant. Goats are resistant to a number of natural toxicants, including pyrrolizidine alkaloids (Goeger et al., 1982) and tannins. Nastis and Malechek (1981) found that diets containing up to 80 percent immature oak with almost 9 percent tannin did not adversely affect goats, although voluntary intake was reduced with the high-oak diets. Nitrogen retention in browsing goats is not adversely affected by consumption of plants with a high phenolic content (Nunez-Hernandez et al., 1989).

Mixed Grazing Programs

The total livestock grazing capacity of shrubby rangelands can often be significantly increased by including goats in a mixed grazing program. Sidahmed et al. (1981, 1983) found that shrubby chaparral species (chamise, scrub oak, and manzanita) were well utilized by goats, although an all-shrub diet is not generally adequate for maintenance. In developing countries, such feedstuffs as cassava, sugarcane residues, molasses, and sorghum may be used as energy supplements (Pfister and Malechek, 1986; Van Eys et al., 1987). However, the much-maligned goat is usually left to fend for itself. Johnson et al. (1986) noted that in Brazil, if supplementary feed were available, farmers would use it for cattle, sometimes for sheep, but almost never for goats.

The potential of goats seems to be unappreciated. But, as noted by NRC (1981), "The goat offers an opportunity, sometimes the only alternative, for deriving value from a vast reservoir of natural resources and unwanted assortments of herbage, shrubs, tree leaves, and plant refuse and by-products."

FEEDING PROGRAMS FOR GOATS

Goats are often raised under circumstances where maximum performance is neither possible nor expected. Goats for meat production can be raised similarly to sheep. In general, goats are not particularly well adapted to intensive lush pastures and prefer rough pastures containing brush for browsing.

Dairy Goats

Dairy goats should be fed a forage (good-quality alfalfa, clover, or grass hay) supplemented with a concentrate. A forage:concentrate ratio of 40:60 is satisfactory. The concentrate mixture, based on grain and a protein supplement, should be formulated using the protein content of the forage and the NRC nutrient requirements for lactation. **By-product feeds,** such as almond hulls, can be successfully used in diets for dairy goats. In California studies, a mixture of almond hulls and urea was used to replace alfalfa meal in goat diets, with no change in milk production (Reed and Brown, 1988). In the tropics, agroindustrial by-products such as banana wastes (stalk, leaf, pseudostem, and peels) can be used as feedstuffs for goats (Poyyamozhi and Kadirvel, 1986). Leaves of tree legumes such as leucaena, gliricidia, and sesbania are useful protein supplements (Van Eys et al., 1986).

Goats are capable of storing considerable amounts of energy in adipose tissue during gestation and mobilizing it during lactation. Does in late gestation should receive some concentrate, approximately 1 lb/day, just before parturition. After lactation begins, the concentrate offered should be increased (gradually) to approximately 2 lb/day. Goats are susceptible to developing **ketosis** in late pregnancy and early lactation. Underfeeding during pregnancy can intensify this problem. Angora goats are also susceptible to abortion if nutrients are restricted during gestation (NRC, 1981). **Stress abortion** is triggered by a low blood glucose level following inadequate energy intake.

Mohair Production

Production of mohair by Angora goats responds to energy intake. Intakes below maintenance cause decreased production (McGregor, 1988). With cashmere goats, production of cashmere fiber is preferentially greater when energy is limiting, whereas with *ad libitum* feeding, production of coarse hair increases (McGregor, 1988). Cashmere is a soft, high-quality down produced by secondary hair follicles, whereas the primary follicles produce the coarse hair.

Nutritional Disorders

Like sheep, goats are susceptible to *Clostridium perfringens* (Type C or D) **enterotoxemia** (overeating disease). With dairy goats, frequent feeding of small amounts of concentrate is less apt to induce enterotoxemia than large meals once or twice a day. Feed changes should be made gradually over a period of several days. Haenlein (1982) suggested that whole sunflower seeds are an excellent source of energy, protein, and fiber for dairy goats. Because much of the energy is from lipid rather than carbohydrate, they provide a means of providing a high-energy intake without inducing enterotoxemia.

There is little information on vitamin and mineral requirements of goats. The NRC values are mainly values recommended for sheep or dairy cattle; it is assumed that goat requirements are similar. Evidence that this is not necessarily the case is the observation that goats are more susceptible than sheep to **nutritional muscular dystrophy** associated with selenium and vitamin E deficiency (Rammell et al., 1989). The National Research Council nutrient requirements for goats are given in Table 16–4.

OTHER RUMINANTS

A variety of other ruminant animals, including the buffalo, yak, camel, and llama, are raised in various parts of the world. Interest in llama production in the United States is increasing. The animals are used as pets, for back-packing, as guard animals for sheep, and for wool production. The llama and other **camelids** are sometimes called pseudoruminants because their stomach anatomy is different from that of the true ruminants. They have three stomach compartments rather than four, with the omasum being absent. The anatomical structure of these compartments is somewhat different from the analogous compartments in ruminants. Camelids have stomach fermentation and regurgitation of ingesta, so for practical purposes they can be considered functional ruminants. Digestion of fibrous feed in camelids is comparable to that in ruminants (Van Soest, 1994).

There is little specific information on the nutritional requirements of llamas. They seem to be similar to sheep in their feeding needs. Warmington et al. (1989) compared the intake and digestibility of camelids (llama and guanaco) to sheep fed a low-quality roughage (ryegrass straw). Per unit of metabolic rate, feed intakes were similar, but water intake of the camelids was approximately 40 percent less. Digestibility values were slightly higher for camelids than for sheep; similar findings were also reported by Hintz et al. (1973). Warmington et al. (1989) concluded that camelids appear to be superior to sheep in their ability to conserve nitrogen and maintain body weight on low-quality diets and might be better suited than sheep as fiber producers on low-quality grasslands.

Junge and Thornburg (1989) reported an incidence of **copper poisoning** in llamas fed a diet containing 64 ppm copper. This sensitivity to copper toxicity is similar to that of sheep. Serum macro and micro mineral concentrations in the llama are similar to those for cattle and sheep (Bechert and Smith, 1996).

Young llamas appear to be unusually sensitive to disturbances of phosphorus metabolism. Llama raisers in California and Oregon have observed numerous cases of **rickets** and signs of phosphorus deficiency in crias (baby llamas). Signs include poor growth, excessive time spent lying down, and lameness. Clinically,

TABLE 16–4. Daily Nutrient Requirements of Goats

Body Weight, kg	Feed Energy: TDN, g	Feed Energy: DE, Mcal	Feed Energy: ME, Mcal	Feed Energy: NE, Mcal	Crude Protein: TP, g	Crude Protein: DP, g	Ca, g	P, g	Vita-min A, (1000) IU	Vita-min D, IU	Dry Matter per Animal, 1 kg = 2.0 Mcal ME: Total, kg	Dry Matter per Animal, 1 kg = 2.0 Mcal ME: % of kg BW	Dry Matter per Animal, 1 kg = 2.4 Mcal ME: Total, kg	Dry Matter per Animal, 1 kg = 2.4 Mcal ME: % of kg BW
Maintenance Only (Includes Stable Feeding Conditions, Minimal Activity, and Early Pregnancy)														
10	159	0.70	0.57	0.32	22	15	1	0.7	0.4	84	0.28	2.8	0.24	2.4
20	267	1.18	0.96	0.54	38	26	1	0.7	0.7	144	0.48	2.4	0.40	2.0
30	362	1.59	1.30	0.73	51	35	2	1.4	0.9	195	0.65	2.2	0.54	1.8
40	448	1.98	1.61	0.91	63	43	2	1.4	1.2	243	0.81	2.0	0.67	1.7
50	530	2.34	1.91	1.08	75	51	3	2.1	1.4	285	0.95	1.9	0.79	1.6
60	608	2.68	2.19	1.23	86	59	3	2.1	1.6	327	1.09	1.8	0.91	1.5
70	682	3.01	2.45	1.38	96	66	4	2.8	1.8	369	1.23	1.8	1.02	1.5
80	754	3.32	2.71	1.53	106	73	4	2.8	2.0	408	1.36	1.7	1.13	1.4
90	824	3.63	2.96	1.67	116	80	4	2.8	2.2	444	1.48	1.6	1.23	1.4
100	891	3.93	3.21	1.81	126	86	5	3.5	2.4	480	1.60	1.6	1.34	1.3
Maintenance Plus Low Activity (= 25% Increment, Intensive Management, Tropical Range and Early Pregnancy)														
10	199	0.87	0.71	0.40	27	19	1	0.7	0.5	108	0.36	3.6	0.30	3.0
20	334	1.47	1.20	0.68	46	32	2	1.4	0.9	180	0.60	3.0	0.50	2.5
30	452	1.99	1.62	0.92	62	43	2	1.4	1.2	243	0.81	2.7	0.67	2.2
40	560	2.47	2.02	1.14	77	54	3	2.1	1.5	303	1.01	2.5	0.84	2.1
50	662	2.92	2.38	1.34	91	63	4	2.8	1.8	357	1.19	2.4	0.99	2.0
60	760	3.35	2.73	1.54	105	73	4	2.8	2.0	408	1.36	2.3	1.14	1.9
70	852	3.76	3.07	1.73	118	82	5	3.5	2.3	462	1.54	2.2	1.28	1.8
80	942	4.16	3.39	1.91	130	90	5	3.5	2.6	510	1.70	2.1	1.41	1.8
90	1030	4.54	3.70	2.09	142	99	6	4.2	2.8	555	1.85	2.1	1.54	1.7
100	1114	4.91	4.01	2.26	153	107	6	4.2	3.0	600	2.00	2.0	1.67	1.7
Maintenance Plus Medium Activity (= 50%, Increment, Semiarid Rangeland, Slightly Hilly Pastures, and Early Pregnancy)														
10	239	1.05	0.86	0.48	33	23	1	0.7	0.6	129	0.43	4.3	0.36	3.6
20	400	1.77	1.44	0.81	55	38	2	1.4	1.1	216	0.72	3.6	0.60	3.0
30	543	2.38	1.95	1.10	74	52	3	2.1	1.5	294	0.98	3.3	0.81	2.7
40	672	2.97	2.42	1.36	93	64	4	2.8	1.8	363	1.21	3.0	1.01	2.5
50	795	3.51	2.86	1.62	110	76	4	2.8	2.1	429	1.43	2.9	1.19	2.4
60	912	4.02	3.28	1.84	126	87	5	3.5	2.5	492	1.64	2.7	1.37	2.3
70	1023	4.52	3.68	2.07	141	98	6	4.2	2.8	552	1.84	2.6	1.53	2.2
80	1131	4.98	4.06	2.30	156	108	6	4.2	3.0	609	2.03	2.5	1.69	2.1
90	1236	5.44	4.44	2.50	170	118	7	4.9	3.3	666	2.22	2.5	1.85	2.0
100	1336	5.90	4.82	2.72	184	128	7	4.9	3.6	723	2.41	2.4	2.01	2.0

Maintenance Plus High Activity (= 75% Increments, Arid Rangeland, Sparse Vegetation, Mountainous Pastures, and Early Pregnancy)														
10	278	1.22	1.00	0.56	38	26	2	1.4	0.8	1.50	0.50	5.0	0.42	4.2
20	467	2.06	1.68	0.94	64	45	2	1.4	1.3	252	0.84	4.2	0.70	3.5
30	634	2.78	2.28	1.28	87	60	3	2.1	1.7	342	1.14	3.8	0.95	3.2
40	784	3.46	2.82	1.59	108	75	4	2.8	2.1	423	1.41	3.5	1.18	3.0
50	928	4.10	3.34	1.89	128	89	5	3.5	2.5	501	1.67	3.3	1.39	2.7
60	1064	4.69	3.83	2.15	146	102	6	4.2	2.9	576	1.92	3.2	1.60	2.7
70	1194	5.27	4.29	2.42	165	114	6	4.2	3.2	642	2.14	3.0	1.79	2.6
80	1320	5.81	4.74	2.68	182	126	7	4.9	3.6	711	2.37	3.0	1.98	2.5
90	1442	6.35	5.18	2.92	198	138	8	5.6	3.9	777	2.59	2.9	2.16	2.4
100	1559	6.88	5.62	3.17	215	150	8	5.6	4.2	843	2.81	2.8	2.34	2.3
Additional Requirements for Late Pregnancy (for All Goat Sizes)														
	397	1.74	1.42	0.80	82	57	2	1.4	1.1	213	0.71		0.59	
Additional Requirements for Growth—Weight Gain at 50 g per Day (for All Goat Sizes)														
	100	0.44	0.36	0.20	14	10	1	0.7	0.3	54	0.18		0.15	
Additional Requirements for Growth—Weight Gain at 100 g per Day (for All Goat Sizes)														
	200	0.88	0.72	0.40	28	20	1	0.7	0.5	108	0.36		0.30	
Additional Requirements for Growth—Weight Gain at 150 g per Day (for All Goat Sizes)														
	300	1.32	1.08	0.60	42	30	2	1.4	0.8a	162	0.54		0.45	
Additional Requirements for Milk Production per kg at Different Fat Percentages (Including Requirements for Nursing Single, Twin, or Triplet Kids at the Respective Milk Production Level)														
% Fat														
2.5	333	1.47	1.20	0.68	59	42	2	1.4	3.8	760				
3.0	337	1.49	1.21	0.68	64	45	2	1.4	3.8	760				
3.5	342	1.51	1.23	0.69	68	48	2	1.4	3.8	760				
4.0	346	1.53	1.25	0.70	72	51	3	2.1	3.8	760				
4.5	351	1.55	1.26	0.71	77	54	3	2.1	3.8	760				
5.0	356	1.57	1.28	0.72	82	57	3	2.1	3.8	760				
5.5	360	1.59	1.29	0.73	86	60	3	2.1	3.8	760				
6.0	365	1.61	1.31	0.74	90	63	3	2.1	3.8	760				
Additional Requirements for Mohair Production by Angora at Different Production Levels														
Annual Fleece Yield, kg														
2	16	0.07	0.06	0.03	9	6								
4	34	0.15	0.12	0.07	17	12								
6	50	0.22	0.18	0.10	26	18								
8	66	0.29	0.24	0.14	34	24								

Adapted from National Research Council (1981).

the animals have very low blood phosphorus (0.2–0.3 mg/100 ml; normal is 0.5–0.7). The cause of the **hypophosphatemia** has not been identified, but may be from vitamin D deficiency. Dosing the affected animals with monosodium phosphate (12 mg phosphorus per lb body weight per day) for several weeks is effective in curing the condition. The supplement can be sprinkled on grain, or directly dosed orally.

Van Saun et al. (1996) suggested that **vitamin D deficiency** is the cause of hypophosphatemic rickets. They found low serum vitamin D in affected animals. Treatment with vitamin D increased serum concentrations of vitamin D and phosphorus. They speculated that in the Andes where llamas evolved, they are exposed to intense sunlight at high elevations, providing abundant vitamin D synthesis in the skin. When raised in areas with lower exposure to sunlight, they may be susceptible to vitamin D deficiency and require a dietary source.

Based on serum values for α-tocopherol, vitamin A, and selenium, Dart et al. (1996) indicated that alfalfa hay is an unreliable source of vitamins A and E for llamas and recommended routine dietary supplementation with these nutrients.

The digestive physiology and nutrition of llamas and alpacas have been reviewed by San Martin and Bryant (1989) and Johnson (1989), who concluded that these animals are nutritionally better adapted to harsh environments than domestic ruminants. Until more specific information on their nutrient requirements is available, llamas should be fed diets suitable for sheep.

In spite of the extensive use of camels in many countries, there has been very little research conducted on **camel nutrition.** Bhattacharya et al. (1988) studied protein and energy digestibility of alfalfa hay and barley in young camels; values were similar to those characteristic of sheep. Camels consume considerable browse and will eat thorny and bitter plants rejected by other livestock. They are predominantly browsers, although they consume considerable dry grass in the dry season in Africa when browse species have lost their leaves (Migongo-Bake and Hansen, 1987).

QUESTIONS AND STUDY GUIDE

1. A 4-H member had a small flock of sheep. One day in late winter, one of her largest ewes was weak and incoordinated, and went down in a comalike condition. The ewe's breath had a sweetish, nail-polish remover smell. Despite receiving several injections of antibiotics, the ewe died. She was one of the member's best ewes, having had twins for 3 years in a row. What is your diagnosis? What could have been done to more effectively treat and prevent the condition?
2. Suppose you are the manager of a university farm. You need to allocate the hay crop from a large grass-clover field to the sheep and beef research units. Some of the hay was cut early and has a high content of clover. Some of the field was mainly grass. This part was cut last, and there were several rain showers after the hay was mowed. How would you allocate the hay to these two livestock units? The managers of both of the units insist that their livestock require the highest quality hay available.

3. You are a sheep producer, and you want to flush the ewes for breeding. You have a field of irrigated grass-red clover **aftermath** (lush regrowth after crop harvest) available. Should you use this pasture, or keep the ewes in dry-lot and buy grain to flush them?
4. How suitable is cow's milk or calf milk replacer for feeding orphan lambs? What would be the value of cow's milk with table sugar added to it to increase the energy content?
5. Lambs affected with white muscle disease may recover if injected with a high dose of vitamin E. Explain.
6. Sheep fed dried poultry waste as a nitrogen supplement may do fine for quite a long time, and then suddenly die. For a day or two before death, they have yellowish eyes, and their urine is red. Explain.
7. Sheep that are starved for a period of time (e.g., snowed in by a blizzard for a week on range in North Dakota) may lose their wool several weeks later. The wool is very easily pulled from their bodies and has a pronounced weak point. Explain.
8. Goats and llamas are believed to make more efficient use of low-quality roughage and lignified browse plants than sheep. What factors could account for this?
9. What are some advantages of a mixed grazing system involving beef cattle, sheep, and Angora goats on an irrigated grass-clover pasture? On a brushy semiarid hillside in California, with grass, scrub oak, forbs, and several species of poisonous plants present?
10. Which will require more feed each day: one 1,200-lb beef cow or eight 150-lb ewes? Assume that all are nonpregnant and kept under maintenance conditions. Explain your answer. (Hint: see metabolic body size, Chapter 9).

REFERENCES

Bechert, U.S. and B.B. Smith. 1996. Serum macro and micro element concentrations in the llama. *Vet. Clin. Nutr.* 3:119–127.

Bhattacharya, A.N., S. Al-Mutairi, A. Hashimi, and S. Economides. 1988. Energy and protein utilization of lucerne hay and barley grain by yearling camel calves. *Anim. Prod.* 47:481–485.

Braithwaite, C.D. 1983. Calcium and phosphorus requirements of the ewe during pregnancy and lactation. 1. Calcium. *Brit. J. Nutr.* 50:711–722.

Brown, L.E. and W.L. Johnson. 1985. Intake and digestibility of wheat straw diets by goats and sheep. *J. Anim. Sci.* 60:1318–1323.

Colebrook, W.F. and P.J. Reis. 1969. Relative value for wool growth and nitrogen retention of several proteins administered as abomasal supplements to sheep. *Aust. J. Biol. Sci.* 22:1507–1516.

Dart, A.J., H. Kinde, D.R. Hodgson, Y.R. Peauroi, A.W. Selby, J. Maas, and M.E. Fowler. 1996. Serum α-tocopherol, vitamin A, and blood selenium concentrations, and glutathione peroxidase activity in llamas fed alfalfa hay. *Am. J. Vet. Res.* 57:689–692.

Delisle, J., H.V. Petit, and F. Giguere. 1988. Performance of young lambs fed a clotting or a non-clotting milk replacer. *Can. J. Anim. Sci.* 68:993–996.

DeRosa, G., V. Fedele, F. Napolitano, L. Gubitosi, A. Bordi, and R. Rubino. 1997. Dietary preferences in adult and juvenile goats. *Anim. Sci.* 65:457–463.

Devendra, C. 1981. Potential of sheep and goats in less developed countries. *J. Anim. Sci.* 51:461–473.

Freer, M. and H. Dove, eds. 2002. *Sheep nutrition.* Collingswood VIC, Australia: CSIRO Publishing.

Goeger, D.E., P.R. Cheeke, J.A. Schmitz, and D.R. Buhler. 1982. Toxicity of tansy ragwort (*Senecio jacobaea*) to goats. *Am. J. Vet. Res.* 43:252–254.

Haenlein, G.F. 1982. Feeding sunflowers can prevent enterotoxemia. *Feedstuffs.* 54 August 2:23.

Heaney, D.P., J.N.B. Shrestha, and H.F. Peters. 1982. Performance of lambs fed milk replacers having two levels of fat. *Can. J. Anim. Sci.* 62:837–843.

Heaney, D.P. and J.N.B. Shrestha. 1985. Effects of warm versus cold milk replacers and of free-choice hay postweaning on performance of artificially reared lambs. *Can. J. Anim. Sci.* 65:871–878.

_______ . 1987. Effects of soyflour in milk replacer on the performance of artificially reared lambs. *Can. J. Anim. Sci.* 67:757–763.

Hintz, H.F., H.F. Schryver, and M. Halbert. 1973. A note on the comparison of digestion by New World camels, sheep and ponies. *Anim. Prod.* 16:303–305.

Howe, J.C., T.N. Barry, and A.I. Popay. 1988. Voluntary intake and digestion of gorse (*Ulex europaeus*) by goats and sheep. *J. Agri. Sci.* 11:107–114.

Johnson, L.W. 1989. Nutrition of llamas. *Vet. Clin. North Am.: Food Anim. Pract.* 5:37–54.

Johnson, W.L., J.E. Van Eys, and H.A. Fitzhugh. 1986. Sheep and goats in tropical and subtropical agricultural systems. *J. Anim. Sci.* 63:1587–1599.

Junge, R.E. and L. Thornburg. 1989. Copper poisoning in four llamas. *J. Am. Vet. Med. Assoc.* 195:987–989.

Kingsbury, J.M. 1983. The evolutionary and ecological significance of plant toxins. In *Handbook of Natural Toxins,* Vol. I, *Plant and Fungal Toxins.* R.F. Keeler, and A.T. Tu, eds. pp. 675–706. New York: Marcel Dekker.

Koch, D.W., F.C. Ernst, N.R. Leonard, R.R. Hedberg, T.J. Blenk, and J.R. Mitchell. 1987. Lamb performance on extended grazing of tyfon. *J. Anim. Sci.* 64:1275–1279.

Loerch, S.C., K.E. McClure, and C.F. Parker. 1985. Effects of number of lambs suckled and supplemental protein source on lactating ewe performance. *J. Anim. Sci.* 60:7–13.

Lu, C.D. and J.J. Potchoiba. 1990. Feed intake and weight gain of growing goats fed diets of various energy and protein levels. *J. Anim. Sci.* 68:1751–1759.

McCabe, S.M. and T.N. Barry. 1988. Nutritive value of willow (*Salix* sp.) for sheep, goats and deer. *J. Agr. Sci.* 111:1–9.

McGregor, B.A. 1988. Effects of different nutritional regimens on the productivity of Australian cashmere goats and the partitioning of nutrients between cashmere and hair growth. *Aust. J. Exp. Agric.* 28:459–467.

McGregor, B.A. and Hodge, R.W. 1989. Influence of energy and polymer-encapsulated methionine supplements on mohair growth and fibre diameter of Angora goats fed at maintenance. *Aust. J. Exp. Agri.* 29:179–181.

Migongo-Bake, W. and R.M. Hansen. 1987. Seasonal diets of camels, cattle, sheep and goats in a common range in Eastern Africa. *J. Range Manage.* 40:76–79.

Nastis, A.S. and J.C. Malechek. 1981. Digestion and utilization of nutrients in oak browse by goats. *J. Anim. Sci.* 53:283–290.

National Research Council. 1981. *Nutrient Requirements of Goats: Angora, Dairy and Meat Goats in Temperate and Tropical Countries.* Washington, DC: National Academy Press.

_______ . 1985. *Nutrient Requirements of Sheep.* Washington, DC: National Academy Press.

Nunez-Hernandez, G., J.L. Holechek, J.D. Wallace, M.L. Galyean, A. Tembo, R. Valdez, and M. Cardenas. 1989. Influence of native shrubs on nutritional status of goats: Nitrogen retention. *J. Range Manage.* 42:228–232.

Pfister, J.A. and J.C. Malechek. 1986. The voluntary forage intake and nutrition of goats and sheep in the semi-arid tropics of northeastern Brazil. *J. Anim. Sci.* 63:1078–1086.

Pfister, J.A., J.C. Malechek, and D.F. Balph. 1988. Foraging behaviour of goats and sheep in the *Caatinga* of Brazil. *J. Appl. Ecol.* 25:379–388.

Poyyamozhi, V.S. and R. Kadirvel. 1986. The value of banana stalk as a feed for goats. *Anim. Feed Sci. Tech.* 15:95–100.

Quick, T.C. and B.A. Dehority. 1986. A comparative study of feeding behavior and digestive function in dairy goats, wool sheep and hair sheep. *J. Anim. Sci.* 63:1516–1526.

Rammell, C.G., K.G. Thompson, G.R. Bentley, and M.W. Gibbons. 1989. Selenium, vitamin E and polyunsaturated fatty acid concentrations in goat kids with and without nutritional myodegeneration. *NZ Vet. J.* 37:4–6.

Reed, B.A. and D.L. Brown. 1988. Almond hulls in diets for lactating goats: Effects on yield and composition of milk, feed intake and digestibility. *J. Dairy Sci.* 71:530–533.

Ryder, M.L. and S.K. Stephenson. 1968. *Wool Growth.* New York: Academic Press.

Russel, A.J.F, J.M. Doney, and R.G. Gunn. 1969. Subjective assessment of body fat in live sheep. *J.Ag Sci.* 72:451–454.

San Martin, F. and F.C. Bryant. 1989. Nutrition of domesticated South American llamas and alpacas. *Small Rumin. Res.* 2:191–216.

SID. 2002. *Sheep Production Handbook.* Centennial, CO: American Sheep Industry Association.

Sidahmed, A.E., J.G. Morris, L.J. Koong, and S.R. Radosevich. 1981. Contribution of mixtures of three chaparral shrubs to the protein and energy requirements of Spanish goats. *J. Anim. Sci.* 53:1391–1400.

Sidahmed, A.E., J.G. Morris, S.R. Radosevich, and L.J. Koong. 1983. Seasonal changes in composition and intake of chaparral by Spanish goats. *Anim. Feed Sci. Tech.* 8:47–61.

Treacher, T.T. and G. Caja. 2002. Nutrition during lactation. In *Sheep Nutrition,* M. Freer and H. Dove, eds. pp. 213–236. Collingswood VIC, Australia: CSIRO Publishing.

Van Eys, J.E., I.W. Mathius, P. Pongsapan, and W.L. Johnson. 1986. Foliage of the tree legumes gliricidia, leucaena, and sesbania as supplement to napier grass diets for growing goats. *J. Agri. Sci.* 107:227–233.

Van Eys, J.E., H. Pulungan, M. Rangkuti, and W.L. Johnson. 1987. Cassava meal as supplement to napier grass diets for growing sheep and goats. *Anim. Feed Sci. Tech.* 18:197–207.

Van Saun, R.J., B.B. Smith, and B.J. Watrous. 1996. Evaluation of vitamin D status of llamas and alpacas with hypophosphatemic rickets. *J. Am. Vet. Med. Assoc.* 209:1128–1133.

Van Soest, P.J. 1994. *Nutritional Ecology of the Ruminant,* Ithaca, NY: Cornell University Press.

Warmington, B.G., G.F. Wilson, and T.N. Barry. 1989. Voluntary intake and digestion of ryegrass straw by llama x guanaco crossbreds and sheep. *J. Agr. Sci.* 113:87–91.

Wilkinson, S.C. and D.M.B. Chestnutt. 1988. Effect of level of food intake in mid and late pregnancy on the performance of breeding ewes. *Anim. Prod.* 47:411–419.

Wilson, A.D. 1977. The digestibility and voluntary intake of the leaves of trees and shrubs by sheep and goats. *Aust. J. Agr. Res.* 28:501–508.

CHAPTER 17

Feeding and Nutrition of Dairy Cattle

H. G. Bateman, II

Objectives

1. To introduce the factors and their integration that determine nutrient requirements for the different classes of dairy cattle
2. To describe the factors that determine if a given diet will supply the required nutrients for milk production and growth by dairy cattle
3. To describe situations where nutritional management can impact other management areas of a dairy farm
4. To describe unique aspects of dairy nutrition

Nutritional management of the lactating dairy cow may be the most complex system worked with by livestock nutritionists. In 2002, the average dairy cow in the United States produced approximately 27.7 kg (60 lb) of milk daily. If one assumes that this milk contained 3.6 percent fat and 3.0 percent protein, over 800 g of protein and 19 Mcal of NE_L must be provided daily just to meet the milk requirements. These nutrient concentrations will leave nothing for maintenance, gestation, or growth. To further complicate matters, we must maintain a minimum level of fiber in the diet to sustain proper rumen health and to reduce or even prevent metabolic disorders (Fig. 17.1). Technological advances have allowed nutritional management of the lactating dairy cow to greatly advance. As recently as 1988, protein nutrition of the lactating dairy cow was based on crude protein (CP). However, the most recent set of nutrient requirements (NRC, 2001) places emphasis on meeting the amino acid requirements of the cow. As knowledge advances and technology increases, our understanding of the nutritional management of the dairy cow will also change. All students are advised to evaluate these changes for their particular situation and implement the ones that will have a positive impact. It is beyond the scope of this text to "teach you everything there is to know about dairy nutrition." It is the hope that the material presented will empower students with knowledge and understanding and prepare them to carefully and objectively evaluate and implement strategies to meet the needs on their particular farm.

NUTRIENT REQUIREMENTS OF MATURE COWS

The nutrient requirements for dairy cattle were revised in 2001 (NRC, 2001). After an extensive review and evaluation, the NRC proceeded with quite a complicated system for predicting nutrient requirements. For the first time, nutrient requirements of the animal were somewhat dependent upon the diet being fed. This was based on the theory that not all feedstuffs were fermented and digested

FIGURE 17.1 Feeding dairy cattle requires an integrated knowledge of nutrition, housing, and management skills.

equally and that some feedstuffs could be complementary to other feedstuffs. There has been a lot of criticism of the system because a computer is necessary for proper and full evaluation of rations, thus resulting in a decreased ability of nutritionists to "troubleshoot" on a farm. However, the advances and accuracy that the current system should provide to the researcher, field person, and nutritionist must not be overlooked.

Water

Water intake is often overlooked and misunderstood when feeding most species of livestock. Dairy cattle are no exception. However, water intake is extremely important to the dairy cow if one considers that milk is approximately 87 percent water (Winchester and Morris, 1956). Obviously, limiting water intake by lactating cows will have a negative impact on milk production. Water requirements can be met through three sources: drinking water, metabolic water, and water in feed. Water generated through metabolic processes in the body is a minor component of the total water "intake" and should not be relied upon to meet requirements. Water in feed is dependent upon the diet and, particularly, the forages fed. Diets with high amounts of fermented feeds will have greater water content than diets based on dry forages. However, even when feeding diets highly dependant upon fermented feeds, free drinking water is needed to meet a cow's water requirement.

Many factors will influence the cow's desire to consume water. As has already been mentioned, milk production requires large amounts of water and does play a role in determining water intake. Similarly, dry-matter intake (DMI) appears to positively influence water intake. Whether this is due to an increased need for water as solvent for metabolic reactions, to solvent for microbial fermentation, or to increased losses of water through increased fecal and urinary turnover is not completely clear, but the relationship between DMI and water intake is quite strong. Environmental factors such as temperature, humidity, and wind speed are also factors in determining water intake. The physiochemical properties of water make it both a good insulator during periods of cold stress and an immensely important conductor for heat loss during periods of heat stress; therefore, water intake should increase during periods of temperature extremes. Humidity levels will be negatively correlated to the amount of water the animal will lose through evaporation. In periods of high humidity (similar to summers in the southeastern United States), evaporative losses of water are minimized. However, in periods of low humidity (similar to the southwestern United States), evaporative losses will increase; therefore, increased water intake must occur to prevent dehydration. **Water quality** also affects water intake by dairy cows (Fig. 17.2). Unfortunately, the scientific literature concerning the impact of quality on water intake by dairy cattle is quite limited, and precise recommendations are hard to make and defend. The National Research Council (2001) provides estimates of the upper limits for many of the common contaminates of drinking water, and these must be considered the best recommendations for current production practices in the United States.

FIGURE 17.2 Providing ample clean water is paramount for optimal milk production.

Dry-Matter Intake (DMI)

It is often joked that the three most important words in real estate are "location, location, and location." A similar case can be made for feeding dairy cattle, the three most important words being "DMI, DMI, and DMI." The DMI of a cow or group of cows determines the concentrations of nutrients that must be in their ration to meet their needs. Field nutritionists commonly use a factor of 2 lb of milk for each lb of DMI as a benchmark to evaluate the farm's nutrition program. The question can be asked, "Does DMI drive milk production or does milk production drive DMI?" There is not a good answer to this question. In early lactation, milk production outpaces DMI, but in late lactation DMI will outpace milk production. If cows consume more DMI from a given diet, they produce more milk, but if DMI from a given diet is limited, milk production will decline. Obviously, milk production and DMI are intricately intertwined.

Prediction of DMI by lactating dairy cows is not easy. In fact, there is no one equation that is widely accepted by the industry. However, the NRC (2001) predicts DMI of lactating cows by using a function of milk yield (corrected to a constant 4 percent milk fat), body weight, and week of lactation. Although other factors such as environmental temperature, dietary fiber level, dietary DM content, forage to concentrate ratio, and dietary fat concentrations are known to influence DMI, no attempt to predict these influences has been made. There is currently a lack of controlled data available to accurately model these influences, but nutritionists are cautioned to buffer their predictions if extremes in these factors are occurring on the particular farm with which they are working.

Feeding management practices are known to influence DMI by lactating cows. Availability of feed bunk space will play a role in determining DMI. Social dominance of cattle is widely observed. If feed bunk space is limited, aggressive cows ("boss cows") will prevent less-dominant or timid cows from having adequate access to feed for optimizing and maximizing DMI. **Frequency of feeding** seems positively correlated with DMI. However, the increase in feeding frequency can be as simple as moving feed already present back to the area accessible to the cow (Fig. 17.3).

Energy

After water, **energy** is probably the most important nutrient to provide for milk production. Each kg of milk (corrected to 4 percent milk fat) produced requires approximately 0.74 Mcal of NE_L (NRC, 1989). The average cow in the United States in 2002 produced 8441.4 kg of milk or 27.7 kg/d during the 10-month (305 day) lactation. Therefore, this cow needed 20.5 Mcal of NE_L/d or more just to meet the energy requirements for milk production. Using an old system for predicting nutrient requirements (NRC, 1989), this requirement could be met through the consumption of 24.7 lb of corn grain. However, when the requirements for maintenance, reproduction, and growth are added to the lactation requirement, the amount of grain needed to meet the total requirement becomes excessive and is not recommended for long-term stability of the cow's health.

Numerous factors are known to affect the energy value of feeds. The current system of predicting nutrient requirements (NRC, 2001) exploits this knowledge by discounting the energy value of a diet when other nutrients, most notably

FIGURE 17.3 Moving feed into areas that are readily accessible by dairy cattle often will stimulate increased intake and result in increased milk production.

protein, are marginal or limiting. The current system (NRC, 2001) predicts the energy value for a feedstuff at maintenance based on the chemical composition of that feed and then adjusts that value based on any processing that has been done to the feed. Because efficiency of fermentation and absorption decreases as DMI increases, these values are then further discounted based on DMI. This system has been criticized for being unsuitable for on-farm use. However, with the widespread availability of portable computer equipment, the system can easily be transported to the farm and used to troubleshoot nutritional problems. In addition, the advances that this system have made in explaining variation in milk yield are useful.

Protein

Until recently, when discussing **ruminant protein nutrition,** crude protein was the unit of choice. This was based on the understanding that fermentation in the rumen would modify the proteins fed by converting them into microbial protein. Therefore, dietary composition was believed to have little impact on the amino acids that were digested by the cow. In the late 1980s, dairy nutritionists became aware that microbial protein could not supply enough amino acids to meet the requirements of a high-producing dairy cow and the RUP/RDP (RUP = ruminally undegradable protein; RDP = ruminally degradable protein) system began to get more usage. Although this system worked well in the field, it was not able to fully optimize amino acid nutrition of the cow because it still treated the rumen as a "black box" that one protein went into and a different protein came out. As computer technology advanced, so did the ability of the nutritionist to predict flow of proteins and amino acids at the small intestine. This allows for better fine-tuning of amino acid nutrition in the modern dairy ration.

Advances in the understanding of protein nutrition of the dairy cow made between the 1989 and 2001 editions of the *Nutrient Requirements of Dairy Cattle* publications were enormous. This has resulted in a markedly different way of formulating diets. In the 1989 edition, protein requirements were based on **microbial crude protein (MCP)** and **ruminally undegradable protein (RUP)** (NRC, 1989). The MCP was predicted based on NE_L, and RUP was predicted based on constants for the degradability of proteins in feeds. Time and experience have proven that neither of these assumptions is true. Research has shown that microbial protein synthesis is better related to organic matter fermentation in the rumen than to NE_L intake, which is reflective of both ruminal fermentation and postruminal digestion. The current publication (NRC, 2001) predicts MCP production based on total digestible nutrient intake but requires that **ruminally degradable protein (RDP)** is adequate for optimal fermentation. It has also been recognized that RUP values for a feed cannot be static across all feeding situations and diets. The current NRC publication (2001) attempts to predict the RUP content of feeds based on their fermentability in a given diet at a given DMI. As with energy values, these changes appear to better explain variation in milk production.

Lysine and methionine have been identified as the first two limiting amino acids for milk production under most production settings (Schwab, 1996). Because of this observation, efforts to predict the flow and absorption of these amino acids at the small intestine along with methods to increase their flow and absorption have become a part of dairy nutrition consideration. The NRC (2001) computer model provides requirements and predicts the flow of nine essential amino acids to the small intestine. The computer model takes advantage of differing ruminal degradability for different feeds and allows for some manipulation of this intestinal flow. Additionally, there are commercial products available that claim to prevent ruminal degradation of amino acids but allow for intestinal absorption of those same amino acids. Research data evaluating the claims of the different products are conflicting (Block and Jenkins, 1994; Papas et al., 1984a, b, Robinson et al., 1998; Rogers et al., 1989). However, many nutritionists observe positive milk production responses when using these products in the field and their potential impact should not be underestimated.

Minerals

Although they constitute a small proportion of most diets, minerals are rapidly becoming more important in dairy cattle nutrition as advances in our understanding of their biological roles unfold. Probably the largest impact of proper mineral nutrition is on the incidence of the metabolic disorder **parturient paresis** or milk fever. Incidence of parturient paresis in dairy herds has ranged from zero to over 20 percent (Kelton et al., 1998). **Calcium** metabolism is the controlling factor for developing parturient paresis. Abnormal Ca metabolism at or around parturition will predispose cows to developing the disorder. Historically, milk fever was prevented by feeding diets with minimal Ca contents to "perk up" the normal Ca homeostatic regulatory mechanism. However, this is almost impossible under modern feeding practices in the United States (NRC, 2001). As a result, many nutritionists are now recommending the use of **anionic salts** along with Ca supplementation during the dry period to alter the cow's **dietary cation-anion balance (DCAB)** and induce a mild metabolic acidosis. The DCAB can be estimated based on the dietary concentrations of sodium, potassium, chlorine, and sulfur. It may be represented as dietary cation-anion difference (DCAD), and most computer formulation software will calculate the value. In all cases, negative values or values close to zero are desired during the dry period. This will cause the cow to excrete Ca in the urine and increase her metabolic drive for Ca absorption from the gut and reabsorption from bone. Feeding the salts is discontinued after parturition, but Ca supplementation is continued, thus providing the Ca necessary for milk synthesis and preventing parturient paresis. **Hypomagnesemia** should not be confused with parturient paresis. Both conditions exhibit similar symptoms; however, traditional treatment for parturient paresis (supplemental Ca either via oral drench or intravenous) may exaggerate hypomagnesemia. As a consequence, cows that are marginal in either Ca or Mg status must be monitored closely during the transition period to prevent either of their plasma concentrations from dropping to dangerously low levels. Supplementing Mg during the dry period and throughout lactation may be warranted to prevent onset of metabolic disorders such as parturient paresis or hypomagnesemia.

There are reports in the scientific literature that increased phosphorous may positively impact reproduction (Morrow, 1969). However, these and similar data actually evaluated supplemental P in times of dietary deficiency. There has been concern about feeding excess P and the potential impact of this practice on the environment. The NRC (2001) was unable to conclude that feeding P above recommended levels had a positive impact on milk production or reproduction and therefore recommended that excess P be avoided. Many **trace minerals** are required in the diets of dairy cattle. Special attention is usually placed on Zn and Se at particular times during the lactation cycle. **Zinc** has many metabolic roles; however, it appears to play a major role in formation of keratinzed tissue formation in dairy cattle (Miller, 1970; Miller et al., 1989). Keratin is a fibrous protein that forms the basis for hair and horns. In addition, it forms the plug for the teat canal after milking. Therefore, keratin is important in locomotion of the dairy cow and also in preventing intermammary infection (mastitis). The form that Zn is supplemented in has a direct influence on its digestibility. Organic forms of Zn (Zn attached to amino acids)have been developed and are marketed to increase the bioavailability of Zn above inorganic sources and have a positive impact on the nutrition and immune status of the cow. However, reports in the scientific literature have both supported (Spain, 1999) and contradicted (Whitaker et al., 1997) this idea. When Zn status of the cow is marginal, there is probably a benefit to added Zn. If supplemental Zn as an inorganic source is already present in the diet, providing additional Zn from the organic sources should improve the absorption and therefore have a positive impact on the immune status of the cow. Similar to Zn, **selenium** has been associated with immune function in dairy cattle (Foster and Sumar, 1997; Smith et al., 1997; Smith and Hogan, 1988). Additionally, Se has been associated with reproductive performance of dairy cattle (Ammerman and Miller, 1984; Ceballos and Wittwer, 1996; Oldfield, 2002). Cows with marginal Se status at parturition are more susceptible to having retained fetal membranes. Supplemental Se during the transition period (either via feeding or injecting) may provide some protection against this disorder, which will have lasting benefits throughout the subsequent lactation. The impact of Se on immune status of the dairy cow is not unique. Reports of the role of Se in the immune response can be located for many species. As with the other species, the impact of Se on the immune response appears to be tied with that of vitamin E. In dairy cattle, Se has been shown to interact with vitamin E to reduce the incidence of mastitis (Smith et al., 1997; Smith and Hogan, 1988). However, students are cautioned that under current United States law supplementation of Se above 0.3 mg/kg of diet is prohibited. Literature data evaluating the potential impact of the other trace minerals on dairy nutrition are not as readily available. However, limited information indicates that Cu, Fe, Mo, and Co are needed in the diet for optimal production by dairy cattle. Copper and, to a smaller extent Fe are used by the immune system to ward off infectious organisms. Molybdenum is used as a cofactor for many metabolic reactions in the body. Cobalt is an integral component of vitamin B_{12} and therefore is needed both for the metabolic processes of the cow and for optimizing ruminal fermentation. For practical purposes, the requirements of dairy cattle for these trace minerals can be met by providing a trace mineralized salt for free-choice consumption.

Vitamins

Historically, supplementation of water-soluble vitamins to ruminant animals has not been needed. Advances and understanding of dairy nutrition have suggested that, under some conditions, supplemental B vitamins may provide some health and production benefits (Girard, 1998). Positive responses to supplemental biotin, choline, and niacin have been reported in the literature (Duffield, 2000; Girard, 1998; Linden, 2001). Supplemental biotin (20mg/head/day) improves hoof health and reduces lameness in both intensively-managed (Midla et al., 1998) and pasture fed (Fitzgerald et al., 2000) dairy cattle. Supplementing B vitamins must be evaluated on a case-by-case basis, and no blanket recommendations can be made at this point. When considering the use of B vitamins in diets fed to dairy cattle, the cost-to-benefit ratio should be examined and decisions based on benefits as compared to the cost of incorporation.

Unlike water-soluble vitamins, **fat-soluble vitamins** are usually supplemented to dairy cattle. Weiss (1998) reviewed the scientific literature and concluded that feeding approximately twice the requirement (NRC, 1989) of **vitamin A** was warranted. In the most recent update for nutrient requirements of dairy cattle (NRC, 2001), an increase in the requirements was adopted, although it was not quite twice that of the 1989 edition. This increase is based on information that supplemental vitamin A (above 1989 requirements) may improve immunity and therefore reduce incidence of mastitis and increase milk production. Very little data exist concerning **vitamin D** requirements of dairy cattle in the scientific literature. After thorough review, Weiss (1998) and the NRC (2001) both recommend 30 IU/kg of BW. They do note that total confinement conditions may warrant increased supplementation levels. However, in practice most dairy cattle in the United States have some access to sunlight and therefore will experience some vitamin D synthesis in the body. Weiss (1998) does comment that there is potential for improved milk production and reproductive efficiency from cows fed up to 1.8 times the requirement and that decisions to supplement at these levels should be based on a cost-to-benefit analysis. As efforts to reduce the incidence of mastitis have increased, researchers and field nutritionists have investigated supplemental **vitamin E** as a potential prophylactic measure. Literature data have indicated that supplemental vitamin E may reduce the incidence of mastitis (Smith et al., 1997; Smith and Hogan, 1988). However, the response to vitamin E appears to be highly related to Se supplementation strategy. This is probably due to the interaction of vitamin E and Se to reduce cell wall destruction. Based on the reduction in mastitis, the current recommendation for vitamin E in diets fed to lactating dairy cattle is 20 IU/kg DMI, which equates to approximately 500 IU/d.

FEEDING MANAGEMENT

A successful nutritional management plan must be in place for optimal utilization of nutrients by lactating dairy cattle. An obvious scheme for making nutritional plans would be to synchronize delivery of nutrients with changes in nutrient requirements throughout the **lactation cycle.** In fact, this idea is widely practiced in the dairy industry. By breaking the lactation cycle into distinct phases, better management of the overall nutritional program can be accomplished.

Different nutritionists will argue for different numbers of periods of the lactation cycle. However, six distinct periods can be distinguished. Some of these periods can be combined under many management schemes, while others may be further divided in other management schemes (Fig. 17.4). The periods include fresh cow (0 to 14 d postpartum), peak milk (14 to 90 d postpartum), peak DMI (90 to 200 d postpartum), tail end (200 d postpartum until dry), far-off dry cow (60 to 21 d prepartum), and close-up dry cow (21 d prepartum until calving). Varied metabolic processes are dominating during each of these periods, and this can be exploited to optimize nutritional management of the dairy cow.

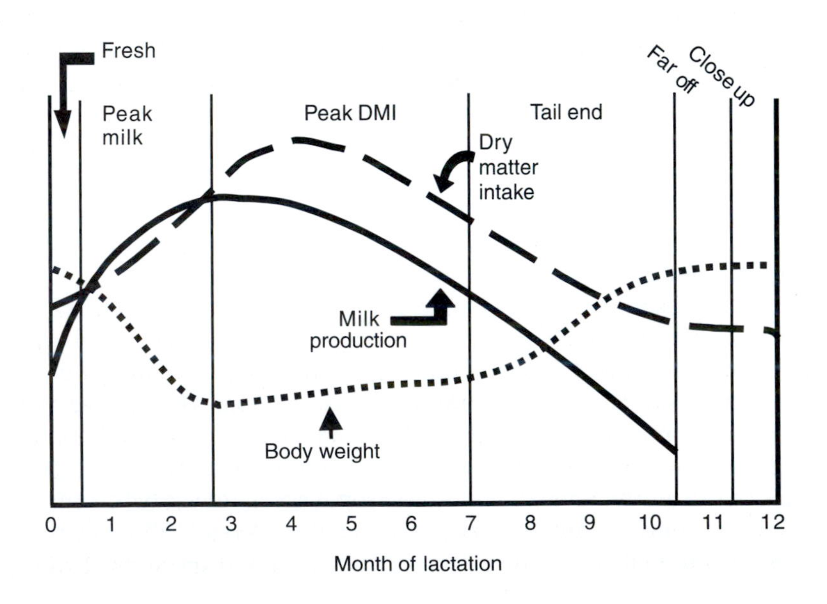

FIGURE 17.4 The lactation cycle can be broken into distinct periods or "phases" based on production, management, and physiological circumstances that the cow encounters. (Adapted from *Feeding the Dairy Herd,* North Regional Extension Publication 346, p. 18. Courtesy of Mike Hutjens, University of Illinois.)

During **early lactation** (fresh cows), hormonal changes following parturition are occurring, DMI is increasing, and lactation is beginning. The combination of these events creates a large amount of metabolic stress for the cow that is compounded by social stress from being moved from a group of nonlactating animals into a different group of lactating animals and by physical stress that occurred during parturition. There is usually a rapid change in diet that accompanies parturition, and this can result in an increased incidence of metabolic disorders. During the fresh period, emphasis should be placed on first maintaining and then increasing DMI. Fresh cows that decrease DMI, especially below what they consumed before calving, should be evaluated to determine the cause. Decreases in DMI by fresh cows are usually an indication of an underlying disease or disorder that, if caught early, may be more easily treated than if allowed to continue. Although the energy and protein requirements of the cow will be rapidly increasing due to the increase in milk production, it is ill advised to alter the diet to meet these increased requirements. Long forage should remain a part of the diet to promote optimal ruminal health and fermentation. During this period, an increased effort to monitor and treat metabolic disorders such as ketosis, displaced abomasums, metritis, retained placenta, and parturient paresis can produce large economic returns throughout the remaining lactation.

During the **peak milk** period, intake will lag behind milk production. This will result in a loss of body weight and is considered normal for dairy cows. Although cows are expected to use body stores to partially meet their requirements for protein and energy during this period, they should not be allowed to lose excessive weight. A simple method for estimating body stores is called **body condition scoring.** Body condition refers to the amount of fat stores that an animal carries and is quantified in dairy cattle on a scale of 1 to 5 (Fig. 17.5) (Edmonson et al., 1989). On this scale, the optimal body condition score for cows entering lactation is between 3.5 and 4. Cows that lose more than two points of body condition score during lactation should be evaluated to determine the underlying cause. Cows that do not lose body condition during the peak milk period should also be evaluated to determine if they are producing at their genetic potential. Although cows are in an energy deficit during the peak milk period, there are some modifications that can be made to minimize the weight loss that occurs. Lowering the **forage to concentrate ratio** (increasing concentrate feeds) will result in a higher energy concentration in the diet. Care must be taken to maintain adequate fiber for proper rumen function. Low levels of fiber are associated with decreased ruminal pH. This decrease in ruminal pH has a depressing effect on fermentation and can result in a metabolic acidosis. The metabolic acidosis can, in turn, have detrimental effects on hoof health. However, this will not be observed on-farm for approximately 6 months following the insult. **Buffers** such as sodium bicarbonate or sodium sesquacarbonate may be useful during the peak milk period. These chemicals assist in maintaining an acceptable ruminal pH and therefore inhibit the effects of low forage to concentrate ratio. This is the period of lactation that has the greatest potential for producing economic returns to **added fat sources.** Substituting fats for other feeds with a lower energy density will help minimize the energy deficit that the cow experiences during this period. Anything that can be done to increase DMI during this period will result in greater milk production. Frequent feeding, clean feeding environments, and flavors and other additives can be used to increase

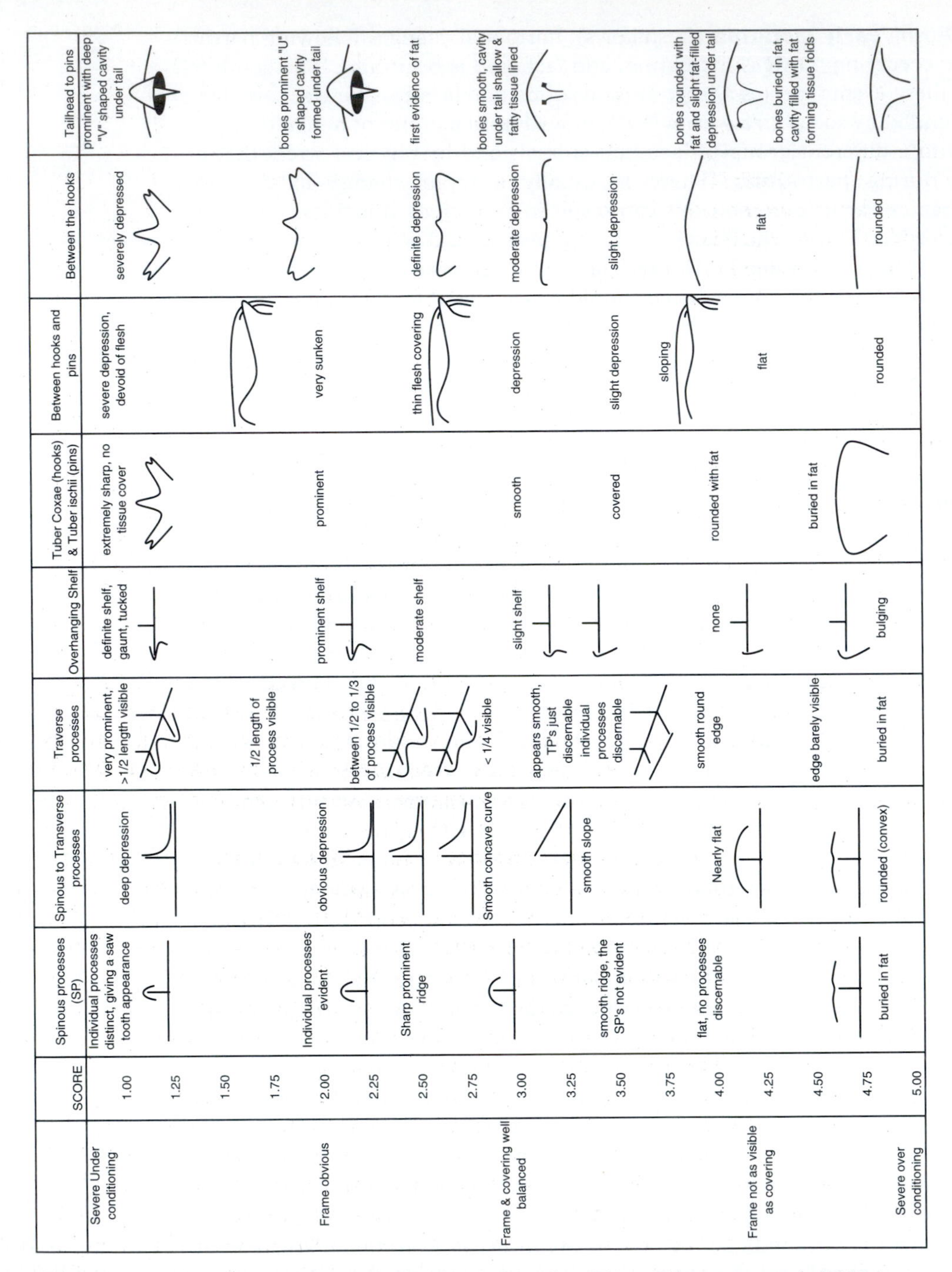

FIGURE 17.5 Body condition scoring chart. (Reprinted from Edmonson et al., 1989.)

DMI. If several choices of feeds are available to a farm, it may be profitable to selectively feed those that have the greatest palatability to cows in peak milk since they will stimulate DMI.

Although milk production will eventually decline, DMI will continue to increase until approximately 4.5 to 5 months after parturition. The combination of increased DMI and declining milk production allows for nutrients to be dedicated to reproduction without continuing the loss of body weight. Increased DMI also supports milk production, allowing for a greater persistency of lactation. Metabolic demands remain high for the cow during this period; however, neither gain nor loss of body weight should be occurring to any great extent. Many production settings will not require changes to the diet from what was being fed during the peak milk period.

During the **tail end of lactation,** milk production continues to decline, but because the cow is expected to be pregnant, fetal development occurs. It is during this period of the lactation cycle that the cow is most efficient at replacing her body stores for use during the next lactation. However, care must be taken not to overfeed the cow. An optimal body condition score at the time of **drying off** is between 3.5 and 4. Cows that are allowed to gain excess body weight will be predisposed to dystocia, ketosis, and fatty liver following calving. Cows that do not regain adequate body weight during this period will not have adequate stores to draw upon to meet the requirements for milk production during the subsequent lactation. Traditionally, the tail end of the lactation cycle has been a time that dietary energy and protein concentrations can be decreased as compared to earlier lactation. If cows are in adequate body condition score, fats can be removed and forage-to-concentrate ratio increased (Fig. 17.6). During the tail end period, there is little need for use of buffers and other additives. Recently, however, producers have experimented with the concept of feeding the same ration to tail-end cows as they feed to cows in earlier lactation, but at restricted amounts. Although this practice has merit in that labor for mixing feeds is reduced, it must be carefully monitored. Because intake is limited, the dairy manager must insure that all cows in the group receive adequate feed and do not lose weight. Equally important is to monitor to insure that some cows do not get above their allotment of feed and therefore gain excessive weight. In herds with extremely high production (> 25,000 lb/cow/year), it may be possible to feed a single ration to all lactating cows for maximal DMI.

Historically, the **dry period** for dairy cattle has been considered of little consequence from a nutritional standpoint. The sole objective during this period was to allow the rumen to rest and regenerate tissues that were damaged from the high amounts of fermentation that occurred during the previous lactation. Our current knowledge base indicates that this is far from true. The dry period is a very metabolically active period for dairy cattle. This is when the majority of fetal growth occurs. This period is also when many metabolic disorders that appear postcalving begin. As such, proper nutritional management during this period is essential.

During the first 45 d of the dry period, we essentially are allowing the cow, rumen, and mammary gland to rest and recuperate. To assist in drying the cow off, abrupt changes in diet usually occur. Diets for far-off dry cows usually contain a large percentage of long forage (Fig. 17.7). Because most forage has a lower energy content than a comparable amount of grains, the dietary energy concentration is

(A)

(B)

FIGURE 17.6 Fat stores on a dairy cow are often estimated as body condition and can usually be estimated based on stores in the rear quarters. The three cows represented here are carrying low (A), adequate (B), and high (C) body condition for cows at dry off.

(C)

FIGURE 17.6 continued

FIGURE 17.7 To allow the rumen time for rest and regeneration and to help cease milk production by the mammary gland, long forage is provided as a major component of the diet offered to dry cows.

reduced, which assists in stopping lactation. Long forages also promote a higher pH during ruminal fermentation, which aids in the regeneration of the ruminal papilla. The majority of **fetal growth** occurs during the third trimester. This corresponds to the dry period for dairy cattle. Because of this, care must be taken to insure that adequate protein and minerals are provided to the cow for fetal growth. During the dry period, a minimal amount of protein is required for cow maintenance. However, it is recommended that protein be supplemented above the minimums to maintain adequate RDP for fermentation and to provide the needed amino acids for the rapid fetal growth occurring.

The **close-up dry period** can be one of the most challenging times for nutritional management of dairy cattle. This period should be used to transition the cow from being on a low-energy/high-forage diet toward the diet that she will be offered postcalving. However, to avoid digestive upsets, this transition must be made smoothly and slowly and some long forage must remain in the diet. During this period, feeding to decrease the **dietary cation-anion balance (DCAB)** may prove beneficial in preventing milk fever. Anionic salts along with commercial feeds have been successfully used to modify the DCAB and reduce the incidence of milk fever. However, care must be taken to insure that if these feeds are used the dietary Ca levels remain sufficient to meet the increased demand that occurs. In practice, 150 g of Ca per day has proven sufficient to meet this requirement and prevent a drastic Ca drain from the body stores. In areas where Se is marginal or deficient, supplemental Se at the beginning of the close-up dry period may reduce the incidence of retained placentas, improve uterine involution, decrease metritis, and reduce cystic ovaries (Wichtel, 1998). The close-up dry period is also a time of colostrogenesis (colostrum synthesis). Care should be taken to monitor DMI of the close-up dry cow to insure that adequate nutrient intake is occurring to support colostrum synthesis. Inadequate intake of energy or protein during the close-up dry period may impair colostrum quantity or quality, which would have devastating effects on calf survival and thriftiness.

PASTURE-BASED GRAZING SYSTEMS FOR DAIRY CATTLE

In areas where there is ample grass, pasture-based dairying is a viable alternative to drylot animal housing. The advantages of pasture-based dairying are lower cost for forage harvesting and increased cow comfort. The disadvantages are increased labor for animal movement and limited herd size. For practical purposes, maximum herd size under pasture-based dairying is near 250 lactating cows. Herds larger than this require large land areas, and cow movement to get from pasture to the milking parlor tends to become excessive and counterproductive. Although pasture-based dairying is different from drylot feeding situations, the general nutritional scheme of the two systems is the same. Nutrient requirements are determined in the same manner and are largely based on milk production. Additional energy must be provided to compensate for the increase in activity associated with grazing and movement from pastures to the milk parlor. Under pasture-based systems, it becomes much more important to monitor forage quality and the quantity of forage consumed. During the growing season, it is possible for available forage quality to change daily due to limitations on forage quantity availability. Close monitoring of forage availability and quality will allow the dairy nutritionist to manage pastures for optimal forage intake.

NUTRIENT REQUIREMENTS OF REPLACEMENT HEIFERS

Swanson (1967) has indicated that "The optimum growth pattern for dairy heifers is that regimen which will develop in the heifer her full lactation potential at a desired age and at a minimum of expense." With this goal, there are probably as many management plans as there are dairy producers and all are equally valid. However, there are some generalities that can be made about rearing replacements.

There is a relative lack of information about the nutrient requirements and nutritional management of **replacement heifers** when compared to the knowledge base surrounding mature cows. However, just as with the mature cow, care and planning should be an integral part of the nutrition program for the replacements. Grouping is essential for proper nutritional management of replacement heifers. It is easy to understand that milk-fed calves have radically different nutrient reqiurements and needs than those of bred heifers. It is not as easy to understand that a 3-month-old weaned heifer may not be able to obtain adequate nutrients from the diet fed to a 6-month-old heifer. When grouping replacements, some consideration must also be taken for feeding and housing facilities. If facilities will not properly support twelve groups of heifers, it makes no sense to form that many groups regardless of what the current trend is in the industry. Similarly, if you do not have enough animals to economically utilize an entire facility, it may be more profitable to allow pens to remain empty.

At birth, a calf does not possess an active immunity system. Calves acquire **passive immunity** following consumption of immunoglobulin from **colostrum.** Gut closure occurs when the intestinal mucosa no longer absorbs large molecules such as proteins and immunoglobulin. **Gut closure** is assumed to occur during the first 24 to 36 hours of life. Therefore, it is imperative that newborn dairy heifers be fed good-quality colostrum within 12 hours of birth. Colostrum quality should be measured, and colostrum with less than 17 g IgG/L should not be fed unless supplemented with a source of extra Ig (Arthington et al., 2000; Morin et al., 1997). Currently, there are several of these marketed. Care should be taken to assure that any product being used has been independently tested for its efficacy and digestibility. It is also worth mentioning that these products are sold as supplements to—and not replacements for—colostrum. Often they will not contain the sugars, fats, and minerals that colostrum contains, and, therefore, the young calf will not obtain the energy needed for survival from these products alone. After feeding colostrum, most dairy heifers are reared on **milk replacers.** There has been a lot of investigation into the type and amount of protein needed in milk replacers to optimize growth of the dairy calf. Most commercial milk replacers contain between 12 and 22 percent CP on a dry basis. Blome et al. (2003) reported that increasing the CP content of milk replacer from 16 to 26 percent increased body weight gains of bull calves and that this increase was in structural tissue growth. Protein content of milk replacers should also be evaluated based on the sources. Milk replacers based on whey proteins are usually more digestible than milk replacers based on soy or other plant-based proteins (Davis and Drackley, 1998). Therefore, there may be additional benefits to selecting milk replacers based on milk proteins.

In addition to milk or milk replacers, calves should be offered a high-quality calf starter and fresh, clean water almost from birth. Considerable evidence suggests that they will not consume a lot of these feeds until approximately 3 to 4 weeks of age. However, providing access may stimulate intake of the feeds and encourage early consumption and weaning. **Calf starters** should contain only highly palatable ingredients. In addition, fiber content should be minimized and non-protein nitrogen sources should be avoided since these very young animals will not have a fully functioning rumen.

There is considerable controversy about whether to feed forages to weaned calves. Although many producers do use forages in their **calf rearing programs,** the practice should be avoided until the calf is weaned. Prior to that time, the underdeveloped rumen may limit nutrient intake if forages are fed. Providing high-quality grain-based diets with a minimum of texture appears to promote ruminal development and allow for optimal growth. After weaning, small amounts of forages may be incorporated into rations to stimulate ruminal growth and minimize the cost of gain. As the calf matures and grows, the reliance on forages will increase. It may be possible to easily meet the nutrient requirements of growing heifers solely from stored forage or pasture or with minimal supplementation to provide vitamins and minerals.

As heifers transition into the lactating herd, they deserve some special attention. Although they are expected to compete with mature cows, they are not mature and, in fact, still require substantial nutrients for growth. They can be considered as "teenage mothers." We expect them to act like adults when they are really not yet fully mature. With a little effort on the part of the dairy producer, they can mature into high-producing cows. Because of their lower milk production, they are less likely to develop metabolic disorders such as ketosis or milk fever; however, because they consume less feed they are more likely to develop an acidic rumen and/or a displaced abomasum. Their smaller body size after calving may limit feed intake per lb of milk produced when compared to herd mates. Care should be taken to insure adequate energy intake of first lactation animals to allow them to milk to their genetic potential, conceive, and regain body stores for use during the next lactation. Often, cows milk well during their first lactation but fail to perform during the second lactation. This is not the fault of the cow but the fault of the manager who failed to meet her nutrient needs during the first lactation and caused her to compensate during the second lactation.

UNIQUE TOPICS IN DAIRY NUTRITION

Although many of the topics that follow have been mentioned previously, they deserve special and individualized attention. This list is by no means complete, but does emphasize the areas where reasonably sound recommendations can be made. The reader is advised to investigate each of these concepts fully and evaluate them for their particular setting or management scheme based on current research findings.

Milk Fever

Milk fever, or parturient paresis, is a disorder of Ca (and to a lesser extent Mg) metabolism that occurs near calving. The increased demands of the mammary gland for Ca to support milk production accompanied by decreased DMI at calving results in dangerously low blood Ca levels that become apparent through lack of muscle control of the cow. Although treatment by oral Ca-drench or intravenous Ca-drip are essentially 100 percent effective, prevention of this disorder is also easy and cost-effective (Kirking, 1992). Historically, feeding a low Ca diet prepartum has been recommended to prevent milk fever. However, this has shown to be impractical on-farm due to a relative lack of low Ca forages. More recent research has shown that feeding an acidogenic diet prepartum is an effective method for reducing the incidence of milk fever (Block, 1984; NRC, 2001). Under this system, feeds that provide an excess of anions to the diet are fed and a metabolic acidosis occurs. This acidosis results in increased Ca excretion through the urine, which in turn activates Ca homeostatic mechanisms and increases Ca absorption from the diet. Early attempts to use this technology were cumbersome due to the low palatability of the feeds available. However, there are currently commerical feeds available that combine mineral acids with protein meals to produce highly palatable feeds that minimize depressions in DMI and allow for a wide range of diets to be used during the dry period while minimizing the impact of milk fever.

Ketosis

Ketosis is a metabolic disorder that occurs when fat stores or dietary fat is metabolized at a rate faster than tissues can oxidize it for energy. Ketosis commonly occurs near parturition due to depressed feed intake. This is accompained by increased energy demands from lactation and causes increased breakdown of body stores. Because DMI is limited, there is a relative lack of precursors for efficient use of the fat that is mobilized, and a buildup of **ketone bodies** occurs. Under most practical production settings, ketosis is probably secondary to another disease or disorder. However, there are some nutritional management practices that can be implemented to decrease the susceptibility of the cow. Cows that are obese at calving are at an increased risk for developing ketosis (Guard, 1995). The optimal body condition score for cows at calving is between 3.5 and 4. Cows that are above this range have a higher concentration of intrahepatic fat stores that results in a decrease in the liver's ability to meet the metabolic demands associated with milk production (Bronicki, 2001). Some feed supplements have been reported to reduce the incidence of ketosis (Besong et al., 2001; Duffield et al., 1998; Jaster and Ward, 1990; Stokes and Goff, 2001) when fed during the transition period. Also, increasing cow comfort while eating may decrease the incidence of ketosis (Guard, 1995), presumably by stimulating increased DMI.

Displaced Abomasum

The exact cause of **displaced abomasum (DA)** is unknown (Rebhun, 1995). In this condition, the abomasum is shifted upward and to the left of its normal position (Fig. 17.8). The displacement may cause a twisting (torsion), interfering with movement of digesta through the gut. There is a marked decrease in feed intake and milk production. Surgical intervention may be necessary. However, some nutritional management practices will decrease the incidence of this disorder. Most nutritionists believe that adequate fiber is necessary for prevention of DA. However, obtaining a working definition of adequate fiber is difficult at best. There are tools marketed that have been designed to assist producers in evaluating rations.

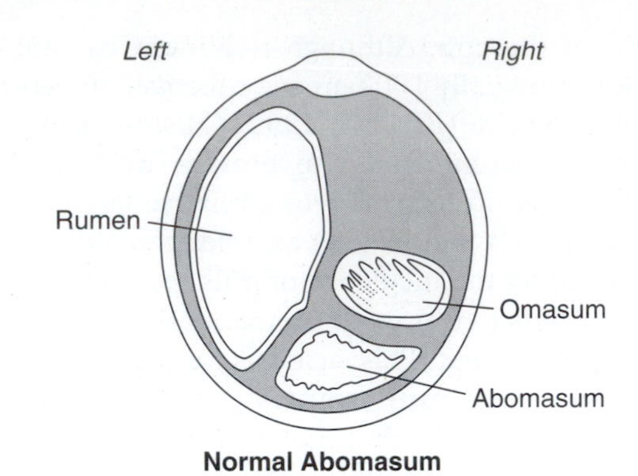

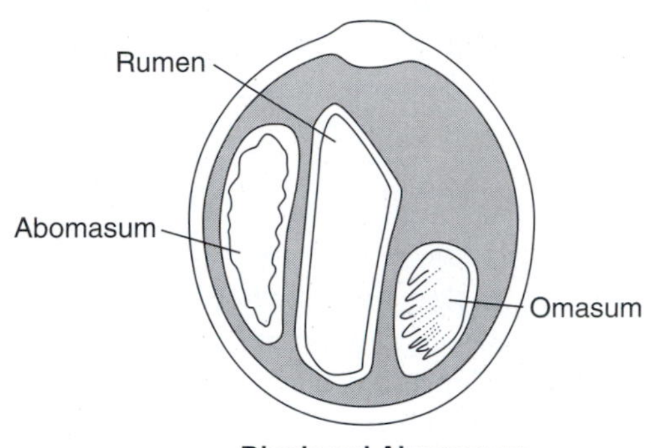

FIGURE 17.8 Normal arrangement of the stomach compartments (top), and displaced abomasum (bottom). The displacement will usually cause twisting of the small intestine, preventing movement of digesta to the intestine.

Readers are encouraged to visit the Web pages of the Penn State University Dairy Cattle Nutrition Group (http://www.das.psu. edu/dcn/index.html) for information about the tool that has become the most accepted for on-farm use. In addition to maintaining adequate fiber, there are other practices that, if implemented, will decrease the incidence of DA. Insuring that feed is readily accessible to cows will encourage intake and thereby decrease the number of DA a herd experiences. This can be accomplished by offering adequate feed to insure that all cows have equal access. In practice this means that a producer must offer enough feed so that 5 to 10 percent of what was offered is removed as unconsumed feed each day. Additionally, antidotal evidence suggests that increased **feeding frequency** may increase intake and thereby decrease the incidence of DA in a herd. However, increasing frequency of feeding may not be profitable or practical in many production settings. In these situations, pushing unconsumed feed closer to the cows has shown to stimulate consumption and may result in the same effect as increased feeding frequency.

Retained Fetal Membranes

Technically, retained fetal membranes (**retained placenta**) are not a nutritional problem. However, if proper nutrition can help prevent this condition it is worthy of discussion. Deficiencies in trace nutrients have been shown to increase the incidence of retained placentas in dairy cattle (LeBlanc et al., 2002). Specifically, deficiencies in vitamin E and Se have been linked to an increase in the number of retained placentas. For many commercial settings a blanket recommendation of supplementing dry cows via injection approximately 21 d prior to calving will alleviate this condition and is cost-effective. Other data suggest that maintaining sound nutritional balance (both energy and protein) during the transition period will reduce the incidence of retained placenta (Drackley, 1999).

Mastitis

Mastitis is an infection of the mammary gland. When individuals are nutritionally stressed the immune system is usually compromised. Therefore, it should be no surprise that good nutrition can help to reduce the incidence of mastitis in any dairy herd. Management of the feed bunk may play a role in decreasing the infection rate of a herd. By insuring that fresh feed is available as cows exit the milking parlor, the producer will encourage them to eat. While cows eat they remain standing. This time allows the teat cistern to fully close prior to the cow lying down and exposing the teat end to mastitis causing organisms. Additionally, nutritional modification may stimulate the immune system and decrease the infection rate. Supplemental vitamin E, Se, Cu, and Zn (Smith et al., 1997; Smith and Hogan, 1988; Spain, 1999) have been reported to reduce the rate of intramammary infections. Recommendations for levels to supplement are not as readily made. Vitamin E is the only one of these nutrients with substantial research data available on which to base recommendations. Current recommendations indicate that providing approximately 500 IU of vitamin E/day may decrease the incidence of mastitis.

Dietary Fat

The final topic that deserves special attention is **dietary fat.** In early lactation, cows cannot physically consume enough feed to meet their energy requirements. Because fats contain 2.5 times the energy as a comparable amount of carbohydrate (grain), adding them to the ration during this period will minimize this deficiency. However, care must be taken since fats are toxic to ruminal microbes and will interfere with fermentation (Jenkins, 1993). Many "rules of thumb" can be found in different recommendations and all are applicable under different circumstances. However, a broad understanding of the scientific principles interacting will allow the nutritionist to better define the optimal inclusion rates for added fats. Unsaturated fatty acids are more toxic than saturated fatty acids (Bateman, II and Jenkins, 1998). Therefore, more fat can be added to rations when it is provided as **saturated fatty acids** (Jenkins, 1993). A free carboxyl group is necessary for the fat to interact with the ruminal microbes (Jenkins, 1993; Jenkins et al., 1996). Therefore, fat sources with the carboxyl group bound to another compound may be added at higher rates than sources with similar fatty acid composition but a free carboxyl group. There are commercial products (**protected fats**) marketed that exploit these concepts and minimize the impact of added fats on ruminal fermentation. The impact of added fats is decreased when dietary fiber is increased (Jenkins et al., 1998; Lewis et al., 1999; Tackett et al., 1996). Jenkins (1997) summarized this information and presented a formula for estimating the maximal amount of unprotected fat to add to rations for optimal conversion of the fat energy into milk energy. This formula recognizes that a minimum amount of fat is always present in the diet and that there is a maximum amount of total fat that needs to be added for optimal efficiency of conversion of the fat energy to milk. The formula uses the ADF content of the diet and the amount of unsaturated fatty acids in the fat source to determine the maximum amount of unprotected fats that should be added to a diet. However, it may not be usable on-farm due to the need for uncommon chemical analysis of the fat sources. For practical purposes, limiting total fat to 8 percent of the diet while minimizing the amount of unprotected fats seems adequate given the current status of our understanding.

QUESTIONS AND STUDY GUIDE

1. Dairy cows in early lactation cannot physically consume adequate feed to meet their energy needs. To meet their needs for energy, they will mobilize body fat. What is the source of the other nutrients (protein, mineral, vitamins) that are required for milk production?
2. During the end of the growing season, your farm started to run short of corn silage. To make sure that you had forage for the entire feeding season (until the next year's crop could be harvested and fed), you decreased the amount of corn silage in the diet to 15 percent of the DM and used soybean hulls to maintain total fiber. Although you maintained dietary NDF at 34 percent of the DM for the entire year, your cows are experiencing an increase in lameness. Could this be nutritionally related? If so, what is the cause?
3. Many managers of high-producing dairy herds have begun feeding the same ration to all lactating cows. Is this a good practice? Why or why not?
4. Calves that do not receive colostrum during the first 12 hours of life tend to have an increased incidence of health problems and an increase in the death rate. What does colostrum provide that helps prevent these problems?
5. How can good nutritional management decrease the amount of health-related problems that are encountered?
6. What is meant by the statement "Dairy heifers are the teenage mothers of the dairy herd"?
7. Many dairy calves have been fed waste milk (mastitic milk or milk from animals that have been given antibiotics) from the lactating herd. Currently, this practice is discouraged. What are the reasons that feeding waste milk to young calves has been discouraged?

REFERENCES

Ammerman, C. B. and S. M. Miller. 1984. Selenium in ruminant nutrition: A review. *J. Anim. Sci.* 58:1561–1577.

Arthington, J. D., M. B. Cattell, J. D. Quigley, III, G. C. McCoy, and W. L. Hurley. 2000. Passive immunoglobin transfer in newborn calves fed colostrum or spray-dried serum protein alone or as a supplement to colostrum of varying quality. *J. Dairy Sci.* 83:2834–2838.

Bateman, H. G., II and T. C. Jenkins. 1998. Influence of soybean oil in high fiber diets fed to nonlactating cows on ruminal unsaturated fatty acids and nutrient digestibility. *J. Dairy Sci.* 81:2451–2458.

Besong, S., J. A. Jackson, D. S. Trammell, and V. Akay. 2001. Influence of supplemental chromium on concentrations of liver triglyceride, blood metabolites and rumen

VFA profile in steers fed a moderately high fat diet. *J. Dairy Sci.* 84:1679–1685.

Block, E. 1984. Manipulating dietary anions and cations for prepartum dairy cows to reduce incidence of milk fever. *J. Dairy Sci.* 67:2939–2948.

Block, S. M. and T. C. Jenkins. 1994. The use of prilled fat to coat and protect amino acids from ruminal degradation. *J. Sci. Food Agric.* 65:441–447.

Blome, R. M., J. K. Drackley, F. K. McKeith, M. F. Hutjens, and G. C. McCoy. 2003. Growth, nutrient utilization, and body composition of dairy calves fed milk replacers containing different amounts of protein. *J. Anim. Sci.* 81:1641–1655.

Bronicki, M. 2001. Pathophysiological conditions of the prophylaxis and treatment of fatty liver in dairy cows. *Medycyna Weterynaryjna* 57:543–546.

Ceballos, M. A. and F. G. Wittwer. 1996. Metabolismo del selenio en ruminantes. *Arch. Med. Vet.* 28:5–18.

Davis, C. L. and J. K. Drackley. 1998. *The Development, Nutrition, and Management of the Young Calf.* Ames: Iowa State University Press.

Drackley, J. K. 1999. Biology of dairy cows during the transition period: The final frontier? *J. Dairy Sci.* 82:2259–2273.

Duffield, T. 2000. Subclinical ketosis in lactating dairy cattle. *Vet. Clin. North Am. Food Anim Pract.* 16:231–253.

Duffield, T. F., D. Sandals, K. E. Leslie, K. Lissemore, B. W. McBride, J. H. Lumsden, P. Dick, and R. Bagg. 1998. Efficacy of monensin for the prevention of subclinical ketosis in lactating dairy cows. *J. Dairy Sci.* 81:2866–2873.

Edmonson, A. J., I. J. Lean, L. D. Weaver, T. Farver, and G. Webster. 1989. A body condition scoring chart for Holstein dairy cows. *J. Dairy Sci.* 72:68–78.

Fitzgerald, T., B. W. Norton, R. Elliott, H. Podlich, and O. L. Svendsen. 2000. The influence of long-term supplementation with biotin on the prevention of lameness in pasture fed dairy cows. *J. Dairy Sci.* 83:338–344.

Foster, L. H. and S. Sumar. 1997. Selenium in health and disease: A review. *Crit. Rev. Food Sci. Nutr.* 37:211–238.

Girard, C. 1998: B-complex vitamins for dairy cows: A new approach. *Can. J. Anim. Sci.* 78:71–90.

Guard, C. 1995. Metabolic diseases: A herd approach. In *Diseases of Dairy Cattle,* W. C. Rebhun, ed, pp. 497–502. Medina, PA: Lippincott Williams & Wilkins.

Jaster, E. H. and N. E. Ward. 1990. Supplemental nicotinic acid or nicotinamide for lactating dairy cows. *J. Dairy Sci.* 73:2880–2887.

Jenkins, T. C. 1993. Lipid metabolism in the rumen. *J. Dairy Sci.* 76:3851–3863.

_______ .C. 1997. Success of fat in dairy rations depends on the amount. *Feedstuffs,* 13 January, 11–12.

Jenkins, T. C., H. G. Bateman, II, and S. M. Block. 1996. Butylsoyamide increases unsaturation of fatty acids in plasma and milk of lactating dairy cows. *J. Dairy Sci.* 79:585–590.

Jenkins, T. C., J. A. Bertrand, and W. C. Bridges, Jr. 1998. Interactions of tallow and hay particle size on yield and composition of milk from lactating Holstein cows. *J. Dairy Sci.* 81:1396–1402.

Kelton, D. F., K. D. Lissemore, and R. E. Martin. 1998. Recommendations for recording and calculating the incidence of selected clinical diseases of dairy cattle. *J. Dairy Sci.* 81:2502–2509.

Kirking, G. A. 1992. Reduce costly milk fevers. *Dairy Herd Workshop.* 22–25.

LeBlanc, S. J., T. F. Duffield, K. E. Leslie, K. G. Bateman, J. TenHag, J. S. Walton, and W. H. Johnson. 2002. The effect of prepartum injection of vitamin E on health in transition dairy cows. *J. Dairy Sci.* 85:1416–1426.

Lewis, W. D., J. A. Bertrand, and T. C. Jenkins. 1999. Interaction of tallow and hay particle size on ruminal parameters. *J. Dairy Sci.* 82:1532–1537.

Linden, J. E. 2001. *The Role of Biotin in Improving the Hoof Condition of Horses.* Basel, Switzerland.

Midla, L. T., K. H. Hoblet, W. P. Weiss, and M. L. Moeschberger. 1998. Supplemental dietary biotin for prevention of lesions associated with aseptic subclinical laminitis (pododermatitis aseptica diffusa) in primaparous cows. *Amer. J. Vet. Res.* 59:733–738.

Miller, W. J. 1970. Zinc nutrition of cattle: A review. *J. Dairy Sci.* 53:1123–1135.

Miller, W. J., H. E. Amos, R. P. Gentry, D. M. Blackmon, R. M. Durrance, C. T. Crowe, A. S. Fielding, and M. W. Neathery. 1989. Long-term feeding of high zinc sulfate diets to lactating and gestating dairy cows. *J. Dairy Sci.* 72:1499–1508.

Morin, D. E., G. C. McCoy, and W. L. Hurley. 1997. Effects of quality, quantity, and timing of colostrum feeding and addition of a dried colostrum supplement on immunoglobulin G1 absorption in Holstein bull calves. *J. Dairy Sci.* 80:747–753.

Morrow, D. A. 1969. Phosphorous deficiency and infertility in dairy heifers. *JAVMA* 154:761–768.

National Research Council. 1989. *Nutrient Requirements of Dairy Cattle,* 6th rev. ed. Washington, DC: National Academy Press.

______ . 2001. *Nutrient Requirements of Dairy Cattle,* 7th rev. ed. Washington, DC: National Academy Press.

Oldfield, J. E. 2002. A brief history of selenium research: From alkali disease to prostrate cancer (from poison to prevention). *J. Anim. Sci.* Online only—accessible at http://www.asas.org/Bios/Oldfieldhist.pdf.

Papas, A. M., C. J. Sniffen, and T. V. Muscato. 1984a. Effectiveness of rumen-protected methionine for delivering methionine postruminally in dairy cows. *J. Dairy Sci.* 67:545–552.

Papas, A. M., J. L. Vicini, J. H. Clark, and S. B. Peirce-Sandner. 1984b. Effect of rumen-protected methionine on plasma free amino acids and production by dairy cows. *J. Nutr.* 114:2221–2227.

Rebhun, W. C. 1995. Abdominal diseases. In *Diseases of Dairy Cattle.* W. C. Rebhun, ed. pp. 106–154. Medina, PA: Lippincott Williams & Wilkins.

Robinson, P. H., W. Chalupa, C. J. Sniffen, W. E. Julien, H. Sato, K. Watanabe, T. Fujieda, and H. Suzuki. 1998. Ruminally protected lysine or lysine and methionine for lactating dairy cows fed a ration designed to meet requirements for microbial and postruminal protein. *J. Dairy Sci.* 81:1364–1373.

Rogers, J. A., S. B. Peirce-Sandner, A. M. Papas, C. E. Polan, C. J. Sniffen, T. V. Muscato, C. R. Staples, and J. H. Clark. 1989. Production responses of dairy cows fed various amounts of rumen protected methionine and lysine. *J. Dairy Sci.* 72:1800–1817.

Schwab, C. G. 1996. Amino acid requirements of the dairy cow: Current status. In *Cornell Nutrition Conference for Feed Manufacturers.* Ithaca, N.Y. pp. 184–197.

Smith, L. and J. S. Hogan. 1988. Vitamin E and selenium can help lower incidence of mastitis. *Large Anim. Vet.* 20–24.

Smith, L., J. S. Hogan, and W. P. Weiss. 1997. Dietary vitamin E and selenium affect mastitis and milk quality. *J. Anim. Sci.* 75:1659–1665.

Spain, J. N. 1999. Mastitis and Nutrition. A better understanding of the big picture. *Agribusiness Dairyman* 18:10–17.

Stokes, S. R. and J. P. Goff. 2001. Case study: Evaluation of calcium propionate and propylene glycol administered into the esophagus of dairy cattle at calving. *Prof. Anim. Sci.* 17:115–122.

Swanson, E. W. 1967. Optimum growth patterns for dairy cattle. *J. Dairy Sci.* 50:244–252.

Tackett, V. L., J. A. Bertrand, T. C. Jenkins, F. E. Pardue, and L. W. Grimes. 1996. Interaction of dietary fat and acid detergent fiber in diets of lactating dairy cows. *J. Dairy Sci.* 79:270–275.

Weiss, W. P. 1998. Requirements of fat-soluble vitamins for dairy cows: A review. *J. Dairy Sci.* 81:2493–2501.

Whitaker, D. A., H. F. Eayres, K. Aitchison, and J. M. Kelly. 1997. No effect of a dietary zinc proteinate on clinical mastitis, infection rate, recovery rate and somatic cell count in dairy cows. *Vet. J.* 153:197–204.

Wichtel, J. J. 1998. A review of selenium deficiency in grazing ruminants. Part 1: New roles for selenium in ruminant metabolism. *New Zealand Vet. J.* 46:47–52.

Winchester, C. F. and M. J. Morris. 1956. Water intake rates of cattle. *J. Anim. Sci.* 15:722–740.

CHAPTER 18

Feeding and Nutrition of Horses

Larry Lawrence and P.R. Cheeke

Objectives

1. To describe the general principles of equine nutrition
2. To describe feeding programs for horse production
3. To describe feeding programs for athletic horses
4. To discuss the causes of important nutritional problems of horses including colic, laminitis, and bone development disorders
5. To briefly describe the nutritional characteristics of other equines such as the donkey

The vast majority of the research published on horse nutrition has appeared in scientific journals in the last 30 years. Nutritional studies on most other livestock species have been published in significant numbers for over 75 years. Although excellent progress has been made in horse nutrition, there is still a great deal of work to be done. Many of the published requirements are extrapolated from studies in other species or developed from the limited research available in horses. Research in horses is complicated by the need to know the nutritional requirements for a variety of athletic endeavors. Exercise changes the nutrient requirements for energy, protein, calcium, phosphorus, magnesium, trace minerals, and vitamins. Requirements for production (i.e., reproduction and growth) must also be fine-tuned.

Production agriculture is certainly important to the horse industry; however, horse production and general livestock production diverge somewhat when the ultimate goals are considered. The goal of most horse owners is to produce horses capable of exceptional athletic performance over an extended period of time. Rapid weight gains in growing animals are known to be counterproductive to the long-term soundness and athleticism of horses. Horse producers also often view least-cost rations and other common livestock management systems with caution. Horses are marketed and treated as individuals throughout much of their lives. The value of individual horses becomes of primary importance. Horses that are shown or raced usually increase in value with additional training, and such horses generally have value as companions. The real or perceived value of horses often leads owners to make management decisions about feed and supplements based on the idea of "protecting their investments." Supplement manufacturers direct marketing concepts toward enhancing performance or protecting the health of the horse. Some of the supplements are useful, some are useless. One important aspect of an education in applied animal sciences is the ability to sort out the advantages and disadvantages of the vast array of nutritional supplements available for horses.

THE DIGESTIVE SYSTEM AND FEEDING MANAGEMENT

Horses are **nonruminant herbivores** and continuous grazers. Management of many horses includes keeping them in stalls or drylots (small enclosures with little or no vegetation) and feeding them two or three times a day. The digestive system of the horse evolved to digest low-nutrient forages in an animal that grazes for 16 to 18 hours per day. When placed in management situations where they must conform to human schedules, horses are often fed higher-energy diets twice a day. The result is reduced exercise and boredom along with digestive disturbances and the development of behavioral vices or abnormalities such as cribbing and wood chewing.

Microbial activities of the hindgut of the horse are similar to ruminal fermentation. The principal microorganisms populating the hindgut of horses are similar to those found in the rumen with some variation in relative percentages. Bacteria, protozoa, and fungi constitute the majority of microbes (Hintz and Cymbaluk, 1994). These microbes are responsive to diet and need to be maintained in a delicate balance. The microbes have the ability to break down the structural carbohydrates cellulose and hemicellulose through fermentation. The products of this fermentation process are volatile fatty acids and methane gas. The **volatile fatty acids** acetate, propionate, and butyrate can provide a significant proportion of the energy needs of the horse. The remaining energy needs must come from digestible nonstructural carbohydrates, protein, and fats in forages or grains.

Changing rapidly from a medium- to low-quality hay diet to either a lush pasture or a high-grain diet will upset the microbial balance of the hindgut. The most serious situation would be a horse that eats large amounts of grain accidently. **Grain overload** can dramatically alter the normal population of gram-negative rods or cocci. Fast-growing spring and fall pastures or pastures containing more than 25 percent clover can also cause soluble and highly fermentable carbohydrate overloads. The overload of nonstructural carbohydrates causes a proliferation of gram-positive bacteria such as *Streptococcus bovis* and *Lactobacillus* spp., often causing **laminitis** (see Hoof Disorders, this chapter). These bacteria produce lactic acid, which can lower the pH of the hindgut. Acid-sensitive gram-negative bacteria are killed at lower pH levels, releasing endotoxins from the cell walls and into the circulatory system. The combination of lactic acid, decreasing pH, and endotoxins contributes to laminitis.

Horses suspected of grain overload need immediate veterinary attention. The laminitis process can take from 18 to 28 hours before obvious signs of the condition are seen. Once in the latter stages, there is little that can be done to stop the cascade of events that eventually leads to the separation of the sensitive and insensitive lamina of the hoof and rotation of the coffin bone.

When adapting horses to pasture, grain, or hay, changes must be made slowly. Horses should be allowed access to pasture for short periods of time while continuing to be fed hay when confined. When changing the diet, intake of any feedstuff should not be altered by more than 25 percent by weight per day. The rule of thumb is to allow a minimum of a week to make changes in feeding programs.

NUTRIENT REQUIREMENTS

The National Research Council (NRC) published minimum requirements in *Nutrient Requirements of Horses* in 1989 (see Table 18–1). In practice, many feed companies adjust these requirements when formulating diets to meet the optimum requirements for growth, reproduction, and work. To take full advantage of the nutritional information available, owners should know the weight and body condition score of their horses and the weight of hay and grain being fed. Knowledge of the estimated contributions of pasture throughout the year is also an important component of understanding the nutrition of horses.

Energy

The energy needs are the top priority of nutritional requirements for most classes of horses. Other nutrients are no less important, but the bulk of the feed offered is to supply energy and that drives ration formulation. Energy requirements for maintenance are fairly low for most horses; however, growth, lactation, and exercise increase energy requirements significantly. Intense exercise such as race training and early lactation can cause energy requirements to increase up to 50 percent over maintenance needs.

Nonstructural carbohydrates (NSC) are mainly digested and absorbed in the small intestine. Pancreatic and intestinal enzymes digest these soluble carbohydrates to varying degrees. One of the factors affecting NSC is the amount of forage in the diet. Hintz et al. (1971) found that when horses were fed a high-alfalfa diet, pre-cecal NSC digestibility was 46 percent. However, when fed a high-grain diet, apparent NSC digestibility values were 71 percent.

Plasma glucose concentrations reflect dietary energy sources. Plasma glucose concentrations are lower in horses than in simple-stomach omnivores but higher than in ruminants (Evans, 1971). Plasma glucose and insulin have received much attention recently. Glucose metabolism is influenced by energy partitioning in the diet. Diets high in NSC increase glycemic indexes when compared to diets high in structural carbohydrates, fat, and/or protein. The **glycemic index** is determined by feeding a set amount of soluble carbohydrate and collecting blood over several hours to measure glucose uptake. Blood glucose is absorbed into muscle and liver tissue and is stored as glycogen. Pancreatic insulin is responsible for clearing glucose from the blood. Insulin is thought to affect additional hormones such as growth hormone, insulin-like growth factors, and other hormones. Current interest in nutrition and endocrinology suggest that different nutrients may affect metabolic pathways not only by enzymatic effects but also by altering hormonal profiles.

Certain metabolic problems associated with high-grain diets such as **tying-up** and **developmental orthopedic disease** (DOD) have led to the creation of **low-glycemic diets** (Valberg et al., 2001; Pagan et al., 2001). These diets contain highly digestible sources of fiber such as beet pulp and soybean hulls (not the outside hull but the seed coating). The energy density of the feed is maintained by adding vegetable oil or high-fat rice bran. The fat is usually included at 10 percent or less of the grain portion of the diet. Unsaturated oils are typically the source of lipids added to horse diets. Soybean oil and corn oil make up the majority of supplemented oil.

TABLE 18–1. Nutrient Concentrations in Total Diets for Horses and Ponies (Dry-Matter Basis)

	Digestible Energy[a]		*Diet Proportions*		*Crude*						*Vitamin A*	
	Mcal/kg	*Mcal/lb*	*Conc., %*	*Hay, %*	*Protein, %*	*Lysine %*	*Ca, %*	*P, %*	*Mg, %*	*K, %*	*IU/kg*	*IU/lb*
Mature Horses												
Maintenance	2.00	0.90	0	100	8.0	0.28	0.24	0.17	0.09	0.30	1830	830
Stallions	2.40	1.10	30	70	9.6	0.34	0.29	0.21	0.11	0.36	2640	1200
Pregnant mares												
9 months	2.25	1.00	20	80	10.0	0.35	0.43	0.32	0.10	0.35	3710	1680
10 months	2.25	1.00	20	80	10.0	0.35	0.43	0.32	0.10	0.36	3650	1660
11 months	2.40	1.10	30	70	10.6	0.37	0.45	0.34	0.11	0.38	3650	1660
Lactating mares												
Foaling to 3 months	2.60	1.20	50	50	13.2	0.46	0.52	0.34	0.10	0.42	2750	1250
3 months to weaning	2.45	1.15	35	65	11.0	0.37	0.36	0.22	0.09	0.33	3020	1370
Working horses												
Light work[b]	2.45	1.15	35	65	9.8	0.35	0.30	0.22	0.11	0.37	2690	1220
Moderate work[c]	2.65	1.20	50	50	10.4	0.37	0.31	0.23	0.12	0.39	2420	1100
Intense work[d]	2.85	1.30	65	35	11.4	0.40	0.35	0.25	0.13	0.43	1950	890
Growing Horses												
Weanling, 4 months	2.90	1.40	70	30	14.5	0.60	0.68	0.38	0.80	0.30	1580	720
Weanling, 6 months												
Moderate growth	2.90	1.40	70	30	14.5	0.61	0.56	0.31	0.08	0.30	1870	850
Rapid growth	2.90	1.40	70	30	14.5	0.61	0.61	0.34	0.08	0.30	1630	740
Yearling, 12 months												
Moderate growth	2.80	1.30	60	40	12.6	0.53	0.43	0.24	0.08	0.30	2160	980
Rapid growth	2.80	1.30	60	40	12.6	0.53	0.45	0.25	0.08	0.30	1920	870
Long yearling, 18 months												
Not in training	2.50	1.15	45	55	11.3	0.48	0.34	0.19	0.08	0.30	2270	1030
In training	2.65	1.20	50	50	12.0	0.50	0.36	0.20	0.09	0.30	1800	820
Two-year-old, 24 months												
Not in training	2.45	1.15	35	65	10.4	0.42	0.31	0.17	0.09	0.30	2640	1200
In training	2.65	1.20	50	50	11.3	0.45	0.34	0.20	0.10	0.32	2040	930

Adapted from National Research Council (1989).

[a] Values assume a concentrate feed containing 3.3 Mcal/kg and hay containing 2.00 Mcal/kg of dry matter.

[b] Examples are horses used in Western and English pleasure, bridle path hack, equitation, etc.

[c] Examples are horses used in ranch work, roping, cutting, barrel racing, jumping, etc.

[d] Examples are race training, polo, etc.

Fatty acids are classified by the position of the first double bond. The essential fatty acids **omega-3** (linolenic) and **omega-6** (linoleic) compete for the same enzymes in their conversion pathways to **eicosapentaenoic acid** (EPA), **docosahexaenoic acid** (DHA), and **arachidonic acid.** In general, the derivatives of omega-3 (EPA and DHA) decrease inflammation, and the derivatives of omega-6 (arachidonic acid) increase inflammation. Both of these processes and many other functions of the derivatives are important to overall health. Work in other species suggests that a ratio of omega-6 to omega-3 fatty acids of 4–10:1 would be the most beneficial (Dunnet, 2002).

Protein

Protein makes up a greater proportion of the horse's body than any other constituent with the exception of water. On a dry, fat-free basis, protein makes up 80 percent of the body composition (Robb et al., 1972). Protein is a major component of muscles, organs, enzymes, and blood. Growth and tissue turnover throughout life represent the greatest need for protein. Age and activity determine the protein requirements. The degree to which a particular protein source meets the requirements for horses is based on the digestibility of the protein, the combination of amino acids included in the protein, and the protein-to-energy ratio of the ration.

Variability in **protein digestibility** of pasture forages depends on the season and stage of growth. The stage of growth at which hay is cut will have an important effect on the protein content of hay. As the forage matures, protein digestibility decreases. In certain classes of horses (e.g., young, growing horses, late-gestation mares, and lactating mares), protein quality is very important. High-quality hay such as alfalfa is a good source of protein, but much of the protein in forages will be digested in the hindgut, and absorption of essential amino acids must take place in the small intestine (Gibbs et al., 1988). Lysine is the most limiting essential amino acid in the diets of horses. When feeding growing horses, protein sources with high levels of lysine must be included in the diet. Soybean meal contains 3.2 percent lysine compared to mature coastal bermudagrass hay with 0.3 percent lysine. Alfalfa hay is a good source of protein with levels often reaching 20 percent. Even the best alfalfa hay, however, has a lysine content of less than 1 percent.

Ott and coworkers (1979) and Meakim and coworkers (1981) demonstrated an important relationship between protein and energy, and for growing horses the relationship has been extended to lysine-to-energy ratios. Although microbial synthesis of protein occurs in the hindgut, the significance of this component of protein digestion is not fully known. It appears the microbial protein synthesized in the hindgut is mainly for the microbes. Nonprotein nitrogen (NPN) can be used as the key ingredient of protein nutrition in ruminants, but it is not efficiently utilized by horses and therefore almost never included in the diets of horses. Further discussion about NPN by nonruminant herbivores is provided in Chapter 4.

Dietary **crude protein requirements** vary from approximately 8 percent of the diet for mature horses to 15 percent for foals. Vigorous exercise or work does not increase dietary protein requirements when expressed as a percentage of dietary protein (NRC, 1989; Miller and Lawrence, 1988) because the increased feed intake compensates for any increase in metabolic protein requirements.

Feeding high-protein diets is generally not a problem for adult horses. However, there are exceptions to this rule. **Endurance horses** that race over great distances are not fed high-protein feeds or alfalfa hay. Feeding excessive protein and calcium may increase frequency of urination and adversely affect fluid balance, as elevated levels of degradation products must be filtered through the kidneys. Fluid balance and dehydration are serious concerns for endurance horses and other horses performing hard work in areas of high heat and humidity.

Minerals

Considerable research on **calcium** and **phosphorus** requirements of horses has been conducted because of the importance of these nutrients in bone growth and skeletal development. Deficient or imbalanced intake of these minerals results in inadequate mineralization of the bone matrix, causing lameness, crooked bones, and enlarged joints (see Bone Development Disorders, this chapter). Excess phosphorus relative to calcium may cause **nutritional secondary hyperparathyroidism** (big head), as discussed in Chapter 2. Calcium insufficiency may also be induced by the presence of dietary oxalates, which impair calcium absorption, leading to signs of hyperparathyroidism. Oxalates in tropical grasses are a common cause of this condition (McKenzie et al., 1981). The phytin phosphorus in grains is of higher availability to horses than to other nonruminants because of phytase activity in the hindgut. Because legumes are much higher in calcium than grasses, the mineral supplement should be reformulated when switching from grass hay to alfalfa, or vice versa. A detailed discussion of calcium and phosphorus metabolism and requirements is provided by NRC (1989).

Sodium chloride **(salt)** is deficient in most of the feedstuffs offered to horses. If, however, horses have continual access to loose salt or a salt block, they will usually consume adequate amounts to meet requirements. Most commercial feeds do not contain adequate salt, so a supplemental source should always be available. Care should be taken when providing salt to a horse of unknown salt status because overconsumption may cause excessive water intake, which can then lead to colic. Salt holders should also be drained, as a brine liquid is toxic to horses.

Trace minerals can be supplied by careful supplementation with commercial products. There are elevated requirements for copper and zinc in broodmares and growing horses.

In some areas, selenium deficiency is a problem. In areas known to have selenium-deficient soils, a mineral mix with **selenium** should be used. In areas with high levels of soil selenium, toxicity may occur. Signs of selenium toxicity in horses include loss of hair from the mane and tail, and hoof deformities, including excessive hoof growth and pronounced growth rings. Because of the interrelationships among trace elements in absorption and metabolism, a marked excess of one mineral can induce deficiency of another. Horse owners should refrain from disrupting a balanced commercial diet by feeding supplements of individual minerals.

Equine hyperkalemic periodic paralysis (HYPP), a genetic disease of horses with Quarter Horse breeding, is characterized by muscle spasms and weakness, recumbancy, and high serum **potassium** concentrations. Affected horses are typically heavily muscled. The disease is caused by a defect that disrupts the sodium ion channel, a tiny gateway in the membrane of muscle cells that regulates sodium uptake. The defect allows sodium to flow freely into cells, which in turn incites the aforementioned symptomatology. Horses diagnosed with HYPP should be kept on a low-potassium diet (Naylor, 1994), such as grain and grass hay. Legume hay (e.g., alfalfa), bran, and protein supplements (e.g., soybean meal, canola meal) have high potassium levels. Molasses has high potassium, whereas beet pulp is low. Heavy potassium (potash) fertilization of pastures should be avoided where HYPP animals are kept.

Vitamins

The vitamin requirements of horses have not been well studied. Most of the data available is on vitamins A, D, and E. These are probably the major vitamins of dietary importance, as the B vitamins and vitamin K are synthesized by gut microbes. Horses grazing good-quality pastures generally do not require vitamin supplementation (NRC, 1989). Mineral and vitamin requirements of horses are summarized in Table 18–2.

TABLE 18–2. Other Minerals and Vitamins for Horses and Ponies (on a Dry-Matter Basis)

	Adequate Concentration in Total Rations				
Nutrient	*Maintenance*	*Pregnant and Lactating Mares*	*Growing Horses*	*Working Horses*	*Maximum Tolerance Levels*
Minerals					
Sodium, %	0.10	0.10	0.10	0.30	3[a]
Sulfur, %	0.15	0.15	0.15	0.15	1.25
Iron, mg/kg	40	50	50	40	1,000
Manganese, mg/kg	40	40	40	40	1,000
Copper, mg/kg	10	10	10	10	800
Zinc, mg/kg	40	40	40	40	500
Selenium, mg/kg	0.1	0.1	0.1	0.1	2.0
Iodine, mg/kg	0.1	0.1	0.1	0.1	5.0
Cobalt, mg/kg	0.1	0.1	0.1	0.1	10
Vitamins					
Vitamin A, IU/kg	2,000	3,000	2,000	2,000	16,000
Vitamin D, IU/kg[b]	300	600	800	300	2,200
Vitamin E, IU/kg	50	80	80	80	1,000
Vitamin K, mg/kg	[c]				
Thiamin, mg/kg	3	3	3	5	3,000
Riboflavin, mg/kg	2	2	2	2	
Niacin, mg/kg					
Pantothenic acid, mg/kg					
Pyridoxine, mg/kg					
Biotin, mg/kg					
Folacin, mg/kg					
Vitamin B_{12}, μg/kg					
Ascorbic acid, mg/kg					
Choline, mg/kg					

Adapted from National Research Council (1989).

[a] As sodium chloride.

[b] Recommendations for horses not exposed to sunlight or to artificial light with an emission spectrum of 280–315 nm.

[c] Blank space indicates that data are insufficient to determine a requirement or maximum tolerable level.

Maenpaa et al. (1988a,b) examined the status of fat-soluble vitamins throughout the year in horses in a northern environment (Finland). They found that serum levels of vitamins A, D, and E decreased markedly during the winter, even when good-quality hay and oats were fed along with a mineral-vitamin supplement. They suggested that during periods of prolonged feeding without access to green forage or pasture, vitamin supplementation should receive special attention. Greiwe-Crandell et al. (1993) found that mares fed hay and a vitamin A-free concentrate depleted their **vitamin A** reserves in 2 months. Supplementation with two times the NRC recommended level was not adequate to completely replete stores of vitamin A in mares without access to pasture. However, mares fed the same diets that had access to green pasture had adequate liver stores regardless of vitamin A supplementation. Horses apparently differ from other animals in **vitamin D** metabolism; the normal serum level of 25–OH vitamin D_3 is very low (Smith and Wright, 1984), although it does increase markedly in vitamin D toxicosis (Harrington and Page, 1983).

Low **vitamin E** levels have been reported in horses in training that are confined to stalls and are fed hay (Hall et al., 1991). Hoffman et al. (1999) reported an increase in serum IgG concentrations (indicator of immune status) in foaling mares when mares were increased from 80 IU vitamin E/kg per day to 160 IU vitamin E/kg per day. The suckling foals of the mares fed the 160 IU vitamin E/kg also had higher serum IgG concentrations. Since the IgG concentrations in foals were not different at birth, the inference is that the milk from mares fed vitamin E contained greater IgG concentrations. The form of vitamin E seems to be very important. Wooden and Papas (1991) reported that the d forms of alpha-tocopherol (vitamin E) were more effective in increasing plasma concentration than the dl form. The dl form is found in most synthetic forms of vitamin E, whereas the d form is found in natural sources. Gansen et al. (1995) reported that natural vitamin E elicited serum vitamin E increases equal to that of three times the amount of synthetic E. Because of its importance to the immune system and muscle integrity, more work needs to be done on the best form of vitamin E to use in supplements.

Equine motor neuron disease (EMD) has signs of generalized muscle weakness, muscle atrophy, and degeneration of nerves in the spinal column and brain stem. Affected animals have very low plasma vitamin E levels (Blythe and Craig, 1992).

Vitamin C is produced in the equine liver. However, older horses and horses under stress may benefit from supplementation (Ralston, 1999). Pasture and high-quality hay can be good sources of **B vitamins,** and the microflora of the hindgut produce B vitamins. Bracken fern and *Equisetum* species (horsetails) have thiaminase activity. Horses consuming these toxic plants may develop an induced thiamin deficiency (see Chapter 7). When horses are stalled and in training, supplementation with B vitamins may be necessary because of dietary changes and decreased efficiency of production and absorption from the hindgut.

Water

Horses should have free access to clean water at all times. The water intake is correlated with dry-matter intake and is predicted quite accurately by ADF intake (Cymbaluk, 1989) using the equation $Y = 3.47 + 6.97\ X$, where Y is water intake in liters/day, and X is ADF intake, kg/day. High-roughage (all hay) diets produce a higher water intake than hay-grain diets. Cymbaluk (1989) measured water intake in trials in which 39 different diets were fed. Horses fed grass or legume hays consumed more water than horses fed complete pelleted feeds, but there were no apparent differences in water intake of animals fed grass or legume hays. Hay-fed animals retained more water in the hindgut, due to the water-holding capacity of fibrous material. Adequate water is essential when high-roughage diets are fed, to prevent intestinal impaction. Horses will usually drink within 20 minutes of being fed a meal of grain or hay. Water temperature is also a factor. When water is at freezing temperatures, intake will be decreased by almost one-half.

FEEDING PROGRAMS FOR HORSES

Maintenance Feeding

The majority of horses in the United States are kept for pleasure purposes and are in light work; most of these horses are at maintenance nutrient requirements. Good-quality pasture in summer (about three acres per horse) when available and good-quality grass or grass-legume hay in winter with a salt and a mineral supplement available will usually meet maintenance requirements. If free-choice roughage is not provided or available, concentrate feeds can be fed along with restricted roughage. Roughage is required to maintain normal digestive tract function and to minimize wood chewing and other behavioral vices.

Reproduction

Broodmares (Fig. 18.1) should be in adequate body condition before breeding, as conception rate is influenced by the condition of the mare. A useful **condition-scoring** system (Henneke et al., 1984) quantifies palpable fat cover over the ribs, vertebrae, pelvis, neck, and shoulder to evaluate horses. The system ranks horses with a score of 1 to 9, where 1 is emaciated and 9 is obese. Broodmares should be maintained between 5 and 7, or in moderate to fleshy condition. The ribs should not be visually discernible but should be able to be felt, the back should be level to having a slight crease, and the neck and withers should blend smoothly. As with other livestock, a flushing response may be seen with animals gaining weight at the time of breeding. A good-quality roughage diet is adequate for the nonworking mare during the first two trimesters. During the last trimester and during lactation, a concentrate supplement should be provided. Each horse's diet should be adjusted according to its individual needs, as assessed by body condition. Obesity in horses caused by overfeeding should be avoided. When Gill et al. (1985) restricted mares to 70 percent of protein requirements during pregnancy and lactation, birth weights were not affected, but growth rate of the foal to 90 days was reduced in the restricted group compared to mares fed at NRC requirements.

Calcium and phosphorus requirements increase in the last trimester. Glade (1993) fed mares at NRC-recommended calcium levels or 20 percent below NRC requirements for the last 12 weeks of gestation and 40 weeks of lactation. Ultrasound evaluation of the cannon bones of mares fed NRC levels of calcium showed increased mineralization. Foals from mares receiving levels of calcium 20 percent

FIGURE 18.1 Lactating mares and young foals have high nutrient requirements. (Courtesy of D. W. Holtan.)

below NRC requirements had thinner midcannon circumferences and weaker bones at birth (Pearce et al, 1998), and the differences continued through 40 weeks.

New Zealand researchers demonstrated that a lowered incidence of DOD (see Bone Development Disorders in this chapter) during the first 150 days of life of foals was dependent on the mare passing adequate **copper** to the foal *in utero.* Copper-supplemented mares (250 mg/day) had foals with a lower incidence of physitis and articular cartilage lesions than foals from mares not receiving supplementation. Pearce and coworkers (1998) reported that direct copper supplementation to the foals did not reduce the incidence of DOD.

Feeding for Growth

The nutritional requirements of foals are adequately met by mare's milk during the first 3 to 4 months of life. Managing growing horses to maintain a steady rate of growth from birth through one year of age is critical for avoiding DOD. The DOD syndrome can be precipitated by periods of excessive growth or weight gain. Pagan et al. (2001) conducted a field study of 218 Thoroughbred weanlings. Twenty-five of the 218 weanlings (11.5%) had OCD (see pg. 458) lesions that were treated surgically. The incidence of OCD was significantly higher in foals that averaged 115 percent of the average weight of all the foals studied. On one farm with no incidence of OCD, foals averaged 97 percent of the average weight. After that time, a **creep ration** should be fed to prepare the foal for weaning at 5 to 6 months of age. After weaning, foals should be fed on a high-concentrate diet and good-quality roughage. Balancing the diet to meet the NRC requirements for energy, protein, calcium, and phosphorus is necessary to ensure good bone development. Expected feed consumption of horses of different productive functions, expressed as a percent of body weight, is given in Table 18–3. Suitable diets for various functions are given in Table 18–4.

TABLE 18–3. Expected Feed Consumption by Horses, % Body Weight*

	Forage	*Concentrate*	*Total*
Mature horses			
Maintenance	1.5–2.0	0–0.5	1.5–2.0
Mares, late gestation	1.0–1.5	0.5–1.0	1.5–2.0
Mares, early lactation	1.0–2.0	1.0–2.0	2.0–3.0
Mares, late lactation	1.0–2.0	0.5–1.5	2.0–2.5
Working horses			
Light work	1.0–2.0	0.5–1.0	1.5–2.5
Moderate work	1.0–2.0	0.75–1.5	1.75–2.5
Intense work	0.75–1.5	1.0–2.0	2.0–3.0
Young horses			
Nursing foal, 3 months	0	1.0–2.0	2.5–3.5
Weanling foal, 6 months	0.5–1.0	1.5–3.0	2.0–3.5
Yearling foal, 12 months	1.0–1.5	1.0–2.0	2.0–3.0
Long yearling, 18 months	1.0–1.5	1.0–1.5	2.0–2.5
Two-year-old, 24 months	1.0–1.5	1.0–1.5	1.75–2.5

Adapted from National Research Council (1989).
*Air-dry feed (approximately 90% DM).

TABLE 18–4. Suitable Diets for Horses

Ingredient	*%*	*Ingredient*	*%*	*Ingredient*	*%*
Foal or Creep Diet		*Weanling Diet*		*Yearling-Adult Diet*	
Oats	44.5	Oats	45.9	Oats	53.0
Corn	18.0	Corn	23.0	Corn	25.3
Soybean meal, 44%	15.0	Soybean meal, 44%	18.5	Soybean meal, 44%	9.5
Dried whey	10.0	Alfalfa meal	5.0	Alfalfa	5.0
Alfalfa meal	5.0	Molasses	5.0	Molasses	5.0
Molasses	5.0	Limestone	0.3	Limestone	0.3
Limestone	0.4	Dicalcium phosphate	1.3	Dicalcium phosphate	0.9
Dicalcium phosphate	1.1	Trace mineral salt	0.5	Trace mineral salt	0.5
Trace mineral salt	0.5	Vitamin premix	0.5	Vitamin premix	0.5
Vitamin premix	0.5				
Kind of horse: foals, 100–450 lbs Daily allowance: 1–4 months: 0.5–0.75 lb grain/100 lb BW 5–6 months: 1.0–1.25 lb grain/100 lb BW Roughage: Good pasture; legume or legume-grass hay		Kind of horse: weanlings, 450–750 lb Daily allowance: 1.0–1.5 lb hay/100 lb BW Kind of hay: legume-grass		Kind of horse and daily allowance: Yearling: 700–1000 lb 0.5–1.0 lb grain and 1.0–1.5 lb hay/100 lb BW Adult: pregnant, lactating, maintenance: feed grain as necessary to maintain condition.	

Adapted from Jurgens (1993).

Feeding for Athletic Performance

Energy is the major nutritional consideration for exercise (Fig. 18.2). Exercise for horses varies from light work to intense, and from intense short duration (racing) to intense long duration (endurance racing). Carbohydrates and fats are the major energy-providing nutrients; protein is not an efficient source of energy for intensive muscle work. For prolonged strenuous exercise, muscle and liver **glycogen** stores are important. Attempts to glycogen-load horses by feeding excess carbohydrates, similar to **carbohydrate-loading** practices of human athletes, are not successful

FIGURE 18.2 Athletic performance in horses ranges from light work (top) to intense short duration activities such as racing (bottom). (Courtesy of D. W. Holtan.)

(Hintz, 1994). Horses have a tremendous capacity for physical work. Genetics and breeding have provided certain breeds and lines of horses with inherent abilities for specific events or disciplines. Horses inherit a certain ratio of two main **types of muscle fibers,** fast-twitch and slow-twitch, that influence the types of performance for which they are most suited. Each type of muscle fiber uses a preferential fuel when performing a specific exercise such as sprinting or endurance racing. Slow-twitch fibers are associated with endurance-type activities, and endurance conditioning can increase the ability to deliver and utilize oxygen and energy-rich fatty acids. Speed training can increase glycogen stores in fast-twitch muscles.

Improving performance by dietary manipulation has been studied. Evidence exists that, under some conditions, a specific energy source may provide some benefit. There is a glycogen-sparing effect when **fatty acid sources** are added to diets. Supplementation with oils and fats has been found to increase the capacity for work during treadmill exercise and at the racetrack (Meyers et al., 1989; Oldham et al., 1990; Harkin et al., 1992).

The use of fatty acids to enhance performance in horses is closely related to the type of exercise being performed. Endurance horses are routinely fed supplemental oils and seem to benefit from their inclusion. The benefits are thought to be a glycogen-sparing effect and a reduction in thermal load (Dunnett, 2002). Cutting horses and Thoroughbreds are also reported to benefit from fat supplementation (Julen et al., 1995; Harkin et al., 1992). The amounts of fat required at specific levels of work need to be determined.

The ability of horses to perform well in athletic events is mainly a function of training, genetics, and mental attitude of the horses. **Ergogenic** aids such as carnitine and carnosine have been effective in humans but have not been proven to enhance athletic ability in horses (Hintz, 1994). Supplementation of horses with sodium bicarbonate to buffer lactic acid has in some cases improved racing times and reduced fatigue. However, it is illegal to use "milkshakes" or bicarbonate buffers on the day of a race. Supplemental antioxidants such as vitamin E, vitamin C and lipoic acid may reduce oxidative stress in horses subjected to severe exercise (Williams, 2004; Williams et al, 2004).

GENERAL NUTRITIONAL PROBLEMS OF HORSES

Horses are subject to a number of feed-related disorders. Some of the more important will be discussed in this section.

Colic Problems

Colic is acute abdominal pain in the horse, caused by distention of the stomach or intestines. The distention may be caused by a physical blockage or gases produced by fermentation. Feed-related factors implicated in colic include excessive consumption of lush green feeds or grain, particularly in animals unaccustomed to these feeds. Changes in feed, particularly in the type of hay fed, should be made slowly (Cohen et al, 1999). Consumption of highly fibrous material (e.g., shavings) that can form a blockage may also cause colic. Symptoms of colic may also occur when hay contains blister beetles. These insects often live on alfalfa and may contaminate alfalfa hay. They contain cantharidin (Spanish fly), an extremely potent irritant of mucous membranes.

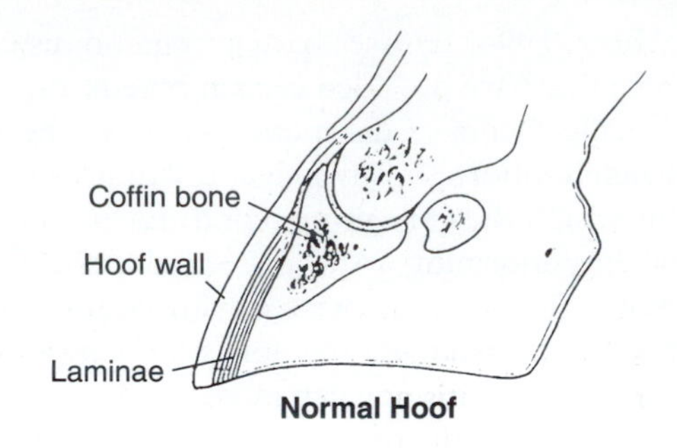

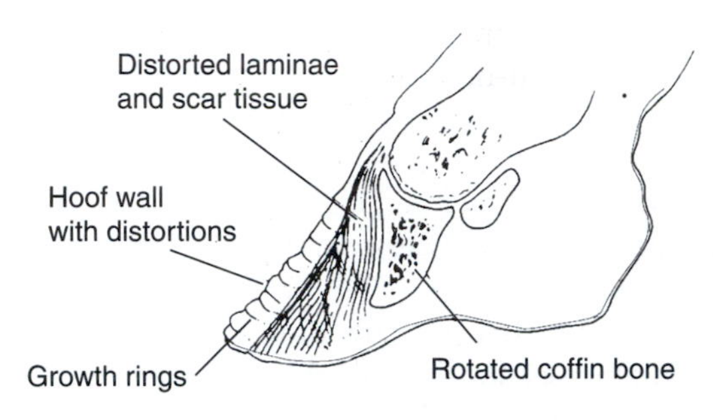

FIGURE 18.3 In the normal hoof, the hoof wall and the laminae are parallel (top). In laminitis, the laminae are distorted, causing rotation of the coffin bone (bottom). Pain, distorted hoof structure, and lameness result.

Hoof Disorders

Founder (also called **laminitis**) is caused by an inflammation of the **laminae** in the hoof and general hoof inflammation. The laminae are leaflike structures that support the coffin bone (Fig. 18.3). When the laminae become inflamed and swell, blood flow through this region of the hoof is impaired. Elevated blood pressure (hypertension) precedes development of symptoms; it results in blood being shunted through a circulatory bypass, denying blood to the hoof. In severe cases, the coffin bone rotates from its normal position. The resulting lameness causes the horse to adopt a "parked out in front" stance to take weight off the front feet.

Laminitis appears to be a result of biochemical changes attributed to lactic acid and bacterial toxins. Common inciting factors are excessive consumption of lush grass (grass founder) or grain (grain founder), and retained placenta with inflammation (metritis). The biochemical causes are not fully understood but seem to involve lactic acid production in the gut due to abnormal fermentation of readily available carbohydrate, acting in association with absorbed bacterial toxins to cause damage to the circulatory system (Garner et al., 1977). The absorbed acid causes a drop in blood pH, electrolyte imbalance, hemoconcentration, and arterial hypertension. In order to study the disorder, Garner et al. (1975) dosed horses orally with cornstarch, causing carbohydrate overload of the hindgut and abnormal fermentation. Horses so treated show elevated blood lactate levels in a few hours and develop all the typical signs of acute laminitis (Garner et al., 1977). Therefore, in grain founder, the probable etiology is that a sudden influx of grain into the digestive tract causes **carbohydrate overload of the hindgut.** This allows proliferation of fast-growing microbes, which produce lactate as their fermentation end product, and encourages growth of coliforms and clostridial species, which produce toxins. The absorbed lactate and microbial toxins cause damage to the circulatory system, particularly to small arterioles in the hoof. Blood circulation to the extremities is impaired and inflammation and damage to the hoof results. Clotting (thrombosis) occurs in the affected blood vessels in the hoof, probably due to activation of platelets by bacterial endotoxins (Weiss et al., 1996). In grass founder, a similar etiology is involved, with the high concentration of soluble carbohydrates and proteins in lush grass causing abnormal fermentation. In laminitis associated with retained placenta, the bacterial toxins would be the primary culprit.

Moore et al. (1981) reviewed the role of bacterial **endotoxins** in founder. Endotoxins are lipopolysaccharide–amino-acid complexes in the cell walls of a number of gram-negative bacteria. The endotoxin is released when the bacteria are lysed in the digestive tract. Elevated blood lactate levels and an acidic cecal pH favor endotoxin absorption, consistent with an interaction between lactate and endotoxin in laminitis. Carbohydrate overload of the equine hindgut causes proliferation of gram-positive bacteria (*S. bovis, Lactobacillus*) that produce lactic acid, a strong acid. This lowers the hindgut pH, killing gram-negative bacteria. As they are lysed, bacteria release cell wall lipopolysaccharides (endotoxins) that, when absorbed, impair circulatory function in the hoof. Absorption of lactic acid, which is normally low from the cecum, is increased by damage to the cecal lining (Kruger et al., 1986).

There are genetic differences in susceptibility to founder; Shetland ponies seem to be particularly susceptible. Hypothyroidism is often suggested as a predisposing factor to laminitis, based on low serum levels of thyroid hormones in affected animals (Hood et al., 1987). However, these authors indicate that the depressed blood hormone levels are a result rather than a cause of the syndrome and that administration of thyroid medication is inappropriate.

Consumption of **black walnut** wood shavings and leaves, and exposure to black walnut pollen, causes laminitis (Galey et al., 1991). The causative agent is probably a derivative of juglone, a toxic component of black walnut (Cheeke, 1998).

Early signs of founder are obviously painful feet, with the animal showing an extreme reluctance to walk. It stands back on its hind feet to remove weight from the sensitive front portion of the hoof. With chronic founder, hoof growth is accelerated, giving rise to elongated hooves with distorted growth rings (Fig. 18.4). The animal walks with its heel landing first, minimizing pain caused by the rotation of the coffin bone. The laminar area suffers permanent damage by formation of scar tissue.

Sudden exposure to lush pasture or grain must be avoided. Animals that have shown a tendency to founder should be kept on dry feed, such as grass hay, and not permitted access to pasture. Rowe et al. (1994) found positive results from oral administration of the antibiotic **virginiamycin** to prevent laminitis induced by grain overload. The antibiotic inhibits the gram-positive bacteria, such as *Streptococcus bovis* and *Lactobacillus* spp., that produce lactic acid. Virginiamycin also reduces formation of amines such as phenylethylamine and isoamylamine, which are vasoactive substances that may play a role in equine laminitis (Bailey et al., 2002). In Australia, the virginiamycin-containing feed additive that Founderguard is used in the nutritional management of horses fed high levels of grain (Rowe et al., 1994).

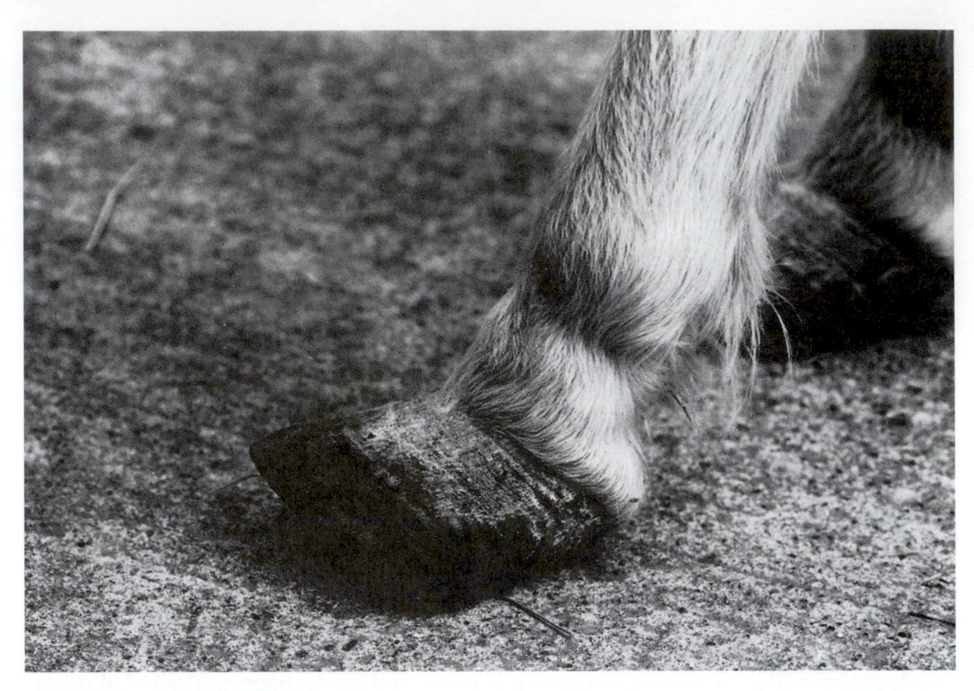

FIGURE 18.4 Hoof of a foundered pony, showing abnormal growth rings and placement of the weight on the heel.

Horses with weak hoof horn, which becomes misshapen and crumbles around the lower parts of the hoof walls, may respond to dietary administration of 10 to 30 mg **biotin** per day (Comben et al., 1984). **Biotin deficiency** in other species, such as swine and poultry, causes foot lesions and dermatitis, supporting a generalized role for this vitamin in formation of keratinized tissue such as the hoof wall.

Cuddeford (1994) reported that replacing half of the grass hay allotment fed to riding school horses with an equal amount of dehydrated alfalfa resulted in favorable effects on hoof growth. Sulfur amino acids, calcium, zinc, other trace elements, and vitamins are nutrients that alfalfa is rich in compared to grass hay; one or more of these nutrients may have had the beneficial effects on hoof growth.

Bone Development Disorders

Bone development disorders are of major concern in young horses. This is particularly true in the thoroughbred racing industry (Fig. 18.5). Horses with abnormal bone development will not reach their racing potential and are susceptible to injury. The term **developmental orthopedic disease** (DOD) is used to describe a complex of bone disorders, including osteochondrosis (OCD), physitis, contracted tendons, cervical vertebral malformation (wobbler syndrome), and similar abnormalities.

Osteochondrosis or OCD is a disease of the growth cartilage at articular or epiphyseal growth regions. It may occur at the growth plate (physis) or in developing the articular cartilage of joints. In the final stages of bone formation, the cartilage cells at the growth plate degenerate, and the region is invaded by osteoprogenitor cells that initiate mineralization of the bone matrix. Disruption of this

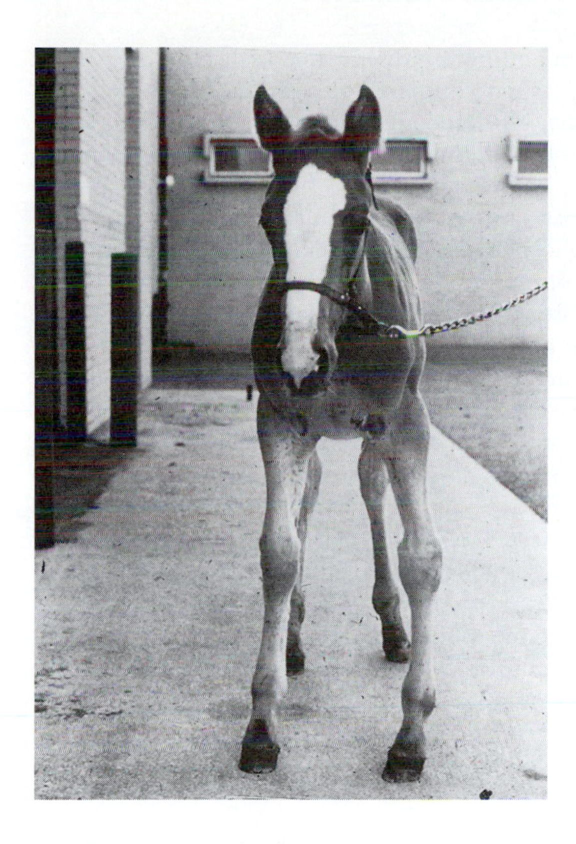

FIGURE 18.5 A foal with angular limb deformity due to nutritional imbalance. (Courtesy of N. F. Cymbaluk.)

process results in thickening of the cartilage, inadequate mineralization, weak joints, fractures, and pain. Extra bone is formed on the sides of the joints, giving a swollen appearance. **Physitis** is another term for this condition, meaning inflammation of the physis (growth plate). Contracted tendons cause a flexure of the leg (knuckling) and may be a secondary effect of osteochondrosis.

Feeding and nutrition influence the DOD syndrome. The most common cause is overfeeding young horses, coupled with lack of exercise. A high-energy intake promotes rapid growth; body weight increase may exceed the rate of normal bone development (Thompson et al., 1988b). Excess energy or protein intake may modify blood hormone levels (e.g., insulin or thyroxin), which in turn may influence bone growth (Glade et al., 1984). Contracted tendons are particularly likely to occur with overfeeding following a period of nutrient restriction. Nutrient imbalance is an important factor in DOD. Protein intake must be adequate because the cartilaginous matrix of bone is proteinaceous. When high-energy diets are used, the protein level must be increased to maintain the optimal protein-to-energy ratio (Thompson et al., 1988b). Savage et al. (1993) fed a control diet that contained digestible energy (DE) at the NRC-recommended level and a diet containing 129 percent of the recommended energy level for foals 2.5 to 6.5 months of age for 16 to 18 weeks. Eleven of the twelve foals on the high-DE diet developed multiple **dyschondroplastic** (DCP) (osteochondrosis) lesions. Only one of the foals on the control diet developed DCP lesions. There were no differences in growth rates between the treatments. The authors suggest there may be an association between high DE and the high incidence of DCP. Schryver et al. (1987) studied the effect of excess dietary protein (20 percent crude protein in the diet) on calcium metabolism and bone growth. The high level of protein had no effect on these parameters when compared to animals fed the recommended protein level, suggesting that NRC recommendations for protein are adequate. Growing horses should not be fed grain diets with more than 16 percent crude protein.

The **calcium-to-phosphorus ratio** is of particular importance in bone growth and development. Excess phosphorus and low calcium is the common situation when animals are overfed on grain and given limited amounts of poor-quality hay. Although excess calcium is well tolerated if adequate phosphorus is present, there is some concern that high calcium intakes may have an adverse effect on bone development (Cymbaluk et al., 1989). The optimal ratio of calcium to phosphorus is 1:1 to 2:1. Because of the high calcium content of alfalfa and the variable, and sometimes low, calcium content of grass and cereal hays (Cymbaluk and Christensen, 1986), alfalfa and grass hays should not be used interchangeably without considering the effect on the balance of calcium and phosphorus.

Adequate trace minerals are necessary for proper bone mineralization of horses. Copper, zinc, iron, manganese, cobalt, and iodine are critical for bone mineralization (Ott and Asquith, 1995).

Thompson et al (1988a) studied the use of creep diets for foals and reported that when a creep diet provided the nutrient levels recommended by the NRC (1978), growth rate and skeletal growth were increased with no detrimental effect on bone quality.

Hyperlipemia

Hyperlipemia means excessive blood lipid. It is a metabolic disease of equines, particularly ponies. It appears similar to ketosis in other species. The condition occurs in animals in negative energy balance, resulting from such factors as pregnancy, lactation, starvation, underfeeding, or transportation. There is a massive mobilization of body fat to provide energy; the blood is opaque or cloudy in appearance because of its high lipid content.

Common clinical signs are dullness and lethargy, progressing to severe depression and coma. Affected animals do not move around much, are anorexic, and rapidly lose condition. The biochemical events have been summarized by Jeffcott and Field (1985). Stress of some type is a precipitating factor, causing stimulation of cortisol secretion into the blood. Elevated plasma cortisol inhibits insulin function in carbohydrate metabolism and lipogenesis. These hormonal changes favor lipid mobilization from adipose tissue, as does inanition from stress. Ponies seem to be somewhat insulin insensitive, which could exacerbate the elevated cortisol condition. A series of metabolic changes occur that cause excessive mobilization of body fat and an increase in the blood lipid concentration. As with other ketotic conditions, therapy involves providing glucose to inhibit fat mobilization and stimulating the animal to eat a highly digestible concentrate feed. Mortality rates of affected animals are high.

Poisonous Plants and Mold Toxins

Poisonous plants are of more concern with horses than with most other livestock for several reasons. With other livestock, the turnover rate is high and the meat animals are marketed at a young age. Horses, however, are often kept until they reach old age. Therefore, chronic toxicities, which might not show up for several years, are of more concern with horses. For example, plants containing pyrrolizidine alkaloids cause irreversible liver damage, which may not result in observable effects for several years, depending upon the amount of toxin consumed. Horses are often kept in overgrazed lots or pastures. Under such conditions, toxic plants may become established because of the lack of competition from forage species and may be consumed by the animals because of boredom or lack of other feed. (As discussed in Chapter 9, most toxic plants are unpalatable because the toxins constitute the plants' chemical defenses against being eaten.) Finally, poisonous plants are of particular concern with horses because of the degree of emotional attachment of the horse owner to the animal. Even though, on a statistical basis, the losses to plant poisonings are low, the death of a particular animal can be devastating to the owner.

Some of the major types of poisonous plants will be briefly described. Further information can be obtained from Cheeke (1998), Kingsbury (1964), and Burrows and Tyrl (2001). A number of plants contain **pyrrolizidine alkaloids.** These include several *Senecio* spp. (e.g., *S. jacobaea,* tansy ragwort; *S. vulgaris,* common groundsel), *Cynoglossum officinale* (hound's tongue), *Crotalaria* spp. (crotalaria, rattle box), and *Symphytum officinale* (comfrey). The alkaloids cause irreversible liver damage, which results in signs of jaundice, poor condition, photosensitization, and in the terminal state, neurological effects (head pressing, or pushing the head into solid objects, walking in a straight line regardless of obstacles). They also cause the liver to accumulate copper (Garrett et al., 1984), which can result in copper toxicity, particularly when horses consume wood treated with copper-containing preservatives (Dewes and Lowe, 1985). Other plants containing alkaloids include the *Astragalus* spp. (locoweeds). Consumption of **locoweeds** may cause neurological effects ("loco" behavior) and birth defects. **Nightshades** (*Solanum* spp.) are common contaminants of alfalfa and oat hay and are common pasture weeds. They cause abdominal irritation and neurological effects. Red clover, when infected with "black patch" fungus, contains a toxic alkaloid called **slaframine** that causes profuse salivation (slobbers) in the horse. **Endophyte-infected tall fescue** contains ergot alkaloids, which cause reproductive problems (prolonged gestation, abortion, thickened placenta, and agalactia) in horses (Cross et al., 1995).

Bracken fern and horsetail (*Equisetum* spp.) contain thiaminase, an enzyme that destroys thiamin (vitamin B_1). These plants are sometimes consumed by horses. A number of plants, including *Senecio* spp., St. John's wort, and buttercups, cause photosensitization. Black locust trees contain lectins in the foliage and bark, causing anorexia and paralysis when consumed by horses. **Black walnut** trees can be toxic to horses. The black walnut contains juglone, a compound that may cause severe allergenic reactions. Outbreaks of toxicity have occurred when black walnut shavings were used as bedding. Signs include depression and laminitis. Other trees that may cause problems are the red maple and Japanese yew. When **red maple** (*Acer rubrum*) leaves are eaten by horses, acute hemolytic anemia may occur, with symptoms including hemoglobinuria (red urine). **Yews** (*Taxus* spp.) are ornamental shrubs and small trees. They contain a number of alkaloids that are cardiac poisons. Only a very small quantity (0.25 to 0.5 lb) of yew leaves can be lethal to a horse. The ornamental shrub **oleander** also contains heart poisons; it is an extremely poisonous plant with just a few leaves being a lethal dose. Finally, **yellow star thistle** and **Russian knapweed** (*Centaurea* spp.) are specifically toxic to horses and not to other livestock. Their consumption causes degeneration of the brain **(equine nigropallidal encephalomalacia).**

Horse owners should learn to identify poisonous plants in their area and take appropriate measures to eliminate them, with herbicides or mechanical removal usually the methods of choice. Horses should not have access to sprayed pastures for 3 to 4 weeks after spraying. Herbicide-treated plants become more palatable as they wilt, due to accumulation of sugars, so there is the potential for animals to consume poisonous plants that they would not normally eat. Also, herbicide residues on sprayed forage should be avoided.

Moldy feeds are undesirable for all livestock, but especially for horses. Mold spores in moldy hay cause dust that can trigger respiratory problems. Many molds produce **mycotoxins** such as aflatoxin, zearalenone, trichothecenes, and so on. Of special concern with horses are **fumonisins,** produced by the mold *Fusarium moniliforme,* which grows particularly on corn. In horses, fumonisins cause **equine leucoencephalomalacia (ELM),** a disorder in which degeneration of the cerebellum occurs. Signs of neurological disturbance occur rapidly after fumonisin-contaminated grain is consumed (Ross et al., 1993; Thiel et al., 1991).

Endophytes in forage grasses produce toxins that can be damaging to horses. As mentioned previously, endophyte-infected tall fescue causes reproductive problems in horses. Endophyte-infected perennial ryegrass contains lolitrem B, a toxin that causes ryegrass staggers (see Chapter 5). Endophyte-infected grasses should never be planted as horse pasture. Mycotoxins in feeds and forages are reviewed comprehensively by Cheeke (1998).

Equine Grass Sickness

Equine grass sickness is a fatal disease of young horses, occurring in England and northern Europe. Intestinal statis occurs and the colon becomes impacted with fecal contents. The etiology of the condition is interesting because it seems to involve a dietary inhibition of gastrointestinal hormones. Regulatory peptides, such as substance P polypeptide and VIP (vasoactive intestinal polypeptide), are released from mucosal cells in the gut and act as hormones. They are neurotransmitters or neuromodulators released at nerve endings of the autonomic nervous system and control the motility of the gut. In equine grass sickness, the peptide-containing nerve ganglia are virtually absent in the distal colon (Bishop et al., 1984). The lack of secretion of regulatory peptides thus results in stasis and impaction of the colon. Bishop et al. (1984) suggest that a dietary neurotoxin may be involved.

Miscellaneous Nutritional Disorders

Woodchewing is sometimes a problem with horses. It often arises from boredom. The main nutritional cause is a diet low in fiber or with fiber of a small particle size. Feeding some coarse roughage is helpful in preventing the condition, instead of feeding only a pelleted diet. Diets inadequate in protein may also result in wood chewing (Schurg et al., 1977).

Protein bumps are hives or welts that appear on the skin. They are caused by allergies to insect bites or chemicals and occasionally to proteins in the feed. Excess dietary protein may be a contributory factor.

Rumensin (see Chapter 8) is a feed additive used for cattle. Horses are unusually sensitive to rumensin and may be poisoned if they have access to rumensin-containing feed or blocks. Signs of toxicity include weakness, staggering, profuse sweating, and head pressing. Liver and kidney damage occur. There is no effective treatment. The minimum lethal intake of rumensin in horses is 1 mg/kg body weight.

NUTRITION OF OTHER EQUINES

Very little work has been done on the nutritional physiology of the **donkey,** including the domestic donkey (*Equus asinus*) and the wild asses (*E. asinus* and *E. hemionus*). These hardy animals can survive in very hostile environments, and donkeys (burros) are still widely used as work and pack animals. According to Schmidt-Nielsen (1964), the donkey has an exceptional tolerance to dehydration and is able to withstand a water loss of 30 percent of its body weight. It also has a large drinking capacity and can consume more than 25 percent of its body weight in water in a few minutes.

Researchers in Israel (Izraely et al., 1989) found that the digestibility of low-quality roughage (wheat straw) was as high in the donkey as in the desert goat. Furthermore, donkeys were able to consume a greater quantity of wheat straw and derive more digestible energy on a straw diet than were goats. Although a direct comparison with horses was not made, Izraely et al. (1989) state that the donkey "is far more efficient in this respect (in digestion of fiber) than the horse." They concluded that the donkey is superior to the horse in its ability to cope with adverse nutritional conditions and is as efficient as such desert ruminants as the Bedouin goat. Similarly, Cuddeford et al. (1995) found that donkeys digested fibrous feedstuffs more efficiently than did horses and ponies, and retained food residues in the hindgut for a longer period. They concluded that donkeys are more efficient at digesting fibrous feeds because they have a higher mean retention time of feed particles in the gut, regardless of intake.

The nutritional needs and digestive efficiencies of wild equids such as zebra and the Przewalski horse have not been studied. Plasma vitamin E levels of free-ranging Przewalski horses in the Ukraine were substantially higher than normal values for domestic horses (Dierenfeld et al., 1997). Nutrient digestibility in ponies is similar to or slightly higher than that in horses (Vermorel et al., 1997).

QUESTIONS AND STUDY GUIDE

1. Are horses more or less susceptible than cattle to urea poisoning? Why?
2. Discuss the value of NPN sources in meeting the protein needs of horses.
3. Explain the importance of the calcium:phosphorus ratio in bone development disorders.
4. Why is water consumption greater with hay than with consumption of an equivalent amount of dry matter from grain? What effect would a diet high in wheat milling by-products have on water consumption?
5. A pleasure horse that is occasionally ridden is kept on a good Kentucky blue-grass pasture containing about 10 percent white clover. Describe a supplementation program for this horse to meet its nutritional needs.
6. Explain why Shetland ponies often founder on grass or clover pasture and have to be kept on dry feed.

7. Name and describe four poisonous plants in your area that horse owners should be able to identify and to which their animals should not have access.
8. Red clover may cause slobbers in horses. What potential use does the alkaloid that causes slobbers have in animal nutrition? (Hint: see Chapter 8.)
9. In what areas of the United States would you anticipate that tall fescue pastures might lead to reproductive problems in horses? (Hint: see Chapter 5.)

REFERENCES

Bailey, S.R., A. Rycroft, and J. Elliott. 2002. Production of amines in equine cecal contents in an in vitro model of carbohydrate overload. *J. Anim. Sci.* 80:2656–2662.

Bishop, A.E., N.P. Hodson, J.H. Major, L. Probert, J. Yeats, G.B. Edwards, J.A. Wright, S.R. Bloom, and J.M. Polak. 1984. The regulatory peptide of the large bowel in equine grass sickness. *Experentia* 40:801–806.

Blythe, L.D. and A.M. Craig. 1992. Equine degenerative myoencephalopathy. 2. Diagnosis and treatment. *Comp. Cont. Edu. Pract. Vet.* 14:1633–1637.

Burrows, G.E. and R.J. Tyrl. 2001. *Toxic Plants of North America.* Ames, IA: Iowa State University Press.

Cheeke, P.R. 1998. *Natural Toxicants in Feeds, Forages, and Poisonous Plants.* Upper Saddle River, NJ: Prentice Hall, Inc.

Cohen, N.D., P.G. Gibbs and A.M. Woods. 1999. Dietary and other management factors associated with colic in horses. *J. Amer. Vet. Med. Assoc.* 215:53–60.

Comben, N., R.J. Clark, and D.J.B. Sutherland. 1984. Clinical observations on the response of equine hoof defects to dietary supplementation with biotin. *Vet. Rec.* 115:642–645.

Cross, D.L., L.M. Redmond, and J.R. Strickland. 1995. Equine fescue toxicosis: Signs and solutions. *J. Anim. Sci.* 73:899–908.

Cuddeford, D. 1994. Artificially dehydrated lucerne for horses. *Vet. Rec.* 135:426–429.

Cuddeford, D., R.A. Pearson, R.F. Archibald, and R.H. Muirhead. 1995. Digestibility and gastro-intestinal transit time of diets containing different proportions of alfalfa and oat straw given to Thoroughbreds, Shetland ponies, Highland ponies and donkeys. *Anim. Sci.* 61:407–417.

Cymbaluk, N.F. 1989. Water balance of horses fed various diets. *Equine Practice* 11:19–24.

Cymbaluk, N.F. and D.A. Christensen. 1986. Nutrient utilization of pelleted and unpelleted forages by ponies. *Can. J. Anim. Sci.* 66:237–244.

Cymbaluk, N.F., G.I. Christison, and D.H. Leach. 1989. Nutrient utilization by limit- and ad libitum-fed growing horses. *J. Anim. Sci.* 67:414–425.

Dewes, H.F. and M.D. Lowe. 1985. Haemolytic crisis associated with ragwort poisoning and rail chewing in two thoroughbred fillies. *NZ Vet. J.* 33:159–160.

Dierenfeld, E.S., P.P. Hoppe, M.H. Woodford, N.P. Krilov, V.V. Klimov, and N.I. Yasinetskaya. 1997. Plasma α-tocopherol, β-carotene, and lipid levels in semi-free-ranging Przewalski horses (*Equus przewalskii*). *J. Zoo Wildlife Med.* 28:144–147.

Dunnett, C.E. 2002. Dietary lipid form and function. In *Proc. 2002 Equine Nutr. Conf.* pp. 1–17. Lexington, KY: Kentucky Equine Research, Inc.

Evans, J.W. 1971. Effect of fasting, gestation, lactation, and exercise on glucose turnover in horses. *J. Anim. Sci.* 33:1001.

Galey, F.D., H.E. Whitley, T.E. Goetz, A.R. Kuenstler, C.A. Davis, and V.R. Beasley. 1991. Black walnut (*Juglans nigra*) toxicosis: A model for equine laminitis. *J. Comp. Path.* 104:313–326.

Gansen, S., A. Linder, and A. Wagener. 1995. Influence of vitamin A supplementation with natural and synthetic vitamin E on serum α-tocopherol content and vitamin A of Thoroughbred horses. *Proc. 1 Equine Nutr. Physiol. Soc. Symp.* 14:68–69.

Garner, H.E., J.R. Coffman, A.W. Hahn, D.P. Hutcheson, and M.E. Tumbleson. 1975.

Equine laminitis of alimentary origin: An experimental model. *Am. J. Vet. Res.* 36:441–444.

Garner, H.E., D.P. Hutcheson, J.R. Coffman, A.W. Hahn, and C. Salem. 1977. Lactic acidosis: A factor associated with equine laminitis. *J. Anim. Sci.* 45:1037–1041.

Garrett, B.J., D.W. Holtan, P.R. Cheeke, J.A. Schmitz, and Q.R. Rogers. 1984. Effects of dietary supplementation with butylated hydroxyanisole, cysteine, and vitamins B on tansy ragwort (*Senecio jacobaea*) toxicosis in ponies. *Am. J. Vet. Res.* 45:459–464.

Gibbs, P.G., G.D. Potter, G.T. Schelling, J.L. Kreider, and C.L. Boyd. 1988. Digestion of hay protein in different segments of the equine digestive tract. *J. Anim. Sci.* 66:400–406.

Gill, R.J., G.D. Potter, J.L. Kreider, G.T. Schelling, W.L. Jenkins, and K.K. Hines. 1985. Nitrogen status and postpartum pH levels of mares fed varying levels of protein. *Proc. Equine Nutr. Physiol. Soc. Symp* 9:84–86.

Glade, M.J. 1993. Equine Osteochondrosis in the 90's. *J. Equine Vet. Sci.* 13:14.

Glade, M.J., S. Gupta, and T.J. Reimers. 1984. Hormonal responses to high and low planes of nutrition in weanling thoroughbreds. *J. Anim. Sci.* 59:658–665.

Griewe-Crandell, K.M., D.S. Kronfeld, G.A. Morrow, and W. Tiego. 1993. Vitamin A depletion in Thoroughbreds: A comparison of pasture and non-pasture feeding regimes. *Proc. Equine Nutr. Physiol. Soc. Symp.* 13:1–2.

Hall, R.R., R.W. Brennan, L.M. Peck, J.P. Lew, and S.E. Duren. 1991. Comparisons of serum vitamin E concentration in yearlings and mature horses. *Proc. Equine Nutr. Physiol. Soc. Symp* 12:263–264.

Harkin, J.D., G.S. Morris, R.T. Tulley, A.G. Nelson, and S.G. Kamerling. 1992. Effect of added dietary fat on racing performance in Thoroughbred horses. *J. Equine Vet. Sci.* 12:123–129.

Harrington, D.D. and E.H. Page. 1983. Acute vitamin D_3 toxicosis in horses: Case reports and experimental studies of the comparative toxicity of vitamins D_2 and D_3. *J. Am. Vet. Med. Assoc.* 182:1358–1369.

Henneke, D.R., G.D. Potter, and J.L. Kreider. 1984. Body condition during pregnancy and lactation and reproductive efficiency of mares. *Theriogenology* 21:897.

Hintz, H.F. 1994. Nutrition and equine performance. *J. Nutr.* 124: 2723S–2729S.

Hintz, H.F. and N.F. Cymbaluk. 1994. Nutrition of the horse. *Ann. Rev. Nutr.* 14:243–267.

Hintz, H.F., E. Hogue, E.F. Walker, Jr., J.E. Lowe, and H.F. Schryver. 1971. Apparent digestion in various segments of the digestive tract of ponies fed diets with varying roughage-grain ratios. *J. Anim. Sci.* 32:245.

Hoffman, R.M., K.L. Morgan, M.P. Lynch, S.A. Zinn, C. Faustman, and P.A. Harris. 1999. Dietary vitamin E supplemented in peripar-turient period influences immunoglobulins in equine colostrum and passive transfer in foals. *Proc. Equine Nutr. and Physiol. Sym.* 16:96–97.

Hood, D.M., D. Hightower, M.S. Amoss, J.D. Williams, S.M. Gremmel, K.A. Stephens, B.A. Stone, and D.A. Grosenbaugh. 1987. Thyroid function in horses affected with laminitis. *Southwest Vet.* 38:85–91.

Izraely, H., I. Choshniak, C.E. Stevens, M.W. Demment, and A. Shkolnik. 1989. Factors determining the digestive efficiency of the domesticated donkey (*Equus asinus asinus*). *Quart. J. Exp. Physiol.* 74:1–6.

Jeffcott, L.B. and J.R. Field. 1985. Current concepts of hyperlipaemia in horses and ponies. *Vet. Rec.* 116:461–466.

Julen, T.R., G.D. Potter, L.W. Greene, G., and G. Stott. 1995. Adaptation to a fat-supplemented diet by cutting horses. *J. Eq. Vet. Sci.* 15:437–440.

Jurgens, M.H. 1993. *Animal Feeding and Nutrition:* Dubuque, IA: Kendall/Hunt.

Kingsbury, J.M. 1964. *Poisonous Plants of the United States and Canada.* Englewood Cliffs, NJ: Prentice Hall, Inc.

Kruger, A.S., D.A. Kinden, H.E. Garner, and R.F. Sprouse. 1986. Ultrastructural study of the equine cecum during onset of laminitis. *Am. J. Vet. Res.* 47:1804–1812.

Maenpaa, P.H., T. Koskinen, and E. Koskinen. 1988a. Serum profiles of vitamins A, E, and D in mares and foals during different seasons. *J. Anim. Sci.* 66:1418–1423.

Maenpaa, P.H., A. Pirhonen, and E. Koskinen. 1988b. Vitamin A, E, and D nutrition in mares and foals during the winter season: Effect of feeding two different vitamin-mineral concentrates. *J. Anim. Sci.* 66:1424–1429.

McKenzie, R.A., R.J.W. Gartner, B.J. Blaney, and R.J. Glanville. 1981. Control of nutritional secondary hyperparathyroidism in grazing horses with calcium and phosphorus supplementation. *Aust. Vet. J.* 57:554–557.

Meakim, D.W., H.F. Hintz, H.F. Schryver, and J.E. Lowe. 1981. The effect of dietary protein on calcium metabolism and growth of the weanling foal. In *Proc. Cornell Nutr. Conf.* p. 95. Ithaca, NY: Cornell University Press.

Meyers, M.C., G.D. Potter, J.W. Evans, L.W. Greene, and S.F. Crouse. 1989. Physiologic and metabolic response of exercising horses to added dietary fat. *J. Equine Vet. Sci.* 9:218–223.

Miller, P.A. and L.M. Lawrence. 1988. The effect of dietary protein level on exercising horses. *J. Anim. Sci.* 66:2185–2192.

Moore, J.N., H.E. Garner, J.E. Shapland, and R.G. Schaub. 1981. Equine endotoxemia: An insight into cause and treatment. *J. Am. Vet. Med. Assoc.* 79:473–477.

National Research Council. 1989 and 1978, *Nutrient Requirements of Horses.* Washington, DC: National Academy Press.

Naylor, J.M. 1994. Equine hyperkalemic periodic paralysis: Review and implications. *Can. Vet. J.* 35:279–285.

Oldham, S.L., G.D. Potter, J.W. Evans, S.B. Smith, T.S. Taylor, and W.S. Barns. 1990. Storage and mobilization of muscle glycogen in exercising horses fed a fat supplemented diet. *J. Equine Vet. Sci.* 10:353–359.

Ott, E.A. and R.L. Asquith. 1995. Trace mineral supplementation of yearling horses. *J. Anim. Sci.* 73:466–471.

Ott, E.A., R.L. Asquith, J.P. Feaster, and F.G. Martin. 1979. Influence of protein level and quality on the growth and development of yearling foals. *J. Anim. Sci.* 49:620–628.

Pearce, S.G., E.C. Firth, N.D. Grace, and P.F. Fennessy. 1998. Effect of copper supplementation on the evidence of developmental orthopedic desease in pasture-fed New Zealand Thouroughbreds. *Equine Vet. J.* 30:211–218.

Pagan, J.D., R.J. Geor, S.E. Caddel, P.B. Pryor, and K.E. Hoekstra. 2001. The relationship between glycemic response and the incidence of OCD in Thoroughbred weanlings: A field study. *Proc. Amer. Assoc. Equine Pract.* 47:322–325.

Ralston, S.L. 1999. Management of geriatric horses. In *Advances in Equine Nutrition II,* J.D. Pagan and R.J. Geor, eds. pp. 393–396. Nottingham, United Kingdom: Nottingham University Press.

Robb, J., R.B. Harper, H.F. Hintz, J.E. Lowe, J.T. Reid, and H.F. Schryver. 1972. Body composition of the horse: Interrelationships among proximate components, energy content, liver and kidney size, and body size and energy value of protein and fat. *Anim. Prod.* 14:25.

Ross, P.F., A.E. Ledet, D.L. Owens, L.G. Rice, H.A. Nelson, G.D. Osweiler, and T.M. Wilson. 1993. Experimental equine leukoencephalomalacia, toxic hepatosis, and encephalopathy caused by corn naturally contaminated with fumonisins. *J. Vet. Diag. Invest.* 5:69–74.

Rowe, J.B., M.J. Lees, and D.W. Pethick. 1994. Prevention of acidosis and laminitis associated with grain feeding in horses. *J. Nutr.* 124: 2742S–2744S.

Savage, C.J., R.N. McCarty, and L.B. Jeffcott. 1993. Effects of dietary energy and protein on induction of dyschondroplasis in foals. *Equine Vet. J. Suppl.* 16:74–79.

Schmidt-Nielsen, K. 1964. *Desert Animals.* London: Oxford University Press.

Schryver, H.F., D.W. Meakim, J.E. Lowe, J. Williams, L.V. Soderholm, and H.F. Hintz. 1987. Growth and calcium metabolism in horses fed varying levels of protein. *Equine Vet. J.* 19:280–287.

Schurg, W.A., D.L. Frei, P.R. Cheeke, and D.W. Holtan. 1977. Utilization of whole corn plant pellets by horses and rabbits. *J. Anim. Sci.* 45:1317–1321.

Smith, B.S.W. and H. Wright. 1984. 25-hyroxy-vitamin D concentrations in equine serum. *Vet. Rec.* 115:579.

Thiel, P.G., G.S. Shephard, E.W. Sydenham, W.F.O. Marasas, P.E. Nelson, and T.M. Wilson. 1991. Levels of fumonisins B_1 and B_2 in feeds associated with confirmed cases of equine leukoencephalomalacia. *J. Agr. Food Chem.* 39:109–111.

Thompson, K.N., J.P. Baker, and S.G. Jackson. 1988a. The influence of supplemental feed on growth and bone development of nursing foals. *J. Anim. Sci.* 66:1692–1696.

Thompson, K.N., S.G. Jackson, and J.P. Baker. 1988b. The influence of high planes of nutrition on skeletal growth and development of weanling horses. *J. Anim. Sci.* 66:2459–2467.

Valberg, S.J., R.J. Goer, and J.D. Pagan. 2001. Muscle disorders: Untying the knots through nutrition. In *Proc. 2001 Equine Nutr. Conf. Feed Manufacturers.* pp. 115–125. Lexington, KY: Kentucky Equine Research Inc.

Vermorel, M. and W. Martin-Rosset. 1997. Concepts, scientific bases, structure and validation of the French horse net energy system (UFC). *Livestock Prod. Sci.* 47:261–275.

Weiss, D.J., R.J. Geor, G. Johnston, and A.M. Trent. 1994. Microvascular thrombosis associated with the onset of acute laminitis in ponies. *Am. J. Vet. Res.* 55:606–612.

Weiss, D.J., L. Monreal, A.M. Angles, and J. Monasterio. 1996. Evaluation of thrombin-antithrombin complexes and fibrin fragment D in carbohydrate-induced acute laminitis. *Res. Vet. Sci.* 61:157–159.

Williams, C.A. 2004. Studies show supplementation with antioxidants may reduce oxidative stress in the exercising horse. *Feedstuffs 76*(13): 11–14. March 29, 2004.

Williams, C.A., D.S. Kronfeld, T.M. Hess, K.E. Saker, J.N. Waldron, K.M. Crandell, R. M. Hoffman, and P.A. Harris. 2004. Antioxidant supplementation and subsequent oxidative stress of horses during an 80-km endurance race. *J. Animal Sci.* 82:588–594.

Wooden, G.R. and A.M. Papas. 1991. Utilization of various forms of vitamin E by horses. *Proc. Equine Nutr. Physiol. Soc. Symp.* 12:265.

CHAPTER 19

Feeding and Nutrition of Rabbits

Objectives

1. To describe the general principles of rabbit nutrition and the role of the enlarged hindgut in digestive processes
2. To describe general nutritional problems of rabbits, such as enteritis, fur chewing, and reproductive disorders

Rabbits are raised for a variety of purposes, including for meat, fur, and wool, and for use as laboratory animals. Large numbers of rabbits are raised for exhibition and showing and many simply as pets. Production of rabbit feed is a minor but significant component of the manufactured feed industry. Rabbit production may have considerable potential in developing countries as a means of converting forages and agricultural by-products to meat (Cheeke, 1986). Detailed reviews of rabbit feeding and nutrition have been given by Cheeke (1987) and deBlas and Wiseman (1998).

Commercial rabbit producers are much concerned with diet cost, whereas raisers of exhibition rabbits are less concerned with cost than with the effects of the diet on animal health and show quality of the animals.

A number of generalizations can be made about rabbit feeds. They should be pelleted because nonpelleted feeds are not well accepted by rabbits and give low growth rates and high feed waste. Perhaps more than with other livestock, the feed has a major impact on the health of the animals. Rabbits are quite susceptible to enteric diseases (enteritis, diarrhea) that have a major dietary component. They are very sensitive to palatability factors and often refuse to consume a batch of feed even though it has the same ingredient specifications as the previous batch that they readily consumed. Per unit of feed produced, feed mills probably have more complaints about rabbit feed than about any other feed.

NUTRIENT REQUIREMENTS

The nutrient requirements of rabbits are influenced by their digestive tract physiology. Rabbits have microbial fermentation in the cecum and consume the cecal contents **(cecotrophy).** Cecotrophy usually occurs once or twice per 24-hour period, generally at night; hence the common name "night feces" for the cecotropes. Consumption of cecotropes provides a source of microbial protein as well as an adequate supply of all of the B vitamins. A summary of nutrient requirements of the rabbit is given in Table 19–1. These requirement figures are provisional; in many cases, they are not well supported by data. Rabbit nutrition research has been quite limited as compared to that of most other domestic species.

Energy levels in typical rabbit diets are quite low, usually being in the range of 2,400 to 2,800 kcal DE/kg diet. **Higher energy diets** tend to promote microbial overgrowth in the cecum and lead to enteric disease (diarrhea). Therefore,

TABLE 19–1. Nutrient Requirements of Meat Rabbits

Nutrient[a]	*Unit*	*Breeding Does*	*Fattening Rabbits*	*Mixed Feed*
Digestible energy	MJ	11.1	10.5	10.5
Metabolizable energy	MJ	10.6	10.0	10.0
NDF[b]	g	31.5 (30.0–34.0)[c]	33.5 (32.0–35.0)	33.0 (32.0–34.0)
ADF	g	16.5 (15.0–18.0)	17.5 (16.0–18.5)	17.0 (16.0–18.0)
Crude fibre	g	13.5 (12.5–14.5)	14.5 (13.5–15.0)	14.0 (13.5–14.5)
ADL	g	5.0	5.5	5.5
Starch	g	18.0 (15.0–21.0)	16.0 (14.5–17.5)	16.0 (15.0–17.0)
Ether extract	g	5.5	Free	Free
Crude protein	g	18.4 (16.3–19.8)	15.3 (14.5–16.2)	15.9 (15.4–16.2)
Digestible protein	g	12.9 (11.4–13.9)	10.7 (10.2–11.3)	11.1 (10.8–11.3)
Lysine				
Total	g	8.4	7.5	8.0
Digestible	g	6.6	5.9	6.3
Sulphur[d]				
Total	g	6.5	5.4	6.0
Digestible	g	5.0	4.1	4.6
Threonine				
Total	g	7.0	6.4	6.8
Digestible	g	4.8	4.4	4.7
Calcium	g	11.5	6.0	11.5
Phosphorus	g	6.0	4.0	6.0
Sodium	g	2.2	2.2	2.2
Chloride	g	2.8	2.8	2.8
Cobalt	mg	0.3	0.3	0.3
Copper	mg	10	6	10
Iron	mg	50	30	50
Iodine	mg	1.1	0.4	1.1
Manganese	mg	15	8	15
Selenium	mg	0.05	0.05	0.05
Zinc	mg	60	35	60
Vitamin A	mIU	10	6	10
Vitamin D	mIU	0.9	0.9	0.9
Vitamin E	IU	50	15	50
Vitamin K_3	mg	2	1	2
Vitamin B_1	mg	1	0.8	1
Vitamin B_2	mg	5	3	5
Vitamin B_6	mg	1.5	0.5	1.5
Vitamin B_{12}	mg	12	9	12
Folic acid	mg	1.5	0.1	1.5
Niacin	mg	35	35	35
Pantothenic acid	mg	15	8	15
Biotin	mg	100	10	100
Choline	mg	200	100	200

(Adapted from deBlas and Wiseman, 1998.)

[a] Amount per kg of diet (90% DM basis).

[b] Long fiber particles (>0.3 mm) higher than 25%.

[c] Range of maximum and minimum values.

[d] Methionine should provide 35% of TSAA.

rabbit diets usually contain alfalfa meal or other fibrous feedstuffs as the main ingredient. As discussed in Chapter 1, rabbits do not digest fiber efficiently. Their digestive strategy is to eliminate fiber rapidly and retain the nonfiber components in the cecum for fermentation, followed by cecotrophy.

The amount and type of **dietary fiber** is a major consideration in rabbit nutrition. Indigestible fiber (cellulose + lignin, i.e., the ADF fraction) has important roles in maintaining adequate gut motility and preventing enteritis. Digestible fiber (hemicellulose + pectins, i.e., NDF-ADF) functions in providing energy and promoting optimal populations of cecal microbes (Gidenne and Bellier, 2000). Digestible fiber from sources such as sugar beet pulp and wheat milling by-products (e.g., wheat bran, wheat mill run) can be used in place of starch sources (cereal grains) to maintain adequate intake of digestible energy while preventing enteritis. Optimal dietary fiber requirements are 15–20 percent ADF, 14–18 percent crude fiber, and a digestible fiber:ADF ratio of 1.3 (Gidenne and Bellier, 2000).

Protein requirements are 16 percent for maximum growth and 18 percent for lactation. **Dietary protein quality** is important, although microbial protein from cecal fermentation does make a significant contribution. Cecal fermentation and cecotrophy result in the ability of the rabbit to use some NPN, but in most cases, the ingredients used (alfalfa meal, wheat middlings, etc.) contain adequate total nitrogen. Dietary protein quality is particularly important for rapidly growing weanling animals.

The rabbit is very efficient in absorption of **calcium.** Excess calcium is excreted in the urine rather than by the fecal route as is typical of most species. Rabbit urine often has a cloudy appearance due to its high content of calcium carbonate. Diets with high (40 to 60 percent) levels of alfalfa meal usually contain excess calcium, which is excreted in the urine.

Both deficiency and toxicity of **vitamin A** cause reproductive problems, including resorbed fetuses, abortion, **hydrocephalus** (enlarged head containing fluid), and small, weak kits at birth (Fig. 19.1). Deficient and toxic dietary vitamin A levels have not been well defined but are in the region of 5,000 and 70,000 IU/kg diet, respectively.

Rabbit diets are typically quite high in alfalfa, which is a good source of β-carotene. Animals convert **β-carotene** to vitamin A, but do not form a toxic level of vitamin A. However, when synthetic vitamin A is added to a diet rich in carotene, a relatively low level of synthetic vitamin A may be enough to induce a toxicity state. Thus, addition of retinyl acetate to commercial rabbit diets can cause severe **reproductive problems** of fetal resorptions, abortions, stillbirths, and weak, nonviable young (DiGiacomo et al., 1992).

Rabbits have very low **vitamin D** requirements (Cheeke, 1987). Excess vitamin D can cause massive mineralization of the soft tissues such as arteries, kidneys, and lungs (Zimmerman et al., 1990).

GENERAL NUTRITIONAL PROBLEMS OF RABBITS

Enteritis

One of the major problems in commercial rabbit production is enteritis. Most cases of enteritis are **enterotoxemia,** meaning that such cases are caused by bacterial toxins elaborated in the gut. The major organisms involved are *Clostridium*

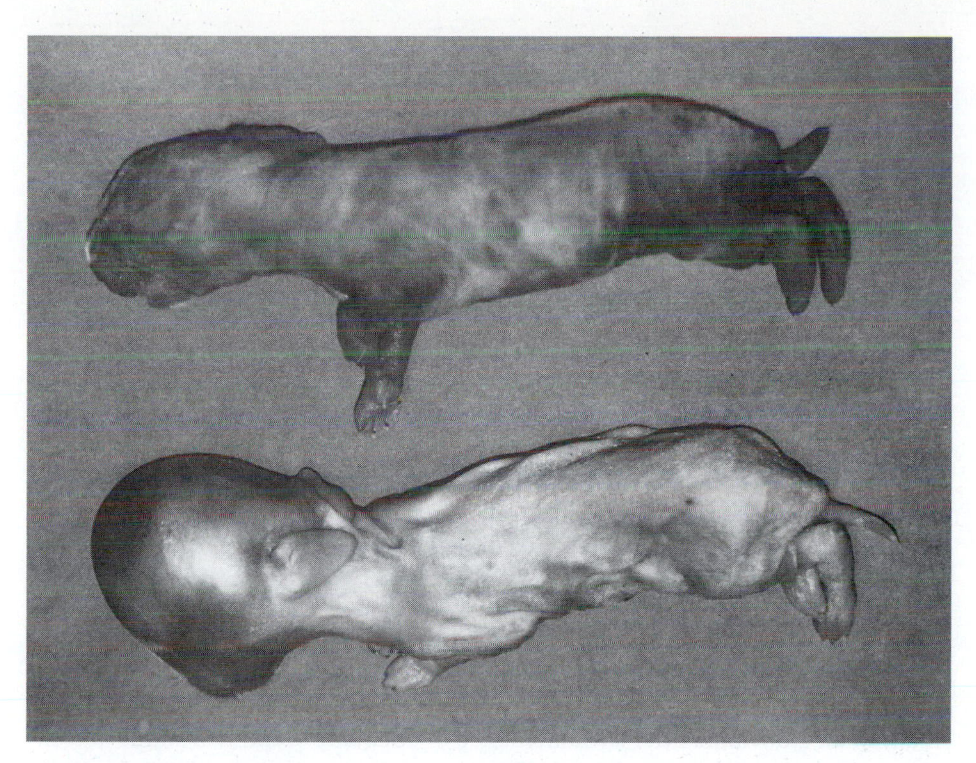

FIGURE 19.1 Hydrocephalus (fluid on the brain) in a newborn rabbit from a doe fed a vitamin A-deficient diet (bottom) compared to a normal kit (top).

spiroforme and toxigenic strains of *E. coli.* Enterotoxemia is of particular importance in weanling rabbits. At weaning, the gut of the young rabbit is susceptible to microbial invasion because the stomach pH (approximately 5) has not dropped to the very acid (pH 1.5 to 2) bacteriocidal level typical of the adult rabbit. If the diet is high in energy (from starch or other readily fermentable carbohydrate), **carbohydrate overload** of the hindgut may occur, with a proliferation of microbes in the cecum. The high quantity of VFA causes a temporary drop in cecal pH, killing many of the normal microbes and allowing proliferation of pathogens such as *C. spiroforme* and *E. coli,* which produce lethal toxins.

Postweaning enteritis can be minimized by feeding a low-energy, high-fiber diet. The crude-fiber content should be 13 to 15 percent, and the diet should contain less than 20 percent cereal grain. **Grain milling by-products,** such as wheat middlings, are an excellent substitute for grain in rabbit diets. The use of high-fiber, low-starch diets reduces the likelihood of carbohydrate overload of the hindgut, and the fiber helps maintain normal gut motility. **Hypomotility,** or lack of normal gut contractions, with low-fiber diets is a contributing factor to enteric problems. Adequate dietary fiber also helps to prevent cecal impaction **(mucoid enteritis).** In this condition, the cecum becomes impacted with fine particulate matter. The goblet cells of the colon mucosa secrete increased quantities of mucus, resulting in large amounts of mucus being excreted with the feces. These and other enteric diseases in rabbits have been reviewed in detail by Cheeke (1987), Lelkes and Chang (1987), and Gidenne (1997).

Other Nutritional Problems

Fur chewing is a common problem in rabbits; animals pull and consume their own hair or that of other animals in the cage. Besides ruining the pelt of market animals, fur chewing leads to **hairballs** in the stomach. Animals with hairballs stop eating and eventually die of inanition. Fur chewing seems to occur mainly when the diet is inadequate in indigestible fiber. Increasing the crude-fiber content of the diet, or providing a source of fiber such as hay or straw to the animals, generally eliminates the problem. Oral administration of a source of proteolytic enzymes, such as bromelain, papain, or raw pineapple juice, results in digestion of the mucus holding the hairball together and allows it to pass through the pyloric sphincter.

Reproduction in rabbits is influenced by nutrition. Inadequate energy intake may result in resorption of the fetuses, or small, weak litters at birth. Excessively fat does will similarly have small, weak litters and poor milking ability. One of the major causes of poor reproduction during the winter is inadequate energy intake of does, often a result of limit-feeding the same amount of feed as provided in the summer. In cold-climate areas, feed intake must be increased during the winter to meet the animals' higher energy requirements. Another common source of reproductive problems is **vitamin A malnutrition.** As mentioned previously, deficient (less than 5,000 IU/kg of diet) or toxic (more than 70,000 IU/kg) levels of vitamin A can result in fetal resorption, abortion, fetal hydrocephalus and small, weak, nonviable litters. Maintenance of adequate vitamin A status is complicated by variability in quality of alfalfa meal, which is usually the main ingredient of rabbit diets.

A common problem with pet rabbits is **urolithiasis,** or urinary and bladder calculi. Rabbits absorb calcium very efficiently and excrete the excess in the urine. When high dietary calcium levels are fed experimentally, the urine can be so loaded with calcium carbonate that it appears as a paste. Rabbit diets are typically high in alfalfa, which is rich in calcium, and usually are supplemented with additional calcium as well. **Pet rabbits** are usually mature, nonreproducing animals kept under maintenance conditions. Typically, they consume much more calcium than needed, and with prolonged excretion of the excess in the urine, urinary tract calculi almost inevitably develop and may require surgical removal. If possible, low calcium (0.5%) diets should be fed to pet rabbits. Rabbit urine is often orange or red in color, particularly in animals excreting large amounts of calcium. The pigmentation is associated with dietary tannins and the animal's own bile pigments, and is accentuated at a high urine pH (Cheeke, 1987).

While there are problems with excess calcium with pet rabbits, there may also be deficiency problems. Harcourt-Brown (1995, 1996) reported that rabbits fed mixtures of unpelleted ingredients such as flaked corn, peas, oats, dried fruit, and so on may select low calcium ingredients and develop calcium deficiency, which is manifested by poor calcification of the skull and teeth. Deformed teeth with little or no enamel may occur. When unpelleted feed mixtures are used, such as those sold in pet stores, care should be taken that all the ingredients are consumed, especially commercial pellets that are usually added to such mixtures. These pellets will contain mineral and vitamin supplements. If the animals specifically select out only the rolled grains, for example, they will then have a severely calcium-deficient diet.

TABLE 19–2. Examples of Satisfactory Diets for Rabbits*

Ingredient	*(%)*	*Ingredient*	*(%)*	*Ingredient*	*(%)*
Diet 1		*Diet 2*		*Diet 3*	
Alfalfa meal	54	Alfalfa meal	40	Alfalfa meal	40
Wheat mill run	36.5	Ground oats	27.5	Ground barley	25
Soybean meal	6	Ground barley	20	Wheat mill run	23.5
Molasses	3	Soybean meal	9	Cottonseed meal	8
Salt (trace mineralized)	0.5	Molasses	3	Molasses	3
		Salt (trace mineralized)	0.5	Salt (trace mineralized)	0.5

*If good-quality, bright-green alfalfa meal is used, a vitamin supplement is unnecessary. If the alfalfa quality is questionable, it is advisable to add a vitamin premix (a swine or poultry vitamin premix is satisfactory).

DIETS FOR RABBITS

Diets for rabbits are based upon low-energy, high-fiber ingredients such as alfalfa meal and wheat milling by-products. Inclusion of grains increases the incidence of enteritis losses and often reduces animal performance. The digestive tract of rabbits is poorly adapted to the use of high-energy diets.

Examples of satisfactory diet formulae are given in Table 19–2.

QUESTIONS AND STUDY GUIDE

1. The digestibility of crude fiber is lower in rabbits than in pigs and rats. This fact seems surprising because rabbits are herbivorous animals and eat fibrous feeds. Explain.
2. Explain how rabbits might derive nutritional benefit from dietary sources of NPN such as urea.
3. A rabbit producer notices that the color of the feed purchased from the local feed company has changed from a bright green color to brown. After a month, there are a lot of small, weak litters born, with many of the kits having enlarged heads. Explain.
4. Discuss the relationship between diet and digestive tract disorders in rabbits.

REFERENCES

Cheeke, P.R. 1986. Potentials of rabbit production in tropical and subtropical agricultural systems. *J. Anim. Sci.* 63:1581–1586.

_______ . 1987. *Rabbit Feeding and Nutrition.* San Diego: Academic Press.

deBlas, C. and J. Wiseman, eds. 1998. *The Nutrition of the Rabbit.* Wallingford, U.K.: CAB International.

DiGiacomo, R.F., B.J. Deeb, and R.J. Anderson. 1992. Hypervitaminosis A and reproductive disorders in rabbits. *Lab. Anim. Sci.* 42:250–254.

Gidenne, T. 1997. Caeco-colic digestion in the growing rabbit: Impact of nutritional factors and disturbances. *Livestock Prod. Sci.* 51:73–88.

Gidenne, T. and R. Bellier. 2000. Use of digestible fibre in replacement to available carbohydrates. Effects on digestion, rate of passage and caecal fermentation pattern during the growth of the rabbit. *Livestock Prod. Sci.* 63:141–152.

Harcourt-Brown, F.M. 1995. A review of clinical conditions in pet rabbits associated with their teeth. *Vet. Rec.* 137:341–346.

_______ . 1996. Calcium deficiency, diet and dental disease in pet rabbits. *Vet. Rec.* 139:567–571.

Lelkes, L. and C.L. Chang. 1987. Microbial dysbiosis in rabbit mucoid enteropathy. *Lab. Anim. Sci.* 37:757–764.

Zimmerman, T.E., W.E. Giddens, Jr., R.F. DiGiacomo, and W.C. Ladiges. 1990. Soft tissue mineralization in rabbits fed a diet containing excess vitamin D. *Lab. Anim. Sci.* 40:212–215.

CHAPTER 20

Feeding and Nutrition of Companion Animals: Dogs and Cats

Julie K. Spears and George C. Fahey, Jr.

Objectives

1. To provide information about the modern-day pet food industry
2. To identify the types of pet foods used today
3. To summarize the nutrient requirements of dogs and cats
4. To outline issues related to nutrient digestion and digestive physiology of dogs and cats
5. To discuss the feeding regimens used in dog and cat nutrition
6. To consider nutritional principles important throughout the life cycle of the dog and cat

THE MODERN-DAY PET FOOD INDUSTRY

In 2002, the global market for pet food and pet care products was valued at U.S. $46 billion, up 2.6 percent from the previous year. This translates into nearly 17 million tons of food produced annually (Crossley, 2003). Sales in North America were responsible for 41 percent of the global market value (39 percentage units of that from the United States alone). Western Europe accounted for a further 30.5 percent of the global market, and Japan another 13 percent (Crossley, 2003). Both Mexico and China are experiencing annual double-digit rates of growth in pet food sales.

The continued expansion of the industry worldwide is driven by increased pet populations, increased spending per pet, and an increased use of prepared pet foods and treats in both developed and emerging nations. Product segmentation strategies also stimulate growth, with brands of food being offered for various life stages, lifestyles, and breeds of pets. "Humanization" of pets has resulted in consumers purchasing foods paralleling human tastes (e.g., gourmet flavors of cat foods, treats for dogs and cats). The popularity of the "pet superstore" has made pet accessories and, more importantly, premium and super-premium dry dog and cat foods (those high in animal protein ingredients and high in digestibility) very accessible to the consumer (Crossley, 2003).

Today, well over 3,000 dog and cat foods are manufactured by over 300 U.S. companies. These numbers are down considerably from those in the 1950s and 1960s, 15,000 brands of dog food alone were marketed by 3,000 different manufacturers ranging in size from small country elevators to large pet food companies. In spite of the relatively large numbers of companies that make pet food yet today, the global market for pet food and pet care products remains concentrated in the hands of two main manufacturers, Mars and Nestlé (Crossley, 2003). Other companies like Iams have benefitted tremendously from the growth surge in premium and super-premium foods, and their products are now available in the supermarket.

Premium dog and cat food sales have experienced dynamic growth throughout the world, but economy and mid-priced foods continue to have a large customer base. Consumers who are cost-conscious regard these foods as being of sufficiently high quality to ensure adequate nutrition of their pets. In addition, many of these foods are formulated based on life stage/lifestyle concepts previously available only with the premium food sector. In terms of volume, sales of these foods are higher than those of premium foods (Crossley, 2003).

As regards animal numbers, it was estimated that 67 million dogs and 65 million cats were owned by Americans in 2001 (Special report: Pet food profiles, 2002). There are over 400 breeds of dogs in the world, with U.S. registering agencies recognizing over 180 different breeds.

TYPES OF PET FOODS

From a marketing perspective, commerical pet foods may be categorized into generic, private label, popular, and premium brands. **Generic foods** carry no brand name and usually are marketed within a rather narrow region of the country. They generally are the lowest cost of any commerical food. Inexpensive ingredients are used in their preparation and little, if any, animal testing is conducted to determine their nutritional efficacy. They may suffer from low palatability by the animal along with low nutrient digestibility and bioavailability (Case et al., 2000).

Private label pet foods carry the name of the store where they are sold, even though they often are produced by the same organization that prepares generic foods and often are of the same general nutritional quality (Case et al., 2000). In many cases, they are generic foods packaged and labeled with the name of the selling entity.

Popular brand pet foods are sold nationally and (or) regionally in grocery store chains. Their name recognition is high as a result of effective advertising. Nutritionally, these foods are often highly palatable to the animal but vary widely in ingredient content among batches depending on ingredient price at the time of manufacture. Nevertheless, the ingredients used are of higher quality than those present in the former two categories, resulting in higher nutrient digestibility and bioavailability (Case et al., 2000).

Premium brand pet foods contain high-quality fixed ingredients that result in excellent palatability and high nutrient digestibility and bioavailability to the animal. These foods are targeted to pets at different life stages and (or) those following different lifestyles. Previously, these foods had limited distribution in pet supply stores and in the offices of the veterinarian, but now are widely available in the supermarket as well. Premium foods are of particular use to pets housed indoors and to pets who have the opportunity to excrete feces and urine only twice daily.

Besides the brands previously listed, there also exist homemade diets and veterinary diets. One popular **homemade diet,** Bones And Raw Food (BARF), consists of a raw meat-based diet. The BARF diet may lead to nutritional problems if not properly balanced for the animal, and feeding raw meat diets may be unsafe for humans due to bacterial contamination. Besides BARF, other cooked homemade diets are popular. To ensure optimal utilization by the pet, homemade diets must be prepared to provide complete and balanced nutrition to the animal. Unfortunately, few homemade diets are tested for nutrient adequacy. Feeding single food items is inadvisable as compared to feeding a prepared diet. **Veterinary diets** are those used for nutritional management of disease states. They contain select types of foods, ingredients, and (or) nutrients. They are sold only through veterinary clinics and are labeled for use under the direction of a veterinarian (Case et al., 2000).

Commercial pet foods also may be categorized according to the processing method used to prepare the food, the methods used to preserve the food, and the moisture content of the resulting foods (Case et al., 2000). **Dry foods** contain between 6 and 10 percent moisture and include kibbles, biscuits, meals, and expanded products. Cereal grains, meat, poultry, or fish meals, some milk products, and vitamin and mineral supplements are used in their preparation. For proper preparation of expanded products, the presence of starch is a requirement.

Kibbles and biscuits are baked products, whereas meals are prepared by mixing dried, flaked, or granular ingredients. **Expanded pet foods** are produced by the extrusion process, and this has essentially replaced use of kibbles and meals. **Extrusion** involves mixing all ingredients together to form a dough that is cooked under conditions of high pressure and increasing temperature. The dough moves through the extrusion machine, mixing all the while. The cooking process generally takes less than 60 seconds and allows for the complete gelatinization of starch with subsequent increases in diet palatability and digestibility At the end of the extruder, dies are used, resulting in selected shapes and sizes of the food. More fat and digest (palatability enhancer) are added to the outside of the expanded food after it is cooled. Moisture content of the final product is reduced to 10 percent or less by hot air drying. Expanded foods are the most common types of dry pet food produced and sold in the United States (Case et al., 2000).

Canned pet foods may supply complete and balanced nutrition or may be a dietary supplement or treat containing canned meats or meat by-products. The former may contain blends of ingredients such as muscle meats, poultry by-products, fish by-products, cereal grains, texturized vegetable protein, and vitamins and minerals. The latter product may contain only meats with no supplemental ingredients added. Meat and fat ingredients are blended with water, after which dry ingredients are added and the mixture heated. Cans are filled on a conveyor line, then sealed and pressure sterilized, a process called *retorting* (250 C; 60 minutes). Canned foods often are more palatable and digestible than dry pet foods and contain a higher proportion of protein (28–50%, dry-matter basis) and fat (20–32%, dry-matter basis). Their water content is extremely high, approximately 75 percent. Canned foods have a long shelf life (Case et al., 2000).

Semimoist pet foods contain fresh or frozen animal tissues, cereal grains, fats, and simple sugars and have a moisture content ranging from 15 to 30 percent. They are softer in texture than dry foods and are highly palatable, largely as a result of the presence of simple sugars. They contain 20 to 28 percent crude protein on a dry-matter basis, and 8 to 14 percent fat. Their appeal to the pet owner lies largely in the fact that they look like ground beef, meat patties, or beef chunks, but this makes little difference to the overall nutrition of the animal (Case et al., 2000).

The last category of pet foods is **snacks and treats,** which are purchased by owners mostly for emotional reasons (Case et al., 2000). Appearance and palatability drive the purchase of the particular snack or treat. Semimoist, biscuit, jerky, and rawhide products are the items available in this food category. Some snacks and treats provide complete and balanced nutrition to the animal, whereas others do not.

NUTRIENT REQUIREMENTS OF DOGS AND CATS

Pets must be fed a diet that provides all of the essential nutrients in their correct quantities and proportions in order to maintain health through all stages of life. As a result of advances made in companion animal nutrition in recent years, nutrient deficiencies are rare in companion animals today. Guidelines of nutrient requirements for dogs and cats are necessary to ensure that commercial pet foods meet the needs of companion animals. A number of governing agencies and organizations regulate the production, marketing, and sales of commercial pet foods in the United States. Two agencies, the Association of American Feed Control Officials **(AAFCO)** and the National Research Council **(NRC),** make recommendations for nutrients needed in pet foods.

AAFCO is instrumental in the regulation of commercial pet foods and ensures that nationally marketed pet foods are uniformly labeled and marketed. Nutrient profiles based on ingredients commonly included in commercial pet foods have been published by AAFCO since the 1990s. The nutrient profiles published by AAFCO are generally believed to be more functional for the commercial pet food industry.

The NRC is a group within the National Academy of Sciences, and for years it has included a Committee on Animal Nutrition that identified problems and needs in animal nutrition. The NRC Subcommittee on Companion Animal Nutrition recently published nutrient requirements of dogs and cats throughout various stages of life (NRC, 2003).

Water

Water has several functions in the body. It acts as a solvent that facilitates intra- and extracellular reactions, a transporter for nutrients and waste products, a regulator of body temperature, and a lubricator of joints and organs. Adequate water intake is required for normal digestion and metabolism of food. Approximately 70 percent of lean body mass is water, and a loss of only 10 percent body water can result in death (Maynard et al., 1979).

All animals experience three major routes of daily **water loss.** The majority of water is lost during urinary excretion. Water excreted with feces accounts for a smaller portion of water loss. The final route of water loss is from the lungs through evaporation. Evaporation is important for dogs and cats for the regulation of body temperature in hot weather. Panting reduces body temperature, but water losses due to evaporation can be large during this time. Daily water consumption must make up for normal water losses.

Three **sources of water** intake exist for dogs and cats: metabolic water, water present in food, and voluntary water consumption. **Metabolic water** is water produced during oxidation of energy-containing nutrients in the body. The metabolism of fat produces the largest amount of metabolic water, whereas protein metabolism produces the smallest amount (Maynard et al., 1979). The amount of metabolic water produced depends on the type of food consumed and the metabolic rate of the animal. The production of metabolic water is small, ranging form 5 to 10 percent of the total water intake of the animal. The amount of water voluntarily consumed by the dog and cat depends on the amount of water present in the food (Cizek, 1959; Seefeldt and Chapman, 1979). Commercially available companion animal diets can range from 7 percent water in dry extruded diets to 84 percent water in canned diets (Anderson, 1981). Some ingredients, such as salt, will result in an increase in water consumption by both dogs and cats (Cowley et al., 1983). High intake of protein and carbohydrate also will result in an increase in water intake by dogs (Golob et al., 1984). Other factors affecting voluntary water intake include environmental temperature, exercise, physiologic state, and overall health.

The water requirement of cats and dogs, expressed in milliliters (mL), is believed to be equal to two to three times the dry-matter intake of food, expressed in grams (Case et al., 2000). Several other formulas exist to calculate the water requirement of companion animals. However, the best method of ensuring adequate water intake is to provide the animal with fresh, clean water at all times, regardless of dry-matter intake.

Energy

Carbohydrates, fats, and proteins supply the energy necessary for support of metabolism of dogs and cats during maintenance, growth, reproduction, lactation, and physical activity (NRC, 2003). Gross energy values are 4.1, 9.4, and 5.7 kcal/g for carbohydrates, fats, and proteins, respectively, in pet foods. Digestible energy (DE) and metabolizable energy (ME) values, particularly the latter, are used to define the energy content of pet foods. **Metabolizable energy** accounts for the energy lost in feces, in urine, and in fermentation gases. However, the latter factor can be neglected for pets (Zentek, 1993). Predictive equations exist allowing determination of ME in dog and cat diets (NRC, 2003). Gross energy content of the food, nutrient digestibility, and an accurate correction for predicted energy losses for protein in urine are required to formulate the equations, which can predict ME with reasonable accuracy only if the digestibility of the diets to which it is applied is close to the digestibility of the diets that were used to write the equation.

As regards energy requirements of companion animals, the **basal metabolic rate (BMR)** is the energy required to maintain homeostasis in an animal in a postabsorptive state, lying down but awake, in a thermoneutral environment to which it has been acclimated (Blaxter, 1989). Values for dogs range from 48 to 114 kcal/kg body weight$^{0.75}$. The BMR of domestic cats and its relationship to body weight or surface area is currently unknown (NRC, 2003). The **resting fed metabolic rate (RFMR)** is measured in animals not in a postabsorptive state but that otherwise meet the criteria for basal metabolism. As heat production increases when food is consumed, the thermic effect of food represents the difference between RFMR and BMR. The RFMR is approximately 15 percent above the BMR in normal adult dogs, and the difference will be greater for growing, gestating, or lactating dogs. The maintenance energy requirement (MER) is the energy required to support energy equilibrium (ME intake = heat production) over a long period of time (Blaxter, 1989), so this value will vary with any factor that affects heat production. Maintenance energy requirements of dogs range from 94 to 250 kcal ME/kg body weight$^{0.75}$. For cats, values range from 22 to 100 kcal/kg body weight.

As regards the basis for establishing energy requirements, allometric considerations of metabolic body weight are of great importance in dogs because of the wide range of body weights that exist among breeds (NRC, 2003). This means that for dogs, energy requirements are not related directly to body weight but to body weight raised to some power (generally accepted to be 0.75). Mature body weights of domestic cats range only from about 2 to 7 kg. The mass exponent often used for cats is 1, given that weights of various breeds of cats do not vary markedly (NRC, 2003).

In terms of physiological state of the dog, growing **puppies** require nearly two times more energy per unit body weight than adult dogs of the same breed (Arnold and Elvehjem, 1939). A decrease to 1.6 times maintenance is recommended when 50 percent of adult body weight has been reached, and 1.2 times maintenance at 80 percent of adult body weight (NRC, 2003). For adult dogs at maintenance, 132 kcal/kg body weight$^{0.75}$ is valid for active young animals kept in kennels (NRC, 2003). Young adult dogs have above average energy requirements, whereas senior dogs have decreased energy requirements. Breed, too, is a factor, as well as activity level (10–20 percent less energy needed by dogs that do not exercise; 90 kcal/kg body weight$^{0.75}$). For **gestation,** Meyer et al. (1985) estimated, using factorial calculations that considered litter weight and puppy body composition, placental weight and composition, and an estimation of extrauterine tissue of the bitch, that the requirements for later gestation (4 weeks after mating until parturition) were 26 kcal/kg body weight$^{0.75}$. This translates into 130 percent of maintenance in a 5 kg bitch and 160 percent of maintenance in a 60 kg bitch (NRC, 2003). NRC (2003) recommends that the energy requirements of pregnant bitches be calculated starting with their individual maintenance energy requirements per kilogram metabolic body weight, then adding 26 kcal/kg body weight rather than calculating percentages of maintenance. For **lactation** (Fig 20.1), milk yield (~8 percent of body weight) and energy content of milk (~1.45 kcal/g wet weight) are important in estimating the energy requirements for milk production (NRC, 2003). The equation for calculating the daily ME requirement (maintenance + milk production) for lactating bitches based on number of puppies and weeks of lactation is the following: ME (kcal) = 145 kcal/kg body weight$^{0.75}$ + body weight × (24n + 12m) × L, where BW = body weight of bitch (kg), n = number of puppies between 1 and 4, m = number of puppies between 5 and 8 (<5 puppies, m = 0), and L = correction factor for stage of lactation (week 1, 0.75; week 2, 0.95; week 3, 1.1; week 4, 1.2). These latter numbers result from the fact that, expressed as a percentage of mean daily milk yield during the first 4 weeks of lactation, milk production is as follows: 75 percent in week 1, 95 percent in week 2, 110 percent in week 3, and 120 percent in week 4.

In terms of physiological state of the cat, newborn **kittens** have an energy requirement of about 20–25 kcal/100 g body weight. For **adult cats** at mainte-

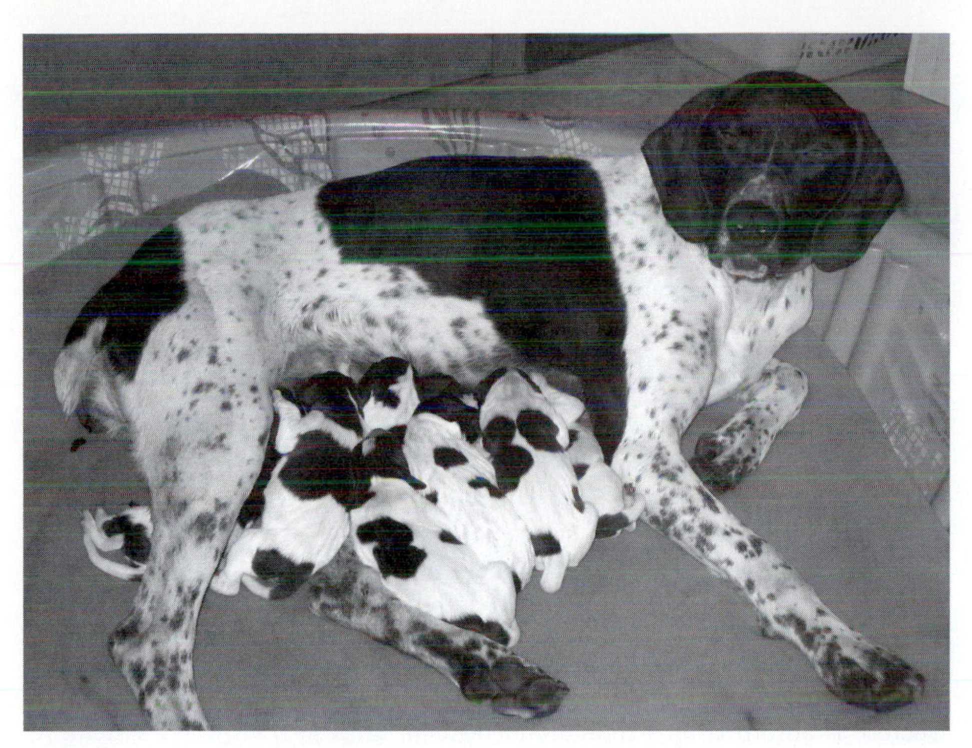

FIGURE 20.1 Lactation greatly increases the nutritional needs of mature dogs. (Courtesy of Howard H. Meyer.)

nance, 31–100 kcal/kg body weight are needed. There is a trend for higher energy requirements for young adult males and for all cats with a body weight lower than 4 kg, suggesting that lean body mass is important as regards energy requirements. Effects of neutering on energy requirements is equivocal (NRC, 2003). For lean cats with a body condition score of 5 (on a scale of 1 to 9), the energy requirement is 70 kcal/kg body weight. For **gestation,** Loveridge (1986) recommended feeding for a 40–50 percent weight increase during pregnancy (140 kcal/kg body weight$^{0.67}$). In **lactation,** queens tend to lose body weight (700–800 g), regardless of their diet. They suckle their kittens for 7 to 9 weeks depending on litter size. Supplemental feeding of kittens may be started at 2.5 weeks of age but by 4 weeks of age at the latest as quantity and nutrient content of queen milk are insufficient for normal development of the kitten (Hendriks and Wamberg, 2000). As is the case for dogs, milk yield (1.5–8 percent of body weight) and energy content of milk (1–1.3 kcal/g wet weight) are important factors for estimating the energy requirements for lactation. The equation for calculating the daily ME requirement of a lactating queen with three to four kittens is as follows: ME (kcal) $= 100 \times$ body weight$^{0.67} + 60 \times$ body weight $\times$ L, where L = factor for stage of lactation from week 1–7 (0.9, 0.9, 1.2, 1.2, 1.1, 1.0, 0.8, respectively). A general rule is that the energy requirement of a lactating queen with more than two kittens is between 2 and 2.5 times the maintenance requirements of the queen at mating (NRC, 2003).

Carbohydrates

Dietary carbohydrates are a heterogeneous group of compounds ranging from simple sugars to complex polysaccharides. Chemically, carbohydrates may be categorized as monosaccharides, disaccharides, oligosaccharides, and polysaccharides. Functionally, they may be categorized as absorbable (monosaccharides), digestible (disaccharides, certain oligosaccharides, nonstructural polysaccharides), fermentable (lactose, certain oligosaccharides, dietary fiber, resistant starch), and nonfermentable or poorly fermented (certain dietary fibers) (NRC, 2003). Carbohydrates sometimes fall into two or more different categories depending on animal species and age, food-processing characteristics, or amount of food ingested. Carbohydrates are not specifically required by dogs and cats, but they play key roles as sources of energy and as functional food components of the diet. Each of the functional categories of carbohydrates will be discussed briefly in the following sections.

Absorbable Carbohydrates Glucose and fructose are the main **monosaccharides** in dog and cat foods and are found primarily in cereal grains and by-products. They are absorbed directly and do not require digestive enzymes to be absorbed. In both dogs and cats, absorption of glucose and galactose across the brush border membrane occurs by an active sodium-dependent transport mechanism as well as sodium-independent simple diffusion. On the other hand, fructose is absorbed by a sodium-independent glucose transporter system, referred to as GLUT–5 in humans, and presumably in dogs and cats. Glucose and galactose are completely absorbed. Glucose provides energy and supplies carbon skeletons for biosynthesis of other compounds in the dog. In the absence of dietary glucose, dogs synthesize glucose from amino acids and glycerol via gluconeogenic processes. Cats are in a constant state of gluconeogenesis, implying that glucose is important for their normal metabolism. But cats do not utilize glucose as rapidly as do dogs and have comparatively longer blood glucose elimination times (NRC, 2003).

Cats are somewhat unique in the manner that they handle glucose. Levinson and Englert (1970) found that increasing concentrations of monosaccharides infused into the cat small intestine resulted in increasing rates of absorption. Buddington et al. (1991) found that cats are unable to up-regulate intestinal glucose or fructose transporter activity in response to changes in dietary carbohydrate concentration. Presumably, very high doses of monosaccharides may exceed the absorptive capacity of the cat small intestine. Cats have lower rates of **glucose utilization** as they have lower glucokinase activity, a hepatic enzyme that in other species is adaptive to diet and blood glucose concentrations. So, even though cats may absorb dietary carbohydrates well, the rate of utilization of the resulting glucose likely is much less efficient.

From a physiological perspective, absorbable carbohydrates have few effects on food intake, transit time of nutrients through the gastrointestinal tract, or fecal characteristics (although glucose- and galactose-containing diets fed to cats resulted in higher fecal dry-matter percentages compared to cats fed carbohydrate-free diets) (NRC, 2003). Whereas pregnant and lactating bitches do not require a dietary source of carbohydrates, they have increased protein requirements when fed a carbohydrate-free diet (Romsos et al., 1981). Currently, there is lack of scientific agreement on how aging affects glucose tolerance in dogs and cats.

Digestible Carbohydrates Lactose, maltose, and sucrose are the common **disaccharides** present in dog and cats foods. Digestible **oligosaccharides** are found in cereals, legumes, and other plant ingredients. Starch is perhaps the major player in this carbohydrate category, with some plants (e.g., rice) having as much as 81 percent starch.

Disaccharides are degraded to their monomeric sugars prior to absorption, and the monomers are absorbed from the jejunum. Diet composition and age of the animal affect disaccharidase activity markedly.

Alpha-amylases degrade the alpha-1, 4 bond of starch constituents (amylose, amylopectin, and maltodextrins). Starch is degraded in the duodenum by pancreatic alpha-amylase to maltose, maltotriose, and branched alpha-dextrins (Gray, 1992). **Amylase activity** is higher in dogs than in cats, and dog alpha-amylase activity is more sensitive to dietary starch concentrations (Kienzle, 1988, 1993). Over 95 percent of the dry pet foods manufactured today are extruded, so the starch present is cooked and rendered nearly 100 percent digestible. Raw starches, however, can be very poorly digested (e.g., ileal digestibility of uncooked potato starch is 0 percent and raw tapioca starch is 65 percent).

As regards physicochemical characteristics of digestible carbohydrates, unabsorbed disaccharides induce fluid secretion into the small intestinal lumen via osmosis. This effect in combination with the rapid colonic fermentation of unabsorbed sugars results in reduced transit time (Washabau et al., 1986). On occasion, fecal pH and fecal moisture concentration can be altered by disaccharide ingestion. Starch hydrates easily and, when heated, gelatinization occurs, resulting in a viscous suspension.

Physiologically, digestible carbohydrates may affect food intake and nutrient digestibility depending on the diet matrix into which they are incorporated and the age of the animal studied. **Source of starch** fed appears to have little influence on growth performance of dogs. Romsos et al. (1981) found that fewer puppies from bitches on a carbohydrate-free diet were alive at birth compared to litters from bitches fed a carbohydrate-containing diet, indicating that pregnant bitches require dietary carbohydrates for optimal reproductive performance and survivability of puppies. As regards glycemic control, large variation in postprandial glucose concentrations and insulin secretory responses to ingestion of different foods have been observed in dogs (Nguyen et al., 1994). The amount of starch ingested is the major determinant of the glycemic response of adult healthy dogs (Nguyen et al., 1998). Postprandial hyperglycemia and insulin secretion depend on the ratio of amylose to amylopectin consumed. A positive linear correlation was found between the starch content of foods and postprandial plasma glucose and insulin responses. Source of grain fed influences postprandial glucose and insulin responses in adult dogs.

Fermentable Carbohydrates This category includes certain **oligosaccharides** (e.g. fructans, mannanoligosaccharides, galactooligosaccharides), **dietary fibers** (e.g., celluloses, hemicelluloses, pectins, lignin, gums, beta-glucans), and **resistant starches** (e.g., entrapped starch components, starch with unique granular structures, cooked and cooled starches, chemically modified starches). They resist hydrolytic digestion in the small intestine but are fermented by microbes present in the large bowel of dogs and cats to **short-chain fatty acids (SCFA),** lactate, and gases (carbon dioxide, hydrogen, methane primarily). Major sources of fermentable carbohydrates in dog and cat foods are cereals, legumes, and other plant ingredients. In addition, certain oligosaccharides are produced synthetically from microbial activity on simple carbohydrate substrates.

Considerable information has been collected in the areas of digestion, absorption, and utilization of fermentable carbohydrates. The major dietary fibers have been examined as components of different types of dog and cat diets. Their effects on nutrient digestibility have been quantified and their fermentation profile delineated. Specific results of these experiments are discussed in detail in NRC (2003).

As regards physicochemical effects of fermentable carbohydrates, the SCFA resulting from their fermentation are weak anions that can exert osmotic pressure in the colon and increase fecal water content. But they are rapidly and nearly completely absorbed, which compensates for any net osmotic action. Indeed, SCFA are probably responsible for promoting water reabsorption from the colonic lumen in dogs (Herschel et al., 1981). Certain of the fermentable carbohydrates (e.g., pectins, gums) have a high water-holding capacity that often is associated with enhanced microbial fermentation. These same carbohydrates tend to have high viscosity that is associated with prolonged gastric emptying, increased intestinal transit time, and slower nutrient absorption.

The diverse physical and chemical properties of fermentable carbohydrates, as well as their concentration in the diet, determine their specific physiological effects. At physiological concentrations in the diet, fermentable carbohydrates generally do not impact food intake, but they nearly always increase volume of feces excreted and fecal moisture content. Their effects on **transit time** through the gastrointestinal tract vary widely. For example, psyllium and guar gum increase, whereas pectin and beet pulp decrease, transit time. Cellulose and wheat bran have no influence on this criterion. Fermentable carbohydrates rarely affect ileal nutrient digestibility but have widely varied effects on total tract nutrient digestion. Oligosaccharides often modify gut microbial populations, favoring growth and activity of more favorable populations at the expense of less favorable populations.

Fermentable carbohydrates affect dogs and cats in various physiological states. For example, high concentrations of fermentable carbohydrates in diets may decrease the growth rate of dogs and cats. In addition, both insoluble and soluble dietary fibers have been used as a means of restricting energy intake by dogs and cats for weight control purposes. Dietary fibers may aid in proper weight maintenance through regulation of appetite, as a result of their physical properties (e.g., water holding capacity, viscosity), and complex hormonal interactions determined by the nature and type of the fiber fed (NRC, 2003). Effects on reproductive performance have not been appropriately quantified to date. And as regards the geriatric animal, fermentable carbohydrates may positively affect bowel health and glycemic control.

Nonfermentable/Poorly Fermentable Carbohydrates Only a few carbohydrates fall into this category, including **purified celluloses** and the natural ingredient **wheat bran.** Both affect laxation and lower nutrient digestibility by dogs and cats.

Fats

Fat serves several functions in a companion animal diet. First, dietary fat provides energy to the animal. Compared with carbohydrates and protein, fat provides the highest amount of metabolizable energy (2.25 times more than either protein or carbohydrate). In addition to having high ME, fat in pet foods is very digestible. The apparent fat digestibility of commercially available diets ranges from 70 to 90 percent for dog food and 85 to 94 percent for cat food (Huber et al., 1986). Due to its high energy content and high digestibility, increasing the concentration of fat in a pet diet will increase the energy density of the diet. Increasing the energy density of pet food is advantageous at life stages where the animal needs to consume a high-calorie food without having to consume a large quantity of dry matter. A high-calorie diet is preferred for growing, gestating, lactating, and exercising/working animals.

Other functions of dietary fat are to act as a carrier of fat-soluble vitamins and provide **essential fatty acids (EFA)** to the animal. Essential fatty acids are required for normal growth and metabolism and cannot be synthesized in adequate amounts by the animal. All EFAs are polyunsaturated fatty acids. The body has a requirement for two families of EFAs: the n-6 and n-3 series. This terminology refers to the location of the first double bond in the fatty acid relative to the methyl end. Plants and algae synthesize EFAs, but the animal can alter the length and degree of saturation of the EFAs after digestion and absorption. Two key enzymes needed to alter EFAs in the body are delta-6-desaturase and delta-5-desaturase. These enzymes are utilized to desaturate and elongate linoleic acid, the parent n-6 fatty acid, to arachidonic acid, and alpha-linolenic acid, the parent n-3 fatty acid, to gamma-linolenic acid. Because all but the parent fatty acids can be synthesized by dogs from other precursors, linoleic acid is regarded as an EFA in the dog. For adult dogs at maintenance, AAFCO (2003) recommends that linoleic acid be provided at 1 percent of the diet (dry matter-basis). The NRC (2003) recommends that **linoleic acid** be provided at 9.5 g/kg dry matter (DM) for adult dogs at maintenance. The cat is unable to synthesize adequate amounts of arachidonic acid from linoleic acid because the cat has low delta-6-desaturase and delta-5-desaturase activities. Because of the low enzyme activities, the cat diet must contain both linoleic and arachidonic acid. The NRC (2003) recommends that a cat at maintenance be supplemented with 0.02 g arachidonic acid/kg DM and 5.5 g linoleic acid /kg DM. The AAFCO (2003) nutrient profiles recommend that a diet for cats at maintenance contain 0.5 percent linoleic acid and 0.02 percent arachidonic acid (dry matter-basis).

A third function of dietary fat is to improve **palatability,** of flavor, of the diet. Fat also improves the texture of the diet. Dogs can adapt to a wide range of fat concentrations in the diet, but studies have shown that they prefer animal fats (Rainbird, 1988; Earle and Smith, 1993). Generally, cats prefer a high-fat diet to a low-fat diet, and they appear to prefer fats of animal origin as well (Greaves, 1965; Kane et al., 1981; Kendall, 1984).

Deficiencies and excesses of fat can occur in the dog and cat. When fat is deficient in the diet, the animal will not be provided with adequate energy and EFAs. Diets that are low in fat will decrease the acceptability of the diet because fat enhances palatability. The inadequate energy and EFA content of a low-fat diet will, therefore, be worsened by low food intakes due to decreased acceptance of the diet. When fat is deficient, the biggest concern is a deficiency in EFA. Due to the high turnover of skin cells, an EFA deficiency will first affect the appearance of the skin. An EFA deficiency will result in a dry hair coat, loss of hair, and skin lesions. Skin lesions will increase the likelihood of infections. Because cats are unable to synthesize other EFAs rapidly from linoleic acid, cats will have reproductive problems as well. Due to the high fat content of commercially available pet foods, EFA deficiencies are rare.

Companion animals are susceptible to excess fat intake as well. Although both dogs and cats can digest high concentrations of dietary fat, an excess amount of fat in the diet will result in diarrhea and increased fat in the stool. In addition, long-term consumption of high-fat diets will lead to weight gain and **obesity.** The weight gain is due to the high energy density of the diet in addition to the increased palatability. Animals that lead sedentary lives are particularly susceptible to weight gain. Obese animals are prone to developing other diseases, such as diabetes and arthritis.

The AAFCO (2003) nutrient profiles and NRC (2003) nutrient requirements for fat and EFAs for dogs and cats during growth and maintenance may be found in Tables 20–1 through 20–4.

Proteins and Amino Acids

Proteins provide amino acids that cannot be made by dogs and cats (essential amino acids) but are required for synthesis of body proteins (NRC, 2003). These include arginine, histidine, isoleucine, leucine, lysine, methionine, phenylalanine, threonine, tryptophan, and valine. Proteins also provide nonessential amino acids needed for dogs and cats in various physiological states (e.g., maintenance, growth, gestation, lactation).

Apparent total tract **digestibility of protein** is similar in dogs and cats for highly digestible proteins (NRC, 2003). Proteins with lower digestibility have higher apparent digestibilities by dogs than cats (Kendall et al., 1982). For dry or canned dog or cat foods, dogs have a protein digestibility approximately 5 to 8 percent higher than that of cats (Kendall et al., 1982). This may be a result of the shorter length of the small intestine relative to body size in cats versus dogs. Animal proteins often have a higher total tract digestibility than do plant proteins (Meyer et al., 1981, 1989), and prolonged heat processing decreases animal protein digestibility by both dogs (Johnson et al., 1998) and cats (Backus et al., 1998). Digestibility of protein varies with size, breed, and age of dogs (e.g., pointers > huskies; young > old). Insoluble fiber content of the diet does not affect protein digestibility, but soluble fiber often decreases it in dogs (Muir et al., 1996) and cats (Sunvold et al., 1995).

Total tract digestibility values are useful in assessing the quality of nutrients in complete foods, but **ileal digestibility** values are better as they are not confounded by the fermentative activity that takes place in the large bowel of the dog and cat. Ileal digestibilities of protein in ingredients contained in typical dog foods are approximately 1 to 20 percentage units lower than apparent total tract digestibility values (63–96 percent vs. 71–98 percent, respectively) (NRC, 2003). Ileal digestibilities of individual amino acids from normal ingredients vary markedly (arginine, 77–87 percent; cystine, 29–66 percent, threonine, 52–78 percent, lysine, 62–84 percent) (Johnson et al., 1998). No similar data exist for cats.

TABLE 20–1. AAFCO Nutrient Profiles for Dogs[1]

Nutrient	*Units (DMB)*[2]	*Growth and Rrproduction (Min)*	*Adult Maintenance (Min)*	*(Max)*
Protein	%	22.00	18.00	
Arginine	%	0.62	0.51	
Histidine	%	0.22	0.18	
Isoleucine	%	0.45	0.37	
Leucine	%	0.72	0.59	
Lysine	%	0.77	0.63	
Met-cys	%	0.53	0.43	
Phe-tyr	%	0.89	0.73	
Threonine	%	0.58	0.48	
Tryptophan	%	0.20	0.16	
Valine	%	0.48	0.39	
Fat[3]	%	8.0	5.0	
Linoleic acid	%	1.0	1.0	
Minerals				
Calcium	%	1.0	0.6	2.5
Phosphorus	%	0.8	0.5	1.6
Ca:P ratio		1:1	1:1	2:1
Potassium	%	0.6	0.6	
Sodium	%	0.3	0.06	
Chloride	%	0.45	0.09	
Magnesium	%	0.04	0.04	0.3
Iron[4]	mg/kg	80.0	80.0	3,000
Copper (extruded)[5]	mg/kg	7.3	7.3	250
Manganese	mg/kg	5.0	5.0	
Zinc	mg/kg	120.0	120.0	1,000
Iodine	mg/kg	1.5	1.5	50
Selenium	mg/kg	0.11	0.11	2
Vitamins				
Vitamin A	IU/kg	5,000.0	5,000.0	250,000
Vitamin D	IU/kg	500.0	500.0	5,000
Vitamin E	IU/kg	50.0	50.0	1,000
Thiamine[6]	mg/kg	1.0	1.0	
Riboflavin	mg/kg	2.2	2.2	
Pantothenic acid	mg/kg	10.0	10.0	
Niacin	mg/kg	11.4	11.4	
Pyridoxine	mg/kg	1.0	1.0	
Folic acid	mg/kg	0.18	0.18	
Vitamin B_{12}	mg/kg	0.022	0.02	
Choline	mg/kg	1200.0	1200.0	

From AAFCO (2003).

[1]Presumes an energy density of 3,500 kcal ME/kg dry matter.

[2]DMB = dry-matter basis.

[3]Although a true requirement for crude fat per se has not been established, the minimum concentration was based on recognition of crude fat as a source of essential fatty acids, as a carrier of fat-soluble vitamins, to enhance palatability, and to supply an adequate caloric density.

[4]Because of very poor bioavailability, iron from carbonate or oxide sources added to the diet should not be considered in determining the minimum nutrient concentration.

[5]Because of very poor bioavailability, copper from oxide sources added to the diet should not be considered in determining the minimum nutrient concentration.

[6]Because processing may destroy up to 90 percent of the thiamine in the diet, allowances in formulation should be made to ensure the minimum nutrient concentration is met after processing.

TABLE 20–2. AAFCO Nutrient Profiles for Cats[1]

Nutrient	*Units (DMB)[2]*	*Growth and Reproduction (Min)*	*Adult Maintenance (Min)*	*(Max)*
Protein	%	30.0	26.0	
Arginine	%	1.25	1.04	
Histidine	%	0.31	0.31	
Isoleucine	%	0.52	0.52	
Leucine	%	1.25	1.25	
Lysine	%	1.20	0.83	
Met-cys	%	1.10	1.10	
Methionine	%	0.62	0.62	1.5
Phe-tyr	%	0.88	0.88	
Phenylalanine	%	0.42	0.42	
Threonine	%	0.73	0.73	
Tryptophan	%	0.25	0.16	
Valine	%	0.62	0.62	
Fat[3]	%	9.0	9.0	
Linoleic acid	%	0.5	0.5	
Arachidonic acid	%	0.02	0.02	
Minerals				
Calcium	%	1.0	0.6	
Phosphorus	%	0.8	0.5	
Potassium	%	0.6	0.6	
Sodium	%	0.2	0.2	
Chloride	%	0.3	0.3	
Magnesium	%	0.08	0.04	
Iron[4]	mg/kg	80	80	
Copper (extruded)[5]	mg/kg	15	5	
Copper (canned)[5]	mg/kg	5	5	
Manganese[6]	mg/kg	7.5	7.5	
Zinc	mg/kg	75	75	2,000
Iodine	mg/kg	0.35	0.35	
Selenium	mg/kg	0.1	0.1	

As protein digestibility and quality increase, the concentration of dietary protein needed to meet the animal's requirement decreases. In addition, the higher the biological value of a protein, less of that protein is needed to meet an animal's essential amino acid requirements (Case et al., 2000).

The **protein requirement** is the minimum intake of dietary protein that promotes optimal performance (Case et al., 2000). Criteria most often used to evaluate performance when determining protein requirements of dogs and cats are **nitrogen balance** and growth rate. Requirement studies with growing animals use maximal positive nitrogen balance or maximal growth rate to indicate an adequate dietary protein concentration. Nitrogen equilibrium, or zero nitrogen balance, is the normal state for healthy adult dogs and cats during maintenance.

TABLE 20–2. continued

Nutrient	*Units (DMB)*[2]	*Growth and Reproduction (Min)*	*Adult Maintenance (Min)*	*(Max)*
Vitamins				
Vitamin A	IU/kg	9,000	5,000	750,000
Vitamin D	IU/kg	750	500	10,000
Vitamin E[7]	IU/kg	30	30	
Vitamin K[8]	mg/kg	0.1	0.1	
Thiamine[9]	mg/kg	5.0	5.0	
Riboflavin	mg/kg	4.0	4.0	
Pantothenic acid	mg/kg	5.0	5.0	
Niacin	mg/kg	60	60	
Pyridoxine	mg/kg	4.0	4.0	
Folic acid	mg/kg	0.8	0.8	
Biotin[10]	mg/kg	0.07	0.07	
Vitamin B_{12}	mg/kg	0.02	0.02	
Choline[11]	mg/kg	2400	2400	
Taurine (extruded)	%	0.10	0.10	
Taurine (canned)	%	0.20	0.20	

From AAFCO (2003).

[1]Presumes an energy density of 4,000 kcal ME/kg dry matter.

[2]DMB = dry-matter basis.

[3]Although a true requirement for crude fat per se has not been established, the minimum concentration was based on recognition of crude fat as a source of essential fatty acids, as a carrier of fat-soluble vitamins, to enhance palatability, and to supply an adequate caloric density.

[4]Because of very poor bioavailability, iron from carbonate or oxide sources added to the diet should not be considered in determining the minimum nutrient concentration.

[5]Because of very poor bioavailability, copper from oxide sources added to the diet should not be considered in determining the minimum nutrient concentration.

[6]If the mean urine pH of cats fed *ad libitum* is not below 6.4, the risk of struvite urolithiasis increases as the magnesium content of the diet increases.

[7]Add 10 IU vitamin E above minimum concentration per gram of fish oil per kilogram of diet.

[8]Vitamin K does not need to be added unless diet contains greater than 25% fish on a dry matter basis.

[9]Because processing may destroy up to 90% of the thiamine in the diet, allowances in formulation should be made to ensure the minimum nutrient concentration is met after processing.

[10]Biotin does not need to be added unless diet contains antimicrobial or antivitamin compounds.

[11]Methionine may be used to substitute for choline as a methyl donor at a rate of 3.75 parts for 1 part choline by weight when methionine exceeds 0.62%.

Long-term protein status can be assessed by the maintenance of serum albumin and lean body mass (NRC, 2003). The basis for determining the amino acids that may be limiting in a diet may be determined by quantifying plasma amino acids.

Dietary factors that may affect nitrogen balance include protein quality and amino acid composition, protein digestibility, and the energy density of the diet. Animal factors include activity level, physiological state, and prior nutritional status (Case et al., 2000).

TABLE 20–3. Minimal Requirements and Adequate Intakes for Dogs at Growth and Maintenance[1]

		Growth[2]		*Maintenance[3]*	
Nutrient	**Units**	*Minimum Requirement[4]*	*Adequate Intake[5]*	*Minimum Requirement[4]*	*Adequate Intake[5]*
Protein	g	140	—	80	—
Arginine[6]	g	5.3	—	2.8	—
Histidine	g	2.0	—	1.5	—
Isoleucine	g	4.0	—	3.0	—
Leucine	g	6.5	—	5.4	—
Lysine	g	5.6	—	2.8	—
Methionine	g	2.1	—	2.6	—
Methionine-cystine	g	4.2	—	5.2	—
Phenylalanine	g	4.0	—	3.6	—
Phenylalnine-tyrosine[7]	g	8.0	—	5.9	—
Threonine	g	5.0	—	3.4	—
Tryptophan	g	1.4	—	1.1	—
Valine	g	4.5	—	3.9	—
Fat	g	—	85	—	40
Linoleic acid	g	—	11.8	—	9.5
α-Linolenic acid[8]	g	—	0.7	—	0.35
Arachidonic acid	g	—	0.3	—	—
Eicosapentaenoic and docosahexanenoic[9]	g	—	0.5	—	0.11
Minerals					
Calcium[10]	g	8.0	—	2	—
Phosphorus[10]	g	—	10	—	3
Potassium	g	—	4.4	—	4.0
Sodium	g	—	2.2	300	—
Chloride	g	—	2.9	—	600
Magnesium	mg	180	—	180	—
Iron[11]	mg	72	—	—	30
Copper[11]	mg	—	11	—	6
Manganese	mg	—	5.6	—	5.0
Zinc	mg	40	—	—	60
Iodine	µg	—	900	700	—
Chromium	µg	—	—	—	700
Selenium	µg	210	—	—	350

Vitamins					
Vitamin A[12]	RE	—	1212	—	1212
Cholecalciferol	μg	—	11.0	—	11.0
α-tocopherol[13]	mg	—	24	—	24
Thiamin	mg	—	1.08	—	1.8
Riboflavin	mg	—	8.4	—	4.2
Pantothenic acid	mg	—	12	—	12
Niacin	mg	—	12	—	12
Pyridoxine	mg	—	1.2	—	1.2
Cobalamin	μg	28	—	—	28
Folic acid	μg	—	216	—	216
Choline	mg	—	1360	—	1360

From National Research Council (2003).

[1]Values expressed as amount/kg DM.

[2]Based on a 5.5-kg 3-month old puppy consuming 1,000 kcal ME/day.

[3]Based on a 15-kg dog consuming 1,000 kcal ME/day.

[4]Minimum requirement defined as the minimal concentration or amount of a maximally bioavailable nutrient that will support a defined physiological state.

[5]Adequate intake defined as the concentration or amount of a nutrient demonstrated to support a defined physiological state when no minimum requirement has been demonstrated.

[6]0.01 g arginine should be added for every gram of crude protein above 180.

[7]The quantity of tyrosine required to maximize black hair color may be about 1.5–2.0 times this quantity.

[8]α-Linolenic acid variable, depending upon linoleic acid content; linoleic acid:α-linolenic acid ratio should be between 2.6 and 16 (2.6 and 26 for growing animals).

[9]For growing animals, eicosapentaenoic acid should not exceed 20 percent of the total amount. For animals at maintenance, 30 percent of the total amount should be eicosapentaenoic acid and 20 percent should be docosahexaenoic acid.

[10]For growing animals, calcium and phosphorus requirements may decrease by up to 20 percent per unit of energy between 60 percent and 100 percent of mature body weight. Values are for giant breeds and may be different in others.

[11]Oxide forms should not be used because of low bioavailability.

[12]Vitamin A requirements are expressed as RE (retinol equivalents). One RE is equal to 1 μg of all-trans retinol, and one IU of vitamin A is equal to 0.3 RE.

[13]Higher concentrations of vitamin E are recommended for high polyunsaturated fatty acid diets.

Table 20–4. Minimal Requirements and Adequate Intakes for Cats at Growth and Maintenance[1]

Nutrient	Units	*Growth*[2]		*Maintenance*[3]	
		Minimum Requirement[4]	*Adequate Intake*[5]	*Minimum Requirement*[4]	*Adequate Intake*[5]
Protein	g	180	—	160	—
Arginine[6]	g	7.7	—	—	7.7
Histidine	g	2.6	—	—	2.6
Isoleucine	g	4.3	—	—	4.3
Leucine	g	10.2	—	—	10.2
Lysine	g	6.8	—	2.7	—
Methionine[7]	g	3.5	—	1.35	—
Methionine-cystine	g	7.0	—	2.7	—
Phenylalanine	g	4.0	—	4.0	—
Phenylalanine-tyrosine[8]	g	15.3	—	15.3	—
Threonine	g	5.2	—	—	5.2
Tryptophan	g	1.3	—	—	1.3
Valine	g	5.1	—	—	5.1
Taurine[9]	g	0.32	—	0.32	—
Fat	g	—	90	—	90
Linoleic acid	g	—	5.5	—	5.5
α-Linolenic acid	g	—	0.2	—	—
Arachidonic acid	g	—	0.02	—	0.02
Eicosapentaenoic and docosahexanenoic[10]	g	—	0.1	—	0.1
Minerals					
Calcium	g	5.2	—	1.6	—
Phosphorus	g	4.8	—	1.4	—
Potassium	g	2.7	—	—	5.2
Sodium	mg	1250	—	650	—
Chloride	mg	760	—		960
Magnesium	mg	160	—	200	—
Iron[11]	mg	70	—	—	80
Copper[11]	mg	4.5	—	—	5.0
Manganese	mg	—	4.8	—	4.8
Zinc	mg	50	—	—	75
Iodine	μg	—	2200	—	2200
Selenium	μg	120	—	—	400

Vitamins					
Vitamin A[12]	RE	—	852	—	852
Cholecalciferol	μg	3.1	—	3.1	—
α-tocopherol[13]	mg	—	30	—	30
Thiamin	mg	4.4	—	—	4.4
Riboflavin	mg	—	3.4	—	3.4
Pantothenic acid	mg	5.0	—	5.0	—
Niacin	mg	—	34	—	34
Pyridoxine	mg	2.0	—	2.0	—
Cobalamin	μg	—	18	—	18
Folic acid	μg	600	—	600	—
Biotin[14]	μg	—	60	—	60
Choline	mg	2040	—	2040	—

From National Research Council (2003).

[1]Values expressed as amount/kg DM.

[2]Based on a 800-g kitten consuming 180 kcal ME/day.

[3]Based on a 4-kg consuming 250 kcal ME/day.

[4]Minimum requirement defined as the minimal concentration or amount of a maximally bioavailable nutrient that will support a defined physiological state.

[5]Adequate intake defined as the concentration or amount of a nutrient demonstrated to support a defined physiological state when no minimum requirement has been demonstrated.

[6]0.02 g arginine should be added for every gram of crude protein above 200.

[7]Methionine is presumed to be one-half the sum of the requirement for methionine + cystine combined.

[8]To maximize black hair color, an equal quantity of tyrosine to that of phenylalanine is required.

[9]The recommended allowances for taurine for dry expanded and canned diets are 1.0 g and 2.0 g/kg diet, respectively.

[10]For growing animals, it is advised that eicosapentaenoic acid not exceed 20 percent of the total eicosapentaenoic acid + docosahexaenoic acid amount. For animals at maintenance, 30 percent of the total amount should be eicosapentaenoic acid and 20 percent should be docosahexaenoic acid.

[11]Oxide forms should not be used because of low bioavailability.

[12]Vitamin A requirements are expressed as RE (retinol equivalents). One RE is equal to 1 μg of all-trans retinol, and one IU of vitamin A is equal to 0.3 RE.

[13]Higher concentrations of vitamin E are recommended for high polyunsaturated fatty acid diets.

[14]A metabolic requirement for biotin has been demonstrated. Diets without raw egg whites should allow for adequate intestinal microbial synthesis.

The cat has a higher dietary requirement for protein than does the dog because of its increased needs for maintenance of normal body tissue, not its increased needs for growth (Case et al., 2000). Approximately 60 percent of the growing kitten's protein requirement is used for maintenance of body tissue, with 40 percent being used for growth. The opposite is true for most other species. The higher **protein requirement for maintenance in the cat** results from the inability of the nitrogen catabolic enzymes in the cat's liver to adapt to changes in dietary protein intake. Also in the cat, enzymes involved in nitrogen catabolism function at relatively high rates of activity, causing the cat to catabolize substantial protein after each meal, regardless of the dietary protein content. The cat has limited capability to conserve nitrogen from the body's general nitrogen pool, so the only alternative that ensures adequate conservation of body protein stores is consistent consumption of a diet containing high concentrations of protein (Case et al., 2000).

Two unique amino acid requirements occur for the cat: (1) the cat does not have the ability to synthesize adequate **arginine** for normal functioning of the urea cycle and protein synthesis, and (2) the cat has a dietary requirement for **taurine,** an amino sulfonic acid. Cats develop severe hyperammonemia within hours of consuming a single arginine-free meal. Clinical signs include vomiting, muscle spasms, ataxia, sensitivity to touch, and tetanic spasms, leading to coma and death. Reasons for this include the fact that the cat is unable to synthesize de novo ornithine, an arginine precursor in the urea cycle. The intestinal mucosal cells of the cat have low concentrations of active pyrroline-5-carboxylate synthase and ornithine aminotransferase, essential enzymes in this pathway. In addition, the cat is unable to synthesize arginine from ornithine for use by extrahepatic tissues, even if dietary ornithine is provided.

Taurine is found as a free amino acid in tissues (brain, heart, skeletal muscle) or as a constituent of small peptides. It is synthesized by most mammals from methionine and cysteine during normal sulfur amino acid metabolism. The mycocardium and retina contain high concentrations of free taurine (100 to 400-fold greater concentrations than those found in plasma). It plays important roles in bile acid conjugation, retinal function, normal functioning of the myocardium, and enhancing excretion of xenobiotics from the body. Cats are able to synthesize only small amounts of taurine because of their low activities of two enzymes essential for taurine synthesis, cysteine dioxygenase and cysteine sulfinic acid decarboxylase. The cat has a high metabolic demand for taurine as it uses only taurine for bile salt formation and cannot convert to conjugation of bile acids with glycine when the taurine supply is limited (Case et al., 2000). **Taurine deficiency** in cats results in feline central retinal degeneration and blindness, dilated cardiomyopathy and heart failure, inadequate immune response, poor neonatal growth, reduced auditory brain stem evoked potential resulting in deafness, poor reproduction resulting in a low number of fetuses, resorptions, abortions, decreased birth weight and low survival rate of kittens, and congenital defects including hydrocephalus and anencephaly.

Requirements for crude protein and amino acids and crude-protein and amino acid profiles may be found in Tables 20–1 through 20–4.

Vitamins

Vitamins are organic components of the diet that are necessary for normal growth and maintenance. They are not used for energy or incorporated into cells. They are most commonly used as enzyme cofactors. The diverse range of abnormalities noted with vitamin deficiencies and excesses reflect the diverse metabolic functions of vitamins. Two classes of vitamins exist: fat-soluble and water-soluble. Although this classification has little to do with their functional roles, it does have relevance as to the route of absorption from the gut.

The AAFCO (2003) nutrient profiles and NRC (2003) nutrient requirements for both fat- and water-soluble vitamins for dogs and cats during growth and maintenance may be found in Tables 20–1 through 20–4.

Fat-Soluble Vitamins Fat-soluble vitamins include vitamins A, D, E, and K. These vitamins are absorbed in the intestine with dietary fat. Therefore, any disturbance in normal lipid metabolism will adversely affect the fat-soluble vitamin status of the animals as well.

Vitamin A Vitamin A has several functions in the body, including vision, bone growth, reproduction, and maintenance of body tissue. Retinol is the most active form of vitamin A in the body; however, all of the vitamin A ingested by animals originates from carotenoids synthesized by plants. Most animals, including the dog, are able to convert other forms of vitamin A, such as β-carotene, to active retinal. The **conversion of β-carotene** to active vitamin A requires the oxidative cleavage of the carotenoid molecule. The cat lacks 15,15'-dioxygenase, the enzyme essential for conversion of β-carotene to retinol. Therefore, the cat must have preformed vitamin A in the diet. Retinol is highly susceptible to oxidation, so retinol is esterified with either acetate or palmitate when added to the diet.

Dogs differ from other domestic animals in their metabolism of vitamin A. Dogs and some other members of the Carnivora order have unusually high concentrations of plasma vitamin A, mainly as retinyl esters (retinyl stearate and palmitate), and excrete retinol and retinyl palmitate in the urine (Schweigert et al., 1990; Schweigert et al., 1991). The use of vitamin A in dermatology therapy for dogs and the high vitamin A content of animal by-products used in canned dog foods suggest that dogs could be exposed to potentially toxic vitamin A levels. However, Cline et al. (1997) fed high levels (225,000 IU/kcal diet) of vitamin A to adult dogs of a variety of genotypes for a one year period. No evidence of vitamin A toxicity was observed, indicating that dogs are resistant to vitamin A toxicity.

Vitamin A toxicity is rare because, with the exception of the cat, animals are capable of regulating absorption of dietary β-carotene. Preformed vitamin A absorption is not regulated in the intestine, so the cat can develop signs of toxicity. Deforming cervical spondylosis is a disorder caused by excess vitamin A consumption by cats that is characterized by the development of bony outgrowths on the cervical vertebrae. **Vitamin A deficiency** is rare in dogs and cats because commercial pet foods contain adequate amounts of the vitamin. Young animals will develop abnormal bones and neurological disorders if diets are deficient in vitamin A. Long-term consumption of vitamin A-deficient diets will cause shortening of the long bones and abnormal skull development (Mellanby, 1938; Hayes, 1971). Adult animals that are deficient in vitamin A will have poor reproduction, vision, and epithelial cell function.

Vitamin D Vitamin D is required for calcium and phosphorus metabolism and homeostasis by increasing circulating calcium and phosphorus. This increase allows for bone remodeling and maintenance of extracellular calcium concentrations. Other functions of vitamin D include maintenance of normal functioning of muscle contraction, blood clotting, nerve conduction, and intracellular signaling. Vitamin D is not the active form of the vitamin, but must undergo two reactions for production of the biologically active form of the vitamin. Most animals, including the dog, are capable of synthesizing vitamin D_3, cholecalciferol, the inactive form of vitamin D, in the skin when exposed to ultraviolet light such as sunlight. Vitamin D_3 is stored in the liver, and it can be converted to active vitamin D in the liver and kidney. The enzyme necessary for vitamin D synthesis in the skin has a low activity in the cat, so cats must have preformed vitamin D in the diet.

Vitamin D toxicity results in calcification of the soft tissue. If the animal consumes high concentrations of vitamin D for a long period of time, skeletal abnormalities can develop in growing animals. **Vitamin D deficiency** in companion animals will result in rickets in growing animals and osteomalacia in adult animals. **Rickets** is caused by insufficient calcium and phosphorus deposition in the bones and is characterized by bowing of the legs. **Osteomalacia** is caused by depletion of calcium and phosphorus in the bones, resulting in bones that are more prone to fractures. The requirement of vitamin D depends on the amount of calcium and phosphorus in the diet, the calcium:phosphorus ratio, and the age of the animal. Adults consuming normal calcium and phosphorus concentrations have a low vitamin D requirement because they can synthesize adequate amounts in the skin. Growing animals have higher vitamin D requirements to aid in normal bone development.

Vitamin E Vitamin E functions in the body as the major lipid-soluble antioxidant that prevents peroxidation of lipids in plasma, red blood cells, and tissues. The active form of vitamin E is α-tocopherol. The amount of vitamin E required in the diet depends on the amount of fat and selenium in the diet. Vitamin E works synergistically with selenium to neutralize free radicals. In many species, the requirement for vitamin E is inversely related to the dietary selenium concentration. As the concentration of fat in the diet increases, the vitamin E requirement increases. To maintain stability of vitamin E during processing of pet foods, esters of α-tocopherol are added to the diet.

Vitamin E toxicity and deficiency are extremely rare in companion animals. Deficiency of vitamin E is associated with skeletal muscle degeneration, poor reproduction, and impaired immune responses (Scott and Sheffy, 1987). Because of the high fat content of diets consumed by cats as compared to dogs, cats are more susceptible to vitamin E deficiencies. A condition called *pansteatitis,* or **yellow fat disease,** occurs in cats fed diets that are very high in unsaturated fat or very low in vitamin E. Diets containing high concentrations of fish, a common ingredient in cat foods, contain very high concentration of unsaturated fats and, therefore, greatly increase vitamin E requirements.

Vitamin K Vitamin K is necessary in the body for normal blood clotting. Bacteria in the intestine can synthesize menadione, the active form of vitamin K, so the dietary requirement is very low. Vitamin K deficiency and toxicity are very rare in companion animals. The dietary requirement of vitamin K is dependent on the extent of intestinal synthesis and the efficiency of absorption. Many commercial pet foods do not contain supplemental vitamin K. Because bacteria in the large intestine synthesize vitamin K, it is not known if supplemental vitamin K is necessary when compounds that interfere with microbial synthesis or absorption are present in the diet. Signs of **vitamin K deficiency** include development of ulcers and an increased time for blood clot formation (Strieker et al., 1996).

Water-Soluble Vitamins Water-soluble vitamins are absorbed either by passive diffusion or sodium-dependent active transport systems. The water-soluble vitamins required for dogs and cats are the nine B vitamins: thiamin, riboflavin, niacin, pyridoxine, pantothenic acid, biotin, folic acid, cobalamin, and choline. These vitamins are involved in the metabolism of energy and tissue synthesis. B vitamins act as coenzymes for specific energy pathways. Deficiencies and toxicities of B vitamins are extremely rare in companion animals.

The AAFCO (2003) nutrient profiles and NRC (2003) nutrient requirements for water-soluble vitamins for dogs and cats during growth and maintenance may be found in Tables 20–1 through 20–4.

Minerals

Minerals are inorganic elements that are essential for metabolism. Minerals play a vital role in maintenance of the skeleton and teeth, nerve impulse transmission, muscle contraction, cell signaling, and maintenance of acid-base balance. Problems with mineral imbalances in companion animal nutrition can greatly affect the availability of other nutrients. Experimental evidence is available for the essentiality of 11 minerals by dogs and cats (NRC, 2003). While nutritional deficiencies and toxicities can be produced experimentally, most minerals are present in adequate concentrations in commercial pet foods. Of practical interest in pet foods are calcium and phosphorus, magnesium, and zinc.

The AAFCO (2003) nutrient profiles and NRC (2003) nutrient requirements for all minerals for dogs and cats during growth and maintenance may be found in Tables 20–1 through 20–4.

Calcium and Phosphorus Calcium and phosphorus are macrominerals necessary for the formation and maintenance of bone. In addition, circulating concentrations of these minerals are involved in body metabolism. Calcium and phosphorus are normally discussed together because their regulation and metabolism in the body are closely interrelated. The availability of calcium and phosphorus in pet food ingredients must be considered to ensure that adequate concentrations of the minerals are available and that a proper calcium:phosphorus ratio is maintained. Calcium and phosphorus toxicities and deficiencies are rare in companion animals today due to properly formulated commercial diets. **Calcium toxicity** can occur in companion animals and is characterized by calcification of soft tissue. When animals are fed diets deficient in calcium, concentrations of parathyroid hormone (PTH) and active vitamin D concentrations in the blood are increased. This increase occurs because elevated blood PTH and vitamin D concentrations increase bone resorption. Resorption of calcium and phosphorus from bone are necessary to maintain normal blood calcium concentrations. Bone demineralization due to deficient calcium intakes will lead to loss of bone mass (Bennett, 1976; Hintz and Schryver, 1987). **Long-term calcium deficiency** will cause the animal to be more prone to long bone fractures. Feeding dogs and cats a diet with an improper calcium:phosphorus ratio can lead to skeletal disorders. A calcium:phosphorus ratio between 1.2:1 and 1.4:1 for dogs and 0.9:1 for cats is recommended (Kealy et al., 1996).

Magnesium Magnesium plays a role in muscle and nervous tissue function and is an important cofactor in several enzymatic reactions. As a cofactor in enzyme function, magnesium is essential for DNA and RNA metabolism and protein synthesis. Approximately 50 percent of the magnesium in the body is found in bone, while the majority of the remaining magnesium is found in the cytoplasm of soft tissues (Favus, 1996). **Magnesium toxicity** is very rare in dogs and cats because excess concentrations are excreted in the urine instead of being absorbed. Excess magnesium has been shown to increase the risk of urolithiasis and **feline lower urinary tract disease (FLUTD)** (Case et al., 2000).

Zinc Zinc has several functions in the body, including nutrient metabolism, maintenance of healthy skin, and immune function. Zinc plays a critical role as a cofactor or catalyst in over 200 reactions in the body. Dietary excesses of zinc are very rare, and dietary deficiencies are more likely to occur (Gross et al., 2000). A dietary **zinc deficiency** will cause impaired growth, poor skin and hair coat, and impaired reproductive performance in dogs (Banta, 1989). Animals fed poorly formulated diets and rapidly growing dogs are most susceptible to zinc deficiencies (Van den Broek and Thoday, 1986; Sousa et al., 1988). Growing dogs that are fed a cereal-based, low-quality pet food that contains high concentrations of antinutritional factors such as phytate, which will bind zinc and impede absorption, can develop a deficiency (Huber et al., 1991). Zinc deficiencies and excesses in cats are very rare.

Other Minerals Companion animals have requirements for other minerals in addition to calcium, phosphorus, magnesium, and zinc. Macrominerals required by dogs and cats in addition to calcium, phosphorus, and magnesium include sodium, potassium, and chloride. Trace minerals required in addition to zinc include iron, copper, manganese, iodine, and selenium. The essentiality of the trace minerals vanadium, molybdenum, boron, and chromium is unknown. Deficiencies and excesses of minerals can be produced experimentally and can occur as clinical problems in dogs and cats that are fed improperly formulated home-prepared diets (NRC, 2003). The current AAFCO (2003) nutrient profiles and NRC (2003) nutrient requirements for all minerals for dogs and cats during growth and maintenance may be found in Tables 20–1 through 20–4.

NUTRIENT DIGESTION AND DIGESTIVE PHYSIOLOGY

AAFCO publishes protocols for measuring dietary ME content of dog and cat foods that have become standards for nutrient digestibility measurement because stool collection and protein digestibility measurement are necessary for both procedures (AAFCO, 2003). Dry-matter and protein digestibility values are used to ensure maintenance of ingredient quality in the nutrition assurance programs developed separately by pet food manufacturers in the United States and Canada to ensure nutritional quality of pet foods through animal testing (Shields, 1993). AAFCO (2003) protocols for dogs and cats recommend a 5-day diet adaptation phase followed by a 5-day period of feces collection to ensure accurate digestibility measurements.

Shields (1993) identified a number of factors that affect **nutrient digestibility.** These include ingredient source, nutrient concentrations in ingredients, bioavailability of nutrients in ingredients, processing effects (e.g., particle size of ingredients, conditions imposed on the processing equipment), feeding management practices, animal factors (breed, age, gender, activity level, physiological state), and housing and environmental factors.

Information on the digestive physiology of nonruminants is covered in Chapter 1, so only features unique to the dog and (or) cat will be mentioned here. Briefly, both dogs and cats have a similar digestive tract except for its length (NRC, 2003). Dogs with a body length of 0.75 m have an intestinal length averaging 4.5 m (small intestine, 3.9 m; large intestine, 0.6m). Cats with a body length of 0.5 m have an intestinal length of approximately 2.1 m (small intestine, 1.7 m; large intestine, 0.4m). **Intestinal length** is one factor that influences the amount of time that food resides in the gut which, in turn, influences the duration of digestion (Maskell and Johnson, 1993). The surface area per centimeter of intestinal length is similar for the dog and the cat (jejunum, 54 and 50 cm^2, ileum, 38 and 36 cm^2, respectively (Maskell and Johnson, 1993). A major difference in the gastrointestinal tracts of dogs and cats is the fact that the dog proximal stomach has a thinner mucous membrane with distinct gastric glands and the distal stomach has a thicker mucous membrane with less distinct glands (NRC, 2003).

An important function of **saliva** in the dog is evaporative cooling. Upon intense parasympathetic stimulation, the parotid gland of the dog secretes saliva at 10 times the rate of the human per gram of gland weight (Argenzio, 1989).

Detailed information about digestive compartments and functions, gastric emptying, small intestinal microbiology, small intestinal transit time, and hormonal aspects of digestion can be found in NRC (2003). Enzyme activity values, electrolyte concentrations, gastric emptying and intestinal transit times, pH values, secretion characteristics and composition of bile, and absorption rates in different parts of the dog and cat gastrointestinal tracts are reported.

FEEDING REGIMENS OF DOGS AND CATS

Domestic dogs and cats have retained certain behavior patterns associated with obtaining food and feeding (Case et al., 2000). Dogs generally eat rapidly and are capable of adapting to a number of different feeding regimens (i.e., portion-controlled feeding, time-controlled feeding, free-choice feeding). Most domestic cats consume their food slowly. If fed free-choice, cats nibble at their food throughout the day (9–16 meals/day; approximately 23 kcal/meal) as opposed to consuming large amounts of food at one time (Case et al., 2000). The cat also is capable of adapting to several types of feeding regimens.

A **free-choice feeding regimen** involves having a surplus amount of food available at all times. The dog and cat must be able to self-regulate food intake in order that energy and nutrient needs are met. Control over the amount of food consumed is primarily the pet's need for energy. Dry food is most suitable for this type of feeding regimen because it will not spoil as quickly as canned food or dry out as easily as semimoist foods (Case et al., 2000). Also, compared with other feeding regimens, free-choice feeding requires the least amount of work and knowledge on the part of the pet owner. The major problem with this feeding regimen is the tendency for some dogs and cats to overeat and become obese. Indeed, most dogs and cats overconsume when first introduced to a free-choice feeding regimen but, over time, they will adjust their intake to meet caloric needs.

Portion size or amount of time that the dog or cat has access to food are controlled in a **meal feeding regimen.** Time-controlled feeding relies, in part, on the animal's ability to regulate its daily energy intake. At mealtime, a surplus of food is provided and the dog or cat is allowed to eat for a predetermined period of time, usually 15 to 20 minutes (Case et al., 2000). Some dogs and cats do not adapt well to this feeding regimen.

The feeding method of choice in most instances is **portion-controlled feeding,** which allows the owner the greatest amount of control over the pet's diet (Case et al., 2000). One or several meals are provided each day that meet the pet's daily caloric and nutrient needs. The pet's growth and weight can be strictly controlled with this method by adjusting either the amount of food or the type of food fed. An underweight condition, an overweight condition, or an inappropriate rate of growth can be corrected at an early age using this feeding regimen.

The best means of determining how much food to provide a particular animal is to first estimate the animal's energy needs and then calculate the amount of a particular pet food that must be fed to meet that need (Case et al., 2000). Age, reproductive status, body condition, level of activity, breed, temperament, environmental conditions, and health all affect a pet's energy requirement. These factors are accounted for by adding or subtracting calories from the quantity of food determined to support the maintenance energy requirement of the adult dog or cat (Case et al., 2000). The **pet food label** also provides information about how much food to provide a dog or cat. All pet foods that carry the "complete and balanced" claim are required to include feeding instructions on the product label (AAFCO, 2003). Estimates of the quantity to feed for several different ranges in body size usually are provided as well.

LIFE CYCLE NUTRITIONAL CONSIDERATIONS

Nutrient requirements of dogs and cats can vary significantly during the lifetime of the animal. While increased demands for energy and nutrients occur during gestation, lactation, growth, and performance, geriatric animals and animals at maintenance generally have a decreased need for energy and nutrients. Proper dietary management of companion animals through all life stages is necessary to maintain optimal health of the animal. The following section outlines changes in nutrient requirements and feeding recommendations for companion animals during all life stages.

Maintenance

The majority of cats and dogs kept as pets today are in the maintenance state. **Maintenance** is defined as an adult animal that is not pregnant, lactating, or working strenuously. Animals in the maintenance state must be provided with a complete and balanced meal to meet their daily needs. The main concern of the animal at maintenance is prevention of **obesity.** Commercial diets are formulated to be highly palatable, and these highly palatable diets, coupled with the less active lifestyle of many pets, have resulted in an increase in obesity. The incidence of obesity in dogs and cats is estimated to be between 25 and 30 percent (Edney and Smith, 1986; Scarlett et al., 1994; Laflamme, 1997). Increasing exercise and monitoring food intake can help prevent obesity. Dry diets are preferred to canned and semimoist foods for companion animals at maintenance. In general, canned and semimoist foods have higher energy densities than dry foods expressed on a dry-matter basis. Consuming a low-energy diet will result in lower food intake by the animal because of physical limitations of the gastrointestinal tract. It is important that low-energy diets be balanced to ensure that the animal may meet its nutrient requirements.

The amount of diet a dog or cat requires can be estimated by either the guidelines printed on the pet food label or by calculations based on the animal's ideal body weight. All pet foods that have a "complete and balanced" claim are required to provide feeding instructions on the label (AAFCO, 2003). Guidelines printed on the label of commercial diets are intended for an animal that is living indoors and provided with moderate exercise. These recommendations may need to be adjusted based on the animal's exercise level and body condition. Calculating the amount of food needed is based on the amount of energy needed to maintain the body's normal metabolism and tissue stores. Equations for determining energy requirements are described in the Energy section of this chapter.

Changing the animal's diet frequently is not recommended. **Frequent diet changes** can result in diarrhea and vomiting. Any changes to the diet should be made gradually. Adding the new diet in increasing amounts to the existing diet over several days will prevent gastrointestinal upset and stress to the animal.

Feeding regimen also must be considered for animals at maintenance. To prevent food overconsumption and weight gain, it is recommended that dogs be fed premeasured meals at regular times as opposed to *ad libitum* feeding. However, cats prefer to eat many small meals during the day (Kanarek, 1975). As a result, many cats can adapt to *ad libitum* feeding and are able to maintain a healthy body weight with this feeding method.

Growth

The most rapid period of growth for both dogs and cats occurs during the first 6 months of life. During this time, body tissue is rapidly accreted and developed, and animals have increased energy and nutrient requirements. Although specific dog and cat breeds do not have unique nutrient requirements for growth, large variation exists in the mature body size of the animals. The time it takes for animals to reach mature body size and weight is dependent on their adult size: small dogs and cats reach their mature size at 8 to 12 months of age; medium size dogs reach their adult size at 12 to 18 months; and large and giant breed dogs reach their mature size at 18 to 24 months of age (Allard et al., 1988; Douglass et al., 1988). Regardless of their mature body size, most dogs and cats have increased their body weight by approximately 50 times by the time they reach mature size. During this time of rapid growth, providing a complete and balanced diet is essential. Of particular concern for companion animals is the energy, protein, and amount and ratio of calcium and phosphorus in the diet (Case et al., 2000).

Energy requirements of puppies and kittens during growth are large. Growing puppies require approximately two times the energy of adult dogs at maintenance. At their greatest period of growth, kittens require 1.5 times more energy than adult cats (Kendall et al., 1983; Lovéridge, 1987). Because energy requirements of growing puppies and kittens are so large, the energy density of the diet is an important factor. Growing animals have a more limited capacity for digesting a meal (Earle, 1993). Due to their small gastrointestinal tract size, growing animals also have a limited amount of food that they are able to consume at one meal. If a growing kitten or puppy is fed a diet that is not energy dense, a large quantity of food must be consumed to meet the animal's nutritional needs. The animal's growth will, therefore, be limited by its food intake. It is best to feed growing animals an energy- and nutrient-dense diet. The animal will not have to consume a large amount of diet to meet its nutritional needs, so the size of the animal's gastrointestinal tract will not limit its capacity for growth.

In addition to an increased energy requirement, growing puppies and kittens have an increased **protein requirement.** This increased requirement is due to the need for protein to build new body tissue. Due to their decreased digestive capacity, dietary protein provided to growing animals should be easily digested. Because puppies and kittens are consuming more food due to their increased energy requirement, they are typically consuming increased concentrations of dietary protein. The total amount of protein in the diet is not as important as the proportion of protein to energy. The optimal concentration of energy that should be provided by dietary protein is between 25 and 29 percent of ME for puppies and 30 and 36 percent of ME for kittens (Case et al., 2000). Providing adequate dietary protein to puppies and kittens is essential for normal growth.

Of particular concern for large breed dogs is the amount of calcium and phosphorus provided in the diet. Diets for growing companion animals should not contain excessive calcium and phosphorus concentrations, or **developmental bone diseases** can occur. Commercially available diets for growing pets will provide more than the adequate calcium and phosphorus recommendation to ensure that all growing animals will be provided with adequate concentrations of these minerals. Commercially available diets that are specifically formulated for large breed dogs will contain lower calcium and phosphorus concentrations. These diets will also contain a slightly decreased energy density. The changes in diet formulation for large breed dogs are to prevent rapid growth. Large breed dogs have a predisposition to skeletal diseases, so diets that promote a slow growth rate and skeletal development are preferred (Dammrich, 1991; Hazewinkel et al.,

1985; Kealy et al., 1992). If large dogs grow at a maximum rate, skeletal diseases such as osteochondrosis and hip dysplasia can develop.

It is recommended that growing animals be fed a diet formulated specifically for growing animals. These growth formulas contain high-quality ingredients that are highly digestible.

It is important to meal-feed growing dogs to prevent rapid growth. Dogs should be fed to attain an average, not maximal, growth rate. The energy needs of the animal will change during the growth period, so the amount fed should be assessed weekly by examining the animal's body condition (Armstrong and Lund, 1996). Dogs should be fed two to three small meals daily throughout their growth period. Dogs should be fed on a portion-controlled basis until they have reached 80 to 90 percent of their adult weight (Malm and Jensen, 1996). In addition to restricting intake to prevent maximal growth, large dogs should not have prolonged exercise. Prolonged stress on growing joints will cause stress on the developing skeletal system.

Growing kittens naturally consume several small meals daily. Most growing cats are capable of regulating their energy intake when fed *ad libitum.* Because of their natural eating patterns, kittens do not experience the same problems as puppies with excessive caloric intakes and growth rates. It is recommended that kittens also be fed a diet formulated for growth. If growing kittens gain excessive weight or do not receive adequate exercise, the animals can be meal-fed (Case et al., 2000).

Gestation and Lactation

Proper care and feeding during reproduction is necessary to ensure the health of both the mother and the offspring. In pregnant bitches, less than 30 percent of fetal growth occurs during the first 6 weeks of pregnancy. The small amount of fetal growth during this time means that the pregnant bitch has only a slight increase in nutritional needs during this time (Case et al., 2000). In the dog, more than 75 percent of fetal growth occurs during the last few weeks of gestation (Moser, 1992). As long as the bitch is healthy at the onset of **gestation,** food intake should not be increased until the sixth week of gestation. After the sixth week, food intake should be gradually increased so that the daily food intake is 25 to 50 percent higher than normal maintenance needs at parturition (Moser, 1992). As the fetuses develop, there is a reduction in abdominal size available for stomach expansion after meal consumption. The bitch will not be able to consume large meals when the abdominal space is reduced. As a result, it is necessary for the bitch to receive several small meals per day to ensure that she is consuming an adequate quantity of food.

Some differences exist between gestating bitches and queens. While bitches gain the majority of their body weight in the last portion of gestation, pregnant queens exhibit a linear increase in weight after the second week of gestation. In addition, queens are able to store body fat during gestation to use during lactation. Dogs lose nearly all of their weight gain at parturition, whereas cats lose only 40 percent of their weight gain at parturition (Loveridge, 1986). The gestating queen's food intake should be gradually increased after the second week of gestation so that the daily food intake is 25 to 50 percent higher than normal maintenance needs at parturition. Queens are capable of *ad libitum* feeding throughout pregnancy to maintain an ideal body weight. However, the queen should be carefully monitored to ensure that she is not losing or gaining excessive weight.

For both the gestating bitch and queen, the best diet to feed is one formulated specifically for animals in this life stage. It is not necessary to supplement the gestating animal with calcium or any other nutrient. Any increase in nutrient requirements is compensated for by the increased intake of their normal diet (Case et al., 2000).

During **lactation,** the most important nutrient consideration is the increased energy requirement. An increased energy intake is necessary for adequate milk production and to prevent the bitch or queen from losing excessive body weight. Peak lactation occurs at 3 to 4 weeks after parturition, so the energy requirement of the lactating animal is highest during this period. Owners should feed 1.5 times the maintenance requirement during the first week of lactation, 2 times the maintenance requirement during the second week, and 2.5 to 3 times the maintenance requirement during the third and fourth week of lactation (Kronfeld, 1975). If a lactating animal is energy deficient, the animal may lose weight and have a decreased milk output.

In addition to an increased energy requirement, the lactating animal will have an increased water requirement. Adequate cool, fresh water must be available to the lactating animal. If the animal is dehydrated, then milk production will be significantly decreased.

It is recommended that the lactating animal be fed a highly digestible, energy-dense diet. A diet that is formulated for performance or high activity is suggested. The quantity of food necessary for the bitch to maintain an adequate body weight may be too large to consume at one meal. Therefore, several small meals may need to be fed during the day, or the animal may be fed *ad libitum.* Puppies and kittens will be introduced to solid food at 3 to 4 weeks. At this time, daily food intake can slowly be reduced. After her offspring are completely weaned from nursing, a highly digestible energy-dense diet should be fed until the queen or bitch has returned to normal body status (Case et al., 2000).

Neonatal Nutrition

The **neonatal period** is defined as the first 2 weeks after birth. During this time, puppies and kittens are rapidly growing and developing and are entirely dependent on the mother for food. The nutritional needs of the neonate are met entirely through nursing. Immediately following parturition, the mother produces colostrum that provides immunity to newborn puppies and kittens. The immune system of animals is not functional immediately after birth. Immunoglobulins and other immune factors produced by the mother are absorbed through the intestinal mucosa of the newborn without being altered by the newborn's digestive enzymes. Absorption of these intact proteins will help prevent infectious disease in the newborn until the animal's immune system is entirely functional. Neonates are able to absorb intact proteins for the first 48 hours after birth (Fisher, 1982). After 48 hours, the gastrointestinal tract is not capable of absorbing intact proteins.

Puppies and kittens will nurse between four and six times per day for the first 4 weeks after birth. After 4 weeks, milk will no longer provide the calories or nutrients needed to continue normal growth. Therefore, at 3 to 4 weeks of age, supplemental food should be introduced to the puppy or kitten. Commercial foods specifically formulated for weaning puppies and kittens are available. A gruel of warm water and the mother's food also can be used for weaning. Puppies and kittens should be offered food for 20 to 30 minutes several times daily. Most animals are completely weaned by 6 weeks of age. Dry pet food that is formulated for puppies and kittens can be fed between the ages of 6 and 10 weeks (Case et al., 2000).

Performance

Endurance Performance Endurance performance of dogs involves prolonged periods of exercise. The best example of dogs involved in **endurance exercise** is sled dogs (Fig.20.2). The Iditarod, perhaps the best-known long-distance sled dog race, is approximately 1,160 miles long. During this race, sled dogs may run for up to 100 miles daily. Other types of **working dogs** include those used in hunting (Fig 20.3) and livestock herding (sheep and cattle dogs).

Early research suggested that the energy requirement of sled dogs increased between 1.5 and 2.5 times the normal maintenance requirement, although recent research has concluded that the energy requirements are much higher than previously estimated. When racing, dogs will expend 460 kcal/kg body weight/day and consume 440 kcal/kg body weight/day (Hinchcliff et al., 1997b).

The two primary metabolic fuels used in muscle are glycogen, the storage form of carbohydrate, and free fatty acids. Dogs performing high-endurance activity rely more on the use of fatty acids as fuel through aerobic metabolism than the use of muscle glycogen through anaerobic metabolism (Hammel et al., 1976).

FIGURE 20.2 Sled dogs (e.g., Those running the Iditarod in Alaska) have high energy requirements met primarily by the metabolism of fatty acids. Fat is the major energy source for endurance performance. (Courtesy of A. M. Craig and Linda L. Blythe.)

FIGURE 20.3 Hunting dogs are endurance performers with high energy requirements. (Courtesy of Howard H. Meyer.)

Research has shown that fat is the preferred fuel for endurance exercise (Reynolds et al., 1994), but sled dogs also must maintain adequate glycogen stores. Some glycogen metabolism is necessary in sled dogs, as they undergo periods of intense anaerobic activity where carbohydrates will be metabolized for energy (Reynolds et al., 1997). A decrease in muscle glycogen stores can result in decreased performance of sled dogs. Feeding sled dogs a diet high in fat will prepare them to utilize free fatty acids as the main energy source and spare muscle glycogen for use in times of anaerobic metabolism during a race.

Protein also plays an important role in the diet of sled dogs. Both training and racing increase the animal's need for protein due to both increased protein synthesis and protein degradation (Adkins and Kronfeld, 1982). Due to training for endurance exercise, dogs will have an increase in lean body mass that must be maintained by increasing dietary protein. During prolonged exercise, protein may be used as well as fat and glycogen for energy. It is estimated that 10 percent of energy used by sled dogs during racing is due to metabolism of gluconeogenic amino acids (Hinchcliff et al., 1993). Providing a high protein diet to dogs undergoing endurance exercise will ensure tissue accretion, prevent tissue loss, and possibly prevent injury.

It is recommended that dogs undergoing endurance exercise be fed a highly digestible energy-dense diet. If their diet is not energy dense, endurance dogs will not be able to consume enough food to meet their energy requirements. Diets formulated for sled dogs should contain 50 to 65 percent of calories from fat, 10 to

15 percent of calories from carbohydrate, and 30 to 35 percent of calories from protein (Case et al., 2000). Supplementing with higher concentrations of protein will cause protein to be utilized for energy instead of fat. Protein is not a desirable fuel for energy during exercise because it is inefficiently metabolized. By providing protein at a concentration of 30 to 35 percent of the calories, the animal will be able to use the dietary protein to maintain tissue body stores. Dogs should be meal-fed twice daily to allow the trainer to regulate meal size and time. A small meal 2 hours before endurance exercise will improve performance (Case et al., 2000). In addition to providing an energy-dense, highly digestible diet, it is important to provide endurance exercise dogs with adequate water. Water losses can increase by 20 times during endurance exercise (Hinchcliff et al., 1997a). Therefore, working dogs should be given frequent opportunities to drink small amounts of water while performing endurance work.

Sprint Racing Performance **Sprint racing** involves brief intense periods of exercise (Fig 20.4). Most racing Greyhounds run short distances between 5/16 and 3/8 of a mile. Due to the short endurance of sprint exercise, fat cannot be mobilized in time to provide energy. Therefore, the requirements of the racing Greyhound differ greatly from those of the sled dog.

Although a sprint racing dog will have an increased requirement for energy, the increase is not nearly as large as is seen in the sled dog. The primary source of energy for racing Greyhounds is the anaerobic metabolism of carbohydrate (Dobson et al., 1988). During a race, blood lactate concentrations increase by nearly

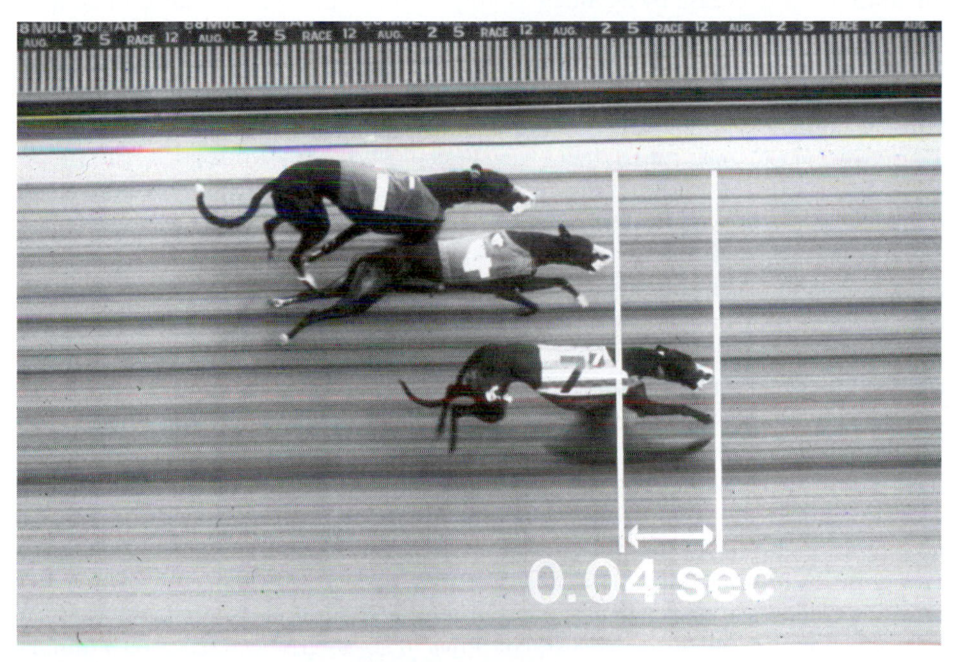

FIGURE 20.4 Greyhounds are sprint racers, having brief intense periods of exercise. Their primary source of energy, in contrast to endurance performers, is the anaerobic metabolism of carbohydrate.(Courtesy of Linda L. Blythe.)

30 times, indicating that the Greyhound is well-adapted to the anaerobic metabolism of glucose for energy (Bjotvedt et al., 1984). Sprinting dogs can rapidly utilize muscle glycogen for anaerobic metabolism.

Sprinting Greyhounds should be fed a highly digestible, energy-dense diet. It is recommended that sprinting dogs be fed a diet that is formulated for performance. The diet should contain moderate concentrations of fat, protein, and carbohydrate. Supplementation of carbohydrate after sprint exercise will help the animals replace depleted muscle glycogen stores (Reynolds et al., 1997).

Geriatrics

Improvements in nutrition and veterinary medicine have resulted in companion animals having an increased life span. It is estimated that greater than 40 percent of dogs and 30 percent of cats are greater than 6 years of age (Case et al., 2000). The breed and adult size of a dog determine when a dog is considered geriatric. Small breed dogs live longer than large and giant breeds (Deeb and Wolf, 1994). Whereas small dogs are considered geriatric at 11.5 years of age, large breed dogs are considered geriatric at 7.5 years. Cats are considered to be geriatric at 12 years of age. Nutritional **goals for geriatric animals** are to slow metabolic changes associated with aging, minimize signs of aging, and increase the life span of the pet.

Several physiological changes are associated with aging, including a decreased RFMR, decreased lean body mass, increased body fat, and decreased digestive efficiency. An animal's RFMR naturally decreases with aging, which is due to changes in the animal's body condition. The total daily energy requirement of an animal can decrease by as much as 40 percent in the last one-third of an animal's life (Mosier, 1989). In addition to a decrease in RFMR, animals naturally experience a decrease in lean body mass and an increase in body fat. The body fat of young dogs is between 15 and 30 percent, whereas geriatric animals have a body fat concentration of between 25 and 30 percent, (Meyer and Stadtfeld, 1980). Young cats have 30 percent body fat, whereas geriatric cats have a body fat concentration of 35 percent (Hayek and Davenport, 1998). As the animal ages, several changes take place in the gastrointestinal tract. Geriatric animals have decreased salivary and gastric secretions, intestinal villus size, cellular turnover, and colonic motility (Venn, 1992). As dogs and cats age, there is a decline in digestive efficiency (Taylor et al., 1995). Although physiological changes associated with aging will not severely affect the health of the animal, these factors must be considered when selecting a diet for geriatric dogs and cats.

A decreased RFMR coupled with reduced exercise results in a decrease in the energy requirement of the geriatric animal. Although elderly dogs and cats have a total reduction in dietary energy needs of 30 to 40 percent, much variation exists in the energy requirements of geriatric animals (Mosier, 1989). Geriatric animals must be closely monitored so that the animals are obtaining an adequate amount of energy without developing obesity.

The decrease in lean body mass noted with geriatric animals is due to a loss of protein reserves that are used in times of stress and illness. Older animals are particularly susceptible to illness and stress. In order to replace losses in lean body mass, geriatric animals should be provided with diets that contain a high-quality protein at a concentration that will adequately replace body losses. Geriatric dogs have a decreased ability to efficiently utilize dietary protein (Wannemacher and McCoy, 1966). Although geriatric dogs and cats have an increased dietary protein requirement, these animals also have a decrease in renal function. Feeding a high-protein diet to animals with kidney disease associated with age may result in chronic renal failure associated with weight loss, muscle wasting, decreased food intake, and decreased intestinal absorption of nutrients (Polzin, 1987). Although geriatric animals are prone to developing kidney disease, it is recommended that healthy animals be offered a high protein diet. Geriatric dogs and cats with renal failure should be fed a diet containing a moderate concentration of protein.

Geriatric companion animals should be fed to maintain optimal health and body condition. Healthy geriatric animals should be fed a high protein, energy dense commercial diet that is formulated for geriatric pets or animals at maintenance. A high-quality protein is preferred, as it is more easily digested by the geriatric animal. It is vital to a geriatric animal's health to maintain a healthy body weight and prevent the onset of obesity. This can be accomplished by observing food intake and maintaining a regular exercise schedule. Because monitoring food intake is essential, it may be necessary to meal-feed geriatric animals. It is recommended that geriatric animals be fed two to three small meals per day to promote improved nutrient utilization (Case et al., 2000). Providing the geriatric dog and cat with a highly digestible diet and maintaining the animal at an ideal body condition will help prolong the life of the animal.

SUMMARY AND CONCLUSIONS

Major advances in the field of companion animal nutrition have been made in recent years, but much remains to be accomplished. Morris and Rogers (1994) indicated that "progress in the nutrition of companion animals requires more precise information on the requirements of cats and dogs at various physiological stages, bioavailability of nutrients, interaction among nutrients and the role of various nutrients in the prevention of disease." This statement succinctly summarizes key areas where more data are required if we are to move beyond the point of using empirical information to formulate the many types of diets currently fed to dogs and cats. Also, as nutrition is an integrative science, more accurate information is needed in the nutrition arena to effectively capitalize on the interactions that can occur between nutrition and other disciplines such as physiology, immunology, pathology, toxicology, and genomics (Fahey, 2003). Companion animal nutrition as a discipline is distinguished by both its unique knowledge base and its mission to apply this knowledge to companion animal health and disease. It is within the realm of possibility that dietary allowances/requirements will one day be tailored to individual genotype for optimal companion animal management, health, and disease prevention (Fahey, 2003). Exciting opportunities exist in companion animal nutrition research and the various other disciplines that impact the nutritional state of the animal.

QUESTIONS AND STUDY GUIDE

1. What advantages do premium brand pet foods offer over generic foods?
2. What are expanded pet foods?
3. How does the extrusion process affect starch utilization by dogs and cats?
4. Most dog foods contain ingredients that have been cooked or extruded. Explain why.
5. What is AAFCO, and what are its contributions to the pet food industry?
6. What are the differences between BMR and RFMR?
7. How and why do dogs and cats differ in their requirements for essential fatty acids? What are the functions of fat in the diet?
8. Why do cats have a high maintenance requirement for protein?
9. Why is taurine an essential nutrient for cats?
10. Cats cannot convert β-carotene to vitamin A. How might this be explained by their evolutionary history?
11. How do the energy requirements of racing Greyhounds differ from those of sled dogs?
12. What are the disadvantages of feeding excess protein to dogs?
13. What is the comparative value of all-meat diets for dogs and cats? Which animal is better suited to all-meat diets?

REFERENCES

AAFCO. 2003. *Official Publication.* Oxford, IN: Association of American Feed Control Officials, Inc.

Adkins, T.O. and D.S. Kronfeld. 1982. Diet of racing sled dogs affects erythrocyte depression by stress. *Can. Vet. J.* 23:260–263.

Allard, R.L., G.M. Douglass, and W.W. Kerr. 1988. The effects of breed and sex on dog growth. *Comp. Anim. Pract.* 2:15–19.

Anderson, R.S. 1981. Water content in the diet of the dog. *Vet. Ann.* 21:171–178.

Argenzio, R.A. 1989. Secretory functions of the gastrointestinal tract. In *Dukes' Physiology of Domestic Animals,* 10th ed., M.J. Swenson, ed. pp. 290–300. Ithaca, NY: Cornell University Press.

Armstrong, P.H. and E.M. Lund. 1996. Changes in body composition and energy balance with aging. *Vet. Clin. Nutr.* 3:83–87.

Arnold, A. and C.A. Elvehjem. 1939. Nutritional requirements of dogs. *J. Am. Vet. Med. Assoc.* 95:187–194.

Backus, R.C., J.G. Morris, S.W. Kim, J.A. O'Donnell, M.A. Hickman, C.A. Kirk, J.A. Cooke, and Q.R. Rogers. 1998. Dietary taurine needs of cats vary with dietary protein quality and concentration. *J. Vet. Clin. Nutr.* 5:18–22.

Banta, C.A. 1989. The role of zinc in canine and feline nutrition. In *Nutrition of the Dog and Cat,* I.H. Burger and J.P.W. Rivers, eds. pp. 317–327. New York: Cambridge University Press.

Bennett, D. 1976. Nutrition and bone disease in the dog and cat. *Vet. Rec.* 98:313–320.

Bjotvedt, G., C.W. Weems, and K. Foley. 1984. Strenuous exercise may cause health hazards for racing Greyhounds. *Vet. Med.* 79:1481–1487.

Blaxter, K. 1989. *Energy Metabolism in Animals and Man.* Cambridge, UK: Cambridge University Press.

Buddington, R.K., J.W. Chen, and J.M. Diamond. 1991. Dietary regulation of intestinal brush-border sugar and amino acid transport in carnivores. *Am. J. Physiol.* 261:R793–R801.

Case, L.P., D.P. Carey, D.A. Hirakawa, and L. Daristotle. 2000. *Canine and Feline Nutrition: A Resource for Companion Animal Professionals.* 2d ed. St. Louis, MO: Mosby, Inc.

Cizek, L.J. 1959. Long-term observations on relationship between food and water ingestion in the dog. *Am. J. Physiol.* 197:342–346.

Cline, J. L., G.L. Czarnecki-Maulden, J.M. Losonsky, C.R. Sipe, and R.A. Easter. 1997. Effect of increasing vitamin A on bone density in adult dogs. *J. Anim. Sci.* 75: 1045–1056.

Cowley, A.W., M.M. Skelton, D.C. Merrill, E.W. Quillen, and S.J. Switzer. 1983. Influence of daily sodium intake on vasopressin secretion and drinking in dogs. *Am. J. Physiol.* 245:R860–R872.

Crossley, A. 2003. The petfood market: Noteworthy trends. *Petfood Industry* 34(6):14–16.

Dammrich, K. 1991. Relationship between nutrition and bone growth in large and giant dogs. *J. Nutr.* 121:S114–S121.

Deeb B.J. and N.S. Wolf. 1994. Studying longevity and morbidity in giant and small breeds of dogs. *Vet. Med. Suppl.* 7: 702–709.

Dobson, G.B., WS. Parkhouse, J.M. Weber, E. Stuttard, J. Harman, D.H. Snow, and P.W. Hochachka. 1988. Metabolic changes in skeletal muscle and blood of Greyhounds during 800-m track sprint. *Am. J. Physiol.* 255: R513–R519.

Douglass, G.M., E. Kane, and E.J. Holmes. 1988. A profile of male and female cat growth. *Comp. Anim. Pract.* 2:9–12.

Earle, K.E. 1993. Calculations of energy requirements of dogs, cats, and small psittacine birds. *J. Sm. Anim. Pract.* 34:163–183.

Earle, K.E. and P.M. Smith. 1993. Balanced diets for dogs and cats. In *The Waltham Book of Companion Animal Nutrition,* I. Burger, ed. pp. 45–55. Oxford: Pergamon Press.

Edney, A.T.B. and A.M. Smith. 1986. Study of obesity in dogs visiting veterinary practices in the United Kingdom. *Vet. Rec.* 118:391–396.

Fahey, Jr., G.C. 2003. Research needs in companion animal nutrition. In *Petfood Technology,* J. L. Kvamme and T. D. Phillips, eds. pp. 57–61. Mt. Morris, IL: Watt Publishing.

Favus, M. 1996. *Primer on the Metabolic Bone Diseases and Disorders of Mineral Metabolism.* Philadelphia, PA: Lippincott-Raven.

Fisher, E.W. 1982. Neonatal diseases of dogs and cats. *Br. Vet. J.* 138: 277–284.

Golob, P., W.J. O'Connor, and D.J. Potts. 1984. Increase in weight and water retention on overfeeding dogs. *Quart. J. Exp. Physiol.* 69:245–256.

Gray, G. 1992. Starch digestion and absorption in nonruminants. *J. Nutr.* 122:172–177.

Greaves, J.P. 1965. Protein and calorie requirements of the feline. In *Canine and Feline Nutritional Requirements,* O. Graham-Jones, ed. pp. 33–45. London: Pergamon Press.

Gross, K., K. Wedekind, C. Cowell, W. Schoenherr, D. Jewell, S. Zicker, J. Debraekeleer, and R. Frey. 2000. Nutreints. pp. 66–80. In *Small Animal Clinical Nutrition.* M. Hand, C. Thatcher, R. Remillard, and P. Roudebush, eds. Marceline, MD: Walsworth Publishing.

Hammel, E.P., D.S. Kronfeld, V.K. Ganjam, and H.L. Dunlap, Jr. 1976. Metabolic responses to exhaustive exercise in racing sled dogs fed diets containing medium, low, and zero carbohydrate. *Am. J. Clin. Nutr.* 30:409–418.

Hayek, M.G. and G.M. Davenport. 1998. Nutrition and aging in companion animals. *J. Anti-Aging Med.* 1:117–123.

Hayes, K.C. 1971. On the pathophysiology of vitamin A deficiency. *Nutr. Rev.* 29:3–6.

Hazewinkel, H.A.W., S.A. Goedegebuure, P.W. Poulos, and W.A. Wolvekamp. 1985. Influences of chronic calcium excess on the skeletal development of growing Great Danes. *J. Am. Anim. Hosp. Assoc.* 21:377–391.

Hendriks, W.H. and S. Wamberg. 2000. Milk intake of suckling kittens remains relatively constant from one to four weeks of age. *J. Nutr.* 130:77–82.

Herschel, D.A., R.A. Argenzio, M. Southworth, and C.E. Stevens. 1981. Absorption of volatile fatty acids and water by the colon of the dog. *Am. J. Vet. Res.* 42:1118–1124.

Hinchcliff, K.W., J. Olson, C. Crusberg, J. Kenyon, R. Long, W. Royle, W. Weber, and J. Burr. 1993. Serum biochemical changes in dogs competing in a long-distance sled race. *J. Am. Vet. Med. Assoc.* 202:401–405.

Hinchcliff, K.W., G.A. Reinhart, J.R. Burr, C. J. Schreier, and R.A. Swenson. 1997a. Effect of racing on serum sodium and potassium concentration and acid: Base status of Alaskan sled dogs. *J. Am. Vet. Med. Assoc.* 210: 1615–1618.

Hinchcliff, K.W., G.A. Reinhart, J.R. Burr, C.J. Schreier, and R.A. Swenson. 1997b. Metabolizable energy intake and sustained energy expenditure of Alaskan sled dogs during heavy exertion in the cold. *Am. J. Vet. Res.* 58:1457–1462.

Hintz, H.F. and H.F. Schryver. 1987. Nutrition and bone development in dogs. *Comp. Anim. Pract.* 1:44–47.

Huber T., D. Laflamme, L. Medleau, K. Comer, and P. Rakich. 1991. Comparison of procedures for assessing adequacy of dog foods. *J. Am. Vet. Med. Assoc.* 199:731–734.

Huber, T.L., R.C. Wilson, and S.A. McGarity. 1986. Variations in digestibility of dry dog food with identical label guaranteed analyses. *J. Am. Anim. Hosp. Assoc.* 22: 571–575.

Johnson, M.L., C.M. Parsons, G.C. Fahey, Jr., N.R. Merchen, and C.G. Aldrich. 1998. Effects of species raw material source, ash content, and processing temperature on amino acid digestibility of animal by-product meals by cecectomized roosters and ileally cannulated dogs. *J. Anim. Sci.* 76:1112–1122.

Kanarek, R.B. 1975. Availability and caloric density of diet as determinants of meal patterns in cats. *Physiol. Behav.* 15:611–618.

Kane, E., J.G. Morris, and Q.R. Rogers. 1981. Acceptability and digestibility by adult cats of diets made with various sources and levels of fat. *J. Anim. Sci.* 53:1516–1523.

Kealy, R.D., D.F. Lawler, and J.M. Ballam. 1996. Dietary calcium:phosphorus ratio for adult cats. *Vet. Clin. Nutr.* 3:28.

Kealy, R.D., S.E. Olsson, K.L. Monti, D.F. Lawler, D.N. Biery, R.W. Helms, G. Lust, and G.K. Smith. 1992. Effects of limited food consumption on the incidence of hip dysplasia in growing dogs. *J. Am. Vet. Med. Assoc.* 201:857–863.

Kendall, P.T. 1984. The use of fat in dog and cat diets. In *Fats in Animal Nutrition,* J. Wiseman, ed. pp. 383. Boston, MA: Butterworths.

Kendall, P.T., S.E. Blaza, and P.M. Smith. 1983. Comparative digestible energy requirements of adult beagles and domestic cats for body weight maintenance. *J. Nutr.* 113:1946–1955.

Kendall, P.T., D.W. Holme, and P.M. Smith. 1982. Comparative evaluation of net digestive and absorptive efficiency in dogs and cats fed a variety of contrasting diet types. *J. Sm. Anim. Pract.* 23:577–587.

Kienzle, E. 1988. Enzymeaktivitaet in pancreas, darmwant und chymus des hundes in abhangigkeit von alter und futterart. *J. Anim. Physiol. Anim. Nutr.* 60:276–288.

_______ . 1993. Carbohydrate metabolism in the cat. 1. Activity of amylase in the gastrointestinal tract of the cat. *J. Anim. Physiol. Anim. Nutr.* 69:92–101.

Kronfeld, D.S. 1975. Nature and use of commercial dog foods. *J. Am. Vet. Med. Assoc.* 166:487–493.

LaFlamme, D. 1997. Development and validation of a body condition score system for dogs. *Canine Pract.* 22:10–15.

Levinson, R.A. and E. Englert, Jr. 1970. Small intestinal absorption of simple sugars and water in the cat. *Experentia* 26:262–263.

Loveridge, G.G. 1986. Body weight changes and energy intakes of cats during gestation and lactation. *Anim. Technol.* 37:7–15.

Loveridge, G.G. 1987. Factors affecting growth performance in male and female kittens. *Anim. Tech.* 38:9–18.

Malm, K. and P. Jensen. 1996. Weaning in dogs: Within and between litter variation in milk and solid food intake. *Appl. Anim. Behav. Sci.* 49:223–235.

Maskell, I.E. and J.V. Johnson. 1993. Digestion and absorption. In *The Waltham Book of Companion Animal Nutrition,* I. Burger, ed. pp. 25–44. Oxford, UK: Pergamon Press.

Maynard, L.A., J.K. Loosli, and H.F. Hintz. 1979. *Animal Nutrition* 7th ed. New York: McGraw-Hill.

Mellanby, E. 1938. The experimental production of deafness in young animals by diet. *J. Physiol.* 94:316–321.

Meyer, H., J. Arndt, T. Behfeld, H. Elbers, and C. Schunemann. 1989. Praecaecale und postileale Verdaulichkeit verschiedener Eiweisse. *Adv. Anim. Physiol. Anim. Nutr.* 19:59–77.

Meyer, H., C. Dammers, and E. Kienzle. 1985. Body composition of newborn puppies and nutrient requirements of pregnant bitches. *Adv. Anim. Physiol. Anim. Nutr.* 16:7–25.

Meyer, H., P.J. Schmitt, and E. Heckotter. 1981. Nutritional content and digestibility of food stuffs for dogs. *Anim. Nutr.* 9:71–104.

Meyer, H. and G. Stadtfeld. 1980. Investigation on the body and organ structures of dogs. In *Nutrition of the Dog and Cat,* R.S. Anderson, ed. pp. 15–30. Oxford, UK: Pergamon Press.

Morris, J.G. and Q.R. Rogers. 1994. Assessment of the nutritional adequacy of pet foods through the life cycle. *J. Nutr.* 124:2520S–2534S.

Moser, D. 1992. Feeding to optimize canine reproductive efficiency. *Prob. Vet. Med.* 4:545–550.

Mosier, J.E. 1989. Effect of aging on body systems of the dog. *Vet. Clin. N. Am. Sm. Anim. Pract.* 19:1–13.

Muir, H.E., S.M. Murray, G.C. Fahey, Jr., N.R. Merchen, and G.A. Reinhart. 1996. Nutrient digestion by ileal cannulated dogs as affected by dietary fibers with various fermentation characteristics. *J. Anim. Sci.* 74:1641–1648.

NRC. 2003. *Nutrient Requirements of Dogs and Cats.* Washington, DC: The National Academy Press.

Nguyen, P., H. Dumon, V. Biourge, and E. Pouteau. 1998. Glycemic and insulinemic responses after ingestion of commercial foods in healthy dogs: Influence of food composition. *J. Nutr.* 128:2654S–2658S.

Nguyen, P., H. Dumon, P. Buttlin, L. Martin, and A. Gouro. 1994. Composition of meal influences changes in postprandial incremental glucose and insulin in healthy dogs. *J. Nutr.* 2707S–2711S.

Polzin, D.J. 1987. Topics in general medicine: General nutrition; the problems associated with renal failure. *Vet. Med.* 82:1027–1035.

Rainbird, A.L. 1988. A balanced diet. In *The Waltham Book of Canine Nutrition,* 2d ed., A.T.B. Edney, ed. pp. 57–74. Oxford, UK: Pergamon Press.

Reynolds, A.J., D.P. Carey, G.A. Reinhart, R.A. Swenson, and F.A. Kallfelz. 1997. Effect of postexercise carbohydrate supplementation on muscle glycogen repletion in trained sled dogs. *Am. J. Vet. Res.* 60:789–795.

Reynolds, A.J., L. Fuhrer, H.L. Dunlap, M.D. Finke, and F.A. Kallfelz. 1994. Lipid metabolite responses to diet and training in sled dogs. *J. Nutr.* 124:2754S–2759S.

_______ . 1995. Effect of diet and training on muscle glycogen storage and utilization in sled dogs. *J. Appl. Physiol.* 79:1601–1607.

Romsos, D.R., H.J. Palmer, K.L. Muiruri, and M.R. Bennink. 1981. Influence of a low carbohydrate diet on performance of pregnant and lactating dogs. *J. Nutr.* 111:678–689.

Scarlett, J.M., S. Donoghue, J. Saidla, and J. Wills. 1994. Overweight cats: Prevalence and risk factors. *Int. J. Obesity.* 18:S22–S28.

Schweigert, F.J., O.A. Ryder, W.A. Rambeck, and H. Zucker. 1990. The majority of vitamin A is transported as retinyl esters in the blood of most carnivores. *Comp. Biochem. Physiol.* 95A:573–578.

Schweigert, F.J., E. Thomann, and H. Zucker. 1991. Vitamin A in the urine of carnivores. *Int. J. Vit. Nutr. Res.* 61:110–113.

Scott, D.W. and B.E. Sheffy. 1987. Dermatosis in dogs caused by vitamin E deficiency. *Comp. Anim. Pract.* 1:42–46.

Seefeldt, S.L. and T.E. Chapman. 1979. Body water content and turnover in cats fed dry and canned rations. *Am. J. Vet. Res.* 40:183–185.

Shields, R.G. 1993. Digestibility and metabolizable energy measurement in dogs and cats. In *Proc. Petfood Forum 93.* pp. 21–35. Mt. Morris, IL: Watt Publishing.

Sousa, C.A., A.A. Stannard, and P.J. Ihrke. 1988. Dermatosis associated with feeding generic dog food: 13 cases (1981–1982). *J. Am. Vet. Med. Assoc.* 192:676–680.

Strieker, M.J., J.G. Morris, and B.F. Feldman. 1996. Vitamin K deficiency in cats fed commercial fish-based diets. *J. Sm. Anim. Pract.* 37:322–326.

Sunvold, G.D., G.C. Fahey, Jr., N.R. Merchen, L.D. Bourquin, E.C. Titgemeyer, L.L. Bauer, and G.A. Reinhart. 1995. Dietary fiber for cats: *In vitro* fermentation of selected fiber sources by cat fecal inoculum and *in vivo* utilization of diets containing selected fiber sources and their blends. *J. Anim. Sci.* 73:2329–2339.

Taylor, E.J., C. Adams, and R. Neville. 1995. Some nutritional aspects of ageing in dogs and cats. *Proc. Nutr. Soc.* 54:645–656.

Van den Broek, A.H.M. and K.L. Thoday. 1986. Skin disease in dogs associated with zinc deficiency: A report of five cases. *J. Sm. Anim. Pract.* 27:313–323.

Venn, A. 1992. Diets for geriatric patients. *Vet. Times.* May.

Wannemacher, R.W. and J.R. McCoy. 1966. Determination of optimal dietary protein requirements of young and old dogs. *J. Nutr.* 88:66–74.

Washabau, R.J., D.R. Strombeck, C.A. Buffington, and D. Harrold. 1986. Evaluation of intestinal carbohydrate malabsorption in the dog by pulmonary hydrogen gas excretion. *Am. J. Vet. Res.* 47:1402–1406.

Zentek, J. 1993. Studies on the effect of feeding on the microbial metabolism in the intestinal tract of dogs. Habilitation thesis, Tieraerzliche Hochschule, Hanover, Germany.

Zentek, J. and H. Meyer. 1993. Digestibility and faecal excretion of water—A comparison between Great Danes and Beagles. *Kleintierpraxis* 38:311–318.

CHAPTER 21

Feeding and Nutrition of Fur-Bearing Animals

A. Skrede and P. R. Cheeke

Objectives

1. To describe the principles of feeding carnivorous fur-bearing animals such as mink and foxes
2. To describe the hair-growth cycle and the role of nutrition in hair growth
3. To discuss nutritional deficiencies associated with the feeding of animal by-products to fur animals, including deficiencies of biotin, iron, and thiamin

The principal animal grown commercially for its fur is the **mink,** which is produced mainly in Denmark and other Scandinavian countries, the Netherlands, China, the United States, Canada, and Russia. To a lesser extent, polar **foxes** (blue foxes) and silver foxes are raised for their pelts, mainly in Finland, Norway, China, Russia, and Poland. The production of other fur-bearing species such as nutria and chinchilla is very minor. Mink and foxes are by nature carnivores and their natural diets are rich in animal protein and fat, and low in carbohydrate. The farmed mink and foxes are usually fed diets containing animal sources such as fish by-products, either raw or converted into fish meal or fish silage, and by-products from slaughterhouses. These products are mixed with cereals, animal or vegetable fats, vitamins, minerals, and water to a moist diet in the consistency of raw hamburger. There is a trend to reduced levels of protein and increased use of fats and precooked cereal grains to produce high-energy diets for mink and fox feeding. Carnivores have a short intestinal tract, rapid rate of passage of ingesta, and minor microbial action in the gut, so they require easily digestible nutrient sources and a high content of metabolizable energy in the diet.

HAIR GROWTH

The only economically important product of fur animals is the **pelt,** so it is appropriate to briefly consider the physiology of hair growth. The physiology of wool growth is very similar; these comments on hair growth are relevant to sheep and Angora goats as well.

Hair **follicles** develop as extensions of the outer layer of the skin. At birth, an animal has its full complement of primary hair follicles; subsequent development of the hair coat depends mainly on nutritional and environmental factors. At the base of the follicle is a highly vascular layer called the *papilla,* which provides the developing hair with nutrients. The growth of the hair occurs from the matrix plate directly above the papilla. The hair cells become cornified shortly after their formation, so the hair fiber, as it emerges from the skin, consists of dead cornified cells.

There are two main types of hair follicles, the primary and the secondary follicles. The **primary follicles** produce the longest type of hair, the **guard hair,** and the shorter, intermediate guard hairs grow from lateral primary follicles, while the underfur grows from secondary follicles. The primary follicles are arranged as a triad, with a central follicle and two laterals. The primary follicles have an arrector pili muscle, which allows erection of the guard hair during fright or anger and, under cold conditions, increases insulatory properties by trapping air in the coat. The **secondary follicles** lack the arrector muscle. The primary follicles have a sebaceous gland that secretes lipid, giving the fur its glossy appearance and providing water-shedding properties. Hairs in fur animals are grouped into bundles containing a guard hair and several underfur hairs. Hairs in a bundle emerge from a common opening to the skin surface. In some bundles the guard hair is so tiny that the difference in structure from underfur hairs is not visible to the naked eye. In mink winter fur there are about 15 underfur hairs per bundle, whereas the winter fur of the blue fox contains about 40 underfur hairs per bundle (Valtonen et al. 1995; Blomstedt, 1998). The dense winter coat is a result of an increase in the number of secondary follicles that can branch to form new follicles. These derived follicles are temporary structures. The development of the winter coat is due to the action of **melatonin,** a hormone produced by the pineal gland. Melatonin production and secretion are regulated by day length (photoperiod). The metabolic effects of melatonin on hair growth are mediated by its effects on prolactin synthesis and secretion (Rose et al., 1984; Worthy et al., 1986).

The hair coat undergoes a growth cycle (Fig. 21.1). At one stage in the cycle, it reaches a state referred to as the **prime** condition. The pelt must be harvested when the fur is prime. If the pelt is not prime, the garment manufactured from it will shed hair and be unattractive. A hair undergoes three distinct growth phases in which periods of active growth alternate with resting periods. These periods are the anagen, catagen, and telogen phases. During the **anagen phase,** the

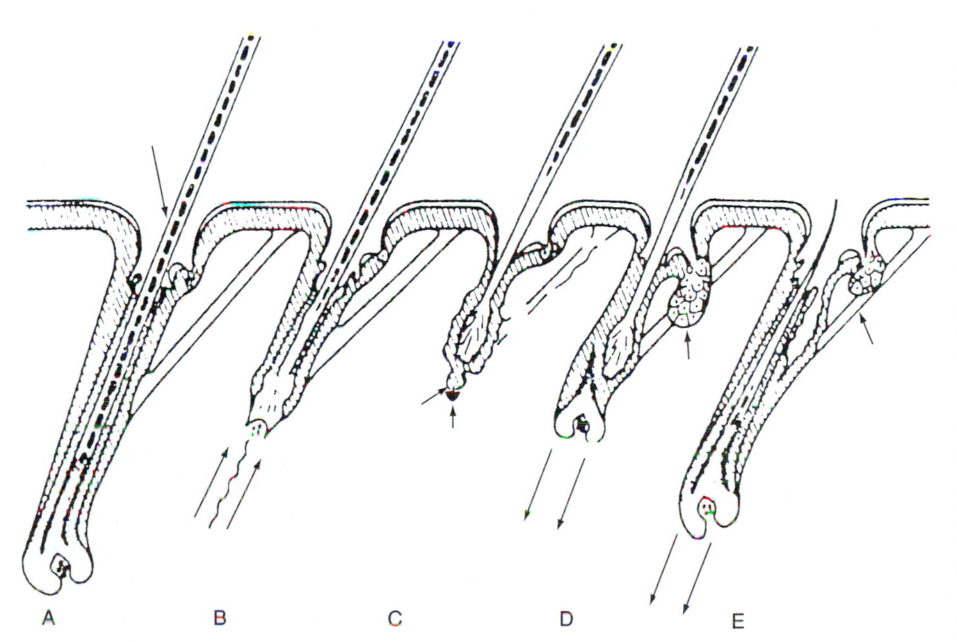

FIGURE 21.1 The hair growth cycle. The growth of a new hair, or anagen phase (A), is followed by the catagen phase (B), during which the lower part of the follicle atrophies. Melanin pigments move up the hair fiber from the dermis to the epidermal area. When this occurs, the underside of the pelt changes from dark to white (Figure 21.2), indicating that the hair is prime (C, D). In the telogen phase (D), a new hair follicle is formed and a new hair begins growth (E), which causes the previous hair to be shed.

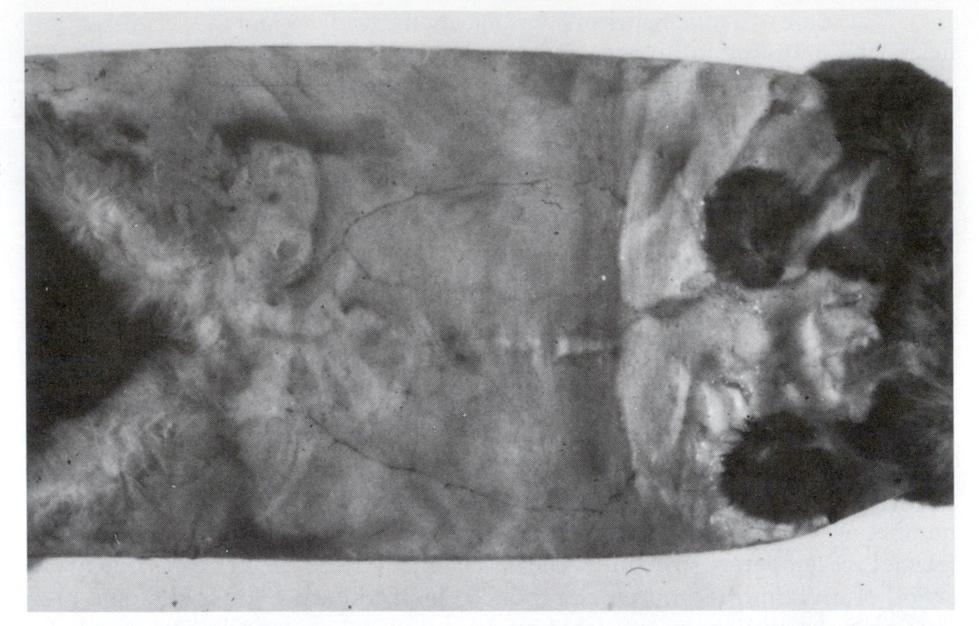

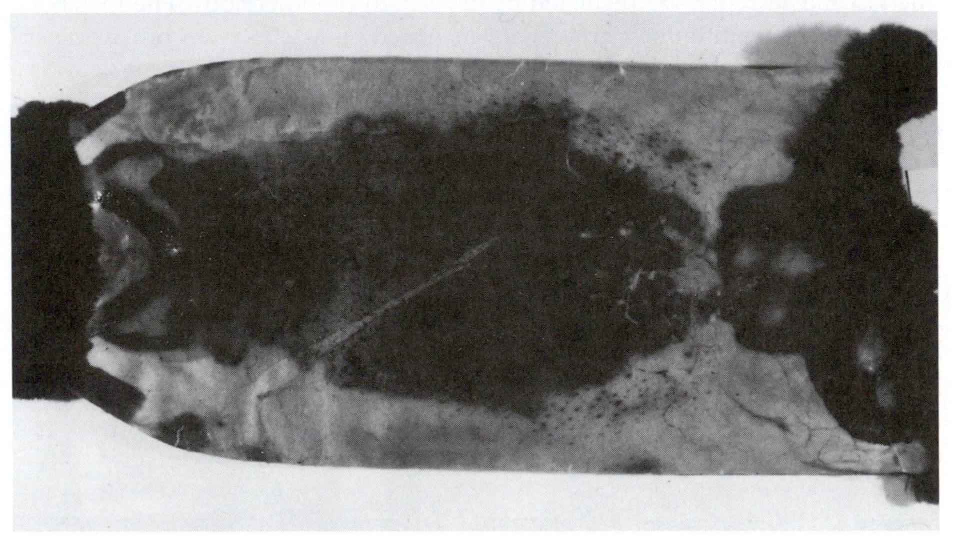

FIGURE 21.2
Examples of prime (top) and non-prime (bottom) pelts turned inside out. The dark areas on the skin of the nonprime pelt correspond to A or B in Figure 21.1 and consist of the melanin pigment granules that have not yet moved up the hair shaft. (Courtesy of N.P. Johnston.)

hair shaft undergoes rapid growth. The follicle is embedded deep in the skin. Pigments are actively synthesized in this phase and move up the hair shaft as it grows. The hair pigment cells, called **melanocytes,** can be seen as dark-colored areas on the leather side of an unprime pelt (Fig. 21.2). As the growth cycle proceeds, the hair follicle shortens until its base is located just below the sebaceous gland in the upper layer of the skin, and the papilla atrophies. These processes occur in the **catagen phase.** During the **telogen phase,** the hair is in a resting state, and the pigment cells have been drawn up into the hair shaft. At this stage, the pelt is prime. The leather side of the prime pelt is creamy white because the pigmented cells are in the hair shaft, and the last grown part of the hair does not contain any color (Fig. 21.3). Following the telogen phase, a new cycle begins with a new hair beginning growth in the same follicle, forcing out the old hair

FIGURE 21.3 A "cotton fur" pelt of a dark mink. The fur is white because of a diet-induced iron deficiency. Iron deficiency impairs the enzymatic conversion of tyrosine to melanin in the hair follicle. Iron is a cofactor for the enzyme involved.

and causing it to be shed (molting). The length of the hair-growth cycle is characteristic of the species. For example, the anagen phase is approximately 3 months in the mink winter fur, 21 to 26 days in the rat, 3 years in the human scalp, and 7 years in the Merino sheep.

Hair is composed of **keratin,** a protein with a high cystine content. Keratin is a polymer of amino acids in a helical coil, where cystine forms sulfhydryl bonds between the coils giving the hair elasticity. Hairs consist of a central core of cells, the medulla, surrounded by the cortex. The cortex is covered with an outer layer of cells called the cuticle; the cuticle cells give water-repellent properties by overlapping like shingles. The medulla contains air spaces. The hair color is determined by its structure and the pigments **melanin** (brown, black) and **pheomelanin** (yellow, red). These pigments are synthesized from tyrosine. The conversion of tyrosine to melanin involves an enzyme, tyrosinase, which requires copper and iron as cofactors. White hair results both from a lack of pigment and reflection of light off the intercellular air spaces. Other factors affecting **hair color** include cuticle structure (rough or smooth) and the degree of glossiness produced by sebaceous gland secretions. Feeding sources of unsaturated fats, such as vegetable oils, increases sebaceous gland secretion and produces a glossy hair coat (sheen, bloom).

Detailed consideration of the chemistry, biosynthesis, and functions of melanin pigments is provided by Jimbow et al. (1986) and Pawelek and Korner (1982).

Mink and foxes have a characteristic annual life cycle. In the Northern Hemisphere, the mink kits are born during a 3-week period from late April to mid May, whereas fox pups are born from early April to late June. The average litter size at birth is about six to seven in mink and silver foxes. In mink, increased litter size may be obtained by applying a flush feeding regime, comprising a 2-week period of moderately restricted feeding, followed by *ad libitum* feeding from 3 to 5 days before mating (Tauson et al., 2000). The polar fox has a remarkable reproductive potential with an average litter size of about 10, and occasional litters with more than 20 liveborn pups. Large litters and rapid growth of the offspring make the lactation period very demanding for breeding mink and fox females. Thus, mobilization of body reserves is usually necessary to meet the high energy requirements during lactation. The offspring are fully grown and the fur is mature in November or December at an age of about 6 months. Both mink and foxes accumulate voluntarily large amounts of body fat during the late phase of growth. Especially the polar fox is by nature adapted to deposit large amounts of body fat during the fall months to survive a harsh winter with limited access to food.

NUTRIENT REQUIREMENTS

The major nutrient requirements of mink and foxes have been tabulated by the NRC (1982). Because the carnivore's digestive tract has a somewhat limited capacity, diets high in metabolizable energy density are needed. Increasing the fat content in feed may benefit feed economy, because fat is a cheap energy source. High levels of fat (up to 30 to 40 percent of dietary dry matter, and 50 to 60 percent of metabolizable energy) can be used in late fall, when muscle growth is declining and fat deposition is intensive (Ahlstrøm and Skrede, 1995). Lower fat levels are used during the reproduction and early growth periods. Fats are of high digestibility in carnivores, although saturated fats are less digestible than unsaturated (Austreng et al., 1979). The rapid muscular development during early growth requires higher protein levels than during later phases of the growth period, when fat accumulation is dominant. The protein requirement for growing mink is 35 percent of ME up 13 weeks of age and 30 to 35 percent of ME thereafter (NRC, 1982). For growing foxes, the protein requirement is 28 to 30 percent of ME up to 23 weeks of age and 25 percent of ME from 23 weeks to maturity (NRC, 1982). Recently, Dahlman et al. (2003) showed that protein levels for polar foxes during late growth can be lowered to about 15 percent of ME, corresponding to 17 percent protein on a dry-matter basis, provided that the methionine requirement is covered. Data on essential amino acids have shown that methionine is commonly the first limiting amino acid in mink and fox diets (Børsting and Clausen, 1996; Dahlman et al., 2003). A high methionine requirement during periods with intensive hair growth is related to the high content of sulfur-containing amino acids, especially cystine, in the fur. Protein sources of good quality are needed to fulfill the amino acid requirements. Fish, fish by-products, eggs, and meat are sources of high-quality protein; however, some by-products, such as chicken feet and low-quality meat-and-bone meal, have a poor amino acid balance and a low digestibility.

Plant protein sources like corn gluten meal, extruded peas, and soybean meal can be used in significant amounts in diets for mink and foxes. Belzile et al. (1986) recommended that dehulled soybean meal be limited to less than 5 percent of the wet diet for mink. This corresponds to about 15 percent of dietary dry matter. At higher levels, reduction in growth rate and pelt size occurred, although fur quality was unaffected. Hydrolyzing the soybean meal with proteolytic enzymes did not improve performance, suggesting the involvement of other factors, such as fiber or phenolic compounds, in soybean meal. Oligosaccharides and fiber disturb digestive processes, and high levels of soybean meal may cause gastrointestinal inflammation in mink (Skrede, 1977).

The evolutionary adaptation to a low carbohydrate intake has not deprived the carnivores of the capability to utilize carbohydrates (Sørensen et al., 1995). Mink and fox diets always contain some plant materials, partly as sources of nutrients, and partly as binders to achieve a suitable consistency of the wet diet and the feces. NRC (1982) recommends that carbohydrates provide 15 to 25 percent of total ME for mink; foxes utilize carbohydrate better than mink. Proper processing of grains with moist heat gelatinizes starch and is desirable to increase digestibility. Lactic acid fermentation of barley and wheat increases carbohydrate digestibility in mink by reducing contents of indigestible soluble β-glucans (Skrede et al., 2001)

The main research efforts on minerals have involved iron. **Induced iron deficiency** occurs with diets containing substantial amounts of raw fish of the cod family, including Pacific hake, Atlantic whiting, and coalfish. Marine fish contain high levels of **trimethylamine oxide** (TMAO) in their tissues. TMAO binds iron by forming insoluble iron oxide hydroxides in the digestive tract, preventing its absorption and inducing severe iron deficiency. This compound is enzymatically converted to **formaldehyde** during cold storage in fish species of the cod family. This is an unusual enzymatic reaction in which the enzyme is activated by freezing. The formaldehyde produced may add to the iron deficiency problem by impairing absorption due to reactions with protein in the intestinal mucosa. Both formaldehyde and TMAO may bind iron in the digestive tract, preventing its absorption and inducing severe iron deficiency. Mink that are fed raw fish of these species become severely anemic, and fur pigmentation is disturbed (Stout et al., 1960). The underfur developed during severe iron deficiency anemia is white or grey, causing the pelt to be almost worthless (Fig. 21.3). Mink ranchers have termed this the **cotton fur syndrome.** Heat treatment of the fish and supplementation with suitable iron sources prevent the condition. Iron deficiency impairs the enzymatic conversion of tyrosine to melanin in the hair follicle. **Copper** is a cofactor also. Aulerich et al. (1982) observed that use of copper sulfate at feed additive levels in mink diets (100 to 200 ppm Cu) resulted in darker fur, presumably from increased melanin synthesis. These workers (Aulerich et al., 1987) also studied effects of dietary fluorine in mink; compared to other livestock, mink are quite tolerant of fluorine toxicity. Shrimp and crab processing by-products may contain sufficient calcium (10 to 15 percent) to reduce the growth rate of mink when these materials are used as feeds (Watkins et al., 1982). These by-products should be used at levels such that the calcium-to-phosphorus ratio does not exceed 2 to 1.

Nursing sickness is a metabolic disorder that may develop in lactating mink when the high demands for lactation require extensive mobilization of body energy reserves (Rouvinen-Watt, 2003). The condition is characterized by lack of feed intake, progressive weight loss, dehydration, and lethargy near the end of the lactation period. The clinical signs point to severe extracellular fluid volume reduction, low levels of sodium and chloride in plasma and urine, and elevated plasma concentrations of glucose, protein, and creatinine (Wamberg et al., 1992). Although dietary salt supplementation may have beneficial effects, it is doubtful that salt deficiency is the cause of nursing sickness. More likely, salt acts as an appetite stimulant to prevent inanition. Recent studies indicate that the etiology of nursing sickness is linked to acquired insulin resistance with three contributing key factors: obesity, n-3 fatty acid deficiency, and high protein oxidation rate (Rouvinen-Watt, 2003).

Problems with **vitamin deficiency** in mink and foxes used to be relatively common, because of the nature of some dietary ingredients. These problems can now be prevented by suitable treatment and storage of the feed ingredients and by vitamin supplementation. Feeding high levels of polyunsaturated fatty acids and oxidized fats can reduce plasma **vitamin E** levels and cause **yellow fat disease** and sudden death in mink and foxes (Ender and Helgebostad, 1975; Brandt et al., 1990). The condition can be prevented by avoiding excessive levels of unsaturated fats, by use of antioxidants to prevent rancidity, and by vitamin E supplementation. Vitamin E is the most important natural antioxidant and sufficient supplementation is very important when high levels of fish fats are fed.

Raw fish of the carp family, Atlantic herring, capelin, and numerous other fish (for list of fish species, see NRC [1982]) contain the enzyme thiaminase in their tissues, which destroys **thiamin** (vitamin B_1). When these fish are fed raw to foxes and mink, thiamin deficiency **(Chastek's paralysis)** may be induced. Symptoms include anorexia, loss of weight, convulsions and lack of muscle coordination, extreme weakness, paralysis, and death. If injected with thiamin, animals recover extremely rapidly (within an hour or less). Thiaminase in fish is destroyed by cooking.

Biotin deficiency in mink can be induced by an inhibitor in raw avian egg white and avian oviducts. Raw eggs contain **avidin,** a glycoprotein that binds biotin and prevents its absorption. Feeding raw eggs, spray-dried eggs, or poultry viscera containing eggs to mink may cause biotin-deficiency symptoms, including grey underfur, "spectacle eyes" (loss of hair around the eyes), foot pad dermatitis, poor growth, and reproductive failure. Turkey eggs have a very high avidin content; the condition was often noted when turkey viscera were fed, leading to the term **"turkey waste greying"** for the syndrome. Heat treatment inactivates avidin (Wehr et al., 1980).

Formic acid is a commonly used preservative in mink and fox diets. Folic acid is a component of the mammalian enzyme that oxidizes formate. Supplemental folic acid may be needed in formate-containing mink diets (Pölönen et al., 1997).

QUESTIONS AND STUDY GUIDE

1. Various nutrient deficiencies can influence hair color. Discuss. Hypothesize an explanation for the greying of hair of older humans and animals. Conduct a literature search to attempt to determine the explanation of the age effect on hair color.
2. Why does the density and sometimes the color of the coat of animals change seasonally?
3. In the fur-dressing process, mink pelts are prepared for garment manufacture. Often the leather is shaved to reduce its thickness to give the garment greater flexibility, particularly with pelts from males, which have thicker skins than females. If the pelt is not fully prime, much of the hair will fall out when the leather is shaved or sanded. Why?

4. The wool of sheep grows continuously from year to year; if not shorn, a sheep will develop a very long fleece. However, with cows and horses, the coat does not increase in length each year. Explain.
5. A mink rancher purchased turkey viscera from a poultry processing plant for mink feed. The first year he used it, the mink produced good pelts. He was very pleased with turkey viscera as feed so he purchased it again the following year. In the second year, his mink were small, had skin lesions and produced very poor-quality grey fur. Explain why the turkey waste apparently caused a problem the second year but not the first.

REFERENCES

Ahlstrøm, Ø. and A. Skrede. 1995. Feed with divergent fat:carbohydrate ratios for blue foxes *(Alopex lagopus)* and mink *(Mustela vison)* in the growing-furring period. *Norw. J. Agric. Sci.* 9:115–126.

Aulerich, R.J., R.K. Ringer, M.R. Bleavins, and A. Napolitano. 1982. Effects of supplemental dietary copper on growth, reproductive performance and kit survival of standard dark mink and the acute toxicity of copper to mink. *J. Anim. Sci.* 55:337–343.

Aulerich, R.J., A.C. Napolitano, S.J. Bursian, B.A. Olson, and J.R. Hochstein. 1987. Chronic toxicity of dietary fluorine to mink. *J. Anim. Sci.* 65:1759–1767.

Austreng, E., A. Skrede, and A. Eldegard. 1979. Effect of dietary fat source on the digestibility of fat and fatty acids in rainbow trout and mink. *Acta Agric. Scand.* 29:119–126.

Belzile, R.J., F. Dauphin, and A.G. Roberge. 1986. Enzymatically prehydrolyzed soybean meal for mink (*Mustela vison*) I. Nutritive value for growth and furring. *Can. J. Anim. Sci.* 66:505–513.

Blomstedt, L. 1998. Pelage cycle in blue-fox *(Alopex lagopus)*: A comparison between animals born early and late in the season. *Acta Agric. Scand., Sect. A, Animal Science* 48:122–128.

Børsting, C. and T. Clausen. 1996. Requirements of essential amino acids for mink in the growing-furring period. *Proc. VIth Int. Sci. Congress on Fur Animal Production. Applied Science Reports* 28:15–24. Polish Society of Animal Production, Warsaw, Poland.

Brandt A., C. Wolstrup, and T. Krogh Nielsen. 1990. The effect of dietary dl-alpha-tocopheryl acetate, sodium selenite and polyunsaturated fatty acids in mink (*Mustela vison*). I. Clinical chemistry and haematology. *J. Anim. Physiol. A Anim. Nutr.* 64:280–288.

Dahlman, T., J. Valaja, and T. Jalava, and A. Skrede. 2003. Growth and fur characteristics of blue foxes *(Alopex lagopus)* fed diets with different protein levels and with or without DL-methionine supplementation in the growing-furring period. *Can. J. Anim. Sci.* 83:239–245.

Ender, F. and A. Helgebostad. 1975. Unsaturated dietary fat and lipoperoxides as etiological factors in vitamin E deficiency in mink—The prophylactic effect of vitamin E and iron compounds. *Acta Vet. Scand.* 55:1–25.

Jimbow, K., T.B. Fitzpatrick, and W.C. Quevedo, Jr. 1986. Formation, chemical composition and function of melanin pigments. In *Biology of the Integument, Vol. 2, Invertebrates.* J. Bereiter-Hahn, A.G. Matoltsy and K.S. Richards, eds. pp. 278–291. Berlin: Springer-Verlag.

National Research Council. 1982. *Nutrient Requirements of Mink and Foxes.* Washington, DC: National Academy Press.

Pawelek, J.M. and A.M. Korner. 1982. The biosynthesis of mammalian melanin. *Am. Sci.* 70:136–145.

Pölönen, I.J., L.T. Vahteristo, and E.J. Tanhuanpaa. 1997. Effect of folic acid supplementation on folate status and formate oxidation rate in mink (*Mustela vison*). *J. Anim. Sci.* 75:1569–1574.

Rose, J., F. Stormshak, J. Oldfield, and J. Adair. 1984. Induction of winter fur growth in mink (*Mustela vison*) with melatonin. *J. Anim. Sci.* 58:57–61.

Rouvinen-Watt, K. 2003. Nursing sickness in the mink—A metabolic mystery or a familiar foe. *Can. J. Vet. Res.* 67:161–168.

Skrede, A. 1977. Soybean meal versus fish meal as protein source in mink diets. *Acta Agric. Scand.* 27:145–155.

Skrede, G., S. Sahlstrøm, A. Skrede, A. Holck, and E. Slinde. 2001. Effect of lactic acid fermentation of wheat and barley whole meal flour on carbohydrate composition and digestibility in mink *(Mustela vison)*. *Anim. Feed Sci. Technol.* 90:199–212.

Sørensen P.G., I.M. Petersen, and O. Sand. 1995. Activites of carbohydrate and amino acid metabolizing enzymes from liver of mink *(Mustela vison)* and preliminary observations on steady state kinetics of the enzymes. *Comp. Biochem. Physiol.* 112B:59–64.

Stout, F.M., J.E. Oldfield, and J. Adair. 1960. Aberrant iron metabolism and the cotton fur abnormality in mink. *J. Nutr.* 72:46–52.

Tauson, A.-H., R. Fink, and A. Chwalibog. 2000. The female mink *(Mustela vison)* as a model for studies on nutrition-reproduction interactions. *Recent Res. Devel. Nutrition* 3:239–263.

Valtonen, M., O. Vakkuri, and L. Blomstedt. 1995. Autumnal timing of photoperiodic manipulation critical via melatonin to winter pelage development in the mink. *Animal Science* 61:589–596.

Wamberg, S., T.N. Clausen, C.R. Olesen and O. Hansen. 1992. Nursing sickness in lactating mink (*Mustela vision*) II. Pathophysiology and changes in body fluid composition. *Can. J. Vet. Res.* 56: 95–101

Watkins, B.E., J. Adair, and J.E. Oldfield, 1982. Evaluation of shrimp and king crab processing byproducts as feed supplements for mink. *J. Anim. Sci.* 55:578–589.

Wehr, N.B., J. Adair, and J.E. Oldfield. 1980. Biotin deficiency in mink fed spray dried eggs *J. Anim. Sci.* 50:877–885.

Worthy, G.A., J. Rose, and F. Stormshak. 1986. Anatomy and physiology of fur growth: The pelage priming process. In *Wild Furbearer Management and Conservation in North America.* M. Novak, J.A. Baker, M.E. Obbard and B. Mallock, eds. pp. 827–841. Toronto: Ministry of Natural Resources.

CHAPTER 22

Feeding and Nutrition of Wild, Exotic, and Zoo Animals

Ellen Dierenfeld and P. R. Cheeke

Objectives

1. To provide an exposure to some of the unique problems in the feeding and nutrition of exotic animals
2. To discuss the roles of feeding behavior and digestive tract physiology in the successful feeding of captive animals

Animal nutritionists may become involved in some aspect of the nutrition of wild or exotic animals. This may arise from requests to furnish feed for starving animals on winter ranges, for provision during translocation or repatriation operations, for zoo animals, or for "nontraditional animals," such as llamas and ostriches, which are increasing in popularity as domestic animals and pets. Because wildlife specialists often do not receive much training in animal nutrition, professional animal nutritionists are a frequent source of information when questions arise in the wildlife field. As quoted in Robbins (1993), wildlife researchers would "do well to obtain the cooperation in their experimental work of experts in animal nutrition." Similarly, animal nutritionists might do well to have some knowledge of nutritional needs of wild animals. Interest in zoo animal nutrition is increasing because it is widely appreciated that the high mortality rates of captive animals, which were once the norm, are no longer acceptable ethically and because of concerns about diminishing wild populations.

Much of the work on the nutrition of wild animals has been done with ruminants such as deer and many African species, as well as numerous African nonruminant herbivores including the rhinoceros, elephant, and hippopotamus. Considerable research has been conducted in Australia with marsupials such as the kangaroo, koala, and possum (Hume, 1999). In North America, research has centered on the large ruminants (e.g., deer, elk, moose, and caribou). There has also been considerable work on small herbivores such as the lagomorphs (rabbits and hares). With zoo animals, research has emphasized the solution of such nutritional problems as mineral metabolism abnormalities in big cats and digestive disturbances and impactions in many herbivorous species, including concentrate-selector ruminants and aboreal monkeys.

Although the literature dealing with the nutrition of wild and exotic species is much less plentiful than for domestic animals, there are a number of comprehensive books and review articles documenting the current state of knowledge. The nutrition of zoo mammals is covered comprehensively in Kleiman et al. (1996), avian species are addressed in Klasing (1998), and Robbins (1993) has reviewed wildlife nutrition in depth. Information is also available from numerous specialty symposia including the *Proceedings of the Dr. Scholl Conferences on the Nutrition of Captive Wild Animals* (see www.lpzoo.org); *Proceedings of the American Zoo and Aquarium's (AZA) Nutrition Advisory Group (NAG)* conferences (www. nagonline.net); *Proceedings of the Comparative Nutrition Society* (www.cns.org); and the *Zoo Animal Nutrition* series published from the European Zoo Nutrition conferences. The intent of this chapter is to provide a brief exposure to some of the major problems that might be encountered in formulating diets for nondomestic species. Because of the tremendous diversity in feeding behavior and digestive-tract physiology of wild animals, only broad guidelines can be given. One must always keep in mind the combination of anatomical, behavioral, and physiologic considerations that must be integrated for applied feeding programs with exotic species; furthermore, external environmental factors play an important role in nutrient utilization and adaptations.

PRINCIPLES OF ZOO ANIMAL NUTRITION

The role of **zoological gardens** is no longer simply the exhibition of exotic animals, but increasingly one of propagation and survival of endangered species. The rapid decimation of the natural environment in tropical areas of Africa, Central America, South America, and Asia imperils the survival of many species. Poaching animals (for various economic reasons) continues to be a critical problem. Environmental changes and human population growth, with consequent habitat alteration, impact nutritional resources of even well-managed lands such that populations of many species may not be viable or sustainable in nature. For some species, intensive management in parks, reserves, and zoos may represent a long-term survival option; this puts a tremendous responsibility on the nutritionists involved, for the survival of a species may rest in their hands. Often, the task must be accomplished without any database(s) on nutritional requirements, nutritional composition of diets, and/or nutritional health assessment criteria. The immensity of the impending problem for wildlife survival can be appreciated by considering projected human population growth in developing countries. (See Table 24–1.) Loss of habitat, disturbance of migration patterns, and direct slaughter of wildlife will be immense; nonetheless, the primary concerns of inhabitants and governments of these regions will be improving the quality of human life. Wildlife preservation will probably be considered a luxury, at least in terms of allocation of resources.

Natural Diets and Feeding Behavior

In developing feeding programs when data is unavailable, it is useful to begin by considering the diet of the animal in the wild state. Even obtaining this information may be difficult, and often only represents a single moment in time. With game animals, digestive contents of freshly killed animals can be examined to identify feeds consumed. However, a list of species consumed provides only partial information necessary for applied feeding programs. We can rarely duplicate the ingredients of any animal's diet in captivity—what we can duplicate are the nutrients consumed. Hence, information on nutrient composition of native foods is a critical component that is often lacking in field observations and reports based solely on feeding ecology. Nonetheless, the **oral and gastrointestinal morphology** provides good clues to the dietary needs of the animals (Hofmann, 1989), with the more chemically complex dietary constituents generally requiring more complex gut anatomy for proper utilization (Stevens and Hume, 1995). The

FIGURE 22.1 Much can be deduced about the feeding and digestive strategies of an animal from its mouth. The wide lips and blunt teeth of the hippopotamus indicate that it is a grazer rather than a browser. (Courtesy of A.G. Hollister.)

oral structure of the black rhinoceros suggests it is a browser (its lack of incisors makes grazing difficult), whereas the white rhinoceros and hippopotamus have wide lips and blunt teeth (Fig. 22.1) designed for grazing (Ullrey and Allen, 1986). Simple observations of **feeding behavior** will indicate whether an animal is a browser or a grazer. The black rhinoceros uses its prehensile upper lip to pull a branch into its mouth and then clip off the branch with its teeth. With a sideways grinding action, it removes the leaves and small shoots (Dierenfeld et al., 1995). In contrast, the white rhino primarily grazes on grass. Along with these external anatomical features, herbivores also display variability in salivary gland size and secretions that alter with diet. Size of the glands varies from 0.05 to 0.2 percent of body mass in grazers versus concentrate selecting mammalian herbivores, and the presence/absence of tannin-binding proteins in the saliva is inducible. Evidence of gut modifications for microbial fermentation may also provide guidance as to the appropriate domestic model and possible physiologic constraints of a given exotic species. Animals with a reticulorumen, or some other foregut modifications, have at least some similarity in nutritional needs to the domestic ruminants, although it is important to bear in mind that many wild ruminants are concentrate-selectors with limited ability to utilize fibrous feeds, whereas the domestic ruminants are grazers or intermediate feeders with a good ability to utilize fiber. Animals with **hindgut fermentation** are likely to be somewhat similar to the horse (large animals) or rabbit (small herbivores) in their nutritional needs. Therefore, considerable extrapolation can be made from domestic species to wild animals that have similar feeding patterns and gut physiology.

With herbivores, selection of appropriate substitute feeds, and problems with utilization of forages and roughages available in the zoo environment often arise, particularly among "browsing" species. Small concentrate-selector ruminants can develop **rumen impaction** when fed fibrous feeds that require long retention times for digestion (Van Soest et al., 1995) and may rely more on hindgut fermentation as a digestive strategy (Conklin-Brittain and Dierenfeld, 1995). Selective feeding for faster digestion and more digestible foods is a principal means of meeting energy needs in these small ungulates, as well as in larger browsers such as giraffe (Clauss et al., 2002c). On the other hand, concentrates containing too little fiber can result in rumen stasis (Willette et al., 2002; Hofmann and Matern, 1988). Even frugivorous browsers routinely consume and digest fibrous diets (Wenniger and Shipley, 2000; Dierenfeld et al., 2002). Further information and understanding of suitable physical as well as chemical characteristics of carbohydrates for zoo herbivores is essential in meeting nutrient needs (Hall et al., 2003). Likewise, many of the arboreal (tree-dwelling) primates have **foregut (pregastric) fermentation.** Langur, proboscis, and colobus monkeys have a four-chambered stomach with rumenlike function (Ensley et al., 1982). However, because they do not ruminate, coarse fibrous material resistant to microbial fermentation may form **phytobezoars** (impacted masses) in their stomachs, leading to death. Ensley et al. (1982) found *Acacia* forage to be especially dangerous in this regard. On the other hand, feeding low-fiber diets to primates with foregut or hindgut fermentation may cause too rapid fermentation, bloat, inflammation of the foregut or hindgut, and diarrhea (Ullrey, 1986; Edwards and Ullrey, 1999). For these primates, combined nutrient requirements of both primates as well as more conventional herbivores (ruminants and equids) may better meet the physiological needs of the species (NRC, 2003). (See Figure 22.2 for a comparison of primate digestive anatomy.)

Some animals have evolved dependence on particular plants; these specialized feeders can be some of the most difficult to feed in captivity. Some of the Australian marsupials will feed only on *Eucalyptus* spp. **Koalas,** for instance, will eat only foliage from certain eucalyptus trees (Fig. 22.3), so their exhibition in zoos is feasible only in areas where the trees can be grown, or suitable browse must be purchased and brought into the facility. Although no single constituent of eucalyptus browse has been shown to consistently correlate with intake in this species, extracted essential oils appear to be useful as feeding stimulants to attract koalas to a manufactured diet (Pahl and Hume, 1990). The koala can absorb several ml of eucalyptus oil daily, a dose that is fatal to humans (McLean et al., 1993). Marsupials detoxify the terpenes in eucalyptus oil by forming highly polar, acidic metabolites (Foley et al., 1995; McLean and Foley, 1997). Another example, the **giant panda,** in the wild eats primarily bamboo, but it can be induced to eat other feed in captivity. The panda is interesting in that it is an herbivorous carnivore! It is a member of the order Carnivora and has a gastrointestinal tract physiology similar to that of other carnivorous species, with a simple stomach, no cecum, and a short colon. It has adapted to a herbivorous diet by having a high feed intake and rapid time of passage, digesting the cell contents and excreting the cell wall material as indigestible residue (Dierenfeld et al., 1982). Nonetheless, pandas appear selective in choice of bamboo species and parts consumed, which must be taken into account for optimal applied feeding programs.

There has been considerable research on the nutrition of **arboreal folivores** (foliage-eating tree dwellers), which has been reviewed by Montgomery (1978). Tree foliage is often fibrous, low in energy, and contains a high concentration of plant secondary compounds such as phenolics, terpenes, essential oils, and cyanogenic glycosides. Most arboreal folivores have either foregut or hindgut

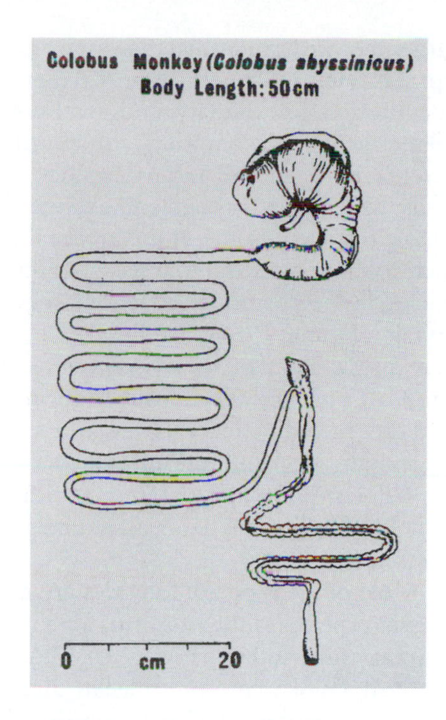

FIGURE 22.2 Colobine monkeys (*Colobus angolensis*)—the primate equivalent of a ruminant (see gut anatomy); nutrient recommendations comprise a combination of primate and ruminant dietary requirements. (NRC, 2003; Stevens and Hume, 1995.)

FIGURE 22.3 The koala (left) is a specialist feeder that will only eat the leaves of a few species of eucalyptus. Koalas coevolved mechanisms, including taste preferences, that overcome the chemical defenses (essential oils) of eucalyptus trees. The giant panda (below), also a specialist feeder, feeds mainly on bamboo in the wild, although it can be induced to eat other feed in captivity. The panda is a vegetarian carnivore!

enlargement with microbial digestion (Langer, 1984a, 1986). Energy-conserving adaptations to a low-energy arboreal diet include small litter size, gliding as a form of locomotion, low activity, and low basal metabolic rate (e.g., sloth). Excretion of plant secondary compounds may utilize energy-requiring reactions, and the difference between DE and ME values may be large when plants contain essential oils, which are absorbed and excreted in the urine. Tree foliage is often low in available nitrogen. Some of the arboreal Australian marsupials have mechanisms similar to those of the rabbit (see Chapter 1) for separation and retention of small particles in the cecum, endogenous recycling of urea into the hindgut, and coprophagy (Hume, 1982). Studies dealing with energy and nitrogen metabolism of arboreal marsupials are presented by Foley (1987) and Foley and Hume (1987a,b). **Foregut fermentation** occurs primarily in folivores that consume a mixed diet including leaves, fruit, and seeds. This allows retention of fiber for microbial digestion while permitting lower-fiber fractions to pass to the intestine (e.g., colobine monkeys). **Hindgut fermentation** with colonic separation of fiber is found primarily in small animals consuming a high-fiber diet, such as the koala (Cork, 1996).

Most of the arboreal folivores are mammals. One of the few obligate folivorous avian species is the **hoatzin,** a neotropical bird that eats green tree leaves as its primary food source (Fig. 22.4). Grajal et al. (1989) reported that the hoatzin is the only known case of an avian digestive system with foregut fermentation. High concentrations of VFA are produced in the enlarged crop. The bird has nu-

FIGURE 22.4 The hoatzin is the only known example of an avian species with foregut fermentation. The diet of this tropical American bird consists of tree leaves. (Courtesy of S. D. Strahl and Chicago Zoological Society.)

merous adaptations for its arboreal life, including functional wing claws and a modified sternum necessary to accommodate the enlarged foregut. Its ability to fly allows it to exploit patches of vegetation not accessible to other herbivorous vertebrates. The hoatzin has a microbial population of bacteria and protozoa in its crop that resembles that of the rumen (Dominguez-Bello et al., 1993 a,b), although digestibilities of tropical forages with hoatzin crop fluid are lower than those with cattle rumen fluid (Jones et al., 2000). The bird avoids leaves with high tannin contents; it selects a high-quality, leafy diet (Jones et al., 2000).

A number of **reptiles** are herbivorous. Based on fossil records of dentition and the presence of large stones in the stomach region of fossils, Farlow (1987) has speculated about the diet and digestive physiology of herbivorous **dinosaurs.** On a more modern level, reptiles such as the **desert tortoise** have a large but simple gut with hindgut fermentation that allows the animal to survive on poor-quality, highly fibrous desert vegetation (Barboza, 1995). It is well adapted to the use of abrasive diets, including the spiny pads of cacti. It has thick, cornified epithelial tissue in the mouth and esophagus that resists sharp spines and copious mucus secretions that facilitate the swallowing and intestinal passage of fibrous and spiney material. The giant **land tortoises** (Galapagos, Aldabra) are also well suited to grazing habits, with a horny beak for clipping forage, and an ileocolic valve to retain foodstuffs in the gut for 10 to 30 days. Native foods are highly fibrous (up to 50 percent NDF), low to moderate in protein content (5 to 20 percent), with a digestibility of <50 percent. Studies recently conducted with tortoises fed differing dietary protein content (14 vs. 19 vs. 30 percent of dry matter) showed a minor impact of protein level on the development of carapace deformities (Wiesner and Iben, 2003). Animals raised in dry conditions, however, produced taller humps compared with those raised under humid conditions, demonstrating that with ectotherms, not only temperature but also humidity can impact nutrient utilization.

Iguanas are highly herbivorous, employing gut valves as well as probable nematode symbionts to assist with digestion. Green iguanas showed better growth with 19–24 percent versus 27 percent NDF diets; digestion ranged from 25 to 60 percent (Baer et al., 1997). Native foods (flowers, fruits, leaves, and seeds) eaten by herbivorous lizards are moderate in protein and fiber content, averaging around 10 percent crude protein and 40 percent NDF (Ward et al., 2003).

There are a number of **aquatic herbivores** such as the manatee (sea cow) and green sea turtle. Both of these species have very high (80 percent) efficiencies of cellulose digestion associated with hindgut and even small intestinal fermentation (Burn, 1986; Bjorndal and Bolten, 1990). In the case of marine mammals, there are no suitable domestic species from which to extrapolate dietary requirements (Bernard and Ullrey, 1989). For these species, knowledge of the diet consumed in the wild is particularly important.

Nutritional Problems

Many nutritional problems with big cats (e.g., lions, tigers, cheetahs) have occurred in zoos, particularly with imbalances in calcium and phosphorus, and were among the first published reports of nutrient imbalance over a century ago (Dierenfeld, 1997). Diets high in meat may induce **juvenile osteoporosis** (nutritional secondary hyperparathyroidism) unless adequate calcium supplementation is provided in the diet (Van Rensburg and Lowry, 1988). In this disorder, the head and paws are enlarged and the skeletal system is soft and decalcified (Fig. 22.5). Although less prevalent in zoos that feed appropriately balanced meat-based diets, or commercial formulations, bone density issues are still a common health problem for large felids as well as rapidly growing carnivorous bird species such as storks. Each kilogram of raw meat fed must be supplemented with 5 g of calcium carbonate and 10 g of dicalcium phosphate (Ullrey and Bernard, 1989), along with appropriate vitamins; this can also be accomplished through the use of commercial supplements. In the wild, carnivores consume small bones of prey, maintaining an adequate calcium-to-phosphorus balance. Bones for chewing should be provided to captive big cats at least once per week to prevent dental plaque and calculus (Ullrey and Bernard, 1989). Wild cats seem to be similar in digestive efficiency to the domestic cat (Barbiers et al., 1982; Crissey et al., 1997) and have similar nutritional peculiarities (see Chapter 20). For example, tigers have a dietary requirement for **taurine** (Pickett et al., 1990). The fact that obligate carnivores such as the felids have unique metabolic adaptations to dietary protein, fatty acids, and vitamins, and a lack of ability to deal with high levels of carbohydrate appears to carry across classes: raptors and crocodilians, for example, also show limited glucokinase upregulation.

Vitamin A and carotenoids are important in zoo animal nutrition. Many of the birds with brilliant plumage, such as the flamingo and scarlet ibis, as well as reptiles and amphibians with brightly colored skin, require dietary carotenoids for this pigmentation. **Canthaxanthin** is included in flamingo diets for this pur-

FIGURE 22.5 Nutritional secondary hyperparathyroidism of a tiger fed meat causing a calcium-to-phosphorus imbalance. (Courtesy of L. Krook.)

pose; β-carotene can also be used because it can be metabolized to canthaxanthin. Natural as well as synthetic carotenoids are used in different regions of the world for coloration, depending on availability and economic factors. Different species have differing abilities to convert various carotenoids; the carmine bee-eater, for example, does not metabolize canthaxanthin and appears rather to utilize lutein for its feather pigmentation. Cats must receive preformed vitamin A in the diet as they cannot convert carotene efficiently. Nonetheless, from recent studies it appears that β-carotene may have some independent and beneficial effects upon reproduction and possibly immune function in carnivores, so it may be more important than previously considered, but studies to confirm this are needed. Very high doses of **vitamin A** given to polar bears have been effective in treating persistent dermatitis (Ullrey and Allen, 1986). **Polar bears** store extremely high levels of vitamin A in the liver, in specialized perisinusoidal cells known as *Ito* or *stellate cells* (Leighton et al., 1988). This ability to sequester vitamin A is important because of the extremely high vitamin A intake from consumption of seals, which are rich in the vitamin. The buildup of vitamin A in the arctic food chain begins with the consumption of carotene-rich plankton by fish. Although the polar bear is resistant to hypervitaminosis A, care to prevent overdosing with the vitamin should be taken with other species.

Dogs and other **carnivores** in the Canidae (canine) and Mustelidae (weasel) families have been reported with very high blood concentrations of **vitamin A,** with levels of 10–50 times higher than for other species (Schweigert et al., 1990, Slifka et al., 1999). The majority of the blood vitamin A in canines and mustelids occurs in the form of **retinyl esters,** primarily as retinyl stearate, bound to lipoproteins. In contrast to other species, the blood vitamin A levels are readily influenced by dietary vitamin A intake (Schweigert et al., 1990). In most species, blood vitamin A exists as retinol bound to retinol-binding protein. Only traces of retinyl esters occur, and blood levels do not reflect intake. In other species, the occurrence of retinyl esters is a sign of severe vitamin A toxicosis. Another difference between canines and other species is that vitamin A is excreted in the urine (Schweigert et al., 1991), mostly as retinyl palmitate/oleate. Schweigert et al. (1990, 1991) speculate that these phenomena allow canines to cope with large amounts of dietary vitamin A in prey—perhaps ecologically on a seasonal basis—without developing **vitamin A intoxication.** Ancecdotal accounts of vitamin A toxicity have been reported in zoo carnivores—primarily snakes—fed whole rodents that have been shown to contain excessive concentrations of preformed vitamin A (Douglas et al., 1994). Vitamin A concentrations measured in free-ranging rodents are about one-tenth the level reported in whole prey raised on commercial laboratory rodent diets. Confirmed vitamin A toxicity when fed at a dietary level considered acceptable for domestic carnivores has been reported for the tamandua, a type of anteater. Because insects as a whole are known to contain low levels of this nutrient (Barker et al., 1998), it is possible that insectivorous species may be more sensitive to dietary excesses, but this has not been investigated systematically.

There are important species differences in **vitamin D** requirements that can lead to unexpected problems in zoo animal nutrition. Vitamin D, in most species, functions in the absorption of calcium. The active form of vitamin D, 1, 25-dihydroxy cholecalciferol (1, 25-OHD_3), regulates the activity of calcium-binding proteins in the intestinal mucosa, which controls the amount of calcium absorbed. In animals that have evolved under conditions of high exposure to uv light, the dietary vitamin D requirement is high in the absence of exposure to high intensity of solar radiation. In contrast, animals that have evolved under conditions of low exposure to uv light have low or, in some cases, no dietary vitamin D requirement (e.g., rabbits, cave-roosting fruit bats). In studies with various lizards, it appears that photoconversion of skin precursors to active metabolites of vitamin D is species-, temperature-, and wavelength-dependent (Allen et al., 1999; Ferguson et al., 2002; Nijboer et al., 2003).

Mixed species exhibits in zoos are becoming increasingly popular. At the Denver Zoo, an exhibit was established with New World primates and two rodent species, pacas and agoutis (Kenny et al., 1993). All animals had access to one diet, which was formulated to meet the requirements of the primates. After several months, paca and agouti mortalities occurred, with extensive soft tissue calcification. Death occurred from kidney failure. The pathological lesions were typical of classic vitamin D toxicity. **New World rodents,** which inhabit the dark forest floor, have evolved physiology that is very efficient in absorbing calcium and phosphorus, with a very low requirement for vitamin D. **New World monkeys,** on the other hand, live in the forest canopy with extensive exposure to solar radiation, so they have evolved with abundant tissue levels of vitamin D. Therefore, they have a high dietary requirement when appropriate light sources are not available, whereas the rodents have a very low requirement. The level of vitamin D supplementation required for the primates is a lethal toxic level for pacas and agoutis.

The **mole-rat** of Africa spends its entire life underground, without ever being exposed to sunlight or a dietary source of vitamin D. Mole-rats have a highly efficient calcium absorption process that is independent of vitamin D regulation (Pitcher et al., 1992). Therefore, the essentiality and toxic levels of vitamin D show great species differences. With the increasing popularity of mixed species exhibits, dietary problems resulting from greatly differing nutrient requirements must be anticipated.

Diets for most **marine mammals** in zoos (e.g., seals, sea lions, porpoises) are based on fish. Frozen fish should be thawed in a refrigerator rather than air thawed. Air thawing may result in a loss of 10 to 15 percent of the moisture; for many marine mammals the only source of fresh water is that consumed with the food. **Thiamin deficiency** is also a potential problem because of the thiaminases present in many fish upon death. The high polyunsaturated fat content of fish enhances the likelihood of vitamin E deficiency. Bernard and Ullrey (1989) recommend that fish-fed marine mammals should receive 25 to 30 mg thiamin and 100 IU vitamin E/kg of fish. Data on composition of whole fish suggest no need for additional vitamin A or D, provided proper storage and handling protocols were incorporated to minimize possible deterioration.

The possible occurrence of toxins in foods offered to zoo animals requires careful consideration. These include thiaminases in raw fish, and a wide array of plant toxins (alkaloids, glycosides, toxic amino acids, etc.). Koalas, for example, have been poisoned by cyanogenic glycosides in *Eucalyptus* forage fed in zoos (Montgomery, 1978).

Known antagonistic nutrient interactions among the fat-soluble vitamins may be precipitated by excesses/imbalances, particularly with the fat-soluble vitamins. Vitamin K deficiency resulting in internal hemorrhage, for example, was likely a result of very high concentrations of vitamin E in fish-based diets fed to pelicans in one zoological collection. Excess vitamin A likely contributed to signs of vitamin K deficiency reported in captive anteaters.

Setchell et al. (1987) found that the poor reproduction of captive cheetahs in zoological parks in North America may be due in part to the effects of **phytoestrogens** in soybean meal used in a commercial zoo feline diet. Phytoestrogens are common in legume species (see clover disease, Chapter 5). Setchell et al. (1987) suggested that because cats have a low activity of liver enzymes involved in metabolism and excretion of toxins, they may be more sensitive than livestock are to plant estrogens. Carnivores would be expected, on an evolutionary basis, to have a lower ability to detoxify plant toxins than have herbivores, which have coevolved with plants heavily defended with toxic chemicals (see Chapter 9). Thus, the inclusion of plant products such as soybean meal in diets for carnivores may have unanticipated detrimental effects.

The significance of **vitamin E** in zoo animal nutrition has been reviewed by Dierenfeld (1989, 1994; Dierenfeld and Traber 1992). Some of the characteristic signs of deficiency include cardiac and skeletal myopathy (white muscle disease) in herbivores, with steatitis (yellow fat), anemia, and poor reproduction, reported more in birds and reptiles, particularly in carnivores. Plasma vitamin E levels in free-ranging animals have been used as a measure of normal status for captives of the same species (Dierenfeld et al., 1988; Dierenfeld, 1989; Dierenfeld and Traber, 1992; Ghebremeskel and Williams 1988), due to concern that the vitamin E status of zoo animals may often be inadequate. According to Dierenfeld (1989) and Dierenfeld and Traber (1992), dietary vitamin E requirements of wild species may be up to ten times higher than the NRC requirements for domestic counterparts due to increased longevity, stresses, and the general low concentrations of this nutrient in zoo diets. To standardize across hoofstock species, Dierenfeld (1989) suggests that plasma cholesterol levels can be used to estimate minimal plasma vitamin E concentrations expected, by dividing normal cholesterol (mg/dl) by 100. For example, a giraffe with a cholesterol value of 40 mg/dl would have an expected minimal plasma vitamin E level of 0.40 mg/ml. Domestic ruminants appear to provide a good comparative physiologic model for vitamin E nutrition of zoo ruminants.

On the contrary, **rhinoceros** and **elephants** appear to have unique vitamin E metabolism, and normally display quite low concentrations of this nutrient (Dierenfeld et al., 1998; Savage et al., 1999; Clauss et al., 2002b). The domestic horse does not appear to be a suitable comparative model for vitamin E nutrition of the pachyderms, although these species share similar gut anatomy. Vitamin E deficient elephants develop cardiac lesions similar to those seen in swine (mulberry heart disease), whereas rhinos develop skeletal and cardiac myopathies, and red blood cell hemolysis. **Vitamin E** concentrations in browse plants consumed in the wild by rhinos and elephants are much higher than in grasses and pelleted diets typically fed to zoo animals (Dierenfeld, 1994).

Neither perissodactyls (horses) nor the pachyderms have a gall bladder. This might be significant in captive animals, which are usually fed meals followed by a need for large amounts of bile for **emulsification of lipids.** Ghebremeskel et al. (1991) suggested that the lower plasma vitamin E levels of captive vs. wild rhinos may reflect a lack of sufficient **bile** in intermittent-fed animals to emulsify and absorb vitamin E efficiently. In contrast, wild rhinos and elephants browse continuously, requiring less bile at any one time than meal-fed animals.

Inadequate bile production in captive animals may be a limiting factor in the absorption of fat-soluble vitamin E, but similar circulating concentrations are seen in both free-ranging and supplemented zoo rhinos (Clauss et al., 2002b); hence, dietary concentrations appear to underlie the previous deficiency issues reported. Also, the nature of the fatty acids in the diets of wild and captive rhinos differs (Ghebremeskel et al., 1991; Wright and Brown, 1997). Ingestion of seeds and kernels favors **linoleic acid** intake, while ingestion of native browse leaves favors **linolenic acid.** Rapid degradation of unsaturated fatty acids occurs in stored feed, while no loss at all would occur with browsed feed in the wild (Grant et al., 2002). These factors could increase the vitamin E requirements of captive rhinos.

The five rhinoceros species still surviving are in great danger of extinction. About 85 percent of the world's rhino population has been lost since 1970. Some species number fewer than 100 surviving animals, and others are under constant armed guard to minimize poaching. **Rhino horn** is in great demand in China as a medicine, (Mainka and Mills, 1995), and in Yemen for dagger handles. The International Rhinoceros Foundation has been formed to help preserve the remaining species. In 1992, an attempt was made to move a small herd of black rhinos from Africa, where they were likely to succumb to poachers in Zimbabwe, to establish a captive herd in Australia. Several animals died, probably of vitamin E deficiency, exacerbated by exposure to creosote in corral timbers (Kelly et al., 1995). **Creosote** is an oxidant that would increase antioxidant (vitamin E) requirements, and black rhinos have been shown to have poor oxidant enzyme activity. This example illustrates the critical need for knowledge of nutritional requirements and peculiarities of wild species, particularly endangered species, where every animal that dies is a critical loss to the gene pool. Other disease problems also occur with black rhinos that may be linked to nutritional status. Captive black rhinos, but not white rhinos, accumulate excessive **iron** as hemosiderin deposits in the liver **(hemosiderosis).** Smith et al. (1995) suggest that in the wild, the black rhino is a browser that has adapted to a diet of low bioavailable iron by increasing the efficiency of iron absorption. In captivity, they are usually fed a diet of grass or alfalfa hay, along with concentrate pellets, which may contain a higher content of available iron than the black rhino's normal diet. Furthermore, captive diets are often low in tannins, which may serve to bind dietary iron in nature. Black (but not white) rhinos have been shown, however, to react to feeding of dietary tannins with increased salivary tannin-binding proteins (Clauss et al., 2002a), hence having differing physiologic mechanisms for dealing with a single nutrient. The role of excessive tissue iron in the suite of health syndromes affecting captive black rhinos is not known. However, iron is one of the most potent stimulators of autooxidation of unsaturated fatty acids, which would increase the vitamin E requirement. It appears from recent summary (Dierenfeld et al., 2004) that dietary iron concentrations are not the sole factor underlying iron status in this species, and rather interrelationships among feed type, feeding behavior, bile secretion, stress, vitamin E, and iron may be involved in the high mortality of captive black rhinos.

Frugivorous species have some specific nutritional problems—identified in zoos—that are related to nutrient imbalances in domestic fruits offered as substitute food items. Primates and fruit bats, for example, can demonstrate calcium deficiencies when unsupplemented fruit-based diets are fed or behaviorally selected. A study of almost 100 native fruits, collected from three continents, showed that indigenous figs contained more than three times as much calcium as nonfig fruits in the same localities (O'Brien et al., 1998). Ficus trees appeared to selectively absorb calcium from soils, at least regionally, and may function as a "keystone" nutrient resource for frugivores. Further, native figs averaged nine times more calcium compared with domestic fruits (including domesticated fig spp.), a nutrient concentration that may not be duplicated without external supplements in captive feeding programs. Vitamin E deficiency has also been documented in fruit-eating bats; in nature, many of these species would obtain this nutrient by browsing on green leafy plants—consuming extracts and obtaining critical nutrients—but avoiding the fibrous components. Hence, evaluating nutrients supplied by native foodstuffs, parts consumed, aspects of food presentation, and behavioral adaptations for eating is a critical part of exotic animal feeding programs (see, e.g., Young, 1997; Silver et al., 2000).

Another serious issue affecting frugivorous primates and birds specifically (but also other herbivores) is the increasing incidence of **hepatic iron storage disease.** Problems could initially be traced to high levels of iron in commercial diets, but other variables including genetic predisposition to enhanced iron uptake (presumably an adaptation for evolution in a low-iron environment) and interactions with other components of mixed diets (other minerals, ascorbic acid, polyphenols, specific sugars, phytates) make this a multifactorial health issue that is receiving much directed research activity.

NUTRITION OF WILD UNGULATES

Information on the nutritional requirements of wild animals is necessary for the assessment of the adequacy and carrying capacity of ranges, for supplemental feeding programs, for game ranching, and for keeping animals in captivity. With increasing human encroachment on wildlife habitats, intervention in the form of wildlife management often becomes essential if the populations are to survive. Accessible databases of native food composition, body condition indices, and other physiologic measures of nutritional status need to be compiled, with contributions from both free-ranging and captive-managed species.

Wild ruminants resemble their domesticated counterparts in digestive physiology and nutritional requirements. Ruminants native to temperate and arctic areas often show a marked **seasonal variation in feed intake,** corresponding to patterns of seasonal availability of feed. Deer, for example, have a high feed intake in the summer and accumulate large fat deposits. In the winter, they may go for many days without eating, meeting their energy needs by mobilization of body fat. This **periodicity in feed intake** is even observed in captive animals with free access to feed (Freudenberger et al., 1994; Parker et al., 1993) and seems to be photoperiodically regulated (Adam et al., 1996; Loudon, 1994; Morgan and Mercer, 1994). In males, there is a second period of **voluntary hypophagia** (lack of feed intake) associated with the autumn mating season or rut. It is correlated with testosterone concentrations (McMillin et al., 1980; West and Nordan, 1976) that show seasonal variations (Whitehead and McEwan, 1973). Hypophagia during the rut period may be an adaptive feature helping to ensure reproductive success.

Seasonal changes in physiology and behavior such as seasonality in breeding, voluntary feed intake, and so on are regulated by **photoperiod,** which is a highly predictable environmental cue (Morgan and Mercer, 1994). The hormone **melatonin** is synthesized by the pineal gland in a precisely regulated pattern, with the neural transmission of a signal from the retina of the eye to the pineal gland.

During the day, low levels of melatonin synthesis and secretion occur. At night, melatonin synthesis and blood melatonin levels are increased in direct proportion to the length of the dark period. Thus, winter and summer photoperiods are reflected in long- and short-duration melatonin signals, respectively.

The **annual cycles** in reproductive behavior and feed intake in wild ruminants have evolved via adaptation mechanisms to ensure reproductive success and winter survival. **Deer** and other temperate ruminants have a number of other physiologic adaptations to enhance their likelihood of winter survival besides the accretion of large fat deposits in the summer. The fasting metabolic rate of white-tailed deer is 30 to 40 percent lower in winter than in summer (McMillin et al., 1980), reducing winter energy requirements. Heart rates and body temperatures are also lower in winter. These changes seem to be mediated by thyroid hormones (McMillin et al., 1980). In **tropical deer,** seasonal patterns of growth, voluntary feed intake, and plasma hormone concentrations do occur, but they are less pronounced than in temperate deer species (Semiadi et al., 1995). Such variability may be important in overall health and body condition of captive hoofstock, but consequences have not been examined in detail in zoo collections. Similar animal responses may be seen *in situ,* even at equatorial latitudes, if one considers seasonality in terms of forage or prey quality and quantities in relation to rainy versus dry environmental conditions.

Supplemental feeding of big game animals is often necessary for a variety of reasons, including emergency feeding during unusually severe winters and enticing wild animals away from haystacks or other sources of privately owned feed. Such feeding is especially important with elk in the western United States. Artificial feeding of deer and elk has been reviewed by Dean (1980). Good-quality alfalfa hay is satisfactory to entice moose and elk to feeding stations. Deer may suffer **rumen and omasum impaction** when fed alfalfa hay or other fibrous feed. Deer should be fed a pelleted concentrate diet containing grain, plant protein supplement, and alfalfa meal.

The nutritional status of wild herbivores such as elk can be assessed by measuring urinary excretion of metabolites deposited in snow **(snow urine).** Analysis of urine spots provides a convenient, noninvasive index of nutrition and nutritional stress or condition in ungulates (White et al., 1997). Trends in urine ratios of metabolites, such as urea:creatinine (UN:C), cortisol:creatinine (Co:C), and **allantoin:creatinine (A:C),** can be used to assess winter nutrient deprivation. The trends in A:C were the most sensitive, interpretable, and consistent with variations in winter severity and digestible energy intake (White et al., 1997). The A:C ratio can also be used to estimate digestible dry-matter intake of free-ranging ungulates (Garrott et al., 1996; Vagnoni et al., 1996), based on the fact that most **urinary allantoin** has as its origin the bacterial nucleic acids absorbed when rumen bacteria are digested (Chen et al., 1997).

Differences between domestic and wild ruminants in their ability to digest and utilize feedstuffs reside mainly in stomach anatomy, which in turn influences **feeding strategy.** As discussed in Chapter 1, ruminants show pronounced variations in reticulo-rumen and omasum anatomy. These differences have been discussed in detail by Hofmann (1973, 1988, 1989) and by Van Hoven and Boomker (1985). A classification for numerous ruminants is given in Table 22–1, and displayed graphically in Figure 22.6. The **concentrate selectors** have a small omasum and the least advanced rumen in terms of its adaptation to fiber digestion. They exhibit a nibbling type of feeding behavior, selecting forage and fruit that is

Table 22–1. Classification of Ruminants According to Feeding Characteristics

Common Name	*Scientific Name*	*Body Weight kg*
Grazers or Roughage Eaters		
American bison	*Bison bison*	800
African buffalo	*Syncerus caffer*	700
Ox	*Bos taurus*	600
Zebu	*Bos indicus*	400
Roan antelope	*Hippotragus equinus*	250
Waterbuck	*Kobus ellipsiprymnus*	220
Wildebeest	*Connochaetes taurinus*	220
Sable antelope	*Hippotragus niger*	200
Oryx	*Oryx gazelle*	180
Hartebeest	*Alcelaphus buselaphus*	150
Topi	*Damalisus lunatus*	120
Uganda kob	*Adenota kob*	90
Nile lechwe	*Kobus megaceros*	80
European sheep	*Ovis aries*	50
Reedbucks	*Redunca* spp.	40
Mouflon	*Ovis musimon*	30
Oribi	*Ourebia ourebi*	16
Intermediate or Adaptable Mixed Feeders		
European bison	*Bison bonasus*	800
Eland	*Taurotragus oryx*	700
Musk Ox	*Ovibos moschatus*	350
Wapiti	*Cervus canadensis*	300
Red deer	*Cervus elaphus*	150
Mule deer	*Odocoileus hemionus*	120
Caribou	*Rangifer tarandus articus*	120
Reindeer	*Rangifer tarandus tarandus*	100
White-tail deer	*Odocoileus virginianus*	100
White sheep	*Ovis dalli*	80
Fallow deer	*Dama dama*	70
Grant's gazelle	*Gazella granti*	60
Impala	*Aepyceros melampus*	60
Pronghorn	*Antilocapra americana*	50
Goat	*Capra hircus*	40
Springbok	*Antidorcas marsupialis*	35
Chamois	*Rupicapra rupricapra*	30
Maasai sheep	*Ovis aries*	30
Thomson's gazelle	*Gaxella thomsoni*	20
Chinese water deer	*Hydropotes inermis*	12
Steinbok	*Raphicerus campestris*	10
Browsers or Concentrate Selectors		
Giraffe	*Giraffa camelopardalis*	800
Moose	*Alces alces*	400
Greater kudu	*Tragelaphus strepsiceros*	250
Bongo	*Taurotragus eurycerus*	200
Lesser kudu	*Tragelaphus imberbis*	90
Bushbuck	*Tragelaphus scriptu*	60
Gerenuk	*Litocranius walleri*	40
Roe deer	*Capreolus capreolus*	20
Muntjacs	*Muntiacus* spp.	20
Red duiker	*Cephalophus harveyi*	16
Grey duiker	*Sylvicapra grimmia*	14
Klipspringer	*Oreotragus oreotragus*	12
Dik-diks	*Madoqua* spp.	5
Suni	*Nesotragus moschatus*	4
Larger mousedeer	*Tragulus napu*	4
Lesser mousedeer	*Tragulus javanicus*	1.5

Adapted from Kay et al. (1980).

high in protein and soluble carbohydrates and low in fiber. Direct absorption of soluble nutrients may occur without the intervention of rumen fermentation. (A detailed description of the anatomical and physiological adaptations of a small concentrate-selector ruminant, the dik-dik, has been provided by Maloiy et al. [1988]). Numerous studies, reviewed by Van Hoven and Boomker (1985), have been made comparing digestibility coefficients in domestic and wild ruminants and nonruminant herbivores. A more recent review article, focusing on browsers/concentrate selectors and intermediate feeders in zoos is available (Clauss et al., 2003), as are the integrative comments of Van Soest (1994).

As discussed in Chapter 9, Hofmann's division of ruminants into three general categories has been very useful in promoting greater understanding of rumen function and feeding strategy. It is now apparent that there is, in fact, little functional difference in digestive function among ruminants that is anatomically driven. Differences between browsers, grazers, and concentrate-feeders

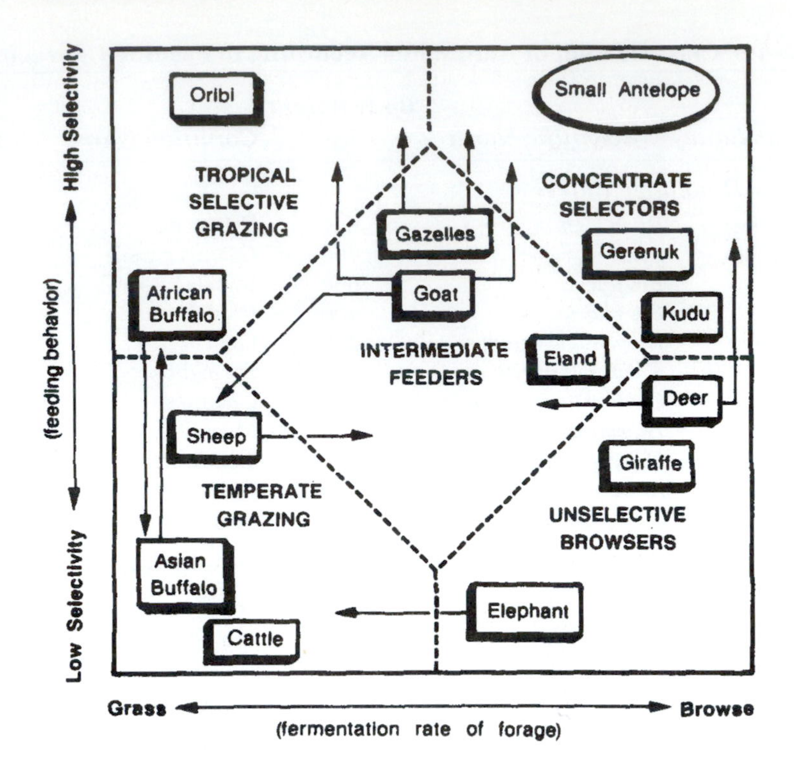

FIGURE 22.6 Herbivorous mammals classified by diet. The axes are the degree of feeding selectivity and the amount of grass versus browse in the diet. The arrows indicate mobility of a species with respect to these axes. (Reprinted with permission from Van Soest, 1994.)

are a consequence of feeding behavior and not digestive tract anatomy (Gordon and Illius, 1994, 1996; Robbins et al., 1995). Nevertheless, the problem of **omasal impaction** by coarse roughage in small, "dainty" ruminants is real.

The microbial population of the rumen varies somewhat according to feeding strategy. The concentrate selectors have a simpler microbial population and may totally lack ciliated protozoa, likely due to short residence time. Grazers and mixed-feeders have a high microbial diversity (Kay et al., 1980). The microbial population may be somewhat specialized if the diet is specialized. In the arctic ruminants, such as caribou and reindeer, lichens containing unique polysaccharides (lichenen and isolichenen) are a major constituent of the diet for much of the year, which may account for the unusual protozoal population characteristic of these animals (Dehority, 1986). The bacterial flora of arctic ruminants are similar to those found in domestic ruminants (Dehority, 1986). The **musk ox,** the only surviving arctic bulk-feeder from the Pleistocene, is adapted to grazing low-quality roughage (Nelleman, 1997), giving it a competitive advantage over caribou and reindeer in many arctic environments. Digestion of low-quality grass hay is higher in musk oxen than in cattle (Adamczewski et al., 1994a). Compared with other ruminants, musk oxen have relatively low maintenance energy requirements and very long rumen retention times (Adamczewski et al., 1994a,b). Their diet is dominated by low-quality, highly fibrous grasses and sedges. They have a high rumen capacity that, combined with a very low rate of digesta passage and low maintenance energy requirements, allows them to thrive on low-quality roughage. Their **low energy requirements** are due to a dense hair coat and a behavioral characteristic of restricting movement and spending a large portion of their time lying down during the winter (Forchhammer and Boomsa, 1995; Schaefer and

Messier, 1996). As with many other wild ungulates, they store large amounts of body fat in the summer and decrease their feed intake and time spent foraging in the winter. Although classed by Kay et al. (1980) as an intermediate feeder (Table 22–1), more recent researchers identify musk oxen as grazers or roughage eaters (Adamczewski et al., 1994a,b).

Other dietary constituents that may influence microbial populations include phenolics, essential oils, and toxic amino acids (e.g., mimosine), and both foregut and hindgut bacterial adaptations have been identified in a variety of wildlife species. Comparative microbiology is an area in need of attention in the field of wildlife nutrition.

Plant secondary compounds (plant toxins) have many interesting interactions with wild herbivores (Palo and Robbins, 1991). In a few cases, mammals have evolved specific detoxification and feeding strategies that have resulted in their becoming **specialist feeders.** For example, the **koala** has evolved hepatic enzymes for detoxifying *Eucalyptus* toxins and a specialist feeding behavior, eating only a few species of *Eucalyptus* (McLean and Foley, 1997). The **pygmy rabbit** has an obligate dietary relationship with terpene-containing **sagebrush** (White et al., 1982) and feeds almost entirely on sagebrush. Most herbivores avoid the effects of plant toxins by consuming a generalized diet (McArthur et al., 1993).

The concentration of toxins in plants often varies according to the degree of herbivory pressure and may vary in different parts of the plant. When trees and shrubs have been browsed, the new (juvenile) growth that follows is usually higher in toxins than the more mature vegetation that was browsed. Furstenburg and Van Hoven (1994) in South Africa reported that the browsing of *Acacia* trees by a **giraffe** resulted in an increase in the **tannin** content of the foliage within 2 to 10 minutes of initiation of browsing, causing the tree's foliage to become less palatable. The giraffe then moved to another tree, rather than completely stripping the leaves from any one tree. **Elephants** test small samples of leaves and reject them if they contain high concentrations of saponins, phenolics, and other plant toxins. Young immature leaves of many trees have higher levels of chemical defenses than mature leaves that are above the normal reach of herbivores. Apparently one of the reasons elephants selectively push over trees is to gain access to mature foliage at the top of the tree that has poor chemical defenses (Jachmann, 1989).

Another example of this type is the response of aspen trees to herbivory by **beavers.** When aspen trees are cut down by beavers, the new juvenile shoots emerging from the stumps have high concentrations of defensive chemicals (Basey et al., 1988). Eventually the beavers have to move on, because the new trees replacing the cut trees are unpalatable. Similarly, arctic browse species such as willows, alder, and birch respond to heavy herbivory by hares with high concentrations of phenolics in the new growth (Bryant et al., 1992). Population cycles of **arctic hares** are influenced in part by such changes in plant chemistry in response to herbivory (Krebs et al., 1995).

Many zoological parks maintain forage browse species that are harvested to feed various herbivores. With continued cutting and growth of new shoots (coppicing), the browse could become increasingly unpalatable. Nutritive value of African browse spp. for wild ungulates has been reviewed by Topps (1997). Databases of domestically harvested browses are being compiled and compared within the global zoological community in collaboration with zoo horticultural specialists (Kirschner et al., 1998). Information on species palatability, seasonal usage, and potential toxicities are integrated with data on proximate and mineral composition for a variety of browses fed in facilities in (currently) the United States, the United Kingdom, and Asia.

Nutritional requirements for **antler growth** in deer have received considerable study (Kay and Staines, 1981). Antler growth requires a high intake of energy, protein, and minerals, especially calcium. The organic matrix of antlers is entirely protein, and the hardened antler contains about 40 percent organic matter, so there is a **high protein requirement** for antler formation (Asleson et al., 1996). **Copper deficiency** results in malformed antlers (Gogan et al., 1988). Usually antler growth does not represent a nutritional problem for the wild animal, because it takes place during the summer when feed availability and intake are high. The maturation of antlers, shedding of the velvet, and maintenance of the antlers in the hard, functional condition are dependent on testosterone concentrations, which vary seasonally (West and Nordan, 1976). In New Zealand, deer have been domesticated for production of antlers, which are exported to China to be used for medicinal purposes. Deer farming for meat production has begun in the United States (Buckmaster, 1988), using the fallow deer. An advantage of the fallow deer is that it resembles the bulk and roughage eaters (grazers) more than the concentrate selectors (see Chapter 9). Thus, it thrives on grass and is not susceptible to omasal impaction when fed hay. Nutritional requirements of red deer for meat production have been reviewed by Kay et al. (1984) and Barry and Wilson (1994).

There is relatively little information about mineral deficiencies or toxicities in wild ungulates, although copper deficiency variously associated with mineral interactions, excess dietary protein, and interactions with plants containing pyrrolizidine alkaloids has been documented in dry region (steppe [yak] or savannah [blesbok antelope]) grazers (Clauss and Dierenfeld, 1999). Deer on copper-deficient diets show typical signs of copper deficiency, including ataxia and bone malformations (Audige et al., 1995). **Pronghorn antelope** are native to areas of western North America that tend to have high-selenium soils. They often occur on reclaimed strip-mined areas where selenium in mine spoil becomes oxidized to plant-available forms. The natural diet of pronghorns includes selenium-accumulating plant species. These observations suggest that pronghorn may be somewhat resistant to **selenium toxicity.** Raisbeck et al. (1996), in a feeding trial with high-selenium hay (15 ppm Se), found no evidence of clinical signs of selenium toxicity in captive pronghorns, suggesting that they may be tolerant of selenium toxicity.

NUTRITION OF OTHER WILD ANIMALS

There is a large body of information on the nutritional and feeding requirements of many wild animals and birds. The subject is too diverse to consider further in this book. Entry to the literature of interest may be obtained from a number of the references listed, such as Hume (1982) and Foley and Hume (1987a,b) for marsupials, Montgomery (1978) for a wide variety of tree-dwelling (arboreal) animals, Cheeke (1987) for lagomorphs (rabbits, hares, and pikas), and Robbins (1993) and Fowler (1986) for birds and mammals in general, Klasing (1998) for avian species, and a symposium on "Nutrition of Wild and Captive Wild Animals"

(*Proc. Nutr. Soc.* 56:989–999, 1997). Comparative aspects of digestive physiology are addressed in Stevens and Hume (1995), and a number of summary recommendations can be found on Web sites of the AZA NAG (www.nagonline.net), the European Zoo Nutrition Centre (www.eznc.org), and the Zoo Conservation Outreach Group (www.zcog.org), which focuses on Latin American facilities.

QUESTIONS AND STUDY GUIDE

1. Normally DE and ME values for feedstuffs are quite similar. In koalas consuming eucalyptus foliage and sheep consuming sagebrush, there is a large difference between DE and ME values for these plants. Explain. Both sagebrush and eucalyptus have a high content of essential oils, accounting for their distinctive odors.
2. Discuss the usefulness of β-carotene as a nutrient for lions, tigers, and cheetahs.
3. Name some arboreal folivores and discuss their adaptations to a low-energy diet.
4. Compare winter feeding programs for deer and elk.
5. Discuss metabolic and feeding strategy adaptations of wild herbivores (e.g., mule deer, bison) to seasonal differences in availability of feed.
6. The largest (e.g., elephants) and smallest (e.g., rabbits) herbivores are nonruminants, while ruminants dominate the intermediate body-size range. Explain. (Hint: see Chapter 9.)
7. Why is the black rhino in danger of becoming extinct in the wild? What nutritional factors are involved in the high mortality of rhinos kept in captivity?
8. According to Hofmann (1989), all of the domestic ruminants are either grazers (e.g., cattle, buffalo, sheep) or intermediate feeders (e.g., goat, reindeer). Although concentrate-selectors make up more than 40 percent of all ruminant species, there is not a single domesticated one. What are some possible reasons to account for this?

REFERENCES

Adam, C. L., C. E. Kyle, and P. Young. 1996. Seasonal patterns of growth, voluntary feed intake and plasma concentrations of prolactin, insulin-like growth factor–1, LH and gonadal steroids in male and female prepubertal red deer (*Cervus elaphus*) reared in either natural photoperiod or constant daylight. *Anim. Sci.* 62:605–613.

Adamczewski, J. Z., W. M. Kerr, E. F. Lammerding, and P. F. Flood. 1994a. Digestion of low-protein grass hay by musk oxen and cattle. *J. Wildlife Manage.* 58:679–685.

Adamczewski, J. Z., R. K. Chaplin, J. A. Schaefer, and P. F. Flood. 1994b. Seasonal variation in intake and digestion of a high-roughage diet by musk oxen. *Can. J. Anim. Sci.* 74:305–313.

Allen, M. E., T. C. Chen, M. F. Holick, and E. Merkel. 1999. Evaluation of vitamin D status of the green iguana (*Iguana iguana*): Oral administration vs. UVB exposure. In *Biologic Effects of Light*, M. F. Holick and E. G. Jung, eds. Boston: Kluwer. pp. 99–101.

Asleson, M. A., E. C. Hellgren, and L. W. Varner. 1996. Nitrogen requirements for antler growth and maintenance in white-tailed deer. *J. Wildlife Manage.* 60:744–752.

Audige, L., P. R. Wilson, R. S. Morris, and G. W. Davidson. 1995. Osteochondrosis, skeletal abnormalities and enzootic ataxia associated with copper deficiency in a farmed red deer (*Cervus elaphus*) herd. *N. Z. Vet. J.* 43:70–76.

Baer, D. J., O. T. Oftedal, W. V. Rumpler, and D. E. Ullrey. 1997. Dietary fiber influences nutrient utilization, growth, and dry matter intake of green iguanas (*Iguana iguana*). *J. Nutr.* 127:1501–1507.

Barbiers, R. B., K. M. Vosburgh, P. K. Ku, and D. E. Ullrey. 1982. Digestive efficiencies and maintenance energy requirements of captive wild felidae: Cougar (*Felis concolor*); leopard (*Panthera pardus*); lion (*Panthera leo*); and tiger (*Panthera tigris*). *J. Zoo Anim. Med.* 13:32–37.

Barboza, P. S. 1995. Digesta passage and functional anatomy of the digestive tract in the desert tortoise (*Xerobates agassizii*). *J. Comp. Physiol.* 165B:193–202.

Barker, D., M. P. Fitzpatrick, and E. S. Dierenfeld. 1998. Nutrient composition of selected whole invertebrates. *Zoo Biol.* 17:123–134.

Barry, T. N. and P. R. Wilson. 1994. Venison production from farmed deer. *J. Agr. Sci.* 123:159–165.

Basey, J. M., S. H. Jenkins, and P. E. Busher. 1988. Optimal central-place foraging by beavers: Tree-size selection in relation to defensive chemicals of quaking aspen. *Oecologia* 76:278–282.

Bernard, J. B. and D. E. Ullrey. 1989. Evaluation of dietary husbandry of marine mammals at two major zoological parks. *J. Zoo Wildlife Med.* 20:45–52.

Bjorndal, K. A. and A. B. Bolten. 1990. Digestive processing in a herbivorous freshwater turtle: Consequences of small intestine fermentation. *Physiol. Zoo.* 63:1232–1247.

Bryant, J. P., P. B. Reichardt, and T. P. Clausen. 1992. Chemically mediated interactions between woody plants and browsing mammals. *J. Range Manage.* 45:18–24.

Buckmaster, R. 1988. Deer farming: A new veterinary frontier. *Large Animal Vet.* 43:20–24.

Burn, D. M. 1986. The digestive strategy and efficiency of the West Indian manatee, *Trichechus manatus. Comp. Biochem. Physiol.* 85A:139–142.

Cheeke, P. R. 1987. *Rabbit Feeding and Nutrition.* San Diego: Academic Press.

Chen, X. B., T. Fujihara, K. Nakamura, P. O. Mawuenyegah, M. F. Franklin, and D. J. Kyle. 1997. Response of urinary and plasma purine derivatives to various rates and infusion patterns of purines in sheep nourished by intragastric infusion. *J. Agr. Sci.* 129:343–352.

Clauss, M. and E. S. Dierenfeld. 1999. Susceptibility of yak (*Bos grunniens*) to copper deficiency. *Vet. Rec.* 145:436–437.

Clauss, M., J. Gehrke, J. Fickel, M. Lechner-Doll, E. J. Flach, E. S. Dierenfeld, and J.-M. Hatt. 2002a. Induction of salivary tannin-binding proteins in captive black rhinoceros (*Diceros bicornis*) by dietary tannins. *Symp. Comp. Nutr. Soc.* 4:119–120. Antwerp, Belgium.

Clauss, M., D. A. Jessup, E. C. Norkus, M. F. Holick, W. J. Streich, and E. S. Dierenfeld. 2002b. Fat soluble vitamins in blood and tissues of free-ranging and captive rhinoceros species. *J. Wildl. Dis.* 38(2): 402–413.

Clauss, M., M. Lechner-Doll, E. J. Flach, J. Wisser, and J.-M. Hatt. 2002c. Digestive tract pathology of captive giraffe (*Giraffa camelopardalis*) a unifying hypothesis. *European Assocation of Zoo and Wildlife Veterinarians 4th Sci meeting,* Heidelberg.

Clauss, M., E. Kienzle, and J.-M. Hatt. 2003. Feeding practice in captive wild ruminants: Peculiarities in the nutrition of browsers/concentrate selectors and intermediate feeders. A review. In *Zoo Animal Nutrition,* Vol II., A. Fidgett, M. Clauss, U. Ganslosser, J.-M. Hatt, J. Nijboer, eds. Fuerth: Filander Verlag.

Clum, N. J., M. P. Fitzpatrick, and E. S. Dierenfeld. 1996. Effects of diet on nutritional content of whole vertebrate prey. *Zoo Biol.* 15:525–537.

Conklin-Brittain, N. L. and E. S. Dierenfeld. 1995. Small ruminants: Digestive capacity differences among four species weighing less than 20 kg. *Zoo Biol.* 15:481–490.

Cork, S. J. 1996. Optimal digestive strategies for arboreal herbivorous mammals in contrasting forest types: Why Koalas and colobines are different. *Aust. J. Ecol.* 21:10–20.

Crawford, M. A., ed. 1968. *Comparative Nutrition of Wild Animals.* London: Academic Press.

Crissey, S. D., J. A. Swanson, B. A. Lintzenich, B. A. Brewer, and K. A. Slifka. 1997. Use of a raw meat-based diet or a dry kibble diet for sand cats (*Felis margarita*). *J. Anim. Sci.* 75:2154–2160.

Dean, R. E. 1980. The nutrition of wild ruminants. In *Digestive Physiology and Nutrition of Ruminants,* Vol. 3, D. C. Church, ed. pp. 278–305. Corvallis, OR: O & B Books.

Dehority, B. A. 1986. Microbes in the foregut of arctic ruminants. In *Control of Digestion and Metabolism in Ruminants*, L. P. Milligan, W. L. Grovum, and A. Dobson, eds. pp. 307–325. Englewood Cliffs, NJ: Prentice Hall.

Dierenfeld, E. S. 1989. Vitamin E deficiency in zoo reptiles, birds, and ungulates. *J. Zoo Wildlife Med.* 20:3–11.

_______ . 1994. Vitamin E in exotics: Effects, evaluation and ecology. *J. Nutr.* 124: 2579S–2581S.

_______ . 1997. Captive wild animal nutrition: A historical perspective. *Proc. Nutr. Soc.* 56:989–999.

Dierenfeld, E. S. and M. G. Traber. 1992. Vitamin E status of exotic animals compared with livestock and domestics. In *Vitamin E in Health and Disease,* L. Packer and J. Fuchs, eds. pp. 345–360. New York: Marcel Dekker, Inc.

Dierenfeld, E. S., R. du Toit, and W. E. Braselton 1995. Nutrient composition of selected browses consumed by black rhinoceros (*Diceros bicornis*) in the Zambezi Valley, Zimbabwe. *J. Zoo Wildlife Med.* 26:220–230.

Dierenfeld, E. S., R. du Toit, and R. E. Miller. 1988. Vitamin E in captive and wild black rhinoceros (*Diceros bicornis*). *J. Wildlife Dis.* 24:547–550.

Dierenfeld, E. S., W. B. Karesh, B. L. Raphael, R. A. Cook, A. M. Kilbourn, E. J. Bosi, and M. Andau. 1998. Circulating α-tocopherol and retinol in free-ranging and zoo ungulates, *Proc. Comp. Nutr. Soc.* 2:42–46.

Dierenfeld, E. S., H. F. Hintz, J. B. Robertson, P. J. Van Soest, and O. T. Oftedal. 1982. Utilization of bamboo by the giant panda. *J. Nutr.* 112:636–641.

Dierenfeld, E. S., P. J. Mueller, and M. B. Hall. 2002. Duikers: Native food composition, micronutrient assessment, and implications for improving captive diets. *Zoo Biol.* 15:185–196.

Dierenfeld, E. S., S. Atkinson, A. M. Craig, K. C. Walker, and M. Clauss. 2004. Mineral concentrations in blood and liver tissue of captive and free-ranging rhinoceros species. *J. Zoo Wildl. Med.* In press.

Dominguez-Bello, M. G., M. Lovera, P. Suarez, and F. Michelangeli. 1993a. Microbial digestive symbionts of the crop of the hoatzin (*Opisthocomus hoazin*): An avian foregut fermenter. *Physiol. Zool.* 66:374–383.

Dominguez-Bello, M. G., M. C. Ruiz, and F. Michelangeli. 1993b. Evolutionary significance of foregut fermentation in the hoatzin (*Opisthocomus hoazin;* Aves: Opisthocomidae). *J. Comp. Physiol.* B163:594–601.

Douglas, T. C., M. Pennino, and E. S. Dierenfeld. 1994. Vitamins E and A, and proximate composition of whole mice and rats used as feed. Comp. *Biochem. Physiol.* 107A:419–424.

Edwards, M. S. and D. E. Ullrey. 1999. Effect of dietary fiber concentrations on apparent digestibility and digesta passage in non-human primates. II. Hindgut- and foregut-fermenting folivores. *Zoo Biol.* 18:537–549.

Ensley, P. K., T. L. Rost, M. Anderson, K. Benirschke, D. Brockman, and D. E. Ullrey. 1982. Intestinal obstruction and perforation caused by undigested *Acacia* sp. leaves in languar monkeys. *J. Am. Vet. Med. Assoc.* 181:1351–1354.

Farlow, J. O. 1987. Speculations about the diet and digestive physiology of herbivorous dinosaurs. *Paleobiology* 13:60–72.

Ferguson, G. W., W. H. Gehrmann, T. C. Chen, E. S. Dierenfeld, and M. F. Holick, 2002. Effects of artificial ultraviolet light exposure on reproductive success of the female panther chameleon (*Furcifer pardalis*) in captivity. *Zoo Biol.* 21:525–537.

Foley, W. J. 1987. Digestion and metabolism in a small arboreal marsupial, the Greater Glider (*Petauroides volans*), fed high-terpene *Eucalyptus* foliage. *J. Comp. Physiol.* 157:355–362.

Foley, W. J. and I. D. Hume. 1987a. Digestion and metabolism of high-tannin *Eucalyptus* foliage by the brushtail possum (*Trichosurus vulpecula*) (*Marsupialia: Phalangeridae*). *J. Comp. Physiol.* 157:67–76.

_______ . 1987b. Nitrogen requirements and urea metabolism in two arboreal marsupials, the greater glider (*Petauroides volans*) and the brushtail possum (*Trichosurus vulpecula*), fed *Eucalyptus* foliage. *Physiol. Zool.* 60:241–250.

Foley, W. J., S. McLean, and S. J. Cork. 1995. Consequences of biotransformation of plant secondary metabolites on acid-base metabolism in mammals—A final common pathway? *J. Chem. Ecol.* 21:721–743.

Forchhammer, M. C. and J. J. Boomsma. 1995. Foraging strategies and seasonal diet optimization of musk oxen in West Greenland. *Oecologia* 104:169–180.

Fowler, M. E., ed. 1986. *Zoo and Wild Animal Medicine*. Philadelphia: W. B. Saunders Co.

Freudenberger, D. O., K. Toyakawa, T. N. Barry, A. J. Ball, and J. M. Suttie. 1994. Seasonality in digestion and rumen metabolism in red deer (*Cervus elaphus*) fed on a forage diet. *Brit. J. Nutr.* 71:489–499.

Furstenburg, D. and W. Van Hoven. 1994. Condensed tannin as anti-defoliate agent against browsing by giraffe (*Giraffa camelopardalis*) in the Kruger National Park. *Comp. Biochem. Physiol.* 107A:425–431.

Garrott, R. A., P. J. White, D. B. Vagnoni, and D. M. Heisey. 1996. Purine derivatives in snow-urine as a dietary index for free-ranging elk. *J. Wildlife Manage.* 60:735–743.

Ghebremeskel, K. and G. Williams. 1988. Plasma retinol and alpha tocopherol levels in captive wild animals. *Comp. Biochem. Physiol.* 89:279–283.

Ghebremeskel, K., G. Williams, R. A. Brett, R. Burek, and L. S. Harbige. 1991. Nutrient composition of plants most favoured by black rhinoceros (*Diceros bicornis*) in the wild. *Comp. Biochem. Physiol.* 98A:529–534.

Gilchrist, F. M. C. and R. I. Mackie, eds. 1984. *Herbivore Nutrition in the Subtropics and Tropics.* Craighall, South Africa: The Science Press.

Gogan, P. J. P., D. A. Jessup, and R. H. Barrett. 1988. Antler anomalies in tule elk. *J. Wildlife Dis.* 24:656–662.

Gordon, I. J. and A. W. Illius. 1994. The functional significance of the browser-grazer dichotomy in African ruminants. *Oecologia* 98:167–175.

_______ . 1996. The nutritional ecology of African ruminants: A reinterpretation. *J. Anim. Ecol.* 65:18–28.

Grajal, A., S. D. Strahl, R. Parra, M. G. Dominguez, and A. Neher. 1989. Foregut fermentation in the hoatzin, a neotropical leaf-eating bird. *Science* 245:1236–1238.

Grant, J. B., D. L. Brown, and E. S. Dierenfeld. 2002. Essential fatty acid profiles differ across diets and browse of black rhinoceros. *J. Wildl. Dis.* 38:132–142.

Hall, M. B., E. S. Dierenfeld, C. C. Kearney, and R. L. Ball. 2003. Implications of carbohydrate feeding for captive herbivore nutrition welfare. *Proc. AZA Nutr. Advisory Group 5th Conf. Zoo Wildlife Nutrition*. pp. 24–25.

Hofmann, R. R. 1973. *The Ruminant Stomach.* Nairobi, Kenya: East African Literature Bureau.

_______ . 1988. Anatomy of the gastro-intestinal tract. In *The Ruminant Animal,* D.C. Church, ed. Englewood Cliffs, NJ: Prentice Hall.

_______ . 1989. Evolutionary steps of ecophysiological adaptation and diversification of ruminants: A comparative view of their digestive system. *Oecologia* 78:443–457.

Hofmann, R. R. and B. Matern. 1988. Changes in gastrointestinal morphology related to nutrition in giraffes (*Giraffa camelopardalis*): A comparison of wild and zoo specimens. *Inter. Zoo Yearbk.* 27:168–176.

Hoppe, P. P. 1984. Strategies of digestion in African herbivores. In *Herbivore Nutrition in the Subtropics and Tropics,* F. M. C. Gilchrist and R. I. Mackie, eds. pp. 222–243. Craighall, South Africa: The Science Press.

Hudson, R. J. and R. G. White, eds. 1985. *Bioenergetics of Wild Herbivores.* Boca Raton, FL: CRC Press.

Hume, I. D. 1982. *Digestive Physiology and Nutrition of Marsupials.* Cambridge: Cambridge University Press.

_______ . 1999. *Marsupial Nutrition*. Cambridge: Cambridge University Press.

Jachmann, H. 1989. Food selection by elephants in the 'Miombo' biome, in relation to leaf chemistry. *Bioch. Syst. Ecol.* 17:15–24.

Jones, R. I., M. A. Garcia-Amado, and M. G. Dominguez-Bello. 2000. Comparison of the digestive ability of crop fluid from the folivorous hoatzin (*Opisthocomus hoazin*) and cow rumen fluid with seven tropical forages. *Anim. Feed Sci. Tech.* 87:287–296.

Kay, R. N. B., W. V. Engelhardt, and R. G. White. 1980. The digestive physiology of wild ruminants. In *Digestive Physiology and Metabolism in Ruminants,* Y. Ruckebusch and P. Thivend, eds. pp. 743–761. Westport, CT: AVI Publishing Co.

Kay, R. N. B., J. A. Milne, and W. J. Hamilton. 1984. Nutrition of red deer for meat production. *Proc. R. Soc. Edinb.* [Biol] 82:231–242.

Kay, R. N. B. and B. W. Staines. 1981. The nutrition of the red deer (*Cervus elaphus*). *Nutr. Abst. Rev.* 51(B):601–622.

Kelly, J. D., D. J. Blyde, and I. S. Denney. 1995. The importation of the black rhinoceros (*Diceros bicornis*) from Zimbabwe into Australia. *Aust. Vet. J.* 72:369–374.

Kenny, D., R. C. Cambre, A. Lewandowski, J. A. Pelto, N. A. Irlbeck, H. Wilson, G. W. Mierau, M. V. Z. Fernando Gaul Sill, and M. V. Z. Alberto Paras Garcia. 1993. Suspected vitamin D_3 toxicity in pacas (*Cuniculus paca*) and agoutis (*Dasyprocta aguti*). *J. Zoo Wildlife Med.* 24:129–139.

Kirschner, A. C., N. A. Irlbeck, and M. M. Moore. 1998. Implications of a zoological database. *Proc. Comp. Nutr. Soc.* 2:108–113.

Klasing, K. C. 1998. *Comparative Avian Nutrition.* Wallingford & New York: CAB International.

Kleiman, D. G., M. E. Allen, K. V. Thompson, and S. Lumpkin, eds. 1996. *Wild Mammals in Captivity.* Chicago: University of Chicago Press.

Krebs, C. J., S. Boutin, R. Boonstra, A. R. E. Sinclair, J. N. M. Smith, M. R. T. Dale, K. Martin, and R. Turkington. 1995. Impact of food and predation on the snowshoe hare cycle. *Science* 269:1112–1115.

Langer, P. 1984a. Anatomical and nutritional adaptations in wild herbivores. In *Herbivore Nutrition in the Subtropics and Tropics,* F. M. C. Gilchrist and R. I. Mackie, eds. pp. 185–221. Craighall, South Africa: The Science Press.

_______ . 1984b. Comparative anatomy of the stomach in mammalian herbivores. *Q. J. Exp. Physiol.* 69:615–625.

_______ . 1986. Large mammalian herbivores in tropical forests with either hindgut- or forestomach-fermentation. *Z. Saugetierkunde* 51:173–187.

Leighton, F. A., M. Cattet, R. Norstrom, and S. Trudeau. 1988. A cellular basis for high levels of vitamin A in livers of polar bears (*Ursus maritimus*): The Ito cell. *Can. J. Zool.* 66:480–482.

Louden, A. S. I. 1994. Photoperiod and the regulation of annual and circannual cycles of food intake. *Proc. Nutr. Soc.* 53:495–507.

Mainka, S. A. and J. A. Mills, 1995. Wildlife and traditional Chinese medicine—Supply and demand for wildlife species. *J. Zoo Wildlife Med.* 26:193–200.

Maloiy, G. M. O and E. T. Clemens. 1991. Aspects of digestion and *in vitro* fermentation in the caecum of some East African herbivores. *J. Zool.* 224:293–300.

Maloiy, G. M. O., B. M. Rugangazi, and E. T. Clemens. 1988. Physiology of the dik-dik antelope. *Comp. Biochem. Physiol.* 91 (A):1–8.

McArthur, C., C. T. Robbins, A. E. Hagerman, and T. A. Hanley. 1993. Diet selection by a ruminant generalist browser in relation to plant chemistry. *Can. J. Zool.* 71:2236–2243.

McLean, S. and W. J. Foley. 1997. Metabolism of *Eucalyptus* terpenes by herbivorous marsupials. *Drug Metab. Rev.* 29:213–218.

McLean, S., W. J. Foley, N. W. Davies, S. Brandon, L. Duo, and A. Blackman. 1993. Metabolic fate of dietary terpenes from *Eucalyptus radiata* in common ringtail possum (*Pseudocheirus peregrinus*). *J. Chem. Ecol.* 19:1625–1643.

McMillin, J. M., U. S. Seal, and P. D. Karns. 1980. Hormonal correlates of hypophagia in white-tailed deer. *Fed. Proc.* 39:2964–2968.

Montgomery, G. G., ed. 1978. *The Ecology of Arboreal Folivores.* Washington, DC: Smithsonian Institute.

Morgan, P. J. and J. G. Mercer. 1994. Control of seasonality by melatonin. *Proc. Nutr. Soc.* 53:483–493.

National Research Council. 2003. *Nutrient Requirements of Nonhuman Primates.* 2d rev. ed. Washington, DC: National Academy Press.

Nelleman, C. 1997. Grazing strategies of musk oxen (*Ovibos moschatus*) during winter in Angujaartorfiup Nunaa in western Greenland. *Can. J. Zool.* 75:1129–1134.

Nijboer, J., H. van Brug, M. A. Tryfonidou, and J. P. T. M. van Leeuwen. 2003. UV-B and vitamin D_3 metabolism in juvenile Komodo dragons (*Varanus komodoensis*). In *Zoo Animal Nutrition,* Vol II, A. Fidgett, M. Clauss, U. Ganslosser, J.-M. Hatt, and J. Nijboer, eds. Fuerth: Filander Verlag. pp. 233–246

O'Brien, T. G., M. F. Kinnaird, E. S. Dierenfeld, N. L. Conklin-Brittain, R. W. Wrangham, and S. C. Silver. 1998. What's so special about figs? *Nature* 392:668.

Pahl, L. I. and I. D. Hume. 1990. Preferences for Eucalyptus species of the New England Tablelands and initial development of an artificial diet for koalas. In *Biology of the Koala,* A. K. Lee, K. A. Handasyde, and G. D. Sanson, eds. Sydney: Surrey Beatty.

Palo, R. T. and C. T. Robbins, eds. 1991. *Plant Defenses Against Mammalian Herbivory.* Boca Raton, FL: CRC Press.

Papas, A. M., R. C. Cambre, S. B. Citino, and R. J. Sokol. 1991. Efficacy of absorption of various vitamin E forms by captive elephants and black rhinoceroses. *J. Zoo Wildlife Med.* 22:309–317.

Parker, K. L., M. P. Gillingham, T. A. Hanley, and C. T. Robbins. 1993. Seasonal patterns in body mass, body composition, and water transfer rates of free-ranging and captive black-tailed deer (*Odocoileus hemionus sitkensis*) in Alaska. *Can. J. Zool.* 71:1397–1404.

Pickett, J. P., R. W. Chesney, B. Beehler, C. P. Moore, S. Lippincott, J. Sturman, and K. L. Ketring. 1990. Comparison of serum and plasma taurine values in Bengal tigers with values in taurine-sufficient and -deficient domestic cats. *J. Am. Vet. Med. Assoc.* 196:342–346.

Pitcher, T., R. Buffenstein, J. D. Keegan, G. P. Moodley, and S. Yahav. 1992. Dietary calcium content, calcium balance and mode of uptake in a subterranean mammal, the Damara mole-rat. *J. Nutr.* 122:108–114.

Raisbeck, M. F., D. O'Toole, R. A. Schamber, E. L. Belden, and L. J. Robinson. 1996. Toxicologic evaluation of a high-selenium hay diet in captive pronghorn antelope (*Antilocapra americana*). *J. Wildlife Dis.* 32:9–16.

Renecker, L. A. and R. J. Hudson. 1990. Digestive kinetics of moose (*Alces alces*), wapiti (*Cervus elaphus*) and cattle. *Anim. Prod.* 50:51–61.

Robbins, C. T. 1993. *Wildlife Feeding and Nutrition.* San Diego: Academic Press.

Robbins, C. T., D. E. Spalinger, and W. Van Hoven. 1995. Adaptations of ruminants to browse and grass diets: Are anatomical-based browser-grazer interpretations valid? *Oecologia* 103:208–213.

Savage, A., K. M. Leong, D. Grobler, J. Lehnhardt, E. S. Dierenfeld, E. F. Stevens, and C. P. Aebischer. 1999. Circulating levels of α-tocopherol and retinol in free-ranging African elephants (*Loxodonta africana*). *Zoo Biol.* 18:319–323.

Schaefer, J. A. and F. Messier. 1996. Winter foraging of musk oxen in relation to foraging conditions. *Ecoscience* 3:147–153.

Schweigert, F. J., O. A. Ryder, W. A. Rambeck, and H. Zucker. 1990. The majority of vitamin A is transported as retinyl esters in the blood of most carnivores. *Comp. Biochem. Physiol.* 95A:573–578.

Schweigert, F. J., E. Thomann, and H. Zucker. 1991. Vitamin A in the urine of carnivores. *Int. J. Vit. Nutr. Res.* 61:110–113.

Semiadi, G., T. N. Barry, and P. D. Muir. 1995. Comparison of seasonal patterns of growth, voluntary feed intake and plasma hormone concentrations in young sambar deer (*Cervus unicolor*) and red deer (*Cervus elaphus*). *J. Agr. Sci.* 125:109–124.

Setchell, K. D. R., S. J. Gosselin, M. B. Welsh, J. O. Johnston, W. F. Balistreri, L. W. Kramer, B. L. Dresser, and M. J. Tarr. 1987. Dietary estrogens: A probable cause of infertility and liver disease in captive cheetahs. *Gastroenterol.* 93:225–233.

Slifka, K. A., P. E. Bowen, M. Stacewicz-Sapuntikis, and S. D. Crissey. 1999. A survey of serum and dietary carotenoids in captive wild animals. *J. Nutr.* 129:380–390.

Silver, S. C., L. E. T. Ostro, C. P. Yeager, and E. S. Dierenfeld. 2000. Phytochemical and mineral components of foods consumed by black howler monkeys (*Alouatta pigra*) at two sites in Belize. *Zoo Biol.* 19:95–109.

Smith, J. E., P. S. Chavey, and R. E. Miller. 1995. Iron metabolism in captive black (*Diceros bicornis*) and white (*Ceratotherium simum*) rhinoceroses. *J. Zoo Wildlife Med.* 26:525–531.

Stevens, C. E. and I. D. Hume. 1995. *Comparative Physiology of the Vertebrate Digestive System.* 2d ed. Cambridge: Cambridge University Press.

Topps, J. H. 1997. Nutritive value of indigenous browse in Africa in relation to the needs of wild ungulates. *Anim. Feed Sci. Tech.* 69:143–154.

Ullrey, D. E. 1986. Nutrition of primates in captivity. In *Primates. The Road to Self-Sustaining Populations,* K. Benirschke, ed. pp. 823–835. New York: Springer-Verlag.

Ullrey, D. E. and M. E. Allen. 1986. Principles of zoo mammal nutrition. In *Zoo and Wild Animal Medicine,* M. E. Fowler, ed. pp. 515–532. Philadelphia: Saunders.

Ullrey, D. E. and J. B. Bernard. 1989. Meat diets for performing exotic cats. *J. Zoo Wildlife Med.* 20:20–25.

Vagnoni, D. B., R. A. Garrott, J. G. Cook, P. J. White, and M. K. Clayton. 1996. Urinary allantoin: Creatinine ratios as a dietary index for elk. *J. Wildlife Manage.* 60:728–734.

Van Hoven, W. and E. A. Boomker. 1985. Digestion. In *Bioenergetics of Wild Herbivores,* R. J. Hudson and R. G. White, eds. pp. 103–120. Boca Raton, FL: CRC Press.

Van Rensburg, I. B. J. and M. H. Lowry. 1988. Nutritional secondary hyperparathyroidism in a lion cub. *J. S. Afr. Vet. Assoc.* 59:83–86.

Van Soest, P. J. 1994. *Nutritional Ecology of the Ruminant.* Ithaca: Cornell University Press.

Van Soest, P. J., E. S. Dierenfeld, and N. L. Conklin. 1995. Digestive strategies and limitations of ruminants. In *Ruminant Physiology: Digestion, Metabolism, Growth and*

Reproduction, Proc. Eighth International Symp. Ruminant Physiol. W. von Englehardt, ed. pp. 581–600. Ferdnand Enke Verlag.

Ward, A. M., J. L. Dempsey, and A. S. Hunt. 2003. A survey of the nutrient content of foods consumed by free-ranging and captive anegada iguanas (*Cyclura pinguis*). *Proc. AZA Nutr. Advis. Group 5th Conf. Zoo Wildlife Nutrition.*

Weisner, C. S. and C. Iben. 2003. Influence of environmental humidity and dietary protein on pyramidal growth of carapaces in African spurred tortoises (*Geochelone sulcata*). *J. Anim. Physiol. Anim. Nutr.* 87:66–74.

Wenniger, P. S. and L. A. Shipley. 2000. Harvesting, ruminantion, digestion and passage of fruit and leaf diets by a small ruminant, the blue duiker. *Oecologia* 123:466–474.

West, N. O. and H. C. Nordan. 1976. Hormonal regulation of reproduction and the antler cycle in the male Columbian black-tailed deer (*Odocoileushemionus Columbianus*). Part I. Seasonal changes in the histology of the reproductive organs, serum testosterone, sperm production, and the antler cycle. *Can. J. Zool.* 54:1617–1636.

White, P. J., R. A. Garrott, and D. M. Heisey. 1997. An evaluation of snow-urine ratios as indices of ungulate nutritional status. *Can. J. Zool.* 75:1687–1694.

White, S. M., B. L. Welch, and J. T. Flinders. 1982. Monoterpenoid content of pygmy rabbit stomach ingesta. *J. Range Manage.* 35:107–109.

Whitehead, P. E. and E. H. McEwan. 1973. Seasonal variation in the plasma testosterone concentration of reindeer and caribou. *Can. J. Zool.* 51:651–658.

Willette, M. M., T. L. Norton, C. L. Miller, and M. G. Lamm. 2002. Veterinary concerns of captive duikers, *Zoo Biol.* 15:197–207.

Wright, J. B. and D. L. Brown. 1997. Identification of 18:3 (n-3) linolenic acid, 18:3 (n-6) linolenic acid and 18:2 (n-6) linoleic acid in Zimbabwean browses preferred by wild black rhinoceroses (*Diceros bicornis*) determined by GC-MS analysis. *Anim. Feed Sci. Tech.* 69:195–199.

Young, R. L. 1997. The importance of food presentation for animal welfare and conservation. *Proc. Nutr. Soc.* 56:1095–1104.

CHAPTER 23

Feeding and Nutrition of Fish and Other Aquatic Organisms

R.A. Swick and P.R. Cheeke

Objectives

1. To describe the basic principles of fish nutrition, emphasizing those aspects that differ from mammalian nutrition
2. To discuss nutrient requirements for fish production
3. To discuss feeding programs for catfish, salmonids, tilapia, and shrimp

AQUACULTURE

There is increasing interest in **aquaculture** (fish farming) as a means of producing high-quality human food. Aquaculture is the fastest growing sector of the world feed market, and consumption of cultured fish and seafood is increasing dramatically.

The main types of commercially farmed fish in North America are carp, catfish, milkfish, rainbow trout, salmon, shrimp, and tilapia (Fig. 23.1). During the 1980s and early 1990s, channel catfish production expanded rapidly in the southern United States, and salmon production expanded in coastal British Columbia. More recently, carp, catfish, milkfish, shrimp, and tilapia production have expanded massively in China and Southeast Asian countries. Cultured shrimp are major export earners in Thailand, Indonesia, and Vietnam. Cultured basa and tra catfish are also a major export earner for Vietnam. Other cultured aquatic animals include various mollusks, crayfish, frogs, freshwater giant prawns, and alligators.

Marketing of fish in contrast to poultry and swine is highly fragmented, as there is a wide array of finfish, mollusks, and crustaceans involving both aquaculture and wild catch. Of the total estimated 142 mmt of fish produced annually (year 2000), the capture fisheries accounted for 102 mmt with 30 mmt of that converted to fish meal and oil. Aquaculture is estimated to have produced nearly 40 mmt in 1999, with over 90 percent of the production in Asia. China is by far the leader with an estimated production of 29 mmt in 2000 (Fig. 23.2). China produces mainly carp and tilapia grown on a sustainable basis in small holdings. Fish are collected and sold in traditional wet markets by third parties. Although Chinese aquaculture lacks capital and organization, the use of processed feed has recently caught the attention of the more progressive growers as a way to increase profit. Aquaculture feed production is expanding rapidly in China.

Shrimp and salmon, because of their relatively high value among the cultured fish species, stand out for their potential to be grown the way broiler chick-

FIGURE 23.1 Various types of aquaculture: Top. Trout production in raceways in Idaho. (Courtesy of R.W. Hardy.) Middle. Salmon farm in Norway. (Courtesy of A. Skrede.) Bottom. Catfish ponds in Mississippi. (Courtesy of C.E. Bond.)

ens are, that is, in vertically integrated businesses. In Taiwan and Southeast Asia, several large feed and poultry companies made significant strides in the late 1980s and early 1990s toward vertical integration in shrimp production. This led the way to similar developments in South and Central America, the Indian subcontinent, and now Africa. In most locations, however, early successes have been punctuated with disease failures. The pathogenic agents, more often than not, have been the consequence of a deteriorated environment caused by the farms themselves and poor practice of biosecurity. Technical advances in disease control, genetics, nutrition, and ecology all have the potential to overcome many of these problems. The value and demand of these food products in the market

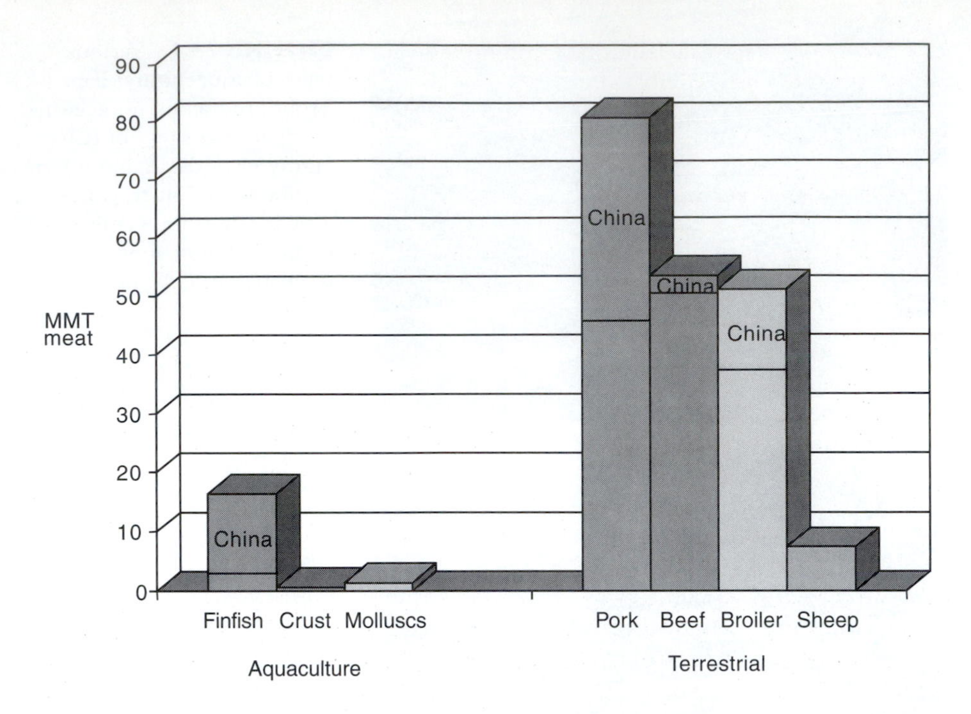

FIGURE 23.2 Annual Production of Meat from Aquaculture and Terrestrial Animals in China and the Rest of the World. (Food and Agriculture Organization, United Nations.)

warrant strong development effort. Business integration and technological development will lead the way for cultured aquatic species to become major protein sources in the human diet.

Various aspects of fish feeding and nutrition have been reviewed comprehensively in the books by Halver (1989) and Lovell (1989), which should be consulted for more detailed information than is presented in this chapter. Scientific journals in which articles on fish nutrition are common include *Aquaculture, Global Aquaculture Advocate, Progressive Fish Culturist, Transactions of the American Fisheries Society, Canadian Journal of Fisheries and Aquatic Sciences, Journal of the World Aquaculture Society,* and *Journal of Nutrition.*

Aquaculture and Fish Feeding

Production of fish in ponds and raceways requires artificial feeding and the preparation of commercial feeds by feed manufacturers. Compared to the traditional livestock species, the nutritional requirements of fish have received little attention. However, the rapid expansion of aquaculture has led to considerable interest in fish nutrition. The nutritional requirements of fish differ substantially from those of mammals. Fish are **poikilothermic,** which means that their body temperature is not homeostatically maintained but fluctuates with water temperature. The body temperature of a resting fish is at or near the environmental water temperature. Thus, the maintenance energy requirements of fish are lower

TABLE 23-1. Comparison of Maintenance Energy Requirements of Fish with Mammals and Birds

Species	*Body Weight g*	*Environmental Temperature, °C*	*ME, $kJ/kg^{0.75}/day$*
Rainbow trout	50–150	7.5	11
		10	24
		15	29
		20	33
Carp	40–90	23	54
Chicken	1500	Thermoneutral	355
Rat	130	22	552
	180–120	30	380

Adapted from Cho and Kaushik (1985).

than for mammals (Table 23–1) because they do not need to expend energy to maintain their body temperature. As a result, feed conversion efficiency is very high, often being one unit or more of weight gain per unit of feed. Another factor contributing to high feed efficiency is that pond-raised fish may consume considerable amounts of **natural foods** in the pond, which are not included in the feed efficiency calculation.

Warmwater fish and shellfish have optimal water temperatures of 25 to 30°C. Examples of common **warmwater fish** are carp, catfish, milkfish, and tilapia; shrimp and prawns are examples of warmwater shellfish. There are both marine and freshwater shrimp species, and the term **prawn** refers to any large shrimp. **Coldwater species,** such as salmon and trout **(salmonids),** have an optimal water temperature range of 10 to 15°C.

In the feeding of fish, it should be appreciated that most fish occupy an ecological niche near the top of the food chain. Most fish have evolved as carnivores; there are few omnivores or herbivores in the natural environment, particularly in the oceans (Cowey and Walton, 1989). The metabolism of fish is adapted to a diet high in protein and low in carbohydrate, as is true for terrestrial carnivores. Therefore, it is not surprising that fish diets are high in protein and low in carbohydrate.

The nutrient levels of fish diets are influenced by the cultural systems used. With fish raised in ponds, insects, plankton, algae, and other **natural food** may provide a significant portion of the nutrient intake and satisfy the requirements for vitamins and pigmenting agents. In some systems, such as carp production, the pond, rather than the fish, is fed. Ponds may be fertilized with manure or inorganic fertilizers to stimulate growth of plankton and algae, and the cultured fish or shrimp feed on these materials, especially in the early growth phases. Fish reared in raceways are entirely dependent on the offered feed, and, therefore, their nutritional requirements are more critical.

In terms of digestive tract physiology, fish are monogastric, although technically some species, such as carp, have no stomach and, therefore, might better be called agastric. Channel catfish, salmonids, and most other fin fishes have true stomachs with hydrochloric acid and pepsinogen secretion. Digestion is accomplished by digestive enzymes similar to those of other animals; the major sources of digestive enzymes are the stomach, intestinal mucosa, pancreas, and pyloric ceca. Digestion of algae cell walls is facilitated by a strongly acid stomach (Hepher, 1988). Some nutrients, such as calcium, are absorbed directly from the water via the gills. **Herbivorous fish,** such as the grass carp, eat large amounts of vegetation. Although they have cellulolytic gut microbes (Hepher, 1988), they derive nutrients mainly from the nonfiber fraction. Grass carp may be added to ponds and waterways to control unwanted vegetation such as water hyacinth. Young grass carp can achieve an excellent growth rate when feeding on aquatic weeds such as *Elodea* spp. (Cai and Curtis, 1989). The intestinal tract in carnivorous fish is shorter than in herbivorous or omnivorous species, as is true with mammals. Compared to mammals, fish have a very short intestinal tract, which is not separated into a small and large intestine. Volatile fatty acid production in herbivorous fish can be nutritionally significant (Clements and Choat, 1995; Kandel et al., 1994).

Feeding and the Environment

Aquaculture, and in particular **shrimp farming,** differs from terrestrial animal farming in that the growout operations are typically located along oceanfront and estuarine areas and are, therefore, under environmental scrutiny. Semi-intensive culture requires large ponds, location on clay soil to reduce seepage, low soil acidity, and low land elevation to reduce pumping costs. Problems with pollution and disease have occurred in shrimp culture areas because of concentrated development without coordination and the use of infected postlarvae. Improved pond management and development according to plans and guidelines will help the situation. In important culture areas such as Thailand, codes of conduct are being developed for shrimp farming to ensure sustainability and provide for environmental, social, and economic benefits for present and future generations. Businesses, organizations, and stakeholders involved in the industry are involved in generating, reviewing, and commenting on the codes of conduct.

Overfeeding of fish may cause nutrient accumulation, fish kills, and eutrophication of water bodies. Feeding may be done by hand, machine blowers, automatic feeders, or demand feeders. Hand feeding is labor intensive but tends to ensure greater awareness by the operator of potential or developing problems. Automatic feeders are labor saving but may waste feed if not carefully monitored. Demand feeders allow the fish to feed themselves. A rod extends from the feeder into the water; fish learn to bump the rod to cause feed to fall into the water. There is little wastage with this system.

Ponds are the primary ecosystems for fish farming and are accountable for >95 percent of world production. The concept of 80:20 **pond fish culture** is to raise several species of fish in ponds (Schmittou, et al., 1998). Approximately 80 percent of the harvest weight is composed of only one feed-taking, high-value species with high consumer demand, and approximately 20 percent is composed of "service species," such as filter feeders that help clean the water and predaceous fishes that control wild fish and other competitors. The 80:20 system typically provides higher yields and higher profit than monoculture or with other relative densities of fish. In intensive feed-based fish production, feed should be formulated to supply all nutritional requirements and should be fresh, clean, and have a water stability of greater than 10 minutes. As a general rule, fish production will increase linearly with feeding rate, whereas water quality deteriorates exponentially with feeding rate. Management techniques to prevent and control water quality deterioration resulting from feed wastes must be based on limiting the feeding amount to a safe level. This level will counter formation of toxic phytoplankton and low dissolved oxygen. In intensive feed-based systems (>10,000 fish/ha), addition of wastes such as fish scraps, vegetable scraps, oilseed cakes, cooked or raw cassava, or corn is strongly discouraged during growout. Any nutritional benefit is outweighed by increased fish stress, disease incidence, and pollution. In addition, adding such wastes may negatively impact the acceptability of the final product in the market. Pond-raised fish should be fed a pelleted sinking feed or an extruded floating feed to 90 percent satiation. The easiest way to do this is to determine 100 percent satiation on day 1 and measure the amount fed. This amount is then fed for the next 10 days. By day 10 the rate will be down to 80 percent satiation with an overall average of about 90 percent satiation. This is then repeated every 10 days. Smaller fish weighing 25 to 100 grams should be fed three times per day, and fish from 100 grams to market are best fed two times a day. Increased daily feedings may increase growth rate but will degrade feed conversion and profit, especially since there is a tendency to overfeed.

Intensive fish culture in low volume (1 to 4 m3) cages at high fish densities (up to 500 individuals or 200 kg of fish per cubic meter) is becoming an important means of expanding production (Schmittou, 1997). Small cages are preferred over the traditional larger cages in intensive culture as more fish can be raised per unit volume. Small cages have greater surface area per volume, which enhances water exchange. Smaller cages, thus, can hold more fish per unit volume and as a result cost less per weight of fish produced. This low-volume high-density (LVHD) design requires that cages be placed individually with no adjacent common sides of any two cages. The LVHD system may be used in lakes, reservoirs, rivers, or ponds. Higher productivity per area of cage and water area is common in open reservoirs, lakes, and rivers, but in many cases this is not sustainable due to lack of carrying capacity of the water body for the number of fish being cultured. Lack of carrying capacity results in nutrient turnover, low dissolved oxygen syndrome, and routine fish kills. Problems with fish kills in open mesotrophic water bodies (80–200 cm visibility) can be avoided by limiting feed input in the total water body to 8 kg/ha/day. Limits in the area where fish are growing should also be imposed. For a 1 ha cage area, a limit of 500 kg/day would be a good starting point. In a 10 ha intensive cage area the limit should be 1500 kg/d (Schmittou, 1997).

The use of **closed biosecure culture systems** with reduced or zero water exchange is an area under intense evaluation at the present time. Such novel production systems have the potential to increase productivity, reduce effluent, and control disease. These systems have already been developed, tested, and proven to work on a prototype basis. At present, however, they are not cost-effective on a commercial scale because of high electricity costs necessary to run aerators and to pump water through raceways and filters (Leung and Moss, 1999). Further refinement of biofilters required to clarify water and metabolize nitrogenous wastes along with improvements in genetics of shrimp may make biosecure systems economically viable in the future. This would allow farms to be located some distance away from the ocean and would reduce the incidence of disease. Much like broiler production on litter floors, several crops of shrimp could be grown in the same water, with time allotted for depopulation between growouts.

TABLE 23-2. Estimated Dietary Protein Requirements of Various Fish

Species	*% Dietary Crude Protein for Maximum Growth*
Rainbow trout	40–46
Carp	31–38
Catfish	32–36
Chinook salmon	40
Eel	44.5
Plaice	50
Shrimp	40

Adapted from National Research Council (1981).

NUTRIENT REQUIREMENTS

Protein Needs

Some typical values of protein contents of fish diets are given in Table 23–2. The values are much higher than for livestock diets. Because fish utilize carbohydrates poorly, the concentration of carbohydrate in the diet is low. Consequently, protein and fat percentages are high. A considerable portion of the dietary protein is used as a source of energy. Protein is used more efficiently as an energy source in fish than in mammals and birds because the nitrogen from amino acid deamination is excreted as ammonia via the gills. In mammals and birds, energy is required to synthesize urea or uric acid from the deaminated nitrogen. On a percent-of-diet basis, the essential amino acid requirements of fish are higher than for swine and poultry.

Because of the high-protein content of fish diets and the high essential amino acid requirements, animal protein sources provide a large part of the protein in fish diets. Fish are quite sensitive to heat-labile toxins such as trypsin inhibitors in soybean products, so proper heat treatment of many **plant protein sources** is necessary. The upper limit for soy protein in diets of carnivorous fish is between 20 and 30 percent, even if heat-labile antinutritional factors are eliminated (Storebakken et al., 2000). With the increasing cost of fish meal due to limited supplies relative to other protein sources, there is greater interest in maximizing use of plant proteins, especially soybean meal (Figure 23.3). Other vegetable protein meals may also be used. For example, lupin seed meal has been found to be satisfactory as a source of up to 30 percent of dietary protein in rainbow trout diets (Higuera et al., 1988). Cottonseed and canola meals are useful protein supplements; they should not exceed dietary levels of 15 percent because of the potential of gossypol and glucosinolate toxicity. **Soybean meal** can be used as a replacement for fish meal in diets for rainbow trout (Refstie et al., 1997), although an adaptation period is required before the feed is well accepted. Similar results were reported by Rumsey et al. (1993).

The use of soybean meal in **shrimp feed** has been studied extensively (Akiyama, 1989). Fiber level of the meal is important as excess fiber reduces pellet stability and reduces energy density of the diet. Levels of 25 to 30 percent dehulled soybean meal have been demonstrated to give good performance in

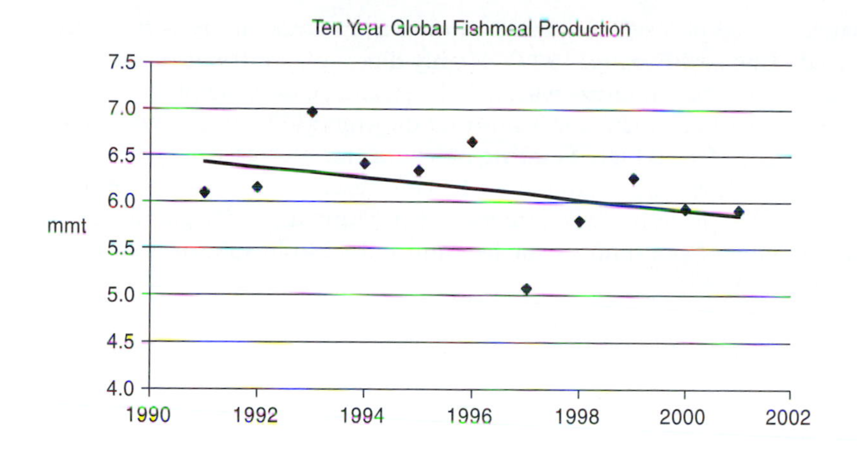

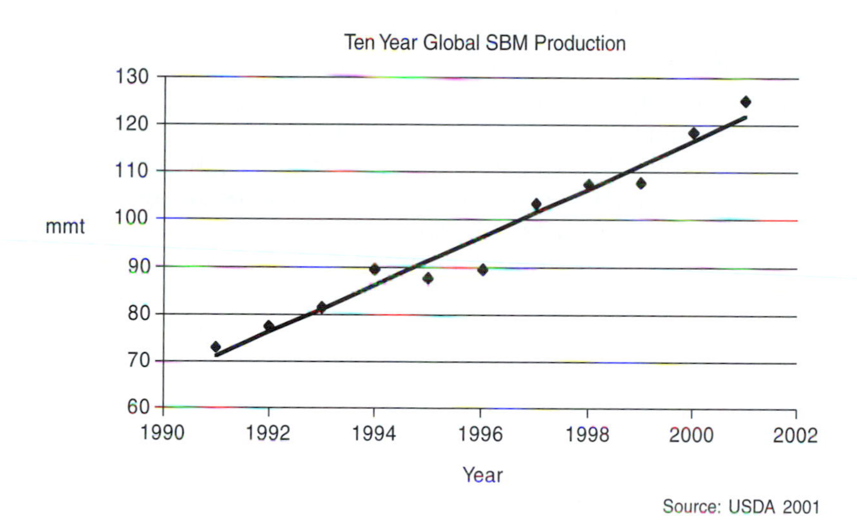

FIGURE 23.3 World production of fishmeal and soybean meal. (USDA 2001.)

shrimp (Lim and Dominy, 1990). Reducing the reliance on fish meal in shrimp feed will reduce cost and lower the possibility of diet-transmitted pathogens. Formulation on a digestible amino acid basis and the use of ideal amino acid ratios will reduce nitrogen loading of pond water and effluent. Use of phytase enzyme will have a beneficial impact on phosphorus levels in water if methods of application can be found to increase enzyme stability and reduce possibility of leaching. Ingredients such as porcine plasma protein, soy protein concentrate, and partially hydrolyzed vegetable protein may have future utility as attractants and nutrient sources.

Fish feeds are typically high in protein content, and modern aquaculture facilities have fish at high stocking densities. These conditions favor the build-up of **nitrogenous wastes** in the water, causing toxic reactions in the fish. Fish excrete excess nitrogen as **ammonia.** Once in the water, ammonia undergoes bacterial oxidation to form **nitrite.** Both ammonia and nitrite are toxic to fish, causing growth inhibition and mortality. They stimulate secretion of stress hormones such as corticosteroid hormones that impair immune function and cause increased susceptibility to disease. The subject of nitrogen toxicity to fish has been reviewed comprehensively by Tomasso (1994).

A relatively large part of the water pollution from aquaculture is from unconsumed feed (Cho and Bureau, 1997). Highly digestible, nutrient-dense diets maximize feed intake and minimize waste. A disadvantage of using plant sources of protein, such as soybean meal, for fish is that they tend to be unpalatable, and the quantity of wasted feed is high. **Feeding attractants,** such as krill hydrolysate or peptides, can increase the palatability of plant-based diets and reduce feed wasteage (Oikawa and March, 1997). Phytase pretreatment of soybean meal improves bioavailability of phosphorus for fish and reduces phosphorus excretion (Cain and Garling, 1995).

Carbohydrate Needs

Fish seem to use most carbohydrate sources poorly. **Starch** in grains has a digestibility of only 20 to 40 percent. Cooking or heat treatment (e.g., extrusion) improves starch digestion considerably. Processing methods for grains could probably be developed to allow greater use of carbohydrates in fish diets, which would offer considerable cost savings. Because wheat contains amylase inhibitors and intestinal amylase activity is low in many fish species, especially carnivorous ones, wheat products should be avoided or used only at low levels. A problem with carbohydrate utilization is that fish may store excess quantities of glycogen in the liver. When **liver glycogen** levels exceed 16 percent, death may occur, apparently due to kidney malfunction and nephrosis. In contrast to most animals, fish do not mobilize liver and muscle glycogen during fasting (Cowey and Walton, 1989). Even in a spawning migration of more than 1,000 km, during which they do not feed, Pacific salmon do not metabolize glycogen stores. During **starvation,** fish mobilize lipid first, then protein, and carbohydrate only during the final stages. Fish may survive a considerable period of time (even years, in some cases) after reaching the "point of no return" from fasting, after which irreversible metabolic changes occur that lead to death.

Lipid Needs

Dietary fats are a major energy source for most fish. Fat metabolism in fish shows considerable differences from that of land animals. The carcass fat has a high content of polyunsaturated fatty acids of 20 to 22 carbon atoms and five and six double bonds, making fish oils highly unsaturated. In land animals, the carcass fatty acids belong to the oleic (omega-9) and linoleic (omega-6) series. In fish, the omega-6 series is replaced by the **omega-3 fatty acid** (linolenic) series. Marine fish have higher omega-3 (n-3) fatty acid levels than freshwater species. The dietary origin of these fatty acids is the phytoplankton at the bottom of the food chain. A dietary deficiency of n-3 fatty acids leads to a number of physiological disorders, including poor growth, fin necrosis, pale fatty liver, edema, heart myopathy, and stress-induced shock. Marine fish have a higher n-3 requirement than freshwater fish, and coldwater species (e.g., salmonids) have a higher requirement than warmwater species (e.g., catfish). More highly unsaturated fatty acids in cell membranes are necessary at coldwater temperatures to maintain normal flexibility and permeability because the melting point of a fat decreases as the degree of unsaturation increases. Channel catfish have a requirement for n-3 fatty acids, of 1.0 to 2.0 percent linolenic acid (18:3:n-3) or 0.5 to 0.75 percent of mixed n-3 fatty acids (Satoh et al., 1989).

Energy Needs

Fish have lower maintenance energy requirements than mammals and birds because they do not have to maintain a constant body temperature, they exert less energy in movement in water than birds and mammals do on land and air, and they excrete ammonia rather than urea or uric acid and thus do not have the energy costs of synthesizing these compounds. A DE requirement of 8 to 9 kcal/g of protein is recommended for warmwater fish as a suitable energy-to-protein ratio.

Vitamin Needs

Fish require the same vitamins as land animals; in addition, many fish require vitamin C (ascorbic acid). **Ascorbic acid** has a metabolic role in the synthesis of connective tissue; it functions in forming cross-links between collagen fibers. Deficiency signs are related to the role of vitamin C in collagen formation and include scoliosis (lateral curvature of spine) and lordosis (vertical curvature). Some species, such as salmon, synthesize ascorbate–2 sulfate, which is a tissue storage form of the vitamin (Halver, 1989). Vitamin C is readily oxidized and much of its activity is lost when catfish feeds are extruded to produce **floating feeds.** Even with ethylcellulose-coated ascorbic acid, the losses are high. A stable form of vitamin C, **L-ascorbyl–2-polyphosphate (AsPP),** is stable during feed processing and is bioactive as a vitamin source in catfish (Wilson et al., 1989). The precision with which some of the vitamin requirements are known is not very high. For example, Woodward and Frigg (1989) found that a biotin supplement not exceeding 0.15 mg biotin/kg dry diet was adequate for rainbow trout; this is 10 times lower than the NRC recommended level. Vitamin requirements are given in Table 23–3 and deficiency signs in Table 23–4.

TABLE 23–3. NRC Vitamin Requirements of Fish, mg/kg diet

		*Warmwater Species**		
Vitamin	*Coldwater Fish*	*Common Carp*	*Catfish*	*Shrimp*
A	2,500 IU	10,000 IU	1,000–2,000 IU	ND
D	2,400 IU	ND	500–1,000 IU	ND
E	30 IU	200–300 mg	30 mg	ND
K	10 IU	ND	ND	ND
Ascorbic acid	100	ND	60	10,000
Thiamin	10	ND	1.0	ND
Riboflavin	20	7.0	9.0	ND
Pyridoxine	10	5–6	3.0	120
Pantothenic acid	40	30–50	10–20	ND
Niacin	150	28	14	ND
Biotin	1	1	ND	ND
Folic acid	5	ND	ND	ND
Vitamin B_{12}	0.02	ND	ND	ND
Choline	3,000	4,000	ND	600
Inositol	400	440	ND	2,000

From National Research Council (1981, 1983).
*Minimum levels required to prevent deficiency.
ND = no data.

TABLE 23–4. Major Vitamin Deficiency Signs in Fish

Vitamin	*Deficiency Sign*
Vitamin A	Xerophthalmia, ascites, poor growth
Vitamin D	Poor growth, low bone ash
Vitamin E	Reduced survival and growth, fatty liver
Vitamin K	Skin hemorrhage
Thiamin	Poor growth, anorexia, loss of equilibrium
Riboflavin	Impaired growth, anorexia, opaque lens
Pyridoxine	Poor growth, anorexia, convulsions, hyperirritability, spiral swimming
Pantothenic acid	Poor growth, clubbed gills
Niacin	Poor growth, tetany, photophobia
Biotin	Skin depigmentation, reduced growth
Folic acid	Anorexia and lethargy
Vitamin B_{12}	Anemia, reduced growth
Choline	Enlarged liver, hemorrhages in kidney and intestine
Ascorbic acid	Lordosis, scoliosis, broken back syndrome

Adapted from Lovell (1989).

Carotenoids are needed for pigmentation of the skin and flesh of colored fish, such as the rainbow trout. Carotenoids, xanthophylls, and similar substances used in the poultry industry can be used as **pigmenting agents** for fish. **Astaxanthin** is absorbed and retained in the flesh more efficiently than **canthaxanthin;** a combination of the two pigments maximizes flesh pigmentation (Torrissen, 1989). Shrimp, daphnia, crabs, and some other herbivorous aquatic animals accumulate phytoplankton carotenoids, mainly astaxanthin, and are good pigmentors when used as feed for salmon and trout.

Mineral Needs

Fish require the same mineral elements as other animals but can often obtain adequate quantities of many minerals from the water. Excretion and absorption of minerals across the gill membranes serve an osmoregulatory function as well as a nutritional role. **Calcium requirements** are usually met by absorption from water, whereas phosphorus is a dietary requirement. For most species, 0.5 to 0.6 percent phosphorus is adequate. Calcium carbonate (lime) is applied to aquaculture ponds to buffer water pH, so calcium in the water is usually present in adequate amounts. Fish require a dietary source of **selenium;** organic selenium compounds such as selenomethionine and selenium in yeast have higher selenium bioavailability than sodium selenite (Wang and Lovell, 1997).

TOXIN PROBLEMS

Fish are sensitive to various toxins in feedstuffs. **Mycotoxins,** and especially aflatoxin, have been of major concern. Rainbow trout are among the most sensitive animals to the carcinogenic effects of aflatoxin. Dietary levels as low as one part per billion (ppb) aflatoxin will cause liver cancer in trout (NRC, 1981). Other dietary toxins, such as gossypol and cyclopropenoid fatty acids in cottonseed meal, may enhance the toxicity or carcinogenicity of aflatoxin. Catfish and most other fish species, except trout, are relatively tolerant to aflatoxin. Field observations have suggested that tilapia may be highly sensitive to T2 toxin.

FEEDING PROGRAMS

Depending upon the rearing facilities used, fish may derive considerable benefit from natural food (bacteria, phytoplankton, zooplankton, insects, crustacea, fish eggs, other fish) or may be entirely dependent on the feed provided by the aquaculturist (**food versus feed**). Fish feed may be of several physical types, including pellets, crumbles, meal, and flakes. **Floating pellets,** prepared by the extrusion process, are often used to aid in the assessment of appetite and the presence of uneaten food. **Sinking feed,** which gradually sinks, may settle on the bottom and remain uneaten, reducing water quality as it decomposes. Crumbles are desirable for small fish, since they may have difficulty eating larger pellets.

Some examples of suitable practical diets for various warmwater fish species are given in Table 23–5 and for coldwater fish in Table 23–6.

TABLE 23–5. Percentage Composition of Practical Diets for Warmwater Fish Species

Ingredient	*32% CP Catfish Diet, with Fish Meal*	*32% CP Catfish Diet, No Fish Meal*	*25% CP Common Carp Diet*	*28% CP Tilapia Diet*	*35% CP Shrimp Diet*
Menhaden fish meal	8.0	—	15.0	5.0	15.0
Soybean meal, 48% CP	48.2	47.5	17.0	45.0 (44% CP)	36.0
Corn	29.2	30.0	—	30.0	—
Rice bran or wheat middlings	10.0	—	—	17.0	12.0
Dicalcium phosphate	1.0	1.25	—	1.0	1.0
Pellet binder (hemicellulose or lignin sulfonate)	2.0	1.5	—	2.0	2.0
Fat (sprayed on after pelleting)	1.5	1.5	—	—	2.0
Trace mineral premix[a]	0.05	0.05	0.5	—	0.5
Vitamin premix[b]	0.05	0.05	0.5	—	0.5
Ascorbic acid, ethylcellulose coated	0.038	0.038	—	—	0.038
Meat and bone meal	—	15.0	—	—	—
Whey, dried	—	2.5	—	—	—
Wheat	—	—	10.0	—	—
Sorghum	—	—	56.85	—	—
DL-methionine	—	—	0.15	—	—
Shrimp waste meal	—	—	—	—	10.0
High gluten wheat flour	—	—	—	—	20.0

Adapted from National Research Council (1983).

[a]Trace mineral mix should provide the following six minerals in approximately these amounts per ton: Zn, 150 g; Fe, 44 g; Mn, 25 g; I, 5 g; Cu, 3 g; Co, 0.05 g; Se, 0.27 g.

[b]Vitamin mix should provide all of the following in approximately the amounts presented below per ton of feed:

Vitamin A	4,000,000 IU	Riboflavin	12 g
Vitamin D_3	2,000,000 IU	Pyridoxine	10 g
Vitamin E	50 g	Thiamine	10 g
Menadione (MPB)	10 g	Pantothenic acid	32 g
Choline chloride (70%)	500 g	Folic acid	2 g
Niacin	80 g	Vitamin B_{12}	8 mg

TABLE 23–6. Examples of Suitable Diets for Coldwater Fishes

Ingredient	*Oregon Moist Pellet*	*Moist Salmon Pellet*	*Dry Salmon Pellet*	*Trout Grower Diet*
Fish meal	28.0	27	50.0	30
Wheat germ meal	9.9	5.0	5.0	—
Cottonseed meal	15.0	—	—	—
Dried whey	5.0	—	10.0	—
Corn distillers dried solids	4.0	5.0	—	—
Trace mineral premix	0.1	—	0.1	0.1
Vitamin premix	1.5	1	—	0.4
Choline chloride, 70%	0.5	0.7	0.58	0.175
Wet fish	30.0	—	—	—
Fish oil	6.0	—	9.0	10.0
Wheat middlings	—	11.3	8.75	17.5
Dried skim milk	—	4	—	—
Brewer's yeast	—	3	—	—
Tuna viscera	—	20	—	—
Turbot	—	10	—	—
Beef liver	—	10	—	—
Crab solubles	—	2	—	—
Cod liver oil	—	1	—	—
Dried blood meal	—	—	10	10
Condensed fish solubles	—	—	3.0	—
Poultry by-product meal	—	—	1.5	—
Ascorbic acid	—	—	0.1	0.075
Pellet binder	—	—	2.0	2.0
Wheat flour	—	—	—	4.75
Soybean meal	—	—	—	25

Adapted from Halver (1989) and Lovell (1989).

FEED FORMULATION AND PROCESSING

Feed ingredients used in the manufacture of fish diets are mainly fish meals, plant proteins, and moderate amounts of cereal grains. Omnivorous or herbivorous species (e.g., carp) can utilize higher grain levels than salmonids and other carnivores. Fish meal should be of good quality, with no putrefaction or oxidized fat. High-ash fish meal, prepared from fish-processing wastes and containing a high percentage of bones, may cause mineral imbalances. High-ash fish meals may cause induced zinc deficiency because of the antagonistic effect of calcium on zinc absorption. A major sign of zinc deficiency is eye cataracts. Fish meals and fish oils are also useful in fish diets as sources of n-3 fatty acids. There is no other practical source.

Some of the particular concerns in manufacturing fish feeds include the following.

1. The particle size must be such that the feed can be easily prehended by the target organism. In the case of feed for larvae, for instance, the particle size must be very small.

2. The nutrients must be stable in the feed and not leached out into the water before the feed is consumed.
3. The feed must have the proper physical properties and water stability for the target organisms. Some species (e.g., salmonids) require a floating feed as they will not eat feed that settles on the bottom. Others, such as carp, are **bottom feeders.** Shrimp prefer sinking feeds, and because they eat very slowly, the feed must have excellent stability in the water.

Moist diets are prepared by mixing dry ingredients with water and a suitable hydrocolloidal binding agent (e.g., carboxymethyl-cellulose, gelatinized starch, gelatin, fresh ground fish tissue) and extruding the mixture to form moist pellets. The best-known feed of this type is the Oregon Moist Pellet, which has been used very extensively in feeding salmonids. **Flaked diets** are prepared by processing moist feed on a rotating drum dryer to form a thin sheet, which then crumbles into flakes. These are most commonly used for aquarium fishes. Most commercial fish feeds are pelleted or extruded dry feeds. **Extruded feeds** are popular because of their floating properties. Special pellet binders (e.g., precooked starch, wheat gluten, guar gum, urea-formaldehyde, wood lignins) are used in pelleted feeds for crustaceans, because the feed must remain intact in the water for several hours.

Feeding may be done by hand, machine blowers, automatic feeders, or **demand feeders.** Hand feeding is labor intensive but tends to ensure greater awareness by the operator of potential or developing problems. Automatic feeders are labor saving but may waste feed if not carefully monitored. Demand feeders allow the fish to feed themselves. A rod extends from the feeder into the water; fish learn to bump the rod to cause feed to fall into the water. There is little wastage with this system.

Feeding Catfish

Catfish production is increasing very rapidly in the southern United States, especially on the Mississippi River floodplain, where large ponds can be readily constructed. Catfish have a rapid rate of growth and are relatively simple to raise compared to other species. Careful feeding management is necessary to avoid overfeeding, which causes a high oxygen demand on the pond. **Oxygen depletion** is a major cause of fish mortality. Floating feeds are useful because they allow the fish farmer to observe the fish feeding and to determine if excessive feed is being offered. On large catfish farms, the feed is spread mechanically; dispersion of feed over a large part of the pond provides maximum feeding opportunities and uniform fish growth. Feeding is usually done once daily. Because catfish have white flesh, feeds with pigmenting agents should be avoided. More than 11 mg of xanthophylls per kg of feed will impart a yellow color to the flesh (Lovell, 1989).

Feeding Salmonids

Salmonid aquaculture is expanding rapidly in North America and Europe in the forms of salmon and trout ranching and farming. **Salmon ranching** refers to the release of artificially reared young salmon into the oceans and harvesting them when they return to the site of release (Isaksson, 1988). During most of their life span, the fish are not artificially fed. In contrast, **salmon farming** involves raising the fish to market weight in pens, usually floating net pens in sheltered areas such as in the fjords of northern Europe and in the long inlets of coastal British Columbia. Intensive salmon farming is intrinsically more advanced than ocean fishing, which might be viewed as analogous to the harvesting of terrestrial animals by hunting rather than by livestock production.

Salmonids include five species of Pacific salmon (Chinook, coho, chum, pink, and sockeye), Atlantic salmon, trouts, and chars. The major species raised in aquacultural systems are coho and Atlantic salmon and **rainbow trout.** In the United States, more than 80 percent of commercial trout culture is in Idaho, where enormous volumes of constant-temperature, fully oxygenated artesian water exist. Rainbow trout in Idaho trout farms are reared in raceways with a continuous flow of large volumes of water (Fig. 23.1).

Salmonid diets are usually either moist or dry pelleted feeds. The Oregon Moist Pellet (Table 23–6) or modifications of it are used quite extensively in Pacific salmon production, especially in hatcheries for young fish. For Atlantic salmon pen culture, dry and semimoist feeds are used. Semimoist diets (12 to 25 percent water) are prepared from ground fish products mixed with dry ingredients. Liquified fish and fish silage (Chapter 4) are also used in semimoist diets. Dry pelleted feeds are generally cheaper than moist diets and are more convenient to use. They contain high levels of fish meal and oilseed meals and low levels of other protein supplements. Young salmon find soybean meal unpalatable, so this ingredient is used only for older fish at levels not exceeding 20 percent of the diet. Sources of pigment are included, to give the skin (for example, rainbow trout) and flesh the desired **pigmentation.** Natural sources include crustacean products (krill, crab waste, shrimp waste) that contain **astaxanthin** or synthetic pigments such as **canthaxanthin.** They are used to provide 40 to 60 mg carotenoid pigment per kg of feed and are fed the last 3 to 6 months before marketing the fish. Rainbow trout develop flesh pigmentation more rapidly than do Pacific or Atlantic salmon (March and MacMillan, 1996).

Feeding Tilapia

Tilapia are tropical fish native to Africa. Tilapia production is expanding rapidly because of the fast growth, use of natural foods, high reproductive rate, and general hardiness characteristic of this species. Large quantities of tilapia are raised in developing countries in fairly simple ponds or lagoons. Often the fish use only natural feeds produced in the pond, but they may receive supplements of byproducts such as rice bran and copra. The ponds are often fertilized with animal excreta. In some countries (e.g., Israel and Taiwan), **polyculture** systems involve tilapia and various carp species to maximize the utilization of natural foods. In Israel, a ratio of 60 tilapia to 40 carp is common (Lovell, 1989). In Southeast Asia, tilapia are cultured in rice paddies. In the United States, tilapia production is small but expanding. Tilapia are the best species for culture where geothermal warm water is available. Commercial catfish diets are suitable for tilapia. A small pellet size is desirable. Tilapia are continuous feeders so frequent feeding, up to four times daily, is recommended.

Feeding Shrimp and Other Crustaceans

Shrimp production is increasing rapidly. Ecuador is the leading producer, followed by Thailand and other Southeast Asian countries. Much of the production, especially from Ecuador, is exported to the United States. Growth of shrimp aquaculture began when successful artificial reproduction methods were developed. Shrimp seem to have nutrient requirements similar to those of finfish. They also require **dietary sterols** for synthesis of sex and molting hormones, and for synthesis of the exoskeleton. The optimal dietary level of **cholesterol** is 0.5 percent (Lovell, 1989). Shrimp may have a dietary need for **phospholipids** (e.g., lecithin), which can be provided by soybean meal and clam meal. Shrimp can convert β-carotene and xanthophylls to the carotenoid pigment **astaxanthin.**

Prawns, shrimp, and other crustaceans have unique lipid requirements. They cannot synthesize linolenic (18:3 n-3) and linoleic (18:2 n-6) acids, which are dietary essential fatty acids for these species (D'Abramo and Sheen, 1994). They cannot synthesize highly unsaturated long chain fatty acids, and they demonstrate a growth response when a source (e.g., shrimp head oil) of these fatty acids is added to the diet. **Phospholipids,** such as lecithin, have growth-promoting activity in crustaceans, which may be related to improved absorption and transport of cholesterol. **Cholesterol** is a dietary requirement and is needed for synthesis of steroid, brain, and molting hormones, and for vitamin D synthesis (Chen, 1993).

Contamination of feed with disease-causing agents from marine meals and live sources is a major concern. Posthatched juvenile shrimp must be fed algae, zooplankton, and other marine-derived ingredients such as live artemia (brine shrimp). Juvenile shrimp are considered to be particulate feeders whether feeding on processed feed or natural sources of plankton. The use of flocculants in pond water has been shown to improve feeding and performance in shrimp as organic particles such as algae and bacteria are aggregated and made available for consumption (Moss, 1995).

Shrimp utilize natural feeds in ponds, such as algae, phytoplankton, and other small organisms. Shrimp feeds are usually dry pellets. Shrimp are slow and selective eaters; they hold the pellet to the mouth and gradually disintegrate it. Therefore, pelleted shrimp feed must be stable in water for several hours to minimize losses from leaching and disintegration. The ingredients should be ground to a particle size of less than 1 mm. Because shrimp do not move large distances, the feed should be distributed over as much of the pond as possible.

In the southern United States, and especially Louisiana, crawfish are a delicacy consumed in large quantities. **Crawfish production** in ponds is increasing in importance. Crawfish consume **detritus** at the bottom of the pond and are raised largely on natural pond foods, using naturally occurring vegetation or planted forages that establish a vegetation base and are then flooded. The decomposing vegetation provides the detritus to begin the food chain consisting of bacteria, protozoa, small crustaceans, worms, mollusks, and insects. Providing supplemental manufactured feeds is not economical. If supplemental feeding is needed, inexpensive fibrous materials such as straw, bagasse, and other fibrous wastes are added to the ponds to initiate detritus production.

OTHER AQUATIC SPECIES

In the southern United States, there is increasing interest in **alligator farming** for production of meat and leather. Alligators are coldblooded carnivores. Young alligators have a dietary requirement for arachidonic acid for maximum growth (Staton et al., 1990a). Alligator diets should contain 45 percent digestible protein. At least 15 percent corn grain can be used as a carbohydrate source (Staton et al., 1990b). Because of their carnivorous nature, alligators are probably not well adapted to high-carbohydrate diets. Staton et al. (1990b) reviewed the current information available on alligator nutrition. There has been some interest in combining alligator farms with swine enterprises, with dead pigs used as alligator feed. This is an ecologically (although not necessarily aesthetically) sound means of disposing of dead animals.

QUESTIONS AND STUDY GUIDE

1. You have a summer job at a fish farm where catfish are being raised in nonaerated ponds. The ponds have been fertilized and have a heavy growth of green algae. After several weeks of clear, sunny weather, there is a cooling trend. When you arrive at the farm one morning to begin feeding the fish, the weather is cloudy and overcast. You go to the ponds and find thousands of dead fish floating on the water, with a few survivors gasping for air at the surface. Explain what happened.
2. It is often observed that weight gains of fish in ponds exceed the weight of feed provided, with feed conversion ratios of 0.8:1 sometimes occurring. How can this be explained?
3. Using the information in Table 23–1, calculate the maintenance ME requirement in kilocalories per unit of metabolic size for a chicken. How does this compare to the Brody or Kleiber values for basal metabolic rate? Perform the same calculation for a carp and compare your answer to the chicken value.
4. Why do fish utilize protein more efficiently as an energy source than do poultry or swine?

REFERENCES

Akiyama, D.M. 1989. The use of soybean meal to replace white fish meal in commercially processed Penaeus monodon Fabricius feeds in Taiwan, R.O.C. *Proc. Third Symp. on Feeding and Nutr. in Fish Toba,* pp. 289–299. Aug. 28-Sept. 1, Japan.

Cai, Z. and L.R. Curtis. 1989. Effects of diet on consumption, growth and fatty acid composition in young grass carp. *Aquaculture* 81:47–60.

Cain, K.D. and D.L. Garling. 1995. Pretreatment of soybean meal with phytase for salmonid diets to reduce phosphorus concentrations in hatchery effluents. *Prog. Fish-Cult.* 57:114–119.

Chen, H.-Y. 1993. Requirements of marine shrimp, *Penaeus monodon,* juveniles for phosphatidyl choline and cholesterol. *Aquaculture* 109:165–176.

Cho, C.Y. and D.P. Bureau. 1997. Reduction of waste output from salmonid aquaculture through feeds and feeding. *Prog. Fish-Cult.* 59:155–160.

Cho, C.Y. and S.J. Kaushik. 1985. Effects of protein intake on metabolizable and net energy values of fish diets. In *Nutrition and Feeding in Fish,* C.B. Cowey, A.M. Mackie, and J.G. Bell, eds. pp. 95–117. New York: Academic Press.

Clements, K.D. and J.H Choat. 1995. Fermentation in tropical marine herbivorous fishes. *Physiol. Zool.* 68:355–378.

Cowey, C.B. and M.J. Walton. 1989. Intermediary metabolism. In *Fish Nutrition,* J.E. Halver, ed. pp. 259–329. San Diego: Academic Press.

Cowey, C.B., A.M. Mackie, and J.G. Bell, eds. 1985. *Nutrition and Feeding in Fish.* New York: Academic Press.

D'Abramo, L.R. and S-S. Sheen. 1994. Nutritional requirements, feed formulation, and feeding practices for intensive production of the freshwater prawn *Macrobrachium rosenbergii. Rev. Fish. Sci.* 2:1–21.

Dupree, H.K. and J.V. Huner, eds. 1984. *Third Report to the Fish Farmers: The Status of Warmwater Fish Farming and Progress in Fish Farming Research.* Washington, DC: U.S. Fish and Wildlife Service.

Gatlin, D.M., W.E. Poe, and R.P. Wilson. 1986. Effects of singular and combined dietary deficiencies of selenium and vitamin E on fingerling Channel catfish (*Ictalurus punctatus*). *J. Nutr.* 116:1061–1067.

Halver, J.E., ed. 1989. *Fish Nutrition.* San Diego: Academic Press.

Hepher, B. 1988. *Nutrition of Pond Fishes.* Cambridge, England: Cambridge University Press.

Higuera, M., M. Garcia-Gallego, A. Sanz, G. Cardenete, M.D. Suarez, and F.J. Moyano. 1988. Evaluation of lupin seed meal as an alternative protein source in feeding of rainbow trout (*Salmo gairdneri*). *Aquaculture* 71:37–50.

Isaksson, A. 1988. Salmon ranching: A world review. *Aquaculture* 75:1–33.

Kandel, J.S., M.H. Horn, and W. Van Antwerp. 1994. Volatile fatty acids in the hindguts of herbivorous fishes from temperate and tropical marine waters. *J. Fish Biol.* 45:527–529.

Leung, Ping Sun and S.M. Moss. 1999. Economic assessment of a prototype biosecure shrimp growout facility. In *Controlled and Biosecure Production Systems,* R.A. Bullis and G.D. Pruder, eds. pp. 97–106. Proceedings of a special session: Integration of shrimp and chickens models. World Aquaculture Society, Sydney, Australia, April 27–30, 1999.

Lim, C. and W. Dominy. 1990. Evaluation of soybean meal as a replacement for marine animal protein in diets for shrimp *(Penaeus vannamei)*. *Aquaculture* 87:53–56.

Lovell, T. 1989. *Nutrition and Feeding of Fish.* New York: Van Nostrand Reinhold.

March, B.E. and C. MacMillan. 1996. Muscle pigmentation and plasma concentrations of astaxanthin in rainbow trout, Chinook salmon, and Atlantic salmon in response to different dietary levels of astaxanthin. *Prog. Fish-Cult.* 58:178–186.

Moss, S.M. 1995. Production of growth-enhancing particles in a plastic-lined shrimp pond. *Aquaculture,* 132:253–260.

National Research Council. 1981. *Nutrient Requirements of Coldwater Fishes.* Washington, DC: National Academy Press.

_______ . 1983. *Nutrient Requirements of Warmwater Fishes and Shellfishes.* Washington, DC: National Academy Press.

National Research Council. 1993. *Nutrient Requirements of Fish.* Washington, DC: National Academy Press.

Oikawa, C.K. and B.E. March. 1997. A method for assessment of the efficacy of feed attractants for fish. *Prog. Fish-Cult.* 59:213–217.

Refstie, S., S.J. Helland, and T. Storebakken. 1997. Adaptation to soybean meal in diets for rainbow trout, *Oncorhynchus mykiss. Aquaculture* 153:263–272.

Rumsey, G.L., S.G. Hughes, and R.A. Winfree. 1993. Chemical and nutritional evaluation of soya protein preparations as primary nitrogen sources for rainbow trout (*Oncorhynchus mykiss*). *Anim. Feed Sci. Tech.* 40:135–151.

Satoh, S., W.E. Poe, and R.P. Wilson. 1989. Effects of dietary *n*-3 fatty acids on weight gain and liver polar fatty acid composition of fingerling channel catfish. *J. Nutr.* 119:23–28.

Schmittou, H.R. 1997. High density fish culture in low volume cages. *Technical Bulletin* AQ41, American Soybean Association.

Schmittou, H.R., J. Zhang, and M.C. Cremer. 1998. Principles and practices of 80:20 pond fish farming. *Technical Bulletin,* American Soybean Association.

Staton, M.A., H.M. Edwards, Jr., I.L. Brisbin, Jr., T. Joanen, and L. McNease. 1990a. Essential fatty acid nutrition of the American alligator (*Alligator mississippiensis*). *J. Nutr.* 120:674–685.

_______ . 1990b. Protein and energy relationships in the diet of the American alligator. (*Alligator mississippiensis*). *J. Nutr.* 120: 775–785.

Storebakken, T., S. Refstie, and B. Ruyter. 2000. Soy products as fat and protein sources in fish feeds for intensive aquaculture. In *Soy in Animal Nutrition,* J.K. Drackley, ed. pp. 127–170. Federation of Animal Science Societies. Savoy, IL.

Tomasso, J.R. 1994. Toxicity of nitrogenous wastes to aquaculture animals. *Rev. Fish. Sci.* 2:291–314.

Torrissen, O.J. 1989. Pigmentation of salmonids: Interactions of astaxanthin and canthaxanthin on pigment deposition in rainbow trout. *Aquaculture* 79:363–374.

Wang, C. and R.T. Lovell. 1997. Organic selenium sources, selenomethionine and selenoyeast, have higher bioavailability than an inorganic selenium source, sodium selenite, in diets for channel catfish (*Ictalurus punctatus*). *Aquaculture* 152:223–234.

Wilson, R.P., W.E. Poe, and E.H. Robinson. 1989. Evaluation of L-ascorbyl–2-polyphosphate (AsPP) as a dietary ascorbic acid source for channel catfish. *Aquaculture* 81:129–136.

Woodward, B. and M. Frigg. 1989. Dietary biotin requirements of young rainbow trout (*Salmo gardneri*) determined by weight gain, hepatic biotin concentration and maximal biotin-dependent enzyme activities in liver and white muscle. *J. Nutr.* 119:54–60.

CHAPTER 24

Epilogue: Livestock Production and Human Welfare

Objectives

1. To provide a perspective on the role of livestock production in world agriculture
2. To provide "food for thought" on controversial issues affecting animal production

T. R. Malthus (1766–1834), a British theologian, is credited with the observation that the human population increases geometrically while the food supply does not, and he predicted that mass starvation was the inevitable result. The Malthusian prospect was postponed by the colonization of the New World, acting as a safety valve for the European population pressures. It is not inconceivable that colonization of space could play a similar role in the future, although it probably will not happen soon enough to relieve the present severe pressures on the environment.

Although most scientists today reject the bleak prediction of Malthus, it is an incontrovertible fact that the world population is increasing rapidly, and before it stabilizes large increases in food production will be necessary to sustain it. The population pressures, now and in the future, are not uniformly distributed. The developed countries (Europe, North America, Australia, Japan) have achieved low growth rates and in some cases virtual stabilization of population size. The developing countries of Latin America, Africa, and Asia are still increasing in population at a very rapid rate. Projected populations at the point of stabilization are, in some cases, of staggering proportion (Table 24–1). Feeding these huge populations will require great increases in total agricultural production and efficiency and likely will be accompanied by massive environmental degradation. Among other adverse consequences, the survival of many wildlife species, both plant and animal, will be seriously jeopardized.

As human population pressures increase, the role of livestock in the food production chain will come under strong scrutiny. As a broad generalization, it can be assumed that where competition between humans and livestock for foodstuffs is high, the use of grains and high-quality protein sources for animal feeding will be curtailed. Under these circumstances, livestock production would probably emphasize those animals that can utilize roughages and by-products unsuitable for direct human consumption. (While this seems intuitively obvious, it is interesting that in countries with high population densities, such as China, livestock production emphasizes poultry and swine rather than ruminants.)

TABLE 24–1. Projected Stabilized Population Sizes of Selected Countries

Place	*Population in 1986, Million*	*Annual Rate of Population Growth, %*	*Size of Population at Stabilization, Million*	*Change from 1986, %*
Slow-Growth Countries				
China	1,050	1.0	1,571	+ 50
Former Soviet Union	280	0.9	377	+ 35
United States	241	0.7	289	+ 20
Japan	121	0.7	128	+ 6
United Kingdom	56	0.2	59	+ 5
West Germany	61	−0.2	52	− 15
Rapid-Growth Countries				
Kenya	20	4.2	111	+455
Nigeria	105	3.0	532	+406
Ethiopia	42	2.1	204	+386
Iran	47	2.9	166	+253
Pakistan	102	2.8	330	+223
Bangladesh	104	2.7	310	+198
Egypt	46	2.6	126	+174
Mexico	82	2.6	199	+143
Turkey	48	2.5	109	+127
Indonesia	168	2.1	368	+119
India	785	2.3	1,700	+116
Brazil	143	2.3	298	+108

Adapted from Brown et al. (1987).

ROLE OF LIVESTOCK IN AGRICULTURE AND WORLD FOOD PRODUCTION

In virtually all human societies that have evolved beyond the hunter-gatherer stage, livestock production has been an integral part of agriculture (Fig. 24.1). Even in the hunter-gatherer stage, of course, meat played a major role in the diet. Domestic animals have provided meat, milk, fiber, work, and fuel, and have served as an essential link in the efficient use of crops by recycling nutrients. In many countries, draft animals (horses, donkeys, oxen, water buffalo) are still very important in providing transportation and farmwork. They are fed on crop residues such as straw. The manure is used as a fertilizer and fuel. In India, for example, **cow dung** is a major fuel used for cooking. Fuel sources are in short supply in many countries, and serious deforestation problems accompany use of wood as fuel. Cattle and buffalo can utilize rice straw as a feedstuff and provide not only work and milk but also dung, which is a more suitable fuel than the original straw.

Ruminant animals provide the major means by which vast areas of rangelands can be used to provide human food. The ability of ruminants to digest fiber allows them to make use of crop residues such as straw, stover, and other agricultural wastes. Additionally, ruminants can utilize nonprotein nitrogen sources such as urea, so they need not be competitive with humans for protein sources.

FIGURE 24.1 Livestock production has many important but sometimes unappreciated roles:
a. Cattle and buffalo in India are very important sources of fuel for cooking. Cattle feed on rice straw and other crop residues; their dung is collected and made into dung patties for fuel. Lack of cattle for fuel production results in increased deforestation as people turn to other fuel sources. (Courtesy of R. A. Leng.)

b. Livestock are important work animals. Water buffalo in Indonesia and other Asian countries convert rice straw and crop residues into mechanical power for growing the next grain crop.

c. In the far north, reindeer harvest the meager growth of lichens and produce food, clothing, and other products for the Laplanders of Finland. (Courtesy of M. Nieminen.)

Livestock production has a number of other attributes, perhaps less obvious than the products of meat, milk, fiber, and work. Production of forages for animal feeding encourages crop rotations involving grasses, clovers, and alfalfa. Forage crops reduce soil erosion, improve soil texture and organic matter content, and in the case of legumes, increase soil fertility by providing nitrogen. A monoculture of cereal grains leads to loss of soil texture and organic matter, reduced water-retaining properties, and increased erosion. Erosion is currently a major problem in the corn and wheat growing areas of the United States. For **sustainable agriculture,** crop rotations involving forages and livestock production are desirable.

There are also sociological reasons why animal production is desirable. Farmers are stewards or caretakers of the land for future generations. It is difficult to be an absentee owner or farmer if a farm produces livestock. Animal production binds farmers to the land, which may contribute to their concern for it and ability to preserve it. Farm children learn positive values such as a sense of responsibility and a respect for life through raising animals. They acquire a perspective of the life cycle and learn to confront dealing with death. Some of the concerns of animal rights activists about killing animals for food may reflect urban childhoods that did not provide exposure to the realities of life cycles.

Livestock production can play a role in the survival of the **family farm,** a sociological issue of considerable current interest and concern (Comstock, 1987). For crop production like corn and wheat, experiences in Hungary and other socialist Eastern European countries have shown that large-scale production can be very efficient. In strict terms of crop production efficiency, it is more efficient to tear out the fence rows, consolidate land in 100,000-acre blocks, and grow corn in a high-technology monoculture than it is to have the 100,000 acres divided up into 200 "family" farms, of 500 acres each, with 200 different sets of management decisions about what crops to plant, how much land to leave for wildlife habitat, and so on. However, as Comstock argues convincingly, there are many other considerations that influence the optimal pattern of land ownership and agricultural systems. The Eastern European experience has shown that megafarms can be successful in crop production, but they are dismal failures in livestock production. Most livestock species thrive best when there is an element of "tender loving care" in their management. Therefore, the family farm is the best setting for the production of livestock, and its sociological benefits are best retained when livestock production is a component of the farming system. Ironically, corporate swine production in the United States is repeating the Eastern European and Soviet experiences with swine megafarms, and getting the same dreary results of massive air and water pollution.

COMMON CRITICISMS OF LIVESTOCK PRODUCTION

Does livestock production, under some conditions, detract from human welfare rather than contribute to it? Orskov (1987) has considered this question in the African context. Animal production is sometimes blamed as a major contributory factor to the **desertification,** the advance of deserts, of the Sahelian region of northern Africa. Certainly overgrazing contributes to desert advancement. The problem, however, is not one of animal production per se but of too many animals, which in turn is a reflection of human overpopulation. Reductions in previously high rates of infant mortality have resulted in very high rates of population growth in Sahelian countries, as in most developing countries. Since it is a cultural tradition and economic necessity for these people to raise cattle and other livestock, an expanding human population means an expanding livestock population. Cattle and other livestock are often owned as **sources of capital,** even by city dwellers who find cattle a safer investment than money in a bank in an unstable economy. As an example of the livestock numbers involved, a survey in Somalia (ILCA, 1988) reported that the average household keeps 25 cattle, 12 camels, 3 donkeys, 62 sheep, and 82 goats. One of the reasons for keeping a mixture of livestock species is to minimize risk of losses in bad years. However, fewer, but more productive, animals could likely provide the products needed with less grazing pressure on the land. Improved reproduction would allow more calves from fewer cows; use of cows as draft animals instead of oxen (steers) would increase efficiency of milk and work production. Improved nutrition, with supplements of bypass protein and other limiting nutrients, could increase efficiency of use of crop residues. Thus, altering the techniques of animal production rather than eliminating them would enhance human welfare.

Many parts of Africa cannot be effectively used for livestock production because of tick and **tsetse fly** (spreader of sleeping sickness) infestations; management of wild ungulates that are resistant to tick- and tsetse fly-borne diseases for food production might be a feasible option. The tsetse fly issue illustrates the complexity of ecological problems. Savory (1988) described his involvement in tsetse fly control programs in which large areas of tropical forest were burned and the large wild animals shot to try to eliminate the tsetse fly. The program was a fiasco because sufficient numbers of small animals remained to serve as blood sources for the flies, and the replacement of trees with grass actually encouraged fly reproduction. Matzke (1983) reviewed the consequences of tsetse fly control programs and suggested that in many cases the best ecological solution was no control program. He concluded that "the push for tsetse eradication arises from its role as a disease vector, its 'non-political' status as a development target and the strong Western bias toward the role of cattle in the modern agricultural mix" (p. 531). The tsetse fly is one of the main factors responsible for the preservation of African forests and wildlife by limiting human encroachment into the infested areas.

Other factors contributing to overgrazing in Africa are forced settlement of nomads by government decree, closure of international boundaries to nomadic stock herders, and use of trees for fuel, thus eliminating them as a source of forage. Common grazing land with no controls on grazing pressure and lack of incentives for individuals to improve it contribute to the problem. Solutions to the severe problem of desertification are largely political.

Desertification is a major ecological problem in Africa, Asia, the Middle East, and Australia. Overgrazing by livestock, especially goats, is widely viewed as a major contributory factor. The effects of livestock on desertification are not as obvious as they might seem to be to the casual observer. The simplistic notion that because overgrazing may contribute to desertification, the removal of livestock should restore the grasslands and prevent further desertification is not valid. Grasslands and large grazing herbivores have coevolved; grasslands depend upon grazing to maintain them. Savory (1988) has documented that the removal of animals from overgrazed rangelands does not restore the range but further accelerates its decline. Establishment of young grass seedlings requires the herd action of large numbers of hooves to break up the soil surface to allow new plants to grow. Further, it appears that under natural conditions, the necessary herd action requires the presence of predators to cause the herd to gallop in a frenzied manner, to break up the soil surface and trample detritus (accumulations of dead plant tissue) into the soil. This effect can be duplicated under controlled grazing practices. Without the necessary congregation to produce the herd effect, livestock and wild herbivores will spread out, place their hooves so as to avoid detritus, and will not break up the soil surface. The importance of this disturbance for the productivity of North American grasslands has been demonstrated by Knapp and Seastedt (1986). Tallgrass prairie grass species declined in growth and productivity when grazing was excluded because of the accumulation of plant litter. Most of the grasslands of the world have evolved under heavy, but not continuous, grazing pressure. The bison and caribou of North America and the wildebeest of Africa are typical examples of large migratory herbivores. They congregate with heavy pressure for a short time and then move on, harassed by predators. Their grazing and trampling action regenerates and invigorates the grassland. The crucial factor is that the grass has a recovery period before the herds arrive again. Therefore, where livestock production has contributed to desertification, it is because of **poor grazing management** by continuous grazing until the plants die. Such a condition has often occurred because of political decisions that restrict movement of nomadic herders across international boundaries, without consideration of ecological consequences.

Various environmentalist groups in the United States are working to eliminate the **western range livestock industry,** particularly that on public lands, which they believe is environmentally undesirable. This position was articulated by Ferguson and Ferguson (1983) in their book *Sacred Cows at the Public Trough.* Slogans such as "cattle-free by ninety-three" and acts of vandalism, including the burning and destruction of ranch property, auction yards, and other livestock-related equipment, illustrate the nature of the opposition to livestock grazing on western U. S. rangelands. This is a simplistic approach to a complex situation. Without grazing, the grasslands will further deteriorate (Savory, 1988). Undoubtedly, grazing management could be improved in many cases, but it must be recognized that whether we like it or not, the rangelands will be grazed by either domestic or wild herbivores. If domestic animals are removed, deer, elk, wild horses, and other wild herbivores will occupy the vacated ecological niche.

Perhaps a look at history gives perspective. Before the development of the western livestock industry, it is estimated that the North American grasslands supported 60 million bison, 40 million white-tailed deer, 40 million pronghorn antelope, 10 million elk, and innumerable prairie dogs, jack rabbits, cottontails, and other small herbivores (Heath and Kaiser, 1985). If these animals did not destroy the grasslands, is it beyond our capabilities to manage livestock so that they do not either?

With the elimination of wolves, grizzly bears, cougars, and other predators, and the conversion of winter ranges into ski resorts and condominium sites, the natural regulation of wild herbivore populations has been irreversibly altered, so that humans are now the main regulators. In the past, populations of these wild herbivores were controlled by predators and winter starvation, whereas today populations of domestic livestock are controlled by marketing for meat. We could allow the populations of wild animals to increase once again until the range is severely damaged and winter starvation controls animal numbers, we could winter-feed wild herbivores, or we could be the predators by hunting. Alternatively, we can continue to raise domestic livestock.

The use of public lands for domestic livestock is becoming increasingly difficult with the prevailing climate of public opinion. It is an issue on which the livestock industry will have to work hard to present its view in the years ahead. The perception of many people is that a Hereford cow on the land is evil, whereas a herd of wild horses on the same land is magnificent and sacred. Ferguson and Ferguson (1983) espouse a common view: "No industry or human activity on earth has destroyed or altered more of nature than the livestock industry"(p. 4).

Livestock, and particularly beef cattle, are now being blamed as one of the chief causes of destruction of **tropical rain forests** (Gradwohl and Greenberg, 1988). Vast areas of rain forest in Brazil and Central America are being converted to cattle pasture. Environmentalists refer to this as the "hamburger connection." Consumption of hamburgers by Americans is blamed for the loss of tropical rain forest because of cattle ranching to produce meat for export (Uhl and Parker, 1986). But there is a misperception here. There is no reason why rain forest should be destroyed to produce meat for American or European consumers. American beef producers can easily supply the American demand for hamburger. Hamburger beef is produced by forage-fed rather than grain-fed cattle. There is great potential to increase American beef production by making efficient use of corn stover, cereal grain straw, and other by-products.

Global warming associated with the greenhouse effects of carbon dioxide and methane emissions is an important ecological concern. Carbon dioxide is a product of combustion and aerobic metabolism while methane is produced by anaerobic fermentation. Major sources of methane include ruminants, termites, rice paddies, and natural gas (methane) emissions from petroleum exploration and processing. McAllister et al. (1996) have reviewed the significance of cattle-generated methane in the global warming scenario. About 7 percent of global methane production is from cattle. This figure could be reduced by use of more intensive animal production practices, such as feeding ionophores, using high-energy diets, and adopting technology (e.g., use of bovine somatotropin) to increase the quantity of animal products per unit of methane emitted.

These and other contemporary issues in animal agriculture have been discussed in detail in a book by Cheeke (2004).

CONCLUDING COMMENTS

As previously discussed, the major increases in human population will occur in the developing countries. The food required by these people will have to be produced largely in their own countries. Agriculture in developing countries generally involves intensive production on small farms, and farm size is getting smaller. Only about 8 percent of the food eaten in developing countries is imported (Raun and Turk, 1983). To maximize food production in developing countries, agriculture must be very intensive. Animal production plays an essential role in utilizing crop residues and by-products. New systems such as alley cropping, in which crops are grown between rows of nitrogen-fixing trees, provide forage. Because of the small farm size and limited amounts of feed per farm, small animals such as sheep, goats, and rabbits are important. These small animals have a rapid generation time and are prolific, and because they are small, refrigerated meat storage is not required.

Animal production techniques in the future are likely to change to accommodate societal concerns, many of which have a nutritional basis. In recent years, **red meat consumption** has been implicated in human health disorders such as heart disease and cancer, because of its cholesterol, fat, and saturated fatty acid contents. These concerns can be addressed by modifications in animal genetics and feeding. The cholesterol content of eggs can be reduced by using certain feed additives. The fat content of meat can be reduced by altering the diet of the animal. Meat will continue to play a major role as a highly nutritious and palatable component of the human diet.

Other factors may modify animal production techniques. Concerns about **animal welfare and animal rights** may result in legislatively mandated changes in animal husbandry techniques. Environmental concerns may lead to legislation to prohibit concentration of large numbers of livestock in feedlots and other intensive facilities. These changes, if they occur, will likely modify how animals are fed. For example, cattle would be fed high-forage diets with much less grain if feedlot feeding is restricted. On the other hand, environmental concerns about agricultural waste disposal practices, such as open field burning, may lead to increased opportunities for animal production by encouraging use of crop residues as feed. Developments in **biotechnology** will affect animal production and feeding. If bovine growth hormone is approved for use with dairy cattle, as it has been in the United States, major changes in dairy cattle feeding will be required to supply cows with the nutrients needed to support a 20 percent increase in milk production.

It is conceivable that artificial fermentation systems could be developed to replace some of the functions of the ruminant animal. Probably the first step would be the development of fermentation systems to convert straw and agricultural wastes to glucose solutions that could be fed to swine and poultry. It is not likely that an acceptable human food to replace meat could be directly produced in such a system. Some of the advantages of the ruminant are difficult to duplicate. The ruminant is a mobile fermentation factory that can travel over rough terrain to harvest plant material, and each year replicates itself to produce a new factory!

Introduction of new technologies may have unexpected consequences. As an example, the **green revolution** in developing countries has involved the introduction of new, high-yielding varieties of wheat and rice. To support the heavier seed heads and to facilitate mechanical harvesting, varieties with short, stiff straw have been selected. This has resulted in reduced quantity and quality of straw for feeding the buffalo and cattle used for working the land. Thus, an improvement in crop production had unexpected adverse effects on animal production.

It is likely that new animals may be domesticated to better utilize certain ecological niches. For example, the **eland** (Fig. 24.2), a large African ruminant, has lower water requirements than cattle and may, under some conditions, be a more suitable food animal than domestic ruminants. Adaptation of the **oryx** (Fig. 10.3) to extremely arid conditions and its potential domestication were described in Chapter 10. Efforts are being made to domesticate the **musk oxen** (Fig. 24.3), which is well adapted to harsh, arctic conditions (Flood, 1989).

Animal production will, like many other things, present a dichotomy of the old and the new. In many parts of the world, water buffalo and oxen are used for

FIGURE 24.2 The eland, an African herbivore with potential for domestication. (Courtesy of R. J. Hudson.)

FIGURE 24.3 The musk ox is well adapted to harsh, arctic conditions. It produces a valuable, fine-textured wool (qiviut) and meat. The animals pictured are part of a musk oxen breeding program at the University of Saskatchewan, Canada.

plowing as they have been for thousands of years, and these traditional techniques will likely continue for many years. In contrast, other areas of animal production will be very sophisticated, incorporating all the advances available in the use of feed additives, modifiers of metabolism, and so on. One thing is certain: domestic animals will continue to play major roles in agriculture and human welfare.

QUESTIONS AND STUDY GUIDE

1. What alterations could be made to livestock diets to produce leaner meat? Which nutrients would be of most importance? Explain.
2. Most students using this text will have 30 to 40 years of their career ahead. Assuming that you are majoring in animal science, how do you view the prospects of animal production and research over the next 40 years?
3. Many countries in tropical Africa, Asia, and Latin America have rapidly increasing human populations. Discuss the potential roles of livestock production in helping to meet the food needs of people in these areas.
4. India has a larger cattle population than does the United States. Cattle in India are not used for meat. It is often suggested by outsiders that India could increase its agricultural efficiency by disposing of its cattle. Discuss this idea, including how it might be done, the effects on crop production and food supplies, and the influence on wildlife habitat and forestry.
5. Discuss ways that biotechnology might influence livestock production and feeding in the future.

REFERENCES

Blaxter, K. 1983. Animal agriculture in a global context. *J. Anim. Sci.* 56:972–978.

Brown, L. R., W. U. Chandler, C. Flavin, J. Jacobson, C. Pollock, S. Postel, L. Starke, and E. C. Wolf. 1987. *State of the World: A Worldwatch Institute Report on Progress Toward a Sustainable Society.* New York: W.W. Norton.

Cheeke, P. R. 2004. *Contemporary Issues in Animal Agriculture.* Upper Saddle River, NJ: Prentice Hall, Inc.

Comstock, G., ed. 1987. *Is There a Moral Obligation to Save the Family Farm?* Ames, IA: Iowa State University Press.

Ferguson, D. and N. Ferguson. 1983. *Sacred Cows at the Public Trough.* Bend, OR: Maverick Publications.

Flood, P. F., ed. 1989. Proceedings of the Second International Muskox Symposium. *Can. J. Zool.* 67:1092–1166; A1–A64.

Gradwohl, J. and R. Greenberg. 1988. *Saving the Tropical Forests.* Washington, DC: Island Press.

Heath, M. E. and C. J. Kaiser. 1985. Forages in a changing world. In *Forages. The Science of Grassland Agriculture,* M. E. Heath, R. F. Barnes, and D. S. Metcalfe, eds. pp. 3–11. Ames, IA: Iowa State University Press.

Int. Livestock Centre Afr. 1988. Small ruminants in central Somalia—Collaborative studies. *Newsletter* 7:1.

Knapp, A. K. and T. R. Seastedt. 1986. Detritus accumulation limits productivity of tallgrass prairie. *Bioscience* 36:662–668.

Matzke, G. 1983. A reassessment of the expected development consequences of tsetse control efforts in Africa. *Soc. Sci. Med.* 17:531–537.

McAllister, T. A., E. K. Okine, G. W. Mathison, and K.-J. Cheng. 1996. Dietary, environmental and microbiological aspects of methane production in ruminants. *Can. J. Anim. Sci.* 76:231–243.

Orskov, E. R. 1987. The role of livestock in Africa: Are livestock occasionally contributing to famine? *Proc. Nutr. Soc.* 46:301–308.

Raun, N. S. and K. L. Turk. 1983. International animal agriculture: History and perspectives. *J. Anim. Sci.* 57:156–170.

Savory, A. 1988. *Holistic Resource Management.* Washington, DC: The Island Press.

Uhl, C. and G. Parker. 1986. Our steak in the jungle. *Bioscience* 36:642.

APPENDIX A

Extensive tables of feed composition have been published by the National Research Council (NRC). Appendix Tables A-1 to A-4 are from United States–Canadian Tables of Feed Composition (NRC, 1982). Because of the voluminous nature of the NRC tables, a single feedstuff, alfalfa, has been chosen for use here to illustrate the type of information available on feeds. The NRC book, or a comparable publication, is essential for those involved in diet formulation.

Several important features of the NRC tables can be appreciated by studying Appendix Tables A-1 to A-4. Note that each feedstuff has an International Feed Number (IFN). This number has been assigned by the International Network of Feed Information Centers to standardize the identification of feedstuffs. For each feedstuff, there may be just one entry or many entries. For example, with alfalfa, there are many different types—fresh alfalfa, alfalfa hay, dehydrated alfalfa, silage—with the material in each case available at different maturities (vegetative, early bloom, midbloom, late bloom, and so forth).

The nutritionist or student should select the category that most closely corresponds to the sample of interest. The values in the table are expressed on a dry-matter basis. Diet formulation on a dry-matter basis is preferred over the as-fed basis because of the variable moisture content of feed ingredients, and because feed intake is correlated with feed dry-matter content. Energy values are expressed both as TDN and on a caloric basis. For ruminants and horses, Mcal are used, whereas for swine and poultry, the values are expressed in kcal. Information on fiber content is given for cell-wall content, cellulose, hemicellulose, lignin, acid detergent fiber, and crude fiber. These values are especially useful in ruminant diet formulation. In Tables A-2 to A-4, values for vitamins, minerals, and amino acids are provided.

It should be recognized that the values in tables of feed composition are average values. These "book values" should be used with good judgment. A certain amount of flexibility is needed in using feed composition tables.

Instructors should make copies of the NRC tables available to students involved in diet formulation, either from the publication itself or on computer diskettes.

TABLE A–1. Composition of Important Feeds: Energy Values, Proximate Analyses, Plant Cell Wall Constituents, and Acid Detergent Fiber, Data Expressed As-Fed and Dry (100% Dry Matter)

				Ruminants					Dairy Cattle	Chickens		
Entry Number	*Alfalfa, Medicago sativa*	*International Feed Number*	*Dry Matter (%)*	*TDN (%)*	*DE (Mcal/kg)*	*ME (Mcal/kg)*	*NE_m (Mcal/kg)*	*NE_g (Mcal/kg)*	*NE_l (Mcal/kg)*	*ME_n (kcal/kg)*	*TME (kcal/kg)*	*NE_p (kcal/kg)*
001	fresh, late	2-00-181	21.0	13.0	0.59	0.50	0.30	0.16	0.30	—	—	—
002	vegetative		100.0	63.0	2.78	2.36	1.39	0.75	1.42	—	—	—
003	fresh, early	2-00-184	23.0	14.0	0.61	0.51	0.30	0.15	0.31	—	—	—
004	bloom		100.0	60.0	2.65	2.22	1.31	0.65	1.35	—	—	—
005	fresh, midbloom	2-00-185	24.0	14.0	0.62	0.52	0.31	0.14	0.32	—	—	—
006			100.0	58.0	2.56	2.13	1.26	0.58	1.30	—	—	—
007	fresh, full bloom	2-00-188	25.0	14.0	0.61	0.50	0.30	0.12	0.31	—	—	—
008			100.0	55.0	2.43	2.00	1.19	0.47	1.23	—	—	—
009	hay, sun-cured,	1-20-681	90.0	47.0	2.06	1.68	1.01	0.32	1.04	—	—	—
010	late bloom		100.0	52.0	2.29	1.87	1.12	0.36	1.15	—	—	—
011	hay, sun-cured,	1-00-071	91.0	46.0	2.01	1.62	0.98	0.25	1.01	—	—	—
012	mature		100.0	50.0	2.21	1.78	1.07	0.28	1.11	—	—	—
013	leaves, sun-cured	1-00-146	89.0	64.0	2.84	2.46	1.46	0.92	1.47	—	—	—
014			100.0	72.0	3.17	2.76	1.64	1.03	1.64	—	—	—
015	meal dehy,	1-00-022	90.0	54.0	2.35	1.97	1.16	0.56	1.20	1,535.0	1,094.0	525.0
016	15% protein		100.0	59.0	2.60	2.18	1.28	0.62	1.33	1,698.0	1,209.0	581.0
017	meal dehy,	1-00-023	92.0	55.0	2.47	2.08	1.22	0.63	1.26	1,504.0	1,393.0	770.0
018	17% protein		100.0	61.0	2.69	2.27	1.33	0.69	1.38	1,640.0	1,519.0	840.0
019	meal dehy,	1-00-024	92.0	57.0	2.51	2.12	1.25	0.66	1.28	1,625.0	1,429.0	1,020.0
020	20% protein		100.0	62.0	2.73	2.31	1.36	0.72	1.40	1,774.0	1,560.0	1,113.0
021	meal dehy,	1-07-851	93.0	62.0	2.74	2.35	1.39	0.82	1.41	1,692.0	1,661.0	1,155.0
022	22% protein		100.0	67.0	2.95	2.53	1.50	0.88	1.52	1,823.0	1,790.0	1,245.0
023	wilted silage,	3-00-216	35.0	21.0	0.92	0.77	0.45	0.23	0.47	—	—	—
024	early bloom		100.0	60.0	2.65	2.22	1.31	0.65	1.35	—	—	—
025	wilted silage,	3-00-217	38.0	22.0	0.97	0.81	0.48	0.22	0.50	—	—	—
026	midbloom		100.0	58.0	2.56	2.13	1.26	0.58	1.30	—	—	—
027	wilted silage,	3-00-218	45.0	25.0	1.09	0.90	0.53	0.21	0.55	—	—	—
028	full bloom		100.0	55.0	2.43	2.00	1.19	0.47	1.23	—	—	—

Horses			Swine				Plant Cell Wall Constituents							
TDN (%)	DE (Mcal/ kg)	ME (Mcal/ kg)	TDN (%)	DE (kcal/ kg)	ME (kcal/ kg)	Crude Pro- tein (%)	Cell Walls (%)	Cell- ulose (%)	Hemi- cell- ulose (%)	Lig- nin (%)	Acid Deter- gent Fiber (%)	Crude Fiber (%)	Ether Extract (%)	Ash (%)
—	—	—	12.0	548.0	502.0	4.3	8.0	5.0	1.0	1.0	6.0	4.9	0.6	2.1
—	—	—	58.0	2,566.0	2,351.0	20.0	38.0	22.0	7.0	7.0	29.0	23.0	2.7	9.8
—	—	—	—	—	—	4.4	9.0	5.0	2.0	2.0	7.0	5.8	0.7	2.2
—	—	—	—	—	—	19.0	40.0	23.0	8.0	7.0	31.0	25.0	3.1	9.5
—	—	—	14.0	631.0	581.0	4.5	11.0	6.0	2.0	2.0	9.0	6.8	0.6	2.1
—	—	—	59.0	2,583.0	2,379.0	18.3	46.0	26.0	10.0	9.0	35.0	28.0	2.6	8.7
—	—	—	—	—	—	3.5	13.0	7.0	3.0	2.0	9.0	7.7	0.7	2.1
—	—	—	—	—	—	14.0	52.0	27.0	13.0	10.0	37.0	31.0	2.8	8.5
—	—	—	—	—	—	12.6	47.0	23.0	11.0	11.0	35.0	28.8	1.6	7.0
—	—	—	—	—	—	14.0	52.0	26.0	12.0	12.0	39.0	32.0	1.8	7.8
42.0	1.68	1.38	—	—	—	11.7	53.0	26.0	12.0	13.0	40.0	34.4	1.2	6.9
46.0	1.84	1.51	—	—	—	12.9	58.0	29.0	13.0	14.0	44.0	37.7	1.3	7.5
52.0	2.04	1.68	—	—	—	20.6	30.0	14.0	5.0	4.0	21.0	15.8	2.7	9.6
58.0	2.29	1.88	—	—	—	23.1	34.0	16.0	6.0	5.0	24.0	17.7	3.0	10.7
46.0	1.83	1.50	31.0	1,372.0	1,293.0	15.6	46.0	26.0	—	11.0	37.0	26.6	2.2	9.1
51.0	2.03	1.66	34.0	1,517.0	1,430.0	17.3	51.0	29.0	—	12.0	41.0	29.4	2.5	10.0
45.0	1.79	1.47	44.0	1,418.0	1,196.0	17.3	41.0	22.0	—	10.0	32.0	24.0	2.7	9.7
49.0	1.95	1.60	48.0	1,546.0	1,304.0	18.9	45.0	24.0	—	11.0	35.0	26.2	3.0	10.6
38.0	1.55	1.27	48.0	2,080.0	1,923.0	20.2	38.0	20.0	—	7.0	28.0	20.6	3.3	10.4
42.0	1.69	1.39	52.0	2,270.0	2,099.0	22.0	42.0	22.0	—	8.0	31.0	22.5	3.7	11.3
27.0	1.14	0.94	49.0	2,186.0	1,855.0	22.2	36.0	19.0	—	7.0	26.0	18.3	4.1	10.2
29.0	1.23	1.01	53.0	2,355.0	1,999.0	23.9	39.0	20.0	—	8.0	28.0	19.8	4.4	11.0
—	—	—	—	—	—	5.9	15.0	8.0	3.0	3.0	11.0	9.7	1.1	2.8
—	—	—	—	—	—	17.0	43.0	23.0	9.0	10.0	33.0	28.0	3.2	8.2
—	—	—	—	—	—	5.9	18.0	9.0	4.0	4.0	13.0	11.4	1.2	3.0
—	—	—	—	—	—	15.5	47.0	24.0	10.0	11.0	35.0	30.0	3.1	7.9
—	—	—	—	—	—	6.3	23.0	11.0	5.0	5.0	17.0	14.9	1.2	3.5
—	—	—	—	—	—	14.0	51.0	25.0	12.0	12.0	38.0	33.2	2.7	7.7

Table A–2. Composition of Important Feeds: Vitamins, Data Expressed As-Fed and Dry (100% Dry Matter)

				Fat-Soluble Vitamins				
Entry Number	***Alfalfa, Medicago sativa***	***International Feed Number***	***Dry Matter (%)***	***Carotene (provitamin A) (mg/kg)***	***Vitamin A (IU/g)***	***Vitamin D_2 (IU/kg)***	***Vitamin E (mg/kg)***	***Vitamin K (mg/kg)***
001	fresh	2–00–196	24.0	45.0	—	46.0	—	—
002			100.0	185.0	—	191.0	—	—
003	hay, sun-cured	1–00–078	90.0	52.0	—	1,417.0	102.0	19.4
004			100.0	58.0	—	1,575.0	113.0	21.6
005	hay, sun-cured,	1–00–050	90.0	181.0	—	1,806.0	—	—
006	early vegetative		100.0	201.0	—	2,007.0	—	—
007	hay, sun-cured,	1–00–054	90.0	181.0	—	—	—	—
008	late vegetative		100.0	202.0	—	—	—	—
009	hay, sun-cured,	1–00–059	90.0	126.0	—	1,796.0	23.0	—
010	early bloom		100.0	140.0	—	1,996.0	26.0	—
011	hay, sun-cured,	1–00–068	90.0	59.0	—	—	—	—
012	full bloom		100.0	65.0	—	—	—	—
013	hay, sun-cured,	1–00–071	91.0	11.0	—	1,287.0	—	—
014	mature		100.0	12.0	—	1,411.0	—	—
015	hay, sun-cured,	1–00–111	91.0	122.0	—	—	60.0	7.8
016	ground		100.0	135.0	—	—	66.0	8.6
017	leaves, sun-cured	1–00–146	89.0	79.0	—	333.0	—	—
018			100.0	88.0	—	373.0	—	—
019	meal dehy,	1–00–022	90.0	74.0	—	—	82.0	9.6
020	15% protein		100.0	82.0	—	—	91.0	10.6
021	meal dehy,	1–00–023	92.0	120.0	—	—	111.0	8.2
022	17% protein		100.0	131.0	—	—	121.0	9.0
023	meal dehy,	1–00–024	92.0	159.0	—	—	151.0	14.2
024	20% protein		100.0	174.0	—	—	165.0	15.5
025	meal dehy,	1–07–851	93.0	235.0	—	—	221.0	11.6
026	22% protein		100.0	253.0	—	—	238.0	12.6
027	silage	3–00–212	41.0	41.0	—	120.0	—	—
028			100.0	99.0	—	289.0	—	—
029	wilted silage	3–00–221	39.0	24.0	—	216.0	—	—
030			100.0	60.0	—	551.0	—	—

Water-Soluble Vitamins									
Biotin (mg/kg)	*Choline (mg/kg)*	*Folic Acid (Folacin) (mg/kg)*	*Niacin (mg/kg)*	*Pantothenic Acid (mg/kg)*	*Riboflavin (mg/kg)*	*Thiamin (mg/kg)*	*Vitamin B_6 (mg/kg)*	*Vitamin B_{12} (μg/kg)*	*Xanthophylli (mg/kg)*
0.12	378.0	0.6	12.0	8.5	3.3	1.4	1.6	—	—
0.49	1,556.0	2.5	49.0	34.9	13.4	5.9	6.7	—	—
0.18	—	3.2	38.0	25.7	12.1	2.7	—	—	33.0
0.20	—	3.6	42.0	28.6	13.4	3.0	—	—	37.0
—	—	—	—	—	—	—	—	—	—
—	—	—	—	—	—	—	—	—	—
—	—	—	—	—	—	—	—	—	—
—	—	—	—	—	—	—	—	—	—
—	—	—	—	—	—	—	—	—	—
—	—	—	—	—	—	—	—	—	—
—	—	—	—	—	—	—	—	—	—
—	—	—	—	—	—	—	—	—	—
—	—	—	—	—	—	—	—	—	—
—	—	—	—	—	—	—	—	—	—
0.29	1,162.0	3.7	38.0	26.3	12.5	3.8	4.0	—	112.0
0.33	1,283.0	4.1	42.0	29.1	13.8	4.2	4.4	—	123.0
0.28	1,062.0	5.8	47.0	29.0	20.6	4.6	—	—	—
0.31	1,189.0	6.5	53.0	32.4	23.1	5.2	—	—	—
0.25	1,573.0	1.6	42.0	20.7	10.6	3.0	6.3	—	171.0
0.28	1,739.0	1.7	46.0	22.9	11.7	3.3	6.9	—	189.0
0.33	1,370.0	4.4	37.0	29.8	12.9	3.4	7.1	—	263.0
0.36	1,494.0	4.8	40.0	32.4	14.1	3.7	7.7	—	287.0
0.35	1,418.0	3.0	48.0	35.5	15.2	5.4	8.8	—	282.0
0.39	1,547.0	3.2	52.0	38.8	16.6	5.9	9.6	—	308.0
0.33	1,605.0	5.1	50.0	39.0	17.6	5.9	8.3	—	330.0
0.36	1,729.0	5.5	54.0	42.9	19.0	6.3	8.9	—	356.0
—	—	—	—	—	—	—	—	—	—
—	—	—	—	—	—	—	—	—	—
—	—	—	—	—	—	—	—	—	—
—	—	—	—	—	—	—	—	—	—

TABLE A–3. Composition of Important Feeds: Amino Acids, Data Expressed As-Fed and Dry (100% Dry Matter)

Entry Number	*Alfalfa, Medicago sativa*	*International Feed Number*	*Dry Matter (%)*	*Crude Protein (%)*	*Arginine (%)*	*Glycine (%)*	*Histidine (%)*	*Isoleucine (%)*
001	hay, sun-cured	1–00–078	90.0	16.4	0.72	0.70	0.30	0.73
002			100.0	18.2	0.81	0.78	0.33	0.81
003	hay, sun-cured, early	1–00–050	90.0	20.7	1.14	1.03	0.50	0.96
004	vegetative		100.0	23.0	1.27	1.14	0.55	1.07
005	hay, sun-cured, late	1–00–054	90.0	17.9	0.84	0.84	0.38	0.79
006	vegetative		100.0	20.0	0.94	0.94	0.42	0.88
007	hay, sun-cured, early	1–00–059	90.0	16.2	0.73	0.68	0.34	0.60
008	bloom		100.0	18.0	0.81	0.75	0.38	0.67
009	hay, sun-cured, full	1–00–068	90.0	13.5	0.67	0.69	0.32	0.61
010	bloom		100.0	15.0	0.74	0.77	0.35	0.68
011	leaves, sun-cured	1–00–146	89.0	20.6	1.16	—	0.36	0.89
012			100.0	23.1	1.30	—	0.40	1.00
013	meal dehy, 15% protein	1–00–022	90.0	15.6	0.59	0.70	0.27	0.64
014			100.0	17.3	0.65	0.78	0.30	0.71
015	meal dehy, 17% protein	1–00–023	92.0	17.3	0.77	0.84	0.33	0.81
016			100.0	18.9	0.84	0.91	0.36	0.88
017	meal dehy, 20% protein	1–00–024	92.0	20.2	0.96	0.98	0.37	0.89
018			100.0	22.0	1.05	1.07	0.41	0.97
019	meal dehy, 22% protein	1–07–851	93.0	22.2	0.96	1.09	0.44	1.06
020			100.0	23.9	1.04	1.18	0.47	1.15

Leucine (%)	*Lysine* (%)	*Methio-nine* (%)	*Cystine* (%)	*Phenyl-alanine* (%)	*Tyrosine* (%)	*Serine* (%)	*Threo-nine* (%)	*Trypto-phan* (%)	*Valine* (%)
1.12	0.75	0.18	0.24	0.69	0.46	0.69	0.63	0.21	0.74
1.25	0.84	0.20	0.27	0.76	0.51	0.76	0.70	0.24	0.82
1.64	1.27	0.36	—	1.07	0.74	0.97	1.08	—	1.22
1.82	1.41	0.40	—	1.19	0.82	1.08	1.20	—	1.35
1.37	0.99	0.27	—	0.83	0.56	0.77	—	—	0.97
1.53	1.10	0.30	—	0.93	0.62	0.86	—	—	1.08
1.07	0.81	0.19	0.31	0.71	0.48	0.65	0.60	—	0.79
1.19	0.90	0.21	0.34	0.78	0.53	0.72	0.66	—	0.88
1.05	0.78	0.20	—	0.68	0.46	0.64	0.55	—	0.77
1.17	0.87	0.22	—	0.75	0.51	0.71	0.61	—	0.86
1.34	0.98	0.36	0.36	0.89	—	—	0.72	0.45	0.98
1.50	1.10	0.40	0.40	1.00	—	—	0.80	0.50	1.10
1.02	0.59	0.22	0.21	0.62	0.41	0.60	0.56	0.38	0.75
1.13	0.66	0.24	0.23	0.69	0.45	0.67	0.62	0.42	0.83
1.28	0.85	0.27	0.29	0.80	0.54	0.71	0.71	0.34	0.88
1.39	0.93	0.29	0.31	0.87	0.59	0.77	0.77	0.37	0.96
1.41	0.90	0.32	0.32	0.94	0.62	0.86	0.81	0.41	1.04
1.54	0.98	0.34	0.35	1.03	0.67	0.94	0.88	0.45	1.13
1.63	0.97	0.34	0.30	1.13	0.64	0.97	0.97	0.49	1.29
1.75	1.05	0.37	0.32	1.22	0.69	1.05	1.04	0.52	1.39

TABLE A–4. Composition of Important Feeds: Mineral Elements, Data Expressed As-Fed and Dry (100% Dry Matter)

Entry Number	*Alfalfa, Medicago sativa*	*International Feed Number*	*Dry Matter (%)*	*Calcium (%)*	*Chlorine (%)*	*Mag-nesium (%)*	*Phos-phorus (%)*	*Potas sium (%)*
001	fresh	2–00–196	24.0	0.48	0.11	0.07	0.07	0.51
002			100.0	1.96	0.47	0.27	0.30	2.09
003	hay, sun-cured	1–00–078	90.0	1.64	0.34	0.27	0.22	2.03
004			100.0	1.82	0.37	0.30	0.24	2.26
005	hay, sun-cured,	1–00–050	90.0	1.62	0.31	0.23	0.32	1.99
006	early vegetative		100.0	1.80	0.34	0.26	0.35	2.21
007	hay, sun-cured,	1–00–054	90.0	1.38	0.31	0.22	0.26	2.29
008	late vegetative		100.0	1.54	0.34	0.24	0.29	2.56
009	hay, sun-cured,	1–00–059	90.0	1.27	0.34	0.29	0.20	2.27
010	early bloom		100.0	1.41	0.38	0.33	0.22	2.52
011	hay, sun-cured,	1–00–063	90.0	1.27	0.34	0.28	0.22	1.54
012	midbloom		100.0	1.41	0.38	0.31	0.24	1.71
013	hay, sun-cured,	1–00–068	90.0	1.13	—	0.28	0.20	1.38
014	full bloom		100.0	1.25	—	0.31	0.22	1.53
015	hay, sun-cured,	1–00–071	91.0	1.03	—	0.24	0.17	1.62
016	mature		100.0	1.13	—	0.27	0.18	1.78
017	leaves, sun-cured	1–00–146	89.0	2.27	0.45	0.36	0.24	1.62
018			100.0	2.54	0.50	0.40	0.27	1.82
019	meal dehy,	1–00–022	90.0	1.24	0.44	0.28	0.22	2.24
020	15% protein		100.0	1.37	0.48	0.31	0.24	2.48
021	meal dehy,	1–00–023	92.0	1.40	0.47	0.29	0.23	2.39
022	17% protein		100.0	1.52	0.52	0.32	0.25	2.60
023	meal dehy,	1–00–024	92.0	1.59	0.47	0.33	0.28	2.50
024	20% protein		100.0	1.74	0.51	0.36	0.30	2.73
025	meal dehy,	1–07–851	93.0	1.69	0.52	0.31	0.30	2.40
026	22% protein		100.0	1.82	0.56	0.33	0.33	2.58
027	wilted silage	3–00–221	39.0	0.52	0.16	0.15	0.12	0.90
028			100.0	1.33	0.41	0.38	0.30	2.29

Sodium (%)	*Sulfur (%)*	*Cobalt (mg/kg)*	*Copper (mg/kg)*	*Iodine (mg/kg)*	*Iron (mg/kg)*	*Manganese (mg/kg)*	*Selenium (mg/kg)*	*Zinc (mg/kg)*
0.05	0.09	0.03	2.0	—	70.0	10.0	—	4.0
0.19	0.37	0.13	10.0	—	286.0	43.0	—	18.0
0.15	0.27	0.21	10.0	—	175.0	28.0	0.49	22.0
0.17	0.30	0.23	11.0	—	195.0	31.0	0.54	24.0
0.20	0.57	0.09	10.0	—	228.0	41.0	—	22.0
0.22	0.63	0.10	11.0	—	253.0	45.0	—	24.0
0.13	0.28	0.08	8.0	—	204.0	30.0	—	25.0
0.15	0.31	0.09	9.0	—	227.0	34.0	—	27.0
0.13	0.25	0.15	10.0	—	173.0	27.0	0.49	22.0
0.14	0.28	0.16	11.0	—	192.0	31.0	0.54	25.0
0.11	0.26	0.32	13.0	—	121.0	25.0	—	21.0
0.12	0.28	0.36	14.0	—	134.0	28.0	—	23.0
0.10	0.25	0.29	13.0	—	135.0	34.0	—	22.0
0.11	0.27	0.33	14.0	—	150.0	37.0	—	25.0
0.08	0.23	0.08	13.0	—	139.0	40.0	—	22.0
0.08	0.25	0.09	14.0	—	153.0	44.0	—	24.0
0.09	—	0.19	10.0	—	319.0	33.0	—	—
0.11	—	0.22	11.0	—	358.0	37.0	—	—
0.07	0.22	0.17	9.0	0.12	280.0	28.0	0.28	19.0
0.08	0.24	0.19	10.0	0.13	309.0	31.0	0.31	21.0
0.10	0.22	0.30	10.0	0.15	405.0	31.0	0.33	19.0
0.11	0.24	0.33	11.0	0.16	441.0	34.0	0.37	21.0
0.12	0.27	0.26	11.0	0.14	380.0	36.0	0.29	20.0
0.14	0.29	0.28	12.0	0.15	415.0	39.0	0.31	22.0
0.12	0.30	0.31	10.0	0.17	355.0	36.0	—	19.0
0.13	0.32	0.34	11.0	0.18	383.0	39.0	—	21.0
0.06	0.14	—	4.0	—	119.0	16.0	—	—
0.15	0.36	—	9.0	—	305.0	40.0	—	—

APPENDIX B

Conversions and Conversion Tables

WEIGHT CONVERSIONS

1 kilogram (kg) = 1000 grams
1 gram (g) = 1000 milligrams
1 milligram (mg) = 1000 micrograms
1 microgram (μg) = 1000 nanograms (ng)
1 pound (lb) = 454 g
1 kg = 2.2 lb
1 U.S. ton = 2000 lb = 908 kg
1 metric tonne = 1000 kg = 2200 lb

USE OF CONVERSION FACTORS

1. Convert grams (g) to pounds (lb)

454 g = 1 lb
Therefore, 1 g = $\frac{1}{454}$ lb
Therefore, W g = $\frac{1}{454} \times$ W lb
e.g., if W = 850 g
850 g = $\frac{1}{454} \times 850 = 1.87$ lb

2. Convert pounds (lb) to grams (g)

1 lb = 454 g
Therefore, W lb = 454 g/lb × W lb
= 454 × W
e.g., if W = 80 lb,
80 lb = 454 × 80 = 36,320 g
= 36.320 kg

3. Calculation of parts per million (ppm)

1 ppm = 1 part per million parts
e.g., 1 g per million g,
1 lb per million lb, etc.

Example:

100 lb of corn contains 90.8 mg of selenium. How many ppm of selenium does the corn contain?

Answer:

100 lb corn contains 90.8 mg Se

Therefore, 1 lb corn has $\frac{90.8}{100} = 0.908$ mg Se

1 lb corn = 454 g

Therefore, 454 g corn contains 0.908 mg Se

Therefore, 1 g corn contains $\frac{0.908}{454} = 0.002$ mg Se

Therefore, 1,000,000 g corn contains

$1{,}000{,}000 \times 0.002$ mg Se

= 2000 mg Se

= 2 g Se

Therefore, 2 g Se in 1 million g corn

= 2 ppm Se

The corn contains 2 ppm selenium.

As a rule of thumb, ppm is the same as mg/kg.

Therefore, 1 kg of the corn contains 2 mg Se.

APPENDIX C

Rumen Microbes

The rumen ecosystem consists of three main types of microbes: bacteria, protozoa, and fungi.

RUMEN BACTERIA

The rumen bacteria can be categorized by their substrates. Two major groups are **cellulolytic** and **amylolytic** bacteria, or cellulose-digesters and starch-digesters, respectively.

Major cellulolytic bacteria are:

Bacteroides succinogenes (−)*
Ruminococcus flavefaciens (+)
Ruminococcus albus (+)
Butyrivibrio fibriosolvens (+)
* Gram + (+) and Gram − (−)

Major amylolytic bacteria are:

Bacteroides amylophilus
Streptococcus bovis (+)
Succinimonas amylolytica
Bacteroides ruminicola (−)

Bacteria can be classified as Gram positive (Gram +) and Gram negative (Gram −). They differ in the structure of the cell wall. The Gram + bacteria have a much thicker cell wall. When a slide of bacteria is treated with Gram stain (crystal violet and iodine), the Gram + are stained purple, and the Gram − are red. The thick cell wall of the Gram + organisms absorbs the purple stain. Because of the differences in their cell walls, Gram + and Gram − bacteria differ in their sensitivity to antibiotics.

RUMEN PROTOZOA

Most rumen protozoa are ciliated, although flagellate species are present. Protozoa feed on rumen bacteria and also engulf small feed particles. *Entodinium* species are a major type, especially on high-starch diets.

RUMEN FUNGI

Various species of anaerobic fungi have been identified in rumen contents. They appear to play a fairly minor role in fiber digestion.

INDEX

C

D

G

H

I

N

O

P

Q

R

T

U

V